Encyclopedia of Knot Theory

Encyclopedia of Knot Theory

Edited by
Colin Adams
Erica Flapan
Allison Henrich
Louis H. Kauffman
Lewis D. Ludwig
Sam Nelson

CRC Press
Taylor & Francis Group
Boca Raton London New York

CRC Press is an imprint of the
Taylor & Francis Group, an **informa** business

A CHAPMAN & HALL BOOK

First edition published 2021
by CRC Press
6000 Broken Sound Parkway NW, Suite 300, Boca Raton, FL 33487-2742

and by CRC Press
2 Park Square, Milton Park, Abingdon, Oxon, OX14 4RN

© 2021 Taylor & Francis Group, LLC

Chapter 58, Satellite and Quantum Invariants: © H. R. Morton
Chapter 87, DNA Knots and Links: © Isabel Darcy

CRC Press is an imprint of Taylor & Francis Group, an Informa business

---

Library of Congress Cataloging-in-Publication Data

Names: Adams, Colin (Colin Conrad), 1956- editor.
Title: Encyclopedia of knot theory / edited by Colin Adams, Erica Flapan,
  Allison Henrich, Louis Kauffman, Lewis D. Ludwig, Sam Nelson.
Description: First edition. | Boca Raton : Chapman & Hall, CRC Press, 2021.
  | Includes bibliographical references and index.
Identifiers: LCCN 2020027796 (print) | LCCN 2020027797 (ebook) | ISBN
  9781138297845 (hardback) | ISBN 9781138298217 (ebook)
Subjects: LCSH: Knot theory--Encyclopedias.
Classification: LCC QA612.2 .E59 2021 (print) | LCC QA612.2 (ebook) | DDC
  514/.2242--dc23
LC record available at https://lccn.loc.gov/2020027796
LC ebook record available at https://lccn.loc.gov/2020027797

ISBN 13: 978-1-138-29784-5 (hbk)
ISBN 13: 978-1-138-29821-7 (ebk)

Contents

Part I

Introduction and History of Knots

Chapter 1

Introduction to Knots

Lewis D. Ludwig, Denison University

1.1 The origins of knot theory

For millennia, humans have been tying knots to bind, lash, and secure. There are numerous knots used for practical purposes by sailors and builders, see *The Ashley Book of Knots* [3] for example, as well as knots used for artistic or even spiritual significance as seen in this Celtic cross and mosaic pattern at the Alhambra in Granada, Spain, Figure 1.1.

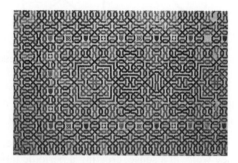

FIGURE 1.1: A Celtic cross adorned with links and a mosaic pattern at the Alhambra.

But what is a mathematical knot? Start with a piece of string, tie a knot in it, then glue the ends to form a knotted loop. To a mathematician, a knot is just a knotted loop of string, except the string has no thickness, its cross-section begins a single point [1].

Why study such objects? In the late 19th century, scientist were trying to develop a model for the atom. At the time, it was believed that an invisible substance called ether pervaded all of space, providing a medium for electricity and magnetism. The prominent German scientist Hermann von Helmholtz (1821-1894) conducted experiments using smoke rings to study vortices in space to possibly explain the nature of ether, see Figure 1.2. In an attempt to explain different types of elements, Lord Kelvin (William Thomson, 1824-1907) hypothesized that atoms were merely knotted vortices in ether. For example, the trivial knot or unknot, the trefoil knot, and the figure-eight knot could represent hydrogen, carbon, and oxygen, respectively, see Figure 1.3.

While this may seem like a fanciful notion compared to the Bohr model of the atom used today, Lord Kelvin had several reasons to believe these knotted vortices may describe the atomic structure of matter. First, vortices in fluid were extremely stable, so each knot could be classified according to its geometrical properties. Second, there are an endless number of knots allowing for a variety of

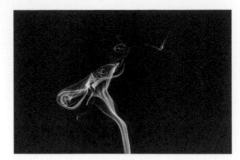

FIGURE 1.2: A smoke ring that could be knotted.

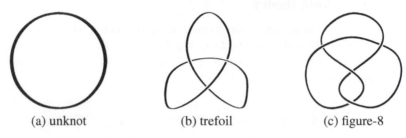

(a) unknot   (b) trefoil   (c) figure-8

FIGURE 1.3: Could knots represent atoms?

chemical elements. Finally, just as smoke rings vibrate, the oscillations of ether vortex tubes might produce atomic spectral lines which can be used to identify atoms and molecules [5].

With the notions that knots may represent the elements, scientists set off to create charts of distinct knots, thinking they were, in a sense, creating a periodic table. This work was largely taken up by a contemporary of Lord Kelvin, the Scottish physicist Peter Guthrie Tait, who in 1877 published the first in a series of papers on knot enumeration. Determining if two knots are distinct, that is, determining it two knots are the same or different, can be challenging. For instance, are the three knots in Figure 1.4 the same?

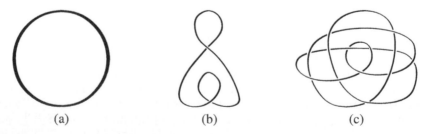

(a)   (b)   (c)

FIGURE 1.4: Are these the same knot?

With the help of others, including American mathematician C.N. Little [4], the enumeration of all knots up to 10 crossings was almost complete by the end of the century. Figure 1.5 is an example of some of Little's work.

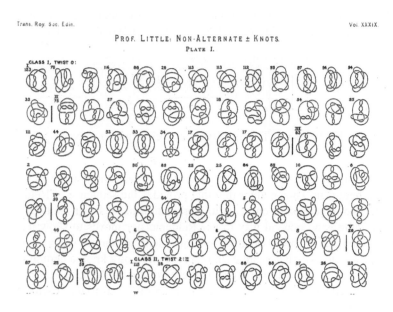

Trans. Roy. Soc. Edin.

Vol XXXIX.

PROF. LITTLE: NON-ALTERNATE ± KNOTS.
PLATE I.

FIGURE 1.5: One of Little's knot tables.

Unfortunately, in 1887, the Michelson-Morley experiment demonstrated that there was no ether to knot. By the end of the century, physicists had developed a new model for the atom and generally lost interest in knots. Fortunately, mathematicians had became intrigued with knots and have been developing mathematical knot theory ever since.

However, more recent advances have rekindled an interest in knots among scientists. In the 1980s biochemists discovered knotted DNA molecules in bacteria. Recent work has also explored the importance of knotting in protein chains. Molecular chemists have created new molecular compounds that are knotted. So a scientific blunder from the late 1800s has turned into an important mathematical field of study with significant applications to chemistry and biology.

1.2 Distinguishing knots

As noted by Figure 1.4, it can be challenging to determine if two knots are equivalent. It turns out that Figure 1.4 illustrates three different projections of the unknot. By projection, we mean the projection of the 3-dimensional knot down to a plane, so that no three or more points on the knot are sent to the same point on the plane and where at double points, which correspond to crossings, it is made clear which strand is the overstrand and which the under. So how can we use projections to determine if two knots are equivalent? An important theorem due to Reidemeister as well as Alexander and Briggs, is that two knots are equivalent if and only if the projection of one knot can be turned in to the projection of the other knot by a sequence of three types of local diagrammatic changes now known as Reidemeister moves and planar isotopy – bending, stretching, or shrinking in the plane [2, 6] . There is a total of three such Reidemeister moves depicted in Figure 1.6.

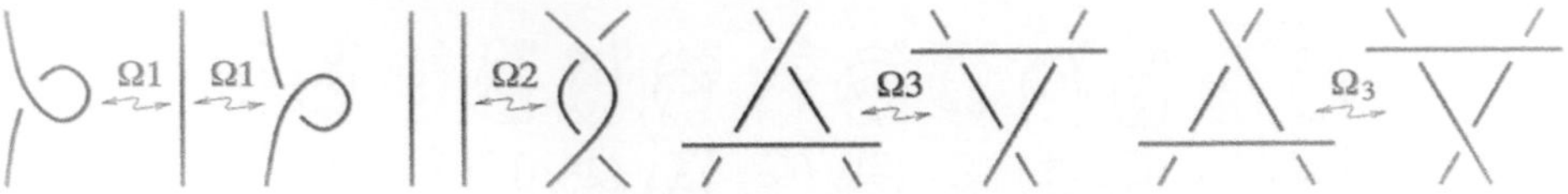

FIGURE 1.6: The three types of Reidemeister moves.

While Reidemeister moves provide a way to determine if two knots are equivalent, given two equivalent knots, there is no known algorithm that provides the sequence of Reidemeister moves needed to transform a projection of one knot to a projection of the other knot. Or given two non-equivalent knots, there is no general way to use Reidemeister moves to show the two knots are distinct. To help determine whether two knots are equivalent, we turn to various knot invariants. A knot invariant is a way to associate, with each representation of a knot, a certain algebraic object – a number, a polynomial – so that this object is preserved under Reidemeister moves and planar isotopy. Later chapters in this volume will provide a wide array of invariants, each with its particular advantages/disadvantages.

1.3 Bibliography

[1] Colin C. Adams. *The Knot Book*. American Mathematical Society, Providence, RI, 2004. An elementary introduction to the mathematical theory of knots. Revised reprint of the 1994 original.

[2] J. W. Alexander and G. B. Briggs. On types of knotted curves. *Ann. of Math. (2)*, 28(1-4):562–586, 1926/27.

[3] Clifford W. Ashley. *The Ashley Book of Knots*. Doubleday, New York, 1944.

[4] C.N. Little. Non-alternate *pm*knots. *Trans. Royal. Soc. Edinburgh*, 39:771–778, 2010.

[5] Mario Livio. Don't bristle at blunders. *Nature*, 497:309–310, 2013.

[6] K. Reidemeister. *Knotentheorie*. Berlin:Springer-Verlag, 1932. Reprint.

Part II

Standard and Nonstandard Representations of Knots

Chapter 2

Link Diagrams

Jim Hoste, Pitzer College

Humans have been drawing pictures of knots for thousands of years—knots appear in Roman mosaics, Central American jade carvings, and the illuminated manuscripts of the Celts, to name just a few examples. So, it is not surprising that pictures of knots were present at the beginning of the mathematical theory of knots. Drawings of knots can be found among Carl Freiderich Gauss's papers and his student Johann Benedict Listing used drawings to depict and discuss knots in his book *Vorstudien zur Topologie*. When Peter Guthrie Tate began to study and enumerate knots, he used drawings of knots as the organizing principle of his tabulation.

A knot is a simple closed curve in space and a *knot diagram* is a projection of the knot into a plane with additional information given so that one can recreate an equivalent knot from the diagram. If the knot is nontrivial, then any projection of it to a plane must contain *singularities*, that is, places where two or more points of the knot project to the same point. In the case of a smooth knot, a particularly nice kind of projection, called a *regular* projection, is one where there are only finitely many singular points, each is the image of exactly two points of the knot, and moreover, at each singularity the tangent vectors to the knot at each preimage point project to linearly independent vectors in the plane. In this case, we call each singular point a *double* point and say that the projection crosses itself *transversely* at each double point. From the regular projection alone, we cannot hope to re-create the knot, because we do not know how "high" each point of the knot lies above its projection. However, if at each double point we indicate which of the two preimage points is "above" the other, we can now create a knot equivalent to the original one. A popular method of indicating which point is higher at each double point is to erase a small piece of the "lower" projected curve, but other methods have been used. The resulting figure is called a *regular diagram* and the double points, now with the added information of which part of the knot is higher, are called *crossings*. An example of a regular knot diagram with five crossings is shown in Figure 2.1.

It can be shown that given any smooth knot, there is some direction in which we may project it to a plane so as to obtain a regular projection. Thus every knot can be represented by a regular knot diagram. Moreover, if the knot is oriented, this additional information can be recorded in the diagram by placing an arrowhead on the projected curve. Regular diagrams can be similarly defined in the case of polygonal knots. We now require a projection that has only finitely many double points, none of which is the image of a vertex. Again, it can be shown that given any polygonal knot, there is some direction in which we may project it so as to obtain a regular projection, which will be a polygonal curve in the plane [5]. As before, diagrams can be decorated with arrowheads so as to depict oriented knots.

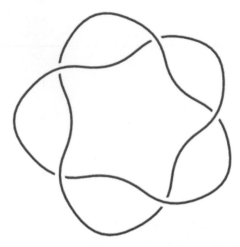

FIGURE 2.1: A regular diagram of a knot with five crossings.

An important theorem due to Reidemeister as well as Alexander and Briggs, is that two regular knot diagrams represent equivalent knots if and only if they are related by a sequence of three types of local diagrammatic changes now known as *Reidemeister* moves [4] [7]. The three types of Reidemeister moves, for unoriented knots, are shown in Figure 2.2.

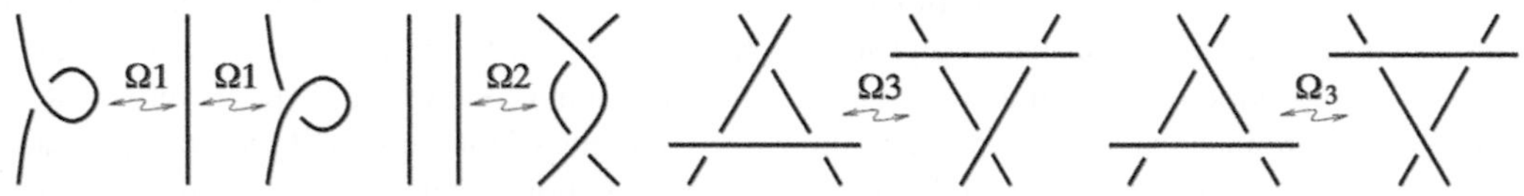

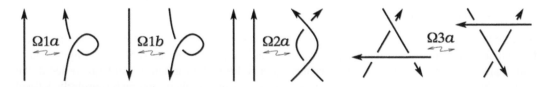

FIGURE 2.2: The three types of Reidemeister moves.

A set of moves sufficient to pass between all oriented diagrams of the same knot can be obtained by introducing orientations in all possible ways. A careful analysis by Ployak shows that the moves depicted in Figure 2.3 form a minimal set of oriented Reidemeister moves sufficient to pass between all diagrams of the same oriented knot [6].

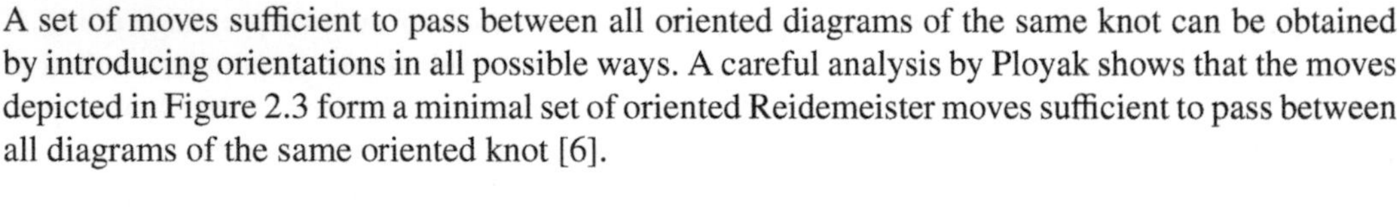

FIGURE 2.3: A generating set of oriented Reidemeister moves.

While knots are really simple closed curves in 3-space, considered up to ambient isotopy, it is possible to approach the entire theory of knots by means of knot diagrams, considered up to Reidemeister moves. Indeed, many important knot invariants are defined in terms of diagrams. For example, the *crossing number* of a knot is defined as the smallest number of crossings in any regular diagram of the knot. Some knot invariants are first defined for a regular diagram and then shown to be preserved by Reidemeister moves, thus being independent of the choice of diagram. Knot diagrams also allow one to define various subclasses of knots consisting of all knots that possess a diagram having certain properties. For example, alternating knots are those that possess an *alternating* diagram—as one traverses the diagram, one passes alternately over and under the knot as each crossing is encountered. Other examples of classes defined by having a diagram of a certain type are *n-bridge* knots and *positive* knots.

While regular diagrams of knots have been the primary type of diagram used in the past, considerable work has been done recently with diagrams allowing multi-crossing points, where three or more strands cross at a point, still pairwise transversely [1] [2]. Figure 2.4 illustrates a diagram of the trefoil knot having two *triple crossings*. For any $n \geq 2$, it has been shown that every knot has a

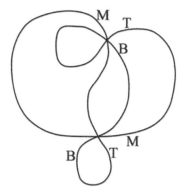

FIGURE 2.4: A diagram of the trefoil knot using only triple crossings. The letters "T," "M," and "B" indicate the top, middle, and bottom level strand at each crossing.

diagram using only n-crossings. In the case of $n = 3$, a set of diagrammatic moves analogous to the Reidemeister moves have been found that allow one to pass between all triple-crossing diagrams of the same knot [3]. Continuing in this vein, it has been shown that every knot has a diagram with only a single n-crossing, for some n [2]. Moreover, one can assume that the diagram appears as the petals of a flower with the single n-crossing in the center. Thus every knot can be encoded by a single permutation used to describe the heights of the strands in the n-crossing.

2.1 Bibliography

[1] C. Adams, Triple crossing number of knots and links. *J. Knot Theory Ramifications*, 22(2):1350006, 2013.

[2] C. Adams, T. Crawford, B. DeMeo, M. Landry, A. T. Lin, M. Montee, S. Park, S. Venkatesh, and F. Yhee, Knot projections with a single multi-crossing. *J. Knot Theory Ramifications*, 24(3):1550011, 2015.

[3] C. Adams, J. Hoste, and M. Palmer, Triple-Crossing Number and Moves on Triple-Crossing Link Diagrams. arXiv:1706.09333, 2017.

[4] J. W. Alexander and G. B. Briggs, On types of knotted curves. *Ann. of Math. (2)*, 28(1-4):562–586, 1926/27.

[5] R.H. Crowell and R.H. Fox, *Introduction to Knot Theory*. New York: Ginn and Co., 1963.

[6] M. Polyak, Minimal generating sets of Reidemeister moves. *Quantum Topol.*, 1(4):399–411, 2010.

[7] K. Reidemeister, *Knotentheorie*. Ergebn. Math. Grenzgeb., Bd.1; Berlin:Springer-Verlag, 1932.

Chapter 3

Gauss Diagrams

Inga Johnson, Willamette University

A knot diagram is a good *visual* representation of a knot, but it is often useful to have a representation that is more *combinatorial* in nature such as a Gauss diagram of a knot. The process to create a Gauss diagram begins with determining the Gauss code of the knot.

The **Gauss code** of a knot diagram with crossings labeled 1 through n is a cyclically ordered, double occurrence list of the integers $1, \ldots, n$ indicating the order of the crossings encountered as the knot is traversed from a selected starting point along the knot. Each integer in the code is preceded with an O or U to indicate whether the curve goes over or under, respectively. For example, the Gauss code for the trefoil in Figure 3.1 is O1U2O3U1O2U3.

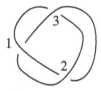

FIGURE 3.1: The Gauss code is O1U2O3U1O2U3.

When creating the Gauss diagram, we simplify our view of the order of the crossings by wrapping the Gauss code around a circle, and we abstract the representation of the crossing by placing a chord of the circle between the two points along the circle that form the crossing. The U's and O's in the Gauss code are dropped and replaced by arrows on the chords that point towards the under-crossing. Last, each oriented chord is decorated with + or − to indicate whether the crossing is positive or negative. An example of the Gauss diagram for the knot 6_3 is given in Figure 3.2.

The representation of crossings as chords in a Gauss diagram makes knot equivalence via Riedemeiseter moves less intuitive to detect visually. To aide our understanding of knot equivalence in the setting of Gauss diagrams we translate the traditional Reidemeister moves to their Gauss diagrammatic equivalences. Figure 3.3 shows a generating set of Reidemeister 1, 2, and 3 moves [1]. Within the equivalences, the dotted portion of the circle could contain several other chords, but the solid portions of the circle contain nothing else.

One observation that is straight-forward to see is that a Gauss diagram without intersecting chords must represent the unknot. This fact follows from noticing that repeated application of the R1 Gauss move on the chords in the Gauss diagram that are exterior to other chords leaves us with a circle that contains no chords.

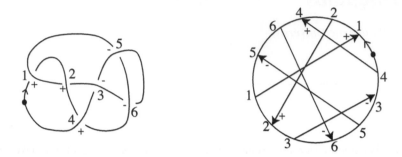

FIGURE 3.2: On the left, 6_3 with Gauss code U1O2U4O6U5O1U2O3U6O5U3O4, and the corresponding Gauss diagram of 6_3 on the right.

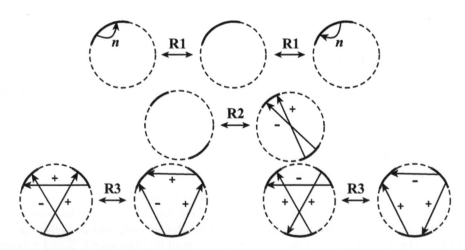

FIGURE 3.3: Reidemeister moves as Gauss diagrams. The n is a variable that may represent $+$ or $-$.

Another interesting feature of Gauss diagrams is that it is easy to draw a diagram for which there is no corresponding classical knot, see Figure 3.4. The theory of this broader collection of Gauss diagrams, for which there is no classical knot diagram, is called the theory of virtual knots and was pioneered independently by L. Kauffman [2] and N. & S. Kamada [3]. We learn more about virtual knots in Chapter 30.

FIGURE 3.4: An example diagram that cannot be realized as the Gauss diagram of a classical knot.

Gauss diagrams are a rich lens through which knots and related objects are studied. They are used to encode knots in a computer recognizable form. In Part VII, we extend the definition of Gauss diagrams to pseudoknots and use them to define invariants. When studying finite type invariants in Part XIII, we learn that any finite type invariant of classical knots is given by a Gauss diagram formula [4].

3.1 Bibliography

[1] M. Polyak, Minimal generating sets of Reidemeister moves, *Quantum Topol.*, **1**, (2010), 399–411.

[2] L. H. Kauffman, Virtual knot theory, *Europ. J. Combinatorics*, 20, (1999), 663–691.

[3] N. Kamada and S. Kamada, Abstract link diagrams and virtual knots, *J. Knot Theory Ramifications*, 9: 1, (2000), 93–106.

[4] M. Goussarov, M. Polyak, O. Viro, Finite-type invariants of classical and virtual knots, *Topology* 39 (2000) 1045–1068.

Chapter 4

DT Codes

Heather M. Russell, University of Richmond

4.1 Codes for Links

Diagrams provide a robust and aesthetically pleasing framework for the study of knots and links, but they are complicated, especially as the number of crossings grows large. To address this issue, there are a variety of notational schemes that compactly summarize the connectivity data of a diagram. Using these knot codes, it is possible to execute computations that would be unreasonable to tackle by hand [4].

In 1898, Tait proposed an early scheme for encoding alternating knot diagrams by labeling crossings and recording the order in which labels are encountered as one walks along the knot [9]. (Tait's work references an even earlier and very different method for encoding diagrams due to Listing [7].) Gauss diagrams were used by Gauss to investigate the crossing sequence problem for plane curves [8]. Gauss codes, which come from these diagrams, closely resemble Tait's notation.

In 1983, Dowker and Thistlethwaite modified Tait's approach with the goal of using computers to tabulate knots [3, 5]. Their codes, which have come to be known as DT codes (or DT sequences), were extended by Doll and Hoste from knots to links [2]. Using computer algorithms to distinguish between realizable vs. nonrealizable and prime vs. composite DT codes together with an arsenal of modern knot invariants, Hoste, Thistlethwaite, and Weeks identified all prime knots up to 16 crossings [4].

There are many other examples of knot codes. For instance, Conway notation, which also helped advance knot tabulation, comes from John H. Conway's work decomposing diagrams into constituent rational tangles [1]. Bar-Natan's PD (planar diagram) codes are used extensively in the Mathematica package Knot Theory [10]. Here, we focus our exploration on DT codes: their definition, examples, features, and limitations.

4.2 Constructing DT Codes

To construct a DT code for a knot diagram, first choose a base point and an orientation. Starting at the base point, travel along the diagram in the direction indicated by the orientation. Beginning with the value 1, label crossings each time you visit them as shown in Figure 4.1. After one trip through the diagram, each crossing will have two labels: one even and one odd.

For an n-crossing diagram, the labeling method thus far has generated n pairs of values in the

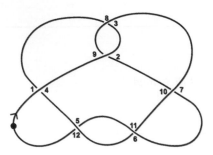

FIGURE 4.1: A diagram with DT code $(4, -8, 12, 10, -2, 6)$.

interval $[1, 2n] = \{1, \ldots, 2n\}$. This data is enough to encode the underlying knot projection, but more information is needed to keep track of crossing information. To do this, add a minus to each even label that corresponds to an overcrossing. For example, the pairs corresponding to the diagram in Figure 4.1 are $(1, 4), (-2, 9), (3, -8), (5, 12), (6, 11)$, and $(7, 10)$.

Since each even number is paired with an odd number, we do not need to record both elements of the pair. Instead, we order the even values according to the odd values with which they are paired, so the value paired with 1 is first followed by the value paired with 3 and so on. This ordered n-tuple of even integers is the DT code for the associated oriented, based knot diagram. Using this construction, we see that the DT code for the diagram in Figure 4.1 is $(4, -8, 12, 10, -2, 6)$. Note that minus signs signify deviation from an alternating crossing pattern, so a code with no minuses (or all minuses) indicates an alternating diagram.

To extend this system from knots to links, first, order the components of the link so that components with smaller numbers of crossing points are first. For example, the central, untwisted component in Figure 4.2 is first because it has four crossing points while the other component has six. Choose a base point and orientation for each component such that traversing the components in order and placing labels at crossings leads to even and odd pairs at each crossing. This is always possible.

As before, modify the even labels with signs, and list them in order according to their odd counterparts. Use vertical bars to separate the code according to the odd labels on each component. For instance, the link diagram in Figure 4.2 has code $(10, 8 \,|\, 2, 4, 6)$ since the first component has odd labels 1 and 3 while the second has odd labels 5, 7 and 9.

4.3   Properties of DT Codes

There are many DT codes for a diagram. Indeed, for an n-crossing knot diagram there are 2 choices of orientation and $2n$ choices for base point. For highly asymmetric diagrams, each of the resulting $4n$ codes can be different [4]. We declare that the *standard* code for a diagram is the minimal one

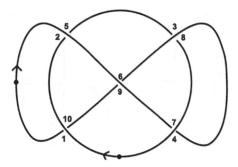

FIGURE 4.2: A diagram with DT code $(10, 8 \mid 2, 4, 6)$.

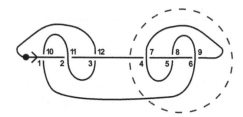

FIGURE 4.3: A composite diagram with DT code $(10, 12, 8, 4, 6, 2)$.

with respect to lexicographic ordering, and the standard code for a link type is the minimal one over all diagrams of that type with minimal crossing number. The standard code for the trefoil is $(4, 6, 2)$, and the standard code for the figure-eight knot is $(4, 6, 8, 2)$.

A diagram is composite if there exists a circle in the plane intersecting the diagram transversely at two points with crossings on both sides. Figure 4.3 shows an example of a composite knot diagram with a dashed circle separating its crossings. A diagram that is not composite is said to be prime. A useful feature of DT codes is that they reveal if a diagram is composite. For an n-crossing diagram, if there exists a proper subinterval of $[1, 2n]$ that is a union of DT code pairs, the diagram is composite. The based, oriented diagram in Figure 4.3 has code $(10, 12, 8, 4, 6, 2)$. Since the subinterval $[4, 9]$ of $[1, 12]$ is the union of pairs $(4, 7)$, $(5, 8)$, and $(6, 9)$, the corresponding diagram must be composite.

DT codes distinguish prime diagrams only up to mirror image. More precisely, if a link has a diagram with a given DT code, its mirror image has a diagram with the same code. This means DT codes cannot directly be used to determine chirality. Figure 4.4 shows an example of based, oriented diagrams for the right- and left-handed trefoils that have the same corresponding DT code. This issue regarding mirror images leads to even more ambiguity in DT codes for composite diagrams.

Finally, we note that not every permutation of the numbers $\pm 2, \pm 4, \ldots, \pm 2n$ yields a realizable DT code. In their paper on DT codes for links, Doll and Hoste share an anecdote that Conway called nonrealizable knot and link DT codes "knits" and "locks," respectively [2]. Nonrealizable codes have inspired the rich theory of virtual links [6]. Dowker and Thistlethwaite foreshadow this at the end of their paper saying that the unrealizable code $(4, 8, 2, 10, 6)$ "can be realized on the torus, and the realization is the projection of an alternating knot in $T^2 \times \mathbb{R}^1$" [3].

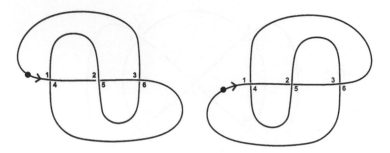

FIGURE 4.4: DT codes cannot detect mirror images.

4.4 Bibliography

[1] J. H. Conway. An enumeration of knots and links, and some of their algebraic properties. In *Computational Problems in Abstract Algebra (Proc. Conf., Oxford, 1967)*, pages 329–358. Pergamon, Oxford, 1970.

[2] Helmut Doll and Jim Hoste. A tabulation of oriented links. *Math. Comp.*, 57(196):747–761, 1991. With microfiche supplement.

[3] C. H. Dowker and Morwen B. Thistlethwaite. Classification of knot projections. *Topology Appl.*, 16(1):19–31, 1983.

[4] Jim Hoste, Morwen Thistlethwaite, and Jeff Weeks. The first 1,701,936 knots. *Math. Intelligencer*, 20(4):33–48, 1998.

[5] I. M. James and E. H. Kronheimer, editors. *Aspects of Topology*, volume 93 of *London Mathematical Society Lecture Note Series*. Cambridge University Press, Cambridge, 1985. In memory of Hugh Dowker: 1912–1982.

[6] Louis H. Kauffman. Virtual knot theory. *European J. Combin.*, 20(7):663–690, 1999.

[7] J. B. Listing. *Vorstudien zur Topologie*. Göttingen Studien, University of Göttingen, Germany, 1848.

[8] R. C. Read and P. Rosenstiehl. On the Gauss crossing problem. In *Combinatorics (Proc. Fifth Hungarian Colloq., Keszthely, 1976), Vol. II*, volume 18 of *Colloq. Math. Soc. János Bolyai*, pages 843–876. North-Holland, Amsterdam-New York, 1978.

[9] P.G. Tait. On knots I, II, III. In *Scientific Papers, Vol. I*, pages 273–347. Cambridge University Press, 1898.

[10] Knot Theory. A knot theory Mathematica package. Available at `http://katlas.math.toronto.edu/wiki/KnotTheory`.

Chapter 5

Knot Mosaics

Lewis D. Ludwig, Denison University

5.1 Introduction

Consider the 11 tiles, labeled T_0 through T_{10}, in Figure 5.1 consisting of five distinct tiles up to rotation. These tiles can be placed in an $n \times n$ grid to form a knot, see Figure 5.2. A *knot mosaic* is the representation of a knot on an $n \times n$ grid composed of a collection of these 11 tiles. Lomonaco and Kauffman [10] used these tiles to create knot mosaics as a model of physical quantum states. They conjectured that knot mosaic theory is equivalent to classic (tame) knot theory. Kuriya and Shehab [7] proved this conjecture, so any knot can be represented by a knot mosaic.

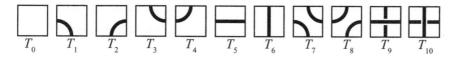

FIGURE 5.1: Tiles T_0-T_{10}.

A *connection point* of a tile is a midpoint of a tile edge that is also the endpoint of a curve drawn on the tile. A mosaic tile is said to be *suitably connected* if each of its connection points touches a connection point of a contiguous tile. Several examples of suitably connected knot mosaics are depicted in Figure 5.2 – the first mosaic is the trefoil knot (3_1), the second mosaic is the Hopf Link, and the third is the composition of two trefoil knots ($3_1 \# 3_1$).

Since knot mosaic theory is equivalent to classic knot theory, usual things like ambient isotopy and Reidemeister moves apply. For a complete list of mosaic planar isotopy moves and mosaic Reidemeister moves see Lomonaco and Kauffman [10]. For example, Figures 5.3 and 5.4 demonstrate the mosaic Reidemeister Type I and Type II moves. Due to this equivalency, we can consider knot invariants such as crossing number (Chapter 26). A *reduced* diagram of a knot is a projection of the knot in which none of the crossings can be reduced or removed. The fourth knot mosaic depicted in Figure 5.2 is an example of a non-reduced trefoil knot diagram. In this example the crossing number of three is not realized because there are two extra crossings that can be easily removed.

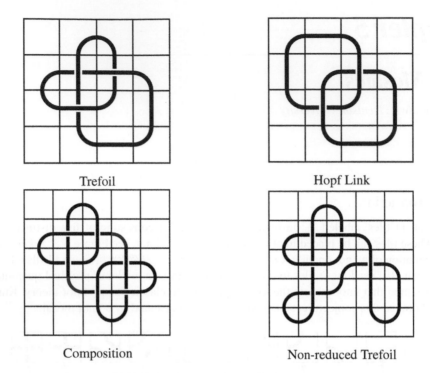

FIGURE 5.2: Examples of link mosaics.

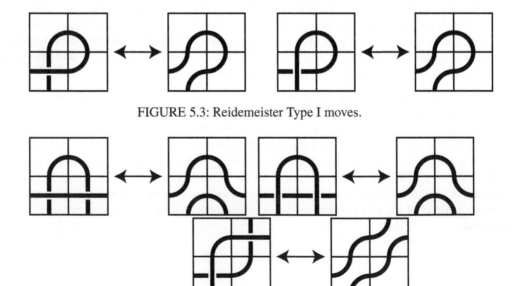

FIGURE 5.3: Reidemeister Type I moves.

FIGURE 5.4: Reidemeister Type II moves.

5.2   The Size of a Knot Mosaic

When considering knot mosaics, one quickly realizes that a knot can be represented as a knot mosaic in many different ways. For example the mosaic could be larger than necessary, with extra loops, bends or twists or unnecessary blank tiles. In this section, we consider "efficient ways" to represent a knot as a knot mosaic. For example, efficient may mean for a given knot K, what is the smallest n for which K can be expressed on an $n \times n$ mosaic? This question is addressed by a knot invariant for knot mosaics known as the *mosaic number*. The mosaic number of a knot K is the smallest integer n for which K can be represented on an $n \times n$ mosaic board. We will denote the mosaic number of a knot K as $m(K)$. For the trefoil, it is an easy exercise to show that the mosaic number for the trefoil is four, or $m(3_1) = 4$. Lee, Ludwig, Paat and Peiffer [9] determined the crossing number for all prime knots with a crossing number of eight or fewer. Interestingly, they showed all four, five, and six crossing prime knots have mosaic number five, except $m(6_3) = 6$. In addition, all seven and eight crossing prime knots have mosaic number six, except $m(7_4) = 6$ (see Figure 5.5).

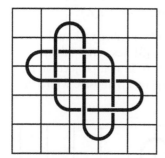
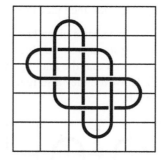

FIGURE 5.5: Knots 6_1 and 7_4.

Notice in Figure 5.5 that although the mosaic number of 6_1 is five, and looks very similar to 7_4, this representation of 6_1 does not have the crossing number six. We say a knot mosaic is *minimal* if the knot mosaic realizes its mosaic number. This led to a result of Ludwig, Evans, and Paat [11] where they produced an infinite family of minimal knot mosaics whose mosaic number is realized in non-reduced projections. That is, this family of mosaic knots realized their mosaic numbers, but not their crossing numbers in those projections. In general, Ludwig et al. created a family of knots whose mosaic number was realized on an $n \times n$ mosaic board with n odd, $n \geq 5$, and whose crossing number was realized on an $(n + 1) \times (n + 1)$ mosaic board.

It is often interesting to see how various knot invariants compare. For example, in 2009, Ludwig, Paat, and Shapiro and Dye et al. [1] showed that the mosaic number and crossing number, $c(K)$, of a knot K can be related in the following way:

$$\lceil \sqrt{c(K)} \rceil + 2 \leq m(K).$$

Howards and Kobin [6] provided another bound on the crossing number based on an odd or even mosaic number:

$$c(K) \leq \begin{cases} (m(K) - 2)^2 - 2 & \text{if } m(K) = 2n + 1 \\ (m(K) - 2)^2 - (m(K) - 3) & \text{if } m(K) = 2n. \end{cases}$$

Using the arc index, in 2014 Lee et al. [8] provided a bound on the mosaic number via the crossing number:

$$m(K) \leq c(K) + 1.$$

This leads to a natural question: Do tighter upper or lower bounds exist on the mosaic number of a knot using the crossing number of the knot?

Here is another possible way to consider efficiency in knot mosaics: what are the fewest non-blank tiles needed to create K? Let $t(K)$ denote the *tile number* of a knot K; the fewest non-T_0 tiles needed to construct K [2],[9]. The *minimal mosaic tile number* of a knot K, denoted $t_m(K)$, is the fewest non-blank tiles needed to construct K on a minimal mosaic. Clearly $t(K) \leq t_m(K)$ for a given knot. Heap and Knowles [2] showed that $t(K) = t_m(K)$ for all prime knots with $m(K) \leq 5$. They further provided upper bounds for the tile number: If a knot or link K with $m(K) = n$ ($n \geq 4$), has no unknotted, unlinked components, then $t(K) \leq n^2 - 4$ for n even and $t(K) \leq n^2 - 9$ for n odd. Further, if K is a prime knot, then they proved the lower bound $5n - 8 \leq t(K)$.

A natural question is whether a given knot could achieve its tile number without being minimal, i.e., without achieving its mosaic number. Heap and Knowles [3] answered this question in the affirmative. They showed that the knot 9_{10} has mosaic number 6. In that projection, $t(9_{10}) = 32$, that is $t_m(K) = 32$. However, they also found a projection of 9_{10} on a 7-mosaic with $t(9_{10}) = 27$ as depicted in Figure 5.6.

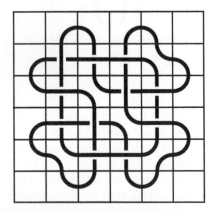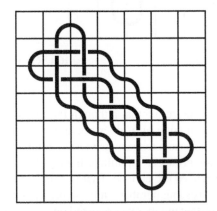

FIGURE 5.6: The 9_{10} knot represented as a 6-mosaic with minimal mosaic tile number 32 and as a 7-mosaic with tile number 27.

5.3   Counting Knot Mosaics

Another question one may consider regarding knot mosaics is, given a mosaic board of a certain size, how many knots or links can be placed on that board up to rotation? In their seminal article [10], Lomonaco and Kauffman defined $\mathbb{K}^n$ as the set of all knot n-mosaics, that is, knots on an $n \times n$ mosaic board, and D_n as the total number of elements of $\mathbb{K}^n$, in the same article they showed $D_3 = 21$. In the summer of 2009, Ludwig, Paat and Shapiro created a computer program that showed $D_4 = 2,594$. This result was independently proven by Lee at al. [5] using combinatorial arguments. Continuing this work, Lee et al. [4] defined $D_{m,n}$ as the total number of all knot (m, n)-mosaics, that is knots on an $m \times n$ mosaic board. In this article, they found the precise values for $4 \leq m \leq n \leq 6$ as below. They used a partition matrix argument and the two-fold rule. The

TABLE 5.1: Number of (m, n)-mosiacs.

$D_{m,n}$	$n = 4$	$n = 5$	$n = 6$
$m = 4$	2,594	54,226	1,144,526
$m = 5$		4,183,954	331,745,962
$m = 6$			101,393,411,126

two-fold rule is an observation that a suitably connected $(m - 2, n - 2)$-mosaic can be extended to exactly two different knot (m, n)-mosaics by augmenting the four sides with a row/column of tiles. Note that state matrices are a generalized version of partition matrices.

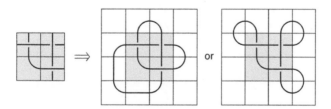

FIGURE 5.7: An example of the two-fold rule.

Using recurrence relations of state matrices, in another article [12] these authors created an algorithm that computes the precise value for $D_{m,n}$.

5.4   Bibliography

[1] J. Alan Alewine, H. A. Dye, David Etheridge, Irina Garduño, and Amber Ramos. Bounds on mosaic knots. *Pi Mu Epsilon J.*, 14(1):1–8, 2014.

[2] Aaron Heap and Douglas Knowles. Tile number and space-efficient knot mosaics, 2018.

[3] Aaron Heap and Douglas Knowles. Space-efficient knot mosaics for prime knots with mosaic number 6, 2019.

[4] Kyungpyo Hong, Ho Lee, Hwa Jeong Lee, and Seungsang Oh. Small knot mosaics and partition matrices. *J. Phys. A*, 47(43):435201, 13, 2014.

[5] Kyungpyo Hong, Seungsang Oh, Ho Lee, and Hwa Jeong Lee. Upper bound on the total number of knot n-mosaics. *J. Knot Theory Ramifications*, 23(13):1450065, 7, 2014.

[6] Hugh Howards and Andrew Kobin. Crossing number bound in knot mosaics, 2018.

[7] Takahito Kuriya and Omar Shehab. The Lomonaco-Kauffman conjecture. *J. Knot Theory Ramifications*, 23(1):1450003, 20, 2014.

[8] Hwa Jeong Lee, Kyungpyo Hong, Ho Lee, and Seungsang Oh. Mosaic number of knots. *J. Knot Theory Ramifications*, 23(13):1450069, 8, 2014.

[9] Hwa Jeong Lee, Lewis Ludwig, Joseph Paat, and Amanda Peiffer. Knot mosaic tabulation. *Involve*, 11(1):13–26, 2018.

[10] Samuel J. Lomonaco and Louis H. Kauffman. Quantum knots and mosaics. *Quantum Inf. Process.*, 7(2-3):85–115, 2008.

[11] Lewis D. Ludwig, Erica L. Evans, and Joseph S. Paat. An infinite family of knots whose mosaic number is realized in non-reduced projections. *J. Knot Theory Ramifications*, 22(7):1350036, 11, 2013.

[12] Seungsang Oh, Kyungpyo Hong, Ho Lee, and Hwa Jeong Lee. Quantum knots and the number of knot mosaics. *Quantum Inf. Process.*, 14(3):801–811, 2015.

Chapter 6

Arc Presentations of Knots and Links

Hwa Jeong Lee, Dongguk University-Gyeongju

6.1 Introduction

An *arc presentation* of a knot or a link L is a special way of presenting L. It is an ambient isotopic image of L contained in the union of finitely many half planes, called *pages*, with a common boundary z-axis, called *binding axis*, in such a way that each half plane contains a properly embedded single simple arc. Figure 6.1 shows an arc presentation of the left-handed trefoil knot. So the minimum number of pages among all arc presentations of a link L is a link invariant. It is called the *arc index* of L and denoted by $\alpha(L)$.

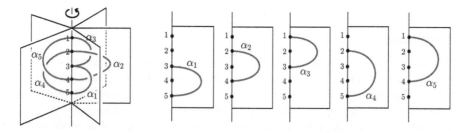

FIGURE 6.1: An arc presentation of the left-handed trefoil knot.

Arc presentations were originally described by Brunn [5] more than 100 years ago, when he proved that any link has a diagram with only one multiple point. Birman and Menasco used arc presentations of companion knots to find braid presentations for some satellites [4]. Cromwell adapted Birman-Menasco's method and used the term "arc index" as an invariant and established some of its basic properties [6]. The theory of arc presentations was developed by Dynnikov [8]. He proved that the decomposition problem of arc presentations is solvable by monotonic simplification. To be more specific, he showed that for any given grid diagram D, one can find a sequence of elementary moves that do not change the topological knot type $D \longrightarrow D_1 \longrightarrow \cdots \longrightarrow D_N$, not including stabilization, such that the final diagram D_N is obtained by the connected sum and distant union operations from diagrams of prime non-trivial non-split links and trivial diagrams. This is a remarkable result in knot theory of which the most important problem is the classification of knots and links.

6.2   Representations of an Arc Presentation

Arc presentations can be represented in various ways [7]. Figure 6.2 shows representations of the arc presentation in Figure 6.1. In the figure, all arcs named and all integers correspond to each other.

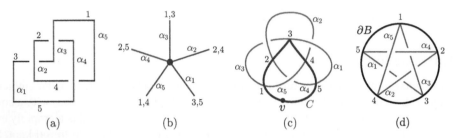

FIGURE 6.2: Representations of the arc presentation: (a) grid diagram, (b) wheel diagram, (c) annotated link diagram, and (d) layered chord diagram.

6.2.1   Grid Diagram

A *grid diagram* of size n is a link diagram which consists only of n vertical and n horizontal line segments in such a way that at each crossing the vertical line segment crosses over the horizontal line segment and no two line segments are colinear. It is also called *"loops and lines" diagram* [6] or *rectangular diagram* [8]. Figure 6.2(a) is a grid diagram of a trefoil knot. Every link diagram is regular isotopic to a grid diagram, where a horizontal overcrossing can be made into a vertical overcrossing as indicated by the ambient isotopy in Figure 6.3. Cromwell [6] showed that a grid diagram can be converted easily to an arc presentation with the number of arcs which is equal to the number of vertical line segments and vice versa as depicted in Figure 6.4.

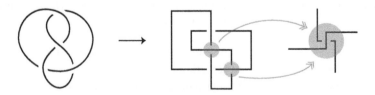

FIGURE 6.3: Every link admits a grid diagram.

6.2.2   Wheel Diagram

In [2], Bae and Park presented an algorithm for constructing arc presentations of a link which is given by edge contractions on a diagram of the link. The resulting diagram is called a *wheel diagram*. Figure 6.2(b) is a wheel diagram of a trefoil knot. Unordered pairs of integers in it indicate z-levels of the end points of the corresponding arcs in the arc presentation of Figure 6.1. A wheel

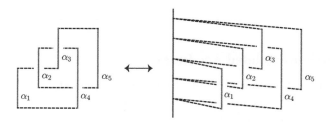

FIGURE 6.4: Grid diagram is an arc presentation.

diagram can be defined as the projection of an arc presentation into the plane perpendicular to the binding axis.

6.2.3 Annotated Link Diagram

Figure 6.2(c) shows how to find an arc presentation on a knot diagram. Let D be a diagram of a link L which lies on a plane P. Suppose that there is a simple closed curve C in P meeting D in $k(> 1)$ distinct points which divide D into k arcs $\alpha_1, \alpha_2, \ldots, \alpha_k$ with the following two properties:

(C1) Each α_i has no self-crossings.

(C2) Let R_I be the inner region bounded by the curve C in the plane P and R_O the outer region. If α_i crosses over α_j at a crossing point in R_I(resp. R_O), then $i > j$(resp. $i < j$) and it crosses over α_j at any other crossing points with α_j.

Then, we call the pair (D, C) an *annotated link diagram* of D with k arcs and C a *dividing circle* of D. In the figure, the thick simple closed curve C is a dividing circle of the diagram. By removing a point v from C away from D, we may identify $C \setminus v$ with the binding axis in Figure 6.1. This shows that (D, C) gives rise to an arc presentation and vice versa.

6.2.4 Layered Chord Diagram

Let B be a 2-dimensional disk. Suppose there are n chords $\alpha_1, \alpha_2, \ldots, \alpha_n$ with the following rules:

(B1) The end points of each α_i lie on the boundary of B.

(B2) If α_i and α_j share a crossing in the interior of B and $i < j$, then α_i underpasses α_j.

If the union of all chords represents a knot K, it is called a *layered chord diagram* of K. Figure 6.2(d) is a layered chord diagram of a trefoil knot. A layered chord diagram can be considered as an annotated link diagram with the property that all arcs are drawn inside of a dividing circle.

6.3 Knot Invariants Related Arc Index

A grid diagram can be transformed into braid forms by flipping all horizontal segments of the grid diagram with the same orientation and by slanting all vertical segments slightly as depiced in Figure 6.5(a), which leads to the following.

Theorem 1 (Cromwell [6]). *Let $\beta(L)$ be the braid index of a link L. Then*

$$\alpha(L) \geq 2\beta(L).$$

In [17], Lee *et al.* used grid diagrams to establish a relation between the mosaic number $m(K)$ and the crossing number $c(L)$ of a link L.

Theorem 2 (Lee-Hong-Lee-Oh [17]). *Let L be a nontrivial knot or link except the Hopf link and 6_3^3 link. Then*

$$\alpha(L) \geq m(L) + 1.$$

Morton and Beltrami gave an explicit lower bound for the arc index of a link L in terms of the Laurent degree of the Kauffman polynomial $F_L(a, z)$. In [14] the reader will find details of the Kauffman polynomial.

Theorem 3 (Morton-Beltrami [24]). *For every link L we have*

$$\alpha(L) \geq \mathrm{spread}_a(F_L) + 2.$$

In [2], Bae and Park showed that the algorithm which transforms a diagram of non-split link L into wheel diagrams leads an upper bound on the arc index in terms of the crossing number, $c(L)$. Jin and Park [12] improved the upper bound for non-alternating prime links.

Theorem 4 (Bae-Park [2]). *Let L be any prime non-split link. Then*

$$\alpha(L) \leq c(L) + 2.$$

Theorem 5 (Jin-Park [12]). *A prime link L is non-alternating if and only if*

$$\alpha(L) \leq c(L).$$

Combining Theorem 3, Theorem 4, and an observation of Thistlethwaite [29] on the Kauffman polynomial of alternating links, we have the following:

Corollary 6. *For a non-split alternating link L,*

$$\alpha(L) = c(L) + 2.$$

In [9], Huh and Oh used the layered chord diagram of a nontrivial knot K to find an upper bound of the stick number $s(K)$ of K.

Theorem 7 (Huh-Oh [9]). *Let K be any nontrivial knot. Then*

$$\alpha(K) \geq \frac{2}{3}s(K) + 1.$$

It is well known that a grid diagram is closely related to front projections of its Legendrian imbedding in contact geometry. In Section 6.5, we will describe that briefly. The Thurston-Bennequinn number is one of the most important invariants for Legendrian knots. Matsuda described a relation between $\overline{\alpha}(K)$ and the maximal Thurston-Bennequinn numbers of a knot K and its mirror K^*, denoted by $\overline{tb}(K)$ and $\overline{tb}(K^*)$.

Theorem 8 (Matsuda [23]).

$$-\alpha(K) \leq \overline{tb}(K) + \overline{tb}(K^*).$$

6.4 Arc Index of Some Knots or Links

The arc index of small knots has been calculated. Beltrami [3] and Ng [25] determined the arc index for prime knots whose crossing number is no greater than 10 and 11, respectively. Nutt [27], Jin *et al.* [10] and Jin and Park [13] identified all prime knots up to arc index 9, 10 and 11, respectively. Recently, Kim [16] identified all prime knots with arc index 12 up to 16 crossings.

As mentioned in Section 6.3, the arc index of alternating knots has been completely determined. So, it is natural to study the arc index of nonalternating knots. Nonalternating knots with specific diagrams have been considered. By refining Bae-Park's algorithm [2], Beltrami [3] determined the arc index of n-semi-alternating links. In [11], Jin and Lee showed that the existence of certain local diagrams indicates that the arc index is strictly less than the crossing number. They also determined the arc index for 364 new knots with 13 crossings and 15 knots with 14 crossings. Matsuda [23] determined the arc index of torus knots using the (maximal) Thurston-Bennequinn number. Finding annotated link diagrams of pretzel knots of type $(-p, q, r)$, Lee and Jin [18] computed the arc index of infinitely many pretzel knots. Recently, Lee and Takioka [19] presented an algorithm, called the *canonical cabling algorithm*, for constructing arc presentations of cable links and Whitehead doubles and determined the arc index of 2-cable links and Whitehead doubles of all prime knots with up to eight crossings. In [20], they also determined the arc index of infinitely many Kanenobu knots.

6.5 Applications of the Grid Diagram

Grid diagrams became more popular in recent years due to a connection with Legendrian links [26, 28] and provide a combinatorial description of knot Floer homology, which is a knot invariant defined in terms of Heegaard Floer homology [21, 22].

Figure 6.5 shows how a grid diagram leads to a braid, a Legendrian link, and a transverse link. In

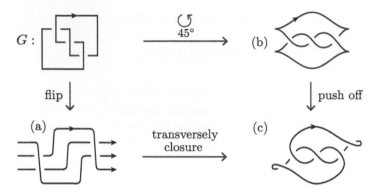

FIGURE 6.5: Grid diagrams are very closely related to (a) braids, (b) the front projection of Legendrian links, and (c) transverse links.

the figure, the braid (a) is obtained by flipping horizontal segments of the grid diagram G going from right to left. The front projection of Legendrian link (b) is naturally transformed from G by rotating the diagram 45 degrees counterclockwise and then smoothing up and down corners and turning right and left corners into cusps. It is well known that transverse links (c) can be obtained by positive push-off of Legendrian links. Using a rotationally symmetric contact structure $\ker(dz - ydx + xdy)$, closed braids can be also transformed into transverse links.

Let $\mathcal{K}, \mathcal{B}, \mathcal{L}$, and $\mathcal{T}$ be the set of all equivalent classes of smooth links in $\mathbb{R}^3$, braids on D^2 modulo conjugation and exchange move, Legendrian links in $(\mathbb{R}^3, \xi_0)$, where $\xi_0 = \ker(dz - ydx)$ is the *standard contact structure*, and *positively oriented* transverse links in $(\mathbb{R}^3, \xi_0)$, respectively. Let $\mathcal{G}$ be the set of all grid diagrams. Khandhawit and Ng [15] proved the commutativity of the maps in Figure 6.6 and Ozsváth-Szabó-Thurston [28] and Ng-Thurston [26] showed that those maps induce bijections. This means that the maps can be understood in grid diagrams. Recently, as an extension of Figure 6.6, An and Lee [1] defined the set of singular grid diagrams $S\mathcal{G}$ which provides a unified description for singular links, singular Legendrian links, singular transverse links, and singular braids. They also classified the complete set of all equivalence relations on $S\mathcal{G}$ which induce the bijection onto each singular object.

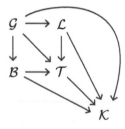

FIGURE 6.6: Commutative diagram of $\mathcal{G}, \mathcal{B}, \mathcal{L}, \mathcal{T}$, and $\mathcal{K}$.

6.6 Bibliography

[1] B.H. An and H.J. Lee. Grid diagram for singular links. *Journal of Knot Theory and Its Ramifications*, 27(4):1850023 (43 pages), 2018.

[2] Y. Bae and C.Y. Park. An upper bound of arc index of links. *Mathematical Proceedings of the Cambridge Philosophical Society*, 129:491–500, 2000.

[3] E. Beltrami. Arc index of non-alternating links. *Journal of Knot Theory and Its Ramifications*, 11(3):431–444, 2002.

[4] J.S. Birman and W.W Menasco. Special positions for essential tori in link complements. *Topology*, 33(3):525–556, 1994.

[5] H. Brunn. Über verknotete Kurven . *Mathematiker-Kongresses Zürich*, pages 256–259, 1897.

[6] P.R. Cromwell. Embedding knots and links in an open book I: Basic properties. *Topology and Its Applications*, 64:37–58, 1995.

[7] P.R. Cromwell. *Knots and Links*. Cambridge University Press, 2004.

[8] I.A. Dynnikov. Arc-presentations of links: Monotonic simplification. *Fundamenta Mathematicae*, 190:29–76, 2006.

[9] Y. Huh and S. Oh. An upper bound on stick number of knots. *Journal of Knot Theory and Its Ramifications*, 20(5):741–747, 2011.

[10] G.T. Jin, H. Kim, G.S. Lee, J. Gong, H. Kim, H. Kim, and S. Oh. Prime knots with arc index up to 10. In *Series on Knots and Everything Book vol. 40*, pages 65–74. World Scientific Pub Co Inc, 2006.

[11] G.T. Jin and H.J. Lee. Prime knots whose arc index is smaller than the crossing number. *Journal of Knot Theory and Its Ramifications*, 21(10):1250103 (33 pages), 2012.

[12] G.T. Jin and W.K. Park. Prime knots with arc index up to 11 and an upper bound of arc index for non-alternating knots. *Journal of Knot Theory and Its Ramifications*, 19(12):1655–1672, 2010.

[13] G.T. Jin and W.K. Park. A tabulation of prime knots up to arc index 11. *Journal of Knot Theory and Its Ramifications*, 20(11):1537–1635, 2011.

[14] L.H. Kauffman. An invariant of regular isotopy. *Transactions of the American Mathematical Society*, 318(2):417–471, 1990.

[15] T. Khandhawit and L. Ng. A family of transversely nonsimple knots. *Algebraic & Geometric Topology*, 10(1):293–314, 2010.

[16] H. Kim. Prime knots with arc index 12 up to 16 crossings. *Masters thesis, KAIST*, 2014.

[17] H.J. Lee, K. Hong, H. Lee, and S. Oh. Mosaic number of knots. *Journal of Knot Theory and Its Ramifications*, 23(13):1450069 (8 pages), 2014.

[18] H.J. Lee and G.T. Jin. Arc index of pretzel knots type of (-p,q,r). *Proceedings of the American Mathematical Society, series B*, 1:135–147, 2014.

[19] H.J. Lee and H. Takioka. On the arc index of cable links and Whitehead doubles. *Journal of Knot Theory and Its Ramifications*, 25(7):1650041 (23 pages), 2016.

[20] H.J. Lee and H. Takioka. On the arc index of Kanenobu knots. *Journal of Knot Theory and Its Ramifications*, 26(4):1750015 (26 pages), 2017.

[21] C. Manolescu, P. Ozsváth, and S. Sarkar. A combinatorial description of knot Floer homology. *Annals of Mathematics*, 169(2):633–660, 2009.

[22] C. Manolescu, P. Ozsváth, Z. Szabó, and D. Thurston. On combinatorial link Floer homology. *Geometry & Topology*, 11:2339–2412, 2007.

[23] H. Matsuda. Links in an open book decomposition and in the standard contact structure. *Proceedings of the American Mathematical Society*, 134(12):3697–3702, 2006.

[24] H.R. Morton and E. Beltrami. Arc index and the Kauffman polynomial. *Mathematical Proceedings of the Cambridge Philosophical Society*, 123(41):41–48, 1998.

[25] L. Ng. On the arc index and maximal Thurston-Bennequin number. *Journal of Knot Theory and Its Ramifications*, 21(4):1250031 (11 pages), 2012.

[26] L. Ng and D. Thurston.

[27] I.J. Nutt. Embedding knots and links in an open book III : On the braid index of satellite links. *Mathematical Proceedings of the Cambridge Philosophical Society*, 126:77–98, 1999.

[28] P. Ozsváth, Z. Szabó, and D. Thurston. Legendrian knots, transverse knots and combinatorial Floer homology. *Geometry & Topology*, 12(2):941–980, 2008.

[29] M.B. Thistlethwaite. Kauffman polynomial and alternating links. *Topology*, 27(3):311–318, 1988.

Chapter 7

Diagrammatic Representations of Knots and Links as Closed Braids

Sofia Lambropoulou, National Technical University of Athens

7.1 Introduction

A way to study knots and links is by studying their regular projections on a plane, called 'diagrams'. One big advancement in low-dimensional topology was the 'discretization' of link isotopy through the well-known moves on knot and link diagrams discovered by K. Reidemeister in 1927 [50], see Fig. 7.1.

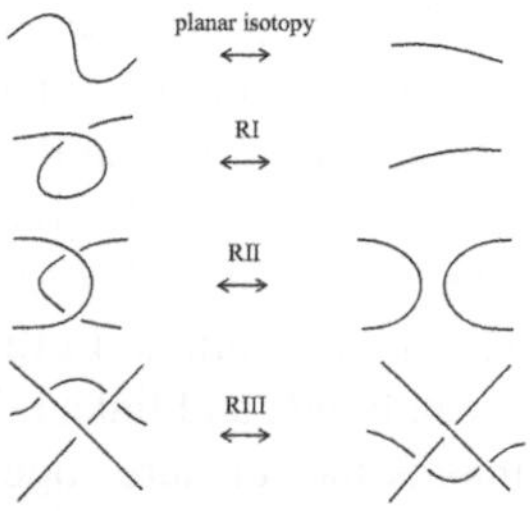

FIGURE 7.1: The Reidemeister moves.

The crucial one is the Reidemeister III move, which in terms of dual planar graphs corresponds to the so-called 'star-triangle relation'. A second advancement was the 'algebraization' of link isotopy, by representing knots and links via braids. Braids are both geometric-topological and algebraic objects; geometric as sets of interwinding, non-intersecting descending strands, algebraic as elements of the so-called Artin braid groups [2, 3]. In Fig. 7.2 middle, we can see an example of a braid on 3 strands. The aim of this chapter is to detail the connection between links and braids, which is marked by two fundamental results: the *Alexander theorem*, whereby links are represented via braids, and the *Markov theorem*, which provides the braid equivalence that reflects link isotopy. See Theorems 1 and 2. In 1984, these theorems played a key role in the discovery of a novel link invariant, the Jones polynomial [29, 30].

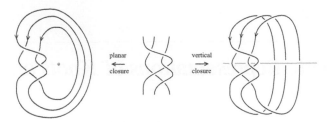

FIGURE 7.2: A braid on 3 strands and two closures.

7.2   From Braids to Links

More rigorously, a classical braid on n strands is, geometrically, a homeomorphic image of n arcs in the interior of a thickened rectangle starting from n top points and running monotonically down to n corresponding points, that is, with no local maxima or minima. Each one of the two sets of endpoints can be assumed colinear and both of them coplanar. So a braid can be identified with the braid diagram on the projection plane of its endpoints, as assumed in the example in the middle illustration of Fig. 7.2. Moreover, two braids are *isotopic* through any isotopy move of its arcs that preserves the braid structure. These isotopies comprise the Reidemeister moves II and III for braids and planar isotopies, which include the change of relative heights for two non-adjacent crossings, as well as small shifts of endpoints that preserve their order. The Reidemeister I move cannot apply as the kink introduces local maxima and minima. In the set of braids, isotopic elements are considered equal. The interaction between braids and links takes place through their diagrams. An operation that connects braids and links is the 'closure' operation. It is realized by joining with simple non-interwinding arcs the corresponding pairs of endpoints of the braid. The result is an oriented link diagram that winds around a specified axis, the *braid axis*, in the same sense, say the counterclockwise. The link orientation is induced by the top-to-bottom direction of the braid. There are many ways of specifying the braid axis. Fig. 7.2 illustrates two of them: in the left-hand illustration the braid axis can be considered to be perpendicular to the braid projection plane, piercing it at one point, its 'trace'; in the right-hand illustration the braid axis is a horizontal line parallel to and behind the projection plane of the braid. The two closures are the *planar closure* and the *vertical closure*, irespectively. Clearly, any two closures of the same braid are isotopic.

7.3   From Links to Braids

Now the following question arises naturally: can one always do the converse? That is, given an oriented link diagram, can one turn it into an isotopic closed braid?

Note that our diagram may already be in closed braid form for some choice of braid axis. Yet, with a different specified axis it may not be braided anymore. Imagine in Fig. 7.2 the axis trace to be placed in some other region of the plane. The answer to the above question is 'yes' and the idea

FIGURE 7.3: Alexander's braiding of an opposite arc.

is quite simple. Indeed, we first specify a point on the plane of the diagram, the trace of the braid axis, and we define a 'good' direction around this point. We then subdivide the arcs of the link diagram into smaller arcs, by marking the transition from 'good' arcs to 'opposite' arcs, according to whether they agree or not with the good direction. The good arcs are left alone.

FIGURE 7.4: Subdividing an opposite arc.

An opposite arc is to be subdivided further if needed (see Fig. 7.4), so that each subarc can be part of the boundary of a 'sliding triangle' across the axis trace. See Fig. 7.3. The sliding triangle of a subarc does not intersect any other arcs of the link diagram. If the subarc lies over other arcs of the diagram, so does its triangle. Similarly, if it lies under other arcs, then its triangle also lies under these arcs. (The notion of the sliding triangle was first introduced by Reidemeister [50] in the general context of his study of isotopy, not of braiding, and he called these isotopy generating moves Δ-*moves*.) So, by an isotopy move we can replace the opposite subarc by two good arcs, the other two sides of the sliding triangle, and the opposite subarc is eliminated. The opposite subarcs are finitely many, so the process terminates with a closed braid. The algorithm outlined above is J.W. Alexander's proof of his homonymous theorem [1]:

Theorem 1 (J.W. Alexander, 1923). *Every oriented link diagram may be isotoped to a closed braid.*

Yet, the idea of turning a knot into a braid goes back to Brunn in 1897 [10], who observed that any knot has a projection with a single multiple point; from this follows immediately (by appropriate small perturbations) that the diagram can be brought to closed braided form. In 1974 Joan Birman made Alexander's braiding algorithm more technical [7] with the purpose of providing a rigorous proof of the Markov theorem (see Section 7.5).

Another proof of the Alexander theorem, very appealing to the imagination, is the one by Hugh Morton ([49], 1986): Instead of having the braid axis fixed and moving by isotopies the arcs of the link diagram for reaching a closed braid form, he considers the diagram fixed and instead he 'threads' the braid axis (in the form of a simple closed curve projected on the plane of the link diagram) through the arcs of the diagram, so that each subarc winds around the axis in the counterclockwise sense. In the same paper, Morton uses a more technical version of his braiding algorithm for proving the Markov theorem.

A novel approach to the Alexander theorem is due to Shuji Yamada ([59], 1987) using the auxilliary concept of 'Seifert circles'. The Seifert circles of a link diagram are created after smoothing all crossings according to the orientations of their arcs. Yamada introduces grouping operations on the system of Seifert circles which correspond to isotopies on the link diagram and which terminate with a braided form.

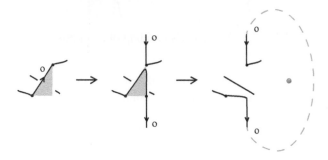

FIGURE 7.5: An *L*-braiding move results in a pair of corresponding braid strands.

An additional value of Yamada's braiding algorithm is that it implies equality between the Seifert number and the braid index of the link. Pierre Vogel gave in 1990 a more sophisticated braiding algorithm based on the one by Yamada, where trees are used for measuring complexity [57]. Vogel's algorithm was then used by Pawel Traczyk for proving the Markov theorem ([55], 1992). A further approach to the Alexander theorem was made by the author in the 1990s [38, 39, 45]. This algorithm results in open braids that one can 'read' directly. Here we mark the local maxima and minima of the link diagram with respect to the height function and the opposite subarcs are now called *up-arcs*, so that we can have for each up-arc an orthogonal sliding triangle of the same type, 'under' or 'over', as the up-arc.

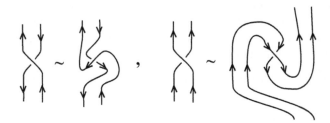

FIGURE 7.6: A half twist and a full twist on crossings.

The elimination of an up-arc is an *L-braiding move* and it consists of cutting the arc at some point, say the upppermost, and then pulling the two ends, the upper upward and the lower downward, and keeping them aligned, so as to obtain a pair of corresponding braid strands, both running entirely *over* the rest of the diagram or entirely *under* it, according to the type of the up-arc. Fig. 7.5 illustrates the abstraction of an 'over' *L*-braiding move and Fig. 7.7 an example of applying the *L*-braiding algorithm. The closure of the resulting tangle is a link diagram, clearly isotopic to the original one, since from the up-arc we created a stretched loop isotopic to the original up-arc.

Any type of closure can apply on the resulting braid. However, if we want to apply the *L*-braiding algorithm to the closure of a braid and obtain the initial braid back, it is convenient to use the vertical closure. Finally, a really elegant version of the above braiding algorithm is given by the

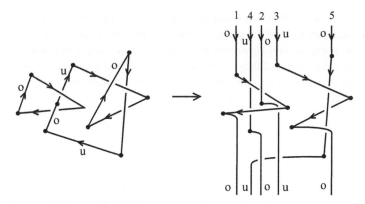

FIGURE 7.7: An example of applying the *L*-braiding algorithm.

author and Louis H. Kauffman ([34], 2004): It uses the basic *L*-braiding move described above for up-arcs with no crossings, but it braids the up-arcs with crossings simply by rotating them on their projection plane, as illustrated in Fig. 7.6.

Each algorithm for proving the Alexander theorem has its own flavour and its own advantages and disadvantages but they are all based on the same idea. Namely, to eliminate one by one the arcs of the diagram that have the wrong sense with respect to the chosen braid axis and the specified right sense, and to replace them with admissible ones.

7.4 The Braid Group

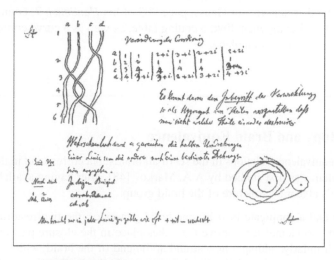

FIGURE 7.8: Gauss' handwritten study of braids.

The systematic theory of braids was introduced by E. Artin in 1925 [2, 3]. He started with the geometric definition and then studied the group structure of the set of equivalence classes of braids on n strands, after analyzing the braid isotopies with respect to the height function. These equivalence classes are also called *braids* and they form the classical *braid group B_n* with the concatenation operation. The group B_n has a presentation with finitely many generators σ_i, $i = 1, \ldots, n - 1$, and a finite set of defining relations [2, 3, 14]:

$$\sigma_i \sigma_{i+1} \sigma_i = \sigma_{i+1} \sigma_i \sigma_{i+1} \quad \text{and} \quad \sigma_i \sigma_j = \sigma_j \sigma_i \quad \text{for} \quad |i - j| > 1.$$

The generators σ_i resemble the elementary transpositions s_i of the symmetric group S_n. They can be viewed as elementary braids of one crossing between the consecutive strands i and $i + 1$, also carrying the topological information of which strand crosses over the other, see Fig. 7.9. The most important braid relation is the braided Reidemeister III move (all arrows down), while the Reidemeister II move is a direct consequence of the fact that the generators σ_i are invertible.

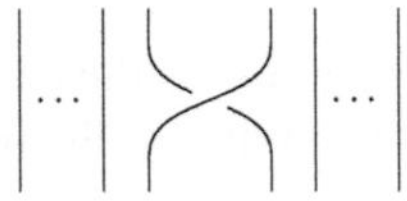

FIGURE 7.9: The braid generator σ_i.

The group B_n surjects on the symmetric group S_n by corresponding the generator σ_i to the elementary transposition s_i. The group S_n has the extra relations $s_i^2 = 1$, $i = 1, \ldots, n - 1$, which are responsible for its finite order. The permutation induced by a braid tells us the number of link components of its closure by counting the disjoint orbits of the permutation. Braid groups play a central role in many areas of mathematics. A complete reference on braid groups and related topics can be found in the book by Kassel and Turaev [32].

It is worth noting that C.F. Gauss was also thinking about the concept of a braid, probably in the frame of studying interwinding curves in space in the context of his theory of electrodynamics. In Fig. 7.8 we see a handwritten note of Gauss, page 283 of his *Handbuch* 7, containing a sketch of a braid that closes to a 3-component link, a coding table, as well as a curve configuration winding around two points on its projection plane.

7.5   Link Isotopy and Braid Equivalence

We next consider equivalence relations in the set of all braids that correspond in the closures to link isotopy. This problem was first studied by A.A. Markov [48], after having available the Alexander theorem and Artin's algebraic structure of the braid group.

Closing a braid a and a conjugate of a by one braid generator, the two resulting links differ by planar isotopy and a Reidemeister II move that takes place in the closure part of the conjugate of a. See Fig. 7.10 . Similarly, adding a new strand at the end of the braid, which crosses once over or under the last strand, corresponds in the closures to a kink, that is, to a Reidemeister I move. This move is called a 'stabilization move'. See Fig. 7.10. Clearly, conjugations and stabilization

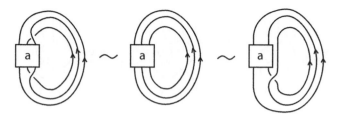

FIGURE 7.10: Closures of braid conjugation and braid stabilization induce isotopy.

moves in the set of all braid groups are apparently independent from each other and must figure in a braid equivalence that reflects link isotopy. The question is whether there are any other hidden moves for capturing the complexity of link isotopy. Yet, this is not the case, as we see from the following:

Theorem 2 (A.A. Markov, 1936). *Two oriented links are isotopic if and only if any two corresponding braids differ by a finite sequence of the braid relations and the moves:*

(i) Conjugation $\sigma_i^{-1} a \sigma_i \sim a$ and (ii) Stabilization $a \sigma_n^{\pm 1} \sim a$ for any $a \in B_n$ and for all $n \in \mathbb{N}$.

The statement of the theorem originally included *three moves*: the two local moves above and another more global one, the exchange move, that generalizes the Reidemeister II move. The sketch of proof by A.A. Markov [48] used Alexander's braiding algorithm. Soon afterwards, N. Weinberg reduced the exchange move to the *two moves* of the statement [58]. The interest in the braid equivalence was rekindled by Joan Birman after following the talk of some unknown speaker. Birman produced a rigorous proof of the theorem, filling in all details, by using a more technical version of Alexander's braiding algorithm [7] . A few years later Daniel Bennequin gave a different proof using 3-dimensional contact topology [6]. In 1984 the Jones polynomial, a new powerful link invariant, was discovered, whose construction used a representation of the braid group in the Temperley–Lieb algebra and the Alexander and Markov theorems [29, 30]. This discovery led to new approaches to the Markov theorem. Hugh Morton gave a new proof using his threading algorithm [49], Pawel Traczyk proved the Markov theorem using Vogel's algorithm [55], and Joan Birman revisited the theorem with William Menasco using Bennequin's ideas [9]. Finally, in the 1990', the author discovered a more geometric braid equivalence move, the L-move, and proved with Colin Rourke a *one-move* analogue of the Markov theorem, whose proof used the L-braiding moves described earlier [38, 39, 45]:

Theorem 3. *Two oriented links are isotopic if and only if any two corresponding braids differ by a finite sequence of the braid relations and the L-moves.*

The *L-move* for braids resembles the L-braiding move: It consists of cutting a braid strand at some point and then pulling the two ends, the upper downward and the lower upward, keeping them aligned, and so as to obtain a new pair of corresponding braid strands, both running entirely *over* the rest of the braid or entirely *under* it, according to the type of the move, denoted L_o or L_u respectively. Fig. 7.11 illustrates an example with both types of L-moves taking place at the same point of a braid. The closure of the resulting braid differs from the closure of the initial one by a stretched loop. View also Fig. 7.13 for an abstract illustration of the similarity of the L-braiding move and the L-move.

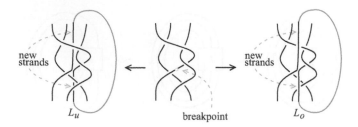

FIGURE 7.11: An L_u-move and an L_o-move at the same point.

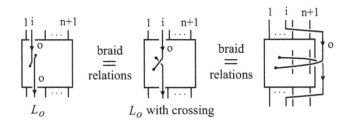

FIGURE 7.12: The L-moves have algebraic expressions.

The L-moves are geometric. However, as we see in the middle illustration of Fig. 7.12, using braid isotopy, an L-move can be also viewed as introducing a crossing inside the braid 'box', so in the closure it creates a stretched Reidemeister I kink. This way of viewing the L-moves shows that they generalize the stabilization moves. It also renders them local and leads to the observation that they have algebraic expressions, as s clear from the right illustration of Fig. 7.12. Furthermore, it follows from Theorem 3 that conjugation can be achieved by braid relations and L-moves.

The Markov theorem or Theorem 3 are not easy to prove. For proving the 'only if' part, one needs first to take any two diagrams of the same link or of two isotopic links and, using some braiding algorithm, produce corresponding braids; then show that the two braids are Markov equivalent (resp. L-equivalent). In practice, this means that any choices made on a given link diagram when applying the braiding algorithm correspond to Markov equivalent (resp. L-equivalent) braids; and that if two link diagrams differ by an isotopy move, the corresponding braids are also Markov equivalent (resp. L-move equivalent). For this analysis it is crucial to have the algorithm choices localized, the isotopy moves local, and the braiding moves independent from each other. In this way one can always assume to have done almost all braiding in the otherwise identical diagrams in question and to be left only with the algorithmic choice or the isotopy move by which they differ. Then the two braid diagrams are directly comparable.

The L-braiding algorithm of [38, 39, 45] is particularly appropriate for proving the Markov theorem or its equivalent Theorem 3, after enhancing it with some extra technicalities for ensuring independence of the sequence of the braiding moves [38, 39, 45]. This is because the L-braiding moves and the L-moves for braids are simple and have a basic symmetric interconnection, as illustrated in Fig. 7.13, so they comprise, in fact, one very fundamental uniform move, enabling one to trace easily how the algorithmic choices and the isotopy moves on diagrams affect the final braids.

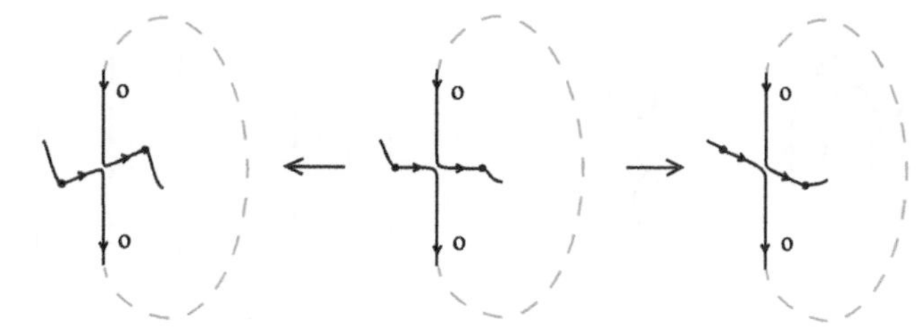

L-braiding move   L-move

FIGURE 7.13: The symmetry of the braiding and the *L*-move.

7.6 Extensions to Other Diagrammatic Settings

Given a diagrammatic knot theory there are deep interrelations between the diagrammatic isotopy, the braid structures and the corresponding braid equivalences in this theory.

More precisely, the isotopy moves that are allowed, but also moves that are forbidden in the setting, determine the corresponding braid isotopy, the way the closure of a braid is realized, and also the corresponding braid equivalence moves. Braid equivalence theorems are important for understanding the structure and classification of knots and links in various settings and for constructing invariants of knots and links using algebraic means, for example via Markov traces on quotient algebras of the corresponding braid groups.

The *L*-braiding moves and the *L*-moves for braids provide a uniform and flexible ground for formulating and proving braiding and braid equivalence theorems for any diagrammatic setting. Indeed, their simple and fundamental nature together with the fact that the *L*-moves are geometric and can be localized are the reasons that they can adapt to all diagrammatic categories where diagrammatic isotopy and the notion of braid are defined. This is particularly useful in settings where algebraic braid structures are not immediately apparent. So, the statements are first formulated geometrically and then they gradually turn algebraic, if algebraic braid structures are available.

The *L*-move techniques were first employed for proving braiding and braid equivalence theorems for knots and links in 3-manifolds with or without boundary; namely in knot complements, in handlebodies, as well as in closed, connected, oriented (c.c.o.) 3-manifolds, which are obtained from the 3-sphere, S^3, via the 'surgery technique'. The idea here is to fix in S^3 a closed braid representation of the 3-manifold and then represent knots and braids in the 3-manifold as *mixed links* and *mixed braids* in S^3, which contain the 'fixed part', representing the 3-manifold, and the 'moving part', representing the knot/braid in the 3-manifold. View Fig. 7.14 for concrete examples.

Then, knot isotopy in the 3-manifold is translated into mixed link isotopy in S^3, which applies only on the moving part. In the case of c.c.o. 3-manifolds we have the isotopy moves for the knot complements, extended by extra isotopy moves, the 'band moves', related to the surgery description of the manifold. The mixed braid equivalence for knot complements comprises *L*-moves which take place only on the moving parts of mixed braids. See Fig. 7.15(a), while in the case of c.c.o. 3-manifolds we have the extra 'braid band moves'. Then the *L*-move braid equivalences turn into

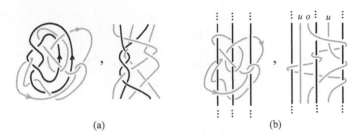

(a)   (b)

FIGURE 7.14: (a) A mixed link and a geometric mixed braid for link complements and c.c.o. 3-manifolds; (b) a mixed link and a geometric mixed braid for handlebodies.

algebraic statements with the use of the algebraic mixed braids [40], see Fig. 7.15(b) for an example.

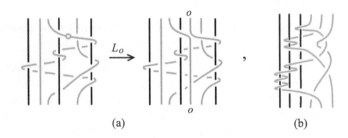

(a)   (b)

FIGURE 7.15: An *L*-move on a mixed braid; (b) an algebraic mixed braid.

In the case of a handlebody we have the same setting as for a knot complement. The handlebody is represented by a fixed identity braid of as many strands as its genus. The difference here is that a knot may not pass beyond its boundary from either end, and this is reflected both in the definition of the closure of a mixed braid as well as in the corresponding braid equivalence. Namely, the closure is realized by simple closing arcs, slightly tilted in the extremes, which run 'over' or 'under' the rest of the diagram, and different choices may lead to non-isotopic closures. Furthermore, in the mixed braid equivalence some 'loop' conjugations are not allowed. Details on the above can be found in [45, 40, 46, 18, 28, 41, 42]. See also [54] and [52].

The next application of the *L*-move methods was in virtual knot theory. Virtual knot theory was introduced by Louis H. Kauffman [33] and it is an extension of classical knot theory. In this extension one adds a 'virtual' crossing that is neither an over-crossing nor an under-crossing. Fig. 7.16 illustrates the diagrammatic moves that contain virtual crossings. In this theory we have the *virtual forbidden moves*, F1 and F2, with two real crossings and one virtual. We also have the virtual braid group [33, 34, 5], which extends the classical braid group (see also Chapter 30). The forbidden moves make it harder to braid a virtual knot diagram, so the idea of rotating crossings that contain up-arcs before braiding a diagram (recall Figure 7.6) comes in handy in this setting. The interpretation of an *L*-move as introducing an in-box crossing proved very crucial in the search of the types of *L*-moves needed in the setting, as they are related to the types of kinks allowed in the given isotopy.

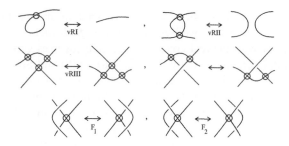

FIGURE 7.16: Virtual moves: allowed and forbidden.

So, we have *L*-moves introducing a real or a virtual crossing facing to the right or to the left of the braid. Moreover, the presence of the forbidden moves in the theory leads to the requirement that the strands of an *L*-move cross the other strands of the virtual braid only virtually, and also to a type of virtual *L*-move coming from a 'trapped' virtual kink. The above led in [35] to formulations of virtual braid equivalence theorems for the virtual braid group, for the welded braid group [19] and for some analogues of these structures, complementing the prior results of Seiichi Kamada in [31], where the more global exchange moves are used.

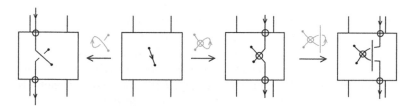

FIGURE 7.17: Types of virtual *L*-moves.

Furthermore, the *L*-move techniques have been used by Vassily Manturov and Hang Wang for formulating a Markov-type theorem for free links [47]. Also, by Carmen Caprau and co-authors for obtaining a braid equivalence for virtual singular braids [11] as well as for virtual trivalent braids [12, 13].

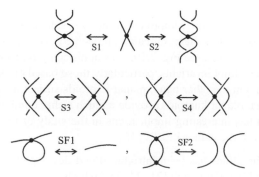

FIGURE 7.18: Singular moves: allowed and forbidden.

Singular knot theory is related to Vassiliev's theory of knot invariants. Fig. 7.18 illustrates the diagrammatic moves in the theory as well as the *singular forbidden moves*, SF1 and SF2. The singular crossings together with the real crossings and their inverses generate the 'singular braid monoid' introduced in different contexts by Baez[4], Birman[8] and Smolin[53] (see also Chapter 38). Braiding a singular knot diagram becomes particularly simple by using the idea of rotating singular crossings that contain up-arcs before braiding (Figure 7.6) and the *L*-braiding moves. An algebraic singular braid equivalence is proved by Bernd Gemein in [20] and, assuming this result, in [41] the *L*-move analogue is formulated. Clearly, there is no *L*-move introducing a singular crossing, as the closure of such a move would contract to a kink with a singular crossing, and this is not an isotopy move in the theory. Also, there is no conjugation by a singular crossing, since this is not an invertible element in the monoid; yet, we can talk about 'commuting' in the singular braid monoid: $ab \sim ba$.

FIGURE 7.19: A knotoid and its two closures to knots.

Another very interesting diagrammatic category is the theory of knotoids and braidoids. The theory of knotoids was introduced by Vladimir Turaev in 2010 [56] (see also Chapter 37). A knotoid diagram is an open curve in an oriented surface, having finitely many self-intersections that are endowed with under/over data and with its two endpoints possibly lying in different regions of the diagram. For an example see middle illustration of Fig. 7.19. The theory of knotoids is a complex diagrammatic theory, and its complexity lies in the *knotoid forbidden moves*, Φ_+ and Φ_-, that prevent the endpoints from slipping under or over other arcs of the diagram. See Fig. 7.20.

FIGURE 7.20: The forbidden moves for knotoids.

The theory of spherical knotoids (i.e., knotoids in the two-sphere) extends the theory of classical knots and also proposes a new diagrammatic approach to classical knots, which arise via the 'overpass' or 'underpass' closures, see Fig. 7.19. This approach promises reduction of the computational complexity of some knot invariants, particularly those based on the number of crossings of a given diagram [56, 15]. On the other hand, planar knotoids surject to spherical knotoids, but do not inject. This means that planar knotoids provide a much richer combinatorial structure than the spherical ones. This fact has interesting implications in the study of proteins, initiated by Dimos Goundaroulis, ex-student of the author [26, 27, 16].

Recently, the theory of braidoids has been introduced and developed, after the suggestion of the author to her student Neslihan Gügümcü [21, 24, 25], which extends the classical braid theory. A 'braidoid' is like a classical braid, but two of its strands terminate at the endpoints. For an example see Fig. 7.21(a). The forbidden moves play a role in the algorithm for turning planar knotoids to

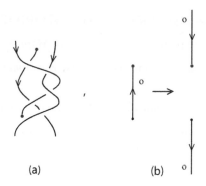

FIGURE 7.21: (a) A braidoid; (b) *L*-braidoiding at an endpoint.

braidoids and they affect the definition of the closure operation on braidoids, in which the endpoints do not participate. Namely, we close corresponding braidoid ends using vertical arcs with slightly tilted extremes, running over or under the rest of the diagram, and this needs to be specified in advance, as different choices may result in non-isotopic knotoids (due to the forbidden moves). For turning a planar knotoid into a braidoid, we use the *L*-braiding moves for up-arcs not containing an endpoint and the analogous moves illustrated in Fig. 7.21(b) (with choice 'o' in the figure) for the ones that contain an endpoint. For a braidoid equivalence we use the *L*-moves that can take place at any point of a braidoid except for the endpoints. We note that for the braidoids we do not have yet an appropriate algebraic structure, so the *L*-equivalence is as good as we can have so far (see also Chapter 37). It is worth adding at this point that in [22], Neslihan Gügümcü and Louis Kauffman give a faithful lifting of a planar knotoid to a space curve, such that the endpoints remain attached to two parallel lines. In [36] Dimitrios Kodokostas and the author make the observation that this interpretation of planar knotoids is related to the knot theory of the handlebody of genus two and these ideas are further explored in [37], where the notion of 'rail knotoid' is introduced.

Further, in [44] the *L*-move techniques are applied to long knots.

Finally, in [51] Nancy Scherich provides a computer-implemented, grid diagrammatic proof of the Alexander theorem, based on the *L*-braiding moves. Another result of analogous flavour is a Markov-type theorem for ribbon torus-links in $\mathbb{R}^4$ by Celeste Damiani [17].

Surveys on many of the above results are included in [41, 42, 23], while a more complete presentation is to appear in [43].

7.7 Bibliography

[1] J. W. Alexander, A lemma on systems of knotted curves, *Proc. Nat. Acad. Sci. U.S.A.* **9** (1923) 93–95.

[2] E. Artin, Theorie der Zöpfe, *Abh. Math. Sem. Hamburg Univ.* **4** (1925) 47–72.

[3] E. Artin, Theory of braids, *Ann. of Math.* (2) **48** (1947) 101–126.

[4] J. Baez, Link invariants and perturbation theory, *Lett. Math. Phys.* **2** (1992) 43–51.

[5] V. G. Bardakov, The virtual and universal braids, *Fundamenta Mathematicae* **184** (2004) 159–186.

[6] D. Bennequin, Entrlacements et équations de Pfaffe, *Asterisque* **107-108** (1983) 87–161.

[7] J.S. Birman, *Braids, Links and Mapping Class Groups*, Annals of Mathematics Studies, Vol. 82 (Princeton University Press, Princeton, 1974).

[8] J. Birman, New points of view in knot theory, *Bull. Amer. Math. Soc. (N.S.)* **28**(2) (1993) 253–287.

[9] J. S. Birman, W. W. Menasco, On Markov's theorem, *J. Knot Theory Ramifications* **11** no.3 (2002), 295–310.

[10] H. Brunn, Über verknotete Kurven, *Verh. des intern. Math. Congr.* **1** (1897), 256–259.

[11] C. Caprau, A. De la Pena, S. McGahan, Virtual singular braids and links, *Manuscripta Math.***151**, No 1 (2016), 147–175.

[12] C. Caprau, A. Dirdak, R. Post, and E. Sawyer, Alexander- and Markov-type theorems for virtual trivalent braids, arXiv:1804.09919 [math.GT], to appear in *J. Knot Theory Ramifications*.

[13] C. Caprau, G. Coloma, M. Davis, The *L*-move and Markov theorems for trivalent braids, arXiv:1807.08095 [math.GT].

[14] W.-L. Chow, On the algebraical braid group, *Ann. of Math.* (2) **49** (1948), 654–658.

[15] K. Daikoku, K. Sakai, and M. Takase, On a move reducing the genus of a knot diagram. In *Indiana University Mathematics Journal* 61.3 (2012), pp. 1111–1127. ISSN: 00222518, 19435258. URL: http://www.jstor.org/stable/24904076.

[16] P. Dabrowski-Tumanski, P. Rubach, D. Goundaroulis, J. Dorier, P. Sulkowski, K.C. Millett, E.J. Rawdon, A. Stasiak, J.I. Sulkowska, KnotProt 2.0: A database of proteins with knots and other entangled structures, *Nucleic Acids Research* **1** (2018). DOI: 10.1093/nar/gky1140.

[17] C. Damiani, Towards a version of Markov's theorem for ribbon torus-links in $\mathbb{R}^4$. In *Knots, Low-dimensional Topology and Applications: Proceedings of the Conference* Knots in Hellas 2016, *Springer Proceedings in Mathematics & Statistics (PROMS)* 284 (2019); S. Lambropoulou et al., Eds.

[18] I. Diamantis, S. Lambropoulou, Braid equivalences in 3-manifolds with rational surgery description, *Topology Applications* (2015), http://dx.doi.org/10.1016/j.topol.2015.08.009.

[19] R. Fenn, R. Rimanyi, C.P. Rourke, The braid permutation group, *Topology* **36** (1997), 123–135.

[20] B. Gemein, Singular braids and Markov's theorem, *J. Knot Theory Ramifications* **6**(4) (1997) 441–454.

[21] N. Gügümcü, On knotoids, braidoids and their applications, PhD thesis, National Technical U. Athens, 2017.

[22] N. Gügümcü and L.H. Kauffman, New invariants of knotoids. *European J. of Combinatorics* (2017), **65C**, 186–229.

[23] N. Gügümcü, L.H. Kauffman, and S. Lambropoulou, A survey on knotoids, braidoids and their applications. In *Knots, Low-dimensional Topology and Applications: Proceedings of the Conference* Knots in Hellas 2016, *Springer Proceedings in Mathematics & Statistics (PROMS)* 284 (2019); S. Lambropoulou et al., Eds.

[24] N. Gügümcü, S. Lambropoulou, Knotoids, braidoids and applications, *Symmetry* 9 (2017), No.12, 315; See http://www.mdpi.com/2073-8994/9/12/315 .

[25] N. Gügümcü, S. Lambropoulou, Bradoids, forthcoming in *Israel J. Math.*.

[26] D. Goundaroulis, J. Dorier, F. Benedetti, and A. Stasiak, Studies of global and local entanglements of individual protein chains using the concept of knotoids. *Sci. Reports*, **2017**, 7, 6309.

[27] D. Goundaroulis, N. Gügümcü, S. Lambropoulou, J. Dorier, A. Stasiak and L. H. Kauffman, Topological models for open knotted protein chains using the concepts of knotoids and bonded knotoids. *Polymers*, *Polymers* Special issue on *Knotted and Catenated Polymers*, D. Racko and A. Stasiak, Eds.; (2017), 9, No.9, 444. http://www.mdpi.com/2073-4360/9/9/444.

[28] R. Häring-Oldenburg, S. Lambropoulou, Knot theory in handlebodies, *J. Knot Theory Ramifications* **11**(6) (2002) 921–943.

[29] V.F.R. Jones, A polynomial invariant for knots via von Neumann algebras, *Bull. Amer. Math. Soc.* **12** No. 1, (1985), 103–111.

[30] V.F.R. Jones, Hecke algebra representations of braid groups and link polynomials, *Ann. of Math.* **126** (1987) 335–388.

[31] S. Kamada, Braid representation of virtual knots and welded knots, *Osaka J. Math.* **44** (2007), no. 2, 441–458. See also *arXiv:math.GT/0008092*.

[32] C. Kassel and V. Turaev, *Braid Groups*, Graduate Texts in Mathematics, Vol. 247 (Springer, 2008).

[33] L.H. Kauffman, Virtual knot theory, *European J. Comb.* **20** (1999) 663–690.

[34] L.H. Kauffman and S. Lambropoulou, Virtual braids, *Fundamenta Mathematicae* **184** (2004) 159–186.

[35] L.H. Kauffman and S. Lambropoulou, Virtual braids and the *L*-move, *J. Knot Theory Ramifications* **15**(6) (2006) 773–811.

[36] D. Kodokostas and S. Lambropoulou, A spanning set and potential basis of the mixed Hecke algebra on two fixed strands. *Mediterranean J. Math.* (2018), **15:192**, https://doi.org/10.1007/s00009-018-1240-7.

[37] D. Kodokostas and S. Lambropoulou, Rail knotoids. See arXiv:1812.09493.

[38] Lambropoulou, S. Short proofs of Alexander's and Markov's theorems. Warwick preprint **1990**.

[39] S. Lambropoulou, A study of braids in 3–manifolds, Ph.D. thesis, Warwick, 1993.

[40] S. Lambropoulou, Braid structures in handlebodies, knot complements and 3-manifolds, in *Proceedings of Knots in Hellas '98*, Series on Knots and Everything Vol. 24 (World Scientific, 2000), pp. 274–289.

[41] S. Lambropoulou, *L*-moves and Markov theorems, *J. Knot Theory Ramifications* **16** no. 10 (2007), 1–10.

[42] Lambropoulou S. Braid equivalences and the *L*-moves. In *Introductory Lectures on Knot Theory: Selected Lectures presented at the Advanced School and Conference on Knot Theory and its Applications to Physics and Biology, ICTP, Trieste, Italy, 11 - 29 May 2009*; L.H. Kauffman, S. Lambropoulou, J.H. Przytycki, S. Jablan, Eds.; Ser. Knots Everything, vol. 46; World Scientific Press, 2011; pp. 281–320. See also ArXiv.

[43] S. Lambropoulou, Braid structures and braid equivalences in different manifolds and in different settings, World Scientific, Series on Knots and Everything, in preparation.

[44] S. Lambropoulou, Braid equivalences for long knots, in preparation.

[45] S. Lambropoulou and C. P. Rourke, Markov's theorem in three-manifolds, *Topology and Its Applications* **78** (1997) 95–122.

[46] S. Lambropoulou and C. P. Rourke, Algebraic Markov equivalence for links in 3-manifolds, *Compositio Math.* **142** (2006) 1039–1062.

[47] V.O. Manturov and H. Wang, Markov theorem for free links, arXiv:1112.4061 [math.GT].

[48] A.A. Markov, Über die freie Äquivalenz der geschlossenen Zöpfe, *Recueil Mathématique Moscou* **1**(43) (1936) 73–78.

[49] H.R. Morton, Threading knot diagrams, *Math. Proc. Cambridge Philos. Soc.* **99** (1986) 247–260.

[50] K. Reidemeister, Elementare Begründung der Knotentheorie, *Abh. Math. Sem. Hamburg Univ.* **5** (1927) 24–32.

[51] N. Scherich, A survey of grid diagrams and a proof of Alexanders theorem. In *Knots, Low-dimensional Topology and Applications: Proceedings of the Conference* Knots in Hellas 2016, *Springer Proceedings in Mathematics & Statistics (PROMS)* 284 (2019); S. Lambropoulou et al., Eds.

[52] R. Skora, Closed braids in 3–manifolds, *Math. Zeitschrift* **211** (1992) 173–187.

[53] L. Smolin, *Knot Theory, Loop Space and the Diffeomorphism Group*, New Perspectives in Canonical Gravity, 245–266, Monogr. Textbooks Phys. Sci. Lecture Notes, **5**, Bibliopolis, Naples, 1988.

[54] P.A. Sundheim, Reidemeister's theorem for 3–manifolds, *Math. Proc. Camb. Phil. Soc.* **110** (1991) 281–292.

[55] P. Traczyk, A new proof of Markov's braid theorem, preprint (1992), Banach Center Publications, Vol. 42, Institute of Mathematics Polish Academy of Sciences, Warszawa (1998).

[56] V. Turaev, Knotoids. *Osaka J. Math.* **49** (2012) 195–223. See also arXiv:1002.4133v4.

[57] P. Vogel, Representation of links by braids: A new algorithm, *Comment. Math. Helvetici* **65** (1990) 104–113.

[58] N. Weinberg, Sur l' equivalence libre des tresses fermée, *Comptes Rendus (Doklady) de l' Académie des Sciences de l' URSS* **23**(3) (1939) 215–216.

[59] S. Yamada, The minimal number of Seifert circles equals the braid index of a link, *Invent. Math.* **89** (1987) 347–356.

Chapter 8

Knots in Flows

Michael C. Sullivan, Southern Illinois University, Carbondale

A solution curve to a differential equation may form a loop. Such cyclic or period behavior is of great interest. The classical Poincaré-Bendixson theorem establishes the existence of a periodic solution curve in vector fields in the plane for suitable hypotheses. It is a mainstay of differential equations courses to this day. If the phase space of interest is three dimensional, periodic solution curves may form knots and different periodic solution curves may be linked. In 1950 Herbert Seifert proved the existence of closed integral curves for certain flows in S^3 and asked if this was always the case for nonsingular continuous flows on S^3 [40]. This became known as Seifert's conjecture. This question drove a great deal of research. It was answered in the negative by Paul Schweitzer in 1974 [39]. His example was C^1. It is now known that there are C^∞ and analytic examples [33].

In 1963 Edward Lorenz published his work examining a 3×3 ODE with very odd behavior [35]. It had many periodic orbits that are knotted and linked in complex ways. In 1983 Joan Birman and Robert Williams developed a systemic framework to analyze this behavior [9]. This was one of the main motivators for the serious study of knots in flows.

We give some technical definitions. Let M be a compact Riemannian 3-manifold. For our purposes a *flow* on M is a smooth map $f : M \times \mathbb{R} \to M$ such that $f(p, 0) = p$ and $f(p, s + t) = f(f(p, s), t)$ for all $p \in M$ and $s, t \in \mathbb{R}$. Often, M will be S^3. For $p \in M$ the *orbit* of p is $O(p) = \{f(p, t) \mid t \in \mathbb{R}\}$. If $O(p) = \{p\}$, then p is a *fixed point* of f. A flow without fixed points is a *nonsingular flow*. If there exists a $T > 0$ such that $f(p, T) = p$, then $O(p)$ is an embedded circle, that is, it is a knot. We say in this case that $O(p)$ is a *periodic orbit* or a *closed orbit*.

As noted above, smooth nonsingular flows on S^3 with no closed orbits exist. As we will see there are flows on S^3 in which every knot and link type is realized by closed orbits [16, 19] and such flows can arise as solutions to 3×3 ODEs in $\mathbb{R}^3$ [17].

8.1   Hyperbolic Flows and Basic Sets

The *chain recurrent set* of a flow f is

$$\mathcal{R} = \{p \in M : \forall \epsilon > 0, \exists \{p_0 = p, p_1, p_2, \ldots, p_k\} \subset M, \exists \{t_1, t_2, \ldots, t_k\} \subset \mathbb{R}^+$$
$$\text{such that } d(f(p_i, t_i), p_{i+1}) < \epsilon, i = 1, \ldots, k-1, \ d(f(p_k, t_k), p_0) < \epsilon\}.$$

The chain recurrent set of a flow includes fixed points and periodic orbits, but also more complicated orbits that come back near themselves over and over.

The chain recurrent set of a flow is said to have a *hyperbolic structure* if the tangent bundle of the manifold can be written as a Whitney sum $T_{\mathcal{R}} = E^u \oplus E^c \oplus E^s$ of sub-bundles invariant under Df where E_p^c is the subspace of TM_p corresponding to the orbit of p and such that there are constants $C > 0$ and $\lambda > 0$ for which $\|Df_t(v)\| \leq Ce^{-\lambda t}\|v\|$ for $v \in E^s$, $t \geq 0$ and $\|Df_t(v)\| \geq 1/Ce^{\lambda t}\|v\|$ for $v \in E^u, t \geq 0$.

Steve Smale showed that when $\mathcal{R}$ is hyperbolic, it is the closure of the periodic orbits of the flow and that it has a finite decomposition into compact invariant sets each containing a dense orbit; he called these the *basic sets* of the flow [41, 15].

We define, respectively, the *stable and unstable manifolds* of an orbit O in a flow f. See Figure 8.1

$$W^s(O) = \{y \in M : d(f(y, t), f(x, t)) \to 0 \text{ as } t \to \infty \text{ for some } x \in O\},$$
$$W^u(O) = \{y \in M : d(f(y, t), f(x, t)) \to 0 \text{ as } t \to -\infty \text{ for some } x \in O\}.$$

That these are manifolds is a classical result of Hirsch and Pugh [29] referred to as the Stable Manifold Theorem. A flow is *structurally stable* if it is *topologically equivalent*, i.e., there is a homeomorphism taking orbits to orbits preserving the flow direction, to flows obtained by small enough perturbations.

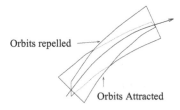

Orbits repelled

Orbits Attracted

FIGURE 8.1: Stable and unstable manifolds.

A flow with hyperbolic chain recurrent set $\mathcal{R}$ satisfies the *transversality condition* if the stable and unstable manifolds of $\mathcal{R}$ always meet transversally. A flow that has a hyperbolic chain recurrent set and satisfies the transversality condition is structurally stable; see [15, Theorem 1.10] for references. The converse — known as the C^1 Stability Conjecture — was proposed by Palis and Smale [37] and was proven by Hu [32] for dimension 3 and for arbitrary dimension by Hayashi [28]; see also [48].

8.1.1 Morse-Smale Flows

If the chain recurrent set of a flow is hyperbolic, consists of a finite collection of periodic orbits and fixed points, and satisfies the transversality condition, we have a *Morse-Smale flow*. Daniel Asimov showed that for $n \neq 3$ all n-manifolds (possibly with boundary), subject to certain obvious Euler characteristic criteria, support nonsingular Morse-Smale flows [4]. John Morgan has characterized which 3-manifolds (possibly with boundary) support nonsingular Morse-Smale flows [36] and Masaaki Wada has determined which labeled links can be realized as the invariant set of a nonsingular Morse-Smale flow on S^3. The components are labeled as attractors, repellers and saddles. See [47]; see also [11]. The simplest nonsingular Morse-Smale flow on S^3 is the Hopf flow which has just two closed orbits, one an attractor, the other a repeller, that form a Hopf link. The essence of Wada's result is that given an allowed labeled link, one can apply certain allowed moves, involving cablings and connected sums, and the resulting labeled link can be realized as the invariant set of a Morse-Smale flow. Then, starting with the labeled link in the Hopf flow, one can generate all other allowed labeled links.

As a possible area for future work, one would like to generalize Wada's theorem to other 3-manifolds, but Wada's proof depends heavily on the triviality of the fundamental group of S^3. Little progress has been made, but see [55].

Wada's links come up in two other types of flows. A certain subset of these links are realized as strands of fixed points for flows on S^3 arising from Bott-integrable Hamiltonian systems and in flows arising from contact structures [18, 45].

8.1.2 Smale Flows

If the chain recurrent set of a flow is hyperbolic, at most one-dimensional and satisfies the transversality condition the flow is known as a *Smale flow*. These were introduced by John Franks who was a student of Smale [15]. Basic sets which are not fixed points or isolated closed orbits are suspensions of nontrivial irreducible *shifts of finite type* (SFTs). (See [34] for definitions of terms for symbolic dynamics, SFTs have infinitely many periodic orbits but rational zeta functions.) These must be saddle sets, referred to as the *chaotic saddle sets*.

These chaotic saddle sets can be modeled by branched 2-manifolds with semi-flows where there is a bijection between any link of closed orbits in the basic set and a link of the same link-type in the semi-flow. These models are referred to as *templates*. Figure 8.2 shows the *Lorenz template*, which is denoted by $L(0, 0)$; its boundary is black, the branch line is blue and the red periodic orbit shown is a trefoil knot. Its periodic orbits were first studied by Joan Birman and Robert (Bob) Williams. [9]. They showed it supports infinitely many distinct knot types as closed orbits, that the set of links that can be realized is a subset of the set of closed positive braids with a full twist. It follows that these knots are fibered and prime. *Lorenz knots*, as they have come to be called, have been studied extensively and we will have more to say about them later.

Figure 8.3 depicts a Smale flow on S^3 with three basic sets, an attracting closed orbit that is a trefoil, a repelling closed orbit that is a meridian of the attractor, and a chaotic saddle set modeled by a Lorenz template. In [44] all possible ways that the Lorenz template can be realized in a nonsingular Smale flow on S^3 with three basic sets are given.

There are many other templates one can construct. In "template theory" one generally asks how a

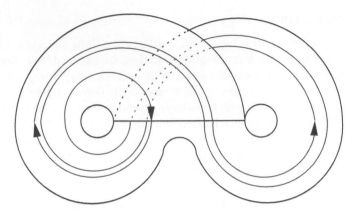

FIGURE 8.2: Lorenz template.

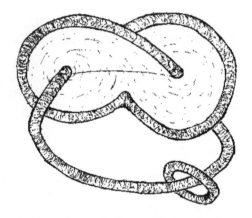

FIGURE 8.3: Smale flow with three basic sets.

given template can be realized in Smale flows, including the knotting and linking of the isolated closed orbits, and what types of knots and links are realized as closed orbits in an embedded template. A Lorenz-like template $L(m, n)$ is like the Lorenz template but there are m and n half twists in the respective bands. The template $L(0, 1)$ has been used to model a suspension of Smale's horseshoe map [10, 31, 30].

Up to homeomorphism there are three types of Lorenz-like templates: $L(0, 0)$, $L(0, 1)$ and $L(1, 1)$. Bin Yu studied how the latter two can be realized as models for Smale flows on S^3 and how each can be realized in Smale flows on certain other 3-manifolds [53]. He later showed that every orientable 3-manifold without boundary supports a nonsingular Smale flow with three basic sets, an attracting closed orbit, a repelling closed orbit and a chaotic saddle set [54]. Elizabeth Haynes and Kamal Adhikari each studied the realization in flows of more complex templates [27, 1].

The templates $L(0, n)$ for $n \geq 0$ (n positive means the band has the same crossing type as the crossing above the branch line) were shown by Robert F. Williams to contain only prime knots [51]. He and Joan Birman in the aforementioned foundational paper [9] conjectured that there would be a bound on the number of prime factors on any template. This was shown to be false in

[42]. Robert Ghrist went further and showed that many templates, including $L(0, -1)$, contain all knot and link types [16, 52]. Templates with this property are called *universal templates*.

Ghrist and Todd Young constructed a family of flows that have a bifurcation from being Morse-Smale flows to Smale flows containing all links [20].

For flows where the hyperbolic chain recurrent set is of dimension two or three, one can "split" along one or two, respectively, periodic orbits and create a one dimensional basic set whose periodic orbits are the same as in the original flow, with one or two exceptions [10, 19]. This provides a connection between Smale flows and Anosov flows.

8.2 Lorenz Knots

The subject of Lorenz knots and links has sparked quite a bit of interest [7, 22]. Birman and Ilya Kofman have characterized Lorenz links, actually a sight generalization, as T-links. A T-link is the closure of a braid of the form

$$(\sigma_1 \sigma_2 \cdots \sigma_{r_1})^{s_1} (\sigma_1 \sigma_2 \cdots \sigma_{r_2})^{s_2} \cdots (\sigma_1 \sigma_2 \cdots \sigma_{r_k})^{s_k}$$

where $1 \leq r_1 \leq r_2 \leq \cdots \leq r_k$, $1 \leq s_i$ for $i = 1, \ldots, k$ and σ_j denotes a positive crossing of the j and $j + 1$ strands of the braid. They make two interesting observations. Following Étienne Ghys they note that of the 1,701,936 prime knots with sixteen or fewer crossings, only twenty are Lorenz knots. Yet, among the 201 simplest hyperbolic knots, at least 107 are Lorenz knots. They then pose the question, "Why are so many geometrically simple knots Lorenz knots?" [8]

Polynomial invariants are very important in knot theory and there are interesting things to be said about polynomial invariants and Lorenz links. In their work [8], Birman and Kofman noted: "The Jones polynomials of Lorenz links are very atypical, sparse with small nonzero coefficients, compared with other links of an equal crossing number." Pierre Dehornoy has shown that the zeroes of the Alexander polynomial of a Lorenz knot all lie in an annulus whose width depends on the genus and the braid index of the knot [12]. This is a deep area of ongoing work.

In one of the most surprising developments Ghys has discovered that Lorenz knots arise in a seemingly unrelated problem. He studied the geodesic flow on the unit tangent bundle of the quotient of the Poincare disk by $PSL(2, \mathbb{Z})$. The periodic orbits are exactly the Lorenz knots excluding the two boundary unknots [21]. See [23] for a visually stunning exposition.

8.3 Partially Hyperbolic Flows and Strange Attractors

Historically, the study of Lorenz knots, and the reason they are called Lorenz knots, started with attempts to understand the "strange attractor" arising in the flow for a 3×3 nonlinear ODE introduced by Edward Lorenz [35].

$$\dot{x}(t) \quad = \quad -10x + 10y$$
$$\dot{y}(t) \quad = \quad 28x - y - xz$$
$$\dot{z}(t) \quad = \quad -8z/3 + xy$$

Figure 8.4 shows three orbits with different initial conditions converging toward the attractor.

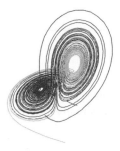

FIGURE 8.4: Lorenz attractor.

It was in this context that Williams introduced a *geometric Lorenz attractor* in the 1970s to study the periodic solution curves for the Lorenz equations [50]. See also [49, 26]; similar ideas were developed independently in the Soviet Union [2]. The model differs from $L(0, 0)$ in two important ways. First, it has a saddle fixed point that cannot be isolated from all the periodic orbits, which implies that it is not hyperbolic. Thus, it was dubbed a *strange attractor*. Second, the two unknotted periodic orbits in the boundary of $L(0, 0)$ are not realized. See Figure 8.5.

Proving that the Lorenz equations actually have a Lorenz attractor is very hard and took many years before this was established by Warwick Tucker in 1999 [46].

Notice that the flow line coming from the left end point of the branch line (blue) next meets the branch line at α, a little to the right of the left end point. Likewise, the flow line coming from the right end point of the branch line next meets the branch line at β, a little to the left of the right end point. It has been shown that the small changes in the parameters of the Lorenz equations shift the location of α and β. Thus, many periodic orbits are created or destroyed in such a perturbation. Hence, the geometric Lorenz attractor is not structurally stable. Yet, the basic form of the attractor remains under small enough perturbations. This phenomenon is called *robustness*. The details of this are beyond the scope of this chapter, but we refer the reader to the books [5] and [56].

Clark Robinson constructed differential equations with an attractor modeled by Lorenz-like templates contained in $L(-1, -1)$ [38]. Here the knots are positive but are not necessarily positive braids. They need not be prime but it is conjectured that there is a bound of two prime factors [43]. Although positive knots need not be fibered, as is the case with positive braids, Ghazwan Alhashimi has shown that the knots in $L(-1, -1)$ are fibered [3].

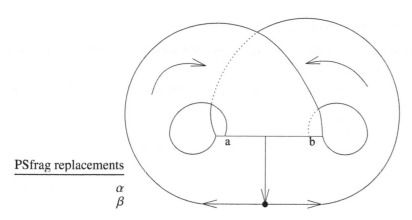

FIGURE 8.5: Geometric Lorenz attractor.

Physicist Robert Gilmore has pioneered the empirical study of strange attractors. He has studied how the existence of certain knotted periodic orbits in flows force the existence of many others. Templates play a key role. With Marc Lefranc he has written a book on this work, *Topology in Chaos* [24]. See also the collection [25] in celebration of his 70th birthday.

8.4 Euler Flows

John Etnyre and Ghrist wrote a series of papers on flows arising in theoretical fluid dynamics. They prove that any steady solution to the C^ω Euler equations on a Riemannian S^3 must possess an unknotted periodic orbit. They employed the relationship between contact structures to show the existence of solutions whose flow lines trace out closed curves of all possible knot and link types by way of a universal template. [13, 14]

8.5 Bibliography

[1] Kamal M. Adhikari and Michael C. Sullivan. Further study of simple smale flows using four band templates, *Top. Proc.* 50 (2017) pp. 21–37.

[2] Afraimovich, V. S., Bykov, V. V. and Shilnikov, L. P. On the origin and structure of the Lorenz attractor, *Akademiia Nauk SSSR Doklady*, vol. 234, May 11, 1977, p. 336–339, and *Soviet Physics Doklady,* vol. 22, May 1977, p. 253–255.

[3] Alhashimi, Ghazwan and Sullivan, Michael C. More on knots in Robinson's attractor, accepted in *Topology and its Applications* with minor revisions.

[4] Asimov, Daniel. Round handles and non-singular Morse-Smale flows, *Ann. of Math.* (2) 102

(1975), no. 1, 41–54. MR0380883 (52 #1780)

[5] Vítor Araújo and Maria José Pacifico. *Three-Dimensional Flows*, A Series of Modern Surveys in Mathematics, Vol. 53, Springer, 2010.

[6] Bendixson, Ivar. Sur les courbes définies par des équations différentielles, *Acta Math.*, Volume 24 (1901), 1–88.

[7] Birman, Joan S. The mathematics of Lorenz knots. *Topology and Dynamics of Chaos: In Celebration of Robert Gilmore's 70th Birthday* , 127–148, World Sci. Ser. Nonlinear Sci. Ser. A Monogr. Treatises, 84, World Sci. Publ., Hackensack, NJ, 2013.

[8] Birman, Joan S.and Kofman, Ilyn. A new twist on Lorenz links. *Journal of Topology*, 2 (2009), 227–248.

[9] J. Birman and R. F. Williams. Knotted periodic orbits in dynamical systems I: Lorenz knots. *Topology* **22** (1983), 47–82.

[10] J. Birman and R. F. Williams. Knotted periodic orbits in dynamical system. II. Knot holders for fibered knots. *Low-Dimensional Topology* (San Francisco, 1981), 1–60, Contemp. Math., 20, Amer. Math. Soc., Providence, RI, 1983.

[11] Casasayas, J., Martinez Alfaro, J., and Nunes, A. Knots and links in integrable Hamiltonian systems. *J. Knot Theory Ramifications* 7 (1998), no. 2, 123–153.

[12] Pierre Dehornoy. On the zeroes of the Alexander polynomial of a Lorenz knot. *Ann. Inst. Fourier, Grenoble* **65**, 2 (2015) 509–548.

[13] J. Etnyre and R. Ghrist, Contact topology and hydrodynamics I: Beltrami fields and the Seifert Conjecture, *Nonlinearity* 13, 441–458.

[14] J. Etnyre and R. Ghrist. Contact topology and hydrodynamics III: Knotted orbits, *Trans. Amer. Math. Soc.*, 352, 5781–5794.

[15] *Homology and Dynamical Systems.* CBMS Regional Conference Series in Mathematics, 49. Published for the Conference Board of the Mathematical Sciences, Washington, D.C.; by the American Mathematical Society, Providence, R. I., 1982. viii+120 pp. Reprinted, with corrections, 1989. MR0669378 (84f:58067)

[16] R. Ghrist. Branched two-manifolds supporting all links. *Topology* 36 (1997), no. 2, 423–448.

[17] R. Ghrist and P. Holmes. An ODE whose solutions contain all knots, *Intl. J. Bifurcation and Chaos*, 6(5), 1999, 779-800.

[18] R. Ghrist and R. Komendarczyk, Topological features of inviscid flows, in *Introduction to the Geometry and Topology of Fluid Flows*, NATO-ASI Series II, vol. 47, Kluwer Press, 183–202.

[19] R. Ghrist, P. Holmes and M. Sullivan. *Knots and Links in Three-Dimensional Flows*, Lecture Notes in Mathematics, Vol. 1654, Springer-Verlag, Berlin, 1997.

[20] R. Ghrist and T. Young. From Morse-Smale to all links, *Nonlinearity*, 11, 1111–1125.

[21] Ghys, Étienne. Knots and dynamics. *International Congress of Mathematicians*. Vol. I, 247277, Eur. Math. Soc., Zürich, 2007.

[22] Ghys, Étienne. The Lorenz attractor, a paradigm for chaos. Chaos, 1–54, *Prog. Math. Phys.*, 66, Birkhäuser/Springer, Basel, 2013.

[23] E. Ghys and J. Leys, Lorenz and modular flows: A visual introduction, Amer. Math. Soc. Feature Column, November 2006, available at http://www.ams.org/featurecolumn/archive/lorenz.html.

[24] Robert Gilmore and Marc Lefranc. *The Topology of Chaos: Alice in Stretch and Squeezeland*, 2nd Ed., Wiley, 2012.

[25] *Topology and Dynamics of Chaos: In Celebration Of Robert Gilmore's 70th Birthday*, 127–148, World Sci. Ser. Nonlinear Sci. Ser. A Monogr. Treatises, 84, World Sci. Publ., Hackensack, NJ, 2013.

[26] Guckenheimer, John; Williams, R. F. Structural stability of Lorenz attractors. *Inst. Hautes Études Sci. Publ. Math.* No. 50 (1979), 59 72.

[27] Elizabeth L. Haynes and Michael C. Sullivan. simple Smale flows with a four band template, *Topology and It's Applications*, Vol. 177, 2014, pages 23–33.

[28] Hayashi, Shuhei. Connecting invariant manifolds and the solution of the C^1 stability and Ω-stability conjectures for flows. *Ann. of Math.* (2) 145 (1997), no. 1, 81–137. Errata: *Ann. of Math.* (2) 150 (1999), no. 1, 353–356. MR1432037 (98b:58096)

[29] Hirsch, Morris W. and Pugh, Charles C. Stable manifolds and hyperbolic sets. 1970 Global Analysis *Proc. Sympos. Pure Math.*, Vol. XIV, Berkeley, Calif., 1968, pp. 133–163, Amer. Math. Soc., Providence, R.I.

[30] Philip Holmes. Knotted periodic orbits in suspensions of Smale's horseshoe: Period multiplying and cabled knots, *Physica D: Nonlinear Phenomena*, Volume 21, Issue 1, August 1986, pages 7–41.

[31] Holmes, Philip and Williams, R. F. Knotted periodic orbits in suspensions of Smale's horseshoe: Torus knots and bifurcation sequences. *Arch. Rational Mech. Anal.* 90 (1985), no. 2, 115–194.

[32] Hu, Sen. A proof of C^1 stability conjecture for three-dimensional flows. *Trans. Amer. Math. Soc.* 342 (1994), no. 2, 753–772.

[33] G. Kuperberg and K. Kuperberg. Generalized counterexamples to the Seifert conjecture, *Ann. of Math.* (2) 143 (1996), no. 3, 547–576.

[34] D. Lind and B. Marcus. *An Introduction to Symbolic Dynamics and Coding*, Cambridge University Press, Cambridge 1995. MR1369092 (97a:58050).

[35] Edward Lorenz. Deterministic non periodic flow, *J. Atmosph. Sci.* 20, 130141 (1963).

[36] Morgan, John W. Nonsingular Morse-Smale flows on 3-dimensional manifolds. *Topology* 18 (1979), no. 1, 41–53. MR0528235 (80j:58043)

[37] Palis, J. and Smale, S. Structural stability theorems. 1970 Global Analysis (*Proc. Sympos. Pure Math.*, Vol. XIV, Berkeley, Calif., 1968) pp. 223–231, Amer. Math. Soc., Providence, R.I. MR0267603 (42 #2505).

[38] Robinson, Clark. Homoclinic bifurcation to a transitive attractor of Lorenz type. *Nonlinearity*, 2:495–518, 1989.

[39] P. A. Schweitzer. Counterexamples to the Seifert conjecture and opening closed leaves of foliations, *Ann. of Math.* (2) 100 (1974), 386–400.

[40] Seifert, Herbert. Closed integral curves in 3-space and isotopic two-dimensional deformations. *Proc. Amer. Math. Soc.* 1 (1950), 287–302.

[41] Smale, S. Differentiable dynamical systems. *Bull. Amer. Math. Soc.* 73 1967 747–817.

[42] Sullivan, Michael C. Composite knots in the Figure-8 knot complement can have any number of prime factors, *Topology Appl.*, **55** (1994) 261–272.

[43] Sullivan, Michael C. Positive knots and Robison's attractor. *Journal of Knot Theory and Its Ramifications*, Vol. 7, No. 1 (1998) 115–121.

[44] Michael C. Sullivan. Visually building Smale flows in S^3. *Topology and Its Applications*, Vol. 106 (2000), no. 1, 1–19.

[45] Michael C. Sullivan. Transverse Foliations to nonsingular Morse-Smale flows on the 3-sphere and Bott-integrable Hamiltonian systems, *Qual. Th. Dyn. Syst.* 7(2):411–415, January 2009.

[46] Tucker, Warwick. The Lorenz attractor exists. *C. R. Acad. Sci. Paris Sér. I Math.* 328 (1999), no. 12, 1197–1202.

[47] Wada, Masaaki. Closed orbits of nonsingular Morse-Smale flows on S^3. *J. Math. Soc. Japan* 41 (1989), no. 3, 405–413. MR0999505 (90g:58059)

[48] Wen, Lan. On the C^1 stability conjecture for flows. *J. Differential Equations* 129 (1996), no. 2, 334–357. MR1404387 (97j:58082)

[49] Williams, R. F. The structure of Lorenz attractors. *Inst. Hautes Études Sci. Publ. Math.* No. 50 (1979), 7399.

[50] Williams, R. F. The structure of Lorenz attractors. *Turbulence Seminar* (Univ. Calif., Berkeley, Calif., 1976/1977), pp. 94–112. Lecture Notes in Math., Vol. 615, Springer, Berlin, 1977.

[51] R. F. Williams. Lorenz knots are prime. *Ergodic Theory and Dynamical Systems*, Volume 4, Issue 1, March 1984, pp. 147–163

[52] R. F. Williams. The universal templates of Ghrist, *Bulletin of the American Mathematical Society*, 35(02), November 1999.

[53] Yu, Bin. Lorenz like Smale flows on three-manifolds, *Topology and Its Applications*, 156 (2009) 2462–2469.

[54] Yu, Bin. Every 3-manifold admits a structurally stable nonsingular flow with three basic sets. *Proc. Amer. Math. Soc.* 144 (2016), no. 11, 4949–4957.

[55] Yu, Bin. Behavior 0 nonsingular Morse Smale flows on S^3. *Discrete Contin. Dyn. Syst.* 36 (2016), no. 1, 509–540.

[56] Elhadj Zeraoulia and Sprott Julien Clinton. *Robust Chaos and Its Applications*, World Scientific, 2012.

Chapter 9

Multi-Crossing Number of Knots and Links

Colin Adams, Williams College

9.1 Introduction

One of the most fundamental ideas in knot theory is that of a crossing. We project a knot to the plane or sphere and see two strands crossing. But what if we allow more than two strands to cross at a crossing?

Definition 9.1.1. An *n-crossing* (also called a *multi-crossing*) is a crossing produced in the projection of a knot or link that has *n* strands passing straight through the singular point.

Each strand passing through the crossing must bisect the crossing, having the same number of strands to its left and right. This idea first appeared in graph theory ([14, 15]), where the authors were allowing so-called degenerate crossings in the projections of spatial graphs to the plane.

Define an *n*-crossing projection to be a projection such that every crossing is an *n*-crossing. The first question one would like to ask is whether every knot and link has an *n*-crossing projection for all $n \geq 2$. The answer turns out to be yes. Thus, we can define the *n*-crossing number $c_n(L)$ of a knot or link L to be the least number of *n*-crossings in any *n*-crossing projection of L. Instead of just one crossing number, each knot has an entire spectrum of crossing numbers associated with it.

We can further ask if each knot has a projection with just a single multi-crossing. In fact, this is true, and such a projection is called an übercrossing projection. Furthermore, there is always such a projection with no nested loops, yielding a petal diagram. Thus, each knot has a well-defined übercrossing number $ü(K)$ and petal number $p(K)$, which are the least possible values of *n* for the single *n*-crossing in each case. In this chapter, we discuss these ideas and what is known about them.

9.2 Triple-Crossing Number

We first consider the case of triple-crossings, as in Figure 9.1. In Figure 9.2, we see triple-crossing projections of the trefoil and the figure-eight knot.

We first prove that every knot or link has a triple-crossing projection. We mention three methods for doing so. But first we need one definition.

FIGURE 9.1: A triple-crossing and a view slightly to the side of that same crossing so we can see how the single triple-crossing resolves into three double crossings. The strands are labelled T, M and B for top, middle and bottom.

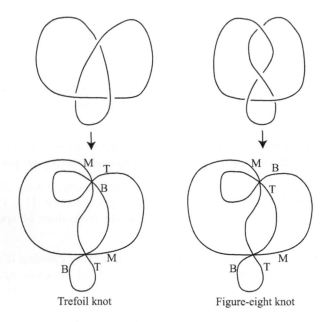

FIGURE 9.2: Triple-crossing projections.

Definition 9.2.1. A collection of disjoint circles C in the projection plane is said to be a *crossing covering collection* for a projection P of a knot K if each circle only intersects P in crossings, and when it does so, it bisects the crossing, with the same number of outgoing strands of the knot to each side of the circle. Moreover, every crossing intersects exactly one of the circles.

When we have such a collection of circles, we can take each circle and choosing the overstrand on one of the crossing through which it passes, we isotope the strand off the over-crossing and around the rest for the circle instead as in Figure 9.3. This move is called a *folding*. It eliminates the one crossing and turns the remaining crossings on the circle into triple-crossings. Repeating this with all the circles yields a triple-crossing projection.

The first proof that every knot and link has a triple-crossing projection relies on Alexander's Theorem ([7]) that every knot or link can be put in braid form. A closed braid obviously possesses a crossing covering collection of circles obtained by taking circles between each of the strings in the braid projection, as in Figure 9.4. Hence, by folding along these circles, we obtain a triple-crossing

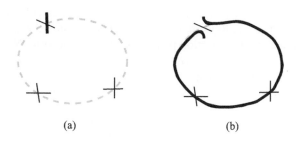

(a)   (b)

FIGURE 9.3: Using a circle to turn double crossings into triple-crossings via a folding.

projection with at most $c - \beta + 1$ triple-crossings where c is the number of crossings in the braid and β is the number of strings in the braid.

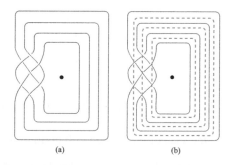

(a)   (b)

FIGURE 9.4: A knot in braid form, and a crossing covering collection for it.

The second proof is due to V. Jones [11]. Take a projection of a knot or link K. Choose a point to fold the link back on itself, as in Figure 9.5 such that as we lay the parallel copy down, each new crossing is an overcrossing. The result is still isotopic to K. Then at the squares created at each of the original crossings, push one strand across via a Type II Reidemeister move to obtain two triple-crossings for each of the original crossings.

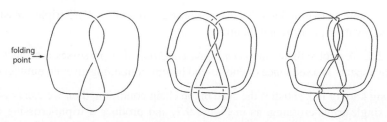

FIGURE 9.5: Turning a double-crossing projection into a triple-crossing projection.

The third proof ([1]), which gives us the best bound on triple-crossing number in terms of the traditional crossing number, is to show that every projection of a knot or link possesses a crossing covering collection of circles. To see this, note that we can always change the crossings in any projection of a knot to yield the trivial knot. Therefore, by ignoring the choice of crossings (that is, treating them as flat), there is a sequence of Reidemeister moves that will take the projection to the trivial projection, which clearly possesses a covering crossing collection of circles, namely an empty one. After showing that Reidemeister moves preserve the existence of covering crossing collections, it follows that every projection possesses such a collection. Note then that if d is the number of circles in the collection, then $c_3(K) \leq c(K) - d$. In particular $c_3(K) \leq c(K) - 1$. In fact, the only knot projections with only crossing covering collections of one circle are the 2-braid knots. So for all other knots and links, $c_3(K) \leq c(K) - 2$. It is still not known whether $c_3(K) = c(K) - 1$ for all 2-braid knots.

From now on, we denote traditional crossing number $c(K)$ as $c_2(K)$. Note that by opening each triple-crossing into three double crossings as in Figure 9.1, we see that $c_3(K) \geq c_2(K)/3$. There are knots which realize this lower bound, as for instance does the 12-crossing knot K that appears in Figure 9.6, which has $c_3(K) = 4$, as can be seen from the diagram.

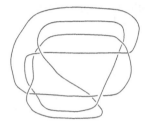

FIGURE 9.6: A knot that realizes $c_3(K) = c_2(K)/3$.

A well-known theorem relates the span of the bracket polynomial to the crossing number.

Theorem 9.2.2. *([10, 12, 16]) span $< K > \leq 4c_2(K)$, and span $< K > = 4c_2(K)$ if and only if K is an alternating knot or link.*

The first part of this theorem was extended to triple-crossing number in [1]:

Theorem 9.2.3. *span $< K > \leq 8c_3(K)$ for any knot or link K.*

These two theorems immediately yield the following corollary:

Corollary 9.2.4. *If K is an alternating knot or link, $c_3(K) \geq c_2(K)/2$.*

In particular, for any alternating knot that can be constructed with $c_2(K)/2$ triple-crossings, we then immediately know its triple-crossing number.

Definition 9.2.5. A knot satisfies the *even bigon condition* if it possesses a projection such that every crossing appears in a sequence of end-to-end bigons containing an even number of crossings.

If an alternating knot or link satisfies the even bigon chain condition, then we can twist every other bigon into a single triple-crossing, as in Figure 9.7, and produce a triple-crossing diagram that realizes this lower bound on triple-crossing number. Using this and related ideas, triple-crossing number can be determined for a variety of knots.

One can also ask if there are moves on triple-crossing projections like the Reidemeister moves between double-crossing projections. See [6] for a set of such moves.

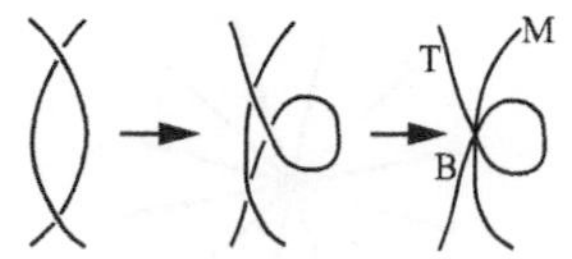

FIGURE 9.7: Twisting a bigon to obtain a triple-crossing.

9.3 Higher Order Multi-Crossings

We can now extend to n-crossing projections for all $n \geq 3$.

Theorem 9.3.1. *Given any integer $n \geq 2$, and any knot or link L, there exists a projection of L with only n-crossings.*

Proof. We already know this to be true for $n = 2$ and 3. For $n = 4$, we take a double-crossing projection, and replace each double crossing with a 4-crossing as in Figure 9.8. A similar construction works to show that there exists an n-crossing projection for any even n.

When n is odd, we first find a triple-crossing projection as we have already described. Then we pull a strand back and forth over each crossing in a manner similar to what we did in Figure 9.8 to obtain a crossing with any odd number of strands passing through it. □

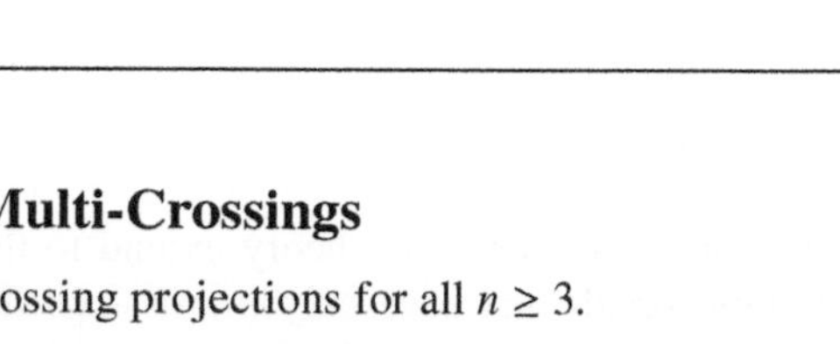

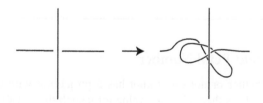

FIGURE 9.8: Converting a double crossing into a quadruple crossing.

Definition 9.3.2. Given any knot or link, the *n-crossing number of K*, denoted $c_n(K)$, is the least number of n-crossings in an n-crossing projection of K.

So now, every knot or link has a spectrum of crossing numbers associated to it. We can keep track of the various types of n-crossings by listing the heights of the strands as they pass through the crossing, read clockwise from the top strand as a permutation of 2 through n (as in Figure 9.9). Thus for a given $n \geq 2$, there are a total of $(n - 1)!$ different n-crossings.

Just as we have a relationship between the span of the bracket polynomial and double- and triple-crossing numbers, we have such a relationship for higher order n-crossing numbers. (See [2] and [5]).

Theorem 9.3.3. *For all $n \geq 3$, $span\langle K \rangle \leq \left(\left\lfloor \frac{n^2}{2} \right\rfloor + 4n - 8 \right) c_n(K)$.*

FIGURE 9.9: A multi-crossing given by the permutation 153264.

See [5] for a list of known n-crossing numbers for low crossing knots.

We can ask how other ideas from traditional knot theory extend to the multi-crossing case. For instance, what can be said about braids that have only n-crossings for a given n? (The idea of investigating this was proposed by Daniel Vitek, who first solved it for a finite number of even n.)

Theorem 9.3.4. *[13]*

1. *For every even $n \geq 2$, every knot and link has an n-crossing braid projection.*

2. *For every odd $n \geq 2$, every multiple component link has an n-crossing braid projection.*

Thus, every knot has a spectrum of braid indices, one for each even $n \geq 2$. And every multiple component link has a spectrum of braid indices, one for each $n \geq 2$. As this theorem is recent, very little is yet known about these higher-order braid indices.

9.4   Übercrossing and Petal Number

A natural question is whether or not every knot has a projection with only one n-crossing. Put another way, given a knot K, is there always a value for n such that $c_n(K) = 1$?

It turns out that the answer is yes.

Definition 9.4.1. An *übercrossing projection* of a knot K is a projection such that there is a single multi-crossing. The übercrossing number of the projection is the number of strands in the single multi-crossing and the übercrossing number of a knot K, denoted $\ddot{u}(K)$, is the least übercrossing number over all übercrossing projections of the knot K.

See, for example, the figure on the left in Figure 9.10, which is an übercrossing number 4 projection of the trefoil knot, and in fact realizes the übercrossing number of the trefoil knot.

Theorem 9.4.2. *([3]) Every knot K has an übercrossing projection and therefore an übercrossing number $\ddot{u}(K)$.*

The proof relies on an algorithm that takes any projection of a knot and turns it into a so-called pre-petal diagram, as in Figure 9.11. Then one can fold the outer loop in to obtain a petal diagram. Note that $\ddot{u}(K) < p(K)$. We can find $\ddot{u}(K)$ and $p(K)$ in certain cases. Let $T_{r,r+1}$ denote the $(r, r + 1)$-torus knot.

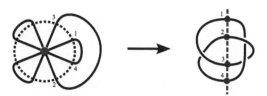

FIGURE 9.10: An übercrossing projection of the trefoil knot and its side view.

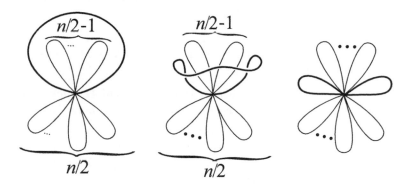

FIGURE 9.11: Turning a pre-petal projection obtained from Theorem 9.4.2 into a petal projection by folding the nesting loop over the middle of the übercrossing projection.

Theorem 9.4.3. *([3, 4])*

    1. $p(T_{r,r+1}) = 2r + 1$ *or all* $r \geq 2$.

    2. $\ddot{u}(T_{r,r+1}) = 2r$ *for all* $r \geq 2$.

If $u(K)$ is the unknotting number, we have the following.

Theorem 9.4.4. *([4])* $u(K) \leq \frac{(p(K)-1)(p(K)-3)}{8}$. *This is an equality if and only if* K *is one of the* $(r, r + 1)$-*torus knots.*

Theorem 9.4.5. *([4])* $c_2(K) \leq \frac{p(K)^2 - 2p(K) - 3}{4}$. *This is an equality if and only if* K *is one of the* $(r, r + 1)$-*torus knots.*

It is clear that $(r, r + 1)$-torus knots are special knots with regard to übercrossing and petal numbers.

Petal projections yield a very simple method for expressing and generating knots. One only needs a permutation on $n - 1$ where n is odd, to describe the levels of the strands going clockwise around the single multi-crossing, starting from the top strand. In [8] and [9], the authors use this as a means to generate random knots and determine properties of those random knots.

There is still much to be done. Choose your favorite concept for double- crossing projections. It could be unknotting number, bridge index, stick number, grid number, mosaic number, Reidemeister moves, Khovanov homology, knot Floer homology, finite type invariants, quantum invariants, Seifert surfaces, etc. See how and if it generalizes to n-crossing, übercrossing and petal projections and numbers. There is lots yet to be investigated.

9.5 Bibliography

[1] C. Adams, Triple crossing number of knots and links, *J. of Knot Theory and Its Ramifications*, Vol. 22, No. 2 (2013), 1350006-1 ˜1350006-17.

[2] C. Adams, Quadruple crossing number of knots and links, *Math. Proceedings of the Cambridge Philosophical Society*, **156** (2014) 241–253.

[3] C. Adams, T. Crawford, B. DeMeo, M. Landry, M. Montee, S. Park, S. Venkatesh, F. Yhee, Knot projections with a single multi-crossing, *Journal of Knot Theory and Its Ramifications* **24** (2015) 1550011 (30 pages).

[4] C. Adams, O. Capovilla-Searle, J. Freeman, D. Irvine, S. Petti, D. Vitek, A. Weber, S. Zhang, Bounds on übercrossing number and petal numbers for knots, *Journal of Knot Theory and Its Ramifications* **24** (2015) 1550012 (16 pages).

[5] C. Adams, O. Capovilla-Searle, J. Freeman, D. Irvine, S. Petti, D. Vitek, A. Weber, S. Zhang, Multicrossing numbers and the span of the bracket polynomial, *Math. Proc. of Cambridge Phil. Soc.* on-line November, 2016, 1–32.

[6] C. Adams, J. Hoste, M. Palmer, Triple-crossing number and moves on triple-crossing diagrams, *Journal of Knot Theory and Its Ramifications*, Vol. 28, No. 11 (2019) 1940001.

[7] J. Alexander, A lemma on a system of knotted curves, *Proc. Natl. Acad. Sci. USA* **9** (1923) 93–95.

[8] H. Even-Zohar, J. Hass, N. Linial, T. Nowik, Invariants of random knots and links, *Discrete Comput. Geom.* **56** (2016) 274–314.

[9] H. Even-Zohar, J. Hass, N. Linial, T. Nowik, The distribution of knots in the petaluma model, i*Alg. Geom. Topology*, Vol. 18, No. 6 (2018) 3647–3667.

[10] L. Kauffman, New Invariants in the theory of knots, *Amer. Math. Monthly* **95** (1988)195–242.

[11] V. Jones, Planar algebras, I, ArXiv 9909027 (1999).

[12] K. Murasugi, Jones polynomials and classical conjectures in knot theory, *Topology* **26** (1987) 187–194.

[13] D. Nishida, Multi-crossing braids, undergraduate thesis, Williams College, 2018.

[14] J. Pach and G. Toth, Degenerate crossing numbers, *Discrete Comput. Geom.*, **41**(3) (2009) 376–384.

[15] H. Tanaka and M. Teragaito, Triple crossing numbers of graphs, ArXiv 1002.4231, (2010).

[16] M. Thistlethwaite, A spanning tree expansion for the Jones polynomial, *Topology* **26** (1987) 297–309.

Chapter 10

Complementary Regions of Knot and Link Diagrams

Colin Adams, Williams College

10.1 Introduction

Diagrams of knots and links play a critical role in the study of knots. They are the primary method of representation of knots. They are used in the tabulation of knots up to a given crossing number. And they yield the means with which to compute many invariants.

Here, we would like to consider diagrams where we ignore the crossing information. Thus, we think of them as 4-valent graphs embedded in the sphere. We would like to understand how the combinatorial structure of the diagram is determined and/or determines the possible knots and links that result.

In particular, we consider the complementary regions of the diagram, each of which is an n-gon bounded by edges and vertices. In this paper, we consider how restricting the possible numbers of edges of faces in diagrams might or might not restrict the knots and links that can result.

10.2 Restrictions on Possible Faces

We assume our diagrams are connected and reduced, so there are no monogonal faces. Let p_i be the number of faces with i edges for all $i \geq 2$. We first note some immediate relations between the possible values for the p_i.

$$e = \frac{2p_2 + 3p_3 + 4p_4 + \ldots}{2} \tag{10.1}$$

$$f = p_2 + p_3 + p_4 + \ldots \tag{10.2}$$

$$e = 4v/2 \tag{10.3}$$

By the Euler characteristic, we can plug this into $v - e + f = 2$ to obtain:

$$2p_2 + p_3 = 8 + p_5 + 2p_6 + 3p_7 + \ldots \qquad\qquad (10.4)$$

In graph theory and geometry, going back to [4], mathematicians have investigated which sequences of values for the $p'_i s$ satisfy Equation 10.4 and correspond to actual 4-valent graphs on the sphere. However, most graph theorists and geometers are interested in these graphs as the 1-skeleta of convex polytopes, and therefore assume that $p_2 = 0$, which we will not usually want to do. Investigations date back to [4].

As an example, Grünbaum ([7]) proved that for any collection of integers with $p_2 = 0$ that satisfies Equation 10.4, there exists a choice of p_4 such that there is a planar 3-connected 4-valent graph realizing these values. This theorem is known as Eberhard's theorem. In [8], this result was extended to show that there is a value for p_4 so that the resulting graph is the projection of a knot, rather than a link.

For our purposes, Equation 10.4 yields several useful pieces of information. First, for any diagram of a knot or link, either p_2 or p_3 is nonzero. In other words, the diagram must contain bigons and/or triangles. Second, the number of 4-gons is not constrained by the Euler characteristic. Third, by considering this equation mod 2, the number of odd-sided regions must always be even.

One of the first investigations of the complementary regions appeared in [5] where the authors define a diagram to be lune-free if there are no bigons, which is to say $p_2 = 0$. They showed that there exist infinitely many lune-free diagrams, and given any number of at least seven vertices, there is such a corresponding diagram.

10.3 Universal Sequences

We are particularly interested in which collections of values of n for the n-gons making up the complementary regions can generate diagrams for all knots and links.

Definition 10.3.1. Given a knot or link K and a strictly increasing sequence of integers $(a_1, a_2, a_3, \ldots)$ with $a_1 \geq 2$, we say the sequence is *realized* by K if there exists a diagram for the knot or link such that each face is an n-gon for some a_n that appears in the sequence. Such a diagram is called an $(a_1, a_2, a_3, \ldots)$-diagram. (Not every a_n must be realized by a face.) We say that a sequence is *universal* if every knot and link has a diagram that realizes the sequence. We sometimes restrict to sequences that are universal just for knots but not for multi-component links.

Note that by Equation 10.4, if a sequence is to be universal, it must start with a 2 or 3.

In [3], it was shown that every knot has a projection with two triangular faces and all of the rest having an even number of edges. This was extended to links in [6], using the fact that every link is carried by a large enough so-called potholder diagram. In particular, this implies that $(2, 3, 4, 6, 8, \ldots)$ is universal for knots and links. In [6], it is also proved that every knot or link has a diagram with two monogonal faces and all the rest having an even number of edges. So $(1, 2, 4, 6, \ldots)$ is universal. The paper [6] also discusses what is known about sequences of immersed curves in the plane that carry all knots, including potholder diagrams, star diagrams, and billiard diagrams, and how efficiently they do so.

In [3], it was further proved that $(3, 5, 7, \ldots)$ and both $(2, n, n + 1, n + 2 \ldots)$ for any integer $n \geq 3$ and $(3, n, n + 1, n + 2, \ldots)$ for any integer $n \geq 4$ are universal for all knots and links.

In [9], the authors proved that every knot has a projection with no lunes, and hence that $(3, 4, 5, \ldots)$ is universal. The authors were particularly interested in moves that eliminated lunes and preserved n-colorings.

In fact, universal sequences need not be infinite. In [3], it is proved the sequence $(3, 4, n)$ is universal for all $n \geq 5$ and that the sequence $(2, 4, 5)$ is universal.

We include here the proof that $(3, 4, 5)$ is universal. A second proof using braids appears in [10], and a third proof using bisected vertex leveling of plane graphs appears in [13]. A projection with only triangular, quadrilateral and pentagonal faces is called a delta diagram. In this last paper, an upper bound on the least crossing number of a delta diagram is given that is quadratic in the crossing number of the knot or link.

Theorem 10.3.2. The sequence $(3, 4, 5)$ is universal for knots and links.

Proof. Take the projection Q that appears in Figure 10.1. Note that its complementary regions are all quadrilaterals except for eight triangles, denoted with the numeral 3. We can choose crossings for Q so that it is the trivial knot. We can also choose such a Q so that the central square grid is as large as we need.

Now let P be a projection of a given knot or link K. We isotope it to a new projection P' so that it lies in the integer lattice in the plane, with all crossings and corners occurring at integer lattice points. Then we superimpose P' on top of Q, so that the vertices of the central grid of Q occur at the half-integer lattice points, and so that the central grid of Q is large enough to entirely contain P' within it. We assume the grid is fine enough that if P' intersects a square of the grid, it does so in a single line segment passing through opposite edges, and there is at least one such square, or in a corner, passing through two adjacent edges of the square or in a crossing, with strands passing through both pairs of opposite edges, as in Figure 10.2. Note that in the new diagram every face is a 3-gon, 4-gon or 5-gon. Now we take a square through which P' passes as a single line segment connecting opposite edges, and we cut open the strand of P', add in a crossing, and cut open a strand of Q and connect the endpoints as in Figure 10.3 to obtain the composition of the knot K with the trivial knot represented by Q. The result is a $(3, 4, 5)$-projection of K.

$\square$

By adding $n - 5$ slightly offset copies of P' to the projection of P' and Q, all with crossings set to trivialize the corresponding knot, and composing them, we obtain the further case that $(3, 4, n)$ is universal for all $n \geq 5$.

It is still an open question as to whether there are any universal sequences of the form (a_1, a_2). Note that if there were such a sequence, then $a_1 = 2$ or 3. And certainly, if (a_1, a_2) is to be universal, there would have to be an infinite set of diagrams that could be realized by (a_1, a_2). In the case of knots, a_1 and a_2 would have to be relatively prime. In [11] (see also [2]), it was proved that there are infinitely many $(2, n)$-knot diagrams for every odd $n \geq 5$ and infinitely many $(3, n)$-knot

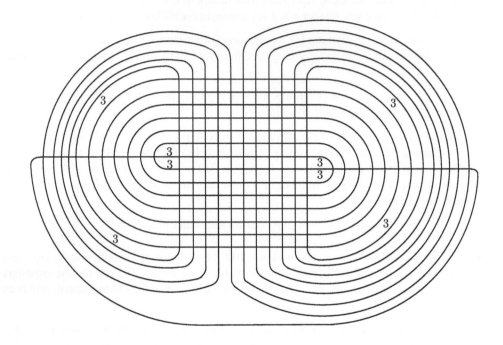

FIGURE 10.1: A $(3, 4)$-projection on the sphere.

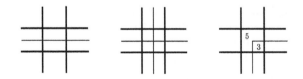

FIGURE 10.2: The three ways P' intersects a square from Q.

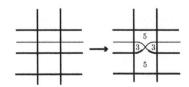

FIGURE 10.3: Composing P' with Q.

diagrams for all n at least 4 and not divisible by 3. Thus, it is possible that any of these two-integer sequences are universal.

Results about universal sequences have also been proved for n-crossing projections. A projection is an n-crossing projection if every crossing has n strands passing directly through it. See the chapter on multi-crossings for more details (Chapter 14). In [1], it was proved that every knot and link has an n-crossing projection. In the ArXiv preprint [12], it was proved that for all $n \geq 3$, $(1, 2, 3, 4)$ is universal for n-crossing projections for all knots and links.

In [14], the ideas here are extended to investigations into universal sequences for spatial graphs.

There is still much work to be done on understanding how the combinatorial patterns of link diagrams influence the types of links that can result.

10.4   Bibliography

[1] C. Adams, Triple crossing number of knots and links, *J. Knot Thy. Ram.* (2013), **22** 1350006.

[2] C. Adams, N. MacNaughton, C. Sia, From doodles to diagrams to knots, *Math. Mag.* (2013), **86** 83–96.

[3] C. Adams, R. Shinjo, K. Tanaka, Complementary regions of knot and link diagrams, *Annals of Combinatorics* (2011), **15** 549–563.

[4] V. Eberhard, Zur morphologie der polyeder, Leipzig, 1891.

[5] S. Eliahou, F. Harary, L. Kauffman, Lune-free knot graphs, *J. Knot Theory Ramifications* **17** (2008), 55–74.

[6] J. Even-Zohar, J. Hass, N. Linial, T. Nowik, Universal knot diagrams, *J. Knot Theory Ramifications* **28** (2019), 1950031.

[7] B. Grünbaum, Some analogues of Eberhard's theorem on convex polytopes, *Israel J. Math.* **6** (1969), 398–411.

[8] D. Jeong, Realizations with a cut-through Eulerian circuit, *Discrete Math.* **137** (1995), 265–275.

[9] S. Jablan, L. Kauffman, P. Lopes, The delunification process and minimal diagrams, *Topology Appl.* **193** (2015), 27–289.

[10] S. Jablan, L. Kauffman, P. Lopes, Delta diagrams, *Jour. Knot Thy. Ram.* **25** (2016), 1641008.

[11] N. MacNaughton, From doodles to diagrams, *Williams College Undergraduate Mathematics Thesis* (2010).

[12] M. Montee, Complementary regions of multi-crossing projections of knots, *ArXiv* **1210.0453** (2012), 1–15.

[13] S. No, S. Oh, H. Yoo, Bisected vertex leveling of plan graphs: Braid index, arc index and delta diagrams, *J. Knot Theory Ramifications* **27** (2018), 1850044.

[14] R. Shinjo, Universal sequences of spatial graphs, *Topology and its Applications* **196** (2015), 868–879.

Chapter 11

Knot Tabulation

Jim Hoste, Pitzer College

The first tables of knots were produced in the late 19th century. Inspired by Lord Kelvin's "Vortex Theory of the Atom," the Scottish physicist, Peter Guthrie Tait, set out to systematically enumerate knots according to their crossing number. Joined later by the English Reverend, Thomas P. Kirkman, and the American mathematician, Charles Newton Little, the trio eventually produced, after an enormous amount of hand-work spread out over a period of about 25 years, a (nearly correct) table of all prime alternating knots to 11 crossings and prime nonalternating knots to 10 crossings [16],[17],[18],[19],[20],[21],[30]. Since 1900, the tables have been considerably expanded, first by hand, and then by computer, due primarily to the work of Burton, Conway, Caudron, Doll, Dowker, Thistlethwaite, Hoste, Weeks, Jablan, Rankin, Flint, and Schermann [2],[4],[5],[8],[9],[10],[13],[14],[27],[28],[29],[31].

Still others, notably Hartley, Perko, and Trotter, were involved in proving that the tabled knots were indeed distinct and prime, or in computing their symmetries [11],[25],[26], [34]. Over the years, important advances in the theory of topology, and especially algebraic and geometric topology, have streamlined the tabulation process. For articles that discuss the history of knot tabulation, see [12],[13].

Knot tables have proven important to the development of knot theory by identifying interesting examples for study and by providing large samples of knots on which to test conjectures and look for patterns. To cite just one example, *KnotInfo* [6], a website that records the values of a variety of knot invariants for each prime knot up to a given crossing number, identifies a number of knots with currently unknown unknotting number. Important examples have also come to light as the result of making the tables. Nontrivial links with trivial Jones polynomials as well as the first examples of amphichiral knots with odd crossing number were first discovered this way by Thistlethwaite.

Knots in the original tables were organized by crossing number, a principle that has continued to this day. The basic idea in making a table of all prime knots with a given crossing number is straightforward. First, adopt some sort of scheme for encoding knot diagrams and use it to enumerate all possible diagrams of prime knots with n crossings. Next, group the diagrams together by knot type. While this process is easy to describe in theory, the sheer number of possible diagrams for anything beyond a relatively small number of crossings, demands a clever encoding scheme and skillful programming. While a number of different methods can be applied in the second step—for example, in the case of hyperbolic knots, the hyperbolic structure provides a complete knot invariant—we quickly reach a practical limit on n, which currently is in the low twenties.

To date, it has been considered easier to first enumerate and classify all prime alternating knots of a given crossing number and then pass from these diagrams to nonalternating diagrams in order to find all possible nonalternating knots. Here we are aided by several important theorems. The first, due to Menasco [22], states that an alternating diagram represents a prime knot if and only if the diagram is prime as a diagram. Thus, when enumerating all possible alternating diagrams of a given crossing number, it is easy to make sure we list only prime knots. A second important result, which came to be known as one of the three "Tait conjectures," and which was proven after the discovery of the Jones polynomial, states that a reduced alternating knot diagram is minimal with respect to the crossing number [15], [24],[32],[33]. Thus the crossing number of alternating knots is readily computable. Finally, the "Tait flyping conjecture," proven by Menasco and Thistlethwaite [23], tells us that that two reduced alternating diagrams represent the same knot if and only if they are related by flypes. Hence, in a sense, we may consider alternating knots to be completely classified. For the knot tabulator, these theorems mean that after producing all possible reduced, prime, alternating diagrams of a given crossing number, they can easily be grouped together by knot type and, moreover, we can be sure of having a complete list of all prime alternating knots with that crossing number.

Several important schemes have been used to encode knot diagrams. The first was introduced by Listing and is now known as the Tait graph. The complementary regions of the knot are first shaded checkerboard fashion after which a vertex is placed in each black region. Next an edge joins two vertices if the corresponding regions share a crossing. Edges may be further labeled with ± 1 to encode the right or left twisting of the crossing, but this data can be safely omitted in the case of an alternating diagram. The sum of the valences of each vertex is equal to twice the number n of crossings. Hence the set of valences provides a partition of $2n$. To enumerate all alternating diagrams with n crossings, one first finds all partitions of $2n$, and then all planar graphs with that set of valences. This is the approach first taken independently by Little and then with the help of Kirkman and Tait.

In the 1960s, Conway introduced a new scheme for encoding knots and used it to tabulate (by hand) all prime knots to 11 crossings and links to 10 crossings (with only a few errors). Conway's idea was to start with a 4-valent planar graph and then to insert *2-string tangles* at each vertex. Only a handful of basic starting graphs are needed to tabulate all knots up to 11 crossings, but as the number of crossings grows, so must the set of starting graphs. This fact deterred Dowker and Thistlethwaite from using Conway's method when they undertook the first computer-aided tabulation in the 1970s. Instead, they introduced an encoding scheme that built on a method first introduced by Gauss and which could be applied to a diagram with any number of crossings. Now known as "DT" code, the method associates to every oriented knot diagram with n crossings a signed sequence of the even integers $2, 4, \ldots, 2n$. A basepoint is first placed somewhere on the diagram and the crossings are then numbered consecutively from 1 to $2n$ as they are encountered while traversing the knot starting from the basepoint. Each crossing receives both an even and odd label and the sequence of even labels associated to the odd sequence $1, 3, 5, \ldots, 2n - 1$ is recorded. If the diagram is alternating, one stops here. If not, even integers that label crossings that must be changed to make the diagram alternate are then negated. One of the advantages of the *DT* sequence is its compactness. But of the $n!$ possible positive *DT* sequences of length n, most do not correspond to a reduced, prime, n-crossing alternating diagram and many encode the same diagram. So, simply listing all possible sequences in order to enumerate the alternating diagrams is inefficient. Nevertheless, both Thistlethwaite and Hoste & Weeks successfully used this basic idea to tabulate all prime knots to 16 crossings.

In 2003, Rankin, Flint, and Schermann created a more sophisticated encoding method that takes advantage of *flype cycles* and *groups of crossings* to enumerate all prime alternating knots to 23 crossings [27],[28]. Their enumeration also proceeded by induction on the number of crossings, using ideas introduced by Calvo [3].

After tabulating the alternating knots, the nonalternating knots can then be derived by changing crossings in all possible ways in every possible alternating diagram. At this point, the number of sheer possibilities to consider is truly overwhelming! Again, some kind of efficient encoding scheme as well as clever programming are absolutely essential. After obtaining a set of diagrams that, in theory, should contain all possible nonalternating diagrams of a given crossing number, it now remains to weed out duplicate knots. Applying a battery of knot invariants, the diagrams can be separated into distinct classes. For example, one could compute skein polynomials, the hyperbolic structure of the complement (if one exists), or count the number of homomorphisms of the knot group into various finite groups (a method put to great use by Thistlethwaite). But for each class of more than one diagram, the problem now becomes quite difficult. Either some new invariant must be found to distinguish a pair of diagrams, or a sequence of Reidemeister moves must be found to confirm that the two diagrams are the same knot. For hyperbolic knots, the hyperbolic structure is a complete invariant and, in theory, solves our problem. Moreover, "most" knots are hyperbolic—of the 352,152,252 nontrivial prime knots with 19 or less crossings, all but 394 are hyperbolic. However, while programs like *SnapPy* [7] can be used to find the hyperbolic structure, when it exists, round-off error can create "close calls" over which the tabulator must now exercise extreme caution.

Recently, Burton [2] has tabulated all prime knots to 19 crossings and all alternating knots to 20 crossings. Like Little, Burton starts with graphs, this time planar quadrangulations generated by the computer program *plantri* [1]. Such a graph is dual to an alternating knot diagram, having one vertex in each region and connecting two vertices if the corresponding regions share an edge. As described before, the graphs are then grouped by flype equivalence class, producing a table of alternating knots, which then leads to all possible nonalternating diagrams by changing crossings. A this point Burton makes extensive use of Reidemeister moves, *SnapPy*, a host of knot invariants, as well as normal surface theory to remove duplicates. Comparison to a "provisional" and unpublished table in the 17–19 crossing range made by Thistlethwaite many years ago turned up a few omissions in Thistlethwaite's work. Thistlethwaite then found and repaired bugs in his program, after which it produced the same numbers of knots as Burton.

Given the huge number of knots involved and the ease with which errors can be made, it should come as no surprise that as each new tabulation was completed, errors were sometimes found in the earlier tables. Remarkably, errors have been few in number. Nevertheless, Table 11.1, which displays the currently known numbers of prime alternating and nonalternating links with a given crossing number and given number of components, should be regarded with healthy skepticism. Presumably all the numbers up to 19 crossings as well as the number of 20-crossing alternating knots are correct, having been checked by more than one independent tabulation. The remaining data are due only to Rankin, Flint, and Schermann.

Finally, the numbers in Table 11.1 represent the number of unoriented links with n crossings and k components up to mirror image. That is, two such links are considered the same if and only if there is a homeomorphism of S^3 (orientation preserving or not) taking one to the other. If L is an oriented link, then reorienting some or all of the components of L and taking the mirror image, or both, could lead to oriented links that are not ambient isotopic. Hence, passing to oriented links considered up to orientation-preserving ambient isotopy will increase the numbers in the table.

TABLE 11.1: Number of alternating and nonalternating, unoriented, prime links with a given number of crossings and components.

Crossing	Number of Components										
	1	2	3	4	5	6	7	8	9	10	11
2a	0	1									
3a	1	0									
4a	1	1									
5a	2	1									
6a	3	3	2								
6n	0	0	1								
7a	7	6	1								
7n	0	2	0								
8a	18	14	6	1							
8n	3	2	4	2							
9a	41	42	12	1							
9n	8	19	9	0							
10a	123	121	43	9	1						
10n	42	64	31	16	2						
11a	367	384	146	17	1						
11n	185	254	138	22	0						
12a	1,288	1,408	500	100	11	1					
12n	888										
13a	4,878	5,100	2,074	341	23	1					
13n	5,110										
14a	19,536	21,854	8,206	1,556	181	13	1				
14n	27,436										
15a	85,263	92,234	37,222	7,193	653	29	1				
15n	168,030										
16a	379,799	427,079	172,678	33,216	3,885	301	16	1			
16n	1,008,906										
17a	1,769,979	2,005,800	829,904	173,549	19,122	1,129	36	1			
17n	6,283,414										
18a	8,400,285	9,716,848	4,194,015	876,173	105,539	8,428	471	19	1		
18n	39,866,181										
19a	40,619,385	48,184,018	21,207,695	4,749,914	599,433	43,513	1,813	43	1		
19n	253,511,073										
20a	199,631,989	241,210,386	110,915,684	25,644,802	3,368,608	282,898	16,613	708	22	1	
21a	990,623,857	1,228,973,463	581,200,584	141,228,387	19,965,215	1,707,147	89,225	2,770	51	1	
22a	4,976,016,485	6,301,831,944	3,091,592,835	786,648,328	115,822,290	10,710,211	673,344	30,671	1016	25	1
23a	25,182,878,921	32,663,182,521	16,547,260,993	4,388,853,201	689,913,117	67,959,962	4,265,763	169,480	4,054	59	1

11.1   Bibliography

[1] G. Brinkmann and B. McKay, *plantri*, a computer program for generating planar graphs, `https://users.cecs.anu.edu.au/~bdm/plantri/`.

[2] B.A. Burton, *The next 350 million knots*, preprint.

[3] J.A. Calvo, *Knot enumeration through flypes and twisted splices*, J. Knot Theory and Its Ramifications **6** (6) (1997), 785–797.

[4] A. Caudron, *Classification des noeuds et des enlacements*, Prepublication Math. d'Orsay, Orsay, France: Universite Paris-Sud.

[5] J.H. Conway, *An enumeration of knots and links*, Computational Problems in Abstract Algebra (Proc.COnf., Oxford, 1967, Leech, ed.) New York: Pergamon Press (1970), 329–358.

[6] J.C. Cha and C. Livingston, *KnotInfo: Table of Knot Invariants*, `http://www.indiana.edu/~knotinfo`, (June 18, 2018).

[7] M. Culler, N.M. Dunfield, M. Goerner, and J.R. Weeks, *SnapPy*, a computer program for studying the geometry and topology of 3-manifolds, `http://snappy.computop.org`.

[8] H. Doll and J. Hoste, *A tabulation of oriented links*, Math. Comp. **57** No.196 (1991), 747–761.

[9] C.H. Dowker and M.B. Thistlethwaite, *On the classification of knots*, C. R. Math. Rep. Acad. Sci. Canada **4** (1982), no. 2, 129–131.

[10] C.H. Dowker and M.B. Thistlethwaite, *Classification of knot projections*, Topol. Appl. **16** (1983), 19–31.

[11] R. Hartley, *Identifying non-invertible knots*, Topology **22** (1983), 137–145.

[12] J. Hoste, *The enumeration and classification of knots and links*, Handbook of Knot Theory, W. Menasco and M.B. Thistlethwaite, eds., Elsevier (2005) 209–232.

[13] J. Hoste, M.B. Thistlethwaite and J. Weeks, *The first 1,701,936 knots*, Math. Intelligencer **20** (4) (1998), 33–48.

[14] S.V. Jablan, *Geometry of links*, Novi Sad Journal of Mathematics 29.3 (1999), 121–139.

[15] L.H. Kauffman, *State models and the Jones polynomial*, Topology **26** (1987), 395–407.

[16] T.P. Kirkman, *The enumeration, description and construction of knots of fewer than ten crossings*, Trans. Roy. Soc. Edinburgh **32** (1885), 281–309.

[17] T.P. Kirkman, *The 364 unifilar knots of ten crossings enumerated and defined*, Trans. Roy. Soc. Edinburgh **32** (1885), 483–506.

[18] C.N. Little, *On knots, with a census of order ten*, Trans. Connecticut Acad. Sci. **18** (1885), 374–378.

[19] C.N. Little, *Non alternate ± knots of orders eight and nine*, Trans. Roy. Soc. Edinburgh **35** (1889), 663–664.

[20] C.N. Little, *Alternate ± knots of order 11*, Trans. Roy. Soc. Edinburgh **36** (1890), 253–255.

[21] C.N. Little, *Non-alternate ± knots*, Trans. Roy. Soc. Edinburgh **39** (1900), 771–778.

[22] W. Menasco, *Closed incompressible surfaces in alternating knot and link complements*, Topology **23** (1) (1984), 37–44.

[23] W. Menasco and M.B. Thistlethwaite, *The classification of alternating links*, Ann. Math. **138** (1993), 113–171.

[24] K. Murasugi, *The Jones polynomial and classical conjectures in knot theory*, Topology **26** (1987), 187–194.

[25] K. Perko, *On the classification of knots*, Proc. Amer. Math. Soc. **45** (1974), 262-266.

[26] K. Perko, *Primality of certain knots*, Topology Proceedings 7 (1982), 109-118.

[27] S. Rankin, O. Flint and J. Schermann, *Enumerating the prime alternating knots, part I*, Journal of Knot Theory and Its Ramifications, **13**, 57–100 (2004)

[28] S. Rankin, O. Flint and J. Schermann, *Enumerating the prime alternating knots, part II*, Journal of Knot Theory and Its Ramifications, **13**, 101–149 (2004)

[29] S. Rankin and O. Flint, *Enumerating the prime alternating links*, J. Knot Theory Ramifications **13**, 151 (2004)

[30] P.G. Tait. *On knots I, II, III*, Scientific Papers, Vol. I, Cambridge: Cambridge University Press, 273–347.

[31] M.B. Thistlethwaite, *Knot tabulations and related topics*, Aspects of Topology in Memory of Hugh Dowker 1912–1982 (James and Kronheimer, eds.) Cambridge, England: Cambridge Press, London Math., Soc. Lecture Note Series 93 (1985), 1–76.

[32] M.B. Thistlethwaite, *A spanning tree expansion of the Jones polynomial*, Topology **26** (3) (1987), 297–309.

[33] M.B. Thistlethwaite, *Kauffman's polynomial and alternating links*, Topology **27** (3) (1987), 311–318.

[34] H. Trotter, *Non-invertible knots exist*, Topology, **2** (1964), 341–358.

Part III

Tangles

Chapter 12

What Is a Tangle?

Emille Davie Lawrence, University of San Francisco

12.1 Introduction

Anyone who has ever hung Christmas lights or brushed their hair has encountered a tangle: some number of strings that are wound in a frustrating mess. However, in mathematics, a tangle, while similar, can be described quite concretely and plays a central role in the study of knots and links. In the late 1960s, John Conway defined, for nonnegative integers m and n, an (m, n)-*tangle* as an embedding of m circles and n disjoint arcs into the 3-ball, B^3, in such a way that the $2n$ endpoints of the arcs are sent to the boundary S^2. There is an example of a $(1, 2)$-tangle and a $(0, 3)$-tangle shown in Figure 12.1.

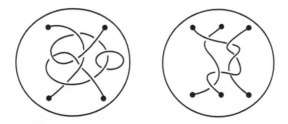

FIGURE 12.1: A $(1, 2)$-tangle and a $(0, 3)$-tangle.

For our purposes here, we will focus on the $(0, 2)$-tangles (referring to them simply as *tangles*) because of the basic symmetry of their endpoints, which keeps them closed under tangle operations. Furthermore, our tangles will be in *general position* having the four endpoints $NE = (0, \frac{1}{\sqrt{2}}, \frac{1}{\sqrt{2}}), SE = (0, \frac{1}{\sqrt{2}}, -\frac{1}{\sqrt{2}}), SW = (0, -\frac{1}{\sqrt{2}}, -\frac{1}{\sqrt{2}})$ and $NW = (0, -\frac{1}{\sqrt{2}}, \frac{1}{\sqrt{2}})$ in the unit ball in $\mathbb{R}^3$. Since all four of these compass points lie in the yz-plane, as we do with knots, we can project the tangle onto the yz-plane with crossing information to obtain what is referred to as a *regular diagram* of the tangle. For simplicity, we will often consider a tangle to be this regular diagram.

Given the choices for each endpoint, the two most basic tangles are the 0-*tangle*, denoted by [0] and the ∞-*tangle*, denoted by [∞]. Each of these tangles has two arcs with no crossings, shown in Figure 12.2. The labels 0 and ∞ give a nod to the slope of the arcs in each tangle. The tangles [0] and [∞] are very similar, and in fact, if we turn our head 90° while looking at one, we see the other.

Hence the question of equivalence arises. What is the notion of equivalence amongst tangles? We say that two tangles T and S are *isotopic* and write $T \sim S$ if there is an orientation-preserving self-homeomorphism $h : (B^3, T) \to (B^3, S)$ that is the identity map on the boundary of B^3. Equivalently, T and S are isotopic if we can transform a diagram of T to a diagram of S through a finite sequence of Reidemeister moves.

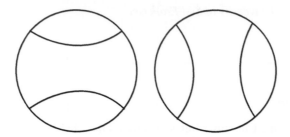

FIGURE 12.2: The 0-tangle and the ∞-tangle.

Right away, we see that the tangles [0] and [∞] are not isotopic because their strands join different pairs of end points. We can also make an infinite family of nonisotopic tangles called the *integer tangles*, denoted by $[n]$, by starting with [0] and twisting the points NE and SE around each other in the boundary of the ball n times. Similarly, the *vertical tangles*, denoted by $\frac{1}{[n]}$, are constructed by starting with [∞] and twisting SE and SW around each other n times. The sign of the integer n in each of these families of tangles gives reference to the sign of the crossings. We use the sign conventions shown below in Figure 12.3.

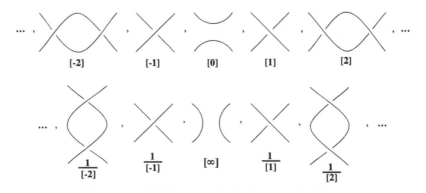

FIGURE 12.3: Integer and vertical tangles.

12.2 Tangle Operations

There are several ways to combine two tangles or to transform one tangle to another. We make the *sum* of two tangles T and S by gluing NE and SE in T to NW and SW in S, and extend the unglued endpoints to the boundary of a "larger" B_3. The resulting tangle is $T + S$. Notice that summing any tangle T and the 0-tangle results in a tangle that is equivalent to T, hence [0] can be regarded as an identity element with respect to this operation. However, summing T with the ∞-tangle could result in an extra component. We can also combine two tangles by the *star product* operation. This operation is performed by placing tangle S underneath tangle T and gluing the NW and NE of S to SW and SE of T to obtain the tangle $T * S$. The tangle $[\infty]$ is an identity element with respect to this operation. The *mirror image* of T, denoted $-T$, is obtained by changing all crossings in T. Taking the mirror image any integer (resp. vertical) tangle creates another integer (resp. vertical) tangle with the same number of twists, but opposite crossings. Thus, $-[n] = [-n]$ and $-\frac{1}{[n]} = \frac{1}{[-n]}$. Finally the *inverse* of T, denoted $\frac{1}{T}$, is obtained by a 90° counterclockwise rotation of T, then taking the mirror image (see Figure 12.4). It is readily seen that $\frac{1}{[0]} = [\infty]$ and $\frac{1}{[\infty]} = [0]$, and in general, the integer tangle $[n]$ is the inverse of the vertical tangle $\frac{1}{[n]}$.

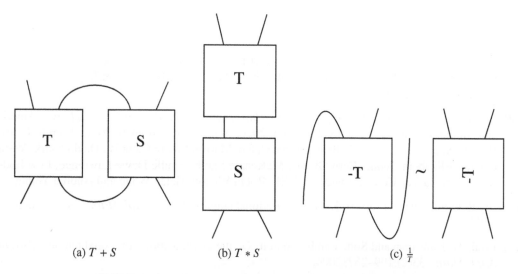

(a) $T + S$ (b) $T * S$ (c) $\frac{1}{T}$

FIGURE 12.4: The sum, star product, and inverse operations.

Intuitively, a tangle is a portion of a knot or link. In fact, every knot or link can be obtained from an (m, n)-tangle. But how? The *numerator* $N(T)$ of a tangle T is obtained by joining the points NE to NW, and SE to SW by simple arcs outside of B_3. The *denominator* $D(T)$ of T is obtained similarly by joining points NE to SE and NW to SW. Figure 12.5 shows how the numerator and denominator of a tangle are formed. Making these closures alleviates the "loose ends" of the tangle. Hence $N(T)$ and $D(T)$ are links.

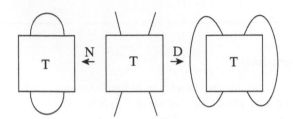

FIGURE 12.5: The numerator and denominator closures of a tangle T.

Every knot or link is the numerator closure of an $(m, 2)$-tangle. It is also the case that two tangles that are not equivalent could have numerators which are equivalent knots. The simplest example of this can be seen with the vertical tangles. Each vertical tangle has numerator closure that is equivalent to the unknot. The general case of this phenomenon will be discussed in Chapter 13 on Rational and non-Rational Tangles. For further information on tangles, see [1], [2], [3], [4], and [5].

12.3   Bibliography

[1] J. H. Conway. An enumeration of knots and links, and some of their algebraic properties. In *Computational Problems in Abstract Algebra (Proc. Conf., Oxford, 1967)*, pages 329–358. Pergamon, Oxford, 1970.

[2] Erica Flapan. *Knots, Molecules, and the Universe: An Introduction to Topology*. American Mathematical Society, Providence, RI, 2016. With contributions by Maia Averett, Lance Bryant, Shea Burns, Jason Callahan, Jorge Calvo, Marion Moore Campisi, David Clark, Vesta Coufal, Elizabeth Denne, Berit Givens, McKenzie Lamb, Emille Davie Lawrence, Lew Ludwig, Cornelia Van Cott, Leonard Van Wyk, Robin Wilson, Helen Wong and Andrea Young.

[3] Jay R. Goldman and Louis H. Kauffman. Rational tangles. *Adv. in Appl. Math.*, 18(3):300–332, 1997.

[4] Louis H. Kauffman and Sofia Lambropoulou. On the classification of rational tangles. *Adv. in Appl. Math.*, 33(2):199–237, 2004.

[5] Kunio Murasugi. *Knot Theory & Its Applications*. Modern Birkhäuser Classics. Birkhäuser Boston, Inc., Boston, MA, 2008. Translated from the 1993 Japanese original by Bohdan Kurpita, Reprint of the 1996 translation [MR1391727].

Chapter 13

Rational and Non-Rational Tangles

Emille Davie Lawrence, University of San Francisco

13.1 Rational Tangles

A special class of $(0, 2)$-tangles can be formed by starting with either the 0-tangle or the ∞-tangle, then adding a sequence of vertical and horizontal twists in an alternating fashion. Any tangle that we can obtain in this manner is called a *rational* tangle. Concretely, here is what we do. Say we start with the 0-tangle. We then add horizontal twists by twisting points NE and SE around each other in B^3 while keeping the other endpoints fixed. Using the sign convention introduced in the previous chapter, these horizontal twists can be encoded by an integer which we call a_1. Next we add some number of vertical twists by twisting points SE and SW around each other while keeping the other endpoints fixed. This gives another integer a_2. Continuing in this way for finitely many iterations, we obtain a rational tangle. If we had instead started with the ∞-tangle, an analogous sequence of steps can be performed only, of course, the first integer would represent some number of vertical twists.

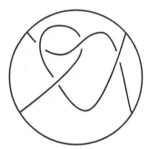

FIGURE 13.1: A rational tangle.

Furthermore given this construction, we can associate to any rational tangle a vector $(a_1, a_2, \ldots, a_n)$. Here, the last entry a_n is where the tangle starts to "unravel" One might wonder how to construct a rational tangle given a vector of integers. How would we know with which tangle to start? To clear up this ambiguity, we adopt the convention that the final entry in this vector represents some number of *horizontal* twists—even if this number is zero. Thus any vector with an odd number of entries must start with the 0-tangle, and any vector with an even number of entries must start with the ∞-tangle. For example, the tangle in Figure 13.1 is represented by the vector $(-2, -1, -1)$. For the sake of convenience, we will denote the tangle represented by the vector $(a_1, a_2, \ldots, a_n)$ by $[a_1, a_2, \ldots, a_n]$. This convention coincides with our notation for horizontal tangles in a previous chapter, and the vertical tangle $\frac{1}{[n]}$ can also be written as $[n, 0]$.

13.2 The Algebra of Rational Tangles

Why do we call these tangles "rational"? It is because to each tangle of this type we can associate a rational number $\frac{p}{q}$. Conway defined the fraction of a rational tangle in [1]. To motivate this idea we must introduce a few more tangle operations. First, observe that any rational tangle can be expressed using the sum and star product operations as:

$$((([a_1] * \frac{1}{[a_2]}) + [a_3]) * \frac{1}{[a_4]}) + \cdots + [a_n],$$

where $a_i \in \mathbb{Z} - \{0\}$ for $i = 1, \ldots, n - 1$, and the tangle $[a_n]$ could be $[0]$. We refer to the expression above as the *standard form* of a rational tangle. For example, the standard form associated to the tangle $[-2, -1, -1]$ in Figure 13.1 is $([-2] * \frac{1}{[-1]}) + [-1]$. Next, we define a tangle isotopy which plays an important role in the theory of rational tangles and in establishing the main result of this chapter, Conway's theorem.

Definition 13.2.1. A *flype* is an isotopy of a tangle applied to a subtangle of the form $[\pm 1] + t$ or $[\pm 1] * t$ shown in Figure 13.2. The endpoints of the subtangle are fixed.

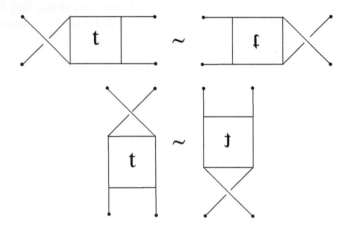

FIGURE 13.2: Flype moves.

The flypes allow us to move a subtangle from right to left or left to right and from top to bottom or bottom to top. The upside-down and backwards letters in Figure 13.2 represent the tangles resulting from a rotation. A *flype* should not be confused with a *flip* which we define as follows.

Definition 13.2.2. A **vertical (resp. horizontal) flip** of a tangle T is a $180°$ rotation of T in its vertical (resp. horizontal) axis.

We will denote the flips of T by T^{vflip} and T^{hflip}. We can readily see that the flype moves give us the isotopies $[\pm 1] + t \sim t^{\text{hflip}} + [\pm 1]$ and $[\pm 1] * t \sim t^{\text{vflip}} * [\pm 1]$. This idea is key in proving the following lemma.

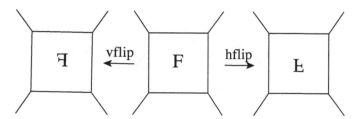

FIGURE 13.3: Vertical and horizontal flips of a tangle F.

Lemma 13.2.3 (Flip Lemma). If T is a rational tangle, then $T \sim T^{\text{hflip}}$ and $T \sim T^{\text{vflip}}$.

The proof of this lemma is by induction on the number of crossings in T and can be found in [2]. As a consequence of the Flip Lemma, if T is a rational tangle,

$$T + [n] \sim [n] + T \quad \text{and} \quad T * \frac{1}{[n]} \sim \frac{1}{[n]} * T.$$

We use flypes and the appropriate flips to show the isotopies above. An example of the first isotopy is shown in Figure 13.4.

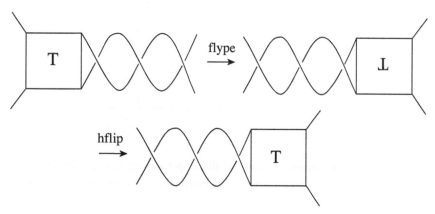

FIGURE 13.4: $T + [3] \sim [3] + T$ whenever $T \sim T^{\text{hflip}}$.

The following lemma gives us an important tangle isotopy.

Lemma 13.2.4. If T is a rational tangle, then T satisfies

$$T * \frac{1}{[n]} = \frac{1}{[n] + \frac{1}{T}}.$$

We now have the necessary tools to motivate the fraction of a rational tangle, and we do so with an example. Consider the tangle $T = [1, 2, -2, 2, 3]$ below in Figure 13.5. The tangle T can be expressed in standard form as

$$((([1] * \frac{1}{[2]}) + [-2]) * \frac{1}{[2]}) + [3].$$

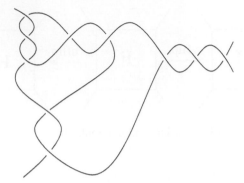

FIGURE 13.5: The tangle $[1, 2, -2, 2, 3]$.

By repeated use of the Flip Lemma and Lemma 13.2.4, we have:

$$((([1] * \tfrac{1}{[2]}) + [-2]) * \tfrac{1}{[2]}) + [3] \sim [3] + ((([1] * \tfrac{1}{[2]}) + [-2]) * \tfrac{1}{[2]})$$

$$\sim [3] + \cfrac{1}{[2] + \cfrac{1}{([1]*\frac{1}{[2]})+[-2]}}$$

$$\sim [3] + \cfrac{1}{[2] + \cfrac{1}{[-2]+([1]*\frac{1}{[2]})}}$$

$$\sim [3] + \cfrac{1}{[2] + \cfrac{1}{[-2]+\cfrac{1}{[2]+\frac{1}{[1]}}}}$$

Thus, after repeated use of Lemma 13.2.4 and the Flip Lemma, we see that the standard form $((([1] * \tfrac{1}{[2]}) + [-2]) * \tfrac{1}{[2]}) + [3]$ transforms to the *continued fraction form* $[3] + \cfrac{1}{[2]+\cfrac{1}{[-2]+\cfrac{1}{[2]+\frac{1}{[1]}}}}$. In

general, the standard form $((([a_1] * \tfrac{1}{[a_2]}) + [a_3]) * \tfrac{1}{[a_4]}) + \cdots + [a_n]$ corresponds to the continued

fraction form $[a_n] + \cfrac{1}{[a_{n-1}]+\cfrac{1}{\cdots+\cfrac{1}{[a_2]+\frac{1}{[a_1]}}}}$. This suggests that to any rational tangle, we can associate an

arithmetic *continued fraction*

$$\frac{p}{q} = a_n + \cfrac{1}{a_{n-1} + \cfrac{1}{\cdots+\cfrac{1}{a_2+\frac{1}{a_1}}}}.$$

For example, the tangle $[-2, -1, -1]$ in Figure 13.1 above has continued fraction $-1 + \cfrac{1}{-1+\frac{1}{-2}} = -\frac{5}{3}$.
It is immediate that the horizontal tangle $[n]$ has fraction n, and the vertical tangle $[n, 0]$ has fraction $\frac{1}{n}$. The ∞-tangle $[0, 0]$ has fraction $0 + \frac{1}{0}$ which we will denote by ∞.

There is more than one way to express a rational number as a continued fraction. In fact, the fraction $-\frac{5}{3}$ above also has the continued fraction expansion $-2 + \frac{1}{3}$. One would certainly hope that

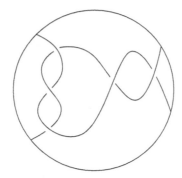

FIGURE 13.6: The tangle $[3, -2]$.

the tangle $[-2, -1, -1]$ shown in Figure 13.1 and the tangle $[3, -2]$ shown in Figure 13.6 are at least related. It turns out that they are, indeed, isotopic tangles.

The isotopy is the result of an R1 type of move involving a second strand followed by an R2 move performed on $[-2, -1, -1]$. The series of pictures in Figure 13.7 shows this isotopy. In fact, the following theorem due to Conway shows that there exists an incredibly nice way to tell if two rational tangles are equivalent.

Theorem 13.2.5 (Conway, 1970). Two rational tangles are isotopic if and only if they have the same continued fractions representing the same rational number.

This is a powerful and surprisingly easy tool to help us determine if two rational tangles are the same or not. As an exercise, show that the tangle $[2, 1, 1, 0]$ and the tangle $[-2, -2, 1]$ are isotopic. A proof of this result can be found in [1].

13.3 Non-rational Tangles

Not all tangles are rational. Recall that any rational tangle can be "unraveled" to a tangle having two straight arcs. Therefore, any tangle which has a knotted strand like the one shown in Figure 13.8 is not rational.

Even the sum of two rational tangles need not be rational. In fact, the sum of two rational tangles is rational only if at least one of them is integral. The sum of two rational tangles can even produce a $(1, 2)$-tangle as in Figure 13.9a. Moreover, we could take additions and multiplications of finitely many rational tangles. The resulting tangle is possibly much more complicated, as in Figure 13.9b.

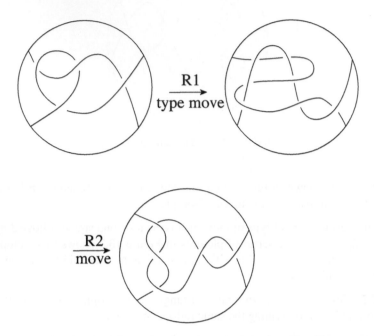

FIGURE 13.7: An isotopy from $[-2, 1, 1]$ to $[3, -2]$.

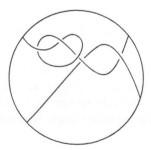

FIGURE 13.8: A non-rational tangle.

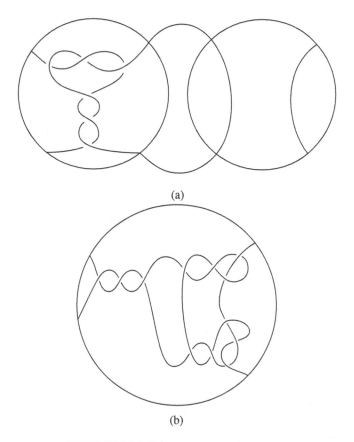

(a)

(b)

FIGURE 13.9: Two non-rational tangles.

13.4 Bibliography

[1] J. H. Conway. An enumeration of knots and links, and some of their algebraic properties. In *Computational Problems in Abstract Algebra (Proc. Conf., Oxford, 1967)*, pages 329–358. Pergamon, Oxford, 1970.

[2] Louis H. Kauffman and Sofia Lambropoulou. On the classification of rational tangles. *Adv. in Appl. Math.*, 33(2):199–237, 2004.

Chapter 14

Persistent Invariants of Tangles

Daniel S. Silver, University of South Alabama

Susan G. Williams, University of South Alabama

14.1 Krebes's Theorem

Imagine a very large and complicated knot diagram in the plane, a diagram with a million crossings. How might we decide whether the knot that it represents is really knotted? Knotting is a holistic phenomenon, one that generally requires consideration of an entire knot diagram, a daunting venture in this case.

If we are fortunate, we might see a "local knot," a nontrivial 2-tangle in our large diagram. A classical theorem of H. Schubert [3], proved in 1949, then comes to our rescue. It assures us that any knot k containing a local knot is itself knotted.

More specifically, Schubert proved that the knot genus of k is at least as great as that of the closure of any nontrivial 2-tangle summand of k. In this sense, the knot genus of the 2-tangle closure is a "persistent invariant."

In his doctoral dissertation of 1997, David Krebes proved a beautiful extension of Schubert's theorem for 4-tangles in knot or link diagrams. In order to state Krebes's theorem we first need to review some terminology.

Recall that a *2n-tangle*, $n \geq 1$, consists of n arcs and a finite number of simple closed curves, pairwise disjoint and properly embedded in the 3-ball. We will require that endpoints of the arcs meet the boundary of the ball along a great circle in $2n$ specified points. Two $2n$-tangles are equivalent if there is an isotopy of the ball, fixing points of the boundary, taking one $2n$-tangle to the other.

We can represent a $2n$-tangle t by a *2n-tangle diagram*, a projection of the tangle to the 2-disk D. We require that the $2n$ endpoints of the arcs lie on the disk boundary. (After isotopy, we may assume that n endpoints are above and n below a horizontal line through the center of the disk.) Such a disk is discovered in a diagram of a link whenever a circle intersects the link diagram in general position. In this case we will say that *t embeds in the link.*

Any 4-tangle diagram can be closed to a link $n(t)$ or $d(t)$ by connecting free ends in pairs of disjoint embedded arcs outside the diagram, as in Figure 14.1. These are referred to as the *numerator* and *denominator* closures of t. A general $2n$-tangle diagram has C_n closures, where C_n is the nth Catalan number $(2n)!/n!(n + 1)!$.

While Schubert's theorem, mentioned above, involves genus, Krebes's theorem considers the *determinant* of a link ℓ. This integer-valued link invariant, denoted here by $\det(\ell)$, can be defined

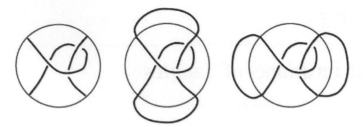

FIGURE 14.1: 4-tangle t (left); numerator closure $n(t)$ (center); denominator closure $d(t)$ (right).

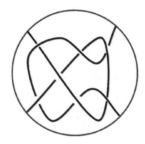

FIGURE 14.2: 4-tangle t with $\det(n(t)) = 5$ and $\det(d(t)) = 10$.

in various equivalent ways. It is the absolute value of the 1-variable Alexander polynomial of ℓ evaluated at -1, and also the absolute value of the Kauffman bracket polynomial evaluated at a primitive eighth root of unity. It is the order of the torsion subgroup of $H_1(M_2(\ell); \mathbb{Z})$, where $M_2(\ell)$ is the 2-fold cover of $\mathbb{S}^3$ branched over each component of ℓ. Finally, the determinant of ℓ can be defined in terms of the spanning trees of a Tait graph corresponding to a diagram of the link, with edges weighted ± 1 according to crossing types (see [3] for details).

Theorem 14.1.1. (Krebes's Theorem) [6, 7] If a 4-tangle t embeds in a link ℓ, then any divisor of both $\det(n(t))$ and $\det(d(t))$ divides $\det(\ell)$.

If this condition is not met, then it is an *obstruction* that prevents a given tangle from being embeddable in a given link. Hence the title of Krebes's original paper.

Corollary 14.1.2. A 4-tangle t cannot embed in the unknot unless $\det(n(t))$ and $\det(d(t))$ are relatively prime.

The corollary is illustrated in Figure 14.2 with a 4-tangle such that $\det(n(t)) = 5$ and $\det(d(t)) = 10$. Consequently, if t embeds in a link ℓ, then $\det(\ell)$ is divisible by 5.

Krebes proved Theorem 14.1.1 by a diagrammatic approach using the Kauffman bracket polynomial and skein theory. His proof was elementary but somewhat involved. Later D. Ruberman gave a short but non-elementary argument by studying the branched 2-fold cover $M_2(\ell)$ mentioned above.

Connections among knots, tangles and graphs abound. (An early, influential paper [4] by J. R. Goldman and L. H. Kauffman discusses all three in the context of electrical networks.) For an elementary proof of Theorem 14.1.1, using spanning trees of a Tait graph, see [11]. It is the shortest proof currently in the literature.

14.2   Generalizations and Extensions

Krebes's theorem admits generalization for $2n$-tangles, with n greater than 2. Moreover, it can be applied in the category of virtual links and tangles.

A *virtual $2n$-tangle* is defined in just the way that a virtual link is defined: it is an equivalence class by generalized Reidemeister moves of a *virtual $2n$-tangle diagram*, a $2n$-tangle diagram in which both classical and virtual crossings are allowed. Endpoints of the diagram should remain fixed when moves are applied.

A virtual $2n$-tangle determines an element of the *virtual Temperley-Lieb algebra* $\mathcal{VTL}_n$, an algebra over the ring $\Lambda = \mathbb{Z}[A, A^{-1}]$ of Laurent polynomials in variable A. It is the free algebra over Λ generated by diagrams D of virtual $2n$-tangles without classical crossings modulo generalized Reidemeister moves and the following two skein relations.

$$(i) \quad \times = A \,)(+ A^{-1} \asymp$$

$$(ii) \quad D \cup \bigcirc = -(A^2 + A^{-2})D.$$

The algebra $\mathcal{VTL}_n$ has a basis $\{U_i\}$ of size $(2n)!/2^n n!$, and any virtual $2n$-tangle diagram determines an element $\sum_i \alpha_i U_i \in \mathcal{VTL}_n$. The coefficients are found by repeated application of relations (i) and (ii). Diagrams that differ by generalized Reidemeister moves other than the first (adding or removing a classical nugatory crossing) yield the same element, while the first move multiplies the sum by $-A^{\pm 3}$.

Denote by $\langle \ell \rangle$ the Kauffman bracket polynomial of a virtual link ℓ. It assigns $-A^2 - A^{-2}$ to an unknot and is well defined up to multiplication by $-A^{\pm 3}$. (See [5].) Finally, let $^- : \Lambda \to \overline{\Lambda}$ denote the projection of Λ onto a quotient ring $\overline{\Lambda}$.

Theorem 14.2.1. [8] Suppose that t is a virtual $2n$-tangle embedded in a virtual link ℓ, and $\sum_i \alpha_i U_i \in \mathcal{VTL}_n$ is associated to some diagram of t. If $\beta \in \overline{\Lambda}$ divides each coefficient $\bar{\alpha}_i$, then β divides $\overline{\langle \ell \rangle}$.

Theorem 14.1.1 is the special case of Theorem 14.2.1 in which $n = 2$, the tangle and link are classical, and $\bar{A}$ is a primitive eighth root of unity.

Following the theme of [10], J. Przytycki and the authors pursued a topological extension of Theorem 14.1.1, replacing the exterior of a classical link and embedded $2n$-tangle with a more general compact 3-manifold and submanifold. Let M be a compact, connected, oriented 3-manifold with boundary and χ an epimorphism from $H_1(M; \mathbb{Z})$ to a free abelian group Π. The epimorphism determines an abelian covering space $\tilde{M}$, and we can consider the 0th elementary divisor $\Delta_0(\tilde{M})$ of $H_1(\tilde{M}; \mathbb{Z})$. Two additional homological invariants $\beta, \tau \in \mathbb{Z}\Pi$ are defined, and Theorem 2.2 of [9] states that if M embeds in another 3-manifold N such that χ extends over $H_1 N$, then the product $\beta\tau$ divides $\Delta_0(\tilde{N})$.

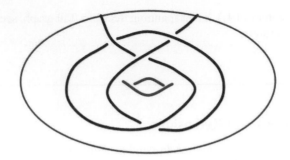

FIGURE 14.3: Does the genus-1 tangle embed in the unknot?

Consider the special case in which M is the exterior of a $2n$-tangle t and N is the exterior of a link ℓ in which t embeds, and assume that Π is infinite cyclic and oriented meridians of t and ℓ map to a generator. Then $\beta\tau$ is a polynomial in a single variable that must divide the 1-variable Alexander polynomial of ℓ. We recover Theorem 14.1.1 by evaluating the variable at -1.

A third direction for extending Theorem 14.1.1 replaces the Kauffman bracket polynomial and virtual Temperley-Lieb algebra with the Homflypt polynomial or the Kauffman 2-variable polynomial and corresponding skein algebras. Details are found in [9].

Yet a fourth approach applies quandles and cocycles to obtain persistent invariants for tangles. This is pursued for 4-tangles in [2].

Finally, a fifth way to extend Theorem 14.1.1 replaces the 3-ball of a $2n$-tangle with other compact bounded submanifolds of $\mathbb{S}^3$. In [1] S. Abernathy defines a *genus-1 tangle* to be a 1-manifold properly embedded in a solid torus with exactly two boundary points. Such a tangle t is said to embed in a link ℓ if ℓ is the union of t and another 1-manifold that meets the solid torus only at the endpoints of the tangle. Homological methods as well as skein theory are used to obtain obstructions for embedding genus-1 tangles in links.

An example, due to Krebes, appears in Figure 14.3. Krebes asked whether it embeds in an unknot. Abernathy's results give restrictions on how such an unknot could intersect the torus complement, but Krebes's question remains open.

Acknowledgments

The authors are grateful for the support of the Simons Foundation.

14.3 Bibliography

[1] S. M. Abernathy, On Krebes's tangle, *Topology Appl.* **160** (2013), 1379–1383.

[2] K. Ameur, M. Elhamdadi, T. Rose, M. Saito and C. Smudde, Tangle embeddings and quandle cocycle invariants, *Experimental Mathematics* **17** (2008), 487–497.

[3] G. Burde, H. Zieschang, *Knots*, Second Edition, Walter de Gruyter and Co., Berlin, 2003.

[4] J. R. Goldman and L. H. Kauffman, Knots, tangles, and electrical networks, *Adv. Appl. Math.* **14** (1993), 267–306.

[5] L. H. Kauffman, Virtual knot theory, *European J. Combin.* **20** (1999), 663–690.

[6] D. A. Krebes, An obstruction to embedding 4-tangles in links. Thesis (Ph.D.) University of Illinois at Chicago. 1997. 48 pp.

[7] D. A. Krebes, An obstruction to embedding 4-tangles in links, *J. Knot Theory and Its Ramifications* **8** (1999), 321–352.

[8] D. A. Krebes, D. S. Silver and S. G. Williams, Persistent invariants of tangles, *J. Knot Theory and Its Ramifications* **9** (2000), 471–477.

[9] J. H. Przytycki, D. S. Silver and S. G. Williams, 3-manifolds, tangles and persistent invariants, *Math. Proc. Camb. Phil. Soc.* **139** (2005), 291–306.

[10] D. Ruberman, Embedding tangles in links, *J. Knot Theory and Its Ramifications* **9** (2000), 523–530.

[11] D. S. Silver and S. G. Williams, Tangles and links: A view with trees, *J. Knot Theory and Its Ramifications* **12** (2018), 1850061, 7 pp.

Part IV

Types of Knots

Chapter 15

Torus Knots

Jason Callahan, St. Edwards University

A torus knot is a knot that lies without self-crossings on an unknotted torus. In other words, a torus knot can be drawn on a bagel without crossing itself. Let p be the number of times the knot goes around the bagel "through" its hole and q the number of times it goes around the hole. On the torus, p is the number of times the knot wraps around in the meridional direction ("the short way around") and q the number of times it wraps around in the longitudinal direction ("the long way around"). Negative values of p or q reverse the direction the knot wraps around; for simplicity we assume $p, q > 0$.

For a curve drawn in this manner to be a knot (i.e., have only one component), p and q must be relatively prime, and the knot is called a (p, q)-torus knot. If p and q are not relatively prime, then the result is a link with $\gcd(p, q)$ components and is called a (p, q)-torus link.

A (p, q)-torus knot is trivial (i.e., equivalent to the unknot) if and only if either p or q is ± 1 (see, for instance, Theorem 2.2.2 in [13]). Figure 15.1 shows $(0, 1)$- and $(1, 0)$-torus knots, both of which are clearly trivial.

FIGURE 15.1: The $(0, 1)$- and $(1, 0)$-torus knots are trivial.

The simplest nontrivial torus knot is the trefoil knot on the left in Figure 15.2. If traced clockwise on a bagel (represented by dashed circles in Figure 15.2) so that the knot wraps around to the bagel's underside at the outer circle (so the knot's undercrossings are on the bagel's underside) and to the topside at the bagel's hole or inner circle (so the knot's overcrossings are on the bagel's topside), the trefoil knot goes around the bagel through its hole thrice and around its hole twice. Thus, on the torus, the trefoil knot wraps around thrice in the meridional direction and twice in the longitudinal direction, so it is a $(3, 2)$-torus knot.

Similarly, a (p, q)-torus knot such as the $(6, 5)$-knot in Figure 15.3 wraps to the underside of the torus and back up through its hole p times, and each time it wraps onto the topside it crosses over $q - 1$ of its strands on the underside before wrapping back to the underside (cf. Section 5.1 of [3]). Thus, a (p, q)-torus knot has a bridge number of at most p, crossing number at most $p(q - 1)$, a closed braid representation with braid word $(\sigma_1 \sigma_2 \cdots \sigma_{q-1})^p$, braid index at most q, and, applying

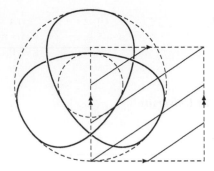

FIGURE 15.2: The trefoil knot is a $(3, 2)$-torus knot.

Seifert's algorithm, a Seifert surface with genus $(p-1)(q-1)/2$, which is minimal by [12] and hence the genus of a (p, q)-torus knot (cf. Theorem 7.5.2 of [19]). Using gauge theory, [14] shows that the slice genus and unknotting number of a (p, q)-torus knot are also $(p-1)(q-1)/2$ (cf. Theorem 7.5.5 of [19]).

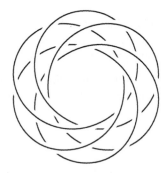

FIGURE 15.3: A $(6, 5)$-torus knot.

The trefoil knot can also be seen on the torus represented by the dashed square with opposite edges identified on the right in Figure 15.2 (cf. Section 3.1 of [16]). If the left and right edges are identified first, followed by the top and bottom, then the trefoil knot wraps around thrice in the meridional direction and twice in the longitudinal direction. Both directions are reversed depending on whether the square is "bent" inward or outward to identify opposite edges, so the trefoil knot is equivalent to both a $(3, 2)$- and a $(-3, -2)$-torus knot. If instead the top and bottom edges are identified first, followed by the left and right, then the trefoil knot wraps around twice in the meridional direction and thrice in the longitudinal direction. Again, both directions are reversed depending on how the square is "bent" to identify opposite edges, so the trefoil knot is also equivalent to both a $(2, 3)$- and a $(-2, -3)$-torus knot.

Such construction and equivalence holds for all torus knots: every torus knot is equivalent to a (p, q)-torus knot for some p and q (see, for instance, Lemma 1 in [26]), and a nontrivial (p, q)-torus knot is equivalent only to the (q, p)-, $(-p, -q)$-, and $(-q, -p)$-torus knots (see, for instance, Theorem 3.29 in [5] or Theorem 2.2.2 in [13]). The $(-p, -q)$-torus knot is the (p, q)-torus knot with reverse orientation, so their equivalence means torus knots are invertible. The $(-p, q)$-torus knot is the mirror image of the (p, q)-torus knot (see, for instance, Proposition 3.27 in [5] or Proposition 7.2.1 in [19]), so their inequivalence means torus knots are chiral.

That torus knots are chiral and invertible was established in [21], which also showed that torus knots have symmetry group D_1 (cf. [7] and Exercise 10.6.4 of [13]). Moreover, torus knots are prime (see, for instance, Corollary 7.11 in [5] or Theorem 5 in [26]) and periodic: the periods of a (p, q)-torus knot are the divisors of p and q (see, for instance, Proposition 14.27 of [5] or Exercise 10.1.9 of [13]).

Since a (q, p)-torus knot has a bridge number of at most q and braid index at most p, the bridge number and braid index of an equivalent (p, q)-torus knot are at most $\min\{p, q\}$ and are in fact both equal to this minimum ([22], Theorem 7.5.3 of [19], and Proposition 10.4.2 of [19]). Likewise, since a (q, p)-torus knot has a crossing number of at most $q(p - 1)$, the crossing number of a (p, q)-torus knot is at most $\min\{p(q - 1), q(p - 1)\}$ and is in fact equal to this minimum by Proposition 7.5 of [18] (cf. Theorem 7.5.4 of [19]). Thus, as listed in [7], up to equivalence, twelve (p, q)-torus knots have sixteen or fewer crossings: $(3, 2)$, $(5, 2)$, $(7, 2)$, $(9, 2)$, $(11, 2)$, $(13, 2)$, $(15, 2)$, $(4, 3)$, $(5, 3)$, $(7, 3)$, $(8, 3)$, and $(5, 4)$.

A famous result of W. Thurston states that every knot in S^3 is either a hyperbolic, satellite, or torus knot (Corollary 2.5 of [27]; see also [7] and Section 5.3 of [3]). To find the fundamental group of the complement of a (p, q)-torus knot $K_{p,q}$ in S^3 (i.e., its knot group), consider $K_{p,q}$ on the standard torus T in S^3 and decompose $S^3 \setminus K_{p,q}$ into two solid tori, U and V, bounded by $T \setminus K_{p,q}$ with intersection an annulus whose generator equals x^p in $\pi_1(U) = \langle x \rangle$ and y^q in $\pi_1(V) = \langle y \rangle$. Then, by the Seifert–van Kampen Theorem,

$$\pi_1(S^3 \setminus K_{p,q}) = \langle x, y \mid x^p = y^q \rangle$$

(see, for instance, Section 3.C of [20], Proposition 6.1.16 of [13], or Proposition 3.28 of [5]). The center of $\pi_1(S^3 \setminus K_{p,q})$ is $\langle x^p \rangle = \langle y^q \rangle$, and torus knots are the only knots whose groups have nontrivial center (Theorem 6.1 in [5]). From this presentation of its group, the Alexander polynomial of $K_{p,q}$ is computed

$$\Delta_{p,q}(t) = \frac{(t^{pq} - 1)(t - 1)}{(t^p - 1)(t^q - 1)}$$

as in Example 9.15 of [5] or Exercise 7.D.8 of [20]. Using Hecke algebras, Proposition 11.9 of [11] finds the Jones polynomial of a (p, q)-torus knot to be

$$V(t) = \frac{t^{(p-1)(q-1)/2}}{1 - t^2}(1 - t^{p+1} - t^{q+1} + t^{p+q}).$$

The commutator subgroup of a torus knot group is a free group of rank twice the knot's genus (Corollary 4.11 of [5]), so the complement of a torus knot in S^3 fibres locally trivially over S^1 with Seifert surfaces of the knot's genus by Theorem 5.6 of [24] (cf. Theorem 5.1 of [5]). Thus, torus knots are fibred knots. By [17] (cf. Remark 9.H.12 of [20]), the only 3-manifolds that result from Dehn surgery on torus knots in S^3 are lens spaces, connected sums of two lens spaces, and Seifert-fibred manifolds with three exceptional fibres. By [23] (cf. Theorem 6.3 of [5]), any fibre of a Seifert fibration of S^3 is a torus knot or the trivial knot, and exceptional fibres are always unknotted.

The $(2, q)$-torus knots are alternating, and the $(3, 4)$- and $(3, 5)$-torus knots are almost alternating: one crossing change yields an alternating knot. In [4], it is proved that a prime almost alternating knot is either a hyperbolic or torus knot and conjectured that the only almost alternating torus knots are the $(3, 4)$- and $(3, 5)$-torus knots. This was proved in [25] using Khovanov homology and in [1] by showing that a knot's dealternating number (the minimum number of crossing changes needed to yield an alternating knot) is at least half the difference of its Rasmussen s-invariant and signature (signatures of torus knots were found in [6], cf. Theorem 7.5.1 of [19], and are nonzero, so torus knots are not slice knots).

Thus, the $(3, 4)$- and $(3, 5)$-torus knots are the only torus knots with dealternating number one. The dealternating number and Turaev genus of $(3, 3n + i)$-torus knots for $i = 1, 2$ are n by [2] and have been determined for torus knots with five or fewer strands either exactly or up to an error of at most two in [9] using knot Floer homology. Finding the dealternating number and Turaev genus for other torus knots remains open.

Several classes of torus knots are also among the few for which the stick number (the least number of straight sticks needed to form a knot) has been found. In [15], the superbridge number and geometric degree of a knot are defined and found to be $\min(q, 2p)$ and $\min(2q, 4p)$, respectively, for a (p, q)-torus knot with $p < q$. Using the superbridge number, [8] shows the stick number of a (p, q)-torus knot is $2q$ if $p < q \leq 2p$, and [10] shows it is $4p$ if $2p < q \leq 3p$ using an algorithm that produces polygonal cable links. Finding the stick number for other torus knots remains open.

15.1   Bibliography

[1] Tetsuya Abe. An estimation of the alternation number of a torus knot. *J. Knot Theory Ramifications*, 18(3):363–379, 2009.

[2] Tetsuya Abe and Kengo Kishimoto. The dealternating number and the alternation number of a closed 3-braid. *J. Knot Theory Ramifications*, 19(9):1157–1181, 2010.

[3] Colin C. Adams. *The Knot Book*. American Mathematical Society, Providence, RI, 2004.

[4] Colin C. Adams, Jeffrey F. Brock, John Bugbee, Timothy D. Comar, Keith A. Faigin, Amy M. Huston, Anne M. Joseph, and David Pesikoff. Almost alternating links. *Topology Appl.*, 46(2):151–165, 1992.

[5] Gerhard Burde and Heiner Zieschang. *Knots*, volume 5. Walter de Gruyter & Co., Berlin, 2003.

[6] C. McA. Gordon, R. A. Litherland, and K. Murasugi. Signatures of covering links. *Canad. J. Math.*, 33(2):381–394, 1981.

[7] Jim Hoste, Morwen Thistlethwaite, and Jeff Weeks. The first 1,701,936 knots. *Math. Intelligencer*, 20(4):33–48, 1998.

[8] Gyo Taek Jin. Polygon indices and superbridge indices of torus knots and links. *J. Knot Theory Ramifications*, 6(2):281–289, 1997.

[9] Kaitian Jin, Adam M. Lowrance, Eli Polston, and Zhen. The turaev genus of torus knots. *J. Knot Theory Ramifications*, 26(14):1750095, 28, 2017.

[10] Maribeth Johnson, Stacy Nicole Mills, and Rolland Trapp. Stick and ramsey numbers of torus links. *J. Knot Theory Ramifications*, 22(7):1350027, 18, 2013.

[11] V. F. R. Jones. Hecke algebra representations of braid groups and link polynomials. *Ann. of Math. (2)*, 126(2):335–388, 1987.

[12] Louis H. Kauffman. Combinatorics and knot theory. In *Low-Dimensional Topology (San Francisco, Calif., 1981)*, volume 20, pages 181–200. Amer. Math. Soc., Providence, RI, 1983.

[13] Akio Kawauchi. *A Survey of Knot Theory*. Birkhäuser Verlag, Basel, 1996.

[14] P. B. Kronheimer and T. S. Mrowka. Gauge theory for embedded surfaces. i. *Topology*, 32(4):773–826, 1993.

[15] Nicolaas H. Kuiper. A new knot invariant. *Math. Ann.*, 278(1-4):193–209, 1987.

[16] Vassily Manturov. *Knot Theory*. Chapman & Hall/CRC, Boca Raton, FL, 2004.

[17] Louise Moser. Elementary surgery along a torus knot. *Pacific J. Math.*, 38:737–745, 1971.

[18] Kunio Murasugi. On the braid index of alternating links. *Trans. Amer. Math. Soc.*, 326(1):237–260, 1991.

[19] Kunio Murasugi. *Knot Theory and Its Applications*. Birkhäuser Boston, Inc., Boston, MA, 1996.

[20] Dale Rolfsen. *Knots and Links*, volume 7. Publish or Perish, Inc., Houston, TX, 1990.

[21] Otto Schreier. Über die Gruppen $a^a b^b = 1$. *Abh. Math. Sem. Univ. Hamburg*, 3(1):167–169, 1924.

[22] Horst Schubert. Über eine numerische Knoteninvariante. *Math. Z.*, 61:245–288, 1954.

[23] H. Seifert. Topologie dreidimensionaler gefaserter Räume. *Acta Math.*, 60(1):147–238, 1933.

[24] John Stallings. On fibering certain 3-manifolds. In *Topology of 3-Manifolds and Related Topics (Proc. The Univ. of Georgia Institute, 1961)*, pages 95–100. Prentice-Hall, Englewood Cliffs, N.J., 1962.

[25] Marko Stoˇsić. On conjectures about positive braid knots and almost alternating torus knots. *J. Knot Theory Ramifications*, 19(11):1471–1486, 2010.

[26] Michael C. Sullivan. Knot factoring. *Amer. Math. Monthly*, 107(4):297–315, 2000.

[27] William P. Thurston. Three-dimensional manifolds, kleinian groups and hyperbolic geometry. *Bull. Amer. Math. Soc. (N.S.)*, 6(3):357–381, 1982.

Chapter 16

Rational Knots and Their Generalizations

Robin T. Wilson, California State Polytechnic University, Pomona

16.1 Rational Tangles and Continued Fractions

Rational knots and links have been recognized as interesting objects of study for some time. They are simple to construct, and they exist in nature. For instance, biologists have discovered them in living organisms as knotted and linked DNA molecules [6]. The association of the term rational with this class of knots is attributed to John Conway in 1967 [4]. Conway took advantage of a clever enumeration scheme for the class of tangles that he called "rational" because they can be completely classified using the non-zero rational numbers [3], [5], [12]. A complete classification of rational knots was first given by Schubert [15] in 1952, however, a simpler combinatorial proof of Conway's classification of rational tangles was given by Goldman and Kauffman in 1996 [7] and of Schubert's classification of rational knots by Kauffman and Lambropoulou in 2003 [9].

FIGURE 16.1: A rational tangle and a rational knot.

A **rational tangle** is a pair of arcs T properly embedded in a 3-dimensional ball B so that the endpoints of each arc lie in the boundary sphere of the 3-ball, and so that there is a homeomorphism $h : (B, T) \to (D^2 \times I, \{x, y\} \times I)$. While this homeomorphism is allowed to move the boundary of the 3-ball, as a set the four compass points $\{NW, NE, SW, SE\}$ must remain fixed. We denote the pair (B, T), and will often use the abbreviation T since all of the tangles we will study are situated in a 3-ball in the same way.

Essentially, this definition says that a rational tangle is any 2-string tangle that can be turned into one of the two tangles in Figure 16.2 by twisting the arcs around each other while keeping the endpoints of the arcs in the boundary of the ball. As a consequence, a rational tangle can be obtained from a finite sequence of vertical and horizontal twists applied consecutively to adjacent endpoints of the 0 tangle or the ∞ tangle. Start with the ∞ tangle and rotate the southern hemisphere of the boundary sphere along the equator, keeping the northern hemisphere fixed and permuting SW and SE results in a **vertical twist**. Similarly, a **horizontal twist** is obtained by rotating the eastern hemisphere while keeping the western hemisphere fixed and transposing the NW and NE endpoints. A twist is **positive** if the over-strands have positive slope, and **negative** otherwise.

The two simplest rational tangles shown in Figure 16.2 are denoted by the 0 tangle, the ∞ tangle. The next simplest rational tangles are the $+1$ tangle, and the -1 tangle, which can be obtained by performing one positive or one negative horizontal twist, respectively, to the 0-tangle. An n-tangle consists of either n vertical or horizontal twists. See Figure 16.3.

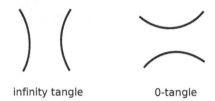

infinity tangle   0-tangle

FIGURE 16.2: The ∞ tangle and 0 tangles.

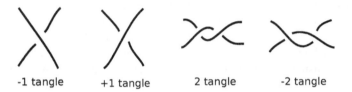

-1 tangle   +1 tangle   2 tangle   -2 tangle

FIGURE 16.3: Some elementary tangles.

This characterization implies that a rational tangle is completely determined by how we perform the finite alternating sequence of vertical and horizontal twists to either the 0 tangle or the ∞ tangle. We can enumerate this list of vertical and horizontal twists as a finite sequence of integers and denote the rational tangle by $T(a_1, a_2, \ldots, a_n)$ as in Figure 16.4. We can ensure that the last twist a_n is a horizontal twist by assuming that if n is odd we begin by performing a horizontal twist to the 0 tangle, and if n is even we begin by performing a vertical twist to the ∞ tangle. This is shown in Figure 16.5 where each box represents a_i half twists.

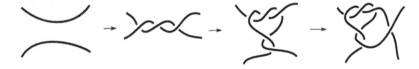

FIGURE 16.4: The rational tangle $T(3, -2, 1)$.

If a projection of a rational tangle shows the sequence of vertical and horizontal twists, we say that the tangle is in **twist form**. From a projection of a rational tangle in twist form we can determine the finite sequence of vertical and horizontal twists used to make the rational tangle. Note that we have the requirement that a_n denotes some number of horizontal twists, which allows for the possibility that $a_n = 0$. However, in general we may assume that all a_i are non-zero since $T(a_1, a_2, \ldots, 0, \ldots, a_n) = T(a_1, a_2, \ldots, a_n)$.

A **continued fraction** is a way of expressing one number as the sum of an integer and the reciprocal of another number, resulting in an expression of the following form:

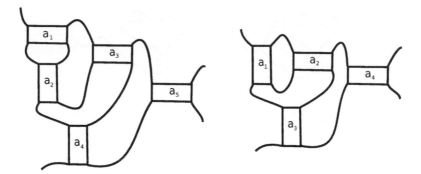

FIGURE 16.5: Rational tangles in twist form with odd and even twists.

$$\frac{337}{104} = 3 + \cfrac{1}{4 + \cfrac{1}{6 + \cfrac{1}{4}}}$$

It is a well-known consequence of the Euclidean algorithm that every rational number p/q has a finite continued fraction expansion and only rational numbers have a finite continued fraction expansion. Thus, every rational tangle $T(a_1, \ldots, a_n)$ corresponds to a rational number of the form

$$\frac{p}{q} = a_n + \cfrac{1}{a_{n-1} + \cfrac{1}{a_{n-2} + \cfrac{1}{\ddots\, a_2 + \cfrac{1}{a_1}}}},$$

denoted by $[a_1, a_2, \ldots, a_n]$, where all a_i are integers. We will assume all $a_i \neq 0$ except possibly a_n.

Theorem 16.1.1 (Conway [4]). Two rational tangles $T(a_1, a_2, \ldots, a_n)$ and $T(b_1, b_2, \ldots, b_m)$ are equivalent if and only if the respective fractions $\frac{p}{q}$ and $\frac{p'}{q'}$ are equal.

While Conway stated this result without proof, there are several well-written expositions of the proof of this theorem in the literature. The first proofs of this theorem used the theory of branched surfaces and the classification of lens spaces [4], [5], [3], [5], [11]. In 1997 Goldman and Kauffman [7] gave the first combinatorial proof of this theorem without the machinery from 3-manifold topology involved in the previous proofs that had previously appeared. Also see [10].

FIGURE 16.6: $T(3, 2, 2)$ and $T(4, -2, 3)$ have the same fraction.

16.2   Rational Knots and Their Classification

Due to the symmetry of the endpoints, there are closing operations that we can perform on a 2-tangle T that result in either a knot or a link. One closing operation is obtained by connecting the NW and NE end points with and arc and then connecting the SW and SE endpoints with an arc giving a knot or link called the **numerator** of T, denoted $N(T)$. Similarly, connecting the SW and SE endpoints, and the NW and NE endpoints gives a knot or link called the **denominator** of T, denoted $D(T)$. The closure of a rational tangle (numerator or denominator) is called a **rational knot** or **rational link**. In what follows we will use the term "link" to refer to both knots and links.

FIGURE 16.7: The rational knot 5_2 is $N(T(3, 2))$ and has fraction $7/3$.

Rational knots and links have been expressed in a handful of different "standard forms." Each arc of a rational tangle (B, T) can be carried via isotopy to an arc in the boundary of the 3-ball in which it is embedded, since this is clearly true for the 0 and ∞ tangles, and adding twists to these two tangles, one after the other, does not chance this fact. We can therefore arrange so that arcs of a rational tangle T lie on ∂B, giving what is sometimes called the **pillowcase** form of the tangle. It also turns out that every rational knot is a 2-bridge knot, meaning it possesses a projection in a place with only two local maxima in some direction (See Chapter 27). Since a rational knot can be decomposed along a 4-punctured 2-sphere S into a rational tangle T and the 0 tangle, we can readily see that each rational tangle is a 2-bridge knot by taking S to be the xy-plane plus infinity and placing the 0-tangle above the plane. The tangle T can then be isotoped into the plane from below, with the arcs winding around each other as in Figure 16.9. This projection is known as **Schubert's normal form**.

While two rational tangles are equivalent if and only if their continued fractions are equal, the theory of rational knots is a bit more subtle. This is because we can find rational tangles that are distinct, but that become equivalent as knots when we take their closure since the endpoints are no

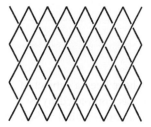

FIGURE 16.8: The rational knot 5_2 in pillow form.

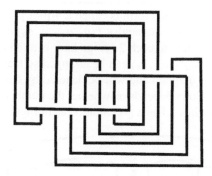

FIGURE 16.9: The rational knot 5_2 in Schubert's normal form.

longer fixed and the isotopies aren't restricted to taking place within the interior of a 3-ball. For example, $T(3, 1, 1)$ has fraction $\frac{7}{4}$ and $T(3, 2)$ has fraction $\frac{7}{2}$, however $N(T(3, 1, 1))$ and $N(T(3, 2))$ are both the knot 5_2. Thus Conway's theorem does not give a classification of rational links.

Theorem 16.2.1 (Conway's Theorem for Rational Links). *Let K_1 and K_2 be rational knots or links. Then K_1 and K_2 are equivalent if their corresponding continued fractions are equal.*

When given a tangle of the form $T(a_1, a_2, \ldots, a_n)$, notice that if all a_i have the same sign then the tangle is alternating. In fact, for any rational number $\frac{p}{q}$, it turns out that we can find a continued fraction expansion in which all of the integer coefficients have the same sign. It follows that given any rational tangle, we can always arrange $T(a_1, a_2, \ldots, a_n)$ so that the twists a_i have the same sign. Thus rational knots are alternating.

Our notation for rational tangles gives us a convenient way to denote rational knots. The knot or link obtained from the numerator/denominator of the tangle $T(a_1, a_2, \ldots, a_n)$ will be denoted by $C(a_1, a_2, \ldots, a_n)$. We call this Conway's notation for rational knots and links. For example, the first six knots of the knot table are rational and have Conway notation $C(3)$, $C(2, 2)$, $C(5)$, $C(3, 2)$, $C(4, 2)$, $C(3, 1, 2)$, and $C(2, 1, 1, 2)$. Conway was interested in a more efficient way to enumerate the knots on the tables created by Kirkman, Little, and Tait in the late 1900s, and he introduced a simple notation that allowed him to re-create the work of Little to enumerate the prime knots of 11 crossings or less by hand in what he says what was an "afternoon's work," although his classification had some errors and was not completed until 1974 [14].

16.3   2-Bridge Knots

As pointed out earlier, rational knots have been a subject of study for some time as 2-bridge knots and 4-plats, and their classification was completed by Schubert in 1956 [15]. He classified this class of knots by taking advantage of an observation of Seifert that the double-branched cover of a 2-bridge knot over S^3 is a lens space and at that time lens spaces had already been classified.

In the image in Figure 16.10, each a_i represents some number of positive or negative twists, depending on the sign of a_i. A knot in this form is called a 4-**plat** presentation. This is readily seen to be the numerator of one of the two the tangles $T(a_1, a_2, \ldots, a_n)$ in Figure 16.5, and since every 2-bridge knot has a 4-plat presentation, each 2-bridge knot is a rational knot. Conversely, the numerator of a rational tangle $T(a_1, a_2, \ldots, a_n)$ can be arranged so that its diagram is in the form of a 4-plat with a standard regular diagram and will be denoted by $C(a_1, a_2, \ldots, a_n)$. Notice that the 4-plat presentation of a rational knot depends on whether the number of twist boxes is even or odd. In this way every 2-bridge knot has a 4-plat presentation as in Figure 16.11.

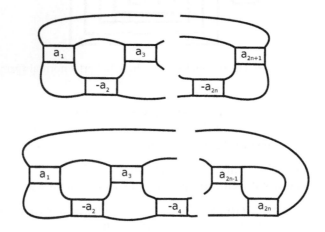

FIGURE 16.10: A rational knot in twist form and as a 4-plat.

FIGURE 16.11: The rational knot 5_2 in a 4-plat presentation.

Just like with rational tangles, given any 2-bridge knot of the form $C(a_1, a_2, \ldots, a_n)$ with $a_i \neq 0$ we can compute a corresponding rational number $\frac{p}{q}$ where $gcd(p, q) = 1$ and $\frac{p}{q} = [a_1, a_2, \ldots, a_n]$. Thus every 2-bridge knot corresponds to a fraction $\frac{p}{q}$. The following theorem of Shubert completely classifies rational knots.

Theorem 16.3.1 (Schubert). Suppose $K(\frac{p}{q})$ and $K(\frac{p'}{q'})$ denote the 2-bridge knots of type $\frac{p}{q}$ and $\frac{p'}{q'}$ respectively. Then $K(\frac{p}{q})$ and $K(\frac{p'}{q'})$ are isotopic if and only if

1. $p = p'$, and

2. either $q \equiv q'$ (mod p) or $qq' \equiv 1$ (mod p).

Notice that it suffices to assume that $0 < q < p$ since we are considering q (mod p). Also, from any rational number $\frac{p}{q}$ where $-p < q < p$ and $p > 0$ we can a construct a unique 2-bridge knot and its 4-plat presentation. Thus, with these conditions, Schubert's theorem provides a complete classification for rational knots.

Another useful fact about 2-bridge knots was noticed by Schubert and his formula for the additivity of the bridge number over connected sums.

Theorem 16.3.2. 2-bridge knots are prime.

This proof relies on Schubert's formula for the additivity of the bridge number $b(K)$:

$$b(K_1 \# K_2) + 1 = b(K_1) + b(K_2)$$

.

To see Theorem 16.3.2, suppose that $b(K) = 2$ and $K = K_1 \# K_2$. Then $b(K) = b(K_1 \# K_2)$, so $b(K) + 1 = b(K_1) + b(K_2)$. Hence $b(K_1) + b(K_2) = 3$. Thus, $b(K_1) = 1$ and $b(K_2) = 2$ or vice versa. Observe that if $b(K_i) = 1$ then K_i is unknotted, hence K must be prime.

Once can also consider the tunnel number, which is the least number of arcs intersecting the knot in their endpoints such that their union with K has a complement, a handlebody. From Figure 16.12, we can see that a 2-bridge knot K has two unknotting arcs τ and τ' so that the union of K with either τ or τ' is isotopic to an unknotted theta graph whose complement is a genus 2 handlebody. These two "unknotting tunnels,", proved to be the only such in [1], show that 2-bridge knots are of tunnel number one.

FIGURE 16.12: Two unknotting arcs for a 2-bridge knot.

Incompressible surfaces in the complement of a 2-bridge link have been classified by Hatcher and Thurston in [8]. In particular, they classify the incompressible, ∂-incompressible surfaces in the complement of a 2-bridge link with fraction $\frac{p}{q}$ in terms of the continued fraction expansion of $\frac{p}{q}$. Their classification uses Thurston's theory of projective lamination spaces and the observation that each incompressible surface is carried by a branched surface that is determined by the continued fraction expansion of $\frac{p}{q}$.

16.4   Montesinos Knots

A Montesinos tangle $T(e; \frac{p_1}{q_1}, \ldots, \frac{p_n}{q_n})$ is the sum of n rational tangles denoted by their fractions $\frac{p_1}{q_1}, \ldots, \frac{p_n}{q_n}$, where the number e is an integer that denotes some number of half twists. The numerator of tangle $T(e; \frac{p_1}{q_1}, \ldots, \frac{p_n}{q_n})$ is called a Montesinos knot or link (also known as star links) and is denoted $M(e; \frac{p_1}{q_1}, \ldots, \frac{p_n}{q_n})$. See Figure 16.13. The 2-bridge knots coincide with the Montesinos knots with $n \leq 2$. In addition, the pretzel knot $(q_1, \ldots, q_n)$ is the Montesinos knot $K(\frac{1}{q_1}, \ldots, \frac{1}{q_n})$. These knots were classified first by Bonahan and Siebenmann in [2]. Proofs of this classification theorem can be found in [3] and [11]. See Chapter 17 for the extension of this classification to the more general class of arborescent (also known as algebraic) links.

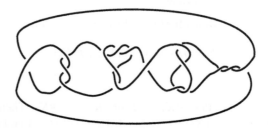

FIGURE 16.13: The Montesinos knot $M(3; \frac{2}{3}, \frac{1}{2}, -\frac{3}{2})$.

Theorem 16.4.1 (Classification of Montesinos Knots). Montesinos links with $n \geq 3$ rational tangles are classified by the ordered set of fractions $(\frac{p_1}{q_1} \pmod 1), \ldots, \frac{p_n}{q_n} \pmod 1)$ together with the number $e_0 = e - \sum_{i=1}^{n} \frac{p_i}{q_i}$. This classification is unique up to cyclic permutations of the rational tangles and reversal of order that each rational tangle appears.

Closed incompressible surfaces in the complements of Montesinos links were classified by Oertel in [13] where it is shown that every incompressible surface in the complement of a Montesinos link is obtained from connecting a finite collection of disjoint incompressible 2-spheres with peripheral tubing operations. Further, in most cases a surface obtained from tubing an incompressible 2-sphere is incompressible if and only if each tube passes through a rational tangle. As a consequence, a Montesinos link is hyperbolic if it is not a torus link, and not equivalent to $K(\frac{1}{2}, \frac{1}{2}, -\frac{1}{2}, -\frac{1}{2})$, $K(\frac{2}{3}, -\frac{1}{3}, -\frac{1}{3})$, $K(\frac{1}{2}, -\frac{1}{4}, -\frac{1}{4})$, $K(\frac{1}{2}, -\frac{1}{3}, -\frac{1}{6})$.

16.5   Bibliography

[1] Colin C. Adams and Alan W. Reid. Unknotting tunnels in two-bridge knot and link complements. *Comment. Math. Helv.*, 71(4):617–627, 1996.

[2] F. Bonahon and L. C. Siebenmann. The characteristic toric splitting of irreducible compact 3-orbifolds. *Math. Ann.*, 278(1-4):441–479, 1987.

[3] Gerhard Burde and Heiner Zieschang. *Knots*, volume 5 of *De Gruyter Studies in Mathematics*. Walter de Gruyter & Co., Berlin, second edition, 2003.

[4] J. H. Conway. An enumeration of knots and links, and some of their algebraic properties. In *Computational Problems in Abstract Algebra (Proc. Conf., Oxford, 1967)*, pages 329–358. Pergamon, Oxford, 1970.

[5] Peter R. Cromwell. *Knots and Links*. Cambridge University Press, Cambridge, 2004.

[6] C. Ernst and D. W. Sumners. A calculus for rational tangles: Applications to DNA recombination. *Math. Proc. Cambridge Philos. Soc.*, 108(3):489–515, 1990.

[7] Jay R. Goldman and Louis H. Kauffman. Rational tangles. *Adv. in Appl. Math.*, 18(3):300–332, 1997.

[8] A. Hatcher and W. Thurston. Incompressible surfaces in 2-bridge knot complements. *Invent. Math.*, 79(2):225–246, 1985.

[9] Louis H. Kauffman and Sofia Lambropoulou. On the classification of rational knots. *Enseign. Math. (2)*, 49(3-4):357–410, 2003.

[10] Louis H. Kauffman and Sofia Lambropoulou. On the classification of rational tangles. *Adv. in Appl. Math.*, 33(2):199–237, 2004.

[11] Akio Kawauchi. *A survey of Knot Theory*. Birkhäuser Verlag, Basel, 1996. Translated and revised from the 1990 Japanese original by the author.

[12] Kunio Murasugi. *Knot Theory and Its Applications*. Birkhäuser Boston, Inc., Boston, MA, 1996. Translated from the 1993 Japanese original by Bohdan Kurpita.

[13] Ulrich Oertel. Closed incompressible surfaces in complements of star links. *Pacific J. Math.*, 111(1):209–230, 1984.

[14] Kenneth A. Perko, Jr. On the classification of knots. *Proc. Amer. Math. Soc.*, 45:262–266, 1974.

[15] Horst Schubert. Knoten mit zwei Brücken. *Math. Z.*, 65:133–170, 1956.

Chapter 17

Arborescent Knots and Links

Francis Bonahon, University of Southern California

Arborescent links form a broad family of knots and links in $S^3 = \mathbb{R}^3 \cup \{\infty\}$, with several convenient features. The main one is that they admit an internal classification, making it easy to decide when two arborescent links are isotopic; the current chapter describes this classification, originally obtained in [7]. They also represent a large fraction of the links with small crossing numbers, for instance close to 90% of those links with at most 10 crossings (although the proportion quickly goes down with the number of crossings afterward). In addition to their classification, the symmetry group of most arborescent links can be explicitly computed, as explained in §17.8.

Another interesting property of arborescent links is that they provide many examples of non-isotopic mutant links. The mutation operation, quickly described in §17.5.4, modifies a link K to give another link K' which traditionally is very difficult to distinguish from the original link, because many combinatorial and hyperbolic invariants of K and K' are equal; see [15] for a survey. Not only do arborescent links provide many examples of nontrivial mutations, but in some sense they even encapsulate the essence of mutations. Indeed, one of the main results of [7] (see also [5]) identifies an "arborescent part" A, well defined up to isotopy respecting the link, in any link K that cannot be simplified as a satellite of another link. This arborescent part A has the property that any nontrivial mutation of K can be realized along a Conway sphere contained in A. Various results of [7] describe all such Conway spheres in A, thereby enumerating all mutations that can be performed on the link K.

Arborescent links were originally identified by Conway [9], who called them "algebraic links" because they occur by some algebraic process in his tangle calculus. Since there is a distinct competing notion of algebraic links, arising from singularities of complex algebraic curves, we prefer the "arborescent" terminology, which reflects the fact that these links are closely associated to certain trees with additional data (and that their double-branched cover is a graph manifold, in Waldhausen's sense [20], whose associated graph is a tree). Arborescent links include Schubert's two-bridge links [17, 21], and Schubert's classification of these links can be seen as a precursor of the classification of all arborescent links described here.

By definition, an arborescent link is a link obtained by gluing together pieces of the type described in Figure 17.3. The gluing of these pieces is specified by a certain operation called plumbing, which is an embedded low-dimensional version of the classical plumbing of disk bundles [13, 11, 12], restricted here to the plumbing of unknotted twisted bands in S^3. The building blocks of the plumbing construction, and the way these blocks fit together, are specified by the combinatorial data of a weighted planar tree: this data consists of a finite tree Γ embedded in the plane and endowed, at each vertex v, with an integer weight assigned to each of the angular sectors delimited in the plane by the edges emanating from v.

Many different weighted planar trees give the same link by plumbing. In §17.3, we describe several moves which modify a weighted planar tree without changing the isotopy class of the resulting link. These moves enable one to simplify the weighted planar tree until it is in a form called "canonical," which is precisely described in §17.4. The Classification Theorem 4 then states that two canonical weighted planar trees describe isotopic links if and only if one can go from one tree to the other by a sequence of particularly simple moves. A paraphrase of this theorem is that, once two links are given such a canonical arborescent presentation, the two links are isotopic if and only if they are obviously isotopic. There is of course much ambiguity in the word "obviously," and subtle phenomena can occur. For instance the two links of Figure 17.1 are isotopic if and only if the number a of half-twists occurring in their central part is odd. However, a precise algorithm described in §17.7.2 makes the Classification Theorem 4 effective. The classification result is also significantly easier to state for the important subfamilies consisting of two-bridge links and Montesinos links; see §17.6.

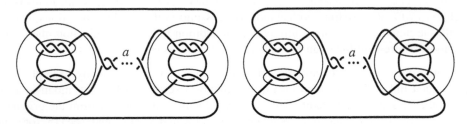

FIGURE 17.1: Two mutant arborescent links.

The classification of arborescent links and their symmetries was made possible by a crucial observation of Montesinos [14], who proved that the double branched covers of arborescent links are graph manifolds in Waldhausen's sense [20]. Montesinos's result rephrased the classification of arborescent links as a classification problem for graph manifolds endowed with certain involutions, for which the techniques of [20] could be used. This was the point of view originally developed in [4], and expanded to the framework of general 3–dimensional orbifolds in [5, 6]. In addition to these theoretical underpinnings, the classification was also made much more practical and effective by the introduction of the combinatorial plumbing calculus, largely developed by L. C. Siebenmann; see also [19] for applications of a similar calculus for graph manifolds. Although inspired by the original point of view, the proofs described in the monograph [7] are more elementary, and are purely centered in S^3 without reference to double branched covers. A different approach to the classification of the subfamily of Montesinos links, and to the determination of their symmetry groups, was developed in [2]; see also [8].

17.1 Arborescent Links

The traditional way to represent a knot or, more generally, a link K in the 3–dimensional sphere $S^3 = \mathbb{R}^3 \cup \{\infty\}$ is by considering a generic projection of this link to a plane in $\mathbb{R}^3$. A *link projection* or, to follow the terminology that seems to have become more widespread in recent years, a *link diagram*, is a family of simple closed curves immersed in the 2–dimensional sphere $S^2 = \mathbb{R}^2 \cup \{\infty\}$, with singularities consisting only of transverse double points, and where each double point in endowed with additional over- and under-crossing information indicating which strand sits above the other one in $\mathbb{R}^3$.

A *Conway circle* for the link diagram $L \subset S^2 = \mathbb{R}^2 \cup \{\infty\}$ is a simple closed curve in S^2 cutting L transversely in exactly four points, away from the crossings. The link diagram $L \subset S^2$ is *arborescent* if there exists a union C of disjoint Conway circles such that, for the closure A of each component of $S^2 - C$, the pair $(A, A \cap L)$ is of one of the three types represented in Figure 17.2.

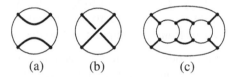

(a)   (b)   (c)

FIGURE 17.2: The building blocks of arborescent link diagrams.

These arborescent link diagrams were first encountered by Conway in his tangle calculus [9] for knots and links. In particular, the pairs of Figure 17.2(a) and (b) correspond to Conway's basic building blocks, the tangles he calls 0 and 1. Conway introduces a "sum" and "product" operation of two tangles, both of which are obtained by suitably plugging the two tangles inside of the two inner Conway circles of the boundary of the pair of Figure 17.2(c). In Conway's terminology, a tangle is *algebraic* if it is obtained by combining a family of tangles 0 and 1 through a series of product and sum operations, and a link diagram is *algebraic* if it is obtained as the "numerator" of an algebraic tangle, an additional operation involving the pair of Figure 17.2(a). Comparing Conway's definitions to ours makes it clear that Conway's algebraic links are precisely the ones we call here arborescent links.

17.2 The Plumbing Construction

Working with a few examples, such as the ones of Figure 17.1, quickly make it clear that the building blocks of Figure 17.2 are too small to be really convenient, as a decomposition of an arborescent link diagram into such blocks typically requires a cumbersomely large number of pieces. This leads us to enlarge our family of building blocks.

17.2.1 Twisted Band Diagrams

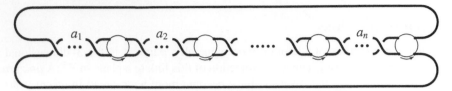

FIGURE 17.3: A twisted band diagram.

The *n–valent twisted band diagram* associated to the integers $a_1, a_2, \ldots, a_n \in \mathbb{Z}$ is the pair (A, L) represented in Figure 17.3. Here A is a connected codimension 0 submanifold of $S^2 = \mathbb{R}^2 \cup \{\infty\}$ delimited by finitely many simple closed curves in S^2, and L is a "partial link diagram" formed by finitely many arcs immersed in A with boundary points in the boundary ∂A and whose only singularities are transverse double points endowed with over- and under-crossing information. The symbol $\mathsf{X}\text{-}\!\overset{a_i}{}\!\mathsf{X}$ represents a band with a_i right-handed half-twists if $a_i \geqslant 0$ and $-a_i$ left-handed half-twists if $a_i \leqslant 0$. For instance, $\mathsf{X}\text{-}\overset{+2}{}\mathsf{X}$ corresponds to XX while $\mathsf{X}\text{-}\overset{-3}{}\mathsf{X}$ stands for XXX. Finally, note that each component C of ∂A meets L in exactly four points, and is endowed with an arrow that runs parallel to C in A and goes from one endpoint of L to the next.

Note that the pair (A, L) can be decomposed into a collection of building blocks of the type of Figure 17.2, by splitting it up along a union of disjoint Conway circles. As a consequence, a link diagram is arborescent if and only if it can be split along a family of disjoint Conway circles to obtain a collection of twisted band diagrams.

The arrows in the definition of twisted band diagrams will be used for a converse construction called plumbing, which glues twisted band diagrams together while carefully keeping track of the underlying combinatorics.

17.2.2 Weighted Planar Trees

The plumbing construction requires the data of a *weighted planar tree* Γ. This is a finite tree Γ endowed with additional data. Recall that a vertex v is *n–valent* if it is adjacent to n edges of Γ. By analogy with chemical terminology, we will refer to the half-edges, or "germs of edges," emanating from v as *bonds*; in particular, an edge connects two bonds, one at each of its ends. The additional data of Γ then consists of

(i) a cyclic order of the bonds at each vertex of Γ;

(ii) an integer $a \in \mathbb{Z}$ sitting between each pair of bonds that are consecutive for the cyclic order of (i).

While we should think of a weighted planar tree Γ as abstract combinatorial data, in practice it is better described by observing that, up to isotopy, there is a unique embedding of Γ in the plane for which the cyclic order of bonds at each vertex corresponds to the clockwise order in the plane. The integer weights of Γ can then be seen as sitting in the angular sectors delimited in the plane by the bonds of a given vertex.

Conversely, a tree embedded in the plane, with an integer weight assigned to each angular sector at its vertices, uniquely determines a weighted planar tree Γ in this way. In practice, it is convenient to omit some of the weights that are equal to 0, so that an angular sector that is blank is understood to carry weight 0. Figure 17.4 offers an example.

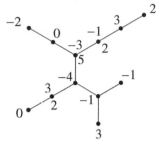

FIGURE 17.4: A weighted planar tree.

This example should clarify one point in the combinatorial definition of weighted planar trees. By convention, a 1–valent vertex of a weighted planar tree Γ is adjacent to a unique angular sector, and in particular it carries an integer weight (possibly omitted in the planar description if equal to 0). We also allow weighted planar trees $\overset{a}{\bullet}$ consisting of a single vertex endowed with an integer weight $a \in \mathbb{Z}$.

17.2.3 Plumbing Twisted Band Diagrams Together

We now use the combinatorial data provided by a weighted planar tree to glue twisted band diagrams together and uniquely determine a link diagram in $S^2 = \mathbb{R}^2 \cup \{\infty\}$, up to isotopy of S^2.

Given a weighted planar tree Γ, we associate to each n–valent vertex v of Γ the n–valent twisted band diagram (A_v, L_v) defined by the weights $a_1, a_2, \ldots, a_n \in \mathbb{Z}$ of the angular sectors of Γ at v, taken in the order defined by the cyclic order, as in Figure 17.3. In particular, A_v has n boundary components, which are in a natural one-to-one correspondence with the bonds of v.

Each edge e of Γ connects two vertices v and v', and singles out one bond at each of these vertices. Let C_v and $C_{v'}$ be the boundary components of A_v and $A_{v'}$ that correspond to the bonds associated to e, each oriented by the boundary orientation of ∂A_v and $\partial A_{v'}$. Up to isotopy, there is a unique orientation-reversing homeomorphism of pairs

$$\theta_e : (C_v, L_v \cap C_v) \to (C_{v'}, L_{v'} \cap C_{v'})$$

sending the point of $L_v \cap C_v$ that is the initial point of the arrow of C_v in A_v to the point of $L_{v'} \cap C_{v'}$ similarly singled out by the arrow of $C_{v'}$ in $A_{v'}$. We can then use this identification θ_e to glue A_v to $A_{v'}$ along C_v and $C_{v'}$, and L_v to $L_{v'}$ along $L_v \cap C_v$ and $L_{v'} \cap C_{v'}$.

After this gluing, the configuration in $A_v \cup A_{v'}$ of the arrows of $C_v = C_{v'}$ in A_v and $A_{v'}$ is then as indicated in Figure 17.5. This property is critical for the plumbing operation to be well-defined.

If we perform this gluing operation for all edges e of the weighted planar tree Γ, the twisted band diagrams (A_v, L_v) are glued into a pair (A, L) where L is a link diagram drawn on a surface A. Since Γ is a tree (namely a graph with no cycle), the surface A is homeomorphic to the sphere; it also carries a natural orientation, induced by the orientation of all $A_v \subset S^2$. Identifying A to S^2 by any orientation-preserving homeomorphism, L now gives a link diagram in S^2, uniquely determined up to isotopy.

FIGURE 17.5: The plumbing rule.

In this way we have associated, up to isotopy of S^2, a link diagram $L \subset S^2$ to each weighted planar tree Γ. This link diagram then determines a link K in S^3.

For instance, the weighted planar tree of Figure 17.4 uniquely determines the link diagram, and the link, of Figure 17.6.

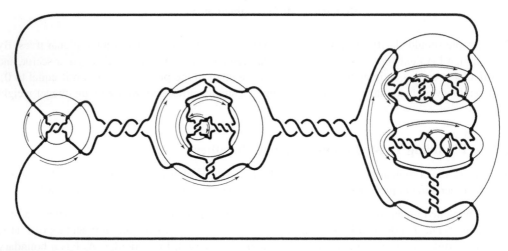

FIGURE 17.6: Plumbing of twisted band diagrams according to the weighted planar tree of Figure 17.4.

When Γ is the single vertex weighted graph $\overset{a}{\bullet}$, (A, L) is the 0–valent twisted band diagram associated to the integer $a \in \mathbb{Z}$, so that the link described by Γ is the boundary of an unknotted twisted band in S^3, with a positive half-twists if $a \geqslant 0$ and $-a$ negative half-twists if $a \leqslant 0$.

17.3 Moves on Weighted Planar Trees

A link can be represented by many different link diagrams, and in fact, an arborescent link can be described by many arborescent link diagrams. We now give a series of moves that modify a weighted planar tree without changing the isotopy class of the associated arborescent link (although they usually change the associated link diagram).

Flips. This first set of moves essentially deals with the cyclic order of weights and bonds around vertices.

(F$_1$) Reverse the cyclic order of bonds and weights at every vertex of the graph.

(F$_2$) Reverse the cyclic order at one vertex v_0, and at each vertex at even distance from v_0 in the tree.

(F$_3$) Replace $- - - \frac{a}{a'} \bullet - - -$ by $- - \frac{a \pm 1}{a' \mp 1} \bullet - -$, where the cyclic order of bonds and weights is reversed at all vertices lying to the right of the vertex shown, and at odd distance from this vertex.

In our description of the flip (F$_3$), only one vertex, one bond, and two (adjacent) weights of the tree are visible. The dashed line $- -$ indicates any continuation to the right, namely a weighted subtree attached to the bond shown. The other dashed line $- - -$ represents any continuation to the left, namely a sequence of integer weights and weighted subtrees attached to the vertex shown. We will use similar conventions in our description of further moves.

Ring move. This move seldom applies, and involves an unknotted component of the link moving freely around one of the twisted bands, whence the name.

(R) Replace $- - \bullet - - - -$ by $- - \bullet - - - -$.

Arithmetic moves. This next set of moves changes the weights of vertices in the tree, and is closely related to arithmetic properties of continued fractions and of the group $SL_2(\mathbb{Z})$. They are labelled by two integers; the first one refers to the weight of the main vertex involved, the second one to its valence. Note the precise location of the weights, which is often critical.

(0.1) Replace $\overset{0}{\bullet}\overset{a}{\longrightarrow}\bullet - - -$ by $\bullet - - - -$.

(0.2) Replace $- - - \overset{a}{\bullet}\overset{0}{\underset{b}{\longrightarrow}}\bullet - -$ by $- - \underset{b}{\overset{a+b}{\bullet}} -$, where the cyclic orders of bonds and weights are preserved in the left-hand side subgraph $- - -$ and reversed anywhere else. In particular, the order of the components of is reversed.

(1.0) Exchange $^{+1}$ and $^{-1}$.

(1.1) Replace $\overset{\pm 1}{\bullet}\overset{a}{\longrightarrow}- - -$ by $\overset{a \mp 1}{\bullet}- - -$.

(1.2) Replace $- - \overset{a}{\bullet}\underset{\pm 1}{\longrightarrow}\overset{b}{\bullet} - - -$ by $- \overset{a \mp 1}{\bullet}\overset{b \mp 1}{\longrightarrow} - -$.

(2.0) Exchange $^{+2}$ and $^{-2}$.

(2.1) Replace $^{+2}\ \ a$ by $^{-2}\ \ a - 1$.

(2.2) Replace $- - \overset{a}{\bullet}\overset{+2}{\underset{b}{\longrightarrow}}\bullet - - -$ by $- \overset{a-1}{\bullet}\overset{-2}{\underset{b-1}{\longrightarrow}}\bullet - - -$.

This list of moves is convenient to work with, but very redundant in the sense that the flips (F_1) and (F_2) easily follow from (F_3), whereas all arithmetic moves are consequences of the three moves (0.1), (0.2) and (1.2).

Proposition 1. If the two weighted planar trees Γ and Γ' are related to each other by a series of the above flip moves, ring move, arithmetic moves, and their inverses, then the arborescent links respectively associated to Γ and Γ' by plumbing are isotopic in S^3.

Proposition 1 is essentially checked by a series of pictures.

As an illustration, consider the weighted planar tree of Figure 17.4, which we rewrite as

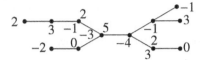

Applying the flip (F_3) to move weights around the vertices, one obtains

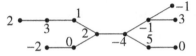

Then, the arithmetic moves (1.2), (0.2), (1.1) and (0.1) give

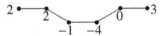

A further application of (0.2) turns this tree into $\overset{2}{\bullet}\!\!-\!\!\overset{2}{\bullet}\!\!-\!\!\overset{-1}{\bullet}\!\!-\!\!\overset{-1}{\bullet}$, which itself can be converted to $\overset{2}{\bullet}\!\!-\!\!\overset{2}{\bullet}\!\!-\!\!\overset{0}{\bullet}$ and finally $\overset{2}{\bullet}$ by (1.1) and (0.2).

What this manipulation proves is that the link of Figure 17.6, associated by plumbing to the weighted planar tree of Figure 17.4, is the same as the link associated to $\overset{2}{}$, namely the Hopf link. This is a good illustration of the power of the calculus of weighted planar trees.

17.4   Canonical Weighted Planar Trees

The moves of the previous section show that an arborescent link can be associated to arbitrarily large weighted planar trees. If these trees are going to be of any use, it therefore appears necessary to impose a certain number of restrictions. The weighted planar trees satisfying these conditions will be called canonical, since we will see that they classify the arborescent pair up to a limited number of moves.

We list the conditions for canonicity.

Weight Condition.

(W) At each vertex of Γ, at most one weight is non-zero.

This property can clearly be attained by a finite sequence of applications of the flip (F_3). We consequently assume it.

The remaining conditions for canonicity concern only the *abstract weighted tree* Γ_0 *underlying* Γ, defined by forgetting the embedding of Γ in the plane. The weights are now assigned to the vertices of Γ_0, by the Weight Condition (W).

Form a weighted graph Γ_0' from Γ_0 by deleting all its vertices of valence $\geqslant 3$ and the bonds (but not the full edges) attached to these vertices. Each component of Γ_0' is a *stick* of Γ_0. It is topologically an interval, and has 0, 1 or 2 free bonds at its ends, which used to be attached to a vertex of valence $\geqslant 3$ in Γ_0. In particular, there are three possible types of sticks: $\overset{a_1}{\bullet}\!-\!\overset{a_2}{\bullet} \cdots -\!\overset{a_n}{\bullet}$, $\overset{a_1}{\bullet}\!-\!\overset{a_2}{\bullet} \cdots -\!\overset{a_n}{\bullet}$ or $-\!\overset{a_1}{\bullet}\!-\!\overset{a_2}{\bullet} \cdots -\!\overset{a_n}{\bullet}$ with $n \geqslant 1$.

Stick Condition.

(S) On every stick of Γ_0, the weights are non-zero and of alternating sign, unless Γ_0 is $\overset{0}{\bullet}$. The end weights a_1 and a_n of a stick are not ± 1, unless Γ_0 is $\overset{\pm 1}{\bullet}$

The last condition for canonicity comes with two options.

Positive Canonicity Condition.

(P) None of the sticks of Γ_0 is $\overset{-1}{\bullet}, \overset{-2}{\bullet}, \overset{-2}{\bullet}\!-\!\bullet$ or $\bullet\!-\!\overset{-2}{\bullet}$.

Negative Canonicity Condition.

(N) None of the sticks of Γ_0 is $\overset{+1}{\bullet}, \overset{+2}{\bullet}, \overset{+2}{\bullet}\!-\!\bullet$ or $\bullet\!-\!\overset{+2}{\bullet}$.

A weighted planar tree is *positively canonical* or (+)–*canonical* if it satisfies the Weight Condition (W), the Stick Condition (S) and the Positive Canonicity Condition (P). It is *negatively canonical* or (−)–*canonical* if it satisfies the Weight Condition (W), the Stick Condition (S) and the Negative Canonicity Condition (N).

The reason to keep these two canonicity conditions is that, if the arborescent link K is associated to the weighted planar tree Γ, applying a mirror image symmetry to K gives the arborescent link $-K$ that is associated to the weighted planar tree $-\Gamma$ obtained by reversing the signs of all the weights of Γ. Clearly Γ is (+)–canonical if and only if $-\Gamma$ is (−)–canonical. This will become particularly important in §17.8, when we consider symmetries of arborescent links that reverse the orientation of S^3.

We would like to show, by a series of moves, that every arborescent link K can be represented by a positively or negatively canonical tree. When trying to do so, one sometimes encounters a situation when a weighted planar tree describing the link is of the form $\overset{0}{\bullet}\!-\!\bullet\,-\,-\,-\,-$ where the central vertex has valence $\geqslant 3$. The arithmetic move (0.1) does not apply here, but the situation reveals a somewhat non-trivial 2–sphere $F \subset S^3$ which meets K in two points, so that the link K splits as a connected sum.

This leads us to introduce an additional move, somewhat troublesome, which replaces a weighted planar graph by a disconnected graph.

Connected Sum Move.

(0.1*) Replace the weighted planar tree $\Gamma = \overset{0}{\bullet}\!\!-\!\!-\!\!-\!\!\bullet - - - -$, where the central vertex has valence $\geqslant 3$, by the family of disjoint weighted planar trees obtained from Γ by erasing the two vertices shown as well as their adjacent edges.

Note that the family associated to Γ by the move (0.1*) consists of at least two weighted planar trees.

Proposition 2. If the family of weighted planar trees $\Gamma_1, \Gamma_2, \ldots, \Gamma_n$ is associated to Γ by the move (0.1*), the arborescent link K associated to the weighted planar tree Γ is a connected sum of the links K_i defined by the Γ_i.

As is well known, the connected sum $K_1 \# K_2$ of two links K_1 and K_2 is in general not uniquely determined, unless:

1. we have chosen the component K_1' of K_1 and the component K_2' of K_2 where the connected sum takes place;

2. in addition, we are given matching orientations for these two components K_1' and K_2', namely orientations of K_1' and K_2' that are defined modulo simultaneous reversal.

Proposition 2 only asserts that K splits as *some* connected sum of the K_i. As a consequence, the move (0.1*) is not reversible.

Proposition 3. There is an effective algorithm which, starting from any weighted planar tree Γ, alters Γ by a sequence of flips, arithmetic moves and their inverses, and connected sum moves, and produces a family of disjoint weighted planar trees which are positively (or negatively) canonical.

The general strategy of the algorithm is fairly simple. First apply the above moves to split and reduce the size of the trees as much as possible. Then arrange for the Stick Condition (S) to be satisfied by applying the inverse of the move (1.2). Finally, use (1.0), (2.0), (2.1), (2.2) or their inverses to guarantee the positive or negative canonicity of the trees.

17.5   The Classification of Prime Arborescent Links

17.5.1   Modified Flips and Ring Moves

The flip (F_3) and the ring move (R) do not apply to canonical weighted planar trees. This leads us to modify these moves as follows.

The flip (F_3) is replaced by the following move, which is an immediate consequence of (F_3).

Modified Flip.

(F_3') Replace $-\,-\,-\,\overset{a}{\bullet}-\,-$ by $-\,-\,-\,-\underset{a}{\bullet}-\,-$ and, if a is odd, reverse the cyclic order of weights and bonds at all vertices of the subgraph $-\,-$ lying at odd distance from the vertex shown.

There are two forms of the modified ring move, one for positively canonical trees and one for negatively canonical trees. These moves are easily deduced from the original ring move by application of the arithmetic move (2.1) and its inverse.

Modified Ring Moves.

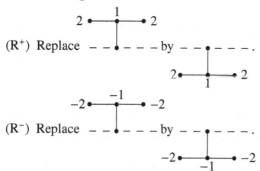

(R⁺) Replace $- - \bullet - - - -$ by $- - \bullet - - - -$.

(R⁻) Replace $- - \bullet - - - -$ by $- - \bullet - - - -$.

17.5.2 The Main Classification Theorem

The following statement is the main classification theorem.

Theorem 4. Let the arborescent links K and K' be respectively associated by plumbing to the (+)–canonical weighted planar trees Γ and Γ'. Then, K is isotopic to K' in S^3 if and only if Γ' can be obtained from Γ by a sequence of flips (F_1), (F_2), (F_3') and modified ring moves (R^+).

There is a similar statement for (−)–canonical trees, replacing the modified ring move (R^+) by (R^-).

As an illustration of Theorem 4, consider the two arborescent links of Figure 17.1. They are respectively associated by plumbing to the (+)–canonical weighted planar trees of Figure 17.7.

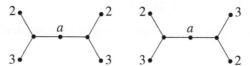

FIGURE 17.7: The weighted planar trees associated to the links of Figure 17.1.

When the weight a is odd, a flip (F_1) followed by a flip (F_3') performed at the central vertex sends the first weighted planar tree to the second one. This explicitly shows that the two links of Figure 17.1 are isotopic when a is odd.

On the other hand, when a is even, we can easily list all weighted planar trees that are obtained from the first tree of Figure 17.7 by a sequence of flips (F_1), (F_2) or (F_3'); note that the modified ring move (R^+) never applies here. There are exactly two such weighted planar trees (if $a \neq 0$, and only one if $a = 0$), neither of which is isomorphic to the second tree of Figure 17.7. Therefore, Theorem 4 proves that the two links of Figure 17.1 are not isotopic when a is even.

17.5.3 Prime Arborescent Links

Theorem 4 provides a classification of those algebraic links which are associated to a positively (or negatively) canonical weighted planar tree. The algorithm of Proposition 3 almost always provides such a canonical tree, except when connected sum moves (0.1*) occur and disconnect the graph.

Recall that a link $K \subset S^3$ is *unsplittable* if there is no 2–dimensional sphere F that is embedded in the complement $S^3 - K$ and separates K. Also, K is *prime* if it does not split as a nontrivial connected sum, namely if every 2–dimensional sphere F in S^3 that cuts K in two points bounds a ball $B \subset S^3$ such that the arc $K \cap B$ is unknotted in B. Note that a prime link is necessarily unsplittable.

Proposition 5. If the arborescent link is associated by plumbing to the positively (or negatively) canonical weighted planar tree Γ, then K is unsplittable and prime unless Γ is the tree $\overset{0}{\underset{\bullet}{\bullet}}$.

Note that the arborescent link associated to $\overset{0}{}$ is the union of two unknots separated by a sphere.

Combining Proposition 5 with Proposition 3 immediately gives a criterion to decide when an arborescent link is unsplittable and/or prime. Observe that, by Theorem 4, the arborescent link associated to a positively canonical weighted planar tree Γ is the unknot exactly when Γ is $\overset{+1}{}$.

Proposition 6. Let the arborescent link K be associated by plumbing to the weighted planar tree Γ, and let $\Gamma_1, \Gamma_2, \dots, \Gamma_n$ be the family of positively canonical weighted planar trees obtained by applying to Γ the algorithm of Proposition 3. Then K is unsplittable if and only if none of the trees Γ_i is $\overset{0}{}$.

If K is unsplittable, it is prime if and only if at most one of the Γ_i is different from $\overset{+1}{}$. In addition, K is then associated by plumbing to this positively canonical tree $\Gamma_i \neq \overset{+1}{}$ if there is such a tree, and is otherwise the trivial knot associated to $\overset{+1}{}$.

There is of course an almost identical statement for negatively canonical trees, obtained by replacing $\overset{+1}{}$ by $\overset{-1}{}$.

Propositions 5 and 6 show that Theorem 4 is essentially a classification theorem for all prime arborescent links. A corollary of Propositions 2, 3, and 5 is that the prime factors of an arborescent link are also arborescent, and are obtained by the algorithm of Proposition 3.

17.5.4 Mutations

The mutation operation modifies a link $K \subset S^3$ by cut-and-paste along a *Conway sphere* for K, namely along a 2–dimensional sphere Σ embedded in S^3 and (transversely) cutting K in exactly four points.

More precisely, consider the *standard Conway sphere* (S^2, P) consisting of the unit sphere $S^2 \subset \mathbb{R}^3$ and of the four points $P = \{(\pm 1, \pm 1, 0)\} \subset S^2$. This pair (S^2, P) is invariant under the action of Klein's *Vierergruppe* $V_4 \cong \mathbb{Z}_2 \times \mathbb{Z}_2$, consisting of the identity together with the rotations of π around the three coordinate axes of $\mathbb{R}^3$.

The link K' is obtained from the link $K \subset S^3 = \mathbb{R}^3 \cup \{\infty\}$ by a *mutation along the Conway sphere* Σ if it results from the following process. First modify K and Σ by an isotopy of S^3 so that the pair $(\Sigma, K \cap \Sigma)$ coincides with the standard Conway sphere (S^2, P); then K' is isotopic to the union of $K - B^3$ and of $\varphi(B^3 \cap K)$ where $B^3 \subset \mathbb{R}^3$ is the unit ball bounded by the sphere S^2 and where φ is one of the four elements of the Vierergruppe V^4. One easily checks that, up to isotopy, there are at most four links that are obtained from K by a mutation along a given Conway sphere Σ, one of which is of course K itself; indeed, to check that the construction is independent of the isotopy moving $(\Sigma, K \cap \sigma)$ to (S^2, P), the key ingredient is that the Vierergruppe V_4 is normal in the group $\pi_0 \, \mathrm{Homeo}(S^2, P)$ of isotopy classes of homeomorphisms of the standard Conway sphere (S^2, P).

Mutant links offer a traditional challenge in knot theory, as they are difficult to distinguish. Indeed, mutant links have the same Alexander polynomial, the same signature, the same Jones polynomial, the same HOMFLY polynomial, the same odd Khovanov homology, the same hyperbolic volume, etc. See the survey [15] for more complete details.

By definition of arborescent links, they exhibit many Conway spheres arising from Conway circles in arborescent link diagrams. For instance, the two links represented in Figure 17.1 are clearly related by a mutation along a Conway sphere.

There is a very simple criterion to decide when two arborescent links K and K' are related by a sequence of mutations along Conway spheres. It involves the abstract weighted tree Γ_0 underlying a canonical weighted planar tree Γ defined, as in §17.4, by forgetting the embedding of Γ in the plane and by considering its integer weights as assigned to the vertices of Γ_0.

Proposition 7. Let the arborescent links K and K' be respectively associated by plumbing to the $(+)$–canonical weighted planar trees Γ and Γ'. Then K' is obtained from K by a sequence of mutations along Conway spheres if and only if, when we forget the embeddings of Γ and Γ' in the plane, the resulting abstract weighted trees Γ_0 and Γ_0' are isomorphic by a graph isomorphism respecting integer weights.

17.6   A Special Case: Montesinos Knots and Links

Montesinos links form an important subfamily of arborescent links. We begin with an even more limited subfamily, that of Schubert's two-bridge links [17].

17.6.1   Two-Bridge Knots and Links

A *two-bridge link* is a link K that can be isotoped in $\mathbb{R}^3$ so that, when we project K to the z–axis, the corresponding height function on K has exactly two nondegenerate local maxima, two nondegenerate local minima, and no other critical point. Note that a two-bridge link can only have 1 or 2 components.

Projecting a two-bridge link to a generic vertical plane easily shows that such a link is arborescent, associated by plumbing to a weighted planar tree consisting of a single stick $\overset{a_1}{\bullet}\!\!-\!\!\overset{a_2}{\bullet}\cdots-\overset{a_n}{\bullet}$.

Schubert [17] associated a two-bridge link $K(\frac{p}{q})$ to each rational number $\frac{p}{q} \in \mathbb{Q} \cup \{\infty\}$ with p and q coprime with $q \geqslant 0$. He then proved the following classification theorem.

Theorem 8 (Schubert [17]). Any two-bridge link is isotopic to a link $K(\frac{p}{q})$ with p coprime to $q \geqslant 0$.

The two-bridge links $K(\frac{p}{q})$ and $K(\frac{p'}{q'})$ are isotopic if and only if $q' = q$ and, either $p \equiv p'$ or $pp' \equiv 1$ mod q.

Conway [9] connected the two viewpoints, and observed that the arborescent link defined by the stick $\overset{a_1}{\bullet}\!\!-\!\!\overset{a_2}{\bullet}\cdots-\overset{a_n}{\bullet}$ is the 2–bridge link $K(\frac{p}{q})$ associated by Schubert to the rational number defined by the continued fraction

$$\frac{p}{q} = \cfrac{1}{a_1 - \cfrac{1}{a_2 - \cfrac{1}{\cdots - \cfrac{1}{a_n}}}}.$$

Theorem 8 can be seen as an immediate consequence of the Classification Theorem 4 and of classical properties of continued fractions, but the converse actually holds: the proof of Theorem 4 for the case of 2–bridge knots that is given in [7] is based on [17].

17.6.2 Montesinos Knots and Links

A *Montesinos link* is an arborescent link obtained by plumbing according to a weighted planar tree that is a star, namely has at most one vertex of valence $\geqslant 3$. In addition, we will require a Montesinos link to be prime, and to not be a two-bridge link. The exclusion of two-bridge links from the family of Montesinos links is not universally shared, but makes the classification result easier to state. We are letting the reader compare this definition to the one given in [21].

The reduction process of Proposition 3 shows that the positively (or negatively) canonical weighted planar tree Γ of a Montesinos link K is a star, namely has exactly one vertex of valence $\geqslant 3$. A convenient consequence of this fact is that, by the flip (F$_3'$), the precise location of the weights in angular sectors is irrelevant; we can here consider the weights as attached to the vertices of Γ.

Consider a Montesinos link K associated to the (+)–canonical weighted planar tree Γ. We saw that Γ has exactly one vertex v_0 of valence $\geqslant 3$. Index the bonds of v_0 from 1 to m, going clockwise around the vertex. The i–th bond leads to a half-open stick $\overset{a_1}{-\!\bullet}\!\!-\!\!\overset{a_2}{\bullet}\cdots-\overset{a_n}{\bullet}$ in Γ, if we neglect the positions of weights in angular sectors (which, as we noted, are irrelevant by appropriate flips). In this case, we associate to the i–th bond the rational number

$$\frac{p_i}{q_i} = \cfrac{1}{a_1 - \cfrac{1}{a_2 - \cfrac{1}{\cdots - \cfrac{1}{a_n}}}}.$$

An easy consequence of the positive canonicity of Γ is that $\frac{p_i}{q_i}$ belongs to the half-open interval $\left]-\frac{1}{2}, +\frac{1}{2}\right]$ and is different from 0.

This associates to the Montesinos link K its *Montesinos data list*

$$M\left(a_0; \frac{p_1}{q_1}, \frac{p_2}{q_2}, \ldots, \frac{p_m}{q_m}\right)$$

where $a_0 \in \mathbb{Z}$ is the weight of the central vertex v_0 in Γ.

Note that, because of the choice of indexing of the bonds of v_0 and because of the flip (F_1), the list of the $\frac{p_i}{q_i}$ above is only defined up to *dihedral permutation*, namely up to the composition of a cyclic permutation and possibly of an order-reversing permutation.

Theorem 9. Two Montesinos links K and K' are isotopic if and only if their respective data lists $M\left(a_0; \frac{p_1}{q_1}, \frac{p_2}{q_2}, \ldots, \frac{p_m}{q_m}\right)$ and $M\left(a'_0; \frac{p'_1}{q'_1}, \frac{p'_2}{q'_2}, \ldots, \frac{p'_{m'}}{q'_{m'}}\right)$ differ only by a dihedral permutation of the $\frac{p_i}{q_i}$.

In addition, any data list $M\left(a_0; \frac{p_1}{q_1}, \frac{p_2}{q_2}, \ldots, \frac{p_m}{q_m}\right)$ is realized by a Montesinos link provided that $m \geqslant 3$, $a_0 \in \mathbb{Z}$ and all $\frac{p_i}{q_i}$ belong to $\mathbb{Q} \cap \left]-\frac{1}{2}, +\frac{1}{2}\right] - \{0\}$.

Theorem 9 is essentially a rephrasing of Theorem 4 in the context of Montesinos pairs. The notation for two-bridge links and Montesinos links provided by Theorems 8 and 9 is certainly more compact than their description by weighted planar trees. Whether it is more convenient depends on the context.

The Montesinos data list is inspired by Seifert's invariants for Seifert fibrations [18]. Indeed, Seifert manifolds occur as double-branched covers of Montesinos links, as was first discovered by Montesinos in [14]. A major difference when $m \geqslant 4$ is that Seifert's invariants can be arbitrarily permuted, whereas the invariants $\frac{p_i}{q_i}$ of a Montesinos link have a well-defined dihedral order.

In fact, Proposition 7 shows that the Montesinos data lists $M\left(a_0; \frac{p_1}{q_1}, \frac{p_2}{q_2}, \ldots, \frac{p_m}{q_m}\right)$ and $M\left(a'_0; \frac{p'_1}{q'_1}, \frac{p'_2}{q'_2}, \ldots, \frac{p'_{m'}}{q'_{m'}}\right)$ differ by a permutation of the $\frac{p_i}{q_i}$ if and only if the associated Montesinos links K and K' are related by a sequence of mutations along Conway spheres, as defined in §17.5.4.

A drawback of the definition of the Montesinos data list $M\left(a_0; \frac{p_1}{q_1}, \frac{p_2}{q_2}, \ldots, \frac{p_m}{q_m}\right)$ is that, when some $\frac{p_i}{q_i}$ are equal to $+\frac{1}{2}$, it does not behave well under reversal of the ambient orientation. Indeed it easily follows from the arithmetic move (2.1) that, if K is the Montesinos link associated to the data $M\left(a_0; \frac{p_1}{q_1}, \frac{p_2}{q_2}, \ldots, \frac{p_m}{q_m}\right)$, its mirror image $-K$ is associated to $M\left(a'_0; \frac{p'_1}{q'_1}, \frac{p'_2}{q'_2}, \ldots, \frac{p'_{m'}}{q'_{m'}}\right)$ where

- $\frac{p_i'}{q_i'} = -\frac{p_i}{q_i}$ is $\frac{p_i}{q_i} \neq +\frac{1}{2}$;

- $\frac{p_i'}{q_i'} = +\frac{1}{2}$ is $\frac{p_i}{q_i} = +\frac{1}{2}$;

- $a_0' = -a_0 + k$, where k is the number of $\frac{p_i}{q_i}$ that are equal to $+\frac{1}{2}$.

For this reason, it is sometimes more convenient to consider the slightly different data list

$$\mathrm{E}\left(e_0; \left[\tfrac{p_1}{q_1}\right], \left[\tfrac{p_2}{q_2}\right], \ldots, \left[\tfrac{p_m}{q_m}\right]\right)$$

where each $\left[\frac{p_i}{q_i}\right]$ is the class of $\frac{p_i}{q_i}$ in $\mathbb{Q}/\mathbb{Z}$, and where

$$e_0 = -a_0 + \sum_{i=1}^{m} \frac{p_i}{q_i} \in \mathbb{Q}.$$

Here E stands for "Euler," as the number $e_0 \in \mathbb{Q}$ can be interpreted as an Euler class for orbifold bundles; see [6].

The Montesinos data $\mathrm{M}\left(a_0; \frac{p_1}{q_1}, \frac{p_2}{q_2}, \ldots, \frac{p_m}{q_m}\right)$ is easily recovered from the Euler data $\mathrm{E}\left(e_0; \left[\frac{p_1}{q_1}\right], \left[\frac{p_2}{q_2}\right], \ldots, \left[\frac{p_m}{q_m}\right]\right)$, and conversely. Theorem 9 then shows that two Montesinos links are isotopic if and only if their Euler data differ only by a dihedral permutation of the $\left[\frac{p_i}{q_i}\right]$. Also, any data list $\mathrm{E}\left(e_0; \left[\frac{p_1}{q_1}\right], \left[\frac{p_2}{q_2}\right], \ldots, \left[\frac{p_m}{q_m}\right]\right)$ is the Euler data of a Montesinos link provided that $n \geqslant 3$, all $\left[\frac{p_i}{q_i}\right]$ belong to $\mathbb{Q}/\mathbb{Z} - \{0\}$ and $[e_0] = +\sum_{i=1}^{m} \left[\frac{p_i}{q_i}\right]$ in $\mathbb{Q}/\mathbb{Z}$.

The main advantage of this Euler data list $\mathrm{E}\left(e_0; \left[\frac{p_1}{q_1}\right], \left[\frac{p_2}{q_2}\right], \ldots, \left[\frac{p_m}{q_m}\right]\right)$ is that replacing K with its mirror image $-K$ replaces its Euler data with $\mathrm{E}\left(-e_0; \left[-\frac{p_1}{q_1}\right], \left[-\frac{p_2}{q_2}\right], \ldots, \left[-\frac{p_m}{q_m}\right]\right)$. Also, this Euler data list avoids the emphasis on $(+)$–canonical, as opposed to $(-)$–canonical, weighted planar trees.

17.7   Detecting Flip and Ring Equivalences

In view of Theorem 4, we are confronted with the following practical problem: Given two $(\pm)$–canonical weighted planar trees Γ and Γ', decide whether or not it is possible to pass from one to the other by a sequence of flips (F_1), (F_2), (F_3') and modified ring moves $(R^{\pm})$.

When the tree is a stick or a star, we observed in the previous section §17.6 that flips are essentially irrelevant.

For trees that are neither sticks nor stars, a simple strategy to decide if Γ' can be obtained from Γ by a sequence of flips and ring moves is to write down the finite list of all weighted planar trees obtained from Γ by a sequence of flips and ring moves, up to weighted planar tree isomorphism, and then to check whether Γ' appears in the list. This naïve method worked well with the example of Figure 17.7, and is often quite sufficient for small trees. However, it ends up being rather clumsy as soon as the tree reaches a reasonable size.

We now describe a more efficient algorithm to solve this practical problem. The mathematical objects that it involves will also provide us with the appropriate tools to describe the symmetry group of an arborescent link in Section 17.8.

17.7.1 Abbreviated Weighted Planar Trees

We first neutralize the ring moves $(R^{\pm})$. These seldom occur (never for a knot, for instance) and are not difficult to handle, but they can be cumbersome.

For this, we introduce a new type of canonical trees. The move (R^+) shows that any subtree $\overset{2}{\bullet}\!\!\underset{1}{\overset{1}{\text{---}}}\!\!\overset{2}{\bullet}$ of a $(+)$–canonical weighted planar tree Γ can be considered as attached to the vertex of Γ that is adjacent to it, since its precise location in the plane is irrelevant by application of the ring move (R^+). The same applies to the subtrees $\overset{2}{\bullet}\!\!\underset{1}{\text{---}}\!\!\overset{2}{\bullet}$ and $\overset{2}{\bullet}\!\!\underset{1}{\text{---}}\!\!\overset{2}{\bullet}$, obtained from it by the flip (F_3'). We will call a subtree of one of the above three types a *positive ring subtree* of Γ.

It is then natural to simplify Γ in the following way: Erase each positive ring subtree that is attached at a vertex v of valence $\geqslant 3$ of Γ together with the correspond bond at v. Then keep track of the number $r \in \mathbb{N}$ of the ring subtrees that were attached to v by making this vertex hollow and inscribing the integer r inside. Thus we get a weighted planar tree Γ_* of a new type, consisting of a combinatorial tree embedded in the plane, of weights $a \in \mathbb{Z}$ located in some of its angular sectors, and of a new weight $r \in \mathbb{N}$ attached to each vertex and called the *ring number* of this vertex; see Figure 17.8. This weighted planar tree Γ_* is the *abbreviated $(+)$–canonical tree asociated to* the $(+)$–canonical weighted planar tree Γ.

FIGURE 17.8: A vertex with ring number r and integer weights $a_i \in \mathbb{Z}$ in its angular sectors.

Since vertices with positive ring numbers are rare, it is convenient to denote each vertex of Γ_* with ring number 0 by a solid black dot of the type we have used so far, and each vertex with ring number 1 by a small hollow (white) dot, with no indication of ring numbers. For instance, the positively canonical tree of Figure 17.9(a) is abbreviated as the tree of Figure 17.9(b).

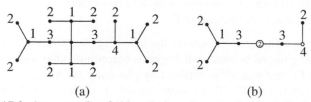

(a)   (b)

FIGURE 17.9: An example of abbreviation of a canonical weighted planar tree.

This construction of Γ_* does not quite make sense when erasing one ring subtree destroys another one. However, this occurs only when Γ is the tree up to a few flips (F_3'). In this case, the abbreviated tree Γ_* is by convention $\overset{2}{\bullet}\!\!\overset{1}{-}\!\!\overset{2}{\bullet}$. Note that this is the only case where a vertex of Γ_* is not canonically associated to a vertex of Γ.

The flips (F_1), (F_2), and (F_3') immediately extend to the framework of abbreviated canonical trees.

The following is essentially a paraphrase of Theorem 4.

Proposition 10. Let the arborescent links K and K' be respectively associated by plumbing to the $(+)$–canonical weighted planar trees Γ and Γ', and let Γ_* and Γ_*' be the abbreviated trees associated to Γ and Γ'. Then K and K' are isotopic if and only if Γ_*' can be obtained from Γ_* by a sequence of flips (F_1), (F_2), (F_3').

These abbreviated weighted planar trees are actually more than a mere combinatorial trick to simplify the calculus of weighted planar trees, and play an important role in the proof of the Classification Theorem 4. Abbreviated weighted planar trees with ring numbers can be used to define more general plumbing constructions, where we associate a *ringed twisted band diagram* of the type shown on Figure 17.10 to each n–valent vertex with ring number r and weights $a_1, a_2, \ldots, a_n \in \mathbb{Z}$ as in Figure 17.8, and we plumb these diagrams together according to the pattern described by the edges of the graph.

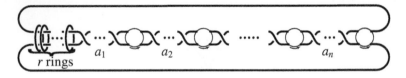

FIGURE 17.10: A ringed twisted band diagram.

The ringed twisted band diagram of Figure 17.10 can be seen as describing a 3–dimensional *ringed twisted band pair* (M_0, K_0), where M_0 is the complement of finitely many disjoint open balls in S_3, and where K_0 is a 1–dimensional submanifold of M_0 such that $\partial K_0 = K_0 \cap \partial M_0$ consists of exactly 4 points in each component of ∂M_0. Then, if the arborescent link K is obtained by plumbing according to the weighted planar tree Γ with ring numbers, the construction provides a decomposition of the pair (S^3, K) into ringed twisted band pairs (M_i, K_i), corresponding to the plumbing blocks associated to the vertices of Γ.

A key step in the proof of Theorem 4 is actually the following. Suppose that the arborescent link K is obtained by plumbing ringed twisted band diagrams according to the abbreviated $(+)$–canonical weighted planar tree Γ_*, and that K' is obtained by plumbing according to the abbreviated $(+)$–canonical weighted planar tree Γ_*'. Then, under suitable hypotheses on Γ, any diffeomorphism $\psi\colon S^3 \to S^3$ sending K to K' can be isotoped, through diffeomorphisms sending K to K', so that it sends each ringed twisted band pair (M_i, K_i) of the plumbing description of K to a ringed twisted band pair (M_j', K_j') of the plumbing description of K'. As a consequence, ψ then induces a combinatorial isomorphism $\varphi\colon \Gamma_c \to \Gamma_c'$ between the combinatorial trees Γ_c and Γ_c' respectively underlying Γ and Γ'.

A second step in the proof of Theorem 4 is that, if ψ is orientation-preserving, this combinatorial isomorphism $\varphi \colon \Gamma_c \to \Gamma'_c$ can be realized by a sequence of flips (F_1), (F_2), (F'_3). Namely, ϕ is what we call a degree $+1$ isogeny in the next section.

To further emphasize the importance of ringed twisted band pairs, we should mention the pioneering work of Montesinos [14], who first observed that a ringed twisted band pair (M_0, K_0) has a unique two-fold branched cover, ramified along K_0, of a very special type. If (M_0, K_0) has $n \geq 0$ boundary components and $r \geq 0$ rings, this two-fold branched cover is a locally trivial bundle with fiber the circle S^1 over the surface that is the connected sum of the n-times punctured sphere with r copies of the projective plane. This fundamental observation provided a key link between the classification problem for arborescent links and Waldhausen's classification of graph manifolds.

17.7.2 An Algorithm

There is a rather efficient algorithm to decide whether two abbreviated $(+)$–canonical trees are flip equivalent or not. The algorithm is based on the following observation: Applying one of the flips (F_1), (F_2), (F'_3) to an abbreviated canonical tree Γ gives another abbreviated canonical tree *together with a preferred isomorphism* $\varphi \colon \Gamma_c \to \Gamma'_c$ between the combinatorial trees Γ_c and Γ'_c respectively underlying Γ and Γ'. Similarly, an isomorphism between two weighted planar trees between Γ and Γ' induces a combinatorial isomorphism $\varphi \colon \Gamma_c \to \Gamma'_c$. We call an isomorphism φ of the above two types an *elementary isogeny* from Γ to Γ'.

A *degree $+1$ isogeny* between the abbreviated trees Γ and Γ' is a combinatorial isomorphism $\varphi \colon \Gamma_c \to \Gamma'_c$ which can be decomposed as a composition of elementary isogenies, namely such that there exists a sequence of abbreviated weighted planar trees $\Gamma = \Gamma^{(0)}$, $\Gamma^{(1)}$, ..., $\Gamma^{(n)} = \Gamma'$ and elementary isogenies $\varphi_i \colon \Gamma_c^{(i-1)} \to \Gamma_c^{(i)}$ such that $\varphi = \varphi_n \circ \cdots \circ \varphi_1$. A notion of degree -1 isogeny will also appear in the next section §17.8, when we determine the symmetries of arborescent links.

By construction, a degree $+1$ isogeny respects weights and ring numbers at each vertex. It also respects, up to orientation-reversal, the cyclic order of the bonds specified by Γ and Γ'. Let us refer to such a cyclic order defined up to orientation-reversal as a *dihedral order* of the bonds at a given vertex.

With this data, the problem of deciding whether two abbreviated $(+)$–canonical trees are flip equivalent can be split into two steps:

1. List all combinatorial isomorphisms $\varphi \colon \Gamma_c \to \Gamma'_c$ that respect weights, ring numbers and dihedral orders of bonds.

2. Given such a $\varphi \colon \Gamma_c \to \Gamma'_c$, decide whether it is a degree $+1$ isogeny or not.

The first step is easily accomplished, at least when the sizes of Γ and Γ' are moderate. We consequently focus attention on the second problem.

The general strategy for this second problem is to progressively modify Γ' by flips, composing φ with the corresponding elementary isogenies, so that φ respects the cyclic order of weights and bonds at as many angular sectors of Γ as possible. The algorithm is then based on the following "magical" fact, stated as Proposition 11 below, which can be paraphrased as follows: *When there* *is no obvious way to go one step further in this construction, there actually is a good reason for this because φ is not a degree +1 isogeny.*

To describe the algorithm, let us start with two abbreviated (+)–canonical weighted planar trees Γ and Γ' and with an isomorphism $\varphi \colon \Gamma_c \to \Gamma'_c$ between their underlying combinatorial trees, respecting weights, ring number and dihedral order of bonds at the vertices.

If all the vertices have valence $\leqslant 2$, a succession of flips (F'_3) readily shows that φ is a degree +1 isogeny. We can therefore assume that Γ contains a vertex v_1 of valence $\geqslant 3$.

Then, list all the vertices as $v_1, v_2, \ldots, v_n$ in such a way that each v_i is adjacent to a v_j with $j < i$.

As a first step, modify Γ' by flips (F_1) and/or (F'_3) if necessary so that φ respects the cyclic order of weights and bonds at v_1.

Now, for $i \geqslant 2$, we assume as an induction hypothesis that we have succeeded in making φ respect the cyclic order of weights and valences at each v_j with $j < i$.

If $i = n + 1$, then φ is a weighted planar tree isomorphism, and is consequently a degree +1 isogeny. Otherwise, we distinguish two cases.

Case 1. *The isomorphism φ respects the cyclic order of bonds at v_i. (This always holds if v_i has valence $\leqslant 2$.)*

In this case, φ induces a unique correspondence between the angular sectors at v_i and those at $\varphi(v_i)$. Making sense of this statement is obvious when the valence of v_i is different from 2. When v_i has valence 2, one needs to use the fact that φ respects the cyclic order of bonds at each v_j with $j < i$, and that there exists at least one such v_j of valence $\geqslant 3$.

By a series of flips (F'_3) performed at v_i, we can then arrange that the weight of $\varphi(v_i)$ in Γ' is located in the angular sector that is the image under φ of the angular sector containing the weight of v_i in Γ. Moreover, by our indexing convention, all v_j with $j < i$ are the vertices of a subtree of Γ, and we can choose the above flips (F'_3) so that they fix this subtree, guaranteeing that the induction hypothesis is still satisfied.

We now have arranged that φ respects the cyclic order of weights and bonds at all vertices v_j with $j \leqslant i$, and we can proceed further in the induction.

Case 2. φ *reverses the cyclic order of bonds at v_i.*

We are stuck. We make another attempt, starting from a different Γ'. Perform on Γ' a flip (F_2) that respects the cyclic order at $\varphi(v_1)$ and reverses it at every vertex at odd distance from $\varphi(v_1)$. This gives a new weighted planar tree $\widehat{\Gamma}'$ and a combinatorial isomorphism $\widehat{\varphi} \colon \Gamma_c \to \widehat{\Gamma}'_c$ obtained by composing φ with the isogeny associated to the flip.

We again try to make $\widehat{\varphi}$ respect the cyclic order of weights and bonds at as many vertices as possible, by modifying $\widehat{\Gamma}'$ (and $\widehat{\varphi}$ accordingly) by a sequence of flips (F'_3) that keep $\widehat{\varphi}(v_1)$ in the subgraph they fix. So, reapply the algorithm right from the beginning, with $\widehat{\Gamma}'$ and $\widehat{\varphi}$ instead of Γ' and φ. This leads to two subcases:

Subcase 2a. *After a certain number of modifications as in Case 1, $\widehat{\varphi}$ eventually respects the cyclic order of weights and bonds at all v_j with $j \leqslant i$.*

Then, replace Γ' by $\widehat{\Gamma'}$ and φ by $\widehat{\varphi}$, and go one step further in the induction.

Subcase 2b. *The algorithm applied to $\widehat{\Gamma'}$ and $\widehat{\varphi}$ again leads to Case 2 for some v_j with $j \leqslant i$.*

We are again stuck. However, the following result, proved in [7], enables us to conclude.

Proposition 11. If Subcase 2b holds at some point, then φ cannot be a degree $+1$ isogeny.

In particular the above process is really an algorithm since, either it explicitly exhibits φ as a degree $+1$ isogeny, or it concludes that φ is not a degree $+1$ isogeny.

17.8   Symmetries of Arborescent Links

The *symmetry group* of a link $K \subset S^3$ is the group $\mathrm{Sym}(K) = \pi_0 \,\mathrm{Diff}(S^3, K)$, consisting of all diffeomorphisms of S^3 sending K to itself and considered modulo isotopy respecting K. The elements of $\mathrm{Sym}(K)$ represented by orientation-preserving diffeomorphisms of S^3 form the *degree $+1$ symmetry group* $\mathrm{Sym}^+(K)$, a subgroup of $\mathrm{Sym}(K)$ of index 1 or 2.

The techniques underlying the proof of the Classification Theorem 4 can also be used to determine the symmetry group of most arborescent links.

Let K be a prime arborescent link. By Propositions 2 and 3, it is associated by plumbing to an abbreviated (+)–canonical weighted planar tree Γ, as defined in §17.7.1. We will determine $\mathrm{Sym}(K)$ in terms of certain symmetries of the tree Γ, called isogenies.

17.8.1   Degree ± 1 Isogenies

We already encountered degree $+1$ isogenies in §17.7.2. Recall that an *elementary degree $+1$ isogeny* $\Gamma \to \Gamma'$ between the abbreviated (+)–canonical weighted planar trees Γ and Γ' is the preferred combinatorial isomorphism $\Gamma_c \to \Gamma'_c$ between their underlying combinatorial trees that arises when Γ and Γ' are related by a weighted planar tree isomorphism or by a flip (F_1), (F_2) or (F'_3). A degree $+1$ isogeny from Γ to Γ' was defined as a composition of such elementary degree $+1$ isogenies.

We now define elementary degree -1 isogenies. If the arborescent link K is obtained by plumbing according to the abbreviated (+)–canonical weighted planar tree Γ, the mirror image $-K$ is obtained by plumbing according to the abbreviated (−)–canonical weighted planar tree $-\Gamma$, obtained by reversing the sign of all weights of Γ. Let Γ' be the abbreviated (+)–canonical weighted planar tree obtained from $-\Gamma$ by performing as many moves (1.0), (2.0), (2.1), (2.2) as necessary. These moves induce a preferred combinatorial isomorphism $\Gamma_c \to \Gamma'_c$ between the combinatorial trees Γ_c and Γ'_c underlying Γ and Γ'. By definition, an *elementary degree -1 isogeny* $\Gamma \to \Gamma'$ is any combinatorial isomorphism $\Gamma_c \to \Gamma'_c$ associated in this way to an abbreviated (+)–canonical weighted planar tree Γ.

We define an *isogeny* $\Gamma \to \Gamma'$ between the (+)–canonical weighted planar trees Γ and Γ' as any composition $\Gamma = \Gamma_0 \to \Gamma_1 \to \cdots \to \Gamma_n = \Gamma'$ of elementary degree ± 1 isogenies $\Gamma_i \to \Gamma_{i+1}$. The isogeny has *degree* $+1$ or -1 according to whether there is an even or odd number of elementary degree -1 isogenies in this decomposition. It is not hard to see that one can always choose the decomposition so that it contains at most one elementary degree -1 isogeny.

An isogeny may exceptionally be of both degree $+1$ and -1. The simplest examples are the isogenies $\Gamma \to \Gamma$ corresponding to the identity of the underlying combinatorial graph Γ_c when Γ is $\overset{1}{\bullet}$,

$\overset{2}{\bullet}$, $\overset{2}{\bullet}\!\!-\!\!\overset{1}{\circ}\!\!-\!\!\overset{2}{\bullet}$ or $\begin{matrix} {}^{2}\!\bullet & {}^{2}\bullet{\diagup}{}^{2} \\ & \times \\ {}_{2}\!\bullet & \bullet{}_{2} \end{matrix}$.

To completely analyze this pathology, let us call a vertex of Γ with integer weight 1 or 2 *active* if an arithmetic move (1.0), (2.0), (2.1), or (2.2) is performed there in the standard transition from $-\Gamma$ to a (+)–canonical weighted planar tree. In particular, an active vertex has ring number 0.

Proposition 12. For an abbreviated (+)–canonical weighted planar tree Γ, the following are equivalent:

1. there exists an isogeny $\Gamma \to \Gamma'$ which has both degree $+1$ and degree -1;

2. the identity isogeny $\Gamma \to \Gamma$ has both degree $+1$ and degree -1;

3. $-\Gamma$ is related to Γ by a series of flips (F_1), (F_2), (F'_3) and of arithmetic moves (1.0), (2.0), (2.1), (2.2);

4. for every vertex v of Γ, either v is active, or it is adjacent to $2n$ active vertices and has weight $n \geqslant 0$.

For a given abbreviated (+)–canonical weighted planar tree Γ, the isogenies $\Gamma \to \Gamma$ are called the *symmetries* of Γ and form a subgroup $\mathrm{Sym}(\Gamma)$ of the group of all isomorphisms of the combinatorial tree underlying Γ. For a given Γ, the symmetry group $\mathrm{Sym}(\Gamma)$ is usually rather easy to calculate explicitly, using the algorithm of Section 17.7.2.

17.8.2 Computing the Symmetries of an Arborescent Link

Let K be the arborescent link associated by plumbing ringed twisted band diagrams according to the abbreviated (+)–canonical weighted planar tree Γ. As indicated in our discussion of the proof of Theorem 4 at the end of §17.7.1, any diffeomorphism $\psi \colon S^3 \to S^3$ can usually be deformed so that it sends each plumbing block (M_i, K_i) to another block (M_j, K_j). As such, ψ induces an isomorphism of the combinatorial tree Γ_c underlying Γ, which can be shown to be an isogeny. We can then attempt to define a group homomorphism $\rho \colon \mathrm{Sym}(S^3, K) \to \mathrm{Sym}(\Gamma) \times \mathbb{Z}_2$ by associating to ψ the isogeny $\varphi \in \mathrm{Sym}(\Gamma)$ so induced, and the degree $\varepsilon(\psi) = \pm 1 \in \mathbb{Z}_2$ of ψ (which is also *a* degree of the isogeny φ).

Theorem 13. Let K be an arborescent link in S^3, associated by plumbing to the abbreviated (+)–canonical weighted planar tree Γ. Assume that Γ is neither $\overset{0}{\bullet}$, nor $\overset{2}{\bullet}\!\!-\!\!\overset{1}{\bullet}\!\!-\!\!\overset{2}{\bullet}$, nor $\begin{matrix} {}^{2}\!\bullet & \bullet{}^{2} \\ & \\ {}_{2}\!\bullet & \bullet{}_{2} \end{matrix}$, nor a three-arm star (with one vertex of valence 3 and all other vertices of valence $\leqslant 2$) where all ring numbers are 0. Then the above group homomorphism

$$\rho\colon \operatorname{Sym}(S^3, K) \to \operatorname{Sym}(\Gamma) \times \mathbb{Z}_2$$

is well defined. The image of ρ consists of all pairs (φ, ε) where the isogeny $\varphi \in \operatorname{Sym}(\Gamma)$ has degree $\varepsilon \in \{\pm 1\} = \mathbb{Z}_2$.

Suppose in addition that all ring numbers in Γ are $\leqslant 1$. If Γ contains at least one vertex of valence 3, the kernel $\mathcal{K}(\Gamma)$ of ρ is isomorphic to $\mathbb{Z}_2$ or to 0. If every vertex of Γ has valence $\leqslant 2$ and Γ contains at least two vertices, $\mathcal{K}(\Gamma)$ is isomorphic to $\mathbb{Z}_2 \times \mathbb{Z}_2$. Finally, if Γ consists of a single vertex with weight $a \in \mathbb{Z}$ (necessarily with ring number 0 and with $a \neq 0, -1, -2$), $\mathcal{K}(\Gamma)$ is isomorphic to $\mathbb{Z}_2 \times \mathbb{Z}_2$ if a is even and to $\mathbb{Z}_2$ if a is odd.

In particular, the group $\operatorname{Sym}(K)$ is finite in all the cases where Theorem 13 computes it.

When all vertices of Γ have valence $\leqslant 2$ and ring number 0, the associated link K is a 2-bridge link, and Theorem 13 is essentially a rephrasing of Horst Schubert's pioneering work in [17].

17.8.3 Hidden Symmetries

When Γ has at least one vertex of valence $\geqslant 3$, Theorem 13 is still incomplete because it only says that the kernel $\mathcal{K}(\Gamma)$ of ρ is isomorphic to $\mathbb{Z}_2$ or to 0. We now make precise this result, namely we determine when the link K has a "hidden symmetry" which acts by the identity on the graph Γ.

It turns out that $\mathcal{K}(\Gamma)$ depends only on the weighted tree Γ_2 that consists of the combinatorial tree Γ_c underlying Γ together with, at each vertex, the class $\bar{a} \in \mathbb{Z}_2$ of the weight $a \in \mathbb{Z}$ of this vertex in Γ. In other words, Γ_2 is obtained from Γ by forgetting the embedding in the plane and the ring numbers, and by keeping only the mod 2 class of the integer weight of vertices. The arithmetic moves (0.2) and (1.2) immediately extend to such $\mathbb{Z}_2$–weighted trees.

Proposition 14. Let the abbreviated (+)–canonical weighted planar tree satisfy the hypotheses of the two parts of Theorem 13, and let Γ_2 be the $\mathbb{Z}_2$–weighted tree associated to Γ as above. Performing arithmetic moves (0.2) *and* (1.2) on Γ_2 sufficiently often, one necessarily reaches a graph Γ_2' for which one of the following holds:

1. some edge of Γ_2' joins two vertices of valence $\geqslant 3$;

2. exactly one vertex of Γ_2' has valence $\geqslant 3$.

Then the kernel $\mathcal{K}(\Gamma)$ of the homomorphism ρ of Theorem 13 is isomorphic to 0 in the first case, and to $\mathbb{Z}_2$ in the second case.

As an example, the arborescent knot K associated to the (+)–canonical tree

is the knot labelled as 8_{17} in [1, 16]. This tree has a degree -1 isogeny, factoring through the weighted planar tree

One easily checks that this is the only non-trivial symmetry of Γ, so that $\mathrm{Sym}(\Gamma) \cong \mathbb{Z}_2$. Proposition 12 shows that no graph symmetry has both degree $+1$ and -1, and the algorithm of §17.7.2 shows that the kernel $\mathcal{K}(\Gamma)$ is trivial. Therefore, by Theorem 13, the symmetry group $\mathrm{Sym}(S^3, K)$ has exactly two elements, the identity and a symmetry reversing the orientation of S^3.

In particular, this shows that 8_{17} is not *invertible*, in the sense that it admits no symmetry that preserves the orientation of S^3 but reverses the orientation of K. This is the first knot in the tables to have this property. This was independently proved by A. Kawauchi [10], using more algebraic techniques.

By inspection, the nontrivial symmetry of 8_{17} reverses both the orientation of S^3 and the orientation of K. Namely, 8_{17} is *negative amphicheiral*.

17.8.4   The Exceptional Cases of Theorem 13

We briefly discuss the exceptions of Theorem 13.

When Γ is $\overset{0}{\underset{\bullet}{\bullet}}$, the link K is the two-component unlink, bounding two disjoint disks in S^3. Its symmetry group $\mathrm{Sym}(K)$ is well known to be $(\mathbb{Z}_2)^3$, the isomorphism being given by considering the action of a symmetry on the components of K and their orientations.

In the excluded case where Γ is $\overset{2}{\bullet}\!-\!\!\overset{1}{\circ}\!\!-\!\overset{2}{\bullet}$, the link K is the Borromean rings and is computed by the following statement.

Proposition 15. When Γ is $\overset{2}{\bullet}\!-\!\!\overset{1}{\circ}\!\!-\!\overset{2}{\bullet}$, so that the corresponding link K is the Borromean rings, consider the group homomorphism $\rho \colon \mathrm{Sym}(K) \to \mathfrak{S}_3 \times \mathbb{Z}_2$ which to a symmetry associates its permutation of the components of K and its degree as a map $S^3 \to S^3$. Then ρ is surjective and its kernel is isomorphic to $\mathbb{Z}_2 \times \mathbb{Z}_2$. In particular, $\mathrm{Sym}(K)$ has order 24.

When Γ is $\overset{2\quad\;\; 2}{\underset{2\quad\;\; 2}{\times}}$, the link K is obtained from the Hopf link by replacing each component with two parallel unlinked copies. In particular, the complement of K is a graph manifold, and the group $\mathrm{Sym}(K)$ is easily computed using [20]. An element of $\mathrm{Sym}(K)$ is determined by its action on the oriented components of K, and $\mathrm{Sym}(K)$ is isomorphic to $(\mathbb{Z}_2)^5$.

When some ring numbers are $\geqslant 2$, the kernel $\mathcal{K}(\Gamma)$ of Theorem 13 includes an additional copy of the braid group B_r for each vertex with ring number $r \geqslant 2$. In particular, it is infinite.

The methods of [7] fail to compute the symmetry group of the link when Γ is a three-arm star where all the ring numbers are 0 (with the exception of a few *ad hoc* examples). However, Boileau and Zimmermann [3] use different techniques to extend the results of Theorem 13 to "most" of these exceptions.

Acknowledgments

While this article was being written, the author's research was partially supported by the grant DMS-1711297 from the U.S. National Science Foundation. The author also owes much gratitude to Larry Siebenmann for their collaboration in [7], which forms the basis of this survey.

17.9 Bibliography

[1] J. W. Alexander, G. B. Briggs, On types of knotted curves, *Annals of Math.* 28 (1926), 562–586.

[2] M. Boileau, L.C. Siebenmann, *A planar Classification of Pretzel Knots and Montesinos Knots*, Publications Mathématiques d'Orsay, Université de Paris-Sud, 1980.

[3] M. Boileau, B. Zimmermann, Symmetries of nonelliptic Montesinos links, *Math. Ann.* 277 (1987), 563–584.

[4] F. Bonahon, *Involutions et fibrés de Seifert dans les variétés de dimension* 3, Thèse de 3ème cycle, University of Paris XI, Orsay, 1979.

[5] F. Bonahon, L.C. Siebenmann, The characteristic toric splitting of irreducible compact 3-orbifolds, *Math. Ann.* 278 (1987), 441–479.

[6] F. Bonahon, L.C. Siebenmann, The classification of Seifert fibred 3-orbifolds, in: *Low-dimensional Topology (Chelwood Gate, 1982)*, 19–85, London Math. Soc. Lecture Note 95, Cambridge University Press, 1985.

[7] F. Bonahon, L.C. Siebenmann, New geometric splittings of classical knots, and the classification and symmetries of arborescent knots, preprint, 2016 version available at `https://dornsife.usc.edu/francis-bonahon/preprints/`

[8] G. Burde, H. Zieschang, *Knots*, de Gruyter Studies in Math. 5, Walter de Gruyter & Co., Berlin, 1985.

[9] J. H. Conway, An enumeration of knots and links, and some of their algebraic properties, in: 1970 *Computational Problems in Abstract Algebra (Proc. Conf., Oxford, 1967)*, 329–358, Pergamon Press, 1970.

[10] A. Kawauchi, The invertibility problem on amphicheiral excellent knots, *Proc. Japan Acad. Ser. A Math. Sci.* 55 (1979), 399–402.

[11] M. A. Kervaire, A manifold which does not admit any differentiable structure, *Comment. Math. Helv.* 34 (1960), 257–270.

[12] M. A. Kervaire and J. W. Milnor, Groups of homotopy spheres. I, *Ann. of Math.* 77 (1963), 504–537.

[13] J. W. Milnor, Differentiable structures on spheres, *Amer. J. Math.* 81 (1959), 962–972.

[14] J.-M. Montesinos, Variedades de Seifert que són recubridores cíclicos de dos hojas, *Bol. Soc. Mat. Mexicana* 18 (1973), 1–32.

[15] H. R. Morton, Mutant knots, in: *New Ideas in Low Dimensional Topology*, 379–412, Ser. Knots Everything 56, World Scientific Publishing, 2015.

[16] D. Rolfsen, *Knots and links*, Mathematics Lecture Series, Publish or Perish, Inc., 1976.

[17] H. Schubert, Knoten mit zwei Brücken, *Math. Z.* 65 (1956), 133–170.

[18] H. Seifert, Topologie dreidimensionaler gefaserter Räume, *Acta Math.* 60 (1933), 147–238.

[19] L. C. Siebenmann, On vanishing of the Rohlin invariant and nonfinitely amphicheiral homology 3-spheres, in: *Topology Symposium, Siegen 1979 (Proc. Sympos., Univ. Siegen, Siegen, 1979)*, 172–222, Lecture Notes in Math. 788, Springer, 1980.

[20] F. Waldhausen, Eine Klasse von 3-dimensionalen Mannigfaltigkeiten, *Invent. Math.* 3 (1967), 308–333; *Invent. Math.* 4 (1967), 87–117.

[21] R. Wilson, Rational knots and their generalizations, this *Encyclopedia*, chapter 16.

Chapter 18

Satellite Knots

Jennifer Schultens, University of California, Davis

18.1 Introduction

Essential tori in knot complements were first studied by Horst Schubert in the 1950s. He described the satellite construction for knots and established several natural properties of this construction. This chapter considers satellite knots and how they fit into a more recent discussion of 3-manifolds, most notably JSJ decompositions and geometrization.

In the 1950s Horst Schubert developed an understanding of essential tori in knot complements. His work can be interpreted as a specialized version and precursor of the work of Jaco-Shalen and Johannson in the 1970s. Specifically, Schubert described JSJ decompositions for knot complements, see [7], and proved that they were finite, see Section 18.7.

Let K be a knot in $\mathbb{S}^3$. An *essential torus* T in the complement of K is, by definition, incompressible and non-peripheral in the complement of K, but necessarily compressible in $\mathbb{S}^3$. Via an exercise in Dehn's lemma and the Schönflies theorem it follows that T bounds a solid torus containing K. Denote this solid torus by V. The complement of V is a knot complement (usually not homeomorphic to the complement of K). This consideration gives rise to the following definition.

18.2 Definition of Satellite Knot

Definition 1. Let J be a nontrivial oriented knot in $\mathbb{S}^3$ and V a closed regular neighborhood of J. Let $\tilde{V}$ be an oriented unknotted closed solid torus in $\mathbb{S}^3$ and $\tilde{K}$ an oriented knot in the interior of $\tilde{V}$. The pair $(\tilde{V}, \tilde{K})$ is called the *pattern*. A meridional disk of $\tilde{V}$ will meet $\tilde{K}$ in a finite subset. The least number of times a meridional disk of $\tilde{V}$ must meet $\tilde{K}$ is called the *wrapping number* of the pattern. Suppose that the wrapping number of the pattern is greater than zero and let $h : (\tilde{V}, \tilde{K}) \to (V, K)$ be an oriented homeomorphism of pairs. The image of $\tilde{K}$ under h, denoted by K, is a knot in $V \subset \mathbb{S}^3$ called a *satellite knot*.

The knot J is called a *companion knot* of K and the torus $T = \partial V$ is called a *companion torus*.

Note that the homeomorphism h in the above definition is unique up to isotopy. The use of the word "satellite" alludes to the fact that a satellite knot "orbits" around its companion knot.

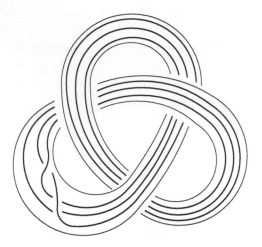

FIGURE 18.1: A satellite knot.

18.3   Case Studies

Three special cases deserve to be pointed out:

Definition 2. A satellite knot with a torus knot as its pattern is called a *cabled knot*. More specifically, a satellite knot with pattern the (p, q)-torus knot and companion J is called a (p, q)-*cable of* J.

See Figure 18.2.

Definition 3. A satellite knot with a pattern of wrapping number 1 is called a *connected sum* of knots. The companion torus of a satellite knot with wrapping number 1 is called a *swallow-follow torus*.

We must reconcile the definition just given with the standard definition, which defines the connected sum of knots $K_1 \subset \mathbb{S}^3$ and $K_2 \subset \mathbb{S}^3$ in terms of the pairwise connected sum $(K, \mathbb{S}^3) = (K_1, \mathbb{S}^3) \# (K_2, \mathbb{S}^3)$. This is accomplished by setting $K_1 = J$ and $K_2 = \tilde{K}$ or vice versa. See Figures 18.3 and 18.4 and Section 18.4 below.

Definition 4. A satellite knot with the pattern pictured in Figure 18.5 is called a *doubled knot*.

18.4   A Theorem of Schubert on Companion Tori

In the 1950s Schubert investigated companion tori and how they lie with respect to each other.

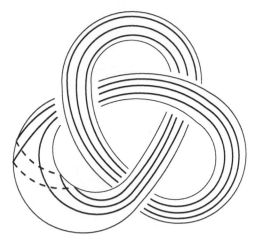

FIGURE 18.2: The $(3, 2)$-cable of the trefoil.

FIGURE 18.3: A swallow-follow torus.

FIGURE 18.4: The connected sum of the figure-8 knot and the trefoil.

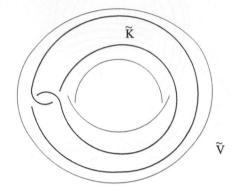

FIGURE 18.5: The pattern for a doubled knot.

FIGURE 18.6: The double of the trefoil.

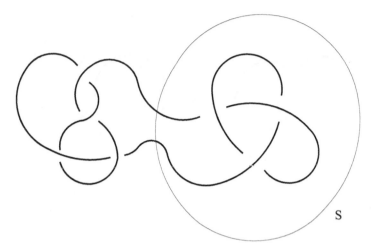

FIGURE 18.7: The connected sum of the figure 8 knot and the trefoil with a decomposing sphere.

Definition 5. A *decomposing sphere* for a knot $K \subset \mathbb{S}^3$ is a 2-sphere S that meets K in exactly two points and separates K into two knotted arcs. (*I.e.*, for B the closure of a 3-ball complementary to S, $K \cap B$ must not be parallel into ∂B.)

Consider a decomposing sphere for the knot K. It is the 2-sphere along which a connected sum of knots

$$(\mathbb{S}^3, K) = (\mathbb{S}^3, K_1) \# (\mathbb{S}^3, K_2)$$

is performed. Given K, a closed regular neighborhood $N(K)$, and the decomposing sphere S, the boundary of the closure of a component of $\mathbb{S}^3 - (N(K) \cup S)$ is a torus. In fact, it is a swallow-follow torus after an isotopy. See Figures 18.3 and 18.8.

For $K = K_1 \# K_2$, denote the swallow-follow torus that swallows K_1 and follows K_2 by T_1 and the swallow-follow torus that swallows K_2 and follows K_1 by T_2. We call T_1 and T_2 *complementary* swallow-follow tori. For $i = 1, 2$, T_i bounds a solid torus containing K which we denote by V_i. If both swallow-follow tori are as in Figure 18.8, then $T_1 \cap T_2 = \emptyset$. Interestingly, V_1 then contains the closure of the complement of V_2 and vice versa.

Unless K_1 or K_2 is the trivial knot, the swallow-follow torus T_1 will not be isotopic to T_2. If T_1 and T_2 are as in Figure 18.8, then they will be disjoint, but we are interested in positioning them as in Figure 18.3. They will then meet in two simple closed curves that are meridians in both T_1 and T_2. Assume that this is the case, then the solid tori $V_1 \cup V_2$ form an unknotted solid torus.

We build a satellite knot L with companion any nontrivial knot and pattern $(K, unknotted\ solid\ torus)$. The images of T_1 and T_2 can no longer be isotoped to be disjoint and isotoping them to meet in two simple closed curves that are meridians in both the image of T_1 and the image of T_2 provides the best positioning of the two tori.

These best possible positionings for tori are described in the theorem below, where Schubert summarizes his findings concerning essential tori in knot complements:

Theorem 1. (Schubert) Let T_1, T_2 be distinct companion tori of a knot K and V_1, V_2 the solid tori

FIGURE 18.8: A swallow-follow torus after an isotopy.

they bound. (*I.e.*, V_1, V_2 each contain K, $T_1 = \partial V_1$ and $T_2 = \partial V_2$.) The solid tori V_1 and V_2 can be isotoped so that (at least) one of the following holds:

- V_1 lies in the interior of V_2;

- V_2 lies in the interior of V_1;

- V_1 contains the closure of the complement of V_2 and V_2 contains the closure of the complement of V_1;

- T_1 meets T_2 in two simple closed curves that are meridians in both solid tori.

For more information on the modern point of view on companion tori in knot complements, that is, as a characteristic submanifold of a JSJ decomposition, see [1].

18.5   The Whitehead Manifold

The double of a knot, discussed in Section 18.3, is also called the Whitehead double. Indeed, note that the complement of $\tilde{K}$ in $\tilde{V}$, for $(\tilde{K}, \tilde{V})$ as pictured in Figure 18.5, is exactly the complement of the Whitehead link, see Figure 18.9.

Definition 6. The *Whitehead manifold* is the 3-manifold M^W obtained by iterating the satellite construction as follows: To begin, we choose $\tilde{K}$ from Figure 18.5 as our companion and also choose $(\tilde{K}, \tilde{V})$ from Figure 18.5 as our pattern to obtain the satellite knot K_1. In Step 2 we choose K_1 as our companion and again choose $(\tilde{K}, \tilde{V})$ from Figure 18.5 as our pattern to obtain the satellite knot K_2. We continue in this fashion to obtain a sequence of satellite knots $K_1, K_2, K_3, \ldots$ each of which is an unknot in $\mathbb{S}^3$. For $i = 1, 2, 3, \ldots$, choose a closed regular neighborhood N_i of K_i such that

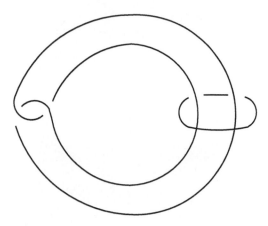

FIGURE 18.9: The Whitehead link.

$N_1 \supset N_2 \supset N_3 \supset \ldots$. We set

$$N_\infty = \cap_{i=1}^\infty N_i$$

and

$$M^W = \mathbb{S}^3 - N_\infty.$$

We can think of the Whitehead manifold as the complement of an iterated satellite knot. Note that for $i = 1, 2, 3, \ldots$ the homomorphism

$$\pi_1(N_{i+1}) \to \pi_1(N_i)$$

is trivial. The Whitehead manifold is interesting for several reasons. Most importantly, it is contractible, yet not homeomorphic to $\mathbb{R}^3$. For more information, see [11] or [5].

18.6 The Genus of a Satellite Knot

It is interesting to consider how invariants of knots behave with respect to the satellite construction.

Definition 7. A *Seifert surface* of a knot K is a compact orientable surface $S \subset \mathbb{S}^3$ such that $\partial S = K$. The *genus* of K is the least possible genus of a Seifert surface of K.

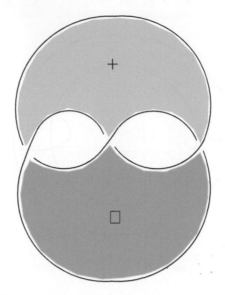

FIGURE 18.10: A Seifert surface for the trefoil.

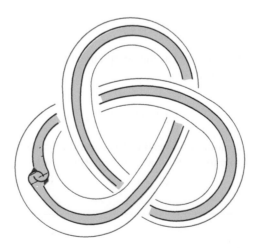

FIGURE 18.11: A Seifert surface for the double of the trefoil.

Figures 18.10 and 18.11 illustrate that the genus of a satellite knot is not necessarily greater than that of its companion. Nevertheless, as we shall see below, the behavior of genus vis-à-vis the satellite construction is completely understood.

Let V be a knotted solid torus in $\mathbb{S}^3$. Then V is homeomorphic to $\mathbb{D}^2 \times \mathbb{S}^1$. Under this homeomorphism, $\mathbf{0} \times \mathbb{S}^1$ is homeomorphic to a curve c called the *core* of V. Since c is a knot in $\mathbb{S}^3$, it has a Seifert surface S. After isotopy, if necessary, the Seifert surface S meets ∂V in a single simple closed curve c'. We leave it as an exercise to show that the isotopy class of c' does not depend on our choice of Seifert surface S.

Definition 8. A simple closed curve in ∂V parallel to c' is called a *preferred longitude*.

We will build a Seifert surface for a satellite knot from a surface inside the companion torus along with Seifert surfaces for the companion torus. The surface inside the companion torus should mimic the features of a Seifert surface:

Definition 9. Let $(\tilde{K}, \tilde{V})$ be a pattern. Orient $\tilde{K}$, $\tilde{V}$ and a meridian disk of $\tilde{V}$. Then the oriented intersection number r between $\tilde{K}$ and the meridian disk of $\tilde{V}$ is called the *winding number* of $(\tilde{K}, \tilde{V})$.

Let $(\tilde{K}, \tilde{V})$ be a pattern with winding number r. A *relative Seifert surface* is a compact orientable surface in $\tilde{V}$ whose interior is disjoint from $\tilde{K}$ and whose boundary consists of $\tilde{K}$ together with r disjoint coherently oriented preferred longitudes.

The *genus* of $(\tilde{K}, \tilde{V})$, denoted by $genus(\tilde{K}, \tilde{V})$, is the smallest possible genus of a relative Seifert surface for $(\tilde{K}, \tilde{V})$.

It is important to distinguish the wrapping number and winding number of a pattern. For instance, the wrapping number of the pattern in Figure 18.5 is two, whereas the winding number is zero.

We obtain a Seifert surface for K from a relative Seifert surface for $(\tilde{K}, \tilde{V})$ by capping off the r preferred longitudes in ∂V with Seifert surfaces for the companion. We thereby obtain the following inequality:

$$genus(K) \le r \cdot genus(J) + genus(\tilde{K}, \tilde{V})$$

Using a delicate argument, Schubert proved that the reverse inequality also holds, thereby establishing the following:

Theorem 2. (Schubert) Let K be a satellite knot with companion J and pattern $(\tilde{K}, \tilde{V})$, then

$$genus(K) = r \cdot genus(J) + genus(\tilde{K}, \tilde{V}).$$

18.7 Bridge Numbers

The bridge number of a knot, denoted $b(K)$, discussed elsewhere in this volume, assigns a natural number to each knot and behaves well with respect to the satellite construction. Specifically, Schubert proves the following (see [8] or [9]):

Theorem 3. (Schubert) Let K be a satellite knot with companion J and pattern $(\tilde{K}, \tilde{V})$ of wrapping number k. Then

$$b(K) \geq k \cdot b(J).$$

If $K = K_1 \# K_2$, then

$$b(K) = b(K_1) + b(K_2) - 1.$$

Recall that in Theorem 1, companion tori are nested unless they are both swallow-follow tori. For a prime knot K, *i.e.*, for a knot that is not a connected sum, there are no swallow-follow tori and hence Theorem 1 tells us that companion tori are nested. Furthermore, if the prime knot K is a satellite knot (with nontrivial companion and pattern), then the wrapping number of the pattern will be at least 2. In particular, Theorem 3 then tells us that each companion knot will have a bridge number strictly lower than that of K. It therefore follows that a prime knot can have only finitely many non-isotopic companion tori. In this manner Schubert established finiteness, in the case of knot complements, for what later became known as JSJ decompositions.

It deserves to be pointed out that decompositions of knots into prime knots, unlike JSJ decompositions, are not canonical. They must always be finite, see [10], but collections of decomposing spheres and swallow-follow tori need not be isotopic when there are more than two prime factors. However, prime factors into which certain maximal collections of decomposing spheres decompose a knot are unique up to homeomorphism. See [6].

Ryan Budney gives a comprehensive survey of JSJ decompositions of knot and link complements, see [1]. His description is formulated in the language of graphs, using what he calls companionship graphs. In particular, he identifies precisely which graphs are companionship graphs of knots and links and what types of knot and link complements occur as basic building blocks. He also describes operations on knots and links such as cabling, connect-sum, Whitehead doubling, and the deletion of a component, and how these operations tie into a description of the JSJ decompositions of knot and link complements.

18.8 Geometrization

Thurston's geometrization conjecture, proved by Grisha Perelman, see [2], [4], [3], tells us that every orientable 3-manifold can be decomposed along a finite collection of 2-spheres and tori into pieces, each of which is geometric, *i.e.*, admits a complete finite volume Riemannian metric and has universal cover isometric to one of the following:

- $\mathbb{H}^3$
- $\mathbb{E}^3$
- $\mathbb{S}^3$
- $\mathbb{H}^2 \times \mathbb{R}$
- $\mathbb{S}^2 \times \mathbb{R}$
- The universal cover of $SL(2, \mathbb{R})$

- *Nil*

- *S olv*

Long before the geometrization conjecture was proved, Thurston established its specialization to knot and link complements. Specifically, he proved that for a prime knot K, one of the following holds:

- K is a torus knot;

- K is a satellite knot;

- the complement of K admits a complete finite volume hyperbolic structure.

18.9   Bibliography

[1] Ryan Budney. JSJ-decompositions of knot and link complements in S^3. *Enseign. Math. (2)*, 52(3-4):319–359, 2006.

[2] Grisha Perelman. The entropy formula for the Ricci flow and its geometric applications. arxiv:0211159.

[3] Grisha Perelman. Finite extinction time for the solutions to the Ricci flow on certain 3-manifolds. arxiv:0307245.

[4] Grisha Perelman. Ricci flow with surgery on 3-manifolds. arxiv:0303109.

[5] Dale Rolfsen. *Knots and Links*, volume 7 of *Mathematics Lecture Series*. Publish or Perish, Inc., Houston, TX, 1990. Corrected reprint of the 1976 original.

[6] Horst Schubert. Die eindeutige Zerlegbarkeit eines Knotens in Primknoten. *S.-B. Heidelberger Akad. Wiss. Math.-Nat. Kl.*, 1949(3):57–104, 1949.

[7] Horst Schubert. Knoten und Vollringe. *Acta Math.*, 90:131–286, 1953.

[8] Horst Schubert. Über eine numerische Knoteninvariante. *Math. Z.*, 61:245–288, 1954.

[9] Jennifer Schultens. Additivity of bridge numbers of knots. *Math. Proc. Cambridge Philos. Soc.*, 135(3):539–544, 2003.

[10] Herbert Seifert. Über das Geschlecht von Knoten. *Math. Ann.*, 110:571–592, 1935.

[11] JHC Whitehead. A certain open manifold whose group is unity. *The Quarterly Journal of Mathematics*, 87(1):43–68, 2013.

Chapter 19

Hyperbolic Knots and Links

Colin Adams, Williams College

19.1   Introduction

In the 1970s and 80s, William Thurston revolutionized low-dimensional topology with his work on hyperbolic 3-manifolds. In particular, with regard to knot theory, he proved that all knots in S^3 are either torus knots, satellite knots (including all composite knots) or hyperbolic knots. These three categories are mutually exclusive. Torus knots sit on the surface of an unknotted torus. Satellite knots are knots that exist in a thickened neighborhood of another "companion" knot in a nontrivial way. That all remaining knots are hyperbolic was a dramatic surprise.

A knot or link L is *hyperbolic* if its complement $S^3 \setminus L$ has a complete metric of constant sectional curvature -1. An alternative equivalent definition is to say that its fundamental group Γ can be realized as a discrete group of torsion-free hyperbolic isometries acting on hyperbolic 3-space $\mathbb{H}^3$. Then the quotient $\mathbb{H}^3/\Gamma$ is a finite-volume hyperbolic manifold homeomorphic to $M = S^3 \setminus L$. The complement then inherits its hyperbolic metric from $\mathbb{H}^3$ via the covering map to $\mathbb{H}^3/\Gamma$. The existence of a hyperbolic metric provides a host of invariants to distinguish the corresponding knots.

Thurston proved that the complement of a nontrivial link $M = S^3 \setminus L$ is hyperbolic if and only if it is non-splittable, with no essential tori or annuli in the complement. In the case of torus knots, the annulus obtained by removing the knot from the torus upon which it sits yields an essential annulus. In the case of satellite knots, the torus boundary of the neighborhood of the companion knot is an essential torus.

But determining whether or not a given knot or link is hyperbolic is often still difficult, since the existence of essential tori and annuli is hard to determine. There are various categories of knots and links known to be hyperbolic. We include here a short and incomplete list of some of those categories.

Hyperbolic links:

1. All but 33 of the 1,701,903 prime knots of 16 or fewer crossings catalogued in [24].

2. Prime non-splittable alternating links that are not 2-braids ([33]).

3. Non 2-braid two-bridge links (follows from the previous item).

4. Generalized augmented alternating links. ([2] and [8]). These are non-splittable prime alter-

nating links with non-parallel trivial components added to a reduced alternating projection, each perpendicular to the projection plane, and each with their pair of intersections with the projection planes in nonadjacent regions.

5. Arithmetic links.

6. Montesinos links. (In [39], it was proved that a Montesinos link $K(\frac{p_1}{q_1}, \ldots, \frac{p_n}{q_n})$ with $q_i \geq 2$, is hyperbolic if it is not a torus link and not equivalent to $K(1/2, 1/2, -1/2, -1/2)$, $K(2/3, -1/3, -1/3)$, $K(1/2, -1/4, -1/4)$, $K(1/2, -1/3, -1/6)$ or the mirror image of these links. In [16] the torus links that are Montesinos links are identified.

7. Many semi-adequate link diagrams (see [22]).

8. Non 2-braid prime links with diagrams having at least six crossings per twist region ([21]). A twist region is a sequence of bigons end-to-end in the diagram.

9. Mutants of a hyperbolic link (see [43]).

Many knot and link complements in the 3-sphere are hyperbolic. Moreover, knot and link complements in other 3-manifolds are often hyperbolic as well. In [38] it was proved that every compact orientable 3-manifold contains a knot such that its complement is hyperbolic. In this sense, hyperbolic knot complements are ubiquitous.

19.2   Volumes of Hyperbolic Links

The most significant advantage to having hyperbolic metrics on knot complements of links comes out of the Mostow-Prasad Rigidity Theorem ([36], [40]). In our context, it says that if a knot or link is hyperbolic, there is only one possible hyperbolic metric on the complement. In particular, the complement has a unique hyperbolic volume associated with it, and any other geometric invariants associated to the metric will be invariants for the link.

Given a hyperbolic knot or link, the SnapPy program, written by Marc Culler, Nathan Dunfield and Matthias Goerner, and based on a kernel written by Jeffrey Weeks, can be utilized to determine the hyperbolic structure (see https://www.math.uic.edu/t3m/SnapPy/). In particular, the program yields the volume of the manifold, the symmetry group, and a variety of invariants that are associated to the cusps of the manifold. The program works by decomposing the complement into ideal tetrahedra (tetrahedra missing their vertices). Each tetrahedron is realized as an ideal tetrahedron in $\mathbb{H}^3$, and is represented by a single complex number, which determines the angles between its faces. The program then attempts to determine the tetrahedra so that they fit together when glued around their edges to yield a compete hyperbolic structure for the link complement. This is achieved by solving a system of n complex polynomial equations in the n complex variables corresponding to the n tetrahedra. It is amazingly effective, and seems to work well even when there are over 50 crossings. If the knot in question is not hyperbolic, the program will fail to find a hyperbolic structure.

Hyperbolic volume is very effective at distinguishing knots, but there are pairs of knots with the same volume. But for low crossing number, they are rare, the first case being the 5-crossing knot 5_2 and the 12-crossing $(-2,3,7)$-pretzel knot.

The set of all volumes of hyperbolic knots and links is a well-ordered set of order type ω^ω. The smallest volume of a hyperbolic link was shown to be 2.0298..., which corresponds to the figure-eight knot ([18]) and is twice the volume of a regular ideal tetrahedron, 1.01494.... Next is 2.828..., for the 5_2 knot and the $(-2, 3, 7)$ -pretzel knot ([23]). In [15], the Whitehead link and the $(-2,3,8)$-pretzel link were shown to be the smallest volume 2-component links, with a volume of 3.6638.... The smallest volume 3-component link is yet to be determined.

A (p, q)-Dehn filling of a link complement M is the procedure of removing an open neighborhood N of a link component and then gluing a new solid torus V along its torus boundary to $\partial(M \setminus N)$ so that the meridian of V goes to a (p, q)-curve on $\partial(M \setminus N)$, yielding a 3-manifold.

Work of Thurston and Jørgensen implies that if M is hyperbolic, then for $p^2 + q^2$ large enough, the resulting manifolds will all be hyperbolic and their volumes will approach the volume of M from below. There has been substantial work on determining how many Dehn fillings can yield non-hyperbolic manifolds. In [30], it was proved that any knot has at most 10 Dehn fillings that are non-hyperbolic, which is the number that occurs for the figure-eight knot.

Considering the set of all volumes of hyperbolic knots and links, we have the least value 2.0298.... Then the volumes increase discretely until they reach the first limit point, which corresponds to the volume of the Whitehead link. As a 2-cusped manifold, $(1, p)$-Dehn filling can be performed on one component to obtain knot complements with volumes approaching 3.6638... from below. A hyperbolic n-component link is known to have volume at least nv_3 where $v_3 = 1.01494...$ is the volume of an ideal regular hyperbolic tetrahedron (cf. [3]). In [29] it was proved that the volumes of the prime non-2-braid alternating links and the augmented prime alternating 2-braid links form a closed subset of $\mathbb{R}^3$.

The operation of mutation preserves volume ([43]). This yields a means to show that arbitrarily many knots can share a volume. Moreover, the operation of twisting a half twist along an essential twice-punctured disk preserves volume as well ([1]), yielding a variety of link complements with identical volumes. There has been much work on finding lower and upper bounds on volume from the combinatorics of the link diagrams.

19.3 Cusps

Given a hyperbolic knot in S^3, one can define a cusp C for the knot to be a neighborhood of the missing knot such that it lifts to a set of horoballs with disjoint interiors in the universal cover $\mathbb{H}^3$. See Figure 19.1.

Topologically, a cusp is homeomorphic to $T \times [0, 1)$, where T is a torus, and its torus boundary inherits a Euclidean structure from the hyperbolic metric. Using hyperbolic geometry, one can show that the hyperbolic volume of a cusp is exactly half the Euclidean area of its boundary.

For any knot, the *maximal cusp* corresponds to expanding a cusp until it touches itself. In the universal cover $\mathbb{H}^3$, the collection of horoballs covering the cusp are equivariantly expanded until two first touch. The volume of the resulting cusp is called the *maximal cusp volume*. It also is an invariant.

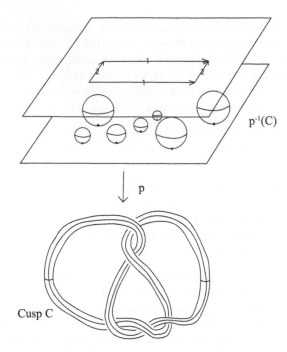

FIGURE 19.1: The cusp of a knot.

In the case of a link complement, we can expand the cusps until they touch themselves and each other, but there is more than one choice as to how to do that. We define the maximal cusp volume to be the largest cusp volume obtained in this manner.

Define the *cusp density* of a knot or link complement to be the ratio of the maximal cusp volume to the total volume for the knot or link complement. Meyerhoff noted in [35] that this number is at most $0.853 = \frac{\sqrt{3}}{2v_3}$ where again, $v_3 = 1.01494\ldots$ is the volume of an ideal regular tetrahedron. The cusp density of the figure-eight knot complement realizes this upper bound. In [13], it is proved that the set of cusp densities for hyperbolic link complements are dense in the interval $[0, 0.853\ldots]$ and for hyperbolic knot complements, it is at least dense in the interval $[0, 0.6826\ldots]$.

The meridian length $|m|$ is the shortest length of a meridian in the maximal cusp boundary. In [14], Agol found examples of knots with $|m|$ approaching 4 from below. It is known that $1 \le |m| \le 6$ (the upper bound coming from the 6-theorem of [14, 29]). The figure-eight knot is the only example of a knot with $|m| = 1$ ([9]). It remains open at this time as to whether or not there are any meridian lengths between 4 and 6.

The volume density $vd(L)$ of a hyperbolic knot or link is given by $vol(S^3 \setminus L)/c(L)$ where $c(L)$ is the crossing number. From [44], it is known that $vd(L)$ is always in the interval $[0, 3.6638\ldots]$. In [19], it was proved that there are links with $vd(K)$ approaching the upper bound. In [12] and [17], it was proved that in fact $vd(L)$ is dense in the interval $[0, 3.6638\ldots]$. Similar types of results for the cusp crossing density, which is the maximal cusp volume divided by crossing number, appear in [13].

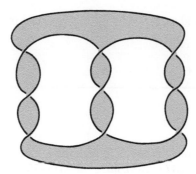

FIGURE 19.2: A (3,3,3)-pretzel knot has a totally geodesic Seifert surface.

19.4 Surfaces

As in Chapter 22, every knot has a Seifert surface, an orientable surface with boundary equal to the knot. Such a surface that is minimal genus is essential. There are three possibilities for such surfaces.

1. The surface is accidental. This means there is a nontrivial curve in the surface that is isotopic to a curve on the boundary of a neighborhood of the link complement.

2. The surface is a virtual fiber. This means that the knot complement can be lifted to a fibered manifold, such that the surface lifts to a fiber.

3. The surface is quasi-Fuchsian.

A quasi-Fuchsian surface is one that lifts to the disjoint union of topological planes in H^3, each with a quasi-circle as the limit set. In the case that the limit set is an actual circle, and the planes are geodesic, we say that the surface is *Fuchsian* or *totally geodesic*. In [10] and [11], examples of knots possessing totally geodesic Seifert surfaces are given. (See Figure 19.2.) But the expectation is that knots with totally geodesic Seifert surfaces are the exception. For instance, in [10], it is proved that 2-bridge knots never have totally geodesic Seifert surfaces.

In [34], it was conjectured that hyperbolic knot complements in S^3 do not contain any closed embedded totally geodesic surfaces. This has been proved for hyperbolic knots that are alternating knots, tunnel number one knots, 2-generator knots and knots of braid index three [34], almost alternating knots [4], toroidally alternating knots [6], Montesinos knots [39], 3-bridge knots and double torus knots [25] and knots of braid index three [31] and four [32].

Many questions remain. Since the set of hyperbolic knots is such a large proportion of knots and links, it is unlikely that we will run out of open questions for far into the foreseeable future. For instance, one direction that shows much promise is the question of whether hyperbolic volume is related to quantum invariants. The *Volume Conjecture* states that the hyperbolic volume for a hyperbolic knot is determined by the asymptotic behavior of the colored Jones polynomial. This relates invariants that on the surface, appear as if they should be independent. The conjecture was originally formulated by [27] and extended by [37]. It has been proved for a handful of knots, but still remains open.

19.5   Bibliography

[1] Adams, C., Thrice-punctured spheres in hyperbolic 3-manifolds, *Transactions of the American Mathematical Society*, **287** (1985), no. 2, pp. 645–56.

[2] Adams, C. Augmented alternating link complements are hyperbolic, London Mathematical Society Lecture Notes Series, 112: *Low Dimensional Topology and Kleinian Groups*, pp. 115–130, Cambridge University Press, Cambridge, 1986.

[3] Adams, C.,Volumes of N-cusped hyperbolic 3-manifolds, *J. London Math. Soc.* (2) **38** (1988), no. 3, 555–565.

[4] Adams, C., Brock, J., Bugbee, J. , Comar, T., Faigin, K., Huston, A., Joseph, A., Pesikoff, D., Almost alternating links, *Topology Appl.* **46** (1992), no. 2, 151–165.

[5] Adams, C. and Reid, A., Quasi-Fuchsian surfaces in hyperbolic knot complements, *J. Austral. Math. Soc. Ser. A* 55 (1993), no. 1, 116–131.

[6] Adams, C., Toroidally alternating knots and links, *Topology* **33**, (1994), no. 2, 353–369.

[7] Adams, C., Cusp densities of hyperbolic 3-manifolds, *Proc. of the Edinburgh Math. Soc.* 45 (2002) 277–284.

[8] Adams, C. Generalized augmented alternating links and hyperbolic volume, *Alg. and Geom. Top.*, 17 (2017) 3375-3397.

[9] Adams, C., Waist size for cusps in hyperbolic 3-manifolds, *Topology* 41 (2002), no.2, 257–270.

[10] Adams, C. and Schoenfeld, E., Totally geodesic Seifert surfaces in hyperbolic knot and link complements I, *Geometriae Dedicata*, Vol. 116 (2005)237–247.

[11] Adams, C. Bennett, H., Davis, C., Jennings, M., Novak, J., Perry, N. and Schoenfeld, E., Totally geodesic surfaces in hyperbolic knot and link complements II, *J. of Diff. Geom.*, Vol. 79, No. 3, May, 2008, 1–23.

[12] Adams, C., Kastner, A., Calderon, A. , Jiang, X., Kehne, G., Mayer, N., Smith, M., Volume and determinant densities of hyperbolic rational links, *J. Knot Theory Ramifications* 26, 1750002 (2017).

[13] Adams, C., Kaplan-Kelly, R., Moore, M., Shapiro, B., Sridhar, S. , Wakefield, J. Densities of hyperbolic cusp invariants, arXiv: 1701.03479 (2017) *Proc. AMS*, 146 (2018) No. 9, 4073-4089.

[14] Agol, I., Bounds on exceptional Dehn filling, *Geom. Topol.* 4 (2000), 431–449.

[15] Agol, I.,The minimal volume orientable hyperbolic 2-cusped 3-manifolds, *Proceedings of the American Mathematical Society* 138(10), March 2008.

[16] Boileau, M. and Siebenmann, L., A planar classification of pretzel knots and Montesinos knots, Orsay preprint (1980).

[17] Burton, S., The spectra of volume and determinant density, Top. and Its Appl., vol. 211 1–56 (2016).

[18] Cao, C. and Meyerhoff, G. R., The orientable cusped hyperbolic 3-manifolds of minimum volume, *Invent. Math.* 146 (2001), no. 3, 451–478.

[19] Champanerkar, A., Kofman, I. , Purcell, J., Geometrically and diagrammatically maximal knots, *Jour. of the London Math. Soc.*, Volume 94, Issue 3, 1 December 2016, Pages 883–908.

[20] Dasbach, O. and Tsvietkova, A., A refined upper bound for the hyperbolic volume of alternating links and the colored Jones polynomial, *Math. Res. Lett.* 22(2015), no. 4, 1047–1060.

[21] Futer, D., Purcell, J., Links with no exceptional surgeries, *Commentarii Mathematici Helvetici* 82 (2007), No. 3, 629–664.

[22] Futer, D., Kalfagianni,E., Purcell, J. Hyperbolic semi-adequate links, *Communications in Analysis & Geometry* 23 (2015), Issue 5, 993–1030.

[23] D. Gabai, R. Meyerhoff, P. Milley, Minimum volume cusped hyperbolic 3-manifolds, *J. Amer. Math. Soc.* 22 (2009), 1157–1215.

[24] Hoste, J., Thistlethwaite, M. and Weeks, J., The first 1,701,936 knots, *Math. Intelligencer* 20 (1998), no. 4, 33–48.

[25] Ichihara, Kazuhiro and Ozawa, Makoto, Hyperbolic knot complements without closed embedded totally geodesic surfaces, *J. Austral. Math. Soc. Ser. A* 68 (2000), no. 3, 379–386.

[26] Kashaev, R. M., A link invariant from quantum dilogarithm, *Modern Phys. Lett. A* 10 (1995), no. 19, 1409–1418.

[27] Kashaev, R. M., The hyperbolic volume of knots from the quantum dilogarithm, *Lett. Math. Phys.*, 39 (1997), no. 3, 269–275.

[28] Lackenby, Marc Word hyperbolic Dehn surgery, *Inventiones Mathematicae*, 140 (2): 243–282.

[29] Lackenby, Marc, The volume of hyperbolic alternating knot complements, with an appendix by Ian Agol and Dylan Thurston. *Proc. London Math. Soc.* (3) 88 (2004), no. 1, 204–224.

[30] M. Lackenby, R. Meyerhoff, The maximal number of exceptional Dehn surgeries, *Inventiones*, February 2013, Volume 191, Issue 2, pp 341–382.

[31] Lozano, M. T., Przytycki, J. H., Incompressible surfaces in the exterior of a closed 3-braid I,*Math. Proc. Cambridge Philos. Soc.* 98 (1985), no. 2, 275–299.

[32] Matsuda, Hiroshi, Complements of hyperbolic knots of braid index four contain no closed embedded totally geodesic surfaces, *Topology Appl.* 119 (2002), no. 1, 1–15.

[33] Menasco, William, Closed incompressible surfaces in alternating knot and link complements, *Topology*, **23**, 1, pp. 37–44, 1984.

[34] Menasco, W. and Reid, A., Totally geodesic surfaces in hyperbolic link complements, *Topology '90 (Columbus, OH, 1990)*, 215–226, Ohio State Univ. Math. Res. Inst. Publ., 1, de Gruyter, Berlin, 1992.

[35] Meyerhoff, G. R., Sphere-packing and volume in hyperbolic 3-space, *Comment. Math. Helv.* 61 (1986), no. 2, 271–278.

[36] Mostow, G.D., Strong rigidity of locally symmetric spaces. *Annals of Mathematics Studies*, **78**, Princeton University Press, Princeton, N.J., 1973.

[37] Murakami, H., Murakami, J., The colored Jones polynomials and the simplicial volume of a knot, *Acta Mathematica*, 186 (1) (2001), 85–104.

[38] Myers, Robert, Simple knots in compact orientable 3-manifolds, *Trans. of A.M.S.* 273 (1982), 75–91.

[39] Oertel, Ulrich, Closed incompressible surfaces in complements of star links, *Pac. J. of Math.*, Vol. 111, No. 1, 1984, 209–230.

[40] Prasad, G. Strong rigidity of Q-rank 1 lattices, *Invent. Math.* 21 (1973), 255–286.

[41] Reid, A., Arithmeticity of knot complements, *J. London Math. Soc.* (2) 43 (1991), no. 1, 171–184.

[42] Reid, A. Totally geodesic surfaces in hyperbolic 3-manifolds, *Proc. Edinburgh Math. Soc.* (2) 34 (1991), no. 1, 77–88.

[43] Ruberman, D., Mutation and volumes of knots in S^3, *Invent. Math.* 90 (1987), no. 1, 189–215.

[44] Thurston, W. The Geometry and Topology of 3-Manifolds, lecture notes, Princeton University, 1978.

Chapter 20

Alternating Knots

William W. Menasco, State University of New York, Buffalo

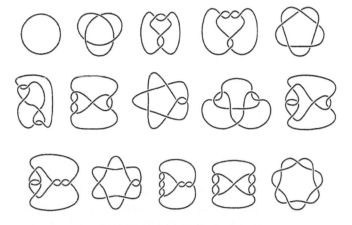

FIGURE 20.1: P.G. Tait's first knot table where he lists all knot types up to 7 crossings. (From reference [6], courtesy of J. Hoste, M. Thistlethwaite and J. Weeks.)

20.1 Introduction

A knot $K \subset S^3$ is *alternating* if it has a regular planar diagram $D_K \subset \mathbb{P}(\cong S^2) \subset S^3$ such that, when traveling around K, the crossings alternate, over-under-over-under, all the way along K in D_K. Figure 20.1 shows the first 15 knot types in P. G. Tait's earliest table and each diagram exhibits this alternating pattern. This simple definition is very unsatisfying. *A knot is alternating if we can draw it as an alternating diagram?* There is no mention of any geometric structure. Dissatisfied with this characterization of an alternating knot, Ralph Fox (1913–1973) asked: "What is an alternating knot?"

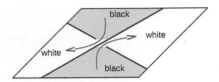

FIGURE 20.2: Going from a black to white region near a crossing.

Let's make an initial attempt to address this dissatisfaction by giving a different characterization of an alternating diagram that is immediate from the over-under-over-under characterization. As with all regular planar diagrams of knots in S^3, the regions of an alternating diagram can be colored in a checkerboard fashion. Thus, at each crossing (see Figure 20.2) we will have "two" white regions and "two" black regions coming together with similarly colored regions being kitty-corner to each other.

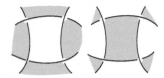

FIGURE 20.3: Black and white regions.

(I will explain my use of quotation marks momentarily.) In the local picture of Figure 20.2, we will make the convention that when standing in a black region and walking to a white region by stepping over the under-strand of the knot; the over-strand is to our left. (Notice this is independent of which side of $\mathbb{P}$ we stand.) Now the feature for an alternating diagram is, if one crossing of a black region satisfies this local scheme, then every crossing of that black region satisfies this scheme. Since both black regions at a crossing satisfy our scheme, this scheme is transmitted to every black region of an alternating diagram. To summarize, a diagram of a knot is *alternating* if we can checkerboard color the regions such that when we stand in any black region, we can walk to an adjacent white region by stepping over the under-strand of a crossing while having the over-strand of that crossing positioned to our left. From the perspective of walking from white regions to black regions, the over-crossing is to our right. Again, one should observe that this coloring scheme is consistent independent of which side of $\mathbb{P}$ one stands.

Coming back now to our previous use of quotation marks—"two" white regions and "two" black regions—we must allow for the possibility that "two" kitty-corner regions may be the same region as shown in Figure 20.4. This possibility forces the occurrence of a *nugatory* crossing in the diagram of D_K. Such a crossing can be eliminated by a π-rotation of K' for half the diagram.

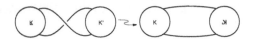

FIGURE 20.4: Eliminating a nugatory crossing.

Observe that such a π-rotation will take an alternating diagram to an alternating diagram with one less crossing. So the elimination of nugatory crossings must terminate and we call a diagram with no nugatory crossings *reduced*. Thus, our study of alternating diagrams is legitimately restricted to ones that are reduced. And, let's add one further restriction, that a diagram D_K be connected, since one that is not connected is obviously representative of a split link. This requirement implies that all regions of a diagram D_K are discs.

With all of the above introductory material in place, and restricting to reduced connected alternating knot diagrams, we now give in to the urge that anyone who has drawn a Möbius band will have. At each crossing we place a half-twisted white band connecting the two white regions; similarly, we place a half-twisted black band connecting the two black regions. This construction gives us a white surface, ω_K, and a black surface, β_K, each one having our knot K as its boundary—they are *spanning surfaces* of K. The white and black bands of each crossing intersect in a single arc that has one endpoint on the under-crossing strand and its other endpoint on the over-crossing strand. Thus, we can reinterpret a checkerboard D_K diagram to be that of a black-and-white surface with each crossing concealing an intersection arc since our eye's viewpoint would be one where we were looking straight down each arc.

Here is where I really wish to start our story of alternating knots. (The story can be expanded to alternating links, but we will leave that for another time.) At the center of this story lies three mysteries, the three conjectures of the Scottish physicist Peter G. Tait (1831–1901). As with all good conjectures, Tait's three point to deeper mathematics. Alternating knots have proven particularly well behaved with respect to tabulation, classification, and computing and topologically interpreting algebraic invariants. Moreover, for hyperbolic alternating knots their alternating diagrams are more closely tied to their hyperbolic geometric structure than other knots. As a collection, alternating knots have supplied researchers with a ready population for experimentation and testing conjectures. And, our two colored surfaces, β_K and ω_K, are rightfully thought of as the heroes the story for they are at the center of establishing the validity of the Tait conjectures and, finally, giving a satisfactory answer to Fox's question.

20.2 The Tait Conjectures

The early efforts of knot tabulation came initially from Tait. He was motivated by Lord Kelvin's program for understanding the different chemical elements as different knotted vortices in ether. Without the aid of any theorems from topology, Tait published in 1876 his first knot table which contained the 15 knot types through seven crossings in Figure 20.1. Specifically, Tait enumerated all possible diagrams up to seven crossings and then grouped together those diagrams that represented the same knot type. For example, the right-handed trefoil has two diagrams, one as a closed 2-braid and one as a 2-bridge knot. Similarly, the figure-eight has two diagrams, one coming from a closed 3-braid and one coming from a 2-bridge presentation. His grouping of these 15 knot types is consistent with modern tables. However, his grouping did contain more than these 15 since he did not possess the notion of prime/composite knots—he included the composite sums of the trefoil with itself and the figure-eight.

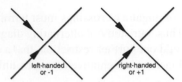

FIGURE 20.5: Left- and right-hand crossings.

Tait also experimented with the assigning of orientations to the knot diagrams—giving a defined direction to the knot along which to travel—which induced a handedness to each crossing. Thus, for right-handed crossings, one could assign a +1 parity and a −1 parity for left-handed crossings. (See Figure 20.5.) The sum of these parities was the *writhe* of a diagram. For example, independent of orientation, the trefoil in Tait's table will have three right-handed crossings yielding a writhe of +3, whereas the figure-eight will always have two right-handed and two left-handed crossings for a writhe of 0.

Tait also observed that one could reduce the crossing number by the elimination of nugatory crossings. Finally, Tait discovered an operation on an alternating diagram—a *flype*—that preserved the number of crossings, the writhe and, more importantly, the knot type. (See Figure 20.6.)

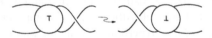

FIGURE 20.6: A flype moves the crossing from right to left.

From the diagrams and their groupings into just 15 knot types, Tait had amassed a sizable about of data. And, from this data Tait proposed a set of conjectures:

The Tait Conjectures

(T1) A reduced alternating diagram has a minimal crossing number.

(T2) Any two reduced alternating diagrams of the same knot type have the same writhe.

(T3) Any two reduced alternating diagrams of a given knot type are related by a sequence of flypes.

One should observe that T1 and T3 imply T2. The solutions to these three conjectures would have to wait until after the work of Vaughan Jones on his knot polynomial in 1984.

20.3   Surfaces and Alternating Knots

During the 20th century, work on understanding the topology and geometry of alternating knots remained active. But, for our story line, we start in the 1980's and describe how our colored surfaces,

β_K and ω_K, control the behavior of closed embedded surfaces in an alternating knot complement $S^3 - K$. Any embedded closed surface, Σ, in S^3 is necessarily orientable and the ones that capture meaningful information about the topology of a knot complement are *incompressible*. That is, every simple closed curve (s.c.c.) $c \subset \Sigma (\subset S^3 - K)$ which bounds an embedded disc in $S^3 - K$, also bounds a sub-disc of Σ. To this end, it is worth observing from the start that $S^3 - (\beta_K \cup \omega_K)$ is the disjoint union of two open 3-balls, which we denote by B_N^3 (N for north) and B_S^3 (S for south).

Given an incompressible surface $\Sigma \subset S^3 - K$, our first task is to put Σ into normal position with respect to β_K and ω_K. Using some basic general position arguments for incompressibility, this means that we can assume: (1) $\beta_K \cap \Sigma$ and $\omega_K \cap \Sigma$ are collections of s.c.c.'s; (2) any curve from either collection cannot be totally contained in a single black or white region used to construct β_K or ω_K—it must intersect some of the half-twisted bands of our colored surfaces—and; (3) $B_N^3 \cap \Sigma$ and $B_S^3 \cap \Sigma$ are collections of open 2-discs which we will call *domes*.

Of particular interest is the behavior of the boundary curves of the domes. Let $\delta \subset B_N^3 \cap \Sigma$ be a dome and consider the curve $\partial(\bar{\delta}) \subset \beta_K \cup \omega_K$. When viewed from inside B_N^3, the s.c.c. $\partial(\bar{\delta})$ will be seen as a union of arcs—an arc in a white region adjoined to an arc in a black region adjoined to an arc in a white region, etc.—where the adjoining of two consecutive arcs occurs within an intersection arc of $\beta_K \cap \omega_K$. Thus, from the viewpoint of B_N^3, $\partial(\bar{\delta})$ respects our earlier scheme of traveling from white to black (resp. black to white) regions having the over-strand to the left (resp. right). One additional normal position condition we can require by general position arguments is that any $\partial(\bar{\delta})$ intersects any intersection arc of $\beta_K \cap \omega_K$ at most once. Now, observe that if we have a curve $\partial(\bar{\delta})$ occurring on one side of an over-strand at a crossing, then we must see a curve $\partial(\bar{\delta}')$ on the other side of the same over-strand. Finally, observe that if $\partial(\bar{\delta}) = \partial(\bar{\delta}')$, then Σ will contain a s.c.c. that is a meridian of K.

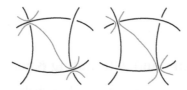

FIGURE 20.7: The view of $\partial(\bar{\delta})$ inside B_N^3. Traveling along $\partial(\bar{\delta})$, we find a crossing to left/right then right/left.

Once placed in the above described normal position, it is readily observed by a nesting argument that Σ must contain a meridian curve of K. If we walk along the boundary curve of any dome, starting in a white region of ω_K and passing into a black region of β_K, our scheme dictates that the over-strand of a crossing be to our left; then traveling through a black region into a white region, our scheme says the over-strand of a crossing is to our right. (See Figure 20.7.) After strolling along all of our boundary curve, if we have not walked on the two sides of the same crossing, then there are at least two additional curves that are boundary curves of differing domes, one to the left side of our first crossing and one to the right side of our second crossing. Walking along either one of these boundary curves and iterating this procedure we will either find a new boundary of a dome that we have not strolled along or we will find a boundary that passes on both sides of the over-strand of a crossing, thus yielding a meridian of K. By the compactness of Σ there are only finitely many domes in B_N^3 so the latter must occur.

The implication of an incompressible Σ always containing a meridian curve are significant. Once a meridian curve has been found we can trade Σ in for a new surface punctured by K, Σ', by compressing Σ along a once-punctured disc that the meridian curve bounds. (See Figure 20.8.) Repeating this meridian surgery whenever a new meridian curve is found, the study of closed incompressible surfaces in the complement of alternating knots becomes the study of incompressible meridionally or *pairwise incompressible* surfaces—if a curve in Σ bounds a disc (resp. once punctured disc) in $S^3 - K$, then it bounds a disc (resp. once-punctured disc) in Σ.

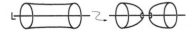

FIGURE 20.8: Pairwise compressing Σ along a meridian disc.

Finally, the above normal position and nesting argument can be enhanced so as to handle the intersection of an incompressible pairwise incompressible surface Σ. Specifically, if K is an alternating composite knot then there exists a twice punctured 2-sphere the intersects β_K and ω_K in single arcs whose union forms a circle that can be thought of as illustrating the composite nature of D_K.

Thus, we have the result that an alternating knot is a composite knot if and only if its diagram is also composite. When combined with William P. Thurston's characterization of hyperbolic knots in S^3, an immediate consequence is that all alternating knots coming from diagrams that are neither composite nor $(2, q)$-torus knots are hyperbolic knots.

20.4   Hyperbolic Geometry and Alternating Knots

In 1975 Robert Riley produced the first example of a hyperbolic knot, the figure eight, by showing that its fundamental group had a faithful discrete representation into $PSL(2, \mathbb{C})$, the group of orientation preserving isometries of hyperbolic 3-space. Inspired by Riley's pioneering work (see page 360 of reference 2), Thurston proved that all knots that are not torus knots and not satellite knots are hyperbolic—their complement has a complete hyperbolic structure. Recalling that a satellite knot is one where there is an incompressible torus that is not a peripheral torus, we now apply the fact that for alternating knots such a torus has a meridian curve. Then pairwise compressing such a torus will produce an annulus illustrating that the knot is a composite knot. As mentioned previously, we then conclude that all prime alternating knots that are knot $(2, q)$-torus knots are hyperbolic.

Similar to Riley, Thurston was able to structure the hyperbolic structures for knot complements. However, he used a geometric approach to see their hyperbolic structures. Specifically, the complement of every hyperbolic knot can be decomposed in a canonical way into ideal tetrahedra, that is, convex tetrahedra in hyperbolic 3-space such that all the vertices are lying on the sphere at infinity.

The most famous example of such a tetrahedron decomposition is Thurston's decomposition of the figure-eight complement into two regular ideal tetrahedra. of our previous two open 3-balls, B_N^3 and B_S^3. Referring to Figure 20.9, if we are viewing the diagram from inside B_N^3 and we try to look into B_N^3, our view is blocked by the non-opaque black and white disc-regions of β_K and ω_K.

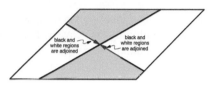

FIGURE 20.9: From the viewpoint of B_N^3, black and white regions are adjoined along an (oriented) edge).

Moreover, when we look at an arc of $\beta_K \cap \omega_K$, from the perspective of B_N^3 it does not appear that there are two black (white) regions adjoined be a half-twisted band. Instead it appears that the black regions are adjoined to the white regions in a manner consistent with our original scheme of going from a white region and black region and having the over-strand to our left. This adjoining scheme for assembling the boundary of B_N^3, and similarly B_S^3 from the disc-regions of our colored surfaces makes sense once one realizes that one's vision in B_N^3 gives us information for only the over-strand at each crossing since we see it in its entirety. To obtain the information for the under-strand of a crossing—to see it in its entirety—we need to view it in B_S^3 where it appears as an over-strand of a crossing.

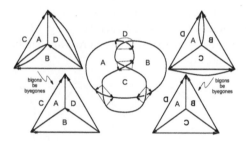

FIGURE 20.10: The upper left (resp. right) configurations are the view of S_N (resp. S_S) from inside B_N^3 (resp. B_S^3).

In Figure 20.10 we show the inside appearance of the 2-sphere boundaries, $S_N = \partial B_N^3$ and $S_S = \partial B_S^3$, coming from this described scheme for assembling of the disc-regions of our colored surface. Notice that S_N and S_S each has four vertices—one for each crossing. When the identification of commonly labeled disc-regions of S_N and S_S is made these 8 vertices will become one vertex and the resulting space will correspond to the space obtained by taking the complement an open tubular neighborhood of the figure eight and coning its peripheral torus to a point. Thus, when we delete this single vertex we have the knot complement. When we place B_N^3 and B_S^3 in hyperbolic 3-space, $\mathbb{H}^3$, so that the vertices of S_N and S_S are in S^∞, the sphere at infinity—they are ideal polyhedra—we have achieved this vertex deletion. A complete hyperbolic structure for the knot complement will come from a tessellation of $\mathbb{H}^3$ by isometric copies of our two ideal polyhedra.

Now, one feature of an ideal polyhedron is that each edge is the unique geodesic line between two points in S^∞. Currently since some edges are boundaries of bigon disc-regions in S_N and S_S we need to collapse them to achieve the needed unique edge—we let bigons be bygones. For the figure eight, once the four bigon disc-regions are collapsed on S_N and S_S, B_N^3 and B_S^3 become Thurston's two tetrahedra decomposition.

The above assembly scheme for producing a ideal polyhedron works for any alternating knot that is not a $(2, q)$-torus or composite knot with only one caveat. If a disc-region is not a triangle, we need to insert additional edges until it is a union of triangles.

___

20.5   Establishing Conjecture T1

In 1985 Vaughan Jones announce the discovery of his polynomial, $V_K(t)$, a Laurent polynomial that is an invariant of oriented knots (and links). His discovery emerged from a study of finite dimensional von Neumann algebras. A number of topologists subsequently reframed the Jones polynomial into more knot theory user friendly settings. Then in 1987 using these reframings of the Jones polynomial, Louis Kauffman, Kunio Murasugi, and Morwen B. Thistlethwaite each independently published proofs of the conjecture T1, *a reduced alternating diagram is a minimal crossing diagram*. Additionally, both Mursugi and Thistlethwaite showed that the writhe was an invariant of a reduced alternating diagram, establishing conjecture T2.

Kauffman's approach using his bracket invariant and state model equation is particularly accessible. Moreover, his approach illustrates the importance of our two colored surfaces. We consider an n-crossing knot diagram, D, where we have given each crossing a label $\{1, \cdots , n\}$. A *state* of D is a choice of smoothly resolving every crossing. The ith crossing can be resolve either positively or negatively as shown in Figure 20.11.

FIGURE 20.11: Two ways of smoothing a crossing.

Thus, a state of D also corresponds to a function $s : \{1, \cdots , n\} \to \{-1, +1\}$ where $s(i) = +1$ (resp. -1) if we smooth the ith crossing in a positive (resp. negative) manner. Applying a state s to D we obtain a new diagram, sD, that is just a collection of disjoint simple closed curves—there are no crossings. We will use $|sD|$ to mean the number of curves in the collection.

Of immediate interest is when D is a reduced alternating diagram and the state is constant, $s_+ \equiv +1$. A constant $+1$ state can be thought of as cutting every half-twisted band of ω_K. Notice then $|s_+D|$ is just equal to the number of disc-regions of ω_K. Similarly, for $|s_-D|$ is a count of the number of disc-regions of β_K. Moreover, using the classical Euler characteristic relation for the sphere, (vertices − edges + faces) = 2, we can derive the equality $|s_+D| + |s_-D| = n + 2$.

We now come to the *Kauffman bracket* of D which is a Laurent polynomial, $\langle D \rangle$, with an indeterminate variable A. Initially Kauffman axiomatically define his bracket in terms of skein relations. He then showed it was an invariant of *regular isotopies*, isotopies that corresponded to a sequence of only Reidemeister type II and III moves. Later, Kauffman derived the following state sum model for his bracket

$$\langle D \rangle = \sum_{s} (A^{\sum_{i=1}^{n} s(i)} (-A^{-2} - A^2)^{|sD|-1}). \tag{20.1}$$

Although the Kauffman bracket is only a regular isotopy invariant, when we insert a term that takes the writhe, $w(D)$, into account we do obtain an oriented knot invariant, the *Kauffman polynomial*:

$$F_K(A) = (-A)^{-3w(D_K)} \langle D_K \rangle. \tag{20.2}$$

As an aside we mention that the Jones polynomial can be obtained from the Kauffman polynomial by a variable substantiation, $F_K(t^{\frac{1}{4}}) = V_K(t)$.

Now since $F_K(A)$ is an oriented knot invariant, the difference between the highest exponent power, $M(F_K)$, and lowest exponent power, $m(F_K)$, of A in $F_K(A)$ is also an invariant. But, since the $(-A)^{-3w(D_K)}$ factor in equation (20.2) is common to both $M(F_K)$ and $m(F_K)$ we see that $M(F_K) - m(F_K) = M(\langle D_K \rangle) - m(\langle D_K \rangle)$, the difference of the highest and lowest power of A in $\langle D_K \rangle$. Now referring back to equation (20.1) with a little arguing one can obtain that $M(\langle D_K \rangle) = n + 2|s_+D| - 2$ and $m(\langle D_K \rangle) = -n - 2|s_-D| + 2$. Then:

$$M(\langle D_K \rangle) - m(\langle D_K \rangle) = 2n + 2(|s_+D| + |s_-D|) - 2.$$

Finally, recalling $|s_+D| + |s_-D| = n + 2$ we obtain $M(\langle D_K \rangle) - m(\langle D_K \rangle) = 4n$.

A variation of the above line of reasoning can establish that for an n-crossing non-alternating prime diagram, D, we have $M(\langle D \rangle) - m(\langle D \rangle) < 4n$. But, since $M(\langle D \rangle) - m(\langle D \rangle)$ is an invariant we conclude that a reduced alternating diagram has minimal number of crossings.

20.6   Establishing Conjectures T2 and T3

Following the successful verification of conjecture T1 in 1987, Thistlethwaite and William W Menasco announced in 1991 a proof verifying conjecture T3, the Tait flyping conjecture. As mentioned before T1 and T3 together imply T2. Their proof of T3 had two components. First, it utilized the new understanding of the interplay between the algebraic invariants coming out of the Jones polynomial revolution and the crossing number of knot diagrams. Second, it utilized enhancements of the elementary geometric technology (previously described) in order to analyze the intersection pattern of a closed incompressible surfaces with the colored surfaces, $\beta \cup \omega$, of an alternating diagram.

Two key lemmas of this enhanced geometric technology are worth mentioning. First, for a reduced alternating diagram the associated colored surfaces, β and ω, are themselves incompressible. Second, for a reduced alternating diagram neither β nor ω contain a meridional simple closed curve.

The overall strategy of the proof is fairly straightforward. Given two reduced alternating diagrams, D_1 and D_2, representing the same knot type, take the associated two sets of incompressible colored surfaces, $\beta_1 \cup \omega_1$ and $\beta_2 \cup \omega_2$, and consider how they will intersect each other. Using nesting arguments that are similar in spirit to the one for closed surfaces, one can establish that intersections are arcs occurring away from $\beta_1 \cap \omega_1$ and $\beta_2 \cap \omega_2$—the portion of our colored surfaces that come from the half-twisted bands. The argument proceeds by trying to "line up" crossings of D_1 with crossings of D_2. An extensive analysis of these arc intersections then leads to the conclusion that if D_1 and D_2 are not isotopic diagrams, there exists a flype. After performing a sufficient number of the flypes we have more crossings lining up. When all the crossings of D_1 line up with those of D_2 we have isotopic diagrams.

With the establishment of T3 we have the structure for a complete classification of alternating knots.

20.7   What is an Alternating Knot?

In 2017 independently Joshua E. Greene and Joshua Howie published differing answers to Fox's fundamental question. Howie's answer starts in the setting of knots in S^3 and supposes that such a knot, K, admits two spanning surfaces, Σ and Σ'. Neither surface need be orientable. Let $i(\partial\Sigma, \partial\Sigma')$ be the intersection number of the the two curves obtained by intersecting $\Sigma \cup \Sigma'$ with a peripheral torus. Then Howie proves that K is an n-crossing alternating knot in S^3 if and only if the following Euler characteristic equation is satisfied:

$$\chi(\Sigma) + \chi(\Sigma') + \frac{1}{2}i(\partial\Sigma, \partial\Sigma') = 2.$$

The "only if" direction of this statement makes sense once one realizes that $\frac{1}{2}i(\partial\beta, \partial\omega) = n$, since each arc of $\beta \cap \omega$ accounts for two intersections on the peripheral torus, and $\chi(\Sigma) + \chi(\Sigma') = (|s_+D| - n) + (|s_-D| - n)$.

Greene's answer starts in a more general setting, knots in a $\mathbb{Z}_2$-homology 3-spheres, M^3. He also supposes that one has two spanning surfaces, Σ and Σ', of a knot, $K \subset M^3$. One then considers the Gordon-Litherland pairing form of the 1st-homology for each surface:

$$\mathcal{F}_\Sigma : H_1(\Sigma) \times H_1(\Sigma) \to \mathbb{Z} \text{ and } \mathcal{F}_{\Sigma'} : H_1(\Sigma') \times H_1(\Sigma') \to \mathbb{Z}.$$

If $\mathcal{F}_\Sigma$ is a positive definite form and $\mathcal{F}_{\Sigma'}$ is a negative definite form then Greene's result states that

$M^3 = S^3$ and K is an alternating knot. Moreover, Σ and Σ' are our colored surfaces—our two heroes—for an alternating diagram, D_K.

20.8 Bibliography

20.8.1 Introduction

[1] P. G. Tait, On knots I, II, III, Scientific Papers, Cambridge University Press, 1898-1900. Including: *Trans. R. Soc. Edin.*, 28, 1877, 35–79. Reprinted by Amphion Press, Washington D.C., 1993.

[2] T. P Kirkman, The enumeration, description and construction of knots of fewer than ten crossings, *Trans. Roy. Soc. Edinburgh* 32 (1885), 281-309.

[3] T. P Kirkman, The 364 unifilar knots of ten crossings enumerated and defined, *Trans. Roy. Soc. Edinburgh 32 (1885)*, 483-506.

[4] C. N. Little, On knots, with a census of order ten, Trans. *Connecticut Acad. Sci.* 18 (1885), 374-378.

[5] C. N. Little, Non alternate ± knots of orders eight and nine, *Trans. Roy. Soc. Edinburgh* 35 (1889), 663-664.

[6] Jim Hoste, Morwen B. Thistlethwaite and Jeffrey Weeks, The first 1, 701,936 knots, *Math. Intelligencer* 20 (4) (1998), 33-48

20.8.2 The Tait Conjectures and Surfaces and Alternating Knots

[1] William W. Menasco, Closed incompressible surfaces in alternating knot and link complements, *Topology* 23 (1) (1984), 37-44.

[2] Menasco, William Determining incompressibility of surfaces in alternating knot and link complements. *Pacific J. Math.* 117 (1985), no. 2, 353-370.

[3] William W. Menasco and M. B Thistlethwaite, Surfaces with boundary in alternating knot exteriors. *J. Reine Angew. Math.* 426 (1992), 47-65.

20.8.3 Hyperbolic Geometry and Alternating Knots

[1] Robert Riley, Discrete parabolic representations of link groups, *Mathematika*, 22 (2) (1975), pp. 141-150.

[2] William P. Thurston, Three dimensional manifolds, Kleinian groups and hyperbolic geometry, *Bull. Amer. Math. Soc.* (NS), 6 (1982), pp. 357-381

[3] William W. Menasco, Polyhedra representation of link complements. Low-dimensional topology (San Francisco, Calif., 1981), 305–325, *Contemp. Math.*, 20, Amer. Math. Soc., Providence, RI, 1983.

[4] Marc Lackenby, The volume of hyperbolic alternating link complements. With an appendix by Ian Agol and Dylan Thurston. *Proc. London Math. Soc.* (3) 88 (2004), no. 1, 204?224.

20.8.4 Establishing Conjecture T1

[1] Louis H. Kauffman, State models and the Jones polynomial, *Topology* 26 (1987), no. 3, 395-407.

[2] Kunio Murasugi, Jones polynomials and classical conjectures in knot theory, *Topology* 26 (1987), no. 2, 187-194.

[3] Morwen B. Thistlethwaite, A spanning tree expansion of the Jones polynomial, *Topology* 26 (1987), no. 3, 297-309.

[4] Vaughan Jones, A polynomial invariant for knots via von Neumann algebras. *Bull. Am. Math. Soc.* 12, 103-111, 1985.

[5] Vaughan Jones, "Hecke algebra representations of braid groups and link polynomials." *Ann. Math.* 126, 335-388, 1987.

[5] W.B.R. Lickorish, An Introduction to Knot Theory, Graduate Texts in Mathematics Series, Springer Verlag, 1997.

20.8.5 Establishing Conjectures T2 and T3

[1] William W. Menasco and M. B Thistlethwaite, The classification of alternating links, *Ann. Math.* 138 (1993), 113-171.

[2] William W. Menasco and M. B Thistlethwaite, The Tait flyping conjecture. *Bull. Amer. Math. Soc.* (N.S.) 25 (1991), no. 2, 403-412.

20.8.6 What is an Alternating Knot?

[1] Joshua A. Howie, A characterisation of alternating knot exteriors. *Geom. Topol.* 21 (2017), no. 4, 2353-2371.

[2] Joshua E. Greene, Alternating links and definite surfaces. With an appendix by Andras Juhasz and Marc Lackenby. *Duke Math. J.* 166 (2017), no. 11, 2133-2151.

Chapter 21

Periodic Knots

Swatee Naik, University of Nevada, Reno

21.1 Introduction

It is naturally interesting to look for patterns and symmetries in the world around us. Knots that remain the same through a rotation of the three-dimensional space around an axis are called periodic or cyclically periodic knots, and they exhibit repeating patterns. The rotation is a map f of the ambient or surrounding space to itself, and the axis is the set of fixed points Fix(f). The map f does not fix any point of the knot, but it rotates or permutes the points within it, thereby leaving the knot invariant as a set. The figure below shows a diagram of the (right-handed) trefoil, or torus knot $T(3,2)$, which remains invariant under a $2\pi/3$ rotation of $\mathbb{R}^3$ about an axis through the indicated point B that is perpendicular to this paper. This order 3 rotation establishes 3-periodicity of the trefoil.

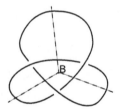

Figure 1

Iterating a rotation f results in f^2, f^3, $\cdots$ which are all distinct maps until we reach a stage when f^q equals the identity map of the space. The set $\{f, f^2, f^3 \cdots\}$ under the operation of composition forms a finite, cyclic group of order q. Therefore the map is also referred to as a group action.

21.2 Definitions and Examples

For reasons of convenience, compactness being one, it is customary in knot theory to consider knots to be subsets of the three-sphere S^3. The three-dimensional Euclidean space $\mathbb{R}^3$ is seen as an embedded subspace that misses only one point of S^3. To avoid certain "wild" cases, smoothness of the map is important, and so we will require that the periodic map of S^3 to itself be a diffeomorphism, that is, a smooth bijection with a smooth inverse. Finally, the axis, which would be a straight line in $\mathbb{R}^3$ is now a circle in S^3 with the "added point at infinity." It is time for

179

a formal definition.

Definition: A knot K in S^3 is said to be periodic if there exist an integer $q > 1$ and an orientation preserving diffeomorphism $f : S^3 \to S^3$ such that

1. $f(K) = K$,

2. $\text{Order}(f) = q$, i.e., $f^q = id$ and $f^n \neq id$, for $0 < n < q$, and

3. $\text{Fix}(f) = B$ is homeomorphic to a circle, and $B \cap K = \phi$.

Any such q is called a period of K, and any such f, a corresponding periodic transformation. The set B is called the axis of periodicity.

Results in [2] show that B is unknotted, and as a consequence the orbit space or quotient space of S^3 under the action of f is itself diffeomorphic to S^3. The image of the knot in this quotient S^3 is called the quotient knot denoted as $\overline{K}$, and the image of B is $\overline{B}$.

In Figure 2 below, we see a period 2 diagram for the trefoil and the quotient knot that captures the repeating pattern. Note the axis B and its image $\overline{B}$, and the similarity in how they go around the knots K and $\overline{K}$, respectively.

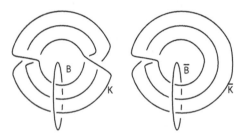

Figure 2

Clearly if a knot has a period q, any factor of q is its period as well. Periods of a torus knot $T(a, b)$ are divisors of a and b.

One may also build a periodic knot starting from a quotient knot and an axis. On the left side in Figure 3, we see a quotient knot (pattern); the knot on the right is obtained by repeating the pattern three times before closing up. The knot K clearly has period 3.

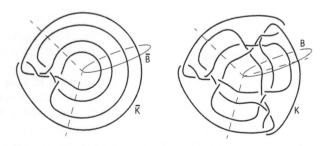

Figure 3

Note that if we repeat the pattern of this $\overline{K}$ twice and then close up, we would get a period 2 *link* instead of a knot. You can see this by drawing the corresponding diagram and checking that it represents two circles, not one!

To understand what happens, we need *linking numbers.*

Definition: The linking number between two disjoint, oriented, simple, closed curves in $\mathbb{R}^3$ or S^3 is obtained by adding all crossing numbers between the two curves according to the following convention and then dividing the sum by 2. In the language of algebraic topology the linking number is the first homology class represented by one curve in the complement of the other in S^3.

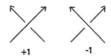

Changing orientation of a curve can change the sign of a linking number, but not its absolute value. The sign is not important for our discussion. In Figure 3, the linking number $\lambda = \text{lk}(K, B) = \text{lk}(\overline{K}, \overline{B}) = \pm 2$ (check, starting first by drawing arrows on K, B, $\overline{K}$ and $\overline{B}$). Starting with this $\overline{K}$, we can build a periodic knot with any odd period q. When we try to imitate the process with an even number, we end up with a *link*. In order to have a period q *knot*, the greatest common divisor $\gcd(\lambda, q)$ has to be 1.

Patterns are not always visible in a given knot diagram, as already seen from the two different pictures for the trefoil. Before we begin playing with a diagram to discover symmetry, it would help to at least know what possible periods that knot may have, or at least to rule out some possibilities. The remainder of our discussion will focus on this.

21.3   Knot Invariants Obstructing Periodicity

It is convenient to associate algebraic objects, such as numbers, polynomials, or groups to a knot in such a way that if a knot diagram is transformed into another by means of diagram moves that do not change the knot type, then the associated algebraic objects are also equivalent under a suitable equivalence relation. Such objects are called knot invariants. Ideally, a knot invariant would not only help distinguish a knot from others, but also reflect some of its properties, including a possible symmetry. If a certain invariant is known to necessarily exhibit a predictable pattern of its own, whenever the knot happens to be periodic, then one can start by examining the invariant to determine potential periods for the knot. Before we illustrate this through an example, we introduce a knot invariant known as the Alexander polynomial.

Definition: An Alexander polynomial is a polynomial Δ with integer coefficients, such that

$$\Delta(t) = t^{\deg(\Delta)}\Delta(t^{-1}) \text{ and } \Delta(1) = \pm 1. \tag{21.1}$$

Every knot has an associated Alexander polynomial that can be computed using a knot diagram, and it closely relates to the topology of the knot embedding. Conversely, any such polynomial is the Alexander polynomial of some knot. The first condition in (21.1) is evident through a symmetric arrangement of the coefficients as in (21.4).

Let K be a q-periodic knot with quotient knot $\overline{K}$ and $\lambda = |\text{lk}(K, B)|$. Let Δ_K and $\Delta_{\overline{K}}$ be the Alexander polynomials of K and $\overline{K}$ respectively. Relating the topology of the 2-component link (K, B) with that of the quotient link $(\overline{K}, \overline{B})$, Kunio Murasugi obtained the following conditions in [7, 8]:

Theorem: With notation as above, and q a prime, we have:

$$\Delta_{\overline{K}}(t) \mid \Delta_K(t) \tag{21.2}$$

$$\Delta_K(t) \equiv \pm t^n \left(\Delta_{\overline{K}}(t)\right)^q \left(1 + t + \cdots + t^{\lambda-1}\right)^{q-1} \bmod q. \tag{21.3}$$

The symbol $\equiv$ stands for congruence (modulo q), n can be any integer, and as discussed earlier, we have $\gcd(\lambda, q) = 1$.

Example: Consider the Alexander polynomial

$$\Delta_K(t) = 3t^6 - 21t^5 + 53t^4 - 71t^3 + 53t^2 - 21t + 3. \tag{21.4}$$

According to KnotInfo [3] this is the polynomial for the knot $K = 12_{a100}$. If K were to have a prime period q, it would have a quotient knot $\overline{K}$, and the polynomials Δ_K and $\Delta_{\overline{K}}$ would satisfy conditions (21.2) and (21.3).

Irreducible factors of this polynomial are $t^3 - 5t^2 + 6t - 3$ and $3t^3 - 6t^2 + 5t - 1$. Since neither of them is symmetric, (21.2) implies that $\Delta_{\overline{K}}$ has to equal 1 or Δ_K. We now examine each case and apply (21.3) with various choices of q.

Case 1: If $\Delta_{\overline{K}} = \Delta_K$, the only way to satisfy (21.3) is to have $n = 0$, $\lambda = 1$, and $\Delta_{\overline{K}} \equiv \Delta_K \equiv 1 \bmod q$. The leading coefficient and the constant term of Δ_K both being 3 quickly eliminates all prime period possibilities other than 3, and using the rest of the coefficients it is easy to see that $\Delta_K \not\equiv 1 \bmod 3$.

Case 2: If $\Delta_{\overline{K}} = 1$, (21.3) $\implies \Delta_K(t) \equiv \pm t^n \left(1 + t + \cdots + t^{\lambda-1}\right)^{q-1} \bmod q$.
This, along with the fact that the degree of Δ_K is 6, severely restricts possible q, λ, n, and it is easy to go through the short list of possibilities one by one.

For instance, if $q = 7$, the degrees on both sides of the congruence ensure that $\lambda = 2$, $n = 0$, and the leading coefficient of Δ_K being 3, we quickly see that $\Delta_K(t) \not\equiv (1 + t)^6 \bmod 7$. This rules out period 7. Period 5 is ruled out in a similar fashion (check), leaving only 2 and 3 as possibilities.

The period 2 condition is satisfied with $\lambda = 7$ and $n = 0$. As period 2 is more difficult to deal with, for the rest of this example we will focus on period 3 only.

Indeed the period 3 conditions are satisfied with $n = 3$, $\lambda = 2$, as $\Delta_K(t) \equiv -t^3(t + 1)^2$ (mod 3). We now call on another invariant, namely the set of Heegaard-Floer correction terms of a knot, defined in [9]. Although the technical details are beyond the scope of this discussion, the following consequence of a theorem of [6] is easy to to state and use:

Theorem: Let K be a period q knot with quotient knot $\overline{K}$, and suppose that q is a prime. If $\Delta_{\overline{K}}(t) = 1$, then all but one of the correction terms of K have values that come in multiples of q.

Here is a table of "5-torsion correction terms" for 12_{a100}:

Value	$-\frac{4}{5}$	$-\frac{2}{5}$	0	$\frac{2}{5}$	$\frac{4}{5}$
Multiplicity	2	6	6	6	4

As the multiplicities 2 (of the correction term value $-\frac{4}{5}$) and 4 (of $\frac{4}{5}$) fail to be divisible by 3, 12_{a100} is not period 3.

For comparison, take a look at a similar table for the period 3 knot 9_{40}, where each value except one, namely $-\frac{1}{12}$, has multiplicity divisible by 3.

Value	$-\frac{9}{10}$	$-\frac{1}{12}$	$-\frac{1}{10}$	$\frac{3}{10}$	$\frac{7}{10}$
Multiplicity	6	1	6	6	6

21.4 A Few Words in Closing

The *Full Symmetry Groups* $S(K)$ for hyperbolic knots K are listed in KnotInfo [3]. Related computer calculations are described in [1].(Note that torus and cable knots are not hyperbolic.) The group $S(K)$ captures symmetries of K; q-periodicity implies the existence of an order q element. However, in the opposite direction, the existence of an order q element does not guarantee q-periodicity, because the corresponding map f may be fixed point free. A *freely periodic* knot is invariant under a group action without any fixed points, and such knots are studied in [4, 5].

In later chapters, you will learn about various other invariants of knots. References listed in [6] provide an extensive list of papers that show how to obstruct periodicity using some of these invariants, such as the Jones polynomial and its 2-variable generalizations, twisted Alexander polynomials, hyperbolic structures on knot complements, homology groups of branched cyclic covers, concordance invariants of Casson and Gordon, Khovanov homology, and link Floer homology.

Acknowledgments

This material is based upon work supported by and while serving at the National Science Foundation. Any opinion, findings, and conclusions or recommendations expressed in this material are those of the authors and do not necessarily reflect the views of the National Science Foundation.

21.5   Bibliography

[1] C. Adams, M. Hildebrand, and J. Weeks. Hyperbolic invariants of knots and links. *Trans. AMS*, 326:1–56, (1991).

[2] H. Bass and J. W. Morgan, editors. *The Smith Conjecture. Papers Presented at the Symposium Held at Columbia University, New York, 1979*, volume 112 of *Pure and Applied Mathematics*. Academic Press, Inc., Orlando, FL, 1984.

[3] Jae Choon Cha and Charles Livingston. Knotinfo: Table of knot invariants, http://www.indiana.edu/~knotinfo, 2018.

[4] Richard Hartley. Knots with free period. *Canad. J. Math.*, 33(1):91–102, 1981.

[5] J. Hillman, C. Livingston, and S Naik. The twisted alexander polynomial and periodicity of knots. *Algebraic and Geometric Topology*, 6:145–169, 2006.

[6] Stanislav Jabuka and Swatee Naik. Periodic knots and Heegaard-Floer correction terms. *Journal of the European Mathematics Society*, 18:1651–1674, 2016.

[7] Kunio Murasugi. On periodic knots. *Comment. Math. Helv.*, 46:162–174, (1971).

[8] Kunio Murasugi. On symmetries of knots. *Tsukuba J. Math.*, 4:331–347, 1980.

[9] Peter Ozsváth and Zoltan Szabó. On the Heegaard Floer homology of branched double-covers. *Adv. Math.*, 194:1–33, 2005.

Part V

Knots and Surfaces

Chapter 22

Seifert Surfaces and Genus

Mark Brittenham, University of Nebraska Lincoln

22.1 Seifert Surfaces

The unknot U can be characterized in many ways; one of these characterizations is that it is the only knot which forms the boundary of a smoothly embedded 2-disk in 3-space. Such a disk is apparent from its zero-crossing diagram, and this disk can be carried along with U under an isotopy to form a disk with boundary equal to any other diagram of U.

While it is not immediately apparent, any knot or link K is the boundary $\partial\Sigma$ of an embedded, orientable surface Σ, known as a *spanning surface* or *Seifert surface* for K. Although he was not the first to demonstrate this [9], Seifert provided an algorithm [49] for building a spanning surface that is quite straightforward to implement. Starting from any diagram for K, orient the components of K, then 'resolve' (i.e., remove) every crossing, using the orientation on K, to create a collection of disjoint oriented *Seifert circles* in the projection plane. Each circle bounds a disk, which we can make disjoint by raising or lowering nested disks. Finally, reintroduce the crossings of the link K by joining the disks with half-twisted bands. These bands will connect the disks in a way that respects a normal orientation given by the right-hand rule along their boundary, resulting in an orientable surface with boundary equal to K. Figure 22.1 illustrates the process.

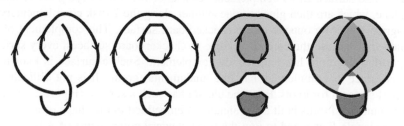

FIGURE 22.1: Seifert's algorithm.

There are, of course, many arbitrary choices in this construction: orientations on the components of the link, raising versus lowering nested disks, and changing the diagram for K at the start of the process. A given knot or link will therefore have many Seifert surfaces.

If there is no nesting among the Seifert circles, the resulting surface is an orientable checkboard surface for the link; it is effectively described by coloring in alternating complementary regions of the link. It is a surprising fact, that has been rediscovered several times [17],[16],[20], that every link has a projection so that Seifert's algorithm results in a checkerboard surface for the link. In fact, every surface built by Seifert's algorithm is isotopic to a checkerboard surface.

22.2   Invariants from Seifert Surfaces

A Seifert surface Σ for a knot provides the backbone to many fundamental constructions in knot theory. If we remove a solid torus neighborhood $N(K)$ of the knot K, to yield the exterior $X(K)$ of K, then Σ meets $\partial X(K)$ in a curve λ which generates the kernel of the inclusion induced map $\iota_* : H_1(\partial X(K)) = Z \oplus Z \rightarrow Z = H_1(X(K))$. This curve, unique up to isotopy, is known as the *longitude* of K. Together with the meridian μ, that is, the boundary of a disk in $N(K)$ meeting K transversely in a single point, the two curves form the standard coordinate system on the boundary torus of a knot exterior for knot-theoretic constructions such as Dehn filling.

From a Seifert surface Σ for K we can construct a *Seifert matrix* V for K. By choosing loops γ_i on Σ representing a basis for the first homology of Σ, $H_1(\Sigma) \cong Z^{2g}$, and pushing the loops normally off of Σ to give loops γ_i^+, this yields the $2g \times 2g$ matrix V of linking numbers $lk(\gamma_i, \gamma_j^+)$. While not itself a knot invariant, V can be used to construct several important knot invariants; the *determinant* $\det(K) = |\det(V + V^T)|$, the *signature* $\sigma(K) = \text{signature}(V + V^T)$ [55], the *Alexander polynomial* $\Delta_K(t) = \det(V^T - tV)$, and the *torsion invariants* of K can all be derived directly from the Seifert matrix V.

22.3   Knot Genera

Seifert's algorithm, however, cannot construct every orientable spanning surface for a knot or link. For a checkerboard surface Σ, its complement consists of the 3-balls lying above and below the projection plane, which are then glued to one another along the 2-disk complementary regions of the knot diagram which lie outside of the checkerboard surface. The complement of Σ is therefore a handlebody. Because they are all isotopic to checkerboard surfaces, every surface built by Seifert's algorithm therefore has handlebody complement. Such a surface is known as a 'free' Seifert surface, since its complement has free fundamental group. It is not difficult to construct non-free Seifert surfaces for knots. Even the unknot has such a surface (Figure 22.2); the example here is a once-punctured torus built as an annulus A embedded in the shape of a trefoil knot, with an untwisted rectangle R attached to join the two boundary components of A into an unknot. An alternate description is as the torus boundary of a neighborhood of the trefoil knot, with a 2-disk removed. This can be grafted onto any Seifert surface to yield a new, non-free, Seifert surface for any knot or link, as in the figure.

The principle invariant distinguishing orientable spanning surfaces is the genus $g(\Sigma)$ = the genus of the closed surface obtained by capping $\partial\Sigma = K$ off with a disk (or disks). This can be used to define an invariant of a knot or link K. The *genus of a knot* of K, denoted $g(K)$, is the minimum of $g(\Sigma)$ over all Seifert surfaces for K. That is, $g(K)$ is the genus of the 'simplest' Seifert surface with boundary K. The unknot, therefore, is the only knot with genus 0; it is the only knot having the 2-disk for a Seifert surface. If we restrict only to free Seifert surfaces, the minimum genus is known as the *free genus* $g_F(K)$ of K; if we restrict only to surfaces built by Seifert's algorithm applied to projections of K, the minimum genus is known as the *canonical genus* $g_c(K)$ of K. Because these

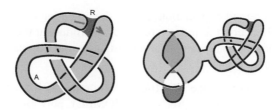

FIGURE 22.2: Non-free unknot.

minima restrict to successively fewer spanning surfaces, we have the basic inequalities $g(K) \leq g_f(K) \leq g_c(K)$.

For many classes of knots (see *Computing knot genus* below) all three invariants are equal, since a $g(K)$-minimizing surface can be built by Seifert's algorithm. In general, though, the three genera can be distinct from one another. Moriah [33] and Livingston [30] showed that doubled knots can have $g_f(K) - g(K)$ arbitrarily large, and Kobayashi and Kobayashi [28] showed, using connected sums, that both $g_c(K) - g_f(K)$ and $g_f(K) - g(K)$ can be simultaneously large. Brittenham [3] showed, using a relationship between $g_c(K)$ and hyperbolic volume, that $g_c(K) - g_f(K)$ can be arbitrarily large for hyperbolic knots, and Nakamura [36] showed the same for fibered knots.

22.4 Properties and Relations to Other Constructions

A spanning surface Σ is *compressible* if there is a simple closed curve γ in Σ which bounds an embedded disk D in the complement of Σ, but does not bound a disk in Σ. Cutting Σ open along γ and gluing in two copies of D yields a new spanning surface with lower genus, and so $g(K)$-minimizing surfaces must be *incompressible*. Every knot therefore has an incompressible Seifert surface. This is important in many cut-and-paste arguments in knot theory, since loops of intersection of a surface with an incompressible spanning surface that are trivial in the knot complement can then be assumed to be trivial on the spanning surface. Incompressible Seifert surfaces need not have minimal genus, however; in fact, there are knots that have incompressible Seifert surfaces of arbitrarily large genus [57].

Knot genus behaves well under many standard constructions. Genus is additive under connected sum: $g(K_1 \sharp K_2) = g(K_1) + g(K_2)$ [46]. Because of this, knots have no inverses under connected sum: you cannot create the unknot by summing two knots together. The free genus g_f is also additive under connected sum [38]. Under the more general operation of band-connected sum $K = K_1 \sharp_b K_2$, where two knots in disjoint 3-balls are connected by an arbitrary embedded rectangle $R = [0, 1] \times [0, 1]$ with $R \cap (K_1 \cup K_2) = [0, 1] \times \{0, 1\}$, the genus is 'superadditive': $g(K) \geq g(K_1) + g(K_2)$ [15]. Another generalization of connected sum is Murasugi sum [50], where two Seifert surfaces Σ_1, Σ_2 are glued together along disks $D_1 \subseteq \Sigma_i$, where $D_i \cap \partial \Sigma_i$ consists of n arcs; this yields a new orientable (Seifert) surface Σ for the knot or link $K = \partial \Sigma$. When $n = 1$ this is connected sum, and when $n = 2$ this is known as a *plumbing*. Gabai [12] showed that if both Σ_1 and Σ_2 are genus-minimizing for their boundaries, then Σ is, as well.

However, genus is not preserved by mutation; that is, the operation of removing a 2-string tangle from K, given as a 3-ball meeting K in two arcs, and reintroducing the tangle after a rotation of the 3-ball. Perhaps the most famous example of mutant knots, the Conway knot and the Kinoshita-Tarasaka knot, have different genera [13],[14].

Immediately after the discovery of the HOMFLY polynomial [10], Morton found an inequality relating the canonical genus and the degree of the HOMFLY polynomial in one of its variables (in his implementation, the z-variable): $\deg_z P_K(v,z) \leq 2g_c(K)$ [34]. This can be shown to be an equality in many cases [54],[51],[37],[4],[22], thereby providing a computation of $g_c(K)$. But there are infinitely many knots for which this inequality is strict [52],[5].

Kobayashi [26] showed that the only knots that have both $g(K) = 1$ and unknotting number one are the Whitehead doubles of knots. Thus knowing that $g(K) = 1$ can be used to prove that the unknotting number of K is greater than one. This, for example, provided the first proof that the knot 9_{25} has unknotting number 2. Scharlemann and Thompson [44] extended Kobayashi's argument, to describe the behavior of $g(K)$ under crossing change. They showed that in a skein relation, that are used to compute the Conway, Jones, and HOMFLY polynomials, where three oriented links $L_+, L_-,$ and L_0 are identical except inside of a 3-ball, where they appear as in Figure 22.3, two of $g(L_-)$, $g(L_+)$ and $g(L_0) - 1$ are equal and the third is not larger. This has the consequence that $g(K)$ is a lower bound for the depth of any skein tree used to compute these polynomial invariants for K.

FIGURE 22.3: Crossing changes.

22.5   Computing Knot Genus

As a minimum of the genera of all Seifert surfaces for K, the genus of K can be, in principle, challenging to compute. But the fact that other knot invariants can be derived from a Seifert matrix can aid in this calculation. If a Seifert surface Σ has genus g, then its Seifert matrix V is a $2g \times 2g$ matrix. Consequently, $|\sigma(K)| \leq 2g$ and $degree(\Delta_K(t)) \leq 2g$. Since this must be true for every Seifert surface for K, these yield the lower bounds $g(K) \geq \frac{1}{2}\sigma(K)$ and $g(K) \geq \frac{1}{2}degree(\Delta_K(t))$. In many cases one can show that a Seifert surface construction matches one of these lower bounds, yielding a determination of $g(K)$. For example, for any torus knot $T(p,q)$ a direct construction shows that Seifert's algorithm applied to the standard diagram of $T(p,q)$ yields a surface with genus $\frac{(p-1)(q-1)}{2} = \frac{1}{2}$(degree of $\Delta_{T(p,q)}(t)$), and so $g(T(p,q)) = \frac{(p-1)(q-1)}{2}$. Murasugi [35] and Crowell [7] showed that for any alternating knot K, the surface built by Seifert's algorithm applied to an alternating diagram has genus $g(K) = \frac{1}{2}$(degree of $\Delta_K(t)$), thereby computing the genera of alternating knots. Similar results have been obtained for the classes of homogeneous knots [6] and alternative knots [25].

The work of Thurston and Gabai in the 1980's established a strong relationship between genus and foliations, which fill a manifold with disjoint surfaces, called leaves, which are locally parallel to one another. Thurston [53] showed that any compact leaf of a taut foliation has least genus in its homology class, while Gabai [11] established the converse, showing that any least genus surface is the leaf of a taut foliation. A Seifert surface can therefore be certified as least genus by constructing a taut foliation around it. Using this, Gabai [13] showed that one can quickly certify a minimal genus Seifert surface for every knot in the standard knot tables, as well as recover the result that Seifert's algorithm will build a least genus spanning surface for an alternating knot. Further constructions [14],[27],[21] have determined the genera of many other classes of knots. The theory of sutured manifolds [11],[45], which is at the foundation of Gabai's constructions, has found wide application in low-dimensional topology.

In recent decades homology theories have been developed for 3- and 4-manifolds, based on gauge theory and symplectic topology, that have yielded powerful tools for exploring the topology and geometry of manifolds. Several of these have a direct bearing in the computation of knot genus. In particular, Kronheimer and Mrowka [29] show that the Thurston norm, which, for a knot exterior, is effectively the knot genus, is determined by the monopole Floer homology. Ozsváth and Szabó [39], on the other hand, show that Heegaard Floer homology can detect the genus of a knot. A combinatorial formulation of knot Floer homology [32], based on a grid diagram for a knot, makes the determination of knot genus computationally feasible.

Using normal surface theory [18], Schubert [47] showed that there is an algorithm to compute the genus of a knot. Extending this work, Hass, Lagarias and Pippenger [19] showed that the computation of knot genus is in the complexity class PSPACE; that is, can be found with an algorithm that uses a polynomial amount of space in the number of crossings of the knot. In the more general setting of knots in 3-manifolds, Agol, Hass and Thurston [1] show that finding the genus of a knot has complexity NP, that is, the assertion $g(K) = n$ can be verified in polynomial time, and they also show that knot genus is NP-hard. The computation of knot genus is therefore an NP-complete problem.

22.6 The Complex of Seifert Surfaces

A given knot K can have many $g(K)$-minimizing Seifert surfaces. There are, in fact, examples of knots which have infinitely many such surfaces [8]. However, for a hyperbolic knot there are only finitely many $g(K)$-minimizing surfaces up to isotopy [48]. Many only have a single such surface up to isotopy [56],[31], including all fibered knots, that is, knots whose exterior fibers over the circle [53]. On the other hand, Roberts [42] has shown that for any positive integer n, there exists an oriented prime knot K in S^3 with at least 2^{2n-1} inequivalent $g(K)$-minimizing Seifert surfaces of genus n.

In [24] Kakimizu introduced a simplicial complex $MS(K)$ to encode the relationships between $g(K)$-minimizing Seifert surfaces for a knot. The vertices are isotopy classes of such Seifert surfaces, and $n + 1$ of them form the vertices of an n-simplex in $MS(K)$ if they have representatives that are all disjoint from one another. For some knots, such as those of Lyon [31], this complex consists of a single point, but there are examples [2] which show that this complex can be locally infinite; that is, one surface is disjoint from infinitely many others. Scharlemann and Thompson [43] showed that two Seifert surfaces Σ and Σ' for a knot K with genera equal to $g(K)$ can be joined by a sequence of Seifert surfaces for K with genus $g(K)$, $\Sigma = \Sigma_1, \Sigma_2, \ldots, \Sigma_n = \Sigma'$, so that, for every i, Σ_i and Σ_{i+1} are disjoint up to isotopy. This implies that the Kakimizu complex is connected. Kakimizu conjectured that $MS(K)$ is always contractible; this was established in 2012 by Przytycki and Schultens [41].

Juhasz [23] has shown that Heegaard Floer homology can constrain the dimension of the Kakimizu complex; if the knot Floer homology has rank less than 2^{n+1} at the $g(K)$-grading, then the dimension of $MS(K)$ is less than n. In particular, if the rank is less than 4, then K has a unique minimal genus Seifert surface, since $MS(K)$ is connected, and so if $MS(K)$ has more than one vertex it must have dimension at least one. For alternating knots, this can be recast in terms of the lead coefficient a_n of the Alexander polynomial $\Delta_K(t)$: if $|a_n| < 2^{n+1}$, then the dimension of $MS(K)$ is less than n.

22.7   Bibliography

[1] I. Agol, J. Hass, and W. Thurston, The computational complexity of knot genus and spanning area, *Trans. Amer. Math. Soc.* **358** (2006) 3821–3850.

[2] J. Banks, On links with locally infinite Kakimizu complexes, *Algebr. Geom. Topol.* **11** (2011) 1445–1454.

[3] M. Brittenham, Free genus one knots with large volume, *Pacific J. Math.* **201** (2001) 61–82.

[4] M. Brittenham and J. Jensen, Canonical genus and the Whitehead doubles of pretzel knots, preprint, arXiv:math.GT/0608765v1.

[5] M. Brittenham and J. Jensen, Families of knots for which Morton's inequality is strict, *Comm. Anal. Geom.* **15** (2007) 971–983.

[6] P. Cromwell, Homogeneous links, *J. London Math. Soc.* **39** (1989) 535–552.

[7] R. Crowell, Genus of alternating link types, *Ann. of Math.* **69** (1959) 258–275.

[8] J. Eisner, Knots with infinitely many minimal spanning surfaces, *Trans. Amer. Math. Soc.* **229** (1977) 329–349.

[9] F. Frankl and L. Pontrjagin, Ein Knotensatz mit Anwendung auf die Dimensionstheorie, *Math. Annalen* **102** (1930) 785–789.

[10] P. Freyd, D. Yetter, J. Hoste, W. Lickorish, K. Millett, and A. Ocneanu, A new polynomial invariant of knots and links, *Bull. Amer. Math. Soc.* **12** (1985) 239–246.

[11] D. Gabai, Foliations and the topology of 3-manifolds, *J. Differential Geom.* **18** (1983) 445–503.

[12] D. Gabai, The Murasugi sum is a natural geometric operation, in Low-dimensional topology (San Francisco, Calif., 1981), *Contemp. Math.* **20** Amer. Math. Soc., Providence, RI, (1983) pp. 131–143.

[13] D. Gabai, Foliations and genera of links, *Topology* **23** (1984) 381–394.

[14] D. Gabai, Genera of the arborescent links, *Mem. Amer. Math. Soc.* **59** (1986) i–viii and 1–98.

[15] D. Gabai, Genus is superadditive under band connected sum, *Topology* **26** (1987) 209–210.

[16] D. Gauld, Simplifying Seifert surfaces, *New Zealand J. Math.* **22** (1993) 61–62.

[17] R. Goodrick, A note on Seifert circles, *Proc. Amer. Math. Soc.* **21** (1969) 615–617.

[18] W. Haken, Theorie der Normalflächen, *Acta Math.* **105** (1961) 245–375.

[19] J. Hass, J. Lagarias, and N. Pippenger, The computational complexity of knot and link problems, *J. ACM* **46** (1999) 185–211.

[20] M. Hirasawa, The flat genus of links, *Kobe J. Math.* **12** (1995) 155–159.

[21] M. Hirasawa and K. Murasugi, Genera and fibredness of Montesinos knots, *Pacific J. Math.* **225** (2006) 53–83.

[22] H. Jang and S. Lee, The canonical genus for Whitehead doubles of a family of alternating knots, *Topology Appl.* **159** (2012) 3563–3582.

[23] A. Juhasz, Knot Floer homology and Seifert surfaces, *Algebr. Geom. Topol.* **8** (2008) 603–608.

[24] O. Kakimizu, Finding disjoint incompressible spanning surfaces for a link, *Hiroshima Math. J.* **22** (1992) 225–236.

[25] L. Kauffman, Combinatorics and knot theory, in Low-dimensional topology (San Francisco, Calif., 1981), *Contemp. Math.* **20** Amer. Math. Soc., Providence, RI, (1983) pp. 181–200.

[26] T. Kobayashi, Minimal genus Seifert surfaces for unknotting number 1 knots, *Kobe J. Math.* **6** (1989) 53–62.

[27] T. Kobayashi, Uniqueness of minimal genus Seifert surfaces for links, *Topology Appl.* **33** (1989) 265–279.

[28] M. Kobayashi and T. Kobayashi, On canonical genus and free genus of knot, *J. Knot Theory Ramifications* **5** (1996) 77–85.

[29] P. Kronheimer and T. Mrowka, Scalar curvature and the Thurston norm, *Math. Res. Lett.* **4** (1997) 931–937.

[30] C. Livingston, The free genus of doubled knots, *Proc. Amer. Math. Soc.* **104** (1988) 329–333.

[31] H. Lyon, Simple knots with unique spanning surfaces, *Topology* **13** (1974) 275–279.

[32] C. Manolescu, P. Ozsváth, and S. Sarkar, A combinatorial description of knot Floer homology, *Ann. of Math.* **169** (2009) 633–660.

[33] Y. Moriah, The free genus of knots, *Proc. Amer. Math. Soc.* **99** (1987) 373–379.

[34] H. Morton, Seifert circles and knot polynomials, *Math. Proc. Cambridge Philos. Soc.* **99** (1986) 107–109.

[35] K. Murasugi, On the genus of the alternating knot. I, II, *J. Math. Soc. Japan* **10** (1958) 94–105 and 235–248.

[36] T. Nakamura, On canonical genus of fibered knot, *J. Knot Theory Ramifications* **11** (2002) 341–352.

[37] T. Nakamura, *On the crossing number of 2-bridge knot and the canonical genus of its Whitehead double*, *Osaka J. Math.* **43** (2006) 609–623.

[38] M. Ozawa, Additivity of free genus of knots, *Topology* **40** (2001) 659–665.

[39] P. Ozsváth and Z. Szabó, Holomorphic disks and genus bounds, *Geom. Topol.* **8** (2004) 311–334.

[40] P. Ozsváth and Z. Szabó, Knot Floer homology, genus bounds, and mutation, *Topology Appl.* **141** (2004) 59–85.

[41] P. Przytycki and J. Schultens, Contractibility of the Kakimizu complex and symmetric Seifert surfaces, *Trans. Amer. Math. Soc.* **364** (2012) 1489–1508.

[42] L. Roberts, On non-isotopic spanning surfaces for a class of arborescent knots, *J. Knot Theory Ramifications* **22** (2013) 16 pp.

[43] M. Scharlemann and A. Thompson, Finding disjoint Seifert surfaces, *Bull. London Math. Soc.* **20** (1988) 61–64.

[44] M. Scharlemann and A. Thompson, Link genus and the Conway moves, *Comment. Math. Helv.* **64** (1989) 527–535.

[45] M. Scharlemann, Sutured manifolds and generalized Thurston norms, *J. Differential Geom.* **29** (1989) 557–614.

[46] H. Schubert, Knoten und Vollringe, *Acta Mathematica* **90** (1953) 131–286.

[47] H. Schubert, Bestimmung der Primfaktorzerlegung von Verkettungen, *Math. Zeit.* **76** (1961) 116–148.

[48] H. Schubert and K. Soltsien, Isotopie von Flächen in einfachen Knoten, *Abh. Math. Sem. Univ. Hamburg* **27** (1964) 116–123.

[49] H. Seifert, Über das Geschlecht von Knoten, *Math. Annalen* **110** (1935) 571–592.

[50] J. Stallings, Constructions of fibred knots and links, in Algebraic and geometric topology, *Proc. Sympos. Pure Math.* **XXXII** Amer. Math. Soc., Providence, R.I. (1978) pp. 55–60.

[51] A. Stoimenow, *On the crossing number of positive knots and braids and braid index criteria of Jones and Morton-Williams-Franks*, *Trans. Amer. Math. Soc.* **354** (2002) 3927–3954.

[52] A. Stoimenow, Knots of (canonical) genus two, *Fund. Math.* **200** (2008) 1–67.

[53] W. Thurston, A norm for the homology of 3-manifolds, *Mem. Amer. Math. Soc.* **59** (1986), no. 339, i–vi and 99–130.

[54] J. Tripp, The canonical genus of Whitehead doubles of a family torus knots, *J. Knot Theory Ramifications* **11** (2002) 1233–1242.

[55] H. Trotter, Homology of group systems with applications to knot theory, *Ann. of Math.* **76** (1962) 464–498.

[56] W. Whitten, Isotopy types of knot spanning surfaces, *Topology* **12** (1973) 373–380.

[57] R. Wilson, Knots with infinitely many incompressible Seifert surfaces, *J. Knot Theory Ramifications* **17** (2008) 537–551.

Chapter 23

Non-Orientable Spanning Surfaces for Knots

Thomas Kindred, University of Nebraska, Lincoln

Given a diagram $D \subset S^2$ of a knot $K \subset S^3$, one can construct two spanning surfaces B and W for K by shading each of the disks of $S^2 \setminus D$ one of two colors (say, dark or light) such that disks of the same color meet only at crossing points, and then modifying the shaded diagram near each crossing point as shown in Figure 23.1. Such B and W are called the *checkerboard surfaces* from D. Figure 23.2 shows an example. At most one of B and W is orientable, since an orientation on a surface induces an orientation on its boundary, but K can be coherently oriented near each crossing in either B or W, but not both.[1]

Starting with this observation, B.E. Clark first laid out several properties of nonorientable spanning surfaces [4]. Thickening any surface F in a 3-manifold and taking the boundary gives a surface

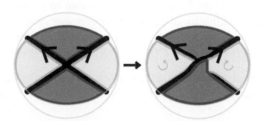

FIGURE 23.1: Modify a shaded knot diagram near each crossing to construct checkerboard surfaces. Locally, exactly one of the two surfaces is coherently oriented, in this case the lightly shaded one.

FIGURE 23.2: Checkerboard surfaces from an alternating diagram of for the 8_{14} knot.

---

[1] More generally, when K is an m-component link and D has n crossings, exactly 2^{m-1} of the 2^n states from K give orientable state surfaces; in particular, both checkerboards may be orientable when $m > 1$.

$V = \partial N(F)$, called the *double* of F, on which projection $N(F) \to F$ restricts to a 2:1 covering map $V \to F$. When F is a nonorientable surface in the knot exterior $E = S^3 \setminus \text{int}(N(K))$, its double V is orientable and connected. Thus, F is *1-sided*, the "side" being the single component of V.[2]

One can view a spanning surface F and its double V as properly embedded surfaces in the knot exterior E, so that ∂F is a curve on $\partial E = \partial N(K)$; then ∂V consists of two curves on ∂E parallel to ∂F. Any of these curves is characterized by its linking number with K, when the curve and K are co-oriented. This number is called the *boundary slope* of F and is given by $\text{slope}(F) = \text{lk}(F \cap \partial N(K), K)$. The curve $F \cap \partial N(K)$ is homologous on $\partial N(K)$ to $\lambda + \text{slope}(F) \cdot \mu$, where λ is a longitude (an essential simple closed curve on $\partial N(K)$ with $\text{lk}(\lambda, K) = 0$) and μ is a meridian curve with $\text{lk}(\lambda, \mu) = +1$. Clark showed for any knot K and any even integer n, there is a spanning surface F for K with $\text{slope}(F) = n$. The idea is that, given any F spanning K, one can change $\text{slope}(F)$ by ± 2 by *attaching a crosscap*. The inverse of this move is pictured in Figure 23.3, right.

Clark defined the *crosscap number* $C(K)$ of a knot K as follows. Given any 1-sided spanning surface F for K, attach a disk D^2 to F along K—either abstractly, or in the closed 3-manifold obtained from the knot exterior $E = S^3 \setminus \mathring{\nu}K$ by Dehn filling along the boundary slope of F.[3] The resulting surface is homeomorphic to a connect sum of projective planes, $F \cup D^2 \cong \#_{i=1}^{m} \mathbb{R}P^2$. The crosscap number $C(K)$ is the minimum number of projective plane summands among all such surfaces coming from K. Equivalently, $C(K)$ is the minimum first betti number $b_1(F) = \text{rank}(H_1(F)) = 1 - \chi(F)$ among all 1-sided surfaces F spanning K; $b_1(F)$ is the number of cuts required to reduce F to a disk. Defining $\Gamma(K)$ in the same way as $C(K)$, but allowing F to be 1- or 2-sided, Murakami-Yasuhara note that $\Gamma(K) = \min(2g(K), C(K))$ [18], where $g(K)$ is the minimal genus of all orientable spanning surfaces for K. Clark notes that every knot K satisfies $2C(K) \leq g(K) + 1$, since one can construct a 1-sided spanning surface by attaching a crosscap to a minimal genus Seifert surface. Also, this inequality and the additivity of genus under connect sum, $g(K_1 \# K_2) = g(K_1) + g(K_2)$, imply that $C(K_1) + C(K_2) - 1 \leq C(K_1 \# K_2) \leq C(K_1) + C(K_2)$. Murakami-Yasuhara show that $C(K_1 \# K_2) = C(K_1) + C(K_2)$ if and only if $C(K_1) = \Gamma(K_1)$ and $C(K_2) = \Gamma(K_2)$.

Crosscap numbers have been computed for several classes of knots. Clark shows that the only knots with $C(K) = 1$ are the $(2, q)$ *cable* and *torus knots*, since if F is a mobius band, then ∂F is a

FIGURE 23.3: Compressing and ∂-compressing a spanning surface.

[2]In an *orientable* 3-manifold, a surface is 1-sided if and only if it is nonorientable; in nonorientable 3-manifolds this is false.
[3]That is, glue E and $D^2 \times S^1$ by a map $\varphi : \partial(D^2 \times S^1) = \partial D^2 \times S^1 \to \partial E = \partial N(K)$ that sends some $\partial D^2 \times \{\text{point}\}$ to $\partial E \cap F$.

cable of the core curve of F.[4] In particular, since $(2, q)$ torus knots $T_{2,q}$ have arbitrarily large genus, $g(K) - 2C(K)$ can be arbitrarily large. Contrast this with $2C(K) - g(K)$, which never exceeds 1. In other words, there exist knots whose simplest 1-sided spanning surface is *much* simpler than their simplest orientable one, but not vice versa. Teragaito determines the crosscap numbers of all *torus knots* $T_{p,q}$, describing these numbers in terms of continued fraction expansions of p/q [21].

Also using continued fraction expansions of p/q, Hatcher-Thurston classify all π_1-injective spanning surfaces for any *2-bridge knot* $K_{p/q}$, up to isotopy [11]. Bessho applies this result to show that $C(K_{p/q})$ equals the minimum length among all such expansions, provided at least one coefficient in the expansion is odd [3]. Hirasawa-Teragaito provide an efficient algorithm for computing these minimum lengths [12]. They also show that $C(K_{p/q}) = 2g(K_{p/q}) + 1$ holds if and only if there is no coefficient ± 2 in the (unique) expansion for K containing only even coefficients.

Ichihara-Mizushima prove that every *pretzel knot* $P(p_1, \ldots, p_n)$ with each $|p_i| > 1$ satisfies $C(K) = n$ if $n, p_1, \ldots, p_n$ are all odd; otherwise, exactly one of $p_1, \ldots, p_n$ is even and $C(K) = n - 1$ [15].

Adams-Kindred show that, given an alternating diagram D of a knot K, any incompressible surface F (1- or 2-sided) spanning K can be isotoped to have a standard half-twist band at some crossing. They apply this crossing-band lemma to show that $C(K)$ is realized by a state surface from D, as are $\Gamma(K)$ and $g(K)$. They also describe an algorithm that finds these particular states and thus computes $C(K)$ and $\Gamma(K)$. Their crossing-band lemma further implies that state surfaces from D, stabilized by attaching tubes or crosscaps, characterize all spanning surfaces for K up to homeomorphism type and boundary slope [1].

Kalfagianni-Lee relate the crosscap numbers of alternating knots to the coefficients of the Jones polynomial (normalized so that $J_{\text{unknot}}(t) = 1$), $J_K(t) = \alpha_K t^n + \beta_K t^{n-1} + \cdots + \beta'_K t^{s-1} + \alpha'_K t^s$. Namely, denoting $T_K := |\beta_K| + |\beta_{K'}|$, and letting $\lceil \cdot \rceil$ denote the function which rounds up to the nearest integer, they show that if K is alternating and hyperbolic (i.e., neither a $(2, q)$ torus link nor a connect sum), then the following bounds hold and are sharp [16]:

$$\left\lceil \frac{T_K}{3} \right\rceil + 1 \le C(K) \le T_K + 1.$$

Garoufalidis's Slopes Conjecture predicts another relationship between the colored Jones polynomials $J_{n,q}(K)$ of a knot K and the properly embedded, π_1-injective surfaces in the exterior E of K [8]: the set of slopes of these surfaces equals the set of cluster points of $\{\frac{2}{n^2} \deg(J_{n,K}(q)) \mid n \in \mathbb{Z}_+\}$.

In general, classifying the essential surfaces in a knot exterior E is a difficult problem, even if one restricts to the case of spanning surfaces. One subtlety is that, for 1-sided surfaces (unlike 2-sided ones), the common algebraic and geometric notions of essentiality are inequivalent. When working exclusively in the class of spanning surfaces, it is natural to define F to be *geometrically essential* if it admits none of the simplifying moves shown in Figure 23.3. This is necessary but insufficient for F to be *algebraically essential*, meaning that F's double V admits no compressing disk or ∂-compressing disk. With the exception of the two mobius bands spanning the unknot, a spanning surface F is *algebraically essential* if and only if it is π_1-injective.

---

[4]Hughes-Kim show that a knot K bounds an *immersed* mobius band, embedded near K, iff K is a $(2p, q)$ or $(p, 2q)$ cable or torus knot [14]. Thus, while a result of Gabai holds that the immersed genus of K always equals $g(K)$ [7], the immersed crosscap number can be less than $C(K)$, and the gap can be arbitrarily large.

These two notions exhibit different sorts of behavior. For example, Ozawa proves that any (generalized) plumbing, also called a Murasugi sum, of algebraically essential spanning surfaces is also algebraically essential [19]; yet, this is not true of geometric essentiality [17]. Since any state surface F is a plumbing of checkerboards, Ozawa's theorem implies that F is algebraically essential as long as its checkerboards are. In particular, adequate state surfaces from alternating diagrams are essential. (For an alternate proof of this fact, see [6].)

As another example, Dunfield showed that the closure of the braid $(\sigma_1\sigma_2\sigma_3\sigma_4)^{13}\sigma_1\sigma_4\sigma_3\sigma_2$ has no algebraically essential spanning surfaces [5]; yet, the following conjecture of Ozawa-Rubinstein remains open to date [20]:

Conjecture 1 (Weakly strong Neuwirth conjecture). Every hyperbolic knot has a geometrically essential, 1-sided spanning surface.

If F spans a knot K, then there is a sequence of spanning surfaces for K, $F = F_0 \rightarrow F_1 \rightarrow \cdots \rightarrow F_k$, each $F_i \rightarrow F_{i+1}$ a geometric compression or ∂-compression move as in Figure 23.3, such that F_k is geometrically essential. Conjecture 1 predicts that there exists no hyperbolic knot for which all such sequences terminate with a Seifert surface.

Interestingly, geometric compression and ∂-compression moves *and their inverses* relate any two surfaces spanning a given knot or link:

Theorem 1 (Gordon-Litherland [9], Yasuhara [22]). Any two surfaces spanning the same link $L \subset S^3$ are related by attaching and deleting tubes and crosscaps.

With $\pi : V \rightarrow F$ denoting the 2:1 covering map from the double of F, Gordon-Litherland define the symmetric, bilinear pairing $\langle \cdot, \cdot \rangle_F$ on $H_1(F)$ by $\langle \alpha, \beta \rangle_V = \mathrm{lk}(\alpha, \pi^{-1}(\beta))$. Under an ordered basis $(\alpha_1, \ldots, \alpha_n)$ for $H_1(F)$, $\langle \cdot, \cdot \rangle$ is described by the *Goeritz matrix* $G = \left(g_{ij}\right)_{i,j=1}^n$, $g_{ij} = \langle \alpha_i, \alpha_j \rangle$ for F.

In particular, when B and W are the checkerboard surfaces from a connected link diagram $D \subset S^2$, the (oriented) boundaries ∂A_i of the disks of $W \setminus B$ generate $H_1(B)$; excluding any one generator gives a basis. The Goeritz matrix $G = \left(g_{ij}\right)$ is then given by:

$$
g_{ii} = \left| A_i \raisebox{0pt}{\includegraphics{}} \right| - \left| A_i \raisebox{0pt}{\includegraphics{}} \right|, \qquad g_{ij} = \left| A_i \raisebox{0pt}{\includegraphics{}} A_j \right| - \left| A_i \raisebox{0pt}{\includegraphics{}} A_j \right|.
$$

Change-of-basis for a Goeritz matrix G from any spanning surface F is given by $G \leftrightarrow R^T G R$, $R \in \mathrm{GL}_n\mathbb{Z}$. After suitable choices of ordered basis for $H_1(F)$, tubing and $\pm$ crosscapping moves on F respectively change G as follows:

$$
\begin{bmatrix} * & \cdots & * \\ \vdots & \ddots & \vdots \\ * & \cdots & * \end{bmatrix}
\leftrightarrow
\begin{bmatrix} * & \cdots & * & \cdot & 0 \\ \vdots & \ddots & \vdots & \vdots & \vdots \\ * & \cdots & * & \cdot & 0 \\ \cdot & \cdots & \cdot & 0 & 1 \\ 0 & \cdots & 0 & 1 & 0 \end{bmatrix}
\qquad
\begin{bmatrix} * & \cdots & * \\ \vdots & \ddots & \vdots \\ * & \cdots & * \end{bmatrix}
\leftrightarrow
\begin{bmatrix} * & \cdots & * & 0 \\ \vdots & \ddots & \vdots & \vdots \\ * & \cdots & * & 0 \\ 0 & \cdots & 0 & \pm 1 \end{bmatrix}
$$

Taking $\sigma(F)$ to denote the signature of this matrix, Gordon-Litherland deduce that the quantity $\sigma(F) - \frac{1}{2}\mathrm{slope}(F)$ is independent of F and therefore equals the signature $\sigma(K)$ of the knot [9]. The quantity $|\det(G)|$ is also independent of F and is called the determinant of the knot. Yasuhara gives an elementary proof of Theorem 1 [22], adapting an argument from Bar Natan-Fulman-Kauffman that all Seifert surfaces for K are related by oriented tubing [2]. We find it instructive to use similar arguments to show:

Theorem 2 ([17]). All checkerboard surfaces from all diagrams of a knot (or link) K are related by attaching and deleting crosscaps. All spanning surfaces for K with equal boundary slopes are related by attaching and deleting tubes.

Proof: Let B and W be the two checkerboards from an n-crossing diagram $D \subset S^2$. To relate B and W by crosscapping moves, let x be any state of D that has a single state circle, and let X be its state surface, which consists of a single disk on one side of S^2 joined to itself by n half-twist bands, one at each crossing. The states b and w underlying B and W have $|\bigcirc|_b$ and $|\bigcirc|_w$ state circles, respectively, with $n + 2$ in total. Attaching $|\bigcirc|_b - 1$ crosscaps to B, one near each crossing where w and x are identical, gives X (see Figure 23.4). Deleting $|\bigcirc|_w - 1$ crosscaps, one near each crossing where b and x are identical, then yields W.

To prove the first claim, it now suffices to show that each of Reidemeister's R-moves is an isotopy or crosscapping move on *at least one of the two* checkerboards. Indeed, the R1 move is an isotopy move on one checkerboard (and a crosscapping move on the other), the R2 move is an isotopy move on one checkerboard (and a tubing move on the other), and the R3 move is a crosscapping move on both checkerboards (see Figure 23.5).

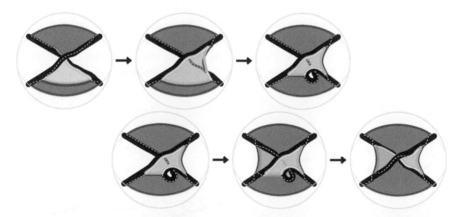

FIGURE 23.4: Re-smoothing a crossing in a Kauffman state attaches or deletes a crosscap from the state surface.

FIGURE 23.5: An R3 move is a crosscapping move on checkerboard surfaces.

For the second claim, isotope an arbitrary spanning surface to look like a disk with bands attached,[5] and then attach a tube anywhere two bands cross. This gives a checkerboard surface. left). Now recycle the above crosscapping argument, with the following modification, illustrated in Figure 23.6, left. Replace each move which attaches a ±-crosscap with a move which attaches one crosscap of each sign (this is equivalent to attaching a tube near K) and which isolates the new $\mp$-crosscap in a tiny ball near K; also replace each move which deletes a ±-crosscap with a move which isolates the same crosscap in a tiny ball near K. At the end of the procedure, the isolated crosscaps will cancel in pairs (using compression moves as shown in Figure 23.6, right), since the initial surfaces had equal boundary slopes. □

Checkerboard surfaces also provide geometric criteria for alternatingness. A spanning surface F is called *positive-definite* if $\langle \alpha, \alpha \rangle_F > 0$ for all nonzero $\alpha \in H_1(F)$, and is called *negative-definite* if $\langle \alpha, \alpha \rangle_F < 0$ for all nonzero $\alpha \in H_1(F)$. Thus F is ±-definite iff $b_1(F) = \pm\sigma(F)$, respectively.

Proposition 1 (Greene [10]). A link diagram $D \subset S^2$ is alternating iff its checkerboard surfaces B and W are definite with opposite signs. Also, D has a *nugatory crossing* iff $\langle \alpha, \alpha \rangle = \pm 1$ for some a in $H_1(B)$ or $H_1(W)$.

Theorem 3 (Greene [10]). A non-split link $L \subset S^3$ is alternating iff it has a positive-definite spanning surface S_+ and a negative-definite one S_-.

Theorem 4 (Howie [13]). A knot $K \subset S^3$ is alternating iff it has spanning surfaces S_+, S_- with $\chi(S_+) + \chi(S_-) + \frac{1}{2}\left|\text{slope}(S_+) - \text{slope}(S_-)\right| = 2$.

A key idea behind Theorems 3 and 4 is that, after S_+, S_- are isotoped to intersect minimally in the link exterior E, they intersect only in arcs, and all of the points of $S_+ \cap S_- \cap \partial E$ have the same sign. This implies that $\nu(S_+ \cup S_-)$ can also be seen as $\nu(S)$ for some closed surface S, on which $\pi : \nu S \to S$ gives a link diagram D of L on S. The other hypotheses imply that S is a sphere.

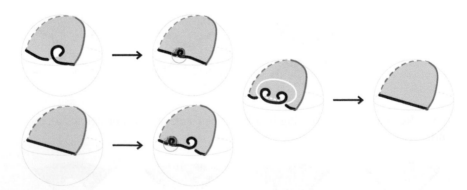

FIGURE 23.6: Tubing moves can replace crosscapping moves by pushing extra crosscaps out near the boundary and cancelling them at the end.

---

[5] In the case of a link, a spanning surface might be disconnected. In that case, one must join the components with tubes before putting the surface in disk-band form.

23.1 Bibliography

[1] C. Adams, C., T. Kindred, A classification of spanning surfaces for alternating links, *Alg. Geom. Topology* 13 (2013), no. 5, 2967–3007.

[2] D. Bar-Natan, J. Fulman, L.H. Kauffman, An elementary proof that all spanning surfaces of a link are tube-equivalent, *J. Knot Theory Ramifications* 7 (1998), no. 7, 873–879.

[3] K. Bessho, Incompressible surfaces bounded by links, master's thesis, Osaka University, 1994 (in Japanese).

[4] B.E. Clark, Crosscaps and knots, Internat. *J. Math. Math. Sci.* 1 (1978), no. 1, 113–123.

[5] N. Dunfield, A knot without a nonorientable essential spanning surface, *Illinois J. Math.* 60 (2016), no. 1, 179–184.

[6] D. Futer, E. Kalfagianni, J. Purcell, Guts of Surfaces and the Colored Jones Polynomial, *Lecture Notes in Mathematics*, 2069. Springer, Heidelberg, 2013.

[7] D. Gabai, Foliations and topology of 3-manifolds, *J. Differential Geometry* 18 (1987), no.3, 445–504.

[8] S. Garoufalidis, The Jones slopes of a knot, *Quantum Topol.* 2 (2011), no. 1, 43–69.

[9] C. McA. Gordon, R.A. Litherland, On the signature of a link, *Invent. Math.* 47 (1978), no. 1, 53–69.

[10] J. Greene, Alternating links and definite surfaces, with an appendix by A. Juhasz, M Lackenby, *Duke Math. J.* 166 (2017), no. 11, 2133–2151.

[11] A. Hatcher, W. Thurston, Incompressible surfaces in 2-bridge knot complements, *Inv. Math.* 79 (1985), 225–246.

[12] M. Hirasawa, M. Teragaito, Crosscap numbers of 2-bridge knots, *Topology* 45 (2006), no. 3, 513–530.

[13] J. Howie, A characterisation of alternating knot exteriors, *Geom. Topol.* 21 (2017), no. 4, 2353–2371.

[14] M. Hughes, S. Kim, Immersed Mobius bands in knot complements, arXiv:1801.00320

[15] K. Ichihara, S. Mizushima, Crosscap numbers of pretzel knots, *Topology Appl.* 157 (2010), no. 1, 193–201.

[16] E. Kalfagianni, C. Lee, Crosscap numbers and the Jones polynomial, *Adv. Math.* 286 (2016), 308–337.

[17] T. Kindred, Checkerboard plumbings, doctoral thesis, University of Iowa, 2018.

[18] H. Murakami, A. Yasuhara, Crosscap number of a knot, *Pacific J. Math.* 171 (1995), no. 1, 261–273.

[19] M. Ozawa, Essential state surfaces for knots and links, *J. Aust. Math. Soc.* 91 (2011), no. 3, 391–404.

[20] M. Ozawa, J.H. Rubinstein, On the Neuwirth conjecture for knots, *Comm. Anal. Geom.* 20 (2012), no. 5, 1019–1060.

[21] M. Teragaito, Crosscap numbers of torus knots, *Topology Appl.* 138 (2004), no. 1–3, 219-238.

[22] A. Yasuhara, An elementary proof that all unoriented spanning surfaces of a link are related by attaching/deleting tubes and Mobius bands, *J. Knot Theory Ramifications* 23 (2014), no. 1, 5 pp.

Chapter 24

State Surfaces of Links

Efstratia Kalfagianni, Michigan State University

24.1 Definitions and Examples

For a link K in S^3, $D = D(K)$ will denote a link diagram, in the equatorial 2–sphere of S^3. We will abuse terminology by referring to the projection sphere using the common term projection plane. In particular, $D(K)$ cuts the projection "plane" into compact regions each of which is a polygon with vertices at the crossings of D.

Given a crossing on a link diagram $D(K)$ there are two ways to resolve it; the A- resolution and the B-resolution as shown on the left of the two diagrams in Figure 24.1. It is borrowed from [13] . Note that if the link K is oriented, only one of the two resolutions at each crossing will respect the orientation of K. A Kauffman state σ on $D(K)$ is a choice of one of these two resolutions at each crossing of $D(K)$ [14]. For each state σ of a link diagram, the *state graph* $\mathbb{G}_\sigma$ is constructed as follows: The result of applying σ to $D(K)$ is a collection $s_\sigma(D)$ of non-intersecting circles in the plane, called *state circles*, together with embedded arcs recording the crossing splice. Next we obtain the *state surface* S_σ, as follows: Each circle of $s_\sigma(D)$ bounds a disk in S^3. This collection of disks can be disjointly embedded in the ball below the projection plane. At each crossing of $D(K)$, we connect the pair of neighboring disks by a half-twisted band to construct a surface $S_\sigma \subset S^3$ whose boundary is K.

FIGURE 24.1: The A-resolution (left), the B-resolution (right) of a crossing, and their contribution to state surfaces.

Example 1. Given an oriented link diagram $D = D(K)$, the Seifert state, denoted by $s(D)$, is the one that assigns to each crossing of D the resolution that is consistent with the orientation of D. The corresponding state surface $S_s = S_s(D)$ is oriented (a.k.a. a Seifert surface). The process of constructing S_s is known a *Seifert's algorithm* [17].

By applying the A–resolution to each crossing of D, we obtain a crossing–free diagram $s_A(D)$. Its state graph, denoted by $\mathbb{G}_A = \mathbb{G}_A(D)$, is called the all–$A$ state graph and the corresponding state surface is denoted by $S_A = S_A(D)$. An example is shown in Figure 24.2, which is borrowed from [10]. Similarly, for the all–B state, the crossing–free resulting diagram is denoted by $s_B(D)$, the state graph is denoted $\mathbb{G}_B$, and the state surface by S_B.

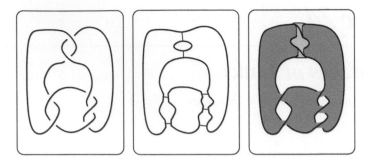

FIGURE 24.2: Left to right: A diagram, the all-A state graph $\mathbb{G}_A$ and the corresponding state surface S_A.

By construction, $\mathbb{G}_\sigma$ has one vertex for every circle of s_σ (i.e. for every disk in S_σ), and one edge for every half–twisted band in S_σ. This gives a natural embedding of $\mathbb{G}_\sigma$ into the surface, where vertices are embedded into the corresponding disks, and edges run through the corresponding half-twisted bands. Hence, $\mathbb{G}_\sigma$ is a spine for S_σ.

Lemma 1. The surface S_σ is orientable if and only if $\mathbb{G}_\sigma$ is a bipartite graph.

Proof. Recall that a graph is bipartite if and only if all cycles (i.e., paths from any vertex to itself) contain an even number of edges.

If $\mathbb{G}_\sigma$ is bipartite, we may assign an orientation on S_σ, as follows: Pick a normal direction to one disk, corresponding to a vertex of $\mathbb{G}_\sigma$, extend over half–twisted bands to orient every adjacent disk, and continue inductively. This inductive process S_σ will not run into a contradiction since every cycle in $\mathbb{G}_\sigma$ has an even number of edges. Thus S_σ is a two–sided surface in S^3, hence orientable. This is the case with the example of Figure 24.2.

Conversely, suppose $\mathbb{G}_\sigma$ is not bipartite, and hence contains a cycle with an odd number of edges. By embedding $\mathbb{G}_\sigma$ as a spine of S_σ, as above, we see that this cycle is an orientation–reversing loop in S_σ.   □

24.2   Genus and Crosscap Number of Alternating Links

The genus of an orientable surface S with k boundary components is defined to be $1 - (\chi(S) + k)/2$, where $\chi(S)$ is the Euler characteristic of S and the *crosscap number* of a non-orientable surface with k boundary components is defined to be $2 - \chi(S) - k$.

Definition 1. Every link in S^3 bounds both orientable and non-orientable surfaces. The *genus* of an oriented link K, denoted by $g(K)$, is the minimum genus over all orientable surfaces S bounded by K. That is we have $\partial S = K$. The *crosscap number* (a.k.a. non-orientable genus) of a link K, denoted by $C(K)$, is the minimum crosscap number over all non-orientable surfaces spanned by K.

For *alternating links* the genus and the crosscap number can be computed using state surfaces of alternating link diagrams. For the orientable case, we recall the following classical result due to Crowell [7] (see also [17]).

Theorem 1. [7] Suppose that D is a connected alternating diagram of a k-component link K. Then the state surface $S_s(D)$ corresponding to the Seifert state of D realizes the genus of K. That is we have $g(K) = 1 - (\chi(S_s(D)) + k)/2$.

In [2], Adams and Kindred used state surfaces to give an algorithm for computing crosscap numbers of alternating links. To summarize their algorithm and state their result, consider a connected alternating diagram $D(K)$ as a 4-valent graph on S^2. Each region in the complement of the graph is an m-gon with vertices at the vertices of the graph.

Lemma 2. Suppose that $D(K)$ is a connected alternating link diagram whose complement has no bigons or 1-gons. Then at least one region must be a triangle.

Proof. Let V, E, F denote the number of vertices, edges and complimentary regions of $D(K)$, respectively. Then, $V - E + F = 2$ and $E = 2V$, which implies that $F > V$. Suppose that none of the F regions is a triangle. Then, $F < 4V/4 = V$ since each region has at least four vertices and each vertex can only be on at most 4 distinct regions. This is a contradiction. □

The Euler characteristic of a surface, corresponding to a state σ, is $\chi(S_\sigma) = v_\sigma - c$, where c is the number of crossings on $D(K)$ and v_σ is the number of state circles in s_σ. Thus to maximize $\chi(S_\sigma)$ we must maximize the number of state circles v_σ. Now we outline the algorithm from [2] that finds a surface of maximal Euler characteristic (and thus of minimum genus) over all surfaces (orientable and non-orientable) spanned by an alternating link.

Adams-Kindred algorithm: Let $D(K)$ be a connected, alternating diagram.

1. Find the smallest m for which the complement of the projection $D(K)$ contains an m-gon.

2. If $m = 1$, then we resolve the corresponding crossing so that the 1-gon becomes a state circle.

   Suppose that $m = 2$. Then some regions of $D(K)$ are bigons. Create one branch of the algorithm for each bigon on $D(K)$. Resolve the two crossings corresponding to the vertices of the bigon so that the bigon is bounded by a state circle. See Figures 1.4 and 24.5.

3. Suppose $m > 2$. Then by Lemma 2, we have $m = 3$. Pick a triangle region on $D(K)$. Now the process has two branches: For one branch we resolve each crossing on the triangle's boundary so that the triangle becomes a state circle. For the other branch, we resolve each of the crossings the opposite way.

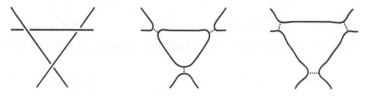

FIGURE 24.3: The two branch of the algorithm for triangle regions. The figure is borrowed from [13].

4. Repeat Steps 1 and 2 until each branch reaches a projection without crossings. Each branch corresponds to a Kauffman state of $D(K)$ for which there is a corresponding state surface. Of all the branches involved in the process, choose one that has the largest number of state circles. The surface S corresponding to this state has maximal Euler characteristic over all the states corresponding to $D(K)$. Note that, *a priori*, more than one branch of the algorithm may lead to surfaces of maximal Euler characteristic.

Theorem 2. [2] Let S be any maximal Euler characteristic surface obtained via above algorithm from an alternating diagram of k-component link K. Then thew following are true:

1. If there is a surface S as above that is non-orientable then $C(K) = 2 - \chi(S) - k$.

2. If all the surfaces S as above are orientable, we have $C(K) = 3 - \chi(S) - k$. Furthermore, S is a minimal genus Seifert surface of K and $C(K) = 2g(K) + 1$.

Example 2. Different choices of branches as well as the order in resolving bigon regions following the algorithm above, may result in different state surfaces. In particular at the end of the algorithm we may have both orientable and non-orientable surfaces that share the same Euler characteristic:

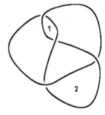 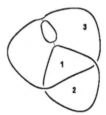

FIGURE 24.4: A diagram of 4_1 with bigon regions 1 and 2 and the result of applying step 2 of the algorithm to bigon 1.

Suppose that we choose the bigon labeled 1 in the left hand side picture of Figure 1.4. Then, for the next step of the algorithm, we have three choices of bigon regions to resolve, labeled 1 and 2 and 3 of the figure.

FIGURE 24.5: Two algorithm branches corresponding to different bigons.

The choice of bigon 1 leads to a non-orientable surface, shown in the left panel of Figure 24.5, realizing the crosscap number of 4_1, which is two. The choice of bigon 2 leads to an orientable surface, shown in the right panel of Figure 24.5, realizing the genus of the knot which is one. Both surfaces realize the maximal Euler characteristic of -1.

24.3   Jones Polynomial and State Graphs

A connected link diagram D defines a 4–valent planar graph $\Gamma \subset S^2$, which leads to the construction of the *Turaev surface* $F(D)$ as follows [8]: Thicken the projection plane to $S^2 \times [-1, 1]$, so that Γ lies in $S^2 \times \{0\}$. Outside a neighborhood of the vertices (crossings) the surface intersect $S^2 \times [-1, 1]$, in $\Gamma \times [-1, 1]$. In the neighborhood of each vertex, we insert a saddle, positioned so that the boundary circles on $S^2 \times \{1\}$ are the components of the A–resolution $s_A(D)$, and the boundary circles on $S^2 \times \{-1\}$ are the components of $s_B(D)$.

When D is an alternating diagram, each circle of $s_A(D)$ or $s_B(D)$ follows the boundary of a region in the projection plane. Thus, for alternating diagrams, the surface $F(D)$ is the projection sphere S^2. For general diagrams, the diagram D is still alternating on $F(D)$.

The surface $F(D)$ has a natural cellulation: the 1–skeleton is the graph Γ and the 2–cells correspond to circles of $s_A(D)$ or $s_B(D)$, hence to vertices of $\mathbb{G}_A$ or $\mathbb{G}_B$. These 2–cells admit a checkerboard coloring, in which the regions corresponding to the vertices of $\mathbb{G}_A$ are white and the regions corresponding to $\mathbb{G}_B$ are shaded. The graph $\mathbb{G}_A$ (resp. $\mathbb{G}_B$) can be embedded in $F(D)$ as the adjacency graph of white (resp. shaded) regions. The *faces* of $\mathbb{G}_A$ (that is, regions in the complement of $\mathbb{G}_A$) correspond to vertices of $\mathbb{G}_B$, and vice versa. Hence the graphs are dual to one another on $F(D)$. Graphs, together with such embeddings in an orientable surface, called *ribbon graphs*, have been studied in the literature [4]. Building on this point of view, Dasbach, Futer, Kalfagianni, Lin and Stoltzfus [8] showed that the ribbon graph embedding of $\mathbb{G}_A$ into the Turaev surface $F(D)$ carries at least as much information as the Jones polynomial $J_K(t)$. To state the relevant result from [8], recall that a *spanning* subgraph of $\mathbb{G}_A$ is a subgraph that contains all the vertices of $\mathbb{G}_A$. Given a spanning subgraph $\mathbb{G}$ of $\mathbb{G}_A$ we will use $v(\mathbb{G})$, $e(\mathbb{G})$, and $f(\mathbb{G})$ to denote the number of vertices, edges and faces of $\mathbb{G}$, respectively.

Theorem 3. [8] For a connected link diagram D, the Kauffman bracket $\langle D \rangle \in \mathbb{Z}[A, A^{-1}]$ is expressed as

$$\langle D \rangle = \sum_{\mathbb{G} \subset \mathbb{G}_A} A^{e(\mathbb{G}_A) - 2e(\mathbb{G})} (-A^2 - A^{-2})^{f(\mathbb{G}) - 1},$$

where $\mathbb{G}$ ranges over all the spanning subgraphs of $\mathbb{G}_A$.

Given a diagram $D = D(K)$, the *Jones polynomial* of K, denoted by $J_K(t)$, is obtained from $\langle D \rangle$ as follows: Multiply $\langle D \rangle$ by $(-A)^{-3w(D)}$, where $w(D)$ is the *writhe* of D, and then substitute $A = t^{-1/4}$ [14, 17].

Theorem 3 leads to formulae for the coefficients of $J_K(t)$ in terms of topological quantities of the state graphs $\mathbb{G}_A$, $\mathbb{G}_B$ corresponding to any diagram of K [8, 9]. These formulae become particularly effective if $\mathbb{G}_A$, $\mathbb{G}_B$ contain no 1-edge loops. In particular, this is the case when $\mathbb{G}_A$, $\mathbb{G}_B$ correspond to an alternating diagram that is *reduced* (i.e., contains no redundant crossings).

Corollary 1. [9] Let $D(K)$ be a reduced alternating diagram and let β_K and β'_K denote the second and penultimate coefficient of $J_K(t)$, respectively. Let $\mathbb{G}'_A$ and $\mathbb{G}'_B$ denote the *simple* graphs obtained by removing all duplicate edges between pairs of vertices of $\mathbb{G}_A(D)$ and $\mathbb{G}_B(D)$. Then,

$$|\beta_K| = 1 - \chi(\mathbb{G}'_B), \quad \text{and} \quad |\beta'_K| = 1 - \chi(\mathbb{G}'_A).$$

24.4   Geometric Connections

To a link K in S^3 corresponds a compact 3-manifold with boundary; namely $M_K = S^3 \smallsetminus N(K)$, where $N(K)$ is an open tube around K. The interior of M_K is homeomorphic to the link complement $S^3 \smallsetminus K$. In the 80's, Thurston [19] proved that link complements decompose canonically into pieces that admit locally homogeneous geometric structures. A very common and interesting case is when the entire $S^3 \smallsetminus K$ has a hyperbolic structure, that is a metric of constant curvature -1 of finite volume. By Mostow rigidity, this hyperbolic structure is unique up to isometry, hence invariants of the metric of $S^3 \smallsetminus K$ give topological invariants of K.

State surfaces obtained from link diagrams $D(K)$ give rise to properly embedded surfaces in M_K. Many geometric properties of state surfaces can be checked through combinatorial and link diagrammatic criteria. For instance, Ozawa [18] showed that the all-A surface $S_A(D)$ is π_1–injective in M_K if the state graph $\mathbb{G}_A(D)$ contains no 1-edge loops. Futer, Kalfagianni and Purcell [10] gave a different proof of Ozawa's result and also showed that M_K is a fiber bundle over the circle with fiber $S_A(D)$, if and only if the simple state graph $\mathbb{G}'_A(D)$ is a *tree*.

State surfaces have been used to obtain relations between combinatorial- or Jones-type link invariants and geometric invariants of link complements. Below we give a couple of sample of such relations. For additional applications the reader is referred to to [1, 5, 10, 11, 15, 16] and references therein. The first result, proven combining [2] with hyperbolic geometry techniques, relates the crosscap number and the Jones polynomial of alternating links. It was used to determine the crosscap numbers of 283 alternating knots of *knot tables* that were previously unknown [6].

Theorem 4. [13] Given an alternating, non-torus knot K, with crosscap number $C(K)$, we have

$$\left\lceil \frac{T_K}{3} \right\rceil + 1 \;\leq\; C(K) \;\leq\; \min\left\{ T_K + 1, \left\lfloor \frac{s_K}{2} \right\rfloor \right\}$$

where $T_K := |\beta_K| + |\beta'_K|$, β_K, β'_K are the second and penultimate coefficients of $J_K(t)$ and s_K is the degree span of $J_K(t)$. Furthermore, both bounds are sharp.

Example 3. For $K = 4_1$ we have $J_K(t) = t^{-2} - t^{-1} + 1 - t + t^2$. Thus $T_K = 2$ and $s_k = 4$ and Theorem 4 gives $C(K) = 2$.

The next result gives a strong connection of the Jones polynomial to hyperbolic geometry as it estimates volume of hyperbolic alternating links in terms of coefficients of their Jones polynomials. The result follows by work of Dasbach and Lin [9] and work of Lackenby [15].

Theorem 5. Let K be an alternating link whose exterior admits a hyperbolic structure with volume $\mathrm{vol}(S^3 \smallsetminus K)$. Then we have

$$\frac{v_{\mathrm{oct}}}{2}(T_K - 2) \leq \mathrm{vol}(S^3 \smallsetminus K) \leq 10 v_{\mathrm{tet}}(T_K - 1),$$

where $v_{\mathrm{oct}} = 3.6638$ and $v_{\mathrm{tet}} = 1.0149$.

To establish the lower bound of Theorem 5 one looks at the state surfaces S_A, S_B corresponding to a reduced alternating diagram $D(K)$: Use $M_K \backslash\backslash S_A$ to denote the complement in M_K of a collar neighborhood of S_A. Jaco-Shalen-Johannson theory [12] implies that there is a canonical way to decompose $M_K \backslash\backslash S_A$ along certain annuli into three types of pieces: (i) I–bundles over subsurfaces of S_A; (ii) solid tori; and (iii) the remaining pieces, denoted by $\text{guts}(M, S)$. On one hand, by work Agol, Storm, and Thurston [3], the quantity $\left| \chi(\text{guts}(M_K, S_A)) \right|$ gives a lower bound for the volume $\text{vol}(S^3 \setminus K)$. On the other hand, [15] shows that this quantity is equal to $1 - \chi(\mathbb{G}'_A)$, which by Corollary 1 is $\left| \beta'_K \right|$. A similar consideration applies to the surface S_B giving the lower bound of Theorem 5. This approach was systematically developed and generalized to non-alternating links in [10].

Acknowledgments

The author is supported in part by NSF grants DMS-1404754 and DMS-1708249.

24.5 Bibliography

[1] C. Adams, A. Colestock, J. Fowler, W. Gillam, and E. Katerman. Cusp size bounds from singular surfaces in hyperbolic 3-manifolds. *Trans. Amer. Math. Soc.*, 358(2):727–741, 2006.

[2] Colin Adams and Thomas Kindred. A classification of spanning surfaces for alternating links. *Algebr. Geom. Topol.*, 13(5):2967–3007, 2013.

[3] Ian Agol, Peter A. Storm, and William P. Thurston. Lower bounds on volumes of hyperbolic Haken 3-manifolds. *J. Amer. Math. Soc.*, 20(4):1053–1077, 2007,. with an appendix by Nathan Dunfield.

[4] Béla Bollobás and Oliver Riordan. A polynomial invariant of graphs on orientable surfaces. *Proc. London Math. Soc. (3)*, 83(3):513–531, 2001.

[5] Stephan D. Burton and Efstratia Kalfagianni. Geometric estimates from spanning surfaces. *Bull. Lond. Math. Soc.*, 49(4):694–708, 2017.

[6] Jae Choon Cha and Charles Livingston. Knotinfo: Table of knot invariants. `http://www.indiana.edu/~knotinfo`, June 14, 2018.

[7] Richard Crowell. Genus of alternating link types. *Ann. of Math. (2)*, 69:258–275, 1959.

[8] Oliver T. Dasbach, David Futer, Efstratia Kalfagianni, Xiao-Song Lin, and Neal W. Stoltzfus. The Jones polynomial and graphs on surfaces. *J. Combin. Theory Ser. B*, 98(2):384–399, 2008.

[9] Oliver T. Dasbach and Xiao-Song Lin. On the head and the tail of the colored Jones polynomial. *Compos. Math.*, 142(5):1332–1342, 2006.

[10] David Futer, Efstratia Kalfagianni, and Jessica Purcell. *Guts of Surfaces and the Colored Jones Polynomial*, volume 2069 of *Lecture Notes in Mathematics*. Springer, Heidelberg, 2013.

[11] David Futer, Efstratia Kalfagianni, and Jessica S. Purcell. Quasifuchsian state surfaces. *Trans. Amer. Math. Soc.*, 366(8):4323–4343, 2014.

[12] William H. Jaco and Peter B. Shalen. Seifert fibered spaces in 3-manifolds. *Mem. Amer. Math. Soc.*, 21(220):viii+192, 1979.

[13] Efstratia Kalfagianni and Christine Ruey Shan Lee. Crosscap numbers and the Jones polynomial. *Adv. Math.*, 286:308–337, 2016.

[14] Louis H. Kauffman. State models and the Jones polynomial. *Topology*, 26(3):395–407, 1987.

[15] Marc Lackenby. The volume of hyperbolic alternating link complements. *Proc. London Math. Soc. (3)*, 88(1):204–224, 2004. With an appendix by Ian Agol and Dylan Thurston.

[16] Marc Lackenby and Jessica S. Purcell. Cusp volumes of alternating knots. *Geom. Topol.*, 20(4):2053–2078, 2016.

[17] W. B. Raymond Lickorish. *An Introduction to Knot Theory*, volume 175 of *Graduate Texts in Mathematics*. Springer-Verlag, New York, 1997.

[18] Makoto Ozawa. Essential state surfaces for knots and links. *J. Aust. Math. Soc.*, 91(3):391–404, 2011.

[19] William P. Thurston. *The Geometry and Topology of Three-Manifolds*. Princeton Univ. Math. Dept. Notes, 1979.

Chapter 25

Turaev Surfaces

Seungwon Kim, Institute for Basic Science

Ilya Kofman, College of Staten Island & CUNY Graduate Center

25.1 Introduction

The two most famous knot invariants, the Alexander polynomial (1923) and the Jones polynomial (1984), mark paradigm shifts in knot theory. After each polynomial was discovered, associating new structures to knot diagrams played a key role in understanding its basic properties. Using the Seifert surface, Seifert showed, for example, how to obtain knots with a given Alexander polynomial. For the Jones polynomial, even the simplest version of that problem remains open: Does there exist a non-trivial knot with trivial Jones polynomial?

Kauffman gave a state sum for the Jones polynomial, with terms for each of the 2^c states of a link diagram with c crossings. Turaev constructed a closed orientable surface from any pair of dual states with opposite markers.

For the Jones polynomial, the Turaev surface is a rough analog to the Seifert surface for the Alexander polynomial. For a given knot diagram, the Seifert genus and the Turaev genus are computed by separate algorithms to obtain each surface from the diagram. The invariants for a given knot K are defined as the minimum genera among all the respective surfaces for K. The Seifert genus is a topological measure of how far a given knot is from being unknotted. The Turaev genus is a topological measure of how far a given knot is from being alternating. (See [26], which discusses alternating distances.) For any alternating diagram, Seifert's algorithm produces the minimal genus Seifert surface. For any adequate diagram, Turaev's algorithm produces the minimal genus Turaev surface. Extending the analogy, we can determine the Alexander polynomial and the Jones polynomial of K from associated algebraic structures on the respective surfaces of K: the Seifert matrix for the Alexander polynomial, and the A–ribbon graph on the Turaev surface for the Jones polynomial.

The analogy is historical, as well. Like the Seifert surface for the Alexander polynomial, the Turaev surface was constructed to prove a fundamental conjecture related to the Jones polynomial. Tait conjectured that an alternating link always has an alternating diagram that has minimal crossing number among all diagrams for that link. A proof had to wait about a century until the Jones polynomial led to several new ideas used to prove Tait's Conjecture [21, 29, 35]. Turaev's later proof in [36] introduced Turaev surfaces and prompted interest in studying their properties.

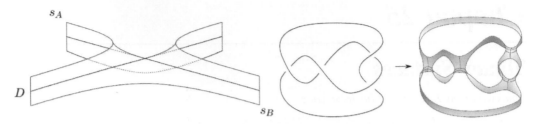

FIGURE 25.1: We put a saddle at each crossing of D to obtain a cobordism between the states s_A and s_B.

25.2 What Is the Turaev Surface?

Let D be the diagram of a link L drawn on S^2. For any crossing ✕, we obtain the A–smoothing as)(and the B–smoothing as ✕. The state s of D is a choice of smoothing at every crossing, resulting in a disjoint union of circles on S^2. Let $|s|$ denote the number of circles in s. Let s_A be the all–A state, for which every crossing of D has an A–smoothing. Similarly, s_B is the all–B state. We will construct the Turaev surface from the dual states s_A and s_B.

At every crossing of D, we put a saddle surface which bounds the A–smoothing on the top and the B–smoothing on the bottom as shown in Figure 25.1. In this way, we get a cobordism between s_A and s_B, with the link projection Γ at the level of the saddles. The *Turaev surface* $F(D)$ is obtained by attaching $|s_A| + |s_B|$ discs to all boundary circles. See Figure 25.1 and [15] for an animation of the Turaev surface for the Borromean link.

The *Turaev genus* of D is the genus of $F(D)$, given by

$$g_T(D) = g(F(D)) = (c(D) + 2 - |s_A| - |s_B|)/2.$$

The Turaev genus $g_T(L)$ of any non-split link L is the minimum of $g_T(D)$ among all diagrams D of L. By [36, 12], L is alternating if and only if $g_T(L) = 0$, and if D is an alternating diagram then $F(D) = S^2$. In general, for any link diagram D, it follows that (see [12]):

1. $F(D)$ is a Heegaard surface of S^3; i.e., an unknotted closed orientable surface in S^3.

2. D is alternating on $F(D)$, and the faces of D can be checkerboard colored on $F(D)$, with discs for s_A and s_B colored white and black, respectively.

3. $F(D)$ has a Morse decomposition, with D and crossing saddles at height zero, and the $|s_A|$ and $|s_B|$ discs as maxima and minima, respectively.

Conversely, in [5], conditions were given for a Heegaard surface with cellularly embedded alternating diagram on it to be a Turaev surface.

25.3 The Turaev Surface and the Jones Polynomial

A diagram D is A–adequate if at each crossing, the two arcs of s_A from that crossing are in different state circles. In other words, $|s_A| > |s|$ for any state s with exactly one B–smoothing. Similarly, we define a B–adequate diagram by reversing the roles of A and B above. If D is both A–adequate and B–adequate it is called *adequate*. If D is neither A–adequate nor B–adequate it is called *inadequate*. A link L is *adequate* if it has an adequate diagram, and is *inadequate* if all its diagrams are inadequate. Any reduced alternating diagram is adequate, hence every alternating link is adequate.

Adequacy implies that s_A and s_B contribute the extreme terms $\pm t^\alpha$ and $\pm t^\beta$ of the Jones polynomial $V_L(t)$, which determine the span $V_L(t) = |\alpha - \beta|$, which is a link invariant. Let $c(L)$ be the minimal crossing number among all diagrams for L. In [36], Turaev proved

$$\operatorname{span} V_L(t) \le c(L) - g_T(L)$$

with equality if L is adequate. If D is a prime non-alternating diagram, then $g_T(D) > 0$. Thus, span $V_L(t) = c(L)$ if and only if L is alternating, from which Tait's Conjecture follows.

Therefore, for any adequate link L with an adequate diagram D (see [1]),

$$g_T(L) = g_T(D) = \frac{1}{2}\left(c(D) - |s_A(D)| - |s_B(D)|\right) + 1 = c(L) - \operatorname{span} V_L(t).$$

So for the connect sum $L\#L'$ of adequate links, $g_T(L\#L') = g_T(L) + g_T(L')$.

Turaev genus and knot homology. Khovanov homology and knot Floer homology categorify the Jones polynomial and the Alexander polynomial, respectively. The *width* of each bigraded knot homology, $w_{KH}(K)$ and $w_{HF}(K)$, is the number of diagonals with non-zero homology. The Turaev genus bounds the width of both knot homologies [10, 27]:

$$w_{KH}(K) - 2 \le g_T(K) \qquad \text{and} \qquad w_{HF}(K) - 1 \le g_T(K). \tag{25.1}$$

For adequate knots, $w_{KH}(K) - 2 = g_T(K)$ [1]. These inequalities have been used to obtain families of knots with unbounded Turaev genus (see [8]).

Ribbon graph invariants. Like the Seifert surface, the Turaev surface provides much more information than its genus. An oriented *ribbon graph* is a graph with an embedding in an oriented surface, such that its faces are discs. Turaev's construction determines an oriented ribbon graph G_A on $F(D)$: We add an edge for every crossing in s_A, and collapse each state circle of s_A to a vertex of G_A, preserving the cyclic order of edges given by the checkerboard coloring (see [8]).

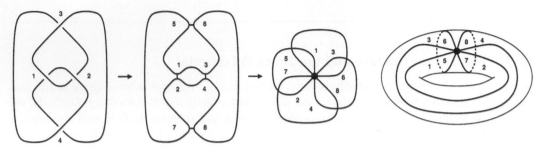

FIGURE 25.2: Ribbon graph G_A for an inadequate diagram of the trefoil.

If L is alternating, then $V_L(t) = T_G(-t, -1/t)$, where $T_G(x, y)$ is the Tutte polynomial [35]. For any L, $V_L(t)$ is a specialization of the Bollobás–Riordan–Tutte polynomial of G_A [12]. These ideas extend to virtual links and non-orientable ribbon graphs [11]. In [13], a unified description is given for all these knot and ribbon graph polynomial invariants.

25.4 Turaev Genus One Links

The Seifert genus is directly computable for alternating and positive links, and has been related to many classical invariants. Moreover, knot Floer homology detects the Seifert genus of knots. In contrast, for most non-adequate links, computing the Turaev genus is an open problem.

The Turaev genus of a link can be computed when the upper bounds in the inequalities (25.1) or those in [26] match the Turaev genus of a particular diagram, which gives a lower bound. So it is useful to know which diagrams realize a given Turaev genus. Link diagrams with Turaev genus one and two were classified in [6, 22].

This classification uses the decomposition of any prime, connected link diagram $D \subset S^2$ into alternating tangles. As in [22], an edge in D is *non-alternating* when it joins two overpasses or two under-passes. If D is non-alternating, we can isotope the state circles in s_A and s_B to intersect exactly at the midpoints of all non-alternating edges of D. In the figure shown to the right, $\alpha \in s_A, \beta \in s_B$. The arc δ joining the points in $\alpha \cap \beta$ is called a *cutting arc* of D ([22]).

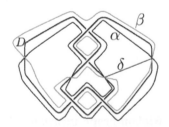

A cutting arc is the intersection of S^2 with a compressing disc of the Turaev surface $F(D)$, which intersects D at the endpoints of δ. The boundary γ of this compressing disc is called a *cutting loop*. Every cutting arc of D has a corresponding cutting loop on $F(D)$, and surgery of D along a cutting arc corresponds to surgery of $F(D)$ along a compressing disc, as shown in the following figure.

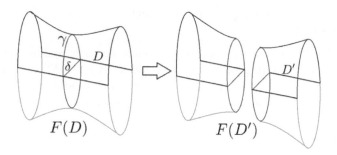

If D' is obtained by surgery from D, the surgered surface is its Turaev surface $F(D')$ with genus $g_T(D') = g_T(D) - 1$. So if $g_T(D) = 1$, then γ is a meridian of the torus $F(D)$, and surgery along all cutting arcs of D cuts the diagram into alternating 2-tangles [22]. Hence, if $g_T(D) = 1$, then D is a cycle of alternating 2-tangles:

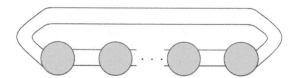

This also implies that for any alternating diagram D on its Turaev surface $F(D)$, if $g_T(D) \geq 1$ there is an essential simple loop γ on F which intersects D twice and bounds a disc in a handlebody bounded by F. Thus, the link on the surface in Example 1.3.1 of [26] cannot come from Turaev's construction. However, this condition is not sufficient; for example, the diagram at right satisfies the condition, but cannot be a Turaev surface because any pla-

nar diagram D for this link has more than four crossings, which would remain as crossings on $F(D)$.

Hayashi [16] and Ozawa [30] considered more general ways to quantify the complexity of the pair (F, D), which has prompted recent interest in *representativity* of knots (see, e.g., [4, 7, 18, 24, 31, 32, 33]).

25.5 Open Problems

Below, we consider open problems in two broad categories:

Question 1. How do you determine the Turaev genus of a knot or link?

Does the Turaev genus always equal the dealternating number of a link? This is true in many cases, and no lower bounds are known to distinguish these invariants (see [26]).

The lower bounds (25.1) vanish for quasi-alternating links. For any $g > 1$, does there exist a quasi-alternating link with Turaev genus g?

The Turaev genus is additive under connect sum for adequate knots, and invariant under mutation if the diagram is adequate [1]. In general, for any K and K', is $g_T(K\#K') = g_T(K) + g_T(K')$? If K and K' are mutant knots, is $g_T(K) = g_T(K')$? The latter question is open even for adequate knots; if D is a non-adequate diagram of an adequate knot K, then for a mutant D' of D, it might be possible that $g_T(K) < g_T(D) = g_T(D') = g_T(K')$.

If K is a positive knot with Seifert genus $g(K)$, then $g_T(K) \leq g(K)$. Is this inequality strict; i.e., is $g_T(K) < g(K)$ for a positive knot? It is known to be strict for $g(K) = 1, 2$ [20, 34] and for adequate positive knots [25].

In general, how do you compute the Turaev genus, which is a link invariant, without using link diagrams? Is it determined by some other link invariants?

Question 2. How do you characterize the Turaev surface?

From the construction in Section 25.2, it is hard to tell whether a given pair (F, D) is a Turaev surface. The existence of a cutting loop implies that the alternating diagram D on F must have minimal complexity; i.e., there exists an essential simple loop on F which intersects D twice. But this condition is not sufficient. What are the sufficient conditions for a given pair (F, D) to be a Turaev surface?

Alternating, almost alternating, and toroidally alternating knots have been characterized topologically using a pair of spanning surfaces in the knot complement [14, 17, 19, 23]. Turaev genus one knots are toroidally alternating, and they contain almost-alternating knots, but they have not been characterized topologically as a separate class of knots. What is a topological characterization of Turaev genus one knots, or generally, of knots with any given Turaev genus?

Any non-split, prime, alternating link in S^3 is hyperbolic, unless it is a closed 2–braid [28]. This result was recently generalized to links in a thickened surface $F \times I$. If the link L in $F \times I$ admits a diagram on F which is alternating, cellularly embedded, and satisfies an appropriate generalization of "prime," then $(F \times I) - L$ is hyperbolic [2, 9, 18]. Now, for a given Turaev surface $F(D)$, let L be a link in $F(D) \times I$ which projects to the alternating diagram on F. It follows that typically the complement $(F(D) \times I) - L$ is hyperbolic, assuming there are no essential annuli. If $g_T(D) = 1$ then $(F(D) \times I) - L$ has finite hyperbolic volume. If $g_T(D) > 1$, then there is a well-defined finite volume if the two boundaries are totally geodesic (see [3]). How do the geometric invariants of $(F(D) \times I) - L$ depend on the original diagram D in S^2?

25.6   Bibliography

[1] Tetsuya Abe. The Turaev genus of an adequate knot. *Topology Appl.*, 156(17):2704–2712, 2009.

[2] C. Adams, C. Albors-Riera, B. Haddock, Z. Li, D. Nishida, B. Reinoso, and L. Wang. Hyperbolicity of links in thickened surfaces. *Topology Appl.*, 256:262–278, 2019.

[3] Colin Adams, Or Eisenberg, Jonah Greenberg, Kabir Kapoor, Zhen Liang, Kate O'connor, Natalia Pacheco-Tallaj, and Yi Wang. TG-Hyperbolicity of virtual links, arXiv:1912.09435 [math.GT], 2019.

[4] Román Aranda, Seungwon Kim, and Maggy Tomova. Representativity and waist of cable knots. *J. Knot Theory Ramifications*, 27(4):1850025, 8, 2018.

[5] Cody Armond, Nathan Druivenga, and Thomas Kindred. Heegaard diagrams corresponding to Turaev surfaces. *J. Knot Theory Ramifications*, 24(4):1550026, 14, 2015.

[6] Cody W Armond and Adam M Lowrance. Turaev genus and alternating decompositions. *Algebraic & Geometric Topology*, 17(2):793–830, 2017.

[7] Ryan Blair and Makoto Ozawa. Height, trunk and representativity of knots. *J. Math. Soc. Japan*, 71(4):1105–1121, 2019.

[8] Abhijit Champanerkar and Ilya Kofman. A survey on the Turaev genus of knots. *Acta Math. Viet.*, 39(4):497–514, 2014.

[9] Abhijit Champanerkar, Ilya Kofman, and Jessica S. Purcell. Geometry of biperiodic alternating links. *J. Lond. Math. Soc. (2)*, 99(3):807–830, 2019.

[10] Abhijit Champanerkar, Ilya Kofman, and Neal Stoltzfus. Graphs on surfaces and Khovanov homology. *Algebr. Geom. Topol.*, 7:1531–1540, 2007.

[11] Sergei Chmutov. Generalized duality for graphs on surfaces and the signed Bollobás-Riordan polynomial. *J. Combin. Theory Ser. B*, 99(3):617–638, 2009.

[12] Oliver T. Dasbach, David Futer, Efstratia Kalfagianni, Xiao-Song Lin, and Neal W. Stoltzfus. The Jones polynomial and graphs on surfaces. *J. Combin. Theory Ser. B*, 98(2):384–399, 2008.

[13] Joanna A. Ellis-Monaghan and Iain Moffatt. *Graphs on surfaces: Dualities, polynomials, and knots*. Springer Briefs in Mathematics. Springer, New York, 2013.

[14] Joshua Evan Greene. Alternating links and definite surfaces. *Duke Mathematical Journal*, 166(11):2133–2151, 2017.

[15] M. Hajij. Turaev Surface Borromean rings - REMIX. [YouTube video]. *http://www.youtube.com/watch?v=j43lionQD9w*, 2010.

[16] Chuichiro Hayashi. Links with alternating diagrams on closed surfaces of positive genus. *Math. Proc. Cambridge Philos. Soc.*, 117(1):113–128, 1995.

[17] Joshua Howie. A characterisation of alternating knot exteriors. *Geometry & Topology*, 21(4):2353–2371, 2017.

[18] Joshua A. Howie and Jessica S. Purcell. Geometry of alternating links on surfaces. *Trans. Amer. Math. Soc.*, 373(4):2349–2397, 2020.

[19] Tetsuya Ito. A characterization of almost alternating knots. *Journal of Knot Theory and Its Ramifications*, 27(01):1850009, 2018.

[20] In Dae Jong and Kengo Kishimoto. On positive knots of genus two. *Kobe J. Math*, 30(1-2):1–18, 2013.

[21] L. Kauffman. State models and the Jones polynomial. *Topology*, 26(3):395–407, 1987.

[22] Seungwon Kim. Link diagrams with low Turaev genus. *Proceedings of the American Mathematical Society*, 146(2):875–890, 2018.

[23] Seungwon Kim. A topological characterization of toroidally alternating knots, arXiv:1608.00521 [math.GT], 2016.

[24] Thomas Kindred. Alternating links have representativity 2. *Algebr. Geom. Topol.*, 18(6):3339–3362, 2018.

[25] Sang Youl Lee, Chan-Young Park, and Myoungsoo Seo. On adequate links and homogeneous links. *Bulletin of the Australian Mathematical Society*, 64(3):395–404, 2001.

[26] Adam Lowrance. Alternating distances of knots, to appear in *A Concise Encyclopedia of Knot Theory*.

[27] Adam M. Lowrance. On knot Floer width and Turaev genus. *Algebr. Geom. Topol.*, 8(2):1141–1162, 2008.

[28] William Menasco. Closed incompressible surfaces in alternating knot and link complements. *Topology*, 23(1):37–44, 1984.

[29] Kunio Murasugi. Jones polynomials and classical conjectures in knot theory. *Topology*, 26(2):187–194, 1987.

[30] Makoto Ozawa. Non-triviality of generalized alternating knots. *Journal of Knot Theory and Its Ramifications*, 15(03):351–360, 2006.

[31] Makoto Ozawa. Bridge position and the representativity of spatial graphs. *Topology and its Applications*, 159(4):936–947, 2012.

[32] Makoto Ozawa. The representativity of pretzel knots. *Journal of Knot Theory and Its Ramifications*, 21(02):1250016, 2012.

[33] John Pardon. On the distortion of knots on embedded surfaces. *Annals of mathematics*, pages 637–646, 2011.

[34] A Stoimenow. Knots of genus one or on the number of alternating knots of given genus. *Proceedings of the American Mathematical Society*, 129(7):2141–2156, 2001.

[35] M. Thistlethwaite. A spanning tree expansion of the Jones polynomial. *Topology*, 26(3):297–309, 1987.

[36] V. G. Turaev. A simple proof of the Murasugi and Kauffman theorems on alternating links. *Enseign. Math. (2)*, 33(3-4):203–225, 1987.

Part VI

Invariants Defined in Terms of Min and Max

Chapter 26

Crossing Numbers

Alexander Zupan, University of Nebraska-Lincoln

26.1 Crossing Number

The crossing number of a link is one of the oldest and best-known link invariants. Given a link K, let $\mathcal{D}(K)$ denote the collection of all diagrams D realizing K. The *crossing number* $c(D)$ of a particular diagram D is defined to be the number of double points of D as an immersed curve in the plane; that is, $c(D)$ is simply the number of crossings contained in D. For any particular diagram, the number $c(D)$ does not necessarily yield much information about K; by performing a sequence of Reidemiester I moves, one can show that for every link K and every integer n, there exists a diagram D realizing K such that $c(D) > n$. Thus, it makes intuitive sense to examine those link diagrams that have relatively few crossings.

An invariant for the link K is obtained by minimizing the number of crossings over all possible diagrams realizing K. In other words,

$$c(K) = \min_{D \in \mathcal{D}(K)} c(D).$$

While some invariants (like colorability and the Jones polynomial) are very good at distinguishing diagrams of different links, the crossing number is—in general—not this type of invariant, since it cannot usually be computed from a single diagram (although some notable exceptions are described below). Instead, the crossing number has historically provided a useful way to organize knots by complexity. Links with a smaller crossing number can be viewed as less complex than links with a larger crossing number, and, unlike some other examples of invariants obtained by minimizing a certain quantity, the crossing number exhibits the following convenient finiteness property:

Proposition 1. For any natural number n, there is a finite collection of links $\mathcal{K}_n$ such that $c(K) \leq n$ for all $K \in \mathcal{K}_n$.

This proposition can be proved in a number of different ways. For instance, observe that for a given integer n, there are only finitely many possible Gauss codes of length $2n$. As a consequence, there is an exhaustive algorithm to produce all link diagrams with a crossing number of at most n, and thus, tabulation of knots and links via the crossing number is a tractable computational problem up to a certain complexity, although the number of links with a given crossing number n grows quickly as n increases. The unknot is the only knot with $c(K) = 0$, the trefoil is the only nontrivial knot with $c(K) \leq 3$, and the figure eight is the only knot with $c(K) = 4$. For $n > 4$, this uniqueness disappears. The first well-known knot table was compiled by Peter Guthrie Tait and published in 1876. Tait produced a list of the prime knots with seven or fewer crossings. At present, Hoste, Thistlethwaite, and Weeks have utilized computer searches to show that there are 1,701,936 prime

n	3	4	5	6	7	8	9	10	11	12	13	14	15
k_n	1	1	2	3	7	21	49	165	552	2176	9988	46927	253293

knots with 16 or fewer crossings [5]. Table 26.1 displays the number k_n of distinct prime knots with a particular crossing number n, computed in [5].

26.1.1   Computing $c(K)$ for Specific Families

In general, it is a difficult problem to determine the crossing number of a given link K. Producing an upper bound is straightforward, since $c(K) \leq c(D)$ for any diagram D realizing K. In general, however, proving minimality is often accomplished by exhaustive argument. For a diagram D conjectured to be a minimal crossing diagram for a link K, one could try to prove that for every possible diagram D' with $c(D') < c(D)$, the realization K' of D' is not equivalent to K. This ad hoc tactic, although useful for producing knot tables, does very little in terms of a general strategy.

Nevertheless, there are infinite families of knots and links for which crossing numbers are completely determined. One example is the collection of torus knots.

Theorem 1. [11, 3] The crossing number of the (p, q)-torus knot K is

$$c(K) = \min\{(|p| - 1)|q|, (|q| - 1)|p|\}.$$

Another family of links for which the crossing number is known is the collection of alternating links. An alternating diagram D is one in which crossings alternate between over- and undercrossings as each link component is traversed, and a link L is *alternating* if it has an alternating diagram.

One might hope that all alternating diagrams realize a minimal crossing number—this is nearly true, and it becomes true when nugatory crossings are removed. A given diagram D is said to have a *nugatory* crossing if there is a simple closed curve in the projection plane that meets the knot only in the crossing. If D has a nugatory crossing, there is an obvious isotopy that decreases $c(D)$. A depiction of a nugatory crossing and the result of this isotopy appears in Figure 26.1. A knot diagram is *reduced* if it does not contain a nugatory crossing. In the late 19th century, Tait—in addition to building one of the first knot tables—made three prominent conjectures. The first of Tait's conjectures took nearly a century to resolve. It asserts that every reduced diagram of an alternating link has a minimal crossing number. The proofs rely heavily on the Jones polynomial.

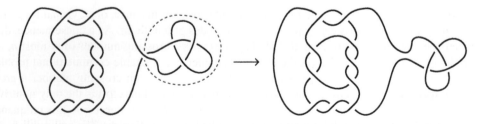

FIGURE 26.1: A nugatory crossing can be eliminated via isotopy.

Theorem 2. [6, 10, 12] If D is a reduced alternating diagram of a link K, then $c(K) = c(D)$.

In fact, Thistlethwaite extended this result to a more general family of links, called adequate links [13]. Given a link diagram D, a *state* of D is a crossing-less diagram obtained from D by making slight modifications, called *smoothings*, in a neighborhood of each crossing of D. The components of a state of D are called *state circles*. Each crossing has two different types of smoothings, the $+$ and $-$ smoothings depicted in Figure 26.2. Adequate diagrams are defined using two particular states, s_+ and s_-, obtained by performing either all $+$ smoothings or all $-$ smoothings, respectively. Finally, we say that a link diagram D is *adequate* if for every crossing in D, the two strands created by the $\pm$ smoothing of the crossing are contained in distinct state circles of $s_\pm$. Equivalently, each smoothing gives rise to a reference arc reflecting the location of the crossing, and D is adequate if every reference arc in the states $s_\pm$ connects distinct state circles. An example appears in Figure 26.3. A link K is *adequate* if K has an adequate diagram.

$$\underset{+}{\overset{\longleftarrow}{\asymp}} \qquad \times \qquad \overset{\longrightarrow}{\underset{-}{)(}}$$

FIGURE 26.2: The positive and negative smoothings of a crossing together with reference arcs.

It is well known that every reduced alternating diagram is adequate, and thus Theorem 2 follows from the next theorem:

Theorem 3. [13] If D is an adequate diagram of a link K, then $c(K) = c(D)$.

As an instance of a practical application of this theorem, there are three non-alternating adequate knots with ten or fewer crossings, 10_{152}, 10_{153}, and 10_{154} in Rolfsen's knot table [2]. The theorem certifies that these diagrams achieve a minimal crossing number. Figure 26.3 verifies that 10_{152} is an adequate knot. Although brute force enumeration also determines the crossing number of these knots, the exhaustive method is (at the present time) only feasible up to twenty or so crossings. Hence, Theorem 3 is a powerful tool for finding the crossing numbers of infinitely many links.

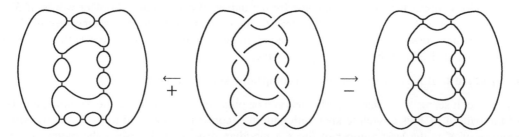

FIGURE 26.3: The states s_+ and s_- of 10_{152} produced from an adequate diagram.

26.1.2 Behavior under connected summation and taking satellites

For an invariant obtained by minimizing a certain geometric quantity, a natural and interesting problem is to determine the behavior of the invariant under standard operations, such at taking connected sums or satellites. Suppose that K_1 and K_2 are knots with minimal diagrams D_1 and D_2; that is, $c(D_i) = c(K_i)$. Then the diagram $D = D_1 \# D_2$ is a diagram for $K_1 \# K_2$ with $c(K_1) + c(K_2)$

crossings, which implies that

$$c(K_1 \# K_2) \le c(K_1) + c(K_2).$$

The diagram $D_1 \# D_2$ for minimal crossing diagrams D_1 of 10_{152} and D_2 of the trefoil 3_1 is shown at right in Figure 26.1. Since the connected sum of adequate diagrams is again adequate, it follows from Theorem 3 that $c(10_{152} \# 3_1) = c(D_1 \# D_2) = 13 = c(10_{152}) + c(3_1)$. In general, the following conjecture is open:

Conjecture 1. For any two knots K_1 and K_2, the crossing number $c(K_1 \# K_2)$ of their connected sum is equal to $c(K_1) + c(K_2)$.

The conjecture is listed as Problem 1.65 in Kirby's famous problem list [7]; some sources claim that it is a hundred-year-old problem [1]. The conjecture is known to be true when both K_1 and K_2 are alternating knots and, more generally, when K_1 and K_2 are adequate knots by Theorems 2 and 3. It has also been proved in the case that both K_1 and K_2 are torus knots (and a broader class of knots, called deficiency zero knots) by Diao [4]. A knot K is said to have *deficiency zero* if $c(K) = br(K) + 2g(K) - 1$, where $br(K)$ is the braid index of K and g is the genus of K.

At the time of the writing of this chapter, the only result that applies to the collection of all knots is the next theorem of Lackenby.

Theorem 4. [8] For any two knots K_1 and K_2, the following inequality holds:

$$c(K_1 \# K_2) \ge \frac{c(K_1) + c(K_2)}{152}.$$

For practical purposes, Lackenby's theorem is perhaps not so useful for actually computing crossing numbers, since the bound is not believed to be sharp. However, it can be regarded as a significant step toward resolving the additivity conjecture.

Another family of knots for which one might wish to predict the crossing number is the collection of satellite knots. This problem is not quite as straightforward as the additivity conjecture, since the "obvious" upper bound does not come from a universal construction. In other words, if K is a satellite knot with companion J and pattern C, the crossing number of K seems very likely to depend on the crossing numbers of both J and C. In general, however, one expects the following conjecture (see Kirby Problem 1.67 [7]) to be true:

Conjecture 2. If K is a satellite knot with companion knot J, then $c(K) \ge c(J)$.

Of course, it seems likely that better bounds can be obtained by considering the wrapping number of the pattern. If K is a satellite knot with pattern C contained in a solid torus V, the *wrapping number* of K is the least number of times a meridian disk for V intersects the pattern C. For the example in Figure 26.4, the satellite knot K is a cable of the trefoil with wrapping number two. The work in [5] verifies that the diagram at right has a minimal crossing number, so that $c(K) = 17$.

In general, an obvious candidate for a minimal diagram of a satellite knot K with wrapping number n and companion J can be constructed from a minimal crossing diagram for the companion J and pattern C, where each crossing of J gives rise to an $n \times n$ grid containing n^2 crossings. This construction informs the next conjecture, which appears in [5]:

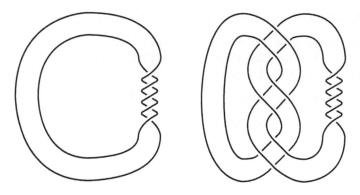

FIGURE 26.4: A pattern at left and satellite knot at right.

Conjecture 3. If K is a satellite knot with wrapping number n and companion knot J, then $c(K) \geq n^2 \cdot c(J)$.

Even less is known about these two conjectures than what is known about the additivity of the crossing number under taking connected sums. Once again, the best result in this direction that holds for all knots comes from work of Lackenby.

Theorem 5. [9] If K is a satellite knot with companion knot J, then

$$c(K) \geq \frac{c(J)}{10^{13}}.$$

While the crossing number is one of the easiest and most natural knot invariants to define, the intractable nature of these conjectures suggests that this invariant is also very difficult to compute and use in practice. Nevertheless, crossing number remains at the heart of a considerable body of research in classical knot theory.

26.2 Bibliography

[1] Colin C. Adams. *The Knot Book*. W. H. Freeman and Company, New York, 1994. An elementary introduction to the mathematical theory of knots.

[2] Yongju Bae and Hugh R. Morton. The spread and extreme terms of Jones polynomials. *J. Knot Theory Ramifications*, 12(3):359–373, 2003.

[3] Peter R. Cromwell. *Knots and Links*. Cambridge University Press, Cambridge, 2004.

[4] Yuanan Diao. The additivity of crossing numbers. *J. Knot Theory Ramifications*, 13(7):857–866, 2004.

[5] Jim Hoste, Morwen Thistlethwaite, and Jeff Weeks. The first 1,701,936 knots. *Math. Intelligencer*, 20(4):33–48, 1998.

[6] Louis H. Kauffman. State models and the Jones polynomial. *Topology*, 26(3):395–407, 1987.

[7] Rob Kirby. Problems in low dimensional manifold theory. In *Algebraic and Geometric Topology (Proc. Sympos. Pure Math., Stanford Univ., Stanford, Calif., 1976), Part 2*, Proc. Sympos. Pure Math., XXXII, pages 273–312. Amer. Math. Soc., Providence, R.I., 1978.

[8] Marc Lackenby. The crossing number of composite knots. *J. Topol.*, 2(4):747–768, 2009.

[9] Marc Lackenby. The crossing number of satellite knots. *Algebr. Geom. Topol.*, 14(4):2379–2409, 2014.

[10] Kunio Murasugi. Jones polynomials and classical conjectures in knot theory. *Topology*, 26(2):187–194, 1987.

[11] Kunio Murasugi. *Knot Theory and Its Applications*. Birkhäuser Boston, Inc., Boston, MA, 1996. Translated from the 1993 Japanese original by Bohdan Kurpita.

[12] Morwen B. Thistlethwaite. A spanning tree expansion of the Jones polynomial. *Topology*, 26(3):297–309, 1987.

[13] Morwen B. Thistlethwaite. On the Kauffman polynomial of an adequate link. *Invent. Math.*, 93(2):285–296, 1988.

Chapter 27

The Bridge Number of a Knot

Jennifer Schultens, University of California, Davis

27.1 Introduction

The bridge number of a knot arose as one of the first numerical knot invariants. This chapter considers the bridge number from a historical perspective, compares it to other knot invariants and reflects on related concepts.

In the 1950s Horst Schubert set out to prove that a given knot has at most a finite number of companion knots. Companion knots are discussed in conjunction with satellite knots elsewhere in this volume. Schubert established this finiteness result with the help of a knot invariant devised for this purpose: The *bridge number*. See [24].

Given a diagram of a knot K, a continuous maximal subarc that includes an overcrossing is called a *bridge*. The number of bridges in a knot diagram is called the *bridge number of the diagram*. The minimum, over all diagrams of K, of the bridge numbers of the diagrams, is called the *bridge number of K* and denoted by $b(K)$.

27.2 The Modern Definition of Bridge Number

In early investigations of the bridge number of a knot, the idea of a knot lying in the plane, with a certain number of bridges venturing out of the plane, informed the discussion. In the 1980s Morse

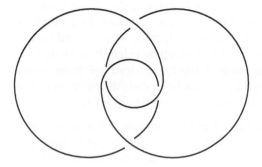

FIGURE 27.1: A diagram of the trefoil with two bridges.

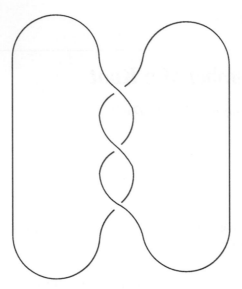

FIGURE 27.2: A representation of the trefoil with two relative maxima.

theoretic considerations and the notion of a height function on both $\mathbb{R}^3$ and $\mathbb{S}^3$ shifted our perspective. A *height function* on $\mathbb{S}^3$ is a Morse function with exactly two critical points: a maximum and a minimum. On $\mathbb{R}^3$ it is a Morse function with no critical points. More concretely, it is projection onto, say, the z-axis.

From this perspective it makes sense to consider the number of relative maxima of the knot K with respect to a height function. We think of the plane used in a knot diagram as the xy-plane and our height function as a projection onto z. Given a diagram of a knot K with b bridges, each subarc of the diagram that is a bridge can be converted into an arc with interior above the plane and exactly one maximum. Subarcs of the diagram that are not bridges can be concatenated and converted into arcs with interior below the plane and exactly one minimum. In this manner we construct a representative of K with exactly b local maxima. As we traverse K, we alternate between traversing arcs above the plane and arcs below the plane. It follows that the representative of K also has exactly b local minima.

Conversely, given a representative of K with exactly b local maxima and b local minima, we can, by raising maxima and lowering minima, if necessary, find a horizontal plane P that divides the representative of K into $2b$ arcs, b of which lie above P and have exactly one local maximum and no other critical points, and b of which lie below P and have exactly one local minimum. By isotoping the arcs that lie below P into P and viewing P from above, we construct a diagram of K with b bridges. See Figure 27.2. It follows that the bridge number of a knot equals the smallest possible number of local maxima for a representative of K.

A plane such as P, that lies above all local minima of K and below all local maxima of K is called a *bridge surface*. When we think of $K \subset \mathbb{S}^3$, then a height function on $\mathbb{S}^3$ decomposes $\mathbb{S}^3$ into level spheres together with one minimum and one maximum. A level sphere that lies above all local minima of the knot K and below all local maxima of K is called a *bridge sphere*.

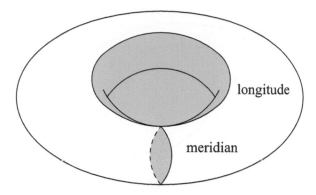

FIGURE 27.3: A meridian/longitude pair.

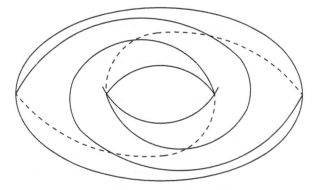

FIGURE 27.4: The torus knot $T(3, 2)$ is the trefoil.

27.3 Bridge Numbers of Torus Knots

One interesting family of knots to consider in the context of bridge number consists of torus knots. A *torus knot* is an isotopy class of knots that are embedded in an unknotted torus T in $\mathbb{S}^3$. The unknotted torus in $\mathbb{S}^3$ is characterized by the existence of two embedded curves called the *meridian* and *longitude* which intersect transversely in one point and bound disks in the complement of T. See Figure 27.3.

An invigorating exercise shows that there is a 1-1 correspondence between the set of isotopy classes of torus knots and pairs (p, q) of relatively prime integers. Given a torus knot K and orientations on K, T, the meridian and the longitude, we take p to be the oriented intersection number of the meridian with K and q to be the oriented intersection number of the longitude with K. The trefoil is an example of a torus knot, see Figures 27.4 and 27.5.

We consider torus knots not just up to isotopy in T but also as knots in $\mathbb{S}^3$. A torus knot will be the unknot in $\mathbb{S}^3$ if and only if $p = \pm 1$ or $q = \pm 1$. Knots in $\mathbb{S}^3$ are considered up to symmetry and homeomorphism. Changing relevant orientations for a torus knots changes p to $-p$ or q to $-q$.

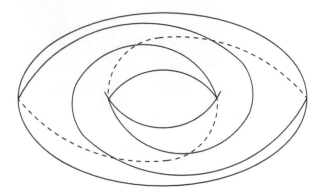

FIGURE 27.5: The torus knot $T(2, 3)$ is also the trefoil.

Interchanging the roles of meridian and longitude exchanges p and q. Thus we need consider only pairs of integers (p, q) such that $0 \leq p < q$ along with $(p, q) = (1, 1)$. Given (p, q) with $0 \leq p < q$ or $(p, q) = (1, 1)$, we denote the corresponding torus knot by $T(p, q)$. In [24], Schubert proved the following theorem:

Theorem 1. (Schubert 1954) The bridge number of $T(p, q)$ equals p as long as $p > 0$.

Recall that $T(p, q)$ is the unknot if $p \leq 1$. Of course, the unknot can be isotoped to lie entirely in a level sphere. However, counting relative maxima only makes sense for simple closed curves that are in general position with respect to a given height function. The unknot therefore, has bridge number 1. Conversely, the only knot with bridge number 1 is the unknot. Notice that this is consistent with the fact that the bridge number of $T(1, q)$ is 1 for every q since $T(1, q)$ is the unknot for every q. The only knot for which $p = 0$ is $T(1, 0)$, which is also the unknot, but rather than having bridge number 0, it has bridge number 1, necessitating the hypothesis $p > 0$ in Schubert's theorem.

By considering Figure 27.5, we see that the bridge number of $T(2, 3)$ is less than or equal to 2. More generally, by drawing an analogous diagram for $T(p, q)$, we see that the bridge number of $T(p, q)$ is at most p. To see that it cannot be strictly less than q requires more work and this work was carried out by Schubert in [24].

Indeed, given a height function, a representative of $T(p, q)$ that realizes bridge number will necessarily lie on an unknotted torus T. However, T could be positioned in an unusual way, folding in on itself, for instance. What Schubert accomplished in [24] was to isotope T into standard position without increasing the bridge number of the representative of $T(p, q)$. For a short Morse-theoretic rendering of Schubert's proof, see [27]. Figures 27.6 and 27.7 illustrate some of the challenges involved in isotoping T into standard position.

Similar reasoning applies in the setting of satellite knots and provides the theorem below:

Theorem 2. (Schubert 1954) Let K be a satellite knot with companion J and pattern with wrapping number k. Then $b(K) \geq k \cdot b(J)$. If K is a cabled knot, *i.e.*, the pattern is a torus knot with wrapping number p, then $b(K) = p \cdot b(J)$.

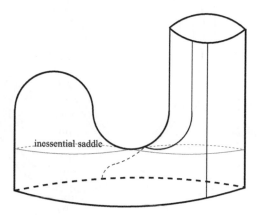

FIGURE 27.6: An inessential saddle in T.

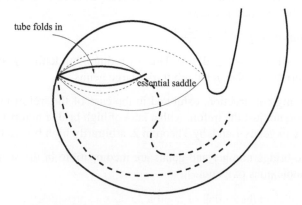

FIGURE 27.7: An essential and nested saddle in T.

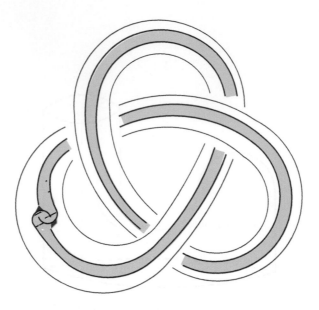

FIGURE 27.8: The double of a trefoil.

27.4 Bridge Number versus Genus

The *genus* of a knot K is the smallest possible genus for a Seifert surface of K. The following theorem is due to Herbert Seifert, see [28]:

Theorem 3. The genus of $T(p, q)$ is $\frac{(p-1)(q-1)}{2}$.

Recall that $b(T(p, q))$, where $q > p \geq 0$, is p. Thus by fixing p but letting q be arbitrarily large, we can have knots with bridge number p but arbitrarily large genus.

Conversely, the doubling construction, exhibited in the case of the trefoil in Figure 27.8, can be performed on any knot, not just the trefoil, with a knot of high bridge number replacing the trefoil. This provides knots with genus 1 and, by Theorem 2, arbitrarily high bridge number.

We conclude that the bridge number and genus are incompatible in the sense that they measure different types of complexities of a knot.

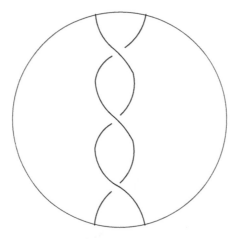

FIGURE 27.9: A twist is a succession of crossings of two strands over each other.

27.5 Bridge Number versus Hyperbolic Volume

A knot K is hyperbolic if its complement supports a complete finite volume hyperbolic structure. For a hyperbolic knot K, the volume of K, denoted by *volume*(K), is the hyperbolic volume of $\mathbb{S}^3 - K$.

A *twist* is a succession of crossings of two strands of a knot over each other that is maximal in the sense that adjacent crossings involve other strands of the knot. See Figure 27.9.

Figure 27.10 schematically exhibits a family of 2-bridge knot diagrams. The boxes labeled with the numbers $t_1, \ldots, t_n$ represent twists with the given number of crossings. The numbers $t_1, \ldots, t_n$ can be chosen so that the knot diagram is alternating. The *twist number* of a knot diagram D, denoted by $t(D)$, is the minimal number of twists in the diagram D. The twist number for the diagrams of the family of 2-bridge knots represented in Figure 27.10 is arbitrarily high. See [26].

In [25], Schubert shows that 2-bridge knots are prime. Furthermore, Allen Hatcher and William Thurston show in [10] that 2-bridge knots are simple, *i.e.*, there are no essential tori in the complements of 2-bridge knots. Some 2-bridge knots will be torus knots, but those with more than one twist will not be torus knots. It follows that the complements of 2-bridge knots that are not torus knots support complete finite volume hyperbolic structures. For details see [10].

Theorem 4. (Lackenby) Let D be a prime alternating diagram of a hyperbolic link K in S^3. Then $v_3(t(D) - 2)/2 \leq volume(K) < v_3(16t(D) - 16)$, where $v_3(\approx 1.01494)$ is the volume of a regular hyperbolic ideal 3-simplex.

Theorem 5. (S) There are 2-bridge knots of arbitrarily large volume.

More generally, Jessica Purcell and Alexander Zupan prove, among other things, the following, see [22]:

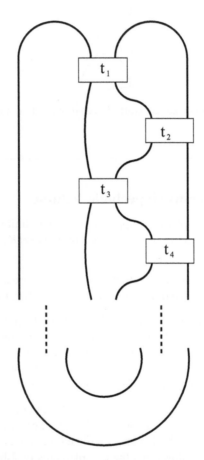

FIGURE 27.10: Schematic of a 2-bridge knot where boxes represent twists.

Theorem 6. (Purcell-Zupan) For any natural number b, there exists a sequence of knots $\{K_n\}$ such that $b(K_n) = b$ but $volume(K_n) \to \infty$ as $n \to \infty$.

They also prove a partial converse:

Theorem 7. (Purcell-Zupan) There is a constant $V > 0$ and a sequence of knots $\{K_n\}$ such that $volume(K_n) < V$ for all n but $b(K_n) \to \infty$ as $n \to \infty$.

We conclude as they do, that the bridge number and volume are incompatible in the sense that they measure different types of complexities of a knot.

27.6 Bridge Number versus Rank and Meridional Rank

Recall that a bridge sphere is a level sphere of a height function that lies above all local minima and below all local maxima of a knot K. Thus a bridge sphere S separates $\mathbb{S}^3$ into two 3-balls, B_1 and B_2, each containing a collection of subarcs of K. If the height function realizes the bridge number b of K, then it has exactly b local maxima and b local minima. Denote the subarcs of K in B_i by $a_1^i, \ldots, a_b^i$. Within B_i, the twists lying above (respectively, below) the bridge surface can be "untwisted" to reveal that $a_1^i, \ldots, a_b^i$ are unknotted in B_i, meaning that there are pairwise disjoint disks $D_1^i, \ldots, D_b^i$ such that ∂D_j^i is partitioned into two subarcs, one lying in ∂B_i and the other equal to a_j^i.

The disks $D_1^i, \ldots, D_b^i$ cut B_i into a 3-ball $\hat{B}_i$. Thus the complement of K in B_i can be constructed from the 3-ball $\hat{B}_i$ by identifying the remnants of the disks $D_1^i, \ldots, D_b^i$ in the boundary of $\hat{B}_i$. This tells us that $B_i - \eta(K)$, for $\eta(K)$ an open regular neighborhood of K, retracts to a wedge of b circles and hence its fundamental group is the free group on b generators.

To compute the fundamental group of the complement of K we choose a basepoint x in the bridge sphere. The bridge sphere retracts to a wedge of $2b$ circles and hence its fundamental group is the free group on $2b$ generators. The generators of the bridge sphere pair up to coincide in $\pi_1(B_i - K)$. It follows that b is an upper bound for the *rank*, *i.e.*, the minimal number of generators needed to generate $\pi_1(\mathbb{S}^3 - K)$:

$$rank(K) := rank(\pi_1(\mathbb{S}^3 - K)) \le b(K).$$

On the other hand, recall that the bridge number of $T(p, q)$ is p. However, regardless of p, q, the complement of $T(p, q)$ is the union of two solid tori along an annulus. It follows that $rank(T(p, q)) = 2$. Thus
$$rank(T(p, q)) < b(K) \; for \; p > 2.$$

Moreover, $rank(T(p, q)) - b(K)$ can be arbitrarily large.

An interesting variation on the rank arises if we restrict our presentations of $\pi_1(\mathbb{S}^3 - K)$ by requiring each generator to be freely homotopic to the meridian. The minimum number of generators required in such a presentation of $\pi_1(\mathbb{S}^3 - K)$ is called the *meridional rank* of K. The above argument involves only meridional generators and hence shows that

$$meridional\ rank(K) \le b(K).$$

Equality holds for several classes of knots, *e.g.*, generalized Montesinos links and iterated torus knots. See [2], [3], [4], [15] and [23]. Whether or not equality holds is currently unknown.

27.7   Recognizing and Computing

In [30], Robin Wilson proved that, subject to certain technical conditions, every bridge sphere is isotopic to a *meridional almost normal* sphere. We will not be interested in a precise definition of meridional almost normal spheres here, suffice it to say that there is an algorithm to detect meridional almost normal spheres. The converse is not true: A meridional almost normal sphere in a knot complement need not be a bridge sphere.

In [11], William Jaco and Jeffrey Tollefson exhibit an algorithm to determine whether or not a given 3-manifold is a 3-ball minus a collection of unknotted arcs. Since meridional almost normal spheres can be detected algorithmically, it is tempting to think that applying Jaco and Tollefson's algorithm to the two components of the complement of the meridional almost normal spheres in a knot complement should provide an algorithm to detect bridge spheres and thereby the bridge number of the knot. However, in toroidal knot complements, *i.e.*, for satellite knots, this method breaks down in the sense that the process need not terminate due to the possible existence of infinitely many meridional almost normal spheres of a given Euler characteristic.

An alternative to the strategy outlined above for computing bridge numbers of knots rests on a result of William Thurston. He proved that for any prime knot K, one of the following holds: 1) K is a torus knot: 2) K is a satellite knot; 3) K is a hyperbolic knot. This trichotomy result proved to be a special case of the geometrization of 3-manifolds. See [19], [21], [20]. In [13], Greg Kuperberg proved a computational analogue of the geometrization of 3-manifolds. Kuperberg's work provides an algorithm to determine whether or not a given knot is a torus knot, a satellite knot or a hyperbolic knot. For torus knots, Schubert's theorem tells us the bridge number. For hyperbolic knots Alex Coward exhibits an algorithm to recognize the bridge number, see [5].

Coward's argument continues a line of investigation begun by Wolfgang Haken in the 1960s to solve recognition problems in low-dimensional topology. However, rather than merely using normal or almost normal surface theory, Coward uses partially flat angled ideal triangulations as described by Marc Lackenby in [14]. Coward's argument also breaks down for knots that are not hyperbolic, again because of the presence of tori. The problem of algorithmically computing bridge number is hence still open.

27.8 Generalized Bridge Number

One can consider the number of relative maxima of the knot K not just with respect to a height function on $\mathbb{S}^3$ but with respect to any self-indexing Morse function on $\mathbb{S}^3$. A level surface of minimal Euler characteristic of such a Morse function is a *bridge surface* for K if it lies above all relative minima and below all relative maxima of K. The *g-bridge number* of K is the least number of relative maxima K will exhibit with respect to a Morse function with bridge surfaces of genus g. Additivity in the sense of Theorem 2 fails. For more on properties of the generalized bridge number, see Helmut Doll's investigation [7].

27.9 Bridge Distance

For any compact connected orientable surface Σ, the *curve complex* of Σ is defined as follows:

- Vertices of $C(\Sigma)$ correspond to isotopy classes of essential simple closed curves in Σ.
- Edges correspond to pairs of vertices admitting disjoint representatives.
- $C(\Sigma)$ is a flag complex.
- The distance between two vertices is the least number of edges in an edge path between the two vertices.

We will be interested in bridge spheres of knots. Let S be a bridge sphere for K and set $\Sigma = S - \eta(K)$. If Σ is a four-times-punctured sphere (in the case where K is a 2-bridge knot) the definition above yields a complex that is disconnected. By convention, the definition of the edges for this curve complex is adjusted (requiring two points of intersection rather than requiring disjointedness), in order to guarantee connectedness.

The bridge sphere S separates $\mathbb{S}^3$ into balls B_1 and B_2. Denote by $\mathcal{D}_i$ the collection of isotopy classes of essential disks in $B_i - \eta(K)$ with a boundary in $S - \eta(K)$. Denote the collection of boundaries of disks in $\mathcal{D}_i$ by ∂_i. The *bridge distance of S*, denoted by $d(S)$, is given by

$$d(S) = \min\{d(c_1, c_2) \mid c_i \in \partial_i\}.$$

The *bridge distance of K*, denoted by $d(K)$, is the greatest possible bridge distance of a bridge sphere of K that meets K in exactly $2b(K)$ points. To understand the subtleties concerning how this gives us a well-defined integer, see [29]. For more on the topic of bridge distance, see [12] and [31]. For a natural generalization of bridge distance, see [16].

27.10   Bridge Number versus Distortion

In [8], Mikhail Gromov studied embeddings of manifolds and defined a notion called *distortion*. In the specialized setting of knots we have the following: Given a knot K in $\mathbb{S}^3$, a smooth representative γ of K and two points $p, q \in \gamma$, the distance between p and q can be measured in two ways: 1) As the distance between p and q in $\mathbb{S}^3$, which we denote by $d_s(p, q)$; 2) As the length of the (shorter) subarc of γ from p to q, which we denote by $d_\gamma(p, q)$. The *distortion* of a knot K, denoted by $\delta(K)$, is then given by:

$$\delta(\gamma) = \sup_{p,q \in \gamma} \frac{d_\gamma(p, q)}{d_s(p, q)}$$

and

$$\delta(K) = \inf_{\gamma \in K} \delta(\gamma).$$

In [6], Elizabeth Denne and John Sullivan proved that the distortion of a nontrivial knot is bounded below by $\frac{5\pi}{3}$. Having observed that the distortion of a knot remains constant under connected sum, Gromov asked in [8] whether there is a universal upper bound on the distortion of a knot. The answer to this question is "no." In [18], John Pardon, building on the work of Makoto Ozawa, see [17], used the bridge number as a tool to prove that torus knots provide a family with arbitrarily high distortion. See also [9].

Theorem 8. (Pardon)

$$\delta(T(p, q)) \geq \frac{1}{160} \min(p, q)$$

Recent work announced by Ryan Blair, Marion Campisi, Scott Taylor and Maggy Tomova suggests that the distortion of a knot is far more closely related to the bridge number than other invariants. See [1].

27.11   Bibliography

[1] Ryan Blair, Marion Campisi, Scott Taylor, and Maggy Tomova. Distortion and the bridge distance of knots. arXiv:1705.08490v1.

[2] Michel Boileau and Heiner Zieschang. Nombre de ponts et générateurs méridiens des entrelacs de Montesinos. *Comment. Math. Helv.*, 60(2):270–279, 1985.

[3] Christopher R. Cornwell. Knot contact homology and representations of knot groups. *J. Topol.*, 7(4):1221–1242, 2014.

[4] Christopher R. Cornwell and David R. Hemminger. Augmentation rank of satellites with braid pattern. *Comm. Anal. Geom.*, 24(5):939–967, 2016.

[5] Alex Coward. Algorithmically detecting the bridge number of hyperbolic knots. arXiv:0710.1262.

[6] Elizabeth Denne and John M. Sullivan. The distortion of a knotted curve. *Proc. Amer. Math. Soc.*, 137(3):1139–1148, 2009.

[7] H. Doll. A generalized bridge number for links in 3-manifolds. *Math. Ann.*, 294(4):701–717, 1992.

[8] Mikhail Gromov. Filling Riemannian manifolds. *J. Differential Geom.*, 18(1):1–147, 1983.

[9] Misha Gromov and Larry Guth. Generalizations of the Kolmogorov-Barzdin embedding estimates. *Duke Math. J.*, 161(13):2549–2603, 2012.

[10] A. Hatcher and W. Thurston. Incompressible surfaces in 2-bridge knot complements. *Invent. Math.*, 79(2):225–246, 1985.

[11] William Jaco and Jeffrey L. Tollefson. Algorithms for the complete decomposition of a closed 3-manifold. *Illinois J. Math.*, 39(3):358–406, 1995.

[12] Jesse Johnson and Yoav Moriah. Bridge distance and plat projections. *Algebr. Geom. Topol.*, 16(6):3361–3384, 2016.

[13] Greg Kuperberg. Algorithmic homeomorphism of 3-manifolds as a corollary of geometrization. arxiv:1508.06720.

[14] Marc Lackenby. An algorithm to determine the Heegaard genus of simple 3-manifolds with nonempty boundary. *Algebr. Geom. Topol.*, 8(2):911–934, 2008.

[15] Martin Lustig and Yoav Moriah. Generalized Montesinos knots, tunnels and N-torsion. *Math. Ann.*, 295(1):167–189, 1993.

[16] Yair N. Minsky, Yoav Moriah, and Saul Schleimer. High distance knots. *Algebr. Geom. Topol.*, 7:1471–1483, 2007.

[17] Makoto Ozawa. Bridge position and the representativity of spatial graphs. *Topology Appl.*, 159(4):936–947, 2012.

[18] John Pardon. On the distortion of knots on embedded surfaces. *Ann. of Math. (2)*, 174(1):637–646, 2011.

[19] Grisha Perelman. The entropy formula for the Ricci flow and its geometric applications. arxiv:0211159.

[20] Grisha Perelman. Finite extinction time for the solutions to the Ricci flow on certain three-manifolds. arxiv:0307245.

[21] Grisha Perelman. Ricci flow with surgery on three-manifolds. arxiv:0303109.

[22] Jessica S. Purcell and Alexander Zupan. Independence of volume and genus g bridge numbers. *Proc. Amer. Math. Soc.*, 145(4):1805–1818, 2017.

[23] Markus Rost and Heiner Zieschang. Meridional generators and plat presentations of torus links. *J. London Math. Soc. (2)*, 35(3):551–562, 1987.

[24] Horst Schubert. Über eine numerische Knoteninvariante. *Math. Z.*, 61:245–288, 1954.

[25] Horst Schubert. Knoten mit zwei Brücken. *Math. Z.*, 65:133–170, 1956.

[26] Jennifer Schultens. Genus 2 closed hyperbolic 3-manifolds of arbitrarily large volume. In *Proceedings of the Spring Topology and Dynamical Systems Conference (Morelia City, 2001)*, volume 26, pages 317–321, 2001/02.

[27] Jennifer Schultens. Additivity of bridge numbers of knots. *Math. Proc. Cambridge Philos. Soc.*, 135(3):539–544, 2003.

[28] Herbert Seifert. Über das geschlecht von knoten. *Math. Ann.*, 110:571–592, 1935.

[29] Maggy Tomova. Multiple bridge surfaces restrict knot distance. *Algebr. Geom. Topol.*, 7:957–1006, 2007.

[30] Robin Todd Wilson. Meridional almost normal surfaces in knot complements. *Algebr. Geom. Topol.*, 8(3):1717–1740, 2008.

[31] Alexander Zupan. Bridge and pants complexities of knots. *J. Lond. Math. Soc. (2)*, 87(1):43–68, 2013.

Chapter 28

Alternating Distances of Knots

Adam Lowrance, Vassar College

28.1   Introduction

A link diagram is *alternating* if its crossings alternate between over and under as one travels along each component of the link, and a link is *alternating* if it has an alternating diagram. Menasco [29] proved that a prime alternating knot is either hyperbolic or a $(2, q)$-torus knot. Kauffman [19], Murasugi [31], and Thistlethwaite [34] used the Jones polynomial to prove Tait's conjecture that an alternating diagram without nugatory crossings has the fewest possible crossings among all diagrams of the link. While the definition of an alternating link is diagrammatic in nature, Greene [15] and Howie [16] gave diagram-free definitions of alternating links in terms of spanning surfaces in the complement.

Non-alternating knots and links that are "nearly alternating" often maintain some of the properties of alternating links. An *alternating distance d* is a non-negatively valued link invariant such that $d(L) = 0$ if and only if L is alternating and such that d is subadditive with respect to connected sums, i.e. $d(L_1 \# L_2) \leq d(L_1) + d(L_2)$ for all links L_1 and L_2 and for any choice of connected sum $L_1 \# L_2$. Alternating distances give a way to measure how far away a link is from alternating and can formalize the idea of a "nearly alternating" link. The alternating distances discussed in this chapter are the dealternating number, alternation number, alternating genus, Turaev genus, and region dealternating number.

In Section 28.2, we define the various different alternating distances and describe some of their properties. In Section 28.3, we compare the different alternating distances by showing the known inequalities they satisfy and by giving examples where some of these inequalities are sharp. In Section 28.4, we give some open questions involving alternating distances.

28.2   Meet the Invariants

The *dealternating number* dalt(D) of a link diagram D is the fewest number of crossing changes needed to transform D into an alternating diagram. The *dealternating number* dalt(L) of a link L is the minimum of dalt(D) taken over all diagrams of L. A link L such that dalt(L) = 1 is called *almost alternating*. Adams et al. [4] proved that a prime almost alternating knot is either a torus knot or a hyperbolic knot, generalizing the corresponding result for alternating knots of Menasco [29] . Abe

[1] later proved that the only almost alternating torus knots are $T_{3,4}$, $T_{3,5}$, and their mirrors. Ito [17] and Kim [23] gave topological characterizations of almost alternating knots.

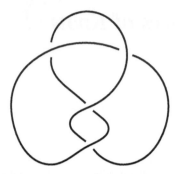

FIGURE 28.1: A diagram D with dalt(D) = 2 and alt(D) = 0.

The *alternation number* alt(D) of a link diagram D is the fewest number of crossing changes needed to transform D into a (possibly non-alternating) diagram of an alternating link. Figure 28.1 shows a non-alternating diagram D of the trefoil with dalt(D) = 2 and alt(D) = 0. The *alternation number* alt(L) of a link L, defined by Kawauchi [20], is the minimum of alt(D) taken over all diagrams of L. An equivalent definition of the alternation number of a link L is the minimum Gordian distance (in the sense of Murakami [30]) between L and any alternating link.

If Σ is a Heegaard surface in S^3, then any link L can be embedded inside a thickening $\Sigma \times [-1, 1]$ of Σ such that projection to $\Sigma = \Sigma \times \{0\}$ yields a link diagram D with only transverse double points on the surface Σ. The *alternating genus* $g_{\text{alt}}(K)$ of a link L is defined to be the minimum genus of a Heegaard surface Σ such that the L has an alternating diagram D on Σ where $\Sigma - D$ is a disjoint union of disks. If $g_{\text{alt}}(L) = 1$, then L is called *toroidally alternating*. Adams [3] proved that prime toroidally alternating knots are either torus knots or hyperbolic knots, again generalizing the corresponding result for alternating knots of Menasco [29]. Kim [23] gave a topological characterization of toroidally alternating knots.

A crossing ⤬ in a link diagram D on S^2 has an *A-resolution*)(and a *B-resolution* ⤬ . The set of simple closed curves resulting from a choice of A-resolution or B-resolution at each crossing of D is a *Kauffman state* of D. When every resolution is an A-resolution, the corresponding state is the *all-A state* of D, and likewise, when every resolution is a B-resolution, the corresponding state is the *all-B state* of D. Construct a cobordism between the all-A and all-B states of D consisting of saddles near the crossings of D and bands away from the crossings of D. Capping off the boundary components of this cobordism with disks forms the *Turaev surface* Σ_D of D. Let $g_T(D)$ denote the genus of the Turaev surface of D, and define the *Turaev genus* $g_T(L)$ of the link L as the minimum of $g_T(D)$ taken over all diagrams D of L. Turaev [35] originally constructed this surface to give an alternate proof that the span of the Jones polynomial is a lower bound on the crossing number of a link. In fact, for a non-split link, he proved that span $V_L(t) \leq c(L) - g_T(L)$. For a comprehensive overview of the Turaev surface and the Turaev genus of a link, see the survey [8] and Chapter 25 of this book.

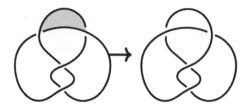

FIGURE 28.2: A region crossing change of the knot 4_1.

A *region* of a link diagram D is a connected component of $S^2 - D$, and a *region crossing change* of D is the operation that changes every crossing incident to a region of D as in Figure 28.2. The *region dealternating number* $\text{dalt}_R(D)$ of the link diagram D is defined as the fewest number of region crossing changes needed to transform D into an alternating diagram. Shimizu [33] proved that any single crossing change on a knot diagram can be realized through a sequence of region crossing changes. Consequently, for any knot diagram D, there is a sequence of region crossing changes transforming D into an alternating diagram, and hence $\text{dalt}_R(D)$ is defined for any knot diagram. For some link diagrams, there is no sequence of region crossing changes taking the diagram to an alternating diagram (see [11, 10]). In this case, we define $\text{dalt}_R(D)$ to be infinite. Kreinbihl et al. [24] define the *region dealternating number* $\text{dalt}_R(L)$ of the link L as the minimum of $\text{dalt}_R(D)$ taken over all diagrams D of L. Despite the fact that $\text{dalt}_R(D)$ can be infinite for some link diagrams, the region dealternating number of a link is always finite, as we will see in Theorem 28.3.1. The reader is invited to verify that all of the preceding invariants are alternating distances.

28.3 Comparisons

Theorem 28.3.1 gives all the known inequalities that hold between our alternating distances. These inequalities are consequences of work found in [2, 12, 24, 27]. The third inequality in the theorem implies that $\text{dalt}_R(L)$ is finite for every link.

Theorem 28.3.1. The following inequalities hold for all links L:

1. $\text{alt}(L) \le \text{dalt}(L)$,

2. $g_{\text{alt}}(L) \le g_T(L) \le \text{dalt}(L)$, and

3. $\text{dalt}_R(L) \le \text{dalt}(L)$.

Proof. Since an alternating diagram is a diagram of an alternating link, $\text{alt}(D) \le \text{dalt}(D)$ for any link diagram D, and thus $\text{alt}(L) \le \text{dalt}(L)$.

The Turaev surface Σ_D of a link diagram D is a Heegaard surface, the diagram D is alternating on Σ_D, and the complement $\Sigma_D - D$ is a disjoint union of disks (see [12, 35]). It follows that $g_{\text{alt}}(L) \le g_T(L)$ for all links L.

When D is a diagram of a non-split link L, the genus of its Turaev surface is $g_T(D) = \frac{1}{2}(2 + c(D) - s_A(D) - s_B(D))$ where $c(D)$ is the number of crossings in D and $s_A(D)$ and $s_B(D)$ are the number of components in the all-A and all-B states of D, respectively. The quantity $s_A(D) + s_B(D)$ can either go up or down by two or stay the same under a crossing change, and thus the genus $g_T(D)$ of the Turaev surface will either go down or up by one or stay the same under a crossing change. Therefore $g_T(D) \leq \text{dalt}(D)$ for every link diagram D, and hence $g_T(L) \leq \text{dalt}(L)$.

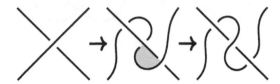

FIGURE 28.3: A Reidemeister 2 move followed by a region crossing change.

Let D be a diagram that minimizes $\text{dalt}(L)$, and suppose that changing crossings $c_1, c_2, \ldots, c_{\text{dalt}(L)}$ of D yields an alternating diagram D_{alt}. Performing a Reidemeister 2 move near each crossing c_i results in a different diagram D' of L, and then performing region crossing changes in the regions formed by those Reidemeister 2 moves (as in Figure 28.3) yields an alternating diagram D'_{alt}. Therefore $\text{dalt}_R(D') = \text{dalt}(D)$, and hence $\text{dalt}_R(L) \leq \text{dalt}(L)$.   Λ

For the remainder of the section, we consider examples where one alternating distance is strictly less than another. The computations of the alternating distances for these examples rely on bounds coming from the Jones polynomial, Khovanov homology, or knot Floer homology [1, 6, 7, 9, 13, 14, 25, 28].

Example 28.3.1. Let L_{grid} be the link in Figure 28.4. Kim [22] proved that this diagram D_{grid} on the torus is not the result of the Turaev surface algorithm, but he left open the possibility that L_{grid} could have some other diagram whose Turaev surface is genus one. Dasbach and Lowrance [14] proved that for any link L, if $g_T(L) = 1$ or $\text{dalt}(L) = 1$, then at least one of the leading or trailing coefficients of the Jones polynomial $V_L(t)$ has absolute value one. Both the leading and trailing coefficients of the $V_{L_{\text{grid}}}(t)$ are -2, and thus $2 \leq g_T(L_{\text{grid}}) \leq \text{dalt}(L_{\text{grid}})$. The diagram in Figure 28.4 implies that $g_{\text{alt}}(L_{\text{grid}}) = 1$.

Example 28.3.2. Let $\widetilde{T}_{4,7}$ be the closure of the braid $(\sigma_1\sigma_2\sigma_3)^6\sigma_1\sigma_2^{-1}\sigma_3$ in the braid group on 4-strands; see Figure 28.4 for a diagram of $\widetilde{T}_{4,7}$. Its Rasmussen s invariant [32] is $s(\widetilde{T}_{4,7}) = 16$, and its signature is $\sigma(\widetilde{T}_{4,7}) = -12$. Abe [1] and Dasbach and Lowrance [13] proved that $|s(K) + \sigma(K)| \leq 2\,\text{alt}(K)$ and $|s(K) + \sigma(K)| \leq 2g_T(K)$, respectively, for any knot K. Therefore $2 \leq \text{alt}(\widetilde{T}_{4,7})$ and $2 \leq g_T(\widetilde{T}_{4,7})$. The diagram of $\widetilde{T}_{4,7}$ in Figure 28.4 shows that $g_{\text{alt}}(\widetilde{T}_{4,7}) = 1$.

Example 28.3.3. Let W_{3_1} be the Whitehead double of the trefoil 3_1 depicted in Figure 28.4. Since W_{3_1} is a prime satellite knot, $1 < g_{\text{alt}}(W_{3_1}) \leq g_T(W_{3_1}) \leq \text{dalt}(W_{3_1})$, and because W_{3_1} has unknotting number one, $\text{alt}(W_{3_1}) = 1$. Example 28.3.2 and this example show there is no inequality relating $g_{\text{alt}}(L)$ and $\text{alt}(L)$ for all links L.

Example 28.3.4. Khovanov and knot Floer homology arguments in [1, 2, 18, 26] show that $\text{alt}(T_{3,7}) = g_T(T_{3,7}) = \text{dalt}(T_{3,7}) = 2$. As observed in [24], a region crossing change in the shaded region of the diagram of $T_{3,7}$ in Figure 28.4 yields an alternating diagram. Thus $\text{dalt}_R(T_{3,7}) = 1$.

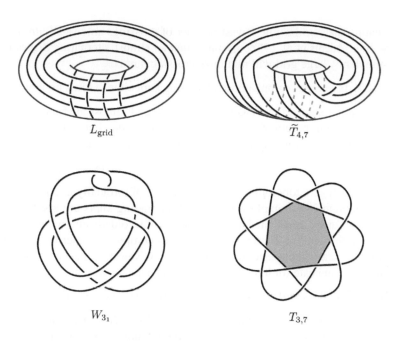

L_{grid}

$\tilde{T}_{4,7}$

W_{3_1}

$T_{3,7}$

FIGURE 28.4: Examples 28.3.1, 28.3.2, 28.3.3, and 28.3.4.

28.4 Open Questions

We conclude with some open questions.

Question 28.4.1. Is there a link L such that $g_T(L) < \text{dalt}(L)$?

The inequality $g_T(D) \leq \text{dalt}(D)$ holds for every link diagram, and one can easily find diagrams where the difference $\text{dalt}(D) - g_T(D)$ is large. For each $n \in \mathbb{N}$, there are standard diagrams D of non-alternating pretzel and Montesinos links such that $\text{dalt}(D) - g_T(D) > n$. However, non-alternating pretzel and Montesinos links have almost alternating diagrams [2, 21], and so $g_T(L) = \text{dalt}(L) = 1$ for all such links. An arbitrary Turaev genus one link can be viewed as a generalization of a non-alternating Montesinos link where every rational tangle is replaced with an alternating tangle, implying that every Turaev genus one link can be obtained by mutations from an almost alternating link [5, 22]. Thus, in order to prove that there is a link with $1 = g_T(L) < \text{dalt}(L)$, one needs link invariants that are sensitive to mutation. One difficulty in answering this question is that every known lower bound for the dealternating number of a link is also a lower bound for the Turaev genus of the link.

Question 28.4.2. Is there an almost alternating diagram that minimizes the crossing number of the link?

Lowrance and Spyropoulos [28] gave sufficient diagrammatic conditions to determine when an almost alternating diagram has the fewest number of crossings among all almost alternating diagrams of the link. Adams et al. [4] conjectured that an almost alternating diagram whose shadow contains a sufficiently large grid will minimize crossing number.

Question 28.4.3. Is a prime region almost alternating knot either a torus knot or a hyperbolic knot?

Toroidally alternating knots, Turaev genus one knots, and almost alternating knots all have this property, and so it seems natural to investigate whether region almost alternating knots do too. A positive answer to the question implies that $\text{dalt}_R(K) \geq 2$ for a prime satellite knot K.

Question 28.4.4. What is a topological characterization of Turaev genus one knots?

Kim [23] and Ito [17] gave topological characterizations of toroidally alternating and almost alternating knots. Theorem 28.3.1 implies that almost alternating knots are a subset of Turaev genus one knots and that Turaev genus one knots are a subset of toroidally alternating knots. Furthermore, the classifications theorem for Turaev genus one links completely describes the alternating tangle decomposition of Turaev genus one diagrams [5, 22]. Given this, it seems that Turaev genus one knots are a natural candidate for having a topological characterization.

Question 28.4.5. Do links with arbitrarily large alternating genus exist?

A positive answer to this question would likely involve discovering a new lower bound for alternating genus. The best known result is that prime satellite knots must have alternating genus at least two.

28.5 Bibliography

[1] Tetsuya Abe. An estimation of the alternation number of a torus knot. *J. Knot Theory Ramifications*, 18(3):363–379, 2009.

[2] Tetsuya Abe and Kengo Kishimoto. The dealternating number and the alternation number of a closed 3-braid. *J. Knot Theory Ramifications*, 19(9):1157–1181, 2010.

[3] Colin C. Adams. Toroidally alternating knots and links. *Topology*, 33(2):353–369, 1994.

[4] Colin C. Adams, Jeffrey F. Brock, John Bugbee, Timothy D. Comar, Keith A. Faigin, Amy M. Huston, Anne M. Joseph, and David Pesikoff. Almost alternating links. *Topology Appl.*, 46(2):151–165, 1992.

[5] Cody W. Armond and Adam M. Lowrance. Turaev genus and alternating decompositions. *Algebr. Geom. Topol.*, 17(2):793–830, 2017.

[6] Marta M. Asaeda and Józef H. Przytycki. Khovanov homology: Torsion and thickness. In *Advances in Topological Quantum Field Theory*, volume 179 of *NATO Sci. Ser. II Math. Phys. Chem.*, pages 135–166. Kluwer Acad. Publ., Dordrecht, 2004.

[7] Abhijit Champanerkar and Ilya Kofman. Spanning trees and Khovanov homology. *Proc. Amer. Math. Soc.*, 137(6):2157–2167, 2009.

[8] Abhijit Champanerkar and Ilya Kofman. A survey on the Turaev genus of knots. *Acta Math. Vietnam.*, 39(4):497–514, 2014.

[9] Abhijit Champanerkar, Ilya Kofman, and Neal Stoltzfus. Graphs on surfaces and Khovanov homology. *Algebr. Geom. Topol.*, 7:1531–1540, 2007.

[10] Zhiyun Cheng. When is region crossing change an unknotting operation? *Math. Proc. Cambridge Philos. Soc.*, 155(2):257–269, 2013.

[11] ZhiYun Cheng and HongZhu Gao. On region crossing change and incidence matrix. *Sci. China Math.*, 55(7):1487–1495, 2012.

[12] Oliver T. Dasbach, David Futer, Efstratia Kalfagianni, Xiao-Song Lin, and Neal W. Stoltzfus. The Jones polynomial and graphs on surfaces. *J. Combin. Theory Ser. B*, 98(2):384–399, 2008.

[13] Oliver T. Dasbach and Adam M. Lowrance. Turaev genus, knot signature, and the knot homology concordance invariants. *Proc. Amer. Math. Soc.*, 139(7):2631–2645, 2011.

[14] Oliver T. Dasbach and Adam M. Lowrance. Invariants of Turaev genus one links. *Comm. Anal. Geom.*, 26(5):1103–1126, 2018.

[15] Joshua Evan Greene. Alternating links and definite surfaces. *Duke Math. J.*, 166(11):2133–2151, 2017. With an appendix by András Juhász and Marc Lackenby.

[16] Joshua Howie. A characterisation of alternating knot exteriors. *Geom. Topol.*, 21(4):2353–2371, 2017.

[17] Tetsuya Ito. A characterization of almost alternating knots. *J. Knot Theory Ramifications*, 27(1):1850009, 13, 2018.

[18] Kaitian Jin, Adam M. Lowrance, Eli Polston, and Yanjie Zheng. The Turaev genus of torus knots. *J. Knot Theory Ramifications*, 26(14):1750095, 28, 2017.

[19] Louis H. Kauffman. State models and the Jones polynomial. *Topology*, 26(3):395–407, 1987.

[20] Akio Kawauchi. On alternation numbers of links. *Topology Appl.*, 157(1):274–279, 2010.

[21] Dongseok Kim and Jaeun Lee. Some invariants of pretzel links. *Bull. Austral. Math. Soc.*, 75(2):253–271, 2007.

[22] Seungwon Kim. Link diagrams with low Turaev genus. *Proc. Amer. Math. Soc.*, 146(2):875–890, 2018.

[23] Seungwon Kim. A topological characterization of toroidally alternating knots. *Comm. Anal. Geom.*, 27(8):1825–1850, 2019.

[24] James Kreinbihl, Ra'jene Martin, Colin Murphy, McKenna Renn, and Jennifer Townsend. Alternation by region crossing changes and implications for warping span. *J. Knot Theory Ramifications*, 25(13):1650073, 14, 2016.

[25] Adam M. Lowrance. On knot Floer width and Turaev genus. *Algebr. Geom. Topol.*, 8(2):1141–1162, 2008.

[26] Adam M. Lowrance. The Khovanov width of twisted links and closed 3-braids. *Comment. Math. Helv.*, 86(3):675–706, 2011.

[27] Adam M. Lowrance. Alternating distances of knots and links. *Topology Appl.*, 182:53–70, 2015.

[28] Adam M. Lowrance and Dean Spyropoulos. The Jones polynomial of an almost alternating link. *New York J. Math.*, 23:1611–1639, 2017.

[29] W. Menasco. Closed incompressible surfaces in alternating knot and link complements. *Topology*, 23(1):37–44, 1984.

[30] Hitoshi Murakami. Some metrics on classical knots. *Math. Ann.*, 270(1):35–45, 1985.

[31] Kunio Murasugi. Jones polynomials and classical conjectures in knot theory. *Topology*, 26(2):187–194, 1987.

[32] Jacob Rasmussen. Khovanov homology and the slice genus. *Invent. Math.*, 182(2):419–447, 2010.

[33] Ayaka Shimizu. Region crossing change is an unknotting operation. *J. Math. Soc. Japan*, 66(3):693–708, 2014.

[34] Morwen B. Thistlethwaite. An upper bound for the breadth of the Jones polynomial. *Math. Proc. Cambridge Philos. Soc.*, 103(3):451–456, 1988.

[35] V. G. Turaev. A simple proof of the Murasugi and Kauffman theorems on alternating links. *Enseign. Math. (2)*, 33(3-4):203–225, 1987.

Chapter 29

Superinvariants of Knots and Links

Colin Adams, Williams College

29.1 Superbridge Index

The first superinvariant for knots was defined by Kuiper in [8]. It is a variant on bridge index. It and all the other superinvariants are physical invariants in the sense that they depend on how conformations of the knots sit in Euclidean 3-space. We begin by stating a definition of bridge index that is equivalent to other definitions as for instance occurring in Chapter 27.

Take a fixed conformation of a knot K in 3-space. For each possible axis in space, project the knot conformation onto the axis, as in Figure 29.1. Relative to that projection, there is some number of local maxima of the knot. The bridge number of the conformation is defined to be the least number of local maxima that can be attained when projecting that conformation to any possible axis. We then minimize this over all conformations of the knot. Note that we need only consider axes through the origin and that we can represent each such axis by a unit vector from the origin ending at a point on the unit sphere S^2.

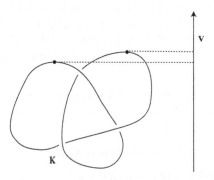

FIGURE 29.1: Counting local maxima when projecting to an axis.

Thus, we can define the bridge index as follows:

Definition 29.1.1. The **bridge index** of a knot is given by

$$b[K] = \min_{K \in [K]} \min_{\vec{v} \in S^2} (\text{\# of local maxima}).$$

Given this formulation of bridge index, it is simple to give Kuiper's variant:

Definition 29.1.2. The **superbridge index** of a knot, denoted $sb(K)$, is given by

$$sb[K] = \min_{K \in [K]} \max_{\vec{v} \in S^2}(\# \text{ of local maxima}).$$

Note that the only difference between the definitions of the bridge index and superbridge index is that for each fixed conformation of the knot, we find the greatest number of local maxima coming from the projection to an axis, rather than the least number. We still then minimize this over all conformations.

From the definitions, it is immediate that $b[K] \leq sb[K]$. In fact, Kuiper proved that for any nontrivial knot, $b[K] < sb[K]$. He further found a relation between the superbridge index and braid index $\beta[K]$:

$$sb[K] \leq 2\beta[K]$$

.

In [4], it was proved that $sb[K] \leq 5b[K] - 3$. In the preprint [2], it is proved that $sb[K] \leq 3b[K] - 1$.

In a difficult tour de force, Kuiper also determined the superbridge index for all torus knots.

Theorem 29.1.3. ([8]) If $2 \leq p < q$, then $sb[T_{p,q}] = min\{2p, q\}$.

This is the only infinite category of knots for which we know the superbridge index. Note that this result implies that for even $n \geq 4$, there are infinitely many knots with $sb[K] = n$.

Since $b[K] < sb[K]$, a nontrivial knot must have the superbridge index of at least 3. In [5], Jeong and Jin showed that there are only finitely many knots with superbridge index 3. The trefoil and figure-eight knots are known to have superbridge index 3. The only other candidates are $5_2, 6_1, 6_2, 6_3, 7_2, 7_3, 7_4, 8_4, 8_7$, and 8_9. Thus, since $sb[K] \leq 2b[K]$, all other 2-bridge knots have superbridge number 4. One would like to show that only the trefoil and figure-eight are superbridge index 3.

It is still unknown how the superbridge index behaves under composition. It is true that $sb[K_1 \# K_2] \leq sb[K_1] + sb[K_2]([2])$, but it is also known that the difference $sb[K_1] + sb[K_2] - sb[K_1 \# K_2]$ can be arbitrarily large([7]).

Jin has conjectured the following for nontrivial knots K_1 and K_2:

Conjecture 29.1.4. $sb[K_1 \# K_2] \leq sb[K_1] + sb[K_2] - 2$.

Conjecture 29.1.5. $sb[K_1 \# K_2] > max\{sb[K_1], sb[K_2]\}$.

The superbridge index plays an important role with respect to stick index $s[K]$. (See Chapter 84.) In [6], Jin noted the fact that when a knot is in its minimal stick conformation, the number of maxima relative to any axis is at most half the stick number, since maxima can only occur at vertices, or along sticks perpendicular to the axis of projection, as in Figure 29.2.

Since the number of vertices is equal to the number of sticks, this implies:

Theorem 29.1.6. $s[K] \geq 2sb[K]$.

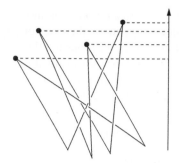

FIGURE 29.2: The number of maxima is at most half the number of vertices and hence sticks.

Considering torus knots $T_{p,q}$ where $2 \leq p < q < 2p$, Theorem 29.1.3 then implies that $s[K] \geq 2sb[K] = 2q$. One can realize such a torus knot with $2q$ sticks as follows. Start with two one-sheeted hyperboloids that open along the z-axis and that intersect in two circles of the same radius, one in the plane $z = 1$ and one in the plane $z = -1$. Each hyperboloid has a distinct waist size. Then the annuli on the hyperboloids between the two circles of intersection together form a torus with crescent-shaped cross section. A one-sheeted hyperboloid is a doubly ruled surface, meaning it is made up of the union of a collection of disjoint lines in two different ways. We can choose a collection of q sticks on the first annulus and q sticks on the second such that the second collection shares endpoints on the top circle with the first. The two sticks sharing an endpoint reach around one meridian on the torus. By adjusting the waist-size of the hyperboloids, we can adjust the slopes of the sticks and therefore make the fraction of longitude this pair of sticks contributes to be p/q. Then, together the $2q$ sticks can be conjoined to generate a (p, q)-torus knot, as in Figure 84.3 from Chapter 84.

The superbridge index is related to another invariant dating back to [3] and [9].

Definition 29.1.7. Let K be a polygonal knot. Traveling along the knot in a choice of the two directions, define the **exterior angles** to be those angles we turn through as we pass through each vertex. The **total curvature** of the conformation of the knot, denoted $tc(K)$, is the sum of these exterior angles. The total curvature of a smooth embedding of a knot is the least upper bound of the total curvatures of the polygons inscribed in it. The **total curvature of the knot type**, denoted $tc[K]$, is the greatest lower bound to the total curvatures of its conformations.

Milnor proved that for a nontrivial knot type $[K]$, $tc[K] = 2\pi b[K]$. Moreover, for any fixed conformation, the total curvature is strictly greater than this lower bound. So, for instance, conformations of the trefoil knot can be constructed with total curvatures arbitrarily close to but slightly greater than 4π.

To see why Milnor's result holds, we consider the *bridge sphere*. Assume K is a polygonal conformation of our knot. Taking a large sphere surrounding the knot, we label each point on the sphere with the number of local extrema we obtain when we project the knot to the axis corresponding to that point on the sphere. Note that antipodal points receive the same label. The largest label must be at least as large as twice the superbridge index of the knot. The smallest label must be at least as large as twice the bridge index of the knot.

Since local extrema generically occur at vertices of the knot, we would like to determine which points on the sphere "see" a given vertex as a local extremum. Let v be a vertex of the knot, e_1 and e_2 the two edges of the knot meeting at v, and α the exterior angle at v. The two planes through the origin that are perpendicular to e_1 and e_2 meet one another at angle α and cut the sphere into 4 regions. The two opposite regions, called lunes, corresponding to the angle α, are the two regions on the sphere where projections to the corresponding axis will generate a local extremum for v. The fraction of the sphere covered by these two lunes is $\frac{2\alpha}{2\pi}$. Thus, the average number of local extrema, over all the choices of axes corresponding to points on the sphere is $\sum_{i=1}^{n} \frac{2\alpha_i}{2\pi}$. Therefore the average number of local maxima is half this, which is $\frac{tc(K)}{2\pi}$.

Thus, we see that for a fixed conformation K, $b(K) \leq \frac{tc(K)}{2\pi} \leq sb(K)$. In fact, since conformations do not exist with only one label on the entire sphere, these are actually strict inequalities. This implies that $tc[K] \geq 2\pi b[K]$ and $tc[K] \leq 2\pi sb[K]$. By applying linear transformations to squeeze a conformation to an axis, we arrive at Milnor's result, that $tc[K] = 2\pi b[K]$.

29.2   Other Superinvariants

In the same manner that the superbridge index comes from the bridge index, we can obtain other superinvariants from certain existing knot invariants. For instance, one can define the crossing index as follows.

Fix a conformation of the knot in space. Consider all possible projections of that conformation to planes passing through the origin (the plane defined by a unit vector perpendicular to the plane). Take the least number of crossings obtained in this manner. Now minimize this over all conformations of the knot.

Definition 29.2.1. The **crossing index** of K, denoted $c[K]$, is given by

$$c[K] = \min_{K \in [K]} \min_{\vec{v} \in S^2} (\# \text{ of crossings}).$$

Once we have defined the crossing number in this manner, it becomes clear how to obtain the "super" version.

Definition 29.2.2. The **supercrossing index** of K, denoted $sc[K]$, is given by

$$sc[K] = \min_{K \in [K]} \max_{\vec{v} \in S^2} (\# \text{ of crossings}).$$

We just maximize the number of crossings we see generated by a given conformation (known as the supercrossing number of that conformation), and then minimize this over all conformations. This is a natural invariant. For a given knot, what is the worst possibility for the number of crossings in a projection of a conformation, if we can minimize this over all possible conformations?

Since the trivial knot can be placed in a plane, we see that its supercrossing index is 0. However for a nontrivial knot, the supercrossing index must be somewhat greater than the crossing index.

Theorem 29.2.3. ([1]) If K is a nontrivial knot, $sc[K] \geq c[K] + 3$.

The idea of the proof is as follows. Travel along the knot in a particular direction. As we travel along, we extend a tangent ray from the point we are at in the direction of travel. Since the knot is nontrivial, there must be a point along the knot when the ray re-intersects the knot somewhere. If we project in the direction of that ray, we obtain a non-regular projection. Slight changes in the projection direction will result in Type I and Type II Reidemeister moves on the corresponding projections. In other words, there will be projections of this conformation near one another that differ by at least three crossings. So every conformation must have a projection that has at least three more crossings than the crossing index for the knot.

However, determining the supercrossing index seems difficult. For instance, since the supercrossing index is at least 3 greater than crossing index, the supercrossing index of any nontrivial knot must be at least 6. In the case of the trefoil, there are conformations such that all of their projections have at most 7 crossings. So for the trefoil knot T, $6 \leq sc(T) \leq 7$.

However, it remains open as to whether 6 or 7 is the supercrossing index of the trefoil knot.

On the other hand, we do know the supercrossing number of the Hopf link. For a link with at least one nontrivial component, we have the same bound as in Theorem 29.2.3. But if all components in a link L are individually trivial, we only know $scr[L] \geq c[L] + 2$. If we take both components of the Hopf link to be perfect circles in their respective planes, then from any view we can have at most four crossings, so the supercrossing number of the Hopf link is 4.

Fixing a conformation of K, we already described how we can create a bridge sphere, with a positive integer label at each point corresponding to the number of local extrema when the conformation is projected onto the axis determined by that point. Similarly, we can create the *crossing sphere* by taking a large sphere around the conformation. For each point on the sphere, we project the conformation to a plane through the origin that is perpendicular to the vector defined by the origin and that point on the sphere, and then label the point with the number of crossings we see. See Figure 29.3 for an example.

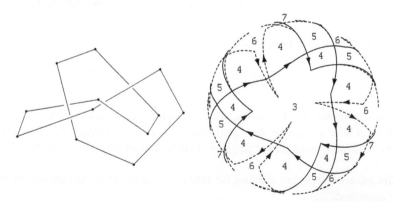

FIGURE 29.3: The crossing sphere for this stick knot, with regions labelled by the number of crossings seen from each point on the sphere. The back side of the sphere is not shown, since antipodal points have the same label.

Note that if K is a fixed conformation of a knot, its supercrossing number will be the largest label that occurs on its crossing sphere. Its average crossing number will be the average of those labels. And the crossing index of the knot must be at most the smallest label that occurs on the crossing sphere. Note that the graph that appears on the crossing sphere that separates regions with a given crossing number consists of points on the sphere corresponding to non-regular projections. These non-regular projections correspond to transitional projections where Reidemeister moves take place. So each edge of the graph either corresponds to a Type I, a Type II or a Type III move. For the crossing sphere, the Type III edges are irrelevant since a Type III move does not change the number of crossings in the projections.

One can also create superinvariants out of other invariants obtained as minima over projections. For instance, one can define the superunknotting index as follows:

Definition 29.2.4. The **superunknotting index** of K, denoted su(K), is given by

$$su(K) = \min_{K \in [K]} \max_{\vec{v} \in S^2} (\text{\# of crossings to trivialize P}).$$

Very little is known about this invariant. Note that there is also an "unknotting sphere" corresponding to this invariant. Each region in the complement of the graph is labelled with the number of crossings that must be changed to unknot the corresponding projections. Here, the Type I edges of the graph become irrelevant.

Similarly, one can define the superbraid index. The braid index can equivalently be defined as follows.

Definition 29.2.5. The **braid index** of a knot or link is given by

$$\beta(K) = \min_{K \in [K]} \min_{\vec{v} \in S^2} (\text{\# of Seifert circles in P}).$$

See [10] for this approach to the braid index. But once we have this definition, the superbraid index is defined as follows.

Definition 29.2.6. The **superbraid index** of a knot or link K is given by

$$s\beta(K) = \min_{K \in [K]} \max_{\vec{v} \in S^2} (\text{\# of Seifert circles in P}).$$

Once again, for a given knot conformation, we can define a "Seifert circle sphere", the largest label of which will be at least as large as the superbraid index and the smallest label of which will be at least as large as the braid index. As of yet, there is little known about this construction.

These superinvariants offer many avenues for further research. There are numerous questions that remain to be answered.

29.3 Bibliography

[1] C. Adams,C. Lefever, J. Othmer, S. Pahk, A. Stier, J. Tripp, An introduction to the super-crossing index of knots and the crossing map, *J. Knot Theory and its Ramifications*, Vol. 11, No. 3(2002) 445-459.

[2] C. Adams, N. Agarwal, R. Allen, T. Khandhawit, A. Simons, R. Winarski, M. Wootters, Superbridge and bridge indices for knots, ArXiv 2008.06483, 2020.

[3] I. Fry, *Sur la courbure totale d'une courbe gauche faisant un noeud*, Bull. Soc. Math. France 77 (1949), 128-138.

[4] E. Furstenberg, J. Lie, and J. Schneider, Stick knots, *Chaos, Solitons, and Fractals* 9(4/5) (1998) 561–568.

[5] J. B. Jeon, G. T. Jin, There are only finitely many 3-superbridge knots, emphKnots in Hellas '98, Vol. 2 (Delphi). *J. Knot Theory Ramifications* 10 (2001), no. 2, 331–343.

[6] G. T. Jin, Polygon indices and superbridge indices of torus knots and links, *J. Knot Theory Ramifications* 6 (1997), no. 2, 281–289.

[7] G. T. Jin, Superbridge index of composite knots, *J. Knot Theory Ramifications* 9 (2000), no. 5, 669–682.

[8] N. Kuiper, A new knot invariant, *Math. Ann.* 278 (1987) 193-209.

[9] J. Milnor, On total curvatures of closed space curves *Math. Scand.* 1 (1953), 289–296.

[10] S. Yamada, The minimal number of Seifert circles equals the braid index of a link, *Invent. Math.* 89 (1987), 347–356.

Part VII

Other Knotlike Objects

Chapter 30

Virtual Knot Theory

Louis H. Kauffman, University of Illinois at Chicago

30.1 Introduction

Virtual knot theory is an extension of classical knot theory to stabilized embeddings of circles into thickened orientable surfaces of genus possibly greater than zero. Classical knot theory is the case of genus zero. There is a diagrammatic theory for studying virtual knots and links, and this diagrammatic theory lends itself to the construction of numerous new invariants of virtual knots as well as extensions of known invariants. Since virtual knot theory extends classical knot theory, there are many relationships with the classical theory. We have emphasized new invariants that arise in studying virtual knots and new questions that have been and can be asked of this theory. In fact, virtual knot theory should be regarded as a beginning of constructive ways to study knots in three manifolds. The case of virtual knots is a success for the manifolds $S_g \times I$, where S_g is an orientable surface of genus g.

This chapter gives a concise introduction and summary of virtual knot theory.

30.2 Virtual Knot Theory

Virtual knot theory studies the embeddings of curves in thickened surfaces of arbitrary genus, up to the addition and removal of empty handles from the surface. Virtual knots have a special diagrammatic theory that makes handling them very similar to the handling of classical knot diagrams. Many structures in classical knot theory generalize to the virtual domain and classical knot theory embeds in virtual knot theory. This makes virtual knot theory a special subject since it includes all of classical knot theory and it contains extensions of that classical theory to the specific class of three manifolds of the form $S_g \times I$, where I is the unit interval and S_g is a surface of genus g. Most invariants of classical links extend to invariants of virtual links, and there are a number of new invariants of virtual links that can handle the new phenomena that appear in this domain.

In the diagrammatic theory of virtual knots, one adds a *virtual crossing* (see Figure 30.1) that is neither an over-crossing nor an under-crossing. A virtual crossing is represented by two crossing segments with a small circle placed around the crossing point.

Moves on virtual diagrams generalize the Reidemeister moves for classical knot and link diagrams. See Figure 30.1. One can summarize the moves on virtual diagrams by saying that the classical crossings interact with one another according to the usual Reidemeister moves, while virtual crossings are artifacts of the attempt to draw the virtual structure in the plane. A segment of diagram consisting of a sequence of consecutive virtual crossings can be excised and a new connection made between the resulting free ends. If the new connecting segment intersects the remaining diagram (transversally) then each new intersection is taken to be virtual. Such an excision and reconnection is called a *detour move*. See Figure 30.2 for an illustration of the detour move. Adding the global detour move to the Reidemeister moves completes the description of moves on virtual diagrams. In Figure 30.1, we illustrate a set of local moves involving virtual crossings. The global detour move is a consequence of moves (B) and (C) in Figure 30.1, and thus, these specific local moves can also be taken to define virtual knot theory (by adding them to the Reidemeister moves). Virtual knot and link diagrams that can be connected by a finite sequence of these moves are said to be *equivalent* or *virtually isotopic*.

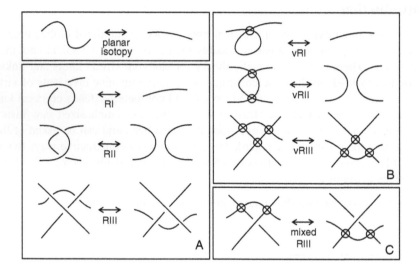

FIGURE 30.1: Moves for virtual diagrams.

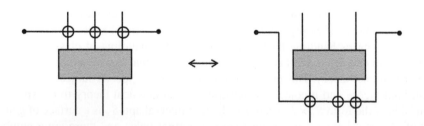

FIGURE 30.2: Detour move.

Another way to understand virtual diagrams is to regard them as representatives for oriented Gauss codes [1, 2, 3] (see Chapter 3). Such codes do not always have planar realizations. An attempt to embed such a code in the plane leads to the production of the virtual crossings. The detour move makes the particular choice of virtual crossings irrelevant. *Virtual isotopy is the same as the*

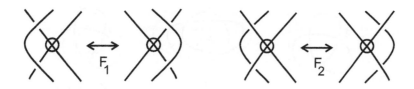

FIGURE 30.3: Forbidden moves.

equivalence relation generated on the collection of oriented Gauss codes by abstract Reidemeister moves on these codes.

The moves illustrated in Figure 30.3 are called the *forbidden moves*. They are not part of the lexicon of allowed moves in virtual knot theory, and indeed, one can show that neither forbidden move is a consequence of the moves in Figure 30.1. It is not hard to construct examples of virtual knots that are non-trivial, but become unknotted on the application of one or both of the forbidden moves. The forbidden moves change the structure of the Gauss code and, if desired, must be considered separately from the virtual knot theory proper. Sam Nelson [4] and Naoko and Seichi Kamada [5] proved that any virtual knot can be unknotted by using the two forbidden moves. Remarkably, if we only allow the "over" forbidden move $F1$ of Figure 30.3, and add this to the Reidemeister and virtual moves of Figure 30.1 we get a non-trivial theory that goes by the name *welded knot theory* (Chapter 33). Welded knot theory was discovered via a corresponding theory of *welded braids* by Rourke, Fenn and Rimiyani [6]. In welded knot theory and virtual knot theory the fundamental group is defined by using a Wirtinger presentation from the diagrams (See Section 30.5). At this writing, it is not known whether this fundamental group detects the unknotted welded knot. We take the definition of welded knots to include both the usual Reidemeister 1 move and the virtual Reidemeister 1 move. One may also consider the *framed welded knots and links* where the classical Reidemeister 1 move is forbidden.

30.3   Interpretation of Virtual Links as Stable Classes of Links in Thickened Surfaces

There is a useful topological interpretation [2, 7] for this virtual theory in terms of embeddings of links in thickened surfaces. Regard each virtual crossing as a shorthand for a detour of one of the arcs in the crossing through a 1-handle that has been attached to the 2-sphere of the original diagram. By interpreting each virtual crossing in this way, we obtain an embedding of a collection of circles into a thickened surface $S_g \times R$ where g is the number of virtual crossings in the original diagram L, S_g is a compact oriented surface of genus g, and R denotes the real line. We say that two such surface embeddings are *stably equivalent* if one can be obtained from another by isotopy in the thickened surfaces, homeomorphisms of the surfaces, and the addition or subtraction of empty handles (i.e., the knot does not go through the handle).

We have the following result.

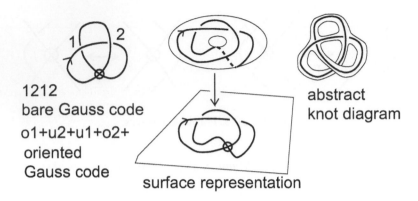

1212
bare Gauss code
o1+u2+u1+o2+
oriented
Gauss code

surface representation

abstract
knot diagram

FIGURE 30.4: Surfaces and virtuals.

Theorem 1 [2, 8, 7, 9]. *Two virtual link diagrams are isotopic if and only if their corresponding surface embeddings are stably equivalent.*

In Figure 30.4, we illustrate some points about this association of virtual diagrams and knot and link diagrams on surfaces. Note that the projection of the knot diagram on the torus to a diagram in the plane (in the center of the figure) has a virtual crossing in the planar diagram, while two arcs that do not form a crossing in the thickened surface project to the same point in the plane. In this way, virtual crossings can be regarded as artifacts of a projection. The same figure shows a virtual diagram on the left and an "abstract knot diagram" [10, 9] on the right. The abstract knot diagram is a realization of the knot on the left in a thickened surface with boundary and it is obtained by making a neighborhood of the virtual diagram that resolves the virtual crossing into arcs that travel on separate bands. The virtual crossing appears as an artifact of the projection of this surface to the plane. The reader will find more information about this correspondence [2, 8] in other papers by the author and in the literature of virtual knot theory.

In [11], Kuperberg shows the uniqueness of the embedding of minimal genus in the stable class for a given virtual link. This is an important result as it shows that each virtual link has a purely topological interpretation in the knot theory of a specific thickened surface that is of minimal genus for this link. See [12] for a description of an algorithm to find the minimal genus surface for a virtual link. The known algorithm for determining this minimal genus is quite complex and uses the three-dimensional topology of the complement of the virtual link in its thickened surface. We can often conjecture the minimal genus of a virtual knot or link and then determine it by using invariants of the link to show that it cannot have smaller genus. See [13]. The Kuperberg result tells us that classical links embed in the theory of virtual links. That is, if two classical links are equivalent as virtual links, then they are equivalent as classical links. The result for knots was proved before Kuperberg's paper by using properties of the fundamental group and peripheral subgroup [2, 1].

30.4 Virtual Braids

Virtual braids are defined by adding new generators to the Artin braid group corresponding to a virtual crossing of adjacent strands (see Chapter 7. The standard braid generators and the new generators are shown in Figure 30.5 for the n-strand virtual braid group VB_n.

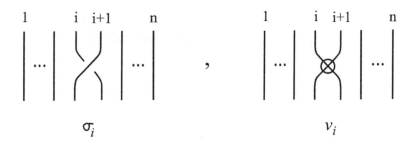

FIGURE 30.5: The Generators of the virtual braid group VB_n.

Among themselves the braid generators satisfy the usual braiding relations:

$$\begin{aligned}
\sigma_i \sigma_{i+1} \sigma_i &= \sigma_{i+1} \sigma_i \sigma_{i+1}, \\
\sigma_i \sigma_j &= \sigma_j \sigma_i, \quad \text{for } j \neq i \pm 1.
\end{aligned}$$

The virtual generators are a presentation for the permutation group S_n, so they satisfy the following *virtual relations:*

$$\begin{aligned}
v_i^2 &= 1, \\
v_i v_{i+1} v_i &= v_{i+1} v_i v_{i+1}, \\
v_i v_j &= v_j v_i, \quad \text{for } j \neq i \pm 1.
\end{aligned}$$

It is worth noting at this point that the virtual braid group VB_n does not embed in the classical braid group B_n, since the virtual braid group contains torsion elements and it is well known that B_n doesn't. The *mixed relations* between virtual generators and braiding generators are as follows:

$$\begin{aligned}
\sigma_i v_j &= v_j \sigma_i, \quad \text{for } j \neq i \pm 1, \\
v_i \sigma_{i+1} v_i &= v_{i+1} \sigma_i v_{i+1}.
\end{aligned}$$

The reader can consult the following references for more information about virtual braids [14, 15, 16, 17, 18, 19, 20, 5].

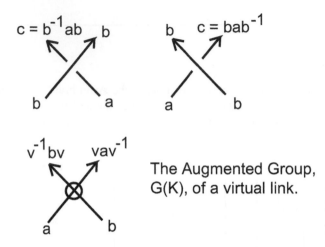

FIGURE 30.6: The augmented fundamental group $G(K)$.

30.5 Fundamental Group and Augmented Fundamental Group for Virtual Knots and Links

One can define a fundamental group for a virtual knot or link by applying the Wirtinger relations [2] at each classical crossing and not using any extra relation at the virtual crossings (see Chapter 47). See [2, 22] for a discussion of this approach and its generalizations using quandles and biquandles (see also Chapter 79). This simple fundamental group for a virtual knot is a weak invariant, but it can be described in terms of the embedding of the virtual knot in a thickened surface. Let F be a surface and $M = F \times I$ where I denotes the unit interval. Let L be a link embedded in the interior of F with a corresponding virtual diagram $D(L)$. Let CM denote the cone on M obtained by taking a cone attached to $F \times 1$. Then the fundamental group $\pi_1(CM - L)$ is isomorphic to the formally defined fundamental group of the virtual diagram as described above.

In this section is given a definition of a more powerful Wirtinger-type fundamental group for virtual knots and links. See Figure 30.6. Here we use Wirtinger relations at the classical crossings, and we have relations at the virtual crossings as shown in the figure. These relations take the form that a group label is conjugated by a special element v when the arc of the diagram goes through a virtual crossing. It is not hard to see that the resulting group $G(L)$ for a virtual diagram L is invariant under virtual isotopy. This group is strong. It has been studied in a different form by Bardakov and Mikhalchishina [17] using a braid group formulation of the invariant. In the form that we have described it, this group first appears in a paper by Manturov [23]. See also [21]. The fundamental group with the standard Wirtinger representation is a weak invariant for virtual knots and even this augmentation of the fundamental group can be trivial (for the augmented group, triviality for a knot is a free group on two generators). For example, the augmented group for the Kishino diagram is easily seen to be trivial. See Figure 30.7 for a presentation of the augmented group $G(K)$ where K

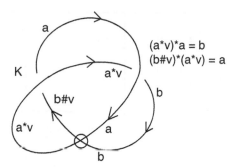

$(a*v)*a = b$
$(b\#v)*(a*v) = a$

The Group $G(K)$ for the
Virtual Trefoil K.
$a*b = b^{-1}ab$
$a\#b = bab^{-1}$

FIGURE 30.7: Augmented Group for virtual trefoil.

is the virtual trefoil diagram. In this case the augmented group is non-trivial while the fundamental group is trivial.

30.6 Biquandles and Quandles

A *biquandle* [22, 24, 25, 26, 7, 27] is an algebra with four binary operations written a^b, a_b, $a^{\bar{b}}$, $a_{\bar{b}}$ together with relations which we will indicate below (see Chapter 79). Biquandles began to be studied primarily for their applications to virtual knot theory, but are now a subject in their own right. We indicate here the definitions and some key applications to virtual knots.

The *fundamental* biquandle is associated with a link diagram and is invariant under the generalized Reidemeister moves for virtual knots and links. The operations in this algebra are motivated by the formation of labels for the edges of the diagram, and by the patterns of relations that first occurred in the Wirtinger presentation for the fundamental group of the complement of a classical knot embedded in three-dimensional space. In Figure 30.8, we have shown the format for the operations in a biquandle. The overcrossing arc has two labels, one on each side of the crossing. There is an algebra element labeling each *edge* of the diagram. An edge of the diagram corresponds to an edge of the underlying plane graph of that diagram.

Let the edges oriented toward a crossing in a diagram be called the *input* edges for the crossing, and the edges oriented away from the crossing be called the *output* edges for the crossing. Let a and b be the input edges for a positive crossing, with a the label of the undercrossing input and b the label on the overcrossing input. In the biquandle, we label the undercrossing output as

$$c = a^b,$$

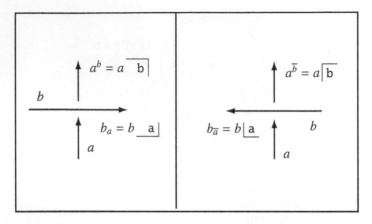

FIGURE 30.8: Biquandle relations at a crossing.

while the overcrossing output is labeled

$$d = b_a.$$

The labeling for the negative crossing is similar using the other two operations.

Another way to write this formalism for the biquandle is as follows

$$a^b = a \ \overline{\ b\ }|$$

$$a_b = a \ \underline{\ b\ }|$$

$$a^{\overline{b}} = a\lceil b$$

$$a_{\overline{b}} = a\lfloor b\ .$$

We call this the *operator formalism* for the biquandle. In Figure 30.8, we show both of these notations. An advantage of the operator notation is that it is always written on a line. Thus,

$$(a^{b^{cd}})^e = a \ b \ \underline{c\ d}\ ||\ \overline{e}|$$

and

$$((a^b)^c)^d = a \ \overline{b}|\ \overline{c}|\ \overline{d}|$$

while

$$a^{b^{cd}} = a \ b \ \overline{c\ d}\ ||.$$

To form the fundamental biquandle, $BQ(K)$, we take one generator for each edge of the diagram and two relations at each crossing (as described above and in Figure 30.8).

These considerations lead to the following definition.

Definition A *biquandle B* is a set with four binary operations indicated above: $a^b, a^{\bar{b}}, a_b, a_{\bar{b}}$. We shall refer to the operations with barred variables as the *left* operations and the operations without barred variables as the *right* operations. The biquandle is closed under these operations and the following axioms are satisfied:

1. Given an element a in B, there exists an x in the biquandle such that $x = a_x$ and $a = x^a$. There also exists a y in the biquandle such that $y = a^{\bar{y}}$ and $a = y_{\bar{a}}$.

2. For any elements a and b in B we have

$$a = a^{b\overline{b_a}}, \; b = b_{\overline{a}a^b}, \; a = a^{\bar{b}b_{\bar{a}}}, \; b = b_{\bar{a}\bar{a}^{\bar{b}}}.$$

3. Given elements a and b in B, there exist elements x, y, z, t such that $x_b = a$, $y^{\bar{a}} = b$, $b^x = y$, $a_{\bar{y}} = x$ and $t^a = b$, $a_t = z$, $z_{\bar{b}} = a$, $b^{\bar{z}} = t$.

   The biquandle is called *strong* if x, y, z, t are uniquely defined, and we then write $x = a_{b^{-1}}, y = b^{\bar{a}^{-1}}, t = b^{a^{-1}}, z = a_{\bar{b}^{-1}}$, reflecting the invertible nature of the elements.

4. For any a, b, c in B, the following equations hold, and the same equations hold when all right operations are replaced in these equations by left operations:

$$a^{bc} = a^{c_b b^c}, \; c_{ba} = c_{a^b b_a}, \; (b_a)^{c_{a^b}} = (b^c)_{a^{c_b}}.$$

These axioms are transcriptions of the Reidemeister moves. The first axiom transcribes the first Reidemeister move. The second axiom transcribes the directly oriented second Reidemeister move. The third axiom transcribes the reverse oriented Reidemeister move. The fourth axiom transcribes the third Reidemeister move. Much ongoing research explores these algebras and their applications to knot theory. Special cases of the biquandle correspond to widely-used algebraic invariants in knot theory and virtual knot theory. When we take the lower operations to be the identity so that

$$a_b = a \underline{\;\; b \;} = a$$

and

$$a_{\bar{b}} = a \underline{\;\; b} = b,$$

then the structure is called a *quandle*, and the axioms simplify to

$$a^a = a$$
$$(a^b)^b = a$$
$$(a^b)^c = (a^c)^{b^c},$$

together with the corresponding equations for the inverse operation.

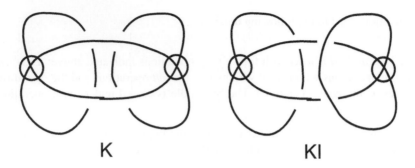

FIGURE 30.9: The knot K and the Kishino diagram KI.

For virtual knots, there is a significant generalization of the quandle and the biquandle that is obtained by adding an extra relation at the virtual crossings, generalizing the construction of the augmented group $G(K)$ that we discussed in the previous section. We will not discuss this aspect of biquandles further in the present chapter.

30.6.1   The Alexander Biquandle

It is not hard to see that the following equations in a module over $\mathbb{Z}[s, s^{-1}, t, t^{-1}]$ give a biquandle structure:

$$a^b = a\ \overline{\boxed{b}} = ta + (1 - st)b, \quad a_b = a\ \underline{\boxed{b}} = sa$$
$$a^{\overline{b}} = a\overline{\lceil b} = t^{-1}a + (1 - s^{-1}t^{-1})b, \quad a_{\overline{b}} = a\underline{\lfloor b} = s^{-1}a.$$

We shall refer to this structure, with the equations given above, as the *Alexander biquandle*.

Just as one can define the Alexander module of a classical knot, we have the Alexander biquandle of a virtual knot or link, obtained by taking one generator for each *edge* of the projected graph of the knot diagram and taking the module relations in the above linear form. Let $ABQ(K)$ denote this module structure for an oriented link K. That is, $ABQ(K)$ is the module generated by the edges of the diagram, factored by the submodule generated by the relations. This module then has a biquandle structure specified by the operations defined above for an Alexander biquandle.

The determinant of the matrix of relations obtained from the crossings of a diagram gives a polynomial invariant (up to multiplication by $\pm s^i t^j$ for integers i and j) of knots and links that we denote by $G_K(s, t)$ and call the *generalized Alexander polynomial. This polynomial vanishes on classical knots, but is remarkably successful at detecting virtual knots and links.* In fact, $G_K(s, t)$ is the same as the polynomial invariant of virtual knots of Sawollek [28, 29] and defined by an alternative method by Silver and Williams [30] and by yet another method by Manturov [23]. It is a reformulation of the invariant for knots in surfaces due to Kauffman, Jaeger and Saleur [31, 32].

We end this discussion of the Alexander biquandle with two examples that show clearly its limitations. View Figure 30.9. In this figure, we illustrate two diagrams labeled K and KI. It is not hard to calculate that both $G_K(s, t)$ and $G_{KI}(s, t)$ are equal to zero. However, the Alexander biquandle of K is non-trivial – it is isomorphic to the free module over $Z[s, s^{-1}, t, t^{-1}]$ generated by elements a and b subject to the relation $(s^{-1} - t - 1)(a - b) = 0$. Thus K represents a non-trivial virtual knot. This shows that it is possible for a non-trivial virtual diagram to be a connected sum of two trivial virtual diagrams. However, the diagram *KI* has a trivial Alexander biquandle. In fact the diagram *KI*, discovered by Kishino [33], is now known to be knotted and its general biquandle is non-trivial. The Kishino diagram has been shown to be non-trivial by a calculation of the three-strand Jones polynomial [33], by the surface bracket polynomial of Dye and Kauffman [34, 35], by the Ξ-polynomial (the surface generalization of the Jones polynomial of Manturov [36]), and its biquandle has been shown to be non-trivial by a quaternionic biquandle representation [37] which we will now briefly describe.

The quaternionic biquandle is defined by the following operations where $i^2 = j^2 = k^2 = ijk = -1$, $ij = -ji = k$, $jk = -kj = i$, $ki = -ik = j$ in the associative, non-commutative algebra of the quaternions. The elements a, b are in a module over the ring of integer quaternions:

$$a^b = a\,\overline{\mathrm{b}} = j \cdot a + (1 + i) \cdot b,$$
$$a_b = a\,\underline{\mathrm{b}} = -j \cdot a + (1 + i) \cdot b,$$
$$a^{\overline{b}} = a\lceil \mathrm{b} = j \cdot a + (1 - i) \cdot b,$$
$$a_{\overline{b}} = a\lfloor \mathrm{b} = -j \cdot a + (1 - i) \cdot b.$$

Amazingly, one can verify that these operations satisfy the axioms for the biquandle.

Equivalently, referring back to the previous section, define the linear biquandle by

$$S = \begin{pmatrix} 1 + i & jt \\ -jt^{-1} & 1 + i \end{pmatrix},$$

where i, j have their usual meanings as quaternions and t is a central variable. Let R denote the ring which they determine. Then as in the Alexander case considered above, for each diagram there is a square presentation of an R-module. We can take the (Study) determinant of the presentation matrix. In the case of the Kishino knot, this is zero. However, the greatest common divisor of the codimension 1 determinants is $2 + 5t^2 + 2t^4$, showing that this knot is not classical.

30.7 Flat Virtual Knots and Links

Every classical knot or link diagram can be regarded as a 4-regular plane graph with extra structure at the nodes. This extra structure is usually indicated by the over- and under-crossing conventions that give instructions for constructing an embedding of the link in three-dimensional space from the diagram. If we take the flat diagram without this extra structure, then the diagram is the shadow of some link in three-dimensional space, but the weaving of that link is not specified. It is well known that if one is allowed to apply the Reidemeister moves to such a shadow (without regard to the types of crossing since they are not specified) then the shadow can be reduced to a disjoint union of circles. This reduction is no longer true for virtual links. More precisely, let a *flat virtual diagram* be a diagram with *virtual crossings* as we have described them and *flat crossings* consisting in undecorated nodes of the 4-regular plane graph. Two flat virtual diagrams are *equivalent* if there is a sequence of generalized flat Reidemeister moves (as illustrated in Figure 30.1) taking one to the other. A generalized flat Reidemeister move is any move as shown in Figure 30.1 where one ignores the over or under crossing structure. Note that in studying flat virtuals, the rules for changing virtual crossings among themselves and the rules for changing flat crossings among themselves are identical. Detour moves as in part C of Figure 30.1 are available for virtual crossings with respect to flat crossings and *not* the other way around. The analogs of the forbidden moves of Figure 30.3 remain forbidden when the classical crossings are replaced by flat crossings.

The theory of flat virtual knots and links is identical to the theory of all oriented Gauss codes (without over or under information) modulo the flat Reidemeister moves. Virtual crossings are an artifact of the realization of the flat diagram in the plane. In Turaev's work [38] flat virtual knots and links are called *virtual strings*. See also recent papers of Roger Fenn [39, 40] for other points of view about flat virtual knots and links.

We shall say that a virtual diagram *overlies* a flat diagram if the virtual diagram is obtained from the flat diagram by choosing a crossing type for each flat crossing in the virtual diagram. To each virtual diagram K there is an associated flat diagram $F(K)$, obtained by forgetting the extra structure at the classical crossings in K. Note that if K and K' are isotopic as virtual diagrams, then $F(K)$ and $F(K')$ are isotopic as flat virtual diagrams. Thus, if we can show that $F(K)$ is not reducible to a disjoint union of circles, then it will follow that K is a non-trivial virtual link. The flat virtual diagrams present a challenge for the construction of new invariants. They are fundamental to the study of virtual knots. A virtual knot is necessarily non-trivial if its flat projection is a non-trivial flat virtual diagram. We wish to be able to determine when a given virtual link is isotopic to a classcal link. The reducibility or irreducibility of the underlying flat diagram is the first obstruction to such an equivalence.

Definition. A *virtual graph* is a flat virtual diagram where the classical flat crossings are not subjected to the flat Reidemeister moves. Thus a virtual graph is a $4-regular$ graph that is represented in the plane via a choice of cyclic orders at its nodes. The virtual crossings are artifacts of this choice of placement in the plane, and we allow detour moves for consecutive sequences of virtual crossings just as in the virtual knot theory. Two virtual graphs are *isotopic* if there is a combination of planar graph isotopies and detour moves that connect them. The theory of virtual graphs is equivalent to the theory of 4-regular graphs on oriented surfaces, taken up to empty handle stabilization, in direct analogy to our description of the theory of virtual links and flat virtual links.

A long knot or link is a $1 - 1$ tangle (see chapters in Part III). It is a tangle with one input end and one output end. In between one has, in the diagram, any knotting or linking, virtual or classical. Classical long knots (one component) carry essentially the same topological information as their closures. In particular, a classical long knot is knotted if and only if its closure is knotted. This statement is false for virtual knots. An example of the phenomenon is shown in Figure 30.11.

The long knots L and L' shown in Figure 30.11 are non-trivial in the virtual category. Their closures, obtained by attaching the ends together are unknotted virtuals. Concomitantly, there can be a multiplicity of long knots associated to a given virtual knot diagram, obtained by cutting an arc from the diagram and creating a $1 - 1$ tangle. *It is a fundamental problem to determine the kernel of the closure mapping from long virtual knots to virtual knots.* In Figure 30.11, the long knots L and L' are trivial as welded long knots (where the first forbidden move is allowed). The obstruction to untying them as virtual long knots comes from the first forbidden move. The matter of proving that L and L' are non-trivial distinct long knots is difficult by direct attack. There is a fundamental relationship between long flat virtual knots and long virtual knots that can be used to see it.

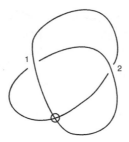

Flat Gauss Code = 1212

FIGURE 30.10: Parity for the virtual trefoil.

Let *LFK* denote the set of long flat virtual knots and let *LVK* denote the set of long virtual knots. We define

$$A : LFK \longrightarrow LVK$$

by letting $A(S)$ be the ascending long virtual knot diagram associated with the long flat virtual diagram S. That is, $A(S)$ is obtained from S by traversing S from its left end to its right end and creating a crossing at each flat crossing so that *one passes under each crossing before passing over that crossing*. Virtual crossings are not changed by this construction. The idea of using the ascending diagram to define invariants of long flat virtuals is exploited in [30].

Long Flat Embedding Theorem. The mapping $A : LFK \longrightarrow LVK$ is well-defined on the corresponding isotopy classes of diagrams, and it is injective. Hence *long flat virtual knots embed in the class of long virtual knots.*

Proof. If two flat long diagrams S, T are virtually isotopic, then $A(S)$ and $A(T)$ are isotopic long virtual knots since flat moves are taken to virtual isotopy moves. To see the injectivity, define

$$Flat : LVK \longrightarrow LFK$$

by letting $Flat(K)$ be the long flat diagram obtained from K by flattening all the classical crossings in K. By definition, $Flat(A(S)) = S$. The map *Flat* takes isotopic long virtual knots to isotopic flat virtual knots. This proves injectivity and completes the proof of the theorem.

Let $Inv(K)$ denote any invariant of long virtual knots. ($Inv(K)$ can denote a polynomial invariant, a number, a group, quandle, biquandle or other invariant structure.) Then we can define

$$Inv(S)$$

for any long flat knot S by the formula

$$Inv(S) = Inv(A(S)),$$

and this definition yields an invariant of long flat knots.

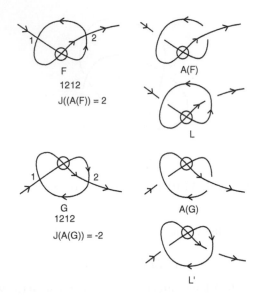

FIGURE 30.11: Ascending map.

30.8   Parity and the Odd Writhe

The Odd Writhe J(K). The *odd writhe* [41] of a virtual knot K (or long virtual knot) is denoted by $J(K)$ (see Chapter 32). It is the sum of the signs of the odd crossings. A crossing is *odd* if it flanks an odd number of symbols in the Gauss code of the diagram. It is easy to check that the odd writhe is invariant under all the Reidemeister moves and the detour move. Hence $J(K)$ is a virtual knot invariant. Classical diagrams have zero odd writhe. If K^* denotes the mirror image of the knot K, then $J(K^*) = -J(K)$.

In Figure 30.10, we illustrate the virtual trefoil K, the simplest virtual knot, and note that it has flat Gauss code 1212 and hence both of its crossings are odd. Thus $J(K) = 2$ and so K is non-classical, non-trivial, and inequivalent to its mirror image.

View Figure 30.11. We show the long flat F and its image under the ascending map, $A(F)$, are non-trivial. In fact, $A(F)$ is non-trivial and non-classical. One computes that $J(A(F))$ is non-zero where $J(K)$ denotes the odd writhe of K. In this case, the flat Gauss code for F is 1212 with both crossings odd. Thus we see from the figure that $J(A(F)) = 2$. Thus $A(F)$ is non-trivial, non-classical, and inequivalent to its mirror image. Once we check that $A(F)$ is non-trivial, we know that the flat knot F is non-trivial, and from this we conclude that the long virtual L is also non-trivial. Note that $J(L) = 0$, so we cannot draw this last conclusion directly from $J(L)$. This same figure illustrates a long flat G that is obtained by reflecting F in a horizontal line. Then, as the reader can calculate from this figure, $J(A(G)) = -2$. Thus F and G are distinct non-trivial long flats. We conclude from these arguments that the long virtual knots L and L' in Figure 30.11 are both non-trivial, and that

L is not virtually isotopic to L' (since such an isotopy would give an isotopy of F with G by the flattening map).

30.9   The Bracket Polynomial for Virtual Knots

In this section, we recall how the bracket state summation model [42] for the Jones polynomial [43, 44] is defined for virtual knots and links (see Chapter 61). In the next section, we give an extension of this model using orientation structures on the states of the bracket expansion. The extension is also an invariant of flat virtual links.

We call a diagram in the plane *purely virtual* if the only crossings in the diagram are virtual crossings. Each purely virtual diagram is equivalent by the virtual moves to a disjoint collection of circles in the plane.

A state S of a link diagram K is obtained by choosing a smoothing for each crossing in the diagram and labeling that smoothing with either A or A^{-1} according to the convention that a counterclockwise rotation of the overcrossing line sweeps two regions labeled A, and that a smoothing that connects the A regions is labeled by the letter A. Then, given a state S, one has the evaluation $< K|S >$ equal to the product of the labels at the smoothings, and one has the evaluation $\|S\|$ equal to the number of loops in the state (the smoothings produce purely virtual diagrams). One then has the formula

$$< K > = \Sigma_S < K|S > d^{\|S\|-1}$$

where the summation runs over the states S of the diagram K, and $d = -A^2 - A^{-2}$. This state summation is invariant under all classical and virtual moves except the first Reidemeister move. The bracket polynomial is normalized to an invariant $f_K(A)$ of all the moves by the formula $f_K(A) = (-A^3)^{-w(K)} < K >$ where $w(K)$ is the writhe of the (now) oriented diagram K. The writhe is the sum of the orientation signs (± 1) of the crossings of the diagram. The Jones polynomial, $V_K(t)$, is given in terms of this model by the formula

$$V_K(t) = f_K(t^{-1/4}).$$

This definition is a direct generalization to the virtual category of the state sum model for the original Jones polynomial. It is straightforward to verify the invariances stated above. In this way, one has the Jones polynomial for virtual knots and links.

In [7], we have the following theorem.

Theorem 3. *To each non-trivial, classical knot diagram of one component, K, there is a corresponding non-trivial, virtual knot diagram Virt(K) with unit Jones polynomial.*

Proof Sketch. This theorem is a key ingredient in the problems involving virtual knots. Here is a sketch of its proof. The proof uses two invariants of classical knots and links that generalize to arbitrary virtual knots and links. These invariants are the *Jones polynomial* and the *involutory quandle* denoted by the notation $IQ(K)$ for a knot or link K.

Given a crossing i in a link diagram, we define $s(i)$ to be the result of *switching* that crossing so that the undercrossing arc becomes an overcrossing arc and vice versa. We define the *virtualization* $v(i)$ of the crossing by the local replacement indicated in Figure 30.12. In this figure we illustrate how, in virtualization, the original crossing is replaced by a crossing that is flanked by two virtual crossings. When we smooth the two virtual crossings in the virtualization we obtain the original knot or link diagram with the crossing switched.

Suppose that K is a (virtual or classical) diagram with a classical crossing labeled i. Let $K^{v(i)}$ be the diagram obtained from K by virtualizing the crossing i while leaving the rest of the diagram just as before. Let $K^{s(i)}$ be the diagram obtained from K by switching the crossing i while leaving the rest of the diagram just as before. Then it follows directly from the expansion formula for the bracket polynomial that

$$V_{K^{s(i)}}(t) = V_{K^{v(i)}}(t).$$

As far as the Jones polynomial is concerned, switching a crossing and virtualizing a crossing look the same. We can start with a classical knot diagram K and choose a subset S of crossings such that the diagram is unknotted when these crossings are switched. Letting $Virt(K)$ denote the virtual knot diagram obtained by virtualizing each crossing in S, it follows that the Jones polynomial of $Virt(K)$ is equal to unity, the Jones polynomial of the unknot. Nevertheless, if the original knot K is knotted, then the virtual knot $Virt(K)$ will be non-trivial. We outline the argument for this fact below.

The involutory quandle [45] is an algebraic invariant equivalent to the fundamental group of the double-branched cover of a knot or link in the classical case (see Chapter 78). In this algebraic system, one associates a generator of the algebra $IQ(K)$ to each arc of the diagram K and there is a relation of the form $c = ab$ at each crossing, where ab denotes the (non-associative) algebra product of a and b in $IQ(K)$. See Figure 30.13. In this figure, we have illustrated the fact that

$$IQ(K^{v(i)}) = IQ(K).$$

As far as the involutory quandle is concerned, the original crossing and the virtualized crossing look the same.

If a classical knot is non-trivial, then its involutory quandle is non-trivial [46]. Hence, if we start with a non-trivial, classical knot and virtualize any subset of its crossings, we obtain a virtual knot that is still non-trivial. There is a subset A of the crossings of a classical knot K such that the knot SK obtained by switching these crossings is an unknot. Let $Virt(K)$ denote the virtual diagram obtained from A by virtualizing the crossings in this subset A. By the above discussion, the Jones polynomial of $Virt(K)$ is the same as the Jones polynomial of SK. Since SK is unknotted, $Virt(K)$ has unit Jones polynomial. The IQ of $Virt(K)$ is the same as the IQ of K, and hence if K is knotted, then so is $Virt(K)$. Thus, we have shown that $Virt(K)$ is a non-trivial virtual knot with unit Jones polynomial.

See Figure 30.14 for an example of a virtualized trefoil, the simplest example of a non-trivial, virtual knot with unit Jones polynomial. More work is needed to prove that the virtual knot, T, in this figure is not classical. In Section 30.10, we will give a proof of this fact by using an extension of the bracket polynomial.

It is an open problem whether there are non-trivial, classical knots with unit Jones polynomial. (There are non-trivial links whose linkedness is unseen [47, 48] by the Jones polynomial.) Can

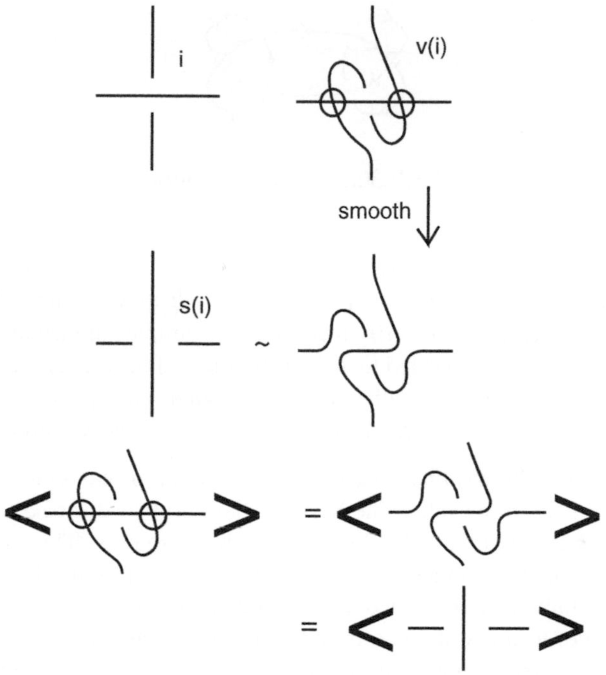

FIGURE 30.12: Switch and virtualize.

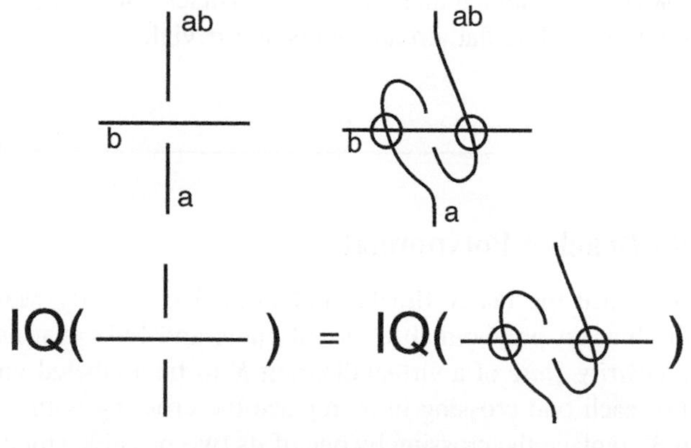

$$IQ\left(\ \Big|\ \right)\ =\ IQ\left(\ \oplus\!\!\!\!\!\!\!\oplus\ \right)$$

FIGURE 30.13: IQ(Virt).

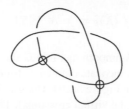

FIGURE 30.14: Virtualized trefoil with unit Jones polynomial.

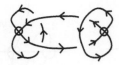

FIGURE 30.15: Kishino diagram.

one of the knots $Virt(K)$, produced as we have described above, be isotopic to a classical knot? Such examples are guaranteed to be non-trivial, but, in fact they are non-classical. This fact is proved in [49]. It is an intricate task to verify that specific examples of virtual knots with unit Jones polynomial are not classical. This has led to an investigation of new invariants for virtual knots. In this way, the search for classical knots with unit Jones polynomial expands to an exploration of the structure of the infinite collection of virtual knots with unit Jones polynomial.

In Figure 30.15, we show the *Kishino diagram K*. This diagram has unit Jones polynomial and its fundamental group is infinite cyclic. The Kishino diagram was discovered by Kishino in [50]. Many other invariants of virtual knots fail to detect the Kishino knot. Thus, it has been a test case for examining new invariants. Heather Dye and the author [51] have used the bracket polynomial defined for knots and links in a thickened surface (the state curves are taken as isotopy classes of curves in the surface) to prove the non-triviality and non-classicality of the Kishino diagram. In fact, we have used this technique to show that knots with unit Jones polynomial obtained by a single virtualization are non-classical. See the problem list by Fenn, Kauffman and Manturov [22] for other problems and proofs related to the Kishino diagram. In the next section, we describe a new extension of the bracket polynomial that can be used to discriminate the Kishino diagram, and, in fact, shows that its corresponding flat virtual knot is non-trivial.

30.10   The Parity Bracket Polynomial

In this section, we introduce the Parity Bracket Polynomial of Vassily Manturov [52]. This is a generalization of the bracket polynomial to virtual knots and links that uses the parity of the crossings. We define a *Parity State* of a virtual diagram K to be a labeled virtual graph obtained from K as follows: For each odd crossing in K, replace the crossing with a graphical node. For each even crossing in K, replace the crossing by one of its two possible smoothings, and label the smoothing site as A or A^{-1} in the usual way. Then, we can define the parity bracket by the state expansion formula

$$\langle K \rangle_P = \sum_S A^{n(S)}[S]$$

where $n(S)$ denotes the number of A-smoothings minus the number of A^{-1} smoothings and $[S]$ denotes a combinatorial evaluation of the state defined as follows: First, reduce the state by Reidemeister 2 moves on nodes, as shown in Figure 30.10. The graphs are taken up to virtual equivalence (planar isotopy plus detour moves on the virtual crossings). Then, regard the reduced state as a disjoint union of standard state loops (without nodes) and graphs that irreducibly contain nodes. With

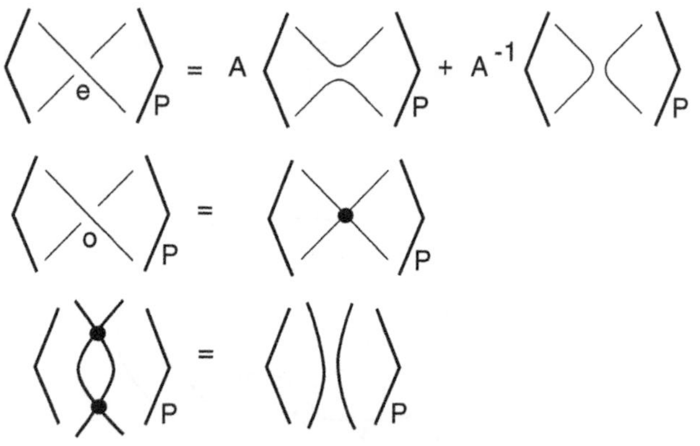

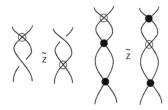

FIGURE 30.16: Parity bracket expansion.

FIGURE 30.17: Z-move and graphical Z-move.

this we write

$$[S] = (-A^2 - A^{-2})^{l(S)}[G(S)]$$

where $l(S)$ is the number of standard loops in the reduction of the state S and $[G(S)]$ is the disjoint union of reduced graphs that contain nodes. In this way, we obtain a sum of Laurent polynomials in A multiplying reduced graphs as the Manturov Parity Bracket. It is not hard to see that this bracket is invariant under regular isotopy and detour moves and that it behaves just like the usual bracket under the first Reidemeister move. However, the use of parity to make this bracket expand to graphical states gives it considerably more power in some situations. For example, consider the Kishino diagram in Figure 30.9. We see that all the classical crossings in this knot are odd. Thus, the parity bracket is just the graph obtained by putting nodes at each of these crossings. The resulting graph does not reduce under the graphical Reidemeister two moves, and so we conclude that the Kishino knot is non-trivial and non-classical. Since we can apply the parity bracket to a flat knot by taking $A = -1$, we see that this method shows that the Kishino flat virtual knot is non-trivial.

In Figure 30.17, we illustrate the *Z-move* and the *graphical Z-move*. Two virtual knots or links that are related by a Z-move have the same standard bracket polynomial. This follows directly from our discussion in the previous section. We would like to analyze the structure of Z-moves using the parity bracket. In order to do this, we need a version of the parity bracket that is invariant under the Z-move. In order to accomplish this, we need to add a corresponding Z-move in the graphical reduction process for the parity bracket. This extra graphical reduction is indicated in Figure 30.17, where we show a graphical Z-move. The reader will note that graphs that are irreducible without the

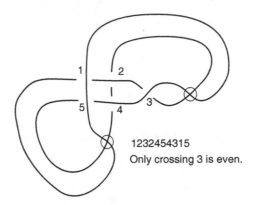

1232454315
Only crossing 3 is even.

FIGURE 30.18: A knot KS with unit Jones polynomial.

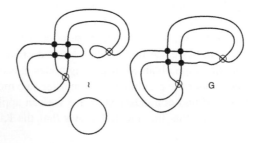

FIGURE 30.19: Parity bracket states for the knot KS.

graphical Z-move can become reducible if we allow graphical Z-moves in the reduction process. For example, the graph associated with the Kishino knot is reducible under graphical Z-moves. However, there are examples of graphs that are not reducible under graphical Z-moves and Reidemister 2 moves. An example of such a graph occurs in the parity bracket of the knot KS shown in Figure 30.18 and Figure 30.19. This knot has one even classical crossing and four odd crossings. One smoothing of the even crossing yields a state that reduces to a loop with no graphical nodes, while the other smoothing yields a state that is irreducible even when the Z-move is allowed. The upshot is that this knot, KS, is not Z-equivalent to any classical knot. Since one can verify that KS has unit Jones polynomial, this example is a counterexample to a conjecture of Fenn, Kauffman and Maturov [22] that suggested that a knot with unit Jones polynomial should be Z-equivalent to a classical knot.

Parity is clearly an important theme in virtual knot theory and will figure in many future investigations of this subject. The type of construction that we have indicated for the bracket polynomial in this section can be varied and applied to other invariants. Furthermore, the notion of describing a parity for crossings in a diagram is also susceptible to generalization. For more on this theme, the reader should consult [53] and [41] for our original use of parity for another variant of the bracket polynomial. See also Chapter 32.

Remark on Free Knots Manturov (See [52]) has defined the domain of *free knots*. A free knot is a Gauss diagram (or Gauss code) without any orientations or signs, taken up to abstract Reidemeister moves. See Chapter 34. The reader will see easily that free knots are the same as flat virtual knots modulo the flat Z-move. Furthermore, the parity bracket polynomial evaluated at $A = 1$ or $A = -1$ is an invariant of free knots. By using it on examples where all the crossings are odd, one obtains infinitely many examples of non-trivial free knots. This is just the beginning of an investigation into free knots that will surely lead to deep relationships between combinatorics and knot theory. Any free knot that is shown to be non-trivial has a number of non-trivial virtual knots overlying it. The free knots deserve to be studied for their own sake. Examples of new invariants of free knots that go beyond Gaussian parity can be found in the papers [54, 55].

30.11 The Arrow Polynomial for Virtual and Flat Virtual Knots and Links

This section describes an invariant for oriented virtual knots and links, and for flat oriented virtual knots and links that we call the *arrow polynomial* [13]. The arrow polynomial is related to constructions of Miyazawa and Kamada. The Miyazawa polynomial [63, 64, 65, 66, 67] is an independently discovered invariant that is equivalent to the arrow polynomial. The arrow polynomial is considerably stronger than the Jones polynomial for virtual knots and links, and it is a very natural extension of the Jones polynomial, using the oriented diagram structure of the state summation. The construction of the arrow polynomial invariant begins with the oriented state summation of the bracket polynomial. This means that each local smoothing is either an oriented smoothing or a *disoriented smoothing* as illustrated in Figure 30.20 and Figure 30.21. In [13] we show how the arrow polynomial can be used to estimate virtual crossing numbers.

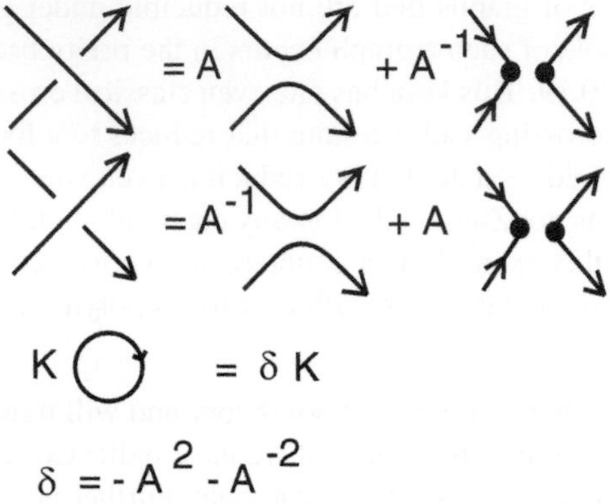

$$K \bigcirc = \delta K$$

$$\delta = -A^2 - A^{-2}$$

FIGURE 30.20: Oriented bracket expansion.

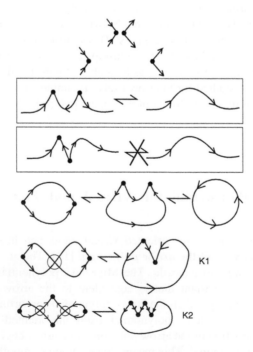

FIGURE 30.21: Reduction relation for the arrow polynomial.

In Figure 30.20, we illustrate the oriented bracket expansion for both positive and negative crossings in a link diagram. An oriented crossing can be smoothed in the oriented fashion or the disoriented fashion, as shown in Figure 30.20. We refer to these smoothings as *oriented* and *disoriented* smoothings. To each smoothing, we make an associated configuration that will be part of the arrow polynomial state summation. The configuration associated to a state with oriented and disoriented smoothings is obtained by applying the reduction rules described below. See Figure 30.21. The arrow polynomial state summation is defined by the formula:

$$\langle\langle K \rangle\rangle = A[K] = \Sigma_S \langle K|S \rangle d^{\|S\|-1}[S]$$

where S runs over the oriented bracket states of the diagram, $\langle K|S \rangle$ is the usual product of vertex weights as in the standard bracket polynomial, and $[S]$ is a product of extra variables $K_1, K_2, \cdots$ associated with the state S. These variables are explained below. Note that we use the designation $\langle\langle K \rangle\rangle$ to indicate the arrow polynomial.

Due to the oriented state expansion, the loops in the resulting states have extra combinatorial structure in the form of paired *cusps*, as shown in Figure 30.20. Each disoriented smoothing gives rise to a cusp pair where each cusp has either two oriented lines going into the cusp or two oriented lines leaving the cusp. We reduce this structure according to a set of rules that yields invariance of the state summation under the Reidemeister moves. The basic conventions for this simplification are shown in Figure 30.21. The reduction move in these figures corresponds to the elimination of two consecutive cusps on the same side of a single loop. Reduced diagrams with $2n$ cusps are denoted by K_n, as in the Figure 30.21.

The phenomenon of cusped states and extra variables, K_n, only occurs for virtual knots and links. Thus, an arrow polynomial with K_n variables ensures that a link is a non-classical virtual link.

There is a state summation for the arrow polynomial

$$A[K] = \Sigma_S < K|S > d^{\|S\|-1} V[S]$$

where S runs over the oriented bracket states of the diagram, $< K|S >$ is the usual product of vertex weights as in the standard bracket polynomial, $\|S\|$ is the number of circle graphs in the state S, and $V[S]$ is a product of the variables K_n.

Theorem. With the above conventions, the arrow polynomial $A[K]$ is a polynomial in A, A^{-1} and the graphical variables K_n (of which finitely many will appear for any given virtual knot or link). $A[K]$ is a regular isotopy invariant of virtual knots and links. The normalized version

$$W[K] = (-A^3)^{-wr(K)} A[K]$$

is a virtual isotopy invariant. If we set $A = 1$ and $d = -A^2 - A^{-2} = -2$, then the resulting specialization

$$F[K] = A[K](A = 1)$$

is an invariant of flat virtual knots and links.

Here is a first example of a calculation of the arrow polynomial invariant. View Figure 30.22. The virtual knot, K, in this figure has two crossings. One can see that this knot is a non-trivial virtual knot by calculating the odd writhe $J(K)$. We have that $J(K) = 2$, proving that K is non-trivial and non-classical. This is the simplest virtual knot, the analog of the trefoil knot for virtual

K
A
A
A
A^-1
A^-1
A
A^-1
A^-1
A^-1
<< >> = A^2 + (2 + A^-2 d) K1

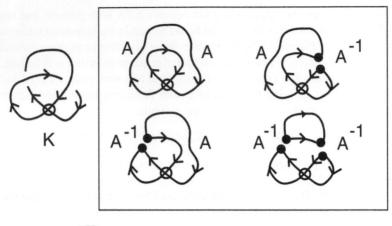

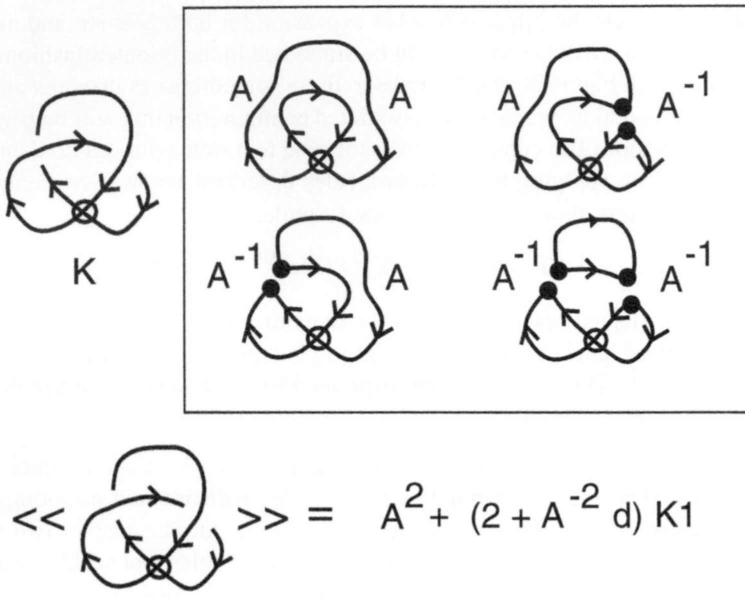

FIGURE 30.22: Arrow polynomial for virtual trefoil.

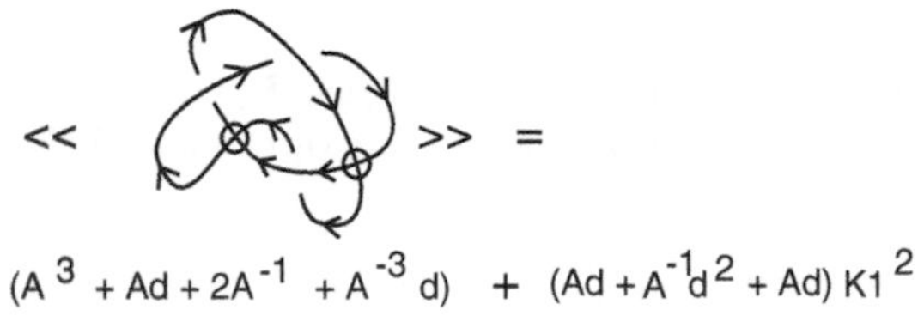

<< >> =
(A^3 + Ad + 2A^-1 + A^-3 d) + (Ad + A^-1 d^2 + Ad) K1^2

FIGURE 30.23: Arrow polynomial for the virtualized trefoil.

knot theory. The arrow polynomial gives an independent verification that K is non-trivial and non-classical.

The next example is given in Figure 30.23. Here, we calculate the arrow polynomial for a non-trivial virtual knot with unit Jones polynomial. Specialization of the calculation to $A = 1$ shows that the corresponding flat knot is non-trivial as well.

Figure 30.24 exhibits the calculation of the arrow polynomial for the Kishino diagram, showing once again that the Kishino knot is non-trivial and that its underlying flat diagram is also non-trivial.

For more information about the arrow polynomial, we refer the reader to [13] where it is proved that the maximal monomial degree in the K_n variables is a lower bound for the virtual crossing number of the virtual knot or link.

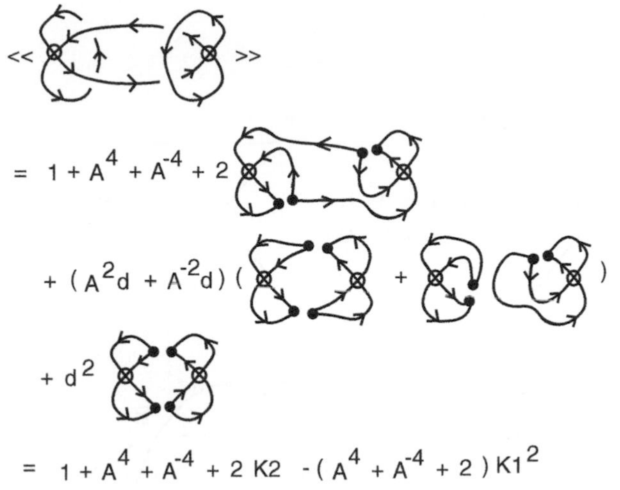

$$= 1 + A^4 + A^{-4} + 2$$

$$+ (A^2d + A^{-2}d) (\qquad + \qquad)$$

$$+ d^2$$

$$= 1 + A^4 + A^{-4} + 2 K2 - (A^4 + A^{-4} + 2) K1^2$$

FIGURE 30.24: Arrow polynomial for the Kishino diagram.

30.12 The Affine Index Polynomial Invariant

We define a polynomial invariant of virtual knots by first describing how to calculate the polynomial. We then justify that this definition is invariant under virtual isotopy. This section follows the definition of the affine index polynomial given in [56] and in [57]. These papers generalize the work in [58, 61] and the affine index polynomial is mathematically equivalent to invariants described independently in [59, 60]. The paper [62] shows that the affine index polynomial can be determined by the Sawollek Polynomial [28].

Calculation begins with a flat oriented virtual knot diagram (the classical crossings in a flat diagram do not have choices made for over or under). An *arc* of a flat diagram is an edge of the 4-regular graph that it represents. That is, an edge extends from one classical node to the next in orientation order. An arc may have many virtual crossings, but it begins at a classical node and ends at another classical node. We label each arc, c, in the diagram with an integer $\lambda(c)$ so that an arc that meets a classical node and crosses to the left increases the label by one, while an arc that meets a classical node and crosses to the right decreases the label by one. See Figure 30.25 for an illustration of this rule. We will prove that such integer labeling can always be done for any virtual or classical link diagram. In a virtual diagram, the labeling is unchanged at a virtual crossing, as indicated in Figure 30.25. One can start by choosing some arc to have an arbitrary integer label, and then proceed along the diagram labelling all the arcs via this crossing rule. We call such an integer labeling of a diagram a *integral coloring* of the diagram.

Given a labeled flat diagram, we define two numbers at each classical node, c: $W_-(c)$ and $W_+(c)$, as shown in Figure 30.25. If we have a labeled classical node with left incoming arc a and right incoming arc b, then the right outgoing arc is labeled $a-1$ and the left outgoing arc is labeled $b+1$, as shown in Figure 30.25. We then define

$$W_+(c) = a - (b + 1)$$

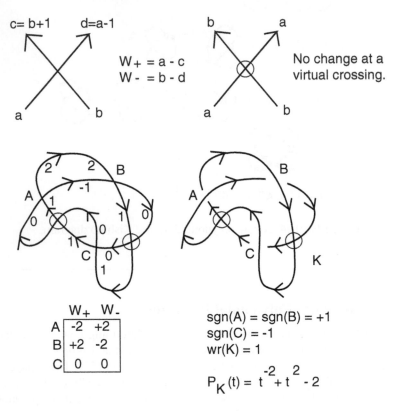

FIGURE 30.25: Affine index polynomial calculation.

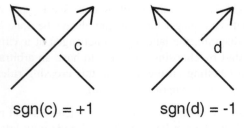

sgn(c) = +1   sgn(d) = -1

FIGURE 30.26: Crossing signs.

and

$$W_-(c) = b - (a - 1).$$

Note that

$$W_-(c) = -W_+(c)$$

in all cases.

Definition. Given a crossing c in a diagram K, we let $sgn(c)$ denote the sign of the crossing. The sign of the crossing is plus or minus one according to the convention shown in Figure 30.26. The *writhe*, $wr(K)$, of the diagram K is the sum of the signs of all its crossings. For a virtual link diagram, labeled in the integers according to the scheme above, and a crossing c in the diagram, define $W_K(c)$ by the equation

$$W_K(c) = W_{sgn(c)}(c),$$

where $W_{sgn(c)}(c)$ refers to the underlying flat diagram for K. Thus $W_K(c)$ is $W_\pm(c)$ according to whether the sign of the crossing is plus or minus. *We shall often indicate the weight of a crossing c in a knot diagram K by $W(c)$ rather than $W_K(c)$.*

Let K be a virtual knot diagram. Define the *affine index polynomial of K* by the equation

$$P_K = \sum_c sgn(c)(t^{W_K(c)} - 1) = \sum_c sgn(c)t^{W_K(c)} - wr(K),$$

where the summation is over all classical crossings in the virtual knot diagram K. We shall prove that the Laurent polynomial P_K is a highly non-trivial invariant of virtual knots.

In Figure 30.25, we show the computation of the weights for a given flat diagram and the computation of the polynomial for a virtual knot K with this underlying shadow. The knot K is an example of a virtual knot with unit Jones polynomial. The polynomial P_K for this knot has the value

$$P_K = t^{-2} + t^2 - 2,$$

showing that this knot is not isotopic to a classical knot.

In order to show the invariance and well-definedness of $P_K(t)$, one must first show the existence of labelings of flat virtual knot diagrams, as described above. This is done by showing that any virtual knot diagram K that overlies a given flat diagram D can be so labeled. One shows that any flat virtual knot diagram has an integral coloring of this type.

Let $\bar{K}$ denote the diagram obtained by reversing the orientation of K and let K^* denote the diagram obtained by switching all the crossings of K. $\bar{K}$ is called the *reverse* of K, and K^* is called the *mirror image* of K.

Theorem. Let K be a virtual knot diagram and $W_\pm(c)$ the crossing weights, as defined above. If α is an arc of K, let $\bar{\alpha}$ denote the corresponding arc of $\bar{K}$, the result of reversing the orientation of K.

1. The polynomial $P_K(t)$ is invariant under oriented virtual isotopy and is, hence, an invariant of virtual knots.

2. Let c be a crossing of K, and let $\bar{c}$ denote the corresponding crossing of $\bar{K}$. Then $W(\bar{c}) = -W(c)$.

3. Consequently, we have

$$P_{\bar{K}}(t) = P_K(t^{-1}).$$

Similarly, we have

$$P_{K^*}(t) = -P_K(t^{-1}).$$

Thus, this invariant changes t to t^{-1} when the orientation of the knot is reversed, and it changes global sign and t to t^{-1} when the knot is replaced by its mirror image.

4. If K is a classical knot diagram, then for each crossing c in K, $W(c) = 0$ and $P_K(t) = 0$.

30.13   Rotational Virtual Knots and Quantum Invariants

There are virtual link invariants corresponding to every quantum link invariant of classical links. However, this must be said with a caveat: We do not assume invariance under the first classical Reidemeister move (the quantum invariants are invariants of regular isotopy), and we do not assume invariance under the flat version of the first Reidemeister move in the list of virtual moves. Otherwise, the usual tensor or state sum formulas for quantum link invariants extend to this generalized notion of regular isotopy invariants of virtual knots and links. See [68] for more information about quantum invariants and virtual links. We define *rotational virtual knots and links* by taking virtual link diagrams up to all the moves in Figure 30.1 except the virtual Reidemeister 1 move ($vR1$). Then it is shown in [68] that every classical quantum link invariant, including those defined in [69, 27] by Hopf algebras, extends to an invariant of rotational virtual links.

See Figure 30.27 for a depiction of the structure of quantum invariants of virtual knots. An operator or matrix is associated to each crossing, including the virtual crossing, and to each maximum or minimum in the diagram. In the partition function (or tensor network) formulation of quantum invariants, one assigns all choices of index from a specified index set to the marked indices in the diagram, and takes the product of all the matrix elements. These products are then added up over all choices of index. See Figure 30.28 for an illustration of examples of non-trivial rotational virtual knots and links.

30.14   Welded Knots and Links

We have already defined welded knots and links as a quotient of virtual knot theory obtained by adding the first forbidden move $F1$ of Figure 30.3. See also Chapter 33. In this section, we describe a remarkable topological representation of welded knots due to Shin Satoh [70]. Satoh defines a

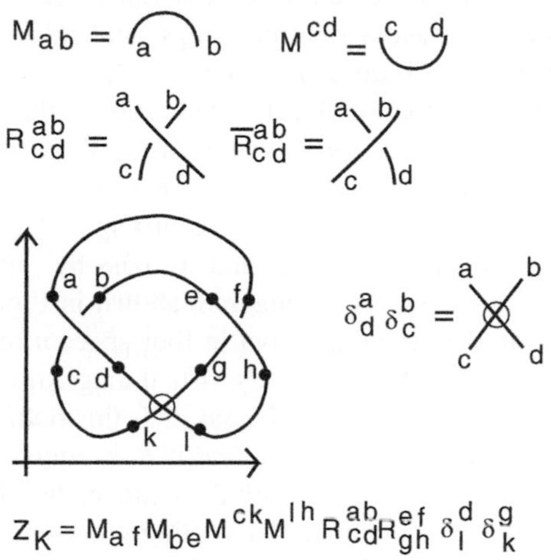

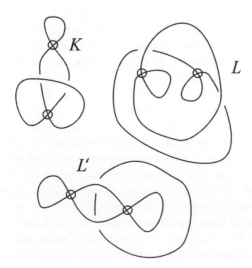

$$M_{ab} = \qquad M^{cd} =$$

$$R^{ab}_{cd} = \qquad \overline{R}^{ab}_{cd} =$$

$$\delta^a_d \, \delta^b_c =$$

$$Z_K = M_{af} M_{be} M^{ck} M^{lh} R^{ab}_{cd} R^{ef}_{gh} \, \delta^d_l \, \delta^g_k$$

FIGURE 30.27: Quantum invariants of virtuals.

K

L

L'

FIGURE 30.28: Non-trivial rotational virtuals.

mapping

$$Tube : WK \longrightarrow RTK$$

where WK denotes the collection of welded knots and links, represented by virtual diagrams, that is, by diagrams with classical and virtual crossings, and RTK denotes a class of "ribbon torus embeddings in R^4." Ribbon tori are embeddings of tori in Euclidean 4-space that can be represented by surface diagrams so that the local singularities in the diagram are all of the form shown in the left-hand part of Figure 30.29. Given a virtual diagram K, one forms the surface diagram for $Tube(K)$ by associating the cross product of a circle with each circular component of the diagram and associating a linking tube in 4-space according to the first part of the Figure 30.29 as in Figure 30.30 for each classical crossing, and a "pass-by" non-interaction of the tubes corresponding to each virtual crossing. In the diagram shown in Figure 30.29, we see one tube apparently over another tube in the virtual part, but in four-space it makes no difference whether they are depicted as over or under. In this way, every virtual diagram K is associated with a ribbon torus link $Tube(K)$ in four-dimensional space. We say that this defines a mapping from welded links to ribbon torus links because one can verify that if K is equivalent to K' via Reidemeister moves, virtual Reidemeister moves and the first forbidden move, then $Tube(K)$ is ambient isotopic to $Tube(K')$ in R^4. Furthermore, the combinatorial Wirtinger presentation fundamental group for any welded diagram K is isomorphic to the topological fundamental group of the complement of $Tube(K)$ in four-space. Thus it is a promising notion that the ribbon torus knots in four-space may be a complete representation for welded knots. At this writing, it is not known if there are distinct welded diagrams K and K' such that $Tube(K)$ and $Tube(K')$ are isotopic surface embeddings in four-space.

Another approach to the torus representation of welded knots was discovered by Colin Rourke [71, 72]. Rourke's approach constructs a toral embedding associated with a diagram K by taking $R^4 = R^2 \times R^2$, where the first factor is the plane for the diagram K. Thus, one views R^4 as a fibration over the diagram plane R^2 such that there is an R^2-fiber over each point p of the diagram plane. Rourke then places a circle in the fiber over each point in the diagram, and two circles in each fiber over a double point of the diagram. In the case when the double point is a virtual crossing, the two circles are adjacent in the fiber plane. In the case when the double point is a classical crossing, the two circles are concentric in the fiber plane with the smaller circle corresponding to the point on the undercrossing arc. It is not hard to see that such an assignment of circles in the fiber planes over the diagram K fit together to form a smoothly embedded torus in $R^4 = R^2 \times R^2$ that is essentially the same as the Satoh tube. See [73] for further discussion of this point. Let $R(K)$ denote this Rourke embedding of a torus in four-space associated with a welded knot diagram K. Rourke shows that *fiberwise locally-trival fiberwise isotopy classes of the tori $R(K)$ are in $1 - 1$ correspondence with framed welded classes of welded knots*. Framed welded knots are the equivalence classes of welded diagrams with neither the virtual, nor classical Reidemeister 1 moves allowed. Locally-trival fiberwise isotopies can always be represented (according to Rourke) by a regular homotopy of the corresponding curve in the diagram plane. In [73], it is shown that by changing the notion of fiberwise isotopy, we can obtain a similar theorem for *rotwelded* classes of welded knot diagrams. Rotwelded knot theory is analogous to rotational virtual knot theory. Rotwelded equivalence includes all the welded moves except the virtual Reidemeister 1 move. These fiberwise theories of welded knots may yield insight into the problem of understanding Satoh's Tube mapping.

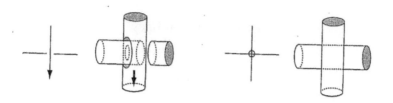

FIGURE 30.29: Satoh representation of welded links.

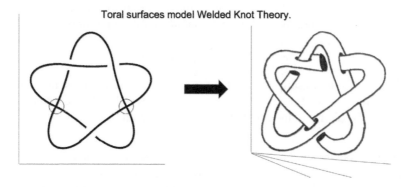

Toral surfaces model Welded Knot Theory.

FIGURE 30.30: Satoh representation of a welded diagram.

30.15 Khovanov Homology

In this section, we describe Khovanov homology (see Chapter 70 along the lines of [74, 75], and we tell the story so that the gradings and the structure of the differential emerge in a natural way. This approach to motivating the Khovanov homology uses elements of Khovanov's original approach, Viro's use of enhanced states for the bracket polynomial [76], and Bar-Natan's emphasis on tangle cobordisms [77].

We begin by working without using virtual crossings, and then we show how to introduce extra structure and generalize the Khovanov homology to virtual knots and links.

Two key motivating ideas are involved in finding the Khovanov invariant. First of all, one would like to *categorify* a link polynomial such as $\langle K \rangle$. There are many meanings to the term categorify, but here the quest is to find a way to express the link polynomial as a *graded Euler characteristic* $\langle K \rangle = \chi_q \langle H(K) \rangle$ for some homology theory associated with $\langle K \rangle$.

The bracket polynomial [42, 45] model for the Jones polynomial [43, 78, 79, 44] is usually described by the expansion

$$\langle \times \rangle = A \langle \asymp \rangle + A^{-1} \langle \rangle \langle \rangle \tag{30.1}$$

and we have

$$\langle K \bigcirc \rangle = (-A^2 - A^{-2})\langle K \rangle \qquad (30.2)$$

$$\langle \diagup\!\!\diagdown \rangle = (-A^3)\langle \smile \rangle \qquad (30.3)$$

$$\langle \diagdown\!\!\diagup \rangle = (-A^{-3})\langle \smile \rangle \qquad (30.4)$$

Letting $c(K)$ denote the number of crossings in the diagram K, if we replace $\langle K \rangle$ by $A^{-c(K)}\langle K \rangle$, and then replace A by $-q^{-1}$, the bracket will be rewritten in the following form:

$$\langle \times \rangle = \langle \asymp \rangle - q\langle \rangle\langle \rangle \qquad (30.5)$$

with $\langle \bigcirc \rangle = (q + q^{-1})$. It is useful to use this form of the bracket state sum for the sake of the grading in the Khovanov homology (to be described below). We shall continue to refer to the smoothings labeled q (or A^{-1} in the original bracket formulation) as *B-smoothings*.

We should further note that we use the well-known convention of *enhanced states* where an enhanced state has a label of 1 or X on each of its component loops. We then regard the value of the loop $q + q^{-1}$ as the sum of the value of a circle labeled with a 1 (the value is q) added to the value of a circle labeled with an X (the value is q^{-1}). We could have chosen the more neutral labels of $+1$ and -1 so that

$$q^{+1} \Longleftrightarrow +1 \Longleftrightarrow 1$$

and

$$q^{-1} \Longleftrightarrow -1 \Longleftrightarrow X,$$

but, since an algebra involving 1 and X naturally appears later, we take this form of labeling from the beginning.

To see how the Khovanov grading arises, consider the form of the expansion of this version of the bracket polynomial in enhanced states. We have the formula as a sum over enhanced states s :

$$\langle K \rangle = \sum_s (-1)^{n_B(s)} q^{j(s)}$$

where $n_B(s)$ is the number of B-type smoothings in s, $\lambda(s)$ is the number of loops in s labeled 1 minus the number of loops labeled X, and $j(s) = n_B(s) + \lambda(s)$. This can be rewritten in the following form:

$$\langle K \rangle = \sum_{i,j} (-1)^i q^j \, dim C^{ij})$$

where we define C^{ij} to be the linear span of the set of enhanced states with $n_B(s) = i$ and $j(s) = j$. Then the number of such states is the dimension $dim(C^{ij})$.

We would like to have a bigraded complex composed of the C^{ij} with a differential

$$\partial : C^{ij} \longrightarrow C^{i+1\,j}.$$

The differential should increase the *homological grading i* by 1 and preserve the *quantum grading j*. Then we could write

$$\langle K \rangle = \sum_j q^j \sum_i (-1)^i dim(C^{ij}) = \sum_j q^j \chi(C^{\bullet\, j}),$$

where $\chi(C^{\bullet\, j})$ is the Euler characteristic of the subcomplex $C^{\bullet\, j}$ for a fixed value of j.

This formula would constitute a categorification of the bracket polynomial. Below, we shall see how *the original Khovanov differential ∂ is uniquely determined by the restriction that $j(\partial s) = j(s)$ for each enhanced state s.* Since j is preserved by the differential, these subcomplexes $C^{\bullet\, j}$ have their own Euler characteristics and homology. We have

$$\chi(H(C^{\bullet\, j})) = \chi(C^{\bullet\, j})$$

where $H(C^{\bullet\, j})$ denotes the homology of the complex $C^{\bullet\, j}$. We can write

$$\langle K \rangle = \sum_j q^j \chi(H(C^{\bullet\, j})).$$

The last formula expresses the bracket polynomial as a *graded Euler characteristic* of a homology theory associated with the enhanced states of the bracket state summation. This is the categorification of the bracket polynomial. Khovanov proves that this homology theory is an invariant of knots and links (via the Reidemeister moves), creating a new and stronger invariant than the original Jones polynomial.

The differential is based on regarding two states as *adjacent* if one differs from the other by a single smoothing at some site. Thus if (s, τ) denotes a pair consisting of an enhanced state s and site τ of that state with τ of type A, then we consider all enhanced states s' obtained from s by resmoothing at such a τ from A to B, and re-labeling only those loops that are affected by the resmoothing. Call this set of enhanced states $S'[s, \tau]$. Then we shall define the *partial differential $\partial_\tau(s)$* as a sum over certain elements in $S'[s, \tau]$, and the differential for the complex by the formula

$$\partial(s) = \sum_\tau (-1)^{c(s,\tau)} \partial_\tau(s)$$

with the sum over all type A sites τ in s. Here $c(s, \tau)$ denotes the number of A-smoothings prior to the A smoothing in s that is designated by τ. Priority is defined by an initial choice of order for the crossings in the knot or link diagram.

In Figure 30.31 we indicate the original forms of the states for the bracket (not labeled yet by 1 or x) and their arrangement as a Khovanov category where the generating morphisms are arrows from one state to another where the domain of the arrow has one more A-state than the target of that arrow. In this figure we have assigned an order to the crossings of the knot, and so the reader can see from it how to define the signs for each partial differential in the complex.

We now explain how to define $\partial_\tau(s)$ so that $j(s)$ is preserved.

Proposition. The partial differentials $\partial_\tau(s)$ are uniquely determined by the condition that $j(s') =$

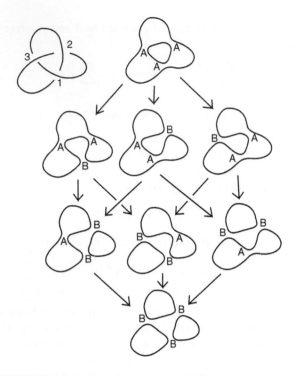

FIGURE 30.31: Bracket states and Khovanov complex.

$j(s)$ for all s' involved in the action of the partial differential on the enhanced state s. This unique form of the partial differential can be described by the following structure of multiplication and comultiplication on the algebra $A = k[X]/(X^2)$ where $k = Z$ for integral coefficients.

1. The element 1 is a multiplicative unit and $X^2 = 0$.

2. $\Delta(1) = 1 \otimes X + X \otimes 1$ and $\Delta(X) = X \otimes X$.

These rules describe the local re-labeling process for loops in a state. Multiplication corresponds to the case where two loops merge to a single loop, while comultiplication corresponds to the case where one loop bifurcates into two loops. (The proof is omitted.)

There is more to say about the nature of this construction with respect to Frobenius algebras and tangle cobordisms. The partial boundaries can be conceptualized in terms of surface cobordisms. The equality of mixed partials corresponds to topological equivalence of the corresponding surface cobordisms, and to the relationships between Frobenius algebras and the surface cobordism category. The proof of invariance of Khovanov homology with respect to the Reidemeister moves (respecting grading changes) will not be given here. See [74, 75, 77]. It is remarkable that this version of Khovanov homology is uniquely specified by natural ideas about adjacency of states in the bracket polynomial.

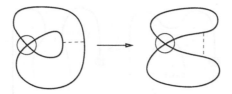

FIGURE 30.32: Single cycle smoothing.

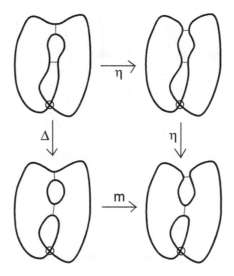

FIGURE 30.33: Khovanov complex for the two-crossing virtual unknot.

30.16   Khovanov Homology for Virtual Links

Extending Khovanov homology to virtual knots for arbitrary coefficients is complicated by the single cycle smoothing as depicted in Figure 30.32. We define a map for this smoothing $\eta : V \longrightarrow V$. In order to preserve the quantum grading $(j(s') = j(s))$ η is the zero map.

Remark. Here we consider only the homology theory related to $\eta(a) = 0$ for all a and thus will still call such a system a Frobenius system and the associated algebra a Frobenius algebra.

Consider the following complex in Figure 30.33 arising from the 2-crossing virtual unknot:

Composing along the top and right we have

$$\eta \circ \eta = 0.$$

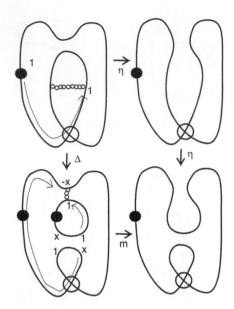

FIGURE 30.34: Transport across virtual crossings.

But composing along the opposite sides we see

$$m \circ \Delta(1) = m(1 \otimes X + X \otimes 1) = X + X = 2X.$$

Hence the complex does not naturally commute or anti-commute.

When the base ring is $Z/2Z$, the definition of Khovanov homology given in the previous section goes through unchanged. Manturov [80] (see also[81], [82]) introduced a definition of Khovanov homology for (oriented) virtual knots with arbitrary coefficients. Dye, Kaestner and Kauffman [49] reformulate Manturov's definition and give applications of this theory, finding a generalization of the Rasmussen invariant and prove that virtual links with all positive crossings have a generalized four-ball genus equal to the genus of the virtual Seifert spanning surface. The method used in these papers to create an integral version of Khovanov homology involves using the properties of the virtual diagram to make corrections in the local boundary maps so that the individual squares commute (or anti-commute).

Here we give a simplified version that obtains the same theory by starting with unoriented diagrams. This definition is much simpler than the definitions given in [80] and [49]. These papers started with oriented states for the Khovanov homology following patterns that began in analysis of the Khovanov-Rozansky $sl(n)$ homology theory [83, 84]. The original formulation of Khovanov homology as in [74] uses unoriented states. It took some time for us to realize that the construction for integral Khovanov homology can be done in this unoriented and simpler manner. This formulation will also be used in a paper of Kauffman and Ogasa that is in preparation [85].

The simplified definition proceeds as follows. We set a base-point on each loop of each state as indicated by the black dots in Figure 30.34. Algebra to be processed by the local boundary maps is placed initially at these base-points. Then it is transferred to the site where the map occurs (either joining two loops at that site, or splitting one loop into two at that site). Taking a path along the diagram from this base-point to the site, one will pass either an even number of virtual crossings or an odd number. If the parity is even, then both x and 1 transport to x and 1 respectively. If the parity is odd, then x is transported to $-x$ and 1 is transported to 1. The local boundary map is performed on the algebra as transported to the site, and then in the image state, the result is transported back to the base-point(s). This transport schema modifies the local boundary maps in the complex so that all squares commute. We then define the signs for the full boundary maps exactly as we have done in the standard classical integral Khovanov homology.

In Figure 30.34 we have illustrated the situation where the top left state is labeled with 1 and in the left vertical column we have $\Delta(1) = 1 \otimes x + x \otimes 1$. We show how the initial element 1 appears at the base-point of the upper left state and how it is transported (as a 1) to the site for the co-multiplication. The result of the co-multiplication is $1 \otimes x + x \otimes 1$ and this is shown at the re-smoothed site. To perform the next local boundary map, we have to transport this algebra to a multiplication site. This result of the co-multiplcation is to be transported back to base-points and then to the new site of multiplication for the next composition of maps. In the figure we illustrate the transport just for $x \otimes 1$. At the new site this is transformed to $(-x) \otimes 1$. Notice that the x in this transport moves through a single virtual crossing. We leave it for the reader to see that the transport of $1 \otimes x$ has even parity for both elements of the tensor product. Thus $1 \otimes x + x \otimes 1$ is transported to $1 \otimes x + (-x) \otimes 1$ at the multiplication site. Upon multiplying we have $m(1 \otimes x + (-x) \otimes 1) = x - x = 0$. Thus we now have that the composition of the left sides of the square is equal to the given zero composition of the right-hand side of the square (which is a composition to two zero single cycle maps). Note that in applying transport for composed maps, we can transport directly from one site to another without going back to the base-point and then to the second site.

Other definitions for virtual Khovanov homology have been given by [86] and [87]. Each of these definitions gives different solutions to handling the difficult morphism square discussed above.

30.17 Virtual Knot Cobordism

The definitions and basic material from this section are from [88, 89, 90]. See also [49, 93] and Chapter 31.

Two oriented knots or links K and K' are *virtually cobordant* if one may be obtained from the other by a sequence of virtual isotopies (Reidemeister moves plus detour moves) plus births, deaths and oriented saddle points, as illustrated in Figure 30.35. A *birth* is the introduction into the diagram of an isolated unknotted cycle. A *death* is the removal from the diagram of an isolated unknotted cycle. A saddle point move results from bringing oppositely oriented arcs into proximity and resmoothing the resulting site to obtain two new oppositely oriented arcs. See Figure 30.35 for an illustration of the process. Figure 30.35 also illustrates the *schema* of surfaces that are generated by the cobordism process. These are abstract surfaces with well-defined genus in terms of the sequence of steps in the cobordism. In Figure 30.35 we illustrate two examples of genus zero, and one example of genus 1. We say that a cobordism has genus g if its schema has that genus. Two knots are *concordant* if there is a cobordism of genus zero connecting them. A virtual knot is said to be a *slice* knot if it is virtually concordant to the unknot, or equivalently if it is virtually concordant to the empty knot. (The unknot is concordant to the empty knot via one death.) Finally we define the *virtual slice genus* (or simply *slice genus* when in the virtual category) of a knot or link to be the minimal genus of all such schema between the knot or link and the unknot.

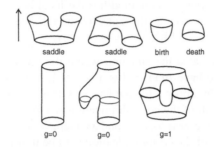

FIGURE 30.35: Saddles, births and deaths.

In Figure 30.36 we illustrate the *virtual Stevedore's knot, VS,* and show that it is a slice knot in the sense of the above definition. We will use this example to illustrate our theory of virtual knot cobordism, and the questions that we are investigating.

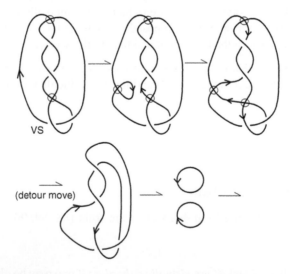

FIGURE 30.36: Virtual Stevedore is slice.

The virtual Stevedore (VS) is an example that illustrates the viability of this theory. One can prove that VS is not classical by showing that it is represented on a surface of genus one and no smaller. The technique for this is to use the bracket expansion on a toral representative of VS and examine the structure of the state loops on that surface. See [88].

30.17.1   Virtual Knot Cobordisms and Seifert Surfaces

It is well-known that every oriented classical knot or link bounds an embedded orientable surface in three-space. A representative surface of this kind can be obtained by the algorithm due to Seifert (See [92]). We have illustrated Seifert's algorithm for a trefoil diagram in Figure 30.37 . The algorithm proceeds as follows: At each oriented crossing in a given diagram K, smooth that crossing in the oriented manner (reconnecting the arcs locally so that the crossing disappears and the connections respect the orientation). The result of this operation is a collection of oriented simple closed curves in the plane, usually called the *Seifert circles*. To form the *Seifert surface* $F(K)$ for the diagram K, attach disjoint discs to each of the Seifert circles, and connect these discs to one another by local half-twisted bands at the sites of the smoothing of the diagram. This process is indicated in the Figure 30.37. In that figure we have not completed the illustration of the outer disc.

It is important to observe that we can calculate the genus of the resulting surface quite easily from the combinatorics of the classical knot diagram K. For purposes of simplicity, we shall assume that we are dealing with a knot diagram (one boundary component) and leave the case of links to the reader.

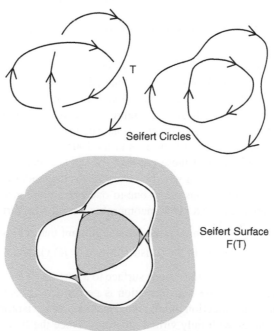

FIGURE 30.37: Classical Seifert surface.

Lemma. Let K be a classical knot diagram with n crossings and r Seifert circles. Then the genus of the Seifert Surface $F(K)$ is given by the formula

$$g(F(K)) = \frac{(-r + n + 1)}{2}.$$

Proof. The proof is omitted.

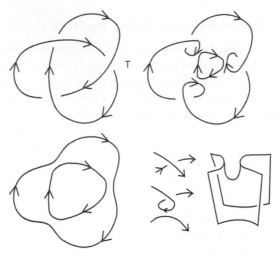

Every classical knot diagram bounds a surface in the four-ball
whose genus is equal to the genus of its Seifert Surface.

FIGURE 30.38: Classical cobordism surface.

We now observe that *for any classical knot K, there is a surface bounding that knot in the four-ball that is homeomorphic to the Seifert surface.* One can construct this surface by pushing the Seifert surface into the four-ball keeping it fixed along the boundary. We will give here a different description of this surface as indicated in Figure 30.38. In that figure we *perform a saddle point transformation at every crossing of the diagram.* The result is a collection of unknotted and unlinked curves. By our interpretation of surfaces in the four-ball obtained by saddle moves and isotopies, we can then bound each of these curves by discs (via deaths of circles) and obtain a surface $S(K)$ embedded in the four-ball with boundary K. As the reader can easily see, the curves produced by the saddle transformations are in one-to-one correspondence with the Seifert circles for K, and it is easy to verify that $S(K)$ is homeomorphic with the Seifert surface $F(K)$. Thus we know that $g(S(K)) = \dfrac{(-r+n+1)}{2}$. In fact, the same argument that we used to analyze the genus of the Seifert surface applies directly to the construction of $S(K)$ via saddles and minima.

Now the stage is set for generalizing the Seifert surface to a surface $S(K)$ for virtual knots K. View Figure 30.39. In this figure a saddle transformation is performed at each classical crossing of a virtual knot K. The result is a collection of unknotted curves that are isotopic (by the first classical Reidemeister move) to curves with only virtual crossings. Once the first Reidemeister moves are performed, these curves are identical with the *virtual Seifert circles* obtained from the diagram K by smoothing all of its classical crossings. One then isotopes these circles into a disjoint collection of circles (since they have no classical crossings) and caps them with discs. The result is a virtual surface $S(K)$ whose boundary is the given virtual knot K. We use the terminology *virtual surface in the four-ball* for this surface schema. In the case of a virtual slice knot, one has that the knot bounds a virtual surface of genus zero. We have the following lemma.

Lemma 1. Let K be a virtual knot diagram, then the virtual Seifert surface $S(K)$ constructed above has genus given by the formula

$$g(S(K)) = \frac{(-r+n+1)}{2}$$

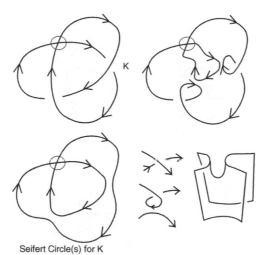

Seifert Circle(s) for K
Every virtual diagram K bounds a virtual orientable surface of
genus g = (1/2)(-r + n +1) where r is the number of Seifert circles,
and n is the number of classical crossings in K.
This virtual surface is the cobordism Seifert surface when K
is classical.

FIGURE 30.39: Virtual cobordism Seifert surface.

where r is the number of virtual Seifert circles in the diagram K and n is the number of classical crossings in the diagram K.

Remark. For a different approach to Seifert surfaces for virtual knots, see [91]. For a graph-theoretic classification of the checker-board colorable virtual diagrams, see [99].

A Rasmussen Invariant for Virtual Knot Cobordisms. Using this version of integral Khovanov homology, we have a generalization of a result of Rasmussen [94] from positive classical knots to positive virtual knots.

Theorem[Generalization of Rasmussen [94]] *If K is a positive virtual knot and $g_s(K)$ is its virtual 4-ball genus, then $g_s(K) = \dfrac{(-r + n + 1)}{2}$ where r is the number of virtual Seifert circles for K and n is the number of classical crossings for K.*

This result is proved in [49].

Remark 1. This theorem generalizes Rasmussen's result, which gave a combinatorial proof of a conjecture by Milnor regarding the genus of torus knots.

This theorem about the 4-ball genus of virtual knots is illustrated by the examples in Figure 30.40 and Figure 30.41. Here we use the fact that the affine index polynomial is a concordance invariant for virtual knots [89]. The knot K_+ is positive and by the above theorem on the genus of positive

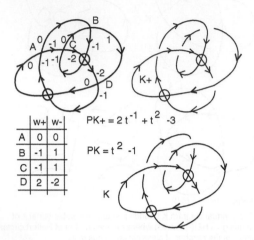

FIGURE 30.40: Polynomial calculation for two knots.

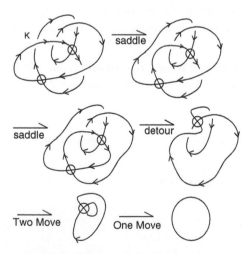

K bounds a virtual surface of genus one.

FIGURE 30.41: The knot K has virtual genus one.

virtual knots, it has genus two. The knot K is obtained from K_+ by switching one crossing. The affine index polynomial shows that it is not slice, and Figure 30.41 shows that K bounds a genus one virtual surface. Thus we know, using the affine index polynomial, that K has genus equal to one.

Finally, an interesting question is: "Can the extension from the category of classical knots to virtual knots lower the slice genus?" While we cannot give a complete answer, for many classes of knots we can say this is false. In particular, since the Rasmussen invariant presented here agrees with the Rasmussen invariant as originally defined for classical knots, any classical knot whose slice genus equals its Rasmussen invariant has the same slice genus in the virtual category. For instance, any (p,q)-torus knot has slice genus $\dfrac{(p-1)(q-1)}{2}$ in the virtual category.

30.18   Relations with Classical Knot Theory

This summary of virtual knot theory has concentrated on the virtual theory itself and the way that theory relates to classical knots embedded in thickened surfaces of genus greater than one. Nevertheless, it is of great interest to ask how virtual knot theory is related to classical knot theory, the case of supporting genus zero. At the present time there are significant relationships with the theory of knotoids originated by Vladimir Turaev [96]. Specific applications of virtual knot theory to knotoids can be found in [97, 98]. Another application occurs in the work of Chrisman and Manturov [100] where they study classical fibered knots by using the virtual knot theory supported by the fiber surface for the fibered knot. Finally, there are many concepts that are in the interface between classical ideas and virtual ideas. These include formulations of Seifert surfaces as we have seen earlier in this exposition, and the intricate role of parity in virtual knot theory that will undoubtedly have its influence in classical theory more widely as time goes on.

30.19   Bibliography

[1] M. Goussarov, M. Polyak, O. Viro. Finite-type invariants of classical and virtual knots. *Topology* 39 (2000), no. 5, 1045–1068.

[2] L. H. Kauffman, Virtual knot theory. *European J. Comb.* **20** (1999), 663–690.

[3] L. H. Kauffman, A survey of virtual knot theory, in *Proceedings of Knots in Hellas '98*, World Sci. Pub. 2000 , 143–202.

[4] S. Nelson. Unknotting virtual knots with Gauss diagram forbidden moves. *J. Knot Theory Ramifications* 10 (2001), no. 6, 931–935.

[5] S. Kamada, Braid presentation of virtual knots and welded knots. *Osaka J. Math.* 44 (2007), no. 2, 441–458.

[6] R. Fenn, R. Rimanyi and C. Rourke, The braid permutation group. *Topology*, Vol. 36, No. 1 (1997), 123–135.

[7] L. H. Kauffman, Detecting virtual knots, in *Atti. Sem. Mat. Fis. Univ. Modena Supplemento al Vol. IL*, 241–282 (2001).

[8] L. H. Kauffman, Knot diagrammatics, in *Handbook of Knot Theory*, edited by Menasco and Thistlethwaite, 233–318, Elsevier B. V., Amsterdam, 2005. math.GN/0410329.

[9] J. S. Carter, S. Kamada and M. Saito, Stable equivalences of knots on surfaces and virtual knot cobordisms. Knots 2000 Korea, Vol. 1, (Yongpyong), *JKTR* **11** (2002) No. 3, 311–322.

[10] N. Kamada and S. Kamada, Abstract link diagrams and virtual knots. *JKTR* **9** (2000), No. 1, 93–106.

[11] G. Kuperberg, What is a virtual link? *Algebr. Geom. Topol.* **3** (2003), 587–591.

[12] V. Manturov. *Knot Theory*. Chapman & Hall, CRC, Boca Raton, FL, 2004.

[13] H. A. Dye and L. H. Kauffman, Virtual crossing number and the arrow polynomial, *JKTR*, Vol. 18, No. 10 (October 2009), 1335–1357.

[14] L. H. Kauffman and S. Lambropoulou, Virtual braids. *Fund. Math.* 184 (2004), 159–186.

[15] L. H. Kauffman and S. Lambropoulou, Virtual braids and the L-move, *JKTR* **15**, No. 6, August 2006, 773–810.

[16] L. H. Kauffman and S. Lambropoulou. A categorical structure for the virtual braid group. *Communications in Algebra*, Volume 39(12):4679–4704, 2011.

[17] V.G. Bardakov, Y. A. Mikhalchishina and Mikhail V. Neshchadim. Representations of virtual braids by automorphisms and virtual knot groups. *Journal of Knot Theory and Its Ramifications* Vol. 26, No. 1 (2017) 1750003 (17 pages).

[18] V.G. Bardakov, Y. A. Mikhalchishina, M. V. Neshchadim. Groups of the virtual trefoil and Kishino knots. *J. Knot Theory Ramifications* 27 (2018), no. 13, 1842009, 20 pp.

[19] V. G. Bardakov, R. Mikhailov, V. V. Vershinin, J. Wu. On the pure virtual braid group PV3. *Comm. Algebra* 44 (2016), no. 3, 1350–1378.

[20] V. G. Bardakov, R. Mikhailov, V. V. Vershinin, J. Wu. On the pure virtual braid group PV3. *Comm. Algebra* 44 (2016), no. 3, 1350–1378.

[21] H. U. Boden, R. Gaudreau, E. Harper, A. J. Nicas, Lindsay White. Virtual knot groups and almost classical knots. *Fundamenta Mathematicae* 238 (2017), 101–142.

[22] L. H. Kauffman, R. Fenn and V. O. Manturov, Virtual knot theory – unsolved problems, *Fund. Math.* **188** (2005), 293–323, math.GT/0405428.

[23] V. O. Manturov, Multivariable polynomial invariants for virtual knots and links *J. Knot Theory Ramifications* **12**:8 (2003), 1131–1144.

[24] L. H. Kauffman, R. Fenn and M. Jordan-Santana, Biquandles and virtual links. *Topology and Its Applications* 145 (2004), 157–175.

[25] L. H. Kauffman, D. Hrencecin, Biquandles for virtual knots. *J. Knot Theory Ramifications* 16 (2007), no. 10, 1361–1382. 57M27 (57M25).

[26] L. H. Kauffman and Vassily Manturov, Virtual biquandles. *Fundamenta Mathematicae* **188** (2005), 103–146.

[27] L. H. Kauffman and D, E. Radford. Bioriented quantum algebras and a generalized Alexander polynomial for virtual links, in *Diagrammatic Morphisms and Applications* (San Francisco, CA 2000), 113–140. Contemp. Math. 318, AMS, Providence, RI (2003).

[28] Jörg Sawollek. On Alexander-Conway polynomials for virtual knots and links. arXiv:math/991273v2 [math.GT] 6 Jan 2001.

[29] Jörg Sawollek. An orientation-sensitive Vassiliev invariant for virtual knots. *J. Knot Theory Ramifications* 12 (2003), no. 6, 767–779.

[30] D.S. Silver and S.G. Williams, An invariant for open virtual strings, *JKTR* **15**, No. 2 (2006), 143 - 152.

[31] F. Jaeger, L. H. Kauffman and H. Saleur, The Conway polynomial in $\mathbb{R}^3$ and in thickened surfaces: A new determinant formulation. *J. Comb. Theory Ser. B* **61** (1994), 237–259.

[32] L. H. Kauffman and H. Saleur, Free fermions and the Alexander–Conway polynomial *Comm. Math. Phys.* **141** (1991), 293–327.

[33] T. Kishino and S. Satoh, A note on non-classical virtual knots. *J. Knot Theory Ramifications* **13**:7 (2004), 845–856.

[34] H. Dye, Characterizing Virtual Knots, Ph.D. Thesis (2002).

[35] H. Dye and L. H. Kauffman, Minimal surface representations of virtual knots and links *Algebr. Geom. Topol.* **5** (2005), pp. 509–535.

[36] V. O. Manturov, Kauffman-like polynomial and curves in 2-surfaces. *J. Knot Theory Ramifications* **12**:8 (2003), pp. 1145–1153.

[37] A. Bartholomew and R. Fenn, Quaternionic invariants of virtual knots and links. *J. Knot Theory Ramifications* **17**:2 (2008), 231.

[38] V. Turaev, Virtual strings. *Ann. Inst. Fourier (Grenoble)* **54** (2004), no. 7, 2455–2525 (2005).

[39] R. Fenn and V. Turaev. Weyl Algebras and Knots. *Journal of Geometry and Physics* 57 (2007) 1313–1324.

[40] A. Bartholomew, R. Fenn, N. Kamada, S. Kamada. New invariants of long virtual knots. *Kobe J. Math.* 27 (2010), no. 1–2, 21–33.

[41] L. H. Kauffman, A self-linking invariant of virtual knots. *Fund. Math.* **184** (2004), 135–158.

[42] L.H. Kauffman, State models and the Jones polynomial. *Topology* **26** (1987), 395–407.

[43] V. F. R. Jones, A polynomial invariant of links via Von Neumann algebras. *Bull. Amer. Math. Soc.*, 1985, No. 129, 103–112.

[44] E. Witten. Quantum field theory and the Jones polynomial. *Comm. in Math. Phys.* **121** (1989), 351–399.

[45] L. H. Kauffman, *Knots and Physics*. World Scientific, Singapore/New Jersey/London/Hong Kong, (1991, 1994, 2001, 2012).

[46] S. Winker. PhD. Thesis, University of Illinois at Chicago (1984).

[47] M. Thistlethwaite, Links with trivial Jones polynomial, *JKTR* **10**, No. 4 (2001), 641–643.

[48] S. Eliahou, L. Kauffman and M. Thistlethwaite, Infinite families of links with trivial Jones polynomial. *Topology* **42**, 155–169.

[49] H. A. Dye, A. Kaestner, L. H. Kauffman. Khovanov homology, Lee homology and a Rasmussen invariant for virtual knots. *J. Knot Theory Ramifications* 26 (2017), no. 3, 1741001, 57 pp.

[50] T. Kishino, 6 kouten ika no kasou musubime no bunrui ni tsiuti (On classification of virtual links whose crossing number is less than or equal to 6), master's thesis, Osaka City University, 2000.

[51] H. Dye and L. H. Kauffman, Minimal surface representations of virtual knots and links. *Algebr. Geom. Topol.* **5** (2005), 509–535.

[52] V. O. Manturov, Parity in knot theory. *Sbornik: Mathematics* (2010) 201(5), 65–110.

[53] V. O. Manturov, Parity and cobordisms of free knots (to appear in *Sbornik*), arXiv:1001.2827.

[54] L. H. Kauffman, V. O. Manturov. A graphical construction of the sl(3) invariant for virtual knots. *Quantum Topol.* 5 (2014), no. 4, 523–539.

[55] L. H. Kauffman, V. O. Manturov. Graphical constructions for the sl(3), C2 and G2 invariants for virtual knots, virtual braids and free knots. *J. Knot Theory Ramifications* 24 (2015), no. 6, 1550031, 47 pp.

[56] L. H. Kauffman, An affine index polynomial invariant of virtual knots, *J. Knot Theory Ramifications* 22(4) (2013) 1340007.

[57] L. C. Folwaczny, L. H. Kauffman. A linking number definition of the affine index polynomial and applications. *J. Knot Theory Ramifications* 22 (2013), no. 12, 1341004, 30 pp.

[58] Z. Cheng. A polynomial invariant of virtual knots. *Proc. Amer. Math. Soc.* 142 (2014), no. 2, 713–725.

[59] H. A. Dye. Vassiliev invariants from parity mappings. *J. Knot Theory Ramifications* 22 (2013), no. 4, 1340008, 21 pp.

[60] Z. Cheng, H. Gao. A polynomial invariant of virtual links. *J. Knot Theory Ramifications* 22 (2013), no. 12, 1341002, 33 pp.

[61] A. Henrich. A sequence of degree one Vassiliev invariants for virtual knots. *J. Knot Theory Ramifications* 19 (2010), no. 4, 461–487.

[62] B. Mellor, Alexander and writhe polynomials for virtual knots. *J. Knot Theory Ramifications* 25 (2016), no. 8, 1650050, 30 pp.

[63] Y. Miyazawa, Magnetic graphs and an invariant for virtual links *JKTR* **15** (2006), no. 10, 1319–1334.

[64] Y. Miyazawa, A Multivariable polynomial invariant for unoriented virtual knots and links, *JKTR* **17**, No. 11, Nov. 2008, 1311–1326.

[65] N. Kamada, An index of an enhanced state of a virtual link diagram and Miyazawa polynomials. *Hiroshima Math. J.* **37** (2007), no. 3, 409–429.

[66] N. Kamada, Miyazawa polynomials of virtual knots and virtual crossing numbers, in *Intelligence of Low Dimensional Topology*, Knots and Everything Series – Vol. 40, edited by S. Carter, S. Kamada, L. H. Kauffman, A. Kawauchi and T. Khono, World Sci. Pub. Co. (2006), 93–100.

[67] N. Kamada and Y. Miyazawa, A two-variable invariant for a virtual link derived from magnetic graphs. *Hiroshima Math. J.* **35** (2005), 309–326.

[68] L. H. Kauffman, Rotational virtual knots and quantum link invariants. *J. Knot Theory Ramifications* 24 (2015), no. 13, 1541008, 46 pp.

[69] L. H. Kauffman and David E. Radford. Invariants of 3-manifolds derived from finite dimensional Hopf algebras. *Journal of Knot Theory and Its Ramifications*, Vol. 4, No. 1 (1995), pp. 131–162.

[70] Shin Satoh. Virtual knot representation of ribbon torus-knots, *JKTRm* Vol. 9, No. 4 (2000) 531–542.

[71] Colin Rourke. What is a welded link?, corrected version (2018), Mathematics Institute, University of Warwick Coventry CV4 7AL, UK E-mail: cpr@maths.warwick.ac.uk.

[72] Colin Rourke. What is a welded link? *Intelligence of Low Dimensional Topology* 2006, 263–270, Ser. Knots and Everything, 40, World Sci. Publ., Hackensack, NJ (2007).

[73] L. H. Kauffman, E. Ogasa, and J. Schneider, A spinning construction for virtual 1-knots and 2-knots, and the fiberwise and welded equivalence of virtual 1-knots, preprint (2018).

[74] M. Khovanov, A categorification of the Jones polynomial, *Duke Math. J.* **101** (2000), no. 3, 359–426.

[75] D. Bar-Natan, On Khovanov's categorification of the Jones polynomial. *Algebraic and Geometric Topology* **2** (2002), 337–370.

[76] O. Viro, Khovanov homology, its definitions and ramifications. *Fund. Math.* 184 (2004), 317–342.

[77] D. Bar-Natan, Khovanov's homology for tangles and cobordisms, *Geom. Topol.* **9** (2005), 1443–1499.

[78] V. F. R. Jones. Hecke algebra representations of braid groups and link polynomials. *Ann. of Math.*, 126:335–338, 1987.

[79] V. F. R. Jones. On knot invariants related to some statistical mechanics models. *Pacific J. Math.*, 137(2):311–334, 1989.

[80] V. O. Manturov. Khovanov homology for virtual knots with arbitrary coefficients. *J. Knot Theory Ramifications*, 16(3):345–377, 2007.

[81] V. O. Manturov. The Khovanov complex for virtual links. *Journal of Mathematical Sciences*, 144(5):4451–4467, 2007.

[82] V. O. Manturov and D. Ilyutko. *Virtual Knots: The State of the Art.* Series on Knots and Everything. World Scientific Publishing Co, Hackensack, NJ, 51 edition, 2013.

[83] M. Khovanov and L. Rozansky. Matrix factorizations and link homology. *Fund. Math.* 199 (2008), no. 1, 1–91.

[84] M. C. Hughes. A note on Khovanov-Rozansky sl2-homology and ordinary Khovanov homology. *J. Knot Theory Ramifications* 23 (2014), no. 12, 1450057, 25 pp.

[85] L. H. Kauffman and E. Ogasa, On a Steenrod square and Khovanov homotopy type for virtual links (in preparation).

[86] D. Tubbenhauer. Virtual Khovanov homology using cobordisms. *J. Knot Theory Ramifications*, 9(23), 2014.

[87] W. Rushworth. Doubled Khovanov homology. *Canad. J. Math.* 70 (2018), no. 5, 1130–1172.

[88] L. H. Kauffman. Virtual knot cobordism. In L. Kauffman and V. Manturov, editors, *New Ideas in Low Dimensional Topology*, volume 56 of *Series on Knots and Everything*, pages 335–378. World Scientific Publishing Co, 2015.

[89] L. H. Kauffman. Virtual knot cobordism and the affine index polynomial. *J. Knot Theory Ramifications* 27 (2018), no. 11, 1843017, 29 pp.

[90] L. H. Kauffman. Virtual knot theory and virtual knot cobordism. Knots. *Low-Dimensional Topology and Applications*, 67–114, Springer Proc. Math. Stat., 284, Springer, Cham, 2019.

[91] M. W. Chrisman. Virtual Seifert surfaces. *J. Knot Theory Ramifications* 28 (2019), no. 6, 1950039, 33 pp.

[92] L. H. Kauffman. *On Knots*. Princeton University Press, Princeton, NJ, 1987.

[93] H. U. Boden, M. Nagel. Concordance group of virtual knots. *Proc. Amer. Math. Soc.* 145 (2017), no. 12, 5451–5461.

[94] J. Rasmussen. Khovanov homology and the slice genus. *Invent. Math.*, 182(2):419–447, 2010.

[95] H. Dye, L. H. Kauffman, V. O. Manturov. (Chapeter 4) On two categorifications of the arrow polynomial for virtual knots. In M. Banagl and C. Vogel, editors, *The Mathematics of Knots, Theory and Application*, Vol. 1, *Contributions in Mathematical and Computational Sciences*, pp. 95–124, Heidelberg University, Springer-Verlag, 2011. arXiv:0906.3408.

[96] V. Turaev. Knotoids. *Osaka J. Math.* 49 (2012), no. 1, 195–223.

[97] N. Gügümcü and L. H. Kauffman, New invariants of knotoids, *European J. Combin.* 65 (2017) 186–229.

[98] N. Gügümcü, L. H. Kauffman, S. Lambropoulou. A survey on knotoids, braidoids and their applications. *Knots, Low-Dimensional Topology and Applications*, 389–409, Springer Proc. Math. Stat., 284, Springer, Cham, 2019.

[99] Q. Deng, X. Jin, L. H. Kauffman. Graphical virtual links and a polynomial for signed cyclic graphs. *J. Knot Theory Ramifications* 27 (2018), no. 10, 1850054, 14 pp.

[100] M. W. Chrisman, V. O. Manturov. Fibered knots and virtual knots. *J. Knot Theory Ramifications* 22 (2013), no. 12, 1341003, 23 pp.

Chapter 31

Virtual Knots and Surfaces

Micah Chrisman, Ohio State University

Geometric invariants of knots in 3-sphere S^3 can be defined via surfaces. The 3-*genus* of a knot K, for example, is the smallest genus among all Seifert surfaces bounded by K. Another example is the 4-*genus* of K. The 4-genus is the minimum genus among all smooth compact oriented surfaces F embedded in the 4-ball B^4 such that $\partial F = K$. Some geometric invariants can be computed algorithmically. For instance, there exists an algorithm to calculate the 3-genus of a knot [21]. Since the only knot with 3-genus equal to zero is the trivial knot, this also implies there is an algorithm to recognize the unknot.

Algorithms such as these are not practically implementable. Instead, algebraic invariants are used to estimate the value of geometric invariants. For the examples previously mentioned, a lower bound on the 3-genus is half the degree of the Alexander polynomial and a lower bound on the 4-genus is half the absolute value of the knot signature. The algebraic invariants have the additional advantage that they can be calculated using only the crossing data of the knot. Oftentimes the estimates coming from the algebraic invariants are sufficient to calculate the geometric invariant exactly. This leads to the natural question: how does one determine the dependency of a geometric invariant on the crossing data of knots?

The first step in answering such questions is to specify what is meant by "crossing data." As discussed elsewhere in this volume, the configuration of crossings of a knot can be described by the Gauss code (or Gauss diagram) of a knot. Recall that the Gauss diagram records the ordered sequence of over- and under-crossings and local crossing signs as one traverses the underlying curve of the knot diagram. The Reidemeister moves can also be encoded as Gauss diagrams and two Gauss diagrams are said to be equivalent if they may be obtained from one another by a sequence of these Gauss diagram moves. A Reidemeister equivalence of knots can thus be seen as an equivalence of Gauss diagrams. However, an arbitrary equivalence of Gauss codes can introduce configurations of crossings that cannot be realized as a diagram of any knot in S^3. In order to study the crossing data of knots in S^3, it is thus necessary to exit the realm of classical knots and turn to virtual knot theory.

A virtual knot is, by definition, an equivalence class of Gauss diagrams. Although not every Gauss diagram can be realized by a classical knot diagram, every Gauss diagram can be realized by a knot diagram on a closed orientable surface (e.g., the sphere, torus, two-holed torus, etc.). This can be seen by constructing a *Carter surface* [8] for a virtual knot diagram. The Carter surface is defined by a handle decomposition as follows. A small disc (0-handle) is centered at each classical crossing. The 1-handles are untwisted bands attached to the 0-handles along the arcs of the diagram. Virtual

FIGURE 31.1: A diagram of 6.87548 (left) and its Carter surface (right).

crossings are ignored. The closed surface Σ obtained by attaching discs (2-handles) to the boundary is the Carter surface of the diagram. The virtual knot diagram now appears as a knot diagram on Σ. The knot diagram on Σ is a projection of a knot in the thickened surface $\Sigma \times [0, 1]$. An example of a Carter surface for the virtual knot 6.87548[1] is depicted in Figure 31.1

Thus, to investigate the dependency of geometric invariants of classical knots on their crossing data, we study how the same geometric invariants of knots in thickened surfaces behave under the equivalence relation for Gauss diagrams and virtual knots. Before returning to the 3-genus and 4-genus of classical knots, we will discuss a fundamental geometric invariant of virtual knots themselves.

31.1 Planarity and the Minimal Supporting Genus

The *minimal supporting genus* of a virtual knot v is the smallest genus Carter surface of all virtual knot diagrams equivalent to v. A Gauss diagram (or virtual knot diagram) is said to be *planar* if it is realizable as the diagram of a classical knot in S^3. A virtual knot admits a planar diagram if and only if its minimal supporting genus is zero, which occurs if and only if it is a classical knot.

While it is difficult to compute the minimal supporting genus of a virtual knot, determining whether or not a given diagram is planar is straightforward. Here we describe a solution to the planarity problem that is due to Cairns-Elton [7]. Suppose K is an oriented virtual knot diagram and let $\mathfrak{D}$ denote the immersed curve of K on its Carter surface Σ. Let $x_1, \ldots, x_n$ denote the classical crossings of K. For a crossing x_i of $\mathfrak{D}$, performing the oriented smoothing at x_i gives two immersed curves on Σ (see Figure 31.2, left). Let $\mathfrak{D}_i$ be the component that lies to the right of the crossing after the smoothing. Let $\mathfrak{D}_0$ be the immersed curve of K itself. From these curves we compute an $(n + 1) \times (n + 1)$ matrix B, indexed by $0, 1, \ldots, n$. The i, j entry $b_{i,j}$ is given by $b_{i,j} = [\mathfrak{D}_i] \cdot [\mathfrak{D}_j]$, where $[\mathfrak{D}_x]$ denotes the first integral homology class of $\mathfrak{D}_x$ on Σ and $\cdot$ denotes the intersection pairing. To determine the integer $[\mathfrak{D}_i] \cdot [\mathfrak{D}_j]$, travel along the curve $\mathfrak{D}_i$ and count the left-to-right crossings of $\mathfrak{D}_j$ as -1 and the right-to-left crossings of $\mathfrak{D}_j$ as $+1$. Then $[\mathfrak{D}_i] \cdot [\mathfrak{D}_j]$ is the sum of

---

[1]This notation comes from Green's table [15], where the ones digit indicates the minimum number of classical crossings and the decimal part denotes its position in the table.

these contributions. The Cairns-Elton criterion for planarity states that the crossing configuration of K is planar if and only if the matrix B is identically 0. Figure 31.2, shows the matrix B for the diagram of 6.87548 given on the left in Figure 31.1. As it is non-vanishing, this diagram of 6.87548 is not planar.

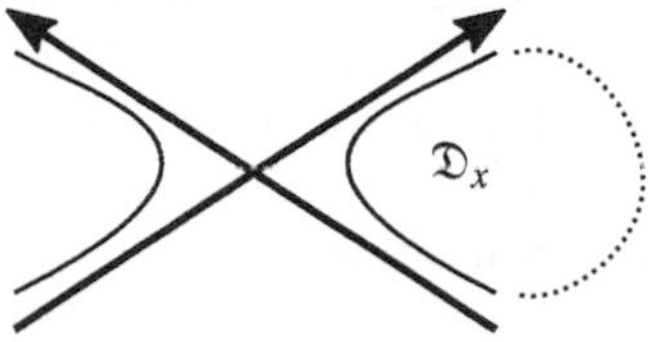

$$B = \begin{bmatrix} 0 & 0 & 0 & 0 & 0 & 0 & 0 \\ 0 & 0 & 1 & -1 & 0 & 0 & 0 \\ 0 & -1 & 0 & 0 & 1 & 0 & 0 \\ 0 & 1 & 0 & 0 & -1 & 0 & 0 \\ 0 & 0 & -1 & 1 & 0 & 0 & 0 \\ 0 & 0 & 0 & 0 & 0 & 0 & 0 \\ 0 & 0 & 0 & 0 & 0 & 0 & 0 \end{bmatrix}$$

FIGURE 31.2: (Left) The right half of a crossing. (Right) The matrix B for the virtual knot diagram of 6.87548 from Figure 31.1.

Given a knot $K \subset \Sigma \times [0, 1]$ representing a virtual knot v, a theorem of Kuperberg [19] states that K can be reduced to a minimal supporting genus representative by removing a sequence of 1-handles from Σ. This representative is unique up to ambient isotopy and orientation preserving diffeomorphisms of surfaces. A number of techniques have been developed to calculate the minimal supporting genus and we will mention a few here. Lower bounds on the minimal supporting genus can be obtained from the *surface bracket* [13], which is a generalization of the Kauffman bracket polynomial to knots in thickened surfaces, and the *arrow polynomial* [14, 1]. If a diagram of a virtual knot v has the minimal number of classical crossings of v, then the minimal supporting genus is the genus of the Carter surface of the diagram [20]. This fact is useful for computing the minimal supporting genus of the virtual knots in Green's table [15]. Carter, Silver, and Williams [9] showed that the minimal supporting genus of a virtual knot v is equal to half the symplectic rank of the covering group of a representative $K \subset \Sigma \times [0, 1]$. The *covering group* $\tilde{\pi}_K$ of K is defined to be the fundamental group $\pi_1(\tilde{\Sigma} \times [0, 1] \setminus \tilde{K})$, where $\tilde{\Sigma}$ is the universal cover of Σ and $\tilde{K}$ is a lift of K to $\tilde{\Sigma} \times [0, 1]$. Lower bounds on the minimal supporting genus can also be obtained from the Alexander invariants of $\tilde{\pi}_K$.

31.2 Almost Classical Knots and the 3-Genus

Given any diagram of a classical knot K, Seifert's algorithm constructs an orientable surface bounded by K. The *canonical 3-genus* of K is the minimum genus Seifert surface that can be constructed from Seifert's algorithm, where the minimum is taken over all diagrams of K. The 3-genus of an alternating knot is equal to the canonical 3-genus [11, 24]. Moriah [23] proved that the 3-genus of K need not be equal to the canonical 3-genus for non-alternating knots (see also Kawauchi [18]). Nonetheless, both the 3-genus and canonical 3-genus are geometric invariants that depend only on the crossing data of knots.

To see why this is true, it is necessary to first define *almost classical knots*. A virtual knot diagram

v is said to have an *Alexander numbering* if the 2-handles of its Carter surface Σ may be numbered so that the face number to the right of each edge of the knot diagram is one less than the face number on the left of the edge [9]. An example is given in Figure 31.3. A virtual knot that admits an Alexander numerable diagram is said to be *almost classical (AC)* . Almost classical knots are precisely those virtual knots to which the definition of 3-genus extends. Indeed, if K is the diagram of v on Σ, the discs $D_1, \ldots, D_m$ are the 2-handles of the Carter surface Σ, and $\alpha_1, \ldots, \alpha_m$ their respective Alexander numbering, then the first integral homology class of K can be recovered as $\partial(\alpha_1 D_1 + \cdots + \alpha_m D_m) = K$. Therefore, every Alexander numerable virtual knot diagram is trivial in the first homology of the Carter surface and hence bounds a compact orientable surface in $\Sigma \times [0, 1]$. The *virtual 3-genus* of an almost classical knot v is defined to be the smallest genus Seifert surface among all its homologically trivial representatives in thickened surfaces.

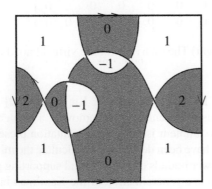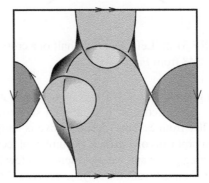

FIGURE 31.3: An Alexander numbering of 6.87548 (left) and a Seifert surface (right) of 6.87548.

The Seifert surface algorithm can be modified to construct Seifert surfaces for homologically trivial knots in thickened surfaces [4] and such surfaces can even be constructed from a Gauss diagram of v [10]. An example is given in Figure 31.3, right. Observe that if the Seifert smoothing is applied at each of the crossings, the result is three discs and one annulus. The main difference between the virtual and classical algorithms is that in the virtual case, the subsurfaces that make up the spanning surface need not be discs. The *virtual canonical 3-genus* is the smallest genus Seifert surface obtained from this virtual Seifert surface algorithm. As the virtual Seifert surface algorithm allows more freedom in the construction of Seifert surfaces, it could potentially lower the canonical 3-genus when applied to a classical knot.

Now, Boden et al. [4] proved that the virtual 3-genus of an almost classical knot v can be realized by a Seifert surface $F \subset \Sigma \times [0, 1]$, where Σ has the minimal supporting genus for v. In particular, if v is classical, this implies that the virtual 3-genus of v is its classical 3-genus. Thus, the 3-genus is a geometric invariant of knots in S^3 that depends on the crossing data. Furthermore, the canonical 3-genus of a classical knot cannot be reduced with the virtual Seifert surface algorithm. It was shown in [10] that the virtual canonical 3-genus of a classical knot coincides with its classical canonical 3-genus. In other words, the canonical 3-genus is itself a geometric invariant that depends only on the crossing data.

31.3 Virtual Knot Cobordism and the 4-Genus

The 4-genus of a knot $K \subset S^3$ has lower bounds derived from the Rasmussen invariant and the Tristam-Levine signature functions. The signature functions actually give lower bounds in the topological category, where the surface $F \subset B^4$ bounded by K is only required to be locally flat. To prove one of these lower bounds is equal to the 4-genus, it is generally necessary to construct a surface realizing the minimum. In the smooth category, a surface can be drawn as a sequence of links in S^3 (or a *movie*) that describes how it changes when passing through a non-degenerate critical point. Each critical point is either a local maximum, a local minimum, or a saddle (see Figure 31.4). Passing through a local minimum, a link K'' changes to a link K' by the addition of an unknotted and unlinked component. The change is called a *birth move*. A local maximum, which is an upside-down local minimum, deletes an unknotted and unlinked component. It is called a *death move*. A *saddle move* either splits a component of K' giving a new link K'' with an additional component, or it merges two components together.

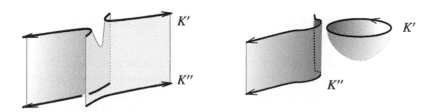

FIGURE 31.4: Saddle (left) and birth/death (right) moves. The vertical coordinate represents time. Time flows downward for the death move and upward for the birth move.

To determine the dependency of the 4-genus on the crossing data, movies as described above are extended to allow virtual knot diagrams [8, 16]. More exactly, two virtual knots are said to be *cobordant* if there is a sequence of extended Reidemeister moves, births, deaths, and saddles taking one virtual knot to the other. The smallest genus cobordism between a virtual knot and the unknot is called the *virtual 4-genus (or virtual slice genus)*. If two virtual knots have a genus zero cobordism between them, then they are said to be *concordant*. A virtual knot that is concordant to the unknot is said to be a *slice*. An example virtual cobordism is depicted in Figure 31.5, which shows a concordance from the Kishino knot to the unknot. In [12], Dye, Kaestner, and Kauffman asked: is the virtual 4-genus of a classical knot equal to its 4-genus? In other words, does the 4-genus of a classical knot depend only on its crossing data?

There is an alternative (but equivalent) definition of virtual knot cobordism, due to Turaev [27], that places this question into a more topological perspective. Let $K \subset \Sigma \times [0, 1]$ be a knot in a thickened surface. Let W be a compact oriented 3-manifold with boundary Σ. Suppose that S is a compact surface properly embedded in the compact 4-manifold $W \times [0, 1]$ such that $\partial S = K$. The virtual slice genus of K is the smallest genus of the surfaces S, minimized over all W with $\partial W = \Sigma$. A positive answer to the question of Dye, Kaestner, and Kauffman then says that if $K \subset S^2 \times [0, 1]$ is a classical knot, a surface realizing the 4-genus in B^4 can be constructed instead in some $W \times [0, 1]$

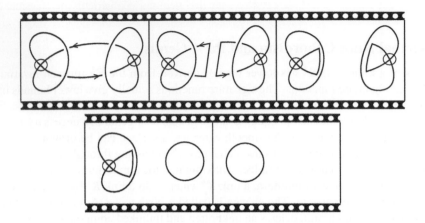

FIGURE 31.5: A concordance between the Kishino knot and the unknot. Time flows left to right. The genus of the cobordism is zero as there is one birth and one death.

with $\partial W = S^2$. Thus, if the Dye, Kaestner, and Kauffman conjecture is true, there is considerable freedom in the construction of cobordisms of classical knots.

Some evidence in support of the conjecture is as follows. Boden and Nagel [5] proved that a classical knot has 4-genus zero if and only if it has virtual 4-genus zero. This proves the conjecture for slice knots in both the smooth and locally flat categories. For the locally flat case, the conjecture is true for all prime knots up to 11 crossings and up to twelve crossings with 15 possible exceptions. In the smooth case, the conjecture is true up to 10 crossings. There are two possible exceptions among the 11 crossing knots and 37 among the 12 crossing knots. For the list of possible exceptions and further discussion on the Dye, Kaestner, and Kauffman conjecture, see [2].

Besides its application to classical knot cobordism, virtual knot cobordism is interesting in its own right. Many of most commonly studied virtual knot invariants are actually concordance invariants. Some examples are the odd writhe, the Henrich-Turaev polynomial, the writhe (or affine index) polynomial, the zero polynomial, and parity projection [3, 2, 17]. Lower bounds on the virtual 4-genus can be obtained from Turaev's graded genus, which is a numerical invariant that can be extracted from the matrix B described in Section 31.1. For smooth cobordisms, bounds can be obtained from categorified invariants, such as the Rasmussen invariant [12] and doubled Khovanov homology [26]. The virtual 4-genus of many of the 92800 virtual knots in Green's table was determined in [3]. For almost classical knots, the existence of Seifert surfaces makes it possible to obtain bounds on the topological virtual 4-genus using a generalization of Tristam-Levine signature functions. These *directed signature functions* [2] behave differently from their classical counterparts. For example, they are not independent of the choice of Seifert surface. This unexpected departure for the classical case was used to determine the slice status of all almost classical knots having at most six classical crossings [2]. Although virtual knot cobordism has its foundation in questions about classical knot cobordism, it offers surprising new phenomena and a bountiful source of future research projects.

The virtual 3-genus and virtual 4-genus are but two geometric invariants of classical knots that have been studied diagrammatically with virtual knots. Other examples of geometric properties are primality [22], periodicity [6], and the bridge number [25]. While virtual knots are typically investigated as combinatorial or algebraic objects, they exhibit interesting geometric behavior that can enrich our understanding of knots and links in the 3-sphere. On the other hand, the geometric properties of virtual knots are often reflected in their combinatorial structure. This can be seen, for example, in the concordance invariance of the Henrich-Turaev and writhe polynomials. The interplay between the combinatorial and geometric properties of virtual knots is just beginning to be explored and much further work will be required before this relationship is completely understood.

31.4 Bibliography

[1] K. Bhandari, H. A. Dye, and L. H. Kauffman. Lower bounds on virtual crossing number and minimal surface genus. In *The Mathematics of Knots*, volume 1 of *Contrib. Math. Comput. Sci.*, pages 31–43. Springer, Heidelberg, 2011.

[2] H. U. Boden, M. Chrisman, and R. Gaudreau. Signature and concordance of virtual knots. *to appear, Indiana University Mathematics Journal*, August 2017.

[3] H. U. Boden, M. Chrisman, and R. Gaudreau. Virtual knot cobordism and bounding the slice genus. *Experimental Mathematics*, 0(0):1–17, 2018.

[4] H. U. Boden, R. Gaudreau, E. Harper, A. J. Nicas, and L. White. Virtual knot groups and almost classical knots. *Fund. Math.*, 238(2):101–142, 2017.

[5] H. U. Boden and M. Nagel. Concordance group of virtual knots. *Proc. Amer. Math. Soc.*, 145(12):5451–5461, 2017.

[6] H. U. Boden, A. J. Nicas, and L. White. Alexander invariants of periodic virtual knots. *Dissertationes Math.*, 530:59, 2018.

[7] G. Cairns and D. M. Elton. The planarity problem for signed Gauss words. *J. Knot Theory Ramifications*, 2(4):359–367, 1993.

[8] J. S. Carter, S. Kamada, and M. Saito. Stable equivalence of knots on surfaces and virtual knot cobordisms. *J. Knot Theory Ramifications*, 11(3):311–322, 2002. Knots 2000 Korea, Vol. 1 (Yongpyong).

[9] J. S. Carter, D. S. Silver, and S. G. Williams. Invariants of links in thickened surfaces. *Algebr. Geom. Topol.*, 14(3):1377–1394, 2014.

[10] M. Chrisman. Virtual Seifert surfaces. *to appear, J. Knot Theory Ramifications*.

[11] R. Crowell. Genus of alternating link types. *Ann. of Math. (2)*, 69:258–275, 1959.

[12] H. A. Dye, A. Kaestner, and L. H. Kauffman. Khovanov homology, Lee homology and a Rasmussen invariant for virtual knots. *J. Knot Theory Ramifications*, 26(3):1741001, 57, 2017.

[13] H. A. Dye and L. H. Kauffman. Minimal surface representations of virtual knots and links. *Algebr. Geom. Topol.*, 5:509–535, 2005.

[14] H. A. Dye and Louis H. Kauffman. Virtual crossing number and the arrow polynomial. *J. Knot Theory Ramifications*, 18(10):1335–1357, 2009.

[15] J. Green. A table of virtual knots. http://www.math.toronto.edu/drorbn/Students/GreenJ, 2004.

[16] L. H. Kauffman. Virtual knot cobordism. In *New Ideas in Low Dimensional Topology*, volume 56 of *Ser. Knots Everything*, pages 335–377. World Sci. Publ., Hackensack, NJ, 2015.

[17] L. H. Kauffman. Virtual knot cobordism and the affine index polynomial. *J. Knot Theory Ramifications*, 27(11):1843017, 29, 2018.

[18] Akio Kawauchi. On coefficient polynomials of the skein polynomial of an oriented link. *Kobe J. Math.*, 11(1):49–68, 1994.

[19] G. Kuperberg. What is a virtual link? *Algebraic and Geometric Topology*, 3:587–591, 2003.

[20] V. O. Manturov. Parity and projection from virtual knots to classical knots. *J. Knot Theory Ramifications*, 22(9):1350044, 20, 2013.

[21] S. Matveev. *Algorithmic Topology and Classification of 3-Manifolds*, volume 9 of *Algorithms and Computation in Mathematics*. Springer, Berlin, second edition, 2007.

[22] S. V. Matveev. Roots and decompositions of three-dimensional topological objects. *Uspekhi Mat. Nauk*, 67(3(405)):63–114, 2012.

[23] Y. Moriah. On the free genus of knots. *Proc. Amer. Math. Soc.*, 99(2):373–379, 1987.

[24] K. Murasugi. On alternating knots. *Osaka Math. J.*, 12:277–303, 1960.

[25] P. Pongtanapaisan. Wirtinger numbers for virtual links. *arXiv e-prints*, page arXiv:1801.02923, Jan 2018.

[26] W. Rushworth. Computations of the slice genus of virtual knots. *ArXiv e-prints*, June 2017.

[27] V. Turaev. Cobordism of knots on surfaces. *J. Topol.*, 1(2):285–305, 2008.

Chapter 32

Virtual Knots and Parity

Heather A. Dye, McKendree University

Aaron Kaestner, North Park University

32.1 Introduction

In Chapter 30, virtual knots and links were defined. Further, in Chapter 3 we learned about Gauss codes for knots. In this chapter we will see how we can utilize the Gauss code to extract additional information about a virtual knot.

To define the parity of a crossing, we can work with a simplified version of the Gauss code by eliminating the over and under information. For example, the trefoil diagram in Figure 32.1a, has a simplified Gauss code of the form *abcabc*. For all classical knot diagrams, each instance of a label in the corresponding Gauss code is separated by an even number of terms. For virtual knots, this is not necessarily true. The virtual trefoil, shown in Figure 32.1b, has a Gauss code of the form

$$abab. \tag{32.1}$$

Geometrically, we can view virtual knots as embeddings of S^1 into a surface cross the unit interval, $F \times I$, modulo the Reidemeister moves, handle cancellation and addition, and Dehn twists. For example, the knot diagram from Figure 32.1b corresponds to the embedded diagram shown in Figure 32.2. From this viewpoint, we can observe that unevenly intersticed symbols in a Gauss code (i.e., odd crossings) occur because the image of the embedding of S^1 in the homotopy of F is non-trivial. For more on this perspective on virtual knots see Chapter 30.

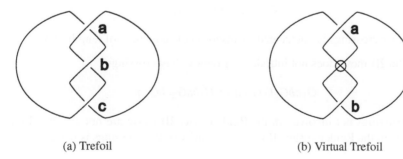

(a) Trefoil   (b) Virtual Trefoil

FIGURE 32.1: Virtual knots.

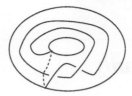

FIGURE 32.2: Embedded virtual trefoil.

32.2   Parity

An *odd* (respectively *even*) classical crossing is unevenly (evenly) intersticed in the Gauss code. This is the *parity* of the crossing.

In terms of the knot diagram, if we view the odd classical crossing as a basepoint, then the knot diagram consists of two loops. As each loop is traversed, an odd number of classical crossings is encountered.

Lemma 1. The Reidemeister moves do not change the parity of existing crossings in a diagram.

Proof. We let $G_1 G_2$ denote the Gauss code obtained from an existing virtual knot diagram. A Reidemeister I move introduces an even crossing. The loop formed by the crossing means that it the Gauss code is altered by the insertion of two adjacent symbols

$$G_1 aaG_2 \leftrightarrow G_1 G_2. \tag{32.2}$$

A Reidemeister II move introduces (or removes) two oppositely signed symbols. This changes the Gauss code:

$$G_1 abG_2 ab \leftrightarrow G_1 G_2 \tag{32.3}$$
$$\text{or} \tag{32.4}$$
$$G_1 abG_2 ba \leftrightarrow G_1 G_2. \tag{32.5}$$

The parity of other crossings is unchanged by insertion of sequences of length two.

The Reidemeister III move does not introduce or remove any crossings,

$$G_1 abG_2 bcG_2 ca \leftrightarrow G_1 baG_2 cbG_3 ac. \tag{32.6}$$

The parity of crossings not involved in the Reidemeister III move are not changed by the move. For the crossings in the Reidemeister III move, the parity of the crossings is unchanged. □

32.3 Odd Writhe

The easiest-to-compute parity invariant is the odd writhe as introduced by Louis Kauffman in [17]. Recall from classical knot theory that the writhe of a knot diagram, $\omega(D)$, is given by adding up the signs of the crossings in the diagram, where the sign of a crossing is as defined in Chapter 20. Given a knot diagram D, let $Odd(D)$ be the collection of odd crossings of D and let $sign(c)$ be the sign of crossing c. The odd writhe, denoted $\omega_o(D)$, is given by

$$\omega_o(D) = \sum_{c \in Odd(D)} sign(c).$$

Note that checking that this is an invariant of virtual knots is rather easy as the Reidemeister II move is the only Reidemeister move which can change the number of odd crossings in a knot diagram. As the Reidemeister II move removes or introduces odd crossings in pairs with opposite signs, it immediately follows that the odd writhe is an invariant of virtual knots.

For example, let T_V be the virtual trefoil in Figure 32.1b. If we orient the diagram we see that both crossings are positive. Since we know both crossings are odd, we have that $\omega_o(T_V) = 2$. Since any classical knot diagram K has only even crossings, we have $\omega_o(K) = 0$. Thus, as $\omega_o(T_V) = 2 \neq 0$, we know that the virtual trefoil is not a classical knot.

While simple to compute, the odd writhe has been shown to be a powerful invariant. Boden, Chrisman and Gadreau [2] and Rushworth [23] have shown that the odd writhe is an obstruction to a knot being virtually slice. (See Chapters 30 and 31 for more on virtual cobordisms.) Kauffman and Folwaczny [3] have shown that under certain conditions the odd writhe is an obstruction to cosmetic crossing change. Kaur, S. Kamada, Kawauchi, and Prabhakar [18] have shown that one-half of the odd writhe is a lower bound on the unknotting number.

32.4 Graphical Parity Bracket Polynomial

Introduced by V.O. Manturov in [22], the graphical parity bracket polynomial, $< D >_P$, is a generalization of the Kauffman bracket polynomial (see Chapter 61), $< D >$, (i.e., the unnormalized Jones polynomial) [15] which, given a knot diagram D, produces a polynomial with graphical coefficients. When normalized for writhe, one gets the graphical Jones polynomial, $J_G(K)$. More precisely, given a knot diagram D, $J_G(D) = (-A)^{-3\omega(D)} < D >_P$. For classical knots, both graphical polynomials are identical to their well-known counterparts. However, for virtual knots with odd crossings, these will frequently, but not always, differ.

The formula for the graphical parity bracket polynomial is as shown in Figure 32.3 where even crossings obey the standard bracket sum, while odd crossings are turned into graphical nodes with a fixed rotational orientation. When a diagram consists only of graphical nodes, and no crossings, we call this a flat knot. As the only Reidemeister move which can remove or introduce odd crossings

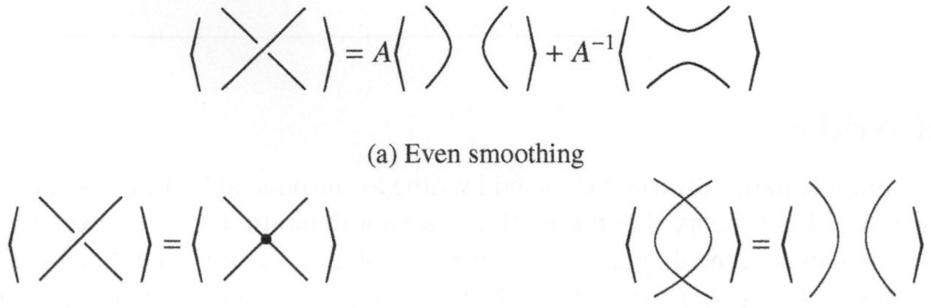

$$\left\langle \times \right\rangle = A \left\langle \right) \left(\right\rangle + A^{-1} \left\langle \smile \atop \frown \right\rangle$$

(a) Even smoothing

$$\left\langle \times \right\rangle = \left\langle \times \right\rangle$$

(b) Odd smoothing

$$\left\langle \bigotimes \right\rangle = \left\langle \right) \left(\right\rangle$$

(c) Bigon Cancellation

FIGURE 32.3: Expansion and reduction for the graphical parity bracket.

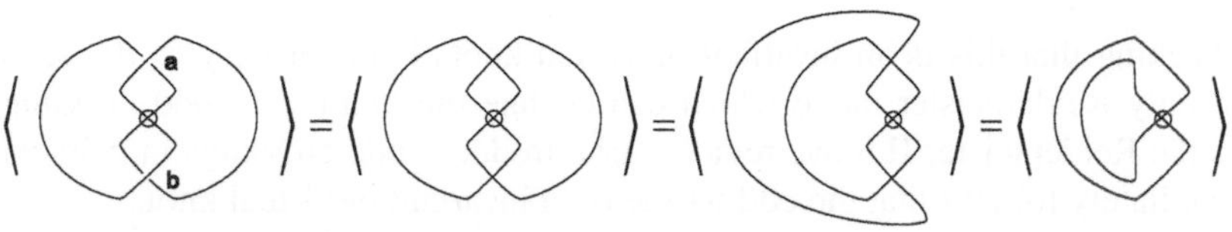

$$\left\langle \bigotimes \right\rangle = \left\langle \bigotimes \right\rangle = \left\langle \bigotimes \right\rangle = \left\langle \bigotimes \right\rangle$$

FIGURE 32.4: Parity bracket applied to the virtual trefoil.

is Reidemeister move II, the flat knots remaining after applying the bracket reductions are invariant up to cancelling of bigons (see Figure 32.3c).

32.5 Examples

Here we will take a look at two examples which illustrate some of the power and weaknesses of the two parity invariants we have previously discussed. Consider once again the virtual trefoil, T_V, as shown in Figure 32.1b. As we saw earlier, $\omega_o(T_V) = 2$ and, hence, the virtual trefoil is non-classical. However, as the graphical parity polynomial ignores signs, $< T_V >_P = < U >_P$, where U is the unknot. This equivalence can be seen in Figure 32.4.

Consider the Kishino knot, K, in Figure 32.5 with corresponding Gauss code *ababcdcd*. Placing an orientation on the diagram we see that this has two positive crossings and two negative crossings.

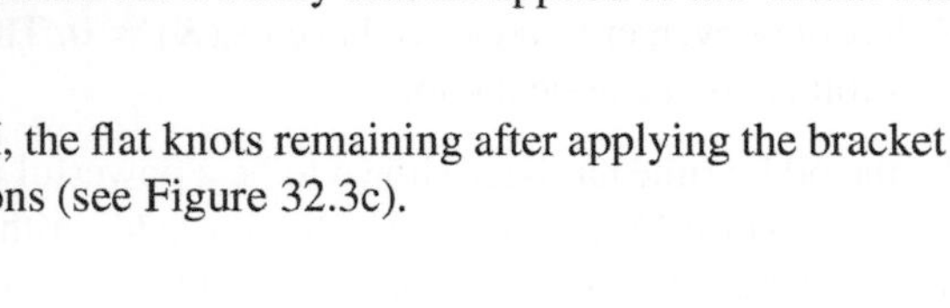

$$\left\langle \bigotimes \overset{a \quad d}{\underset{b \quad c}{\rule{0pt}{0pt}}} \bigotimes \right\rangle = \left\langle \bigotimes \rule{1.5cm}{0pt} \bigotimes \right\rangle$$

FIGURE 32.5: Parity bracket applied to the Kishino knot.

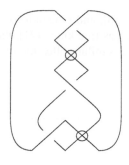 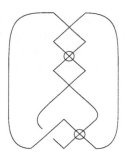

(a) Virtual knot 3.1   (b) Virtual knot 3.1 with graphical nodes

FIGURE 32.6: Virtual knot 3.1 before and after turning odd crossings into nodes.

FIGURE 32.7: Parity bracket applied to virtual knot 3.1.

Hence, $\omega_o(T_V) = 0$ and, thus, is not detected as non-classical by the odd writhe. However, as all four crossings are odd, we have that $< T_V >_P = Flat(K)$, as seen in Figure 32.5.

Note that the Gauss code cannot be reduced by the Reidemeister move II due to the order of the over- and under-crossings. Without this over/under information we would be unable to distinguish the Reidemeister move II from the forbidden move on the Gauss code for the Kishino knot.

Finally, consider the three-crossing virtual knot, $K_{3.1}$, in Figure 32.6a. $K_{3.1}$ has two odd crossings of opposite signs, and thus $\omega_o(K_{3.1}) = 0$. However, if we turn the odd crossings into nodes as in Figure 32.6b, we can see $< K_{3.1} >_P$ contains both graphical and non-graphical coefficients when computing $< K_{3.1} >_P$ as on Figure 32.7. Note that since the graphical nodes have a fixed rotational orientation on the edges, we are prevented from flipping a crossing and transforming the diagram into the unlink.

32.6 Additional Invariants and Parity

The two invariants we have already seen are not the only known parity invariants. In virtual knot theory, generalizations of classical invariants such as Vassiliev invariants based on chord diagrams [1], the Jones polynomial [11] [16], and Khovanov homology [19] [20] to parity invariants [5] [22] [12] play a critical role in helping us understand the full category of (virtual) knots defined by arbitrary Gauss codes. Parity enhancements can even be applied to purely virtual invariants like the arrow polynomial [6] [7] [12], biquandles [8] [9] [10] [13], and biquandle (co) homology [4] [14]. Moreover, if one looks at other families of knots, such as free knots [21], flat knots (see Chapter 34), and welded knots (see Chapter 33), one can find a similar collection of applications of parity to generalize previously known invariants.

32.7   Bibliography

[1] Dror Bar-Natan. On the Vassiliev knot invariants. *Topology*, 34(2):423–472, 1995.

[2] H. U. Boden, M. Chrisman, and R. Gaudreau. Virtual knot cobordism and bounding the slice genus. *ArXiv e-prints*, August 2017.

[3] L. C. Folwaczny and L.H. Kauffman. A linking number definition of the affine index polynomial and applications. *J. Knot Theory and Its Ramifications*, 22(12):1341004, 11 2013.

[4] J. Scott Carter, Mohamed Elhamdadi, and Masahico Saito. Homology theory for the set-theoretic Yang-Baxter equation and knot invariants from generalizations of quandles. *Fund. Math.*, 184:31–54, 2004.

[5] H.A. Dye. Vassiliev invariants from parity mappings. *J. Knot Theory Ramifications*, 22(4):134008, 03 2013.

[6] H.A. Dye and L.H. Kauffman. Virtual crossing number and the arrow polynomial. *J. Knot Theory Ramifications*, 18(10):1335–1357, 2009.

[7] H.A. Dye, L.H. Kauffman, and V.O. Manturov. On two categorifications of the arrow polynomial for virtual knots. In M. Banagl and Vogel. D., editors, *The Mathematics of Knots*, volume 1, pages 95–124. Contributions in Mathematical and Computational Sciences, Berlin, Heidelberg: Springer, 2011.

[8] R. Fenn, M. Jordan-Santana, and L. Kauffman. Biquandles and virtual links. *Topology and Its Applications*, 145:157–175, 2004.

[9] Roger Fenn, Colin Rourke, and Brian Sanderson. An introduction to species and the rack space. In *Topics in Knot Theory (Erzurum, 1992)*, volume 399 of *NATO Adv. Sci. Inst. Ser. C Math. Phys. Sci.*, pages 33–55. Kluwer Acad. Publ., Dordrecht, 1993.

[10] D. P. Ilyutko and V. O. Manturov. Picture-valued biquandle bracket. *ArXiv e-prints*, January 2017.

[11] V.F.R. Jones. A polynomial invariant for links via von Neumann algebras. *Bull. Amer. Math. Soc.*, 129:103–112, 1985.

[12] A. Kaestner and L.H. Kauffman. Parity, skein polynomials and categorification. *J. Knot Theory Ramifications*, 21(13):1240011, 2012.

[13] A. Kaestner and L.H. Kauffman. Parity biquandles. In *Knots in Poland III, part 1*, volume vol. 100 of Banach Center Publications, pages 131–151. Polish Acad. Sci. Inst. Math., Warsaw, 2014.

[14] A. Kaestner, S. Nelson, and L. Selker. Parity biquandle invariants of virtual knots. 209, 07 2015.

[15] L.H. Kauffman. State models and the Jones polynomial. *Topology*, 26(3):395–407, 1987.

[16] L.H. Kauffman. New invariants in the theory of knots. *Amer. Math. Monthly*, 95(3):195–242, 1988.

[17] L.H. Kauffman. A self-linking invariant of virtual knots. *Fund. Math.*, 184:135–158, 2004.

[18] K. Kaur, S. Kamada, A. Kawauchi, and M. Prabhakar. An unknotting index for virtual knots. *ArXiv e-prints*, September 2017.

[19] M. Khovanov. A categorification of the Jones polynomial. *Duke Math. J.*, 101(3):359–426, 2000.

[20] M. Khovanov. A functor-valued invariant of tangles. *Algebr. Geom. Topol.*, 2:665–741, 2002.

[21] O. V. Manturov and V. O. Manturov. Free knots and groups. *J. Knot Theory Ramifications*, 19(2):181–186, 2010.

[22] V.O. Manturov. Parity in knot theory. *Mat. Sb.*, 201(5–6):693–733, 2010.

[23] W. Rushworth. Doubled Khovanov Homology. *ArXiv e-prints*, April 2017.

Chapter 33

Forbidden Moves, Welded Knots and Virtual Unknotting

Sam Nelson, Claremont McKenna College

33.1 The Virtual Forbidden Moves

Virtual knot theory, unlike classical knot theory, is at its core combinatorial rather than geometric. Where classical knot theory starts with geometric objects and represents them combinatorially with diagrams, in virtual knot theory we start with diagrams and define the objects combinatorially as equivalence classes under certain moves. The question is then, which moves?

Let us briefly consider an analogy with linear algebra. A matrix $A \in M_{m,n}(R)$ over a ring R can be understood as a "diagram" representing a subspace of R^n, namely the row space of the matrix. We can ask what operations on the matrix preserve this object, and a bit of thought shows that the three usual row moves do not change the row space. We could even consider an alternative *definition* of subspaces of R^n as "equivalence classes of matrices under the equivalence relation generated by the row moves."

In virtual knot theory, we can start with Gauss diagrams representing oriented knots and links in $\mathbb{R}^3$. We observe that the classical oriented Reidemeister moves have Gauss diagram versions as shown:

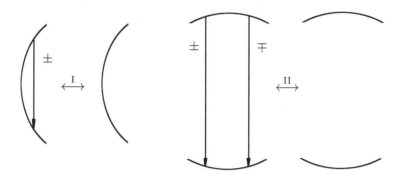

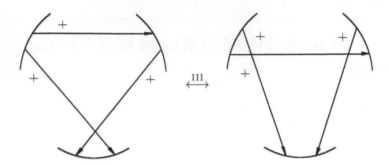

We can then *define* virtual knots as equivalence classes of Gauss diagrams under the equivalence relation generated by these moves. Some Gauss diagrams correspond to classical knot diagrams; to draw the ones that don't, we need to introduce *virtual crossings*. The interaction rules for virtual and classical crossings are defined by the requirement that the virtual moves do not change the Gauss code of the resulting diagram.

We then notice that there are certain operations which look legitimate at first glance but, upon closer inspection, change the object we are trying to represent. In our linear algebra analogy, an example might be multiplying a row by zero or any non-invertible scalar – this changes the span of the rows, even though it looks like a plausible row move.

Similarly, there are certain local moves on virtual knot diagrams which look quite plausible but if allowed, change the underlying object by changing the Gauss code. These are known as *forbidden moves*, and they are not allowed in virtual knot theory.

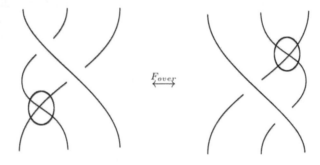

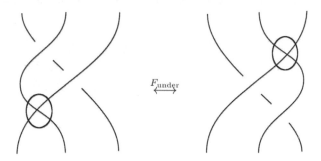

In practice it is very easy to accidentally do these moves when working with virtual knot diagrams. For example, in classical knot theory one frequently employs the *overpass move* where an arc with only overcrossings is moved to a new position with only overcrossings. In virtual knot theory, such a move is allowed only if there are no virtual crossings in the region bounded by before and after positions of the arc.

Why are these moves forbidden? It turns out that if we allow both forbidden moves, then every virtual knot becomes an unknot! This fact was observed in [3], then further explained in [6] and [7]; see also 3. To see why, let us consider the effect of these moves on Gauss diagrams:

An F_{under} move slides an arrowhead past an arrowhead, with no conditions on the tails of the arrows or the crossing signs:

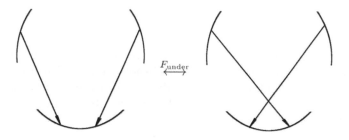

An F_{under} move similarly slides an arrowtail past an arrowtail:

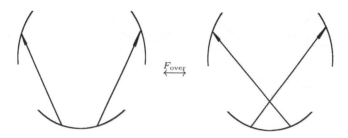

In [6] several other forbidden moves are derived from the first two, including the *mixed forbidden move* F_{mixed}:

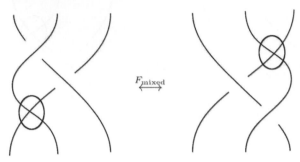

This move is also constructed in [7] using Gauss diagrams. In particular, this move slides an arrowhead past an arrowtail with no conditions on the respective other ends of the arrows.

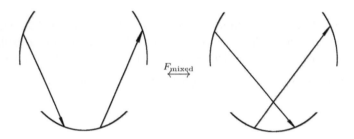

Armed with these forbidden moves, we can simply rearrange the arrows in our Gauss diagram at will, changing any virtual knot into any other virtual knot, and in particular into the unknot. Note that for virtual *links* with multiple components, forbidden moves do not necessarily reduce links to trivial links since multicomponent crossings can be deleted only in pairs with opposite signs. For example, the *virtual Hopf link* (shown with Gauss diagram)

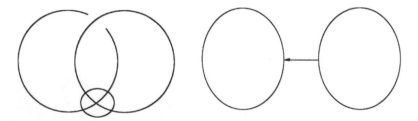

is nontrivial even using both forbidden moves.

33.2 Welded Knots

There are circumstances when we want to allow a move that we would normally forbid. For example, the *knot quandle* of a virtual knot is changed by forbidden move F_{under} but not by forbidden move F_{over}, so for the purposes of studying which properties of a virtual knot are determined by the knot quandle, we may want to allow forbidden move F_{over}. Equivalence classes of Gauss diagrams under the set of usual virtual moves together with the single forbidden move F_{over} are called *welded knots*, first introduced in braid form in [2].

The term "welded" suggests the strands being welded to the page, appropriate since we can move a strand over but not under a welded crossing. Some nontrivial virtual knots are trivial as welded knots, e.g. the *kishino knot* which can be unknotted using the welded move:

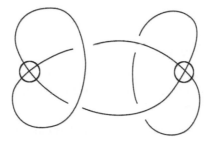

In particular, attempts to reconstruct a knot diagram from a quandle presentation end up being well defined only up to welded moves [8]. For more about welded knots and their invariants, see for example [1, 4, 10].

As observed in [9], welded knots and links are related to *ribbon torus-knots*, a type of knotted torus in $\mathbb{R}^4$. More precisely, we obtain a ribbon torus-knot from a welded knot by interpreting classical crossings as a type of tube-penetrating crossing

 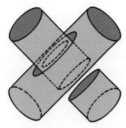

and virtual crossings as classical crossings of tubes, bearing in mind that it doesn't matter which tube is on top since the ambient space is $\mathbb{R}^4$:

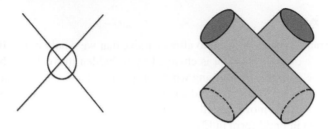

In [9], it is shown that equivalent welded knots yield ambient isotopic ribbon tori.

33.3   Other Forbidden Moves

More generally, any diagrammatic theory comes with its own set of forbidden moves which may look plausible but are disallowed because they change the underlying object represented by the diagram. A very incomplete list includes:

- In classical knot theory, the following moves are forbidden since they also allow unknotting of any knot:

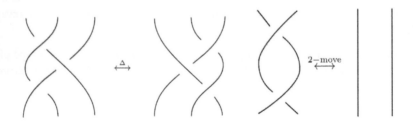

- In *spatial graph theory*, i.e., the study of knotted graphs in $\mathbb{R}^3$, for vertices of degree $n \geq 2$, there are two flavors: *rigid vertex isotopy*, in which the cyclic order of edges around a vertex is fixed, and *non-rigid vertex isotopy* where edges can swivel around a vertex. In the former, the move below is an example of a forbidden move:

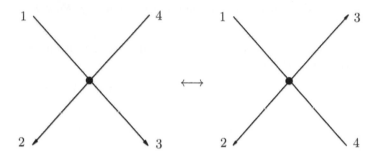

- In trivalent spatial graph theory, the *IH-move*

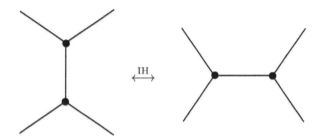

is forbidden. Allowing it yields quotients known as *handlebody-knots*, which correspond to knotted handlebodies in $\mathbb{R}^3$ as shown in [5].

33.4 Bibliography

[1] Benjamin Audoux, Paolo Bellingeri, Jean-Baptiste Meilhan, and Emmanuel Wagner. On usual, virtual and welded knotted objects up to homotopy. *J. Math. Soc. Japan*, 69(3):1079–1097, 2017.

[2] Roger Fenn, Richárd Rimányi, and Colin Rourke. The braid-permutation group. *Topology*, 36(1):123–135, 1997.

[3] Mikhail Goussarov, Michael Polyak, and Oleg Viro. Finite-type invariants of classical and virtual knots. *Topology*, 39(5):1045–1068, 2000.

[4] Young Ho Im, Kyeonghui Lee, and Mi Hwa Shin. Some polynomial invariants of welded links. *J. Korean Math. Soc.*, 52(5):929–944, 2015.

[5] Atsushi Ishii. Moves and invariants for knotted handlebodies. *Algebr. Geom. Topol.*, 8(3):1403–1418, 2008.

[6] Taizo Kanenobu. Forbidden moves unknot a virtual knot. *J. Knot Theory Ramifications*, 10(1):89–96, 2001.

[7] Sam Nelson. Unknotting virtual knots with Gauss diagram forbidden moves. *J. Knot Theory Ramifications*, 10(6):931–935, 2001.

[8] Sam Nelson. Signed ordered knotlike quandle presentations. *Algebr. Geom. Topol.*, 5:443–462, 2005.

[9] Shin Satoh. Virtual knot presentation of ribbon torus-knots. *J. Knot Theory Ramifications*, 9(4):531–542, 2000.

[10] Shin Satoh. Crossing changes, delta moves and sharp moves on welded knots. *Rocky Mountain J. Math.*, 48(3):967–979, 2018.

Chapter 34

Virtual Strings and Free Knots

Nicolas Petit, Emory University

34.1 Virtual Strings

Virtual strings belong to the branch of virtual knot theory (Chapter 30) that considers diagrams not just up to Reidemeister moves, but also up to the crossing change (CC) move, pictured in Fig. 34.1. This move changes a positive crossing into a negative one and vice versa, which can physically be thought of as letting one strand pass through the other in space. In other words, we are considering virtual knots up to homotopy, instead of just isotopy, since at one point through the deformation, we allow the curve to simply be immersed rather than embedded. Since positive and negative crossings are considered to be equivalent for virtual strings, we often draw our virtual knot diagrams in this setting with flat classical crossings (ignoring over-under information).

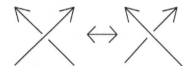

FIGURE 34.1: The crossing change move.

The idea of adding the CC move to the collection of allowable equivalences is not unique to virtual knots; it had been used before for classical knots, giving rise to the theory of flat knots, and it played a big part in the theory of finite-type invariants [2]. Since we can transform any classical knot diagram into the trivial unknot diagram via a sequence of crossing changes and Reidemeister moves (or by flattening all crossings in our diagrams and applying flat Reidemeister moves), there is only one equivalence class of classical knots up to the CC move. However, in the case of virtual knots, that is no longer true. For instance, the knot in Fig. 34.2 is a simple example of a virtual knot that can't be unknotted even if the CC move is allowed [10].

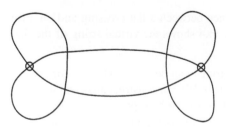

FIGURE 34.2: The flat Kishino knot.

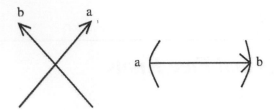

FIGURE 34.3: The way of associating an arrow to a flat crossing and vice versa. The arrow in the picture would be denoted by (a, b).

FIGURE 34.4: The virtual string of the flat Kishino knot.

Virtual knots can also be expressed as equivalence classes of Gauss diagrams (see Chapters 3, 30, 32). It should not come as a surprise that we can approach flat virtual knots from a similar point of view. However, the astute reader familiar with the construction of Gauss diagrams might notice an issue in this generalization: the Gauss diagram construction relies on over-under information to determine the direction the arrow goes in. If which strand of the knot passes under and which passes over at each crossing isn't fixed (since we can simply switch one to the other), how might we represent the crossing in our core circle? To answer this question, various authors (e.g., [3], [7], [14]) have proposed different—albeit ultimately equivalent—solutions. We will focus here on the answer given by Turaev in [14]: given a flat virtual knot diagram, resolve every crossing positively, draw the associated Gauss diagram, and forget all the signs that decorate the arrows. Turaev called this representation of the flat virtual knot diagram its associated virtual string; the reader should be aware that, by abuse of language, the term virtual string is often used to reference the flat virtual knot itself (including in this note). Equivalently, starting from a Gauss diagram of a given virtual knot, we draw the corresponding virtual string by flipping all the negative arrows and throwing away the signs. We typically denote an arrow as an ordered pair (a, b): a is the tail of the arrow, and it corresponds to the strand going up and right in the crossing, while b is the head of the arrow (corresponding to the other strand).

Pictorially, the correspondence between a flat crossing and an arrow in the virtual string is illustrated in Fig. 34.3, and Fig. 34.4 shows the virtual string of the flat Kishino knot, whose diagram is pictured in Fig. 34.2.

Like virtual knots, virtual strings can be represented as curves in orientable surfaces of positive genus (reference Micah's entry). We can construct an orientable surface on which the virtual string (or more generally a virtual knot) can be represented using the *band presentation*. To do so, we assign to every crossing one of the two pictures represented in Fig. 34.5. We then replace the strands connecting the crossings with bands. Finally, we cap everything off with disks until we get a closed surface.

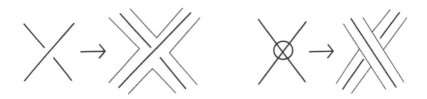

FIGURE 34.5: We associate to each crossing (classical or virtual) a pair of disjoint bands or a pair of crossing bands.

More precisely, the equivalence between knot diagrams and knots on surfaces is achieved via the proxy of abstract link diagrams. The interested reader can consult [8] or Chapter 35 in this book. We now have an orientable surface Σ on which the virtual string is represented as an immersed curve. Using this construction, we can associate to each virtual string an object called a *based matrix*; this algebraic object can naturally be associated to a virtual string, is used to define or give information on invariants (e.g., a lower bound on the slice genus) and can be used to distinguish non-homotopic virtual strings. We briefly summarize the approach to based matrices of [14] in what follows.

Definition 1. A *based matrix* is a triple (G, s, b) where G is a finite set, s is a designated element of G, and b is a skew-symmetric mapping $b \colon G \times G \to \mathbb{Z}$.

By abuse of language, we will often call the "based matrix" the matrix representation of b. Based matrices are interesting objects in their own right, but are mostly of use to us in the following context: if G consists of the homology class in $H_1(\Sigma)$ of the virtual string α (which will correspond to s) and of one homology class in $H_1(\Sigma)$ for each arrow of α (we obtain this via oriented smoothing, see [14] for details), then b is just the restriction of the homology pairing $H_1(\Sigma) \times H_1(\Sigma) \to \mathbb{Z}$ to G.

The entries of this matrix can be easily computed in a combinatorial fashion. As a preamble, given two arrows (c, d) and (e, f), we denote by $cd \cdot ef$ the number of arrows with tail in the interior of the (oriented) arc cd and head in the interior of ef, minus the number of arrows with tail on the interior of ef and head in the interior of cd. The values of the entries in the based matrix are then given by

$$b([g_1], s) = cd \cdot dc \qquad b([g_1], [g_2]) = cd \cdot ef + \varepsilon$$

where $g_1 = (c, d)$, $g_2 = (e, f)$ and $\varepsilon \in 0, \pm 1$ is determined by Fig. 34.6.

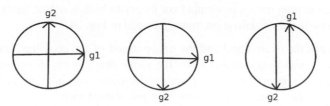

FIGURE 34.6: The pictures have respectively $\varepsilon = 1$, $\varepsilon = -1$, $\varepsilon = 0$.

As an example, the reader may verify that the matrix corresponding to the flat Kishino knot in Fig.

34.2 is given by

$$\begin{pmatrix} 0 & -1 & 1 & -1 & 1 \\ 1 & 0 & 1 & 0 & 0 \\ -1 & -1 & 0 & 0 & 0 \\ 1 & 0 & 0 & 0 & 1 \\ -1 & 0 & 0 & -1 & 0 \end{pmatrix}$$

where the first row/column represents the element s and the other row/columns represent the cross-ings, going counterclockwise starting from the top left. So the second row/column is the top left crossing, the third is the bottom left one, the fourth is the bottom right one, and the last row/column is the top right one.

Notice that this construction depends on the specific flat virtual knot diagram we're considering. To ensure invariance under the Reidemeister moves, we need to introduce a notion of equivalence of based matrices, which Turaev called "homology" of based matrices. We define three basic relations, called elementary extensions: the addition of a row and column of zeroes, the addition of a row and column that coincides with the first row/column, and the addition of two rows and columns that add up to the first row/column; note that all these operations must preserve the skew-symmetry of the matrix. Homology of based matrices is the equivalence relation given by the elementary extensions, their inverses and isomorphism. More details can be found in [14].

Under this notion of homology, the first Reidemeister move corresponds to the first two elementary extensions (depending on how the kink is added to the diagram), the second Reidemeister move corresponds to the third elementary extension, and the third Reidemeister move simply corresponds to an isomorphism. As a consequence, we get the following theorem:

Theorem 1. If two virtual strings are homotopic, then their based matrices are homologous. Conversely, if two virtual strings have non-homologous based matrices, then they must be non-homotopic.

Based matrices have been extended to various other knot categories (namely, singular virtual strings [6], long (singular) virtual strings [12], and virtual n-strings [5]), and they have been used to help distinguish some Vassiliev finite-type invariants, as in [6], [12].

34.2 The Virtualization Move and Free Knots

One of the most fascinating moves in virtual knot theory (which is not an equivalence in the category of virtual knots) is the virtualization move, pictured in Fig. 34.7.

This move exchanges the over-strand and the under-strand of the crossing, keeping the crossing sign intact. At the Gauss diagrammatic level, the move changes the direction of an arrow, and keeps its sign intact. We observe that the virtualization move is an involution, i.e., if you apply it twice to the same crossing, you get back to the knot you started with.

As simple as the idea behind the move is, it is deceptively difficult to get a handle on. For instance, while the virtualization move generally changes the knot type of a virtual knot, the involutory quandle and the Jones polynomial are both invariant under this move [9]. Interestingly, the virtualization

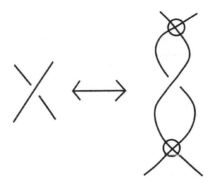

FIGURE 34.7: The virtualization move.

move allows us to easily find infinitely many virtual knots with trivial Jones polynomial, many of which are proven to be non-classical [4], [9], [13]; it is a famous open conjecture whether the only classical knot with trivial Jones polynomial is the unknot.

Since the virtualization move switches the direction of an arrow in the Gauss diagram, but not the sign, while the CC move changes the sign of the arrow and the direction of the arrow, it would seem natural to pair the two moves together. A virtual knot modulo the CC move and the virtualization move is called a *free knot*. Since it is pointless to keep track of the directions or signs of the arrows in a Gauss diagram for a free knot, we typically represent free knots as chord diagrams. We can also draw free knots as flat virtual knots in the plane, but we have to be careful—we are postulating that any two knots related by a virtualization move are the same, which can be a bit awkward for our visual intuition (especially if the reader is used to dealing with virtual knots without the virtualization move). The Reidemeister moves find themselves greatly reduced in number by the CC and virtualization moves; the resulting moves (presented for chord diagrams) are pictured in Fig. 34.8, while an example of a nontrivial flat virtual knot whose underlying free knot is trivial is pictured in Fig. 34.9.

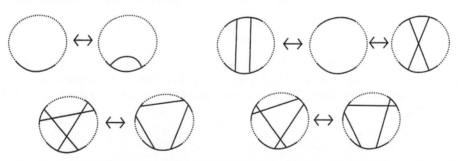

FIGURE 34.8: The Reidemeister-type moves for free knots.

There are currently few combinatorial invariants available for free knots [1], though these knot-like objects can be used to study other knot types. A free knot is often seen as underlying some virtual knot or flat virtual knot, and we'd like to know what information about the (flat) virtual knot the underlying free knot is able to capture. Free knots are also closely connected to the idea of

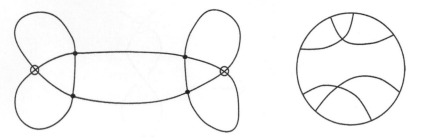

FIGURE 34.9: The free Kishino knot and its chord diagram. Using the second move of Fig. 34.8 twice we can reduce this chord diagram to the empty circle, so the free Kishino knot is trivial [11]. A diagrammatic proof of this fact is also shown in Fig. 34.10.

parity, further developed in Chapter 32 in this book. We present here a couple of important results involving free knots. The interested reader can find the context and details for them in [11].

Definition 2. Let K be a (free/flat) virtual knot diagram, and D its corresponding Gauss diagram/virtual string; a crossing of K is called *even* if the corresponding arrow links with an even number of other arrows, *odd* otherwise. The Gaussian parity is a map gp that assigns to each crossing of K the value 0 if the crossing is even, and the value 1 if the crossing is odd.

More generally, a parity is a map from the set of crossings to some abelian group A that satisfies some conditions related to the second and third Reidemeister move [11].

Theorem 2. If K is a free knot, the Gaussian parity on diagrams of K is the universal parity, i.e., for every other parity p, there is a group homomorphism that transforms the Gaussian parity into p.

Definition 3. The *atom* of a virtual knot diagram K is a connected closed 2-manifold M (not necessarily orientable) on which K is embedded in a way that admits a checkerboard coloring.

Theorem 3. The orientability of the atom of K only depends on the free knot underlying K.

Theorem 4. If a free knot is not slice (i.e., has slice genus > 0), then any virtual knot that it underlies is also not slice (reference Micah's entry).

To conclude, we note that both virtual strings and free knots can be interpreted in a much larger theory, that of the topology of words. Making some special choices in this construction will lead to either virtual strings or free knots. The reader interested in this fascinating theory can learn more about it in [15].

34.3   Bibliography

[1] T. Boothby, A. Henrich, and A. Leaf. Minimal diagrams of free knots. *Journal of Knot Theory and Its Ramifications*, 23:1450032, 05 2014.

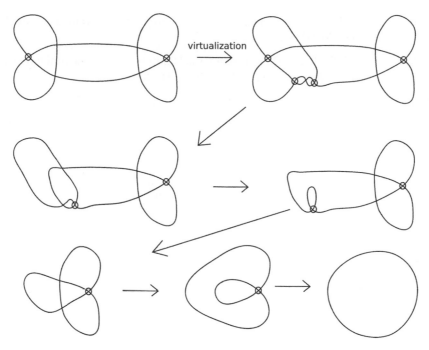

FIGURE 34.10: A diagrammatic proof that the free Kishino knot is trivial.

[2] S. Chmutov, S. Duzhin, and J. Mostovoy. *Introduction to Vassiliev Knot Invariants*. Cambridge University Press, 2012.

[3] H. Dye. Smoothed invariants. *Journal of Knot Theory and Its Ramifications*, 21, 09 2011.

[4] H A Dye and Louis H Kauffman. Minimal surface representations of virtual knots and links. *Algebr. Geom. Topol.*, 5(2):509–535, 2005.

[5] D. Freund. Multistring based matrices. https://arxiv.org/abs/1709.00564.

[6] A. Henrich. A sequence of degree one Vassiliev invariants for virtual knots. *Journal of Knot Theory and Its ramifications*, 19(4):461–487, 2010.

[7] D. Hrencecin and L. H. Kauffman. On filamentations and virtual knots. *Topology and Its Applications*, 134(1):23–52, 2003.

[8] N. Kamada and S. Kamada. Abstract link diagrams and virtual knots. *Journal of Knot Theory and Its Ramifications*, 09(01):93–106, 2000.

[9] L. H. Kauffman. Virtual knot theory. *European J. Comb*, 20(7):663–690, 1999.

[10] L. H. Kauffman. An extended bracket polynomial for virtual knots and links. *Journal of Knot Theory and Its Ramifications*, 18(10):1369–1422, 2009.

[11] V. O. Manturov and D. Ilyutko. *Virtual Knots: The State of the Art*, volume 51 of *Series on Knots and Everything*. World Scientific, 2012.

[12] N.   Petit.   Finite-type   invariants   of   long   and   framed   virtual   knots.
     https://arxiv.org/abs/1610.03825.

[13] D.S. Silver and S. G. Williams. On a class of virtual knots with unit jones polynomial. *Journal
     of Knot Theory and Its Ramifications*, 13(03):367–371, 2004.

[14] V. Turaev. Virtual strings. *Annales de l'Institut Fourier*, 54(7):2455–2525, 2004.

[15] V. Turaev. Topology of words. *Proceedings of the London Mathematical Society*, 95(2):360–
     412, 2007.

Chapter 35

Abstract and Twisted Links

Naoko Kamada, Nagoya City University

An abstract link is an extension of classical links. There is a one-to-one correspondence between virtual links [7] and ordinary abstract links [4]. Through this correspondence, we obtain a one-to-one correspondence between virtual links and stable equivalence classes of links in oriented thickenings of oriented closed surfaces [2, 4]. A twisted virtual link, or simply a twisted link, is a notion generalizing a virtual link so that twisted links correspond to abstract links on (possibly nonorientable) surfaces, and correspond to stable equivalence classes of links in oriented thickenings of closed surfaces [1].

35.1 Abstract Links

An *abstract link diagram* (*ALD*) is a pair (Σ, D) of a compact surface Σ and an oriented link diagram D on Σ such that the underlying 4-valent graph $|D|$ is a deformation retract of Σ [1, 4]. Abstract link diagrams are considered up to homeomorphism of Σ. See an example of an ALD in Figure 35.1.

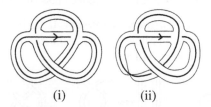

(i)   (ii)

FIGURE 35.1: Example of ALDs.

Precisely speaking, for an ALD (Σ, D), the surface Σ is regarded as a disk band surface, i.e., it is a union of some disks and some bands connecting the disks, such that each disk is a neighborhood of a crossing of D. (Here a disk band surface may have annulus components. This case happens when D has simple loops as its components.)

We assume that the disk neighborhood of each crossing is locally oriented as a disk. The local orientation at a crossing v can be reversed by switching the over/under information of arcs at v, which we call an *inversion* at v. ALDs are considered modulo inversions at crossings.

343

When we draw figures of ALDs, we assume that the local orientations of disk neighborhoods of crossings are counterclockwise. Using this convention, we do not need to specify the local orientations in figures.

An ALD $P = (\Sigma, D)$ is said to be *ordinary* if the surface Σ is oriented and the local orientations of disk neighborhoods of crossings are induced from the orientation.

Let $P = (\Sigma, D)$ be an ALD. When an embedding $f : \Sigma \to F$ into a closed surface F is given, we have a link diagram $f(D)$ on F. It is called a *link diagram realization* of P, denoted by (F, D_F). A link diagram realization always exists. Attach 2-disks to Σ along the boundary to obtain a closed surface. Furthermore, by attaching pipes, we may assume that the closed surface is connected. Figure 35.2 shows examples of link diagram realizations of the ALDs given in Figure 35.1.

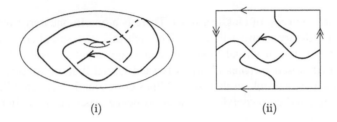

(i)   (ii)

FIGURE 35.2: Link diagram realizations of ALDs in Figure 35.1.

Let $P = (\Sigma, D)$ and $P' = (\Sigma', D')$ be ALDs. We say that P' is obtained from a P by an *abstract Reidemeister move* or an *abstract R-move* (of type I, type II, type III) if there exist embeddings $f : \Sigma \to F$ and $f' : \Sigma' \to F$ to a surface F and there is a disk E on F such that $f(D')$ is obtained from $f(D)$ by a Reidemeister move (of type I, type II, type III) in E on F. Here we assume that E is oriented and that the local orientations of disk neighborhoods of crossings in E are compatible with the orientation of E. See Figure 35.3.

Definition 1. Two ALDs are said to be *equivalent* if they are related by a sequence of abstract Reidemeister moves. An *abstract link* is an equivalence class of ALDs.

Two ordinary ALDs are said to be *equivalent as ordinary ALDs* if they are related by a sequence of abstract Reidemeister moves through ordinary ALDs. An *ordinary abstract link* is an equivalence class, as an ordinary ALD, of ordinary ALDs.

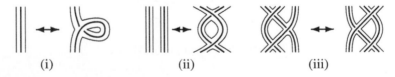

(i)   (ii)   (iii)

FIGURE 35.3: Abstract Reidemeister moves.

Let F be a surface and $I = [-1, 1]$. An *oriented thickening* of F is an oriented 3-manifold M which is an I-bundle over F. That means, there is a surjective map $\pi : M \to F$ called the *projection* such that for each $x \in F$, $\pi^{-1}(x)$ is homeomorphic to I and there is a neighborhood U of x in F equipped with a homeomorphism $\phi : U \times I \to \pi^{-1}(U)$ with $\pi\phi(y, z) = y$ for $y \in U$ and $z \in I$. Then F is called the *base* of the I-bundle. We denote M by $F \widetilde{\times} I$.

Let $P = (\Sigma, D)$ be an ALD. Let $\Sigma \tilde{\times} I$ be an oriented thickening of Σ. We assume that the restriction of $\Sigma \tilde{\times} I$ over the disk neighborhood of each crossing is the product of the disk and the fiber I and the orientation as a 3-manifold coincides with the orientation induced from the local orientation of the disk and the interval I. Then we have a link L in $M = \Sigma \tilde{\times} I$ whose diagram is D. We call the link L in $M = \Sigma \tilde{\times} I$ a *link realization* of $P = (\Sigma, D)$ over Σ.

A *link realization* of $P = (\Sigma, D)$ over a closed surface F is a link L in an oriented thickening $M = F \tilde{\times} I$ of F whose diagram on F is the image of D by an embedding $f : \Sigma \to F$.

For a closed surface F a *stabilization* means attaching a pipe, that is, removing a pair of disks from F and attaching an annulus along the boundary. A *destabilization* is the inverse operation of a stabilization. For an oriented thickening $F \tilde{\times} I$ of F, a *stabilization* (*destabilization*) means a surgery such that it covers a stabilization (destabilization) of F. When we consider links in oriented thickened surfaces, a *stabilization* (*destabilization*) means a stabilization (destabilization) of the ambient thickened surfaces.

Theorem 1 ([1, 2, 4, 5]). The map sending abstract link diagrams to link realizations in oriented thickenings of closed surfaces yields a one-to-one correspondence between abstract links and stable equivalence classes of links in oriented thickened closed surfaces. Moreover, ordinary abstract links correspond to stable equivalence classes of links in oriented thickenings of closed oriented surfaces.

35.2 Abstract Links vs Virtual Links and Twisted Links

A *twisted virtual link diagram* or a *twisted link diagram* is a virtual link diagram possibly with *bars* on arcs [1]. Examples of twisted link diagrams are depicted in Figure 35.4.

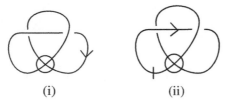

(i) (ii)

FIGURE 35.4: Twisted link diagrams.

The diagram in Figure 35.4 (i) is a virtual link diagram. A virtual link diagram is a twisted link diagram. Namely, the set of virtual link diagrams is a subset of that of twisted link diagrams. Thus the set of classical link diagrams is a subset of twisted link diagrams.

The local moves depicted in Figure 35.5 are called *twisted Reidemeister moves*. Reidemeister moves, virtual Reidemeister moves, and twisted Reidemeister moves are called *extended Reidemeister moves*.

Definition 2. A *twisted link* is an equivalence class of twisted link diagrams under extended Reidemeister moves.

FIGURE 35.5: Twisted Reidemeister moves.

We define a map

$$p : \{\text{twisted link diagrams}\} \to \{\text{abstract link diagrams}\}$$

as depicted in Figure 35.6. For a twisted link diagram D, the image $p(D)$ is called the *ALD associated with D*. For example, the ALDs depicted in Figure 35.1 are the ALDs associated with the twisted link diagrams in Figure 35.4.

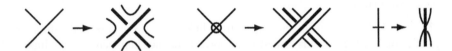

FIGURE 35.6: The correspondence from a twisted link diagrams to an ALD.

Virtual Reidemeister moves yield moves on ALDs depicted as in Figure 35.7 (i)–(iv). Twisted Reidemeister moves yield moves on ALDs depicted as in Figure 35.7 (v)–(vii).

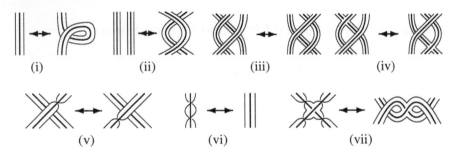

FIGURE 35.7: The correspondence between moves of twisted link diagrams and moves of ALDs.

Theorem 2 ([1, 4, 5]). The map p yields a one-to-one correspondence between twisted links and abstract links. By this correspondence, virtual links correspond to ordinary abstract links.

Now we see that there are one-to-one correspondences among the following families.

- Virtual links,

- Ordinary abstract links,

- Stable equivalence classes of links in oriented thickenings of oriented closed surfaces,

and there are one-to-one correspondences among the following families.

- Twisted links,

- Abstract links,

- Stable equivalence classes of links in oriented thickenings of closed surfaces.

Kuperberg [8] proved that a representative of a stable equivalence class of links in oriented thickenings of oriented closed surfaces is unique. Similarly, a representative of a stable equivalence class of links in oriented thickenings of closed surfaces is unique [1].

Some important notions on virtual links are often defined using abstract links. For example, checkerboard colorability. A virtual or twisted link diagram is *checkerboard colorable* if the associated ALD is checkerboard colorable. A virtual or twisted link is *checkerboard colorable* if there is a checkerboard colorable diagram representing it [1, 3]. It is shown in [3] that Jones polynomials of checkerboard colorable virtual links satisfy a certain property that Jones polynomials of classical links have.

35.3 Virtual Links vs Twisted Links

The inclusion map

$$\iota : \{\text{virtual link diagrams}\} \to \{\text{twisted link diagrams}\}$$

yields a natural map

$$f : \{\text{virtual links}\} \to \{\text{twisted links}\}.$$

For a virtual link diagram D in $\mathbb{R}^2$, let $r(D)$ denote the diagram obtained from D by a reflection along a line on $\mathbb{R}^2$ and switching over/under information on all classical crossings. See Figure 35.8.

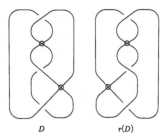

$$D \qquad\qquad r(D)$$

FIGURE 35.8: Example of a virtual link diagram D and $r(D)$.

Theorem 3 ([6]). For any virtual knot diagram D, D and $r(D)$ are equivalent as a twisted knot. Conversely, if two virtual knot diagrams D and D' are equivalent as twisted knots, then D' is equivalent to D or $r(D)$ as virtual knots.

There is a virtual knot diagram D such that D and $r(D)$ are not equivalent as virtual knots. For example, using an invariant defined in [10], we see that D and $r(D)$ in Figure 35.8 are not equivalent. Thus, the map f is not injective. A generalization of Theorem 3 to virtual links is given in [6].

Recall that the set of classical links is a subset of the set of virtual links. It is also a subset of the set of twisted links.

Theorem 4 ([6]). The map f restricted to the set of classical links is injective, i.e., two classical links are equivalent if and only if they are equivalent as twisted links.

In [5], a map

$$d : \{\text{twisted links}\} \to \{\text{virtual links},\}$$

called *the orientation double covering*, is introduced. For a twisted link diagram D, one can easily construct a virtual link diagram $\widetilde{D}$, called an *orientation double covering diagram* of D, such that $d([D]) = [\widetilde{D}]$. See Figure 35.9 for an example of D and $\widetilde{D}$.

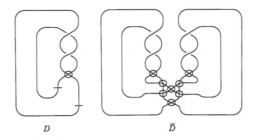

$$D \hspace{5cm} \widetilde{D}$$

FIGURE 35.9: Example of an orientation double covering diagram.

Theorem 5 ([5]). Let D or D' be twisted link diagrams and $\widetilde{D}$ and $\widetilde{D'}$ be orientation double covering diagrams of D and D', respectively. If D and D' are equivalent as twisted links, then $\widetilde{D}$ and $\widetilde{D'}$ are equivalent as virtual links.

Using this theorem, when we have an invariant of virtual links, we can obtain an invariant of twisted links.

Remark 1. The idea of abstract links is due to N. Kamada (cf. [4]). She was inspired by the formal Wirtinger presentation introduced by T. Sakai [9]. In [4], N. Kamada and S. Kamada provided the correspondence between virtual links and ordinary abstract links. J.S. Carter, S. Kamada, and M. Saito provided the correspondence between virtual links and stable equivalence classes in oriented thickenings of oriented closed surfaces. M. Bourgoin extended to nonorientable abstract links and defined twisted links. For link diagrams on closed surfaces presenting links in oriented thickenings of surfaces, the over/under information of classical crossings is discussed in [1, 5] in detail.

35.4   Bibliography

[1] M. O. Bourgoin, Twisted link theory, *Algebr. Geom. Topol.* 8 (2008) 1249–1279.

[2] J. S. Carter, S. Kamada and M. Saito, Stable equivalence of knots on surfaces and virtual knot cobordisms, *J. Knot Theory Ramifications* 11 (2002) 311–322.

[3] N. Kamada, On the Jones polynomials of checkerboard colorable virtual knots, *Osaka J Math.*, 39 (2002), 325–333.

[4] N. Kamada and S. Kamada, Abstract link diagrams and virtual knots, *J. Knot Theory Ramifications* 9 (2000) 93-106.

[5] N. Kamada and S. Kamada, Double coverings of twisted links, *J. Knot Theory Ramifications* 25 (2016) 1641011, 22pages.

[6] N. Kamada and S. Kamada, Virtual links which are equivalent as twisted links, *Proc. Amer. Math. Soc.*, 148 (2020) 2273–2285.

[7] L. H. Kauffman, Virtual knot theory, *European J. Combin.* 20 (1999) 663–690.

[8] G. Kuperberg, What is a virtual link?, *Algebraic and Geometric Topology* 3 (2003) 587–591.

[9] S. Tsuyoshi, Wirtinger presentations and the Kauffman bracket. *Kobe J. Math.*, 17 (2000), 83–98

[10] J. Sawollek, On Alexander-Conway polynomials for virtual knots and links preprint (1999, arXiv:9912173(math.GT)).

Chapter 36

What Is a Knotoid?

Harrison Chapman[1]

Classical knot theory describes circles in space, transformed into a theory of diagrams by Reidemeister's theorem. Chapter 30 on virtual knot theory began with the question, "what if we have knot diagrams on surfaces besides the sphere?" In 2010, Turaev [18] asked a similar question, essentially "what if we have knot diagrams that correspond to open intervals instead of circles," from which the study of knotoids arose.

Although defined simply, there are more knotoids than there are knots, and they provide a new generalization of knots that has nice properties and applications. Every knotoid has a well-defined invariant pair of associated knots. As a result, knot invariants and bounds can be calculated directly from an associated knotoid. In some cases, this has advantages over using the knot itself for computation, especially for calculating knot invariants for which complexity is measured by the number of crossings in a diagram. Additionally, knotoids have their own collection of invariants both similar to knots (such as polynomial invariants) and unique to knotoids (such as the height). They are not just worth studying combinatorially, topologically, and algebraically—they also provide a new way of studying knots in open curves (for instance in biomolecules such as DNA and proteins [9, 11]).

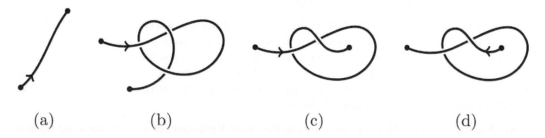

(a)   (b)   (c)   (d)

FIGURE 36.1: Some examples of knotoid diagrams. The knotoid diagrams (a) and (b) are knot-type, while knotoid diagrams (c) and (d) are proper. The knotoid (a) is trivial and the knotoid (b) is the trefoil 3_1. The knotoids (c) and (d) are equivalent as spherical knotoids (they are $k2_1$), but distinct as planar knotoids.

---

[1]The author is now at Google. This chapter was written while the author was at Colorado State University.

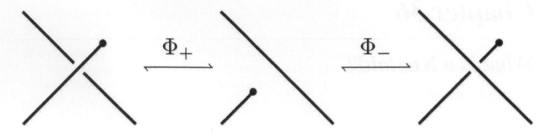

FIGURE 36.2: The forbidden moves Φ_+ and Φ_-.

36.1   What Really is a Knotoid?

Definition 1. If Σ is a surface (the plane, the sphere, the torus, etc.), a *knotoid diagram* K is a generic immersion of the interval $I = [0, 1]$ into the interior of Σ whose only singularities are transverse double points, together with over- and undercrossing information at each double point [18]. Just like with knot diagrams, only the combinatorial data of K is important.

Figure 36.1 shows some examples of knotoid diagrams. The interval has a natural orientation pointing from 0 to 1, so the image of 0 is the *tail* [7] (or *leg*) of a knotoid diagram and the image of 1 is the *head* of the knotoid diagram. We will only consider when the surface Σ is the sphere $S^2 = \mathbb{R}^2 \cup \{\infty\}$ or plane $\mathbb{R}^2$, but in general there is no reason that Σ cannot be any surface.

Knotoids are then defined as equivalence classes of knotoid diagrams under a set of moves: Locally, a knotoid diagram is only really different from a knot diagram at the tail and head endpoints. Everywhere else, the three usual Reidemeister moves $\Omega_1, \Omega_2, \Omega_3$ can still be applied. You might also devise the moves Φ_+, Φ_- in Figure 36.2 that move the endpoints, tucking them under or pushing them over some adjacent edge in the diagram.

What is the equivalence class of knotoid diagrams under the set of the two moves $\{\Phi_+, \Phi_-\}$? If you take the head of *any* knotoid diagram with at least one crossing, you can apply one of these two moves to reduce the number of crossings by one. Eventually you will end up with the trivial knotoid diagram of no crossings in Figure 36.1a. This means that if we ever allow both of the moves Φ_+ and Φ_-, we only have one equivalence class of diagrams, trivializing the problem. This earns the moves Φ_+ and Φ_- the name *forbidden moves*.

If we disallow the forbidden moves and take the usual Reidemeister moves, we wind up with knotoids:

Definition 2. A *knotoid* on a surface Σ is an equivalence class of knotoid diagrams on Σ under the usual Reidemeister moves. The set of all knotoids on Σ is denoted $\mathcal{K}(\Sigma)$. Unless mentioned otherwise, we will use the term *knotoid* to denote a knotoid on S^2.

Intuitively, allowing only the Reidemeister moves forces the tail and head of a knotoid to be pinned rigidly in place, while the rest of the diagram can move freely. The knotoid in Figure 36.1a is the trivial knotoid 0_1. The knotoids depicted by the diagrams in Figure 36.1b–d are nontrivial under the

Reidemeister moves, so the set of all knotoids is nontrivial. Any knot can actually be interpreted as a knotoid, making knotoid theory a generalization of knot theory:

Proposition 3 (Turaev [18]). Taking an oriented knot diagram of a given knot k and cutting any edge into a tail and head produces a knotoid diagram $k^\bullet$. The knotoid equivalence class of $k^\bullet$ depends on neither the initial diagram chosen nor the edge which was cut. This defines an injective map from the set of knots into the set of knotoids. A knotoid in this image is *knot-type*, while those which are not are *proper* [7] (or pure).

For instance, the knotoid in Figure 36.1b is of trefoil type 3_1, while the knotoids in Figure 36.1c–d are proper.

There are likely more proper knotoids than knot-type knotoids. In comparison with the 4 knots of at most 5 crossings, Bartholomew's initial tabulation [3] counts 41 proper spherical knotoids with at most 5 crossings. A later verified tabulation of *prime* proper spherical knotoids with at most 5 crossings by Korbalev, May, and Tarkaev [17] finds 31 such knotoids, including one 5-crossing knotoid not included in Bartholomew's original tabulation! More recently, Goundaroulis, Dorier, and Stasiak tabulated both spherical *and planar* knotoids up to 6 crossings in [10]. These counts are only for small crossing numbers, and the following is more generally not known:

Question 4. How rare are knot-type knotoids? Do they represent an exponentially small fraction of all knotoids?

Here is a reason that knotoids are more peculiar than knots: Even though the theory of knot diagrams on the plane and the sphere are equivalent, there are some knotoids which are equivalent as *spherical* knotoids but different as *planar* knotoids. For example, the knotoids in Figure 36.1c–d are equivalent spherically but distinct in the plane. So there is the (non-injective) surjection,

$$\mathbb{R}^2 \text{ knotoids } \mathcal{K}(\mathbb{R}^2) \twoheadrightarrow S^2 \text{ knotoids } \mathcal{K}(S^2).$$

Güğümcü and Kauffman provide a way to visualize planar knotoids in space [7] as in Figure 36.3. Imagine a knotoid diagram lying flat on a table. Thicken up the diagram in the direction normal to the table, so that you have an embedding of an interval $I \hookrightarrow \mathbb{R}^2 \times [-\epsilon, \epsilon]$. Finally, pin the two ends of the embedded interval in place with rigid lines, forcing the endpoints to stay in the same (x, y) position. Then roughly, planar knotoids are isotopy classes of the string tied to the endpoint lines, making sure that string never passes through itself or the rigid parallel infinite lines. This geometric interpretation of knotoids as space curves was used in [11] to represent protein chain configurations by spherical and planar knotoids, by associating to a protein its dominant knotoid representation. This interpretation of knotoids furthermore shows that the theory of knotoids is connected to the knot theory of the handlebody of genus two [15] and led Kodokostas and Lambropoulou to the topologically equivalent study of *rail knotoids* [16].

36.2 Studying Knotoids with Knots

While knotoid theory is a generalization of classical knot theory, it is possible to use tools built up for classical knots to understand knotoids: Even if a knotoid diagram does not have both its head

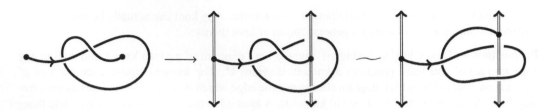

FIGURE 36.3: A knotoid diagram and two isotopic representations as rail knotoids in $\mathbb{R}^3$. The endpoints can slide along their respective lines, but string may never pass through them.

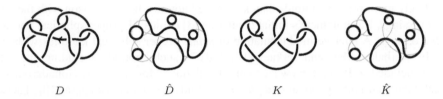

$$D \qquad\qquad \hat{D} \qquad\qquad K \qquad\qquad \hat{K}$$

FIGURE 36.4: A knot diagram D for $11n1$ and a knotoid diagram K which provides a better upper bound on the Seifert genus.

and tail in the same face, there is still a well-defined way to close the knotoid diagram into a knot diagram: For a knotoid diagram K, take any non-self-intersecting path γ connecting the tail and head that intersects K only transversely in double points. The knot diagram K_- (respectively K_+) is the diagram produced by "closing" up K by adding in the path γ so that it is always underneath (respectively over) the original edges of K. The knot K_- is called the underclosure, and K_+ the overclosure, of K. As for which pairs of knots can appear as over- and underclosures of a knotoid, Adams, Henrich, Kearney, and Scoville have shown that the answer is every pair [1]!

For knotoids in the sphere we could have picked any number of different closing paths, but they do not change the knot produced. This means that invariants of K_- (such as the fundamental group) can be computed directly from the knotoid diagram K. As a result, knotoid diagrams K can yield improved insight into the under-closed knots K_-. As an example, consider the Seifert genus $g(k)$ of a knot k (see Chapter 22). Any knot diagram D for a knot k can be coherently smoothed at each crossing into a number $|\hat{D}|$ of circles. This yields the bound $g(k) \leq \frac{1}{2}(\mathrm{cr}(D) - |\hat{D}| + 1)$ by Seifert's algorithm for constructing a Seifert surface. We can do the same for a knotoid diagram K with $K_- = k$. Now, coherently smoothing K produces an interval and a number of circles, where $|\hat{K}|$ is the number of components (that is, the number of circles plus the interval). There is a way to produce a Seifert surface for k from this, which gives the bound that $g(k) \leq \frac{1}{2}(\mathrm{cr}(K) - |\hat{K}| + 1)$. Sometimes this is a better estimate, see Figure 36.4: The knot $11n1$ has a minimal diagram presentation with $\frac{1}{2}(\mathrm{cr}(D) - |\hat{D}| + 1) = 3$, but removing an understrand with two crossings from this diagram yields a knotoid diagram with $\frac{1}{2}(\mathrm{cr}(K) - |\hat{K}| + 1) = 2$ (in fact, $g(11n1) = 2$). This idea has been used by Daikoku, Sakai, and Takase in an algorithm for improving bounds of the Seifert genus of knots [6].

However, a knotoid's associated knots alone do not uniquely determine the knotoid. There are many distinct knotoids with the same underclosure. There are even distinct knotoids with the same

$k2_1$ $k4_6$

FIGURE 36.5: Two knotoids $k2_1$ and $k4_6$ (indexing comes from Bartholomew's table [3]) which have the same under- and overclosure knots.

underclosure and overclosure knots (for instance $k2_1$ and $k4_6$ in Figure 36.5). So knotoids have more structure than the pair of their under- and overclosures alone.

Instead of making the closing arc pass over or under the knotoid diagram, it is also possible to close the knotoid diagram with *virtual crossings*. This connection to virtual knot theory is explored in [7]. Another way to study knotoids with knots was developed by Barbensi, Buck, Harrington, and Lackenby who show [2] that *unoriented* knotoids are in bijection with a specific subset of knot types, their *double-branched covers*.

36.3 Invariants of Knotoids

Just as with classical knots and virtual knots, knotoids have a number of invariants which are left unchanged by Reidemeister moves on their diagrams. Invariants are used for the same reasons as in usual knot theories, including the categorization, distinction, and identification of knotoids. Some invariants, such as the crossing number, are similar to knots:

Definition 5. The *crossing number* of a knotoid K is the minimal number of crossings in a knotoid diagram for K.

Some, like the height, only make sense for knotoids:

Definition 6. The *height* [7] (or *complexity* [18]) of a knotoid diagram D on a surface is the minimal number of edges which a path in the surface connecting the head and tail of D must cross. The *height* of a knotoid K is the minimal height over all knotoid diagrams of K.

Notably, the height distinguishes proper knotoids from knots: The height of a knotoid is 0 if and only if it is a knot-type knotoid [18]. Just as with knots (see Part VI), these invariants defined as minima are difficult to compute and provide little distinction between knotoids: For example, there are seven prime proper knotoids of crossing number 4 [17], *four* of which have height 2.

Just like knots, knotoids have algebraic invariants. There are polynomials defined similarly to knots like the *normalized bracket polynomial* (similar to the Jones polynomial) [18], and some that are based on virtual knot invariants such as the *arrow polynomial* and the *affine index polynomial* discovered by Gügümcü and Kauffman [7]. Beyond polynomials, there are more complex knotoid invariants such as the *biquandle coloring invariants* of Gügümcü and Nelson [14] and Khovanov homology of the virtual closure [7].

For polynomial knotoid invariants we do not know the answer to the analogue of the famous Jones conjecture:

Question 7 (c.f. [7]). Does the normalized bracket polynomial detect the unknot? Is there any polynomial knotoid invariant that detects the unknot?

For knotoids it is not even known if more rich invariants like the virtual closure Khovanov homology detect the trivial knotoid, even though it is known in the case of classical knots!

The arrow and affine index polynomials provide lower bounds on the height of a knotoid. On the other hand, any knotoid diagram of a given knotoid provides an upper bound on the knotoid's height. Since the height of a knotoid is a measurement of its difference from being knot-type, it would be interesting to answer:

Question 8. What is the average height of a knotoid of minimal crossing number n?

It is not even known what the average height of a knotoid diagram itself is. These questions are related to the average distance between faces of quadrangulations of the sphere [4], which is known to grow like $n^{1/4}$.

Surprisingly little is known about knotoids and their invariants, even for knot-type knotoids. In particular, the following question of Turaev [18] remains unanswered:

Question 9. It is known that the minimal crossing number of a knot-type knotoid $k^\bullet$ is at most that of the minimal crossing number of its knot k; are these two quantities equal? That is, is there a knot-type knotoid $k^\bullet$ whose minimal crossing knotoid diagram has positive height?

Like many other questions in knot theory, this is one whose answer may seem obvious but eludes proof. Perhaps there is an invariant that provides the right viewpoint.

36.4   Knotoids and Open Knots

Knotoids have been applied as a new way to study open knotting (see Chapter 86), primarily in proteins [9, 11, 5] where very loose knots essential to the function of the protein can manifest; a study initiated by Goundaroulis. Given an open curve γ embedded in $\mathbb{R}^3$ and a generic normal direction $\vec{n} \in S^2 \subset \mathbb{R}^3$, the diagram projection of γ to $\vec{n}^\perp$ is a knotoid diagram. Of course, the knotoid depends strongly on the direction $\vec{n}$, so to capture an accurate picture of the knotting of γ, it is important to consider all possible directions $\vec{n}$; this is approximated by choosing some large number of directions $\vec{n}$ uniformly randomly. Then the open curve γ is considered, essentially, to be whichever knotoid was most common in the above algorithm. It is also possible to consider the entire "map" of knotoid types for additional detail. For modeling proteins with disulphide bonds, which can exhibit motifs such as *lassos*, the theory of *bonded knotoids* is also introduced in in [11].

This is very similar to how open physical knots are classically studied; it differs mainly in classifying the resulting knotoid diagram K itself rather than simply one of the knot diagrams K_- or K_+. So, it recognizes open curves with rather tight knots to be essentially the same, but provides a finer amount of distinction in general. It will be exciting to see how this technique advances the field!

36.5   More and More Knotoids

There are many more facets of knotoid theory to discover. For example, the theory of braidoids and their connections to knotoids initiated in [13, 12, 8] (see Chapter 37), the connections of knotoids to virtual knot theory [7, 13], as well as the theories of rail knotoids [16] and bonded knotoids [11]. There are other topics—such as multiplication of knotoids, knotoids' connections to theta curves, and flat knotoids—worth learning about. Beyond that, there is still much more to discover for knotoids: There are many more open questions about knotoids awaiting proof.

36.6   Bibliography

[1] Colin Adams, Allison Henrich, Kate Kearney, and Nicholas Scoville. Knots related by knotoids. *The American Mathematical Monthly*, 126(6):483–490, 2019.

[2] Agnese Barbensi, Dorothy Buck, Heather A Harrington, and Marc Lackenby. Double branched covers of knotoids. 2018. http://arxiv.org/abs/1811.09121.

[3] Andrew Bartholomew. Knotoids, 2010. `http://www.layer8.co.uk/maths/knotoids/index.htm`.

[4] J. Bouttier, P. Di Francesco, and E. Guitter. Geodesic distance in planar graphs. *Nuclear Physics B*, 663:535–567, July 2003.

[5] Pawel Dabrowski-Tumanski, Pawel Rubach, Dimos Goundaroulis, Julien Dorier, Piotr Sukowski, Kenneth C Millett, Eric J Rawdon, Andrzej Stasiak, and Joanna I Sulkowska. KnotProt 2.0: a database of proteins with knots and other entangled structures. *Nucleic Acids Research*, 47(D1):D367–D375, 2018.

[6] Kenji Daikoku, Keiichi Sakai, and Masamichi Takase. On a move reducing the genus of a knot diagram. *Indiana University Mathematics Journal*, 61(3):1111–1127, 2012.

[7] Neslihan Gügümcü and LouisăH. Kauffman. New invariants of knotoids. *European Journal of Combinatorics*, 65:186 – 229, 2017.

[8] Neslihan Gügümcü and Sofia Lambropoulou. Knotoids, braidoids and applications. *Symmetry*, 9(12):315, Dec 2017.

[9] Dimos Goundaroulis, Julien Dorier, Fabrizio Benedetti, and Andrzej Stasiak. Studies of global and local entanglements of individual protein chains using the concept of knotoids. *Scientific Reports*, 7(1), Jul 2017.

[10] Dimos Goundaroulis, Julien Dorier, and Andrzej Stasiak. A systematic classification of knotoids on the plane and on the sphere, 2019. https://arxiv.org/abs/1902.07277.

[11] Dimos Goundaroulis, Neslihan Gügümcü, Sofia Lambropoulou, Julien Dorier, Andrzej Stasiak, and Louis Kauffman. Topological models for open-knotted protein chains using the concepts of knotoids and bonded knotoids. *Polymers*, 9(12):444, Sep 2017.

[12] Neslihan Gügümcü and Sofia Lambropoulou. Braidoids. https://arxiv.org/abs/1908.06053.

[13] Neslihan Gügümcü and Sofia Lambropoulou. *On knotoids, braidoids and their applications.* PhD thesis, PhD thesis, National Technical U. Athens, 2017.

[14] Neslihan Gügümcü and Sam Nelson. Biquandle coloring invariants of knotoids. *Journal of Knot Theory and Its Ramifications*, nil(nil):1950029, 2019.

[15] Dimitrios Kodokostas and Sofia Lambropoulou. A spanning set and potential basis of the mixed hecke algebra on two fixed strands. *Mediterranean Journal of Mathematics*, 15(5):192, 2018.

[16] Dimitrios Kodokostas and Sofia Lambropoulou. Rail knotoids. *Journal of Knot Theory and Its Ramifications*, 28(13):1940019, 2019.

[17] Filipp Glebovich Korablev, Yana Konstantinovna May, and Vladimir Viktorovich Tarkaev. Classification of low complexity knotoids. *Sibirskie Èlektronnye Matematicheskie Izvestiya [Siberian Electronic Mathematical Reports]*, 15:1237–1244, 2018.

[18] Vladimir Turaev. Knotoids. *Osaka J. Math.*, 49(1):195–223, 2012. See also arXiv:1002.4133v4.

Chapter 37

What Is a Braidoid?

Neslihan Güğümcü, Izmir Institute of Technology, Turkey

37.1   An Overview

Knotoids, introduced by Turaev [16], are immersions of the oriented unit interval into oriented surfaces, endowed with over/under crossing information at each transversal double point. In other words, knotoid diagrams are open-ended knot diagrams as exemplified in Figure 37.1. A knotoid

FIGURE 37.1: Knotoid diagrams.

diagram in S^2 generalizes the notion of a long knot with two endpoints (called *leg* and *head*) that can lie in any planar region complementary to the knotoid diagram. Knotoid diagrams are considered up to three Reidemeister moves, see Figure 37.2, and the isotopy of the surface they lie in. Moving the endpoints of a knotoid diagram is restricted: We forbid the two moves shown in Figure 37.3. Precisely, it is forbidden to pull/push an endpoint over/under a strand.

The notion of knotoid can be extended to the notion of multi-knotoid. A *multi-knotoid diagram* in an oriented surface Σ is a union of a knotoid diagram and a finite number of oriented knot diagrams in Σ. Multi-knotoid diagrams in Σ are subject to the Reidemeister moves and isotopy of Σ. The induced isotopy classes of multi-knotoid diagrams are called *multi-knotoids*.

Knotoids in $\mathbb{R}^2$ admit a 3-dimensional interpretation through planar projections of open-ended curves in 3-dimensional space whose ends are attached on two parallel lines [3], see Figure 37.4. With this interpretation, knotoids provide a direct and realistic insight for the study of certain entangled physical objects such as proteins [8, 9, 2]. (The reader is referred to the Chapter 36 for further details.) Here we present *braidoids* introduced by the author and Lambropoulou [6, 7]. A braidoid generalizes the notion of a classical braid: It is a system of descending strands interacting with each other at a finite number of points assigned with a specified over and under information. The ends of strands are assumed to be fixed at top and bottom except the two ends that are not necessarily lying at top or bottom and are subject to special topological moves that we explain below. Before going into the precise definition let us view some examples of braidoid diagrams given in Figure 37.5.

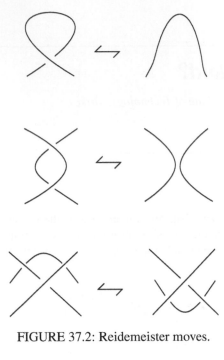

FIGURE 37.2: Reidemeister moves.

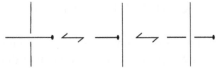

FIGURE 37.3: Forbidden knotoid moves.

37.2   The Precise Definition of a Braidoid

37.2.1   Braidoid Diagrams

Let I denote the unit interval $[0, 1] \subset \mathbb{R}$. We identify $\mathbb{R}^2$ with the xt-plane with the t-axis directed downward. A *braidoid diagram* B is a system of a finite number of arcs lying in $I \times I \subset \mathbb{R}^2$. The arcs are called the *strands* of B. Each strand is naturally oriented downward with no local maxima or minima, and there is only a finite number of intersection points among the strands which are transversal double points endowed with over/under data. Such intersection points are called *crossings* of B.

B has two types of strands. A *classical strand* is like a braid strand connecting two points, one that lies on $I \times \{0\}$ and the other lies on $I \times \{1\}$. Each end of a classical strand is assumed to be fixed. The other type of strands, the so called *free strands*, are of a more flexible nature: One or two ends of a free strand are located at points which are not necessarily on $I \times \{0\}$ or on $I \times \{1\}$, and they are not assumed to be fixed. Such ends of free strands are denoted by graphical nodes to be

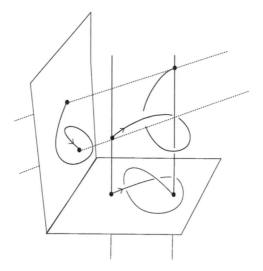

FIGURE 37.4: A space curve and its different planar projections as knotoids.

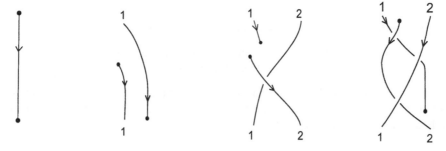

FIGURE 37.5: Some examples of braidoid diagrams.

distinguished from the fixed ends (that might be the ends of either classical or free strands), and are called *endpoints* of B. B has exactly two endpoints. We call an endpoint a *head* if it is the terminal point of a free strand and a *leg* if it is the beginning point of a free strand, in analogy with the leg and the head of a knotoid diagram. Each fixed end of B lying on $I \times \{0\}$ is paired up with a fixed end on $I \times \{1\}$ that is in the same vertical alignment with it. Paired ends are called *corresponding ends*. Each pair of corresponding ends are numbered from left to right, as we show in Figure 37.5.

37.2.2 Isotopy Moves of Braidoid Diagrams

We allow Δ-*moves* on braidoid diagrams that replace a segment of a strand with two segments in a triangular region free of any of the endpoints (see Figure 37.6). A Δ-move on a braidoid diagram preserves the downward orientation of the braidoid strands and respects the crossing information of the strands intersecting the triangular region of the move. The Reidemeister II and III moves in which the downward direction of strands is preserved, are special cases of the Δ-moves on braidoid diagrams. Similar to the forbidden moves of knotoid diagrams, moving the endpoints of a braidoid diagram over/under a strand, as shown in Figure 37.7, are *forbidden moves* for braidoid diagrams. It

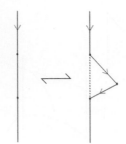

FIGURE 37.6: A Δ-move.

is clear that allowing forbidden moves would cancel any braiding of the free strands. The following

FIGURE 37.7: Forbidden braidoid moves.

moves displacing the endpoints are allowed on braidoid diagrams.

1. *Vertical moves*: An endpoint can be pulled up/down along the vertical direction as long as the forbidden moves are not violated (see Figure 37.8).

2. *Swing moves*: An endpoint can be moved to the right and to the left like a pendulum (see Figure 37.9), as long as the downward orientation on the swinging arc is preserved and the forbidden moves are not violated.

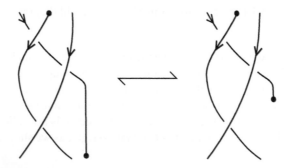

FIGURE 37.8: A vertical move.

Definition 1. A *braidoid* is an isotopy class of braidoid diagrams taken up to the isotopy relation generated by the oriented Reidemeister II and III moves and planar Δ moves together with the swing and vertical moves for the endpoints.

FIGURE 37.9: Swing moves.

37.3   Braidoids in Relation with Knotoids

In 1923 Alexander showed that any classical knot and link can be represented in braided form [1], then in 1936 Markov proved a one-to-one correspondence between the set of classical links considered up to link isotopy and the set of braids considered up to a braid equivalence relation generated by braid isotopy moves, conjugation and stabilization [15]. We have analogues of Alexander's and Markov's theorems that relate braidoids equipped with a special labeling to multi-knotoids in $\mathbb{R}^2$.

37.3.1   Labeled Braidoid Diagrams

Definition 2. A *labeled braidoid diagram* is a braidoid diagram such that each pair of its corresponding ends —not the endpoints—is labeled either with o or u.

Labeled braidoid diagrams are subject to the braidoid Δ-moves, the vertical moves, and the *restricted swing moves*, shown in Figure 37.10, whereby the swinging of an endpoint takes place within the interior of the vertical strip determined by the vertical lines that pass through two neighboring pairs of corresponding ends. During the restricted swing moves, the endpoints cannot surpass the vertical line determined by any pair of corresponding ends. Clearly, a restricted swing move is transformed into a planar isotopy move by the closure defined below. The reason for restricting the swing moves for labeled braidoid diagrams is to avoid any incident of forbidden moves on the multi-knotoid diagram obtained by the closure. See Figure 37.13 that depicts how an unrestricted swing move gives rise to nonequivalent knotoids under the closure.

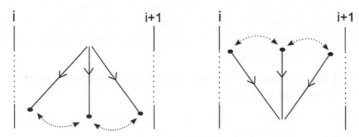

FIGURE 37.10: Restricted swing moves.

Definition 3. *Labeled braidoid isotopy* is generated by the braidoid Δ-moves, vertical moves and the restricted swing moves, preserving at the same time the labeling. Equivalence classes of labeled braidoid diagrams under this isotopy relation are called *labeled braidoids*.

37.3.2　A Closure for Labeled Braidoids

The correspondence between labeled braidoid diagrams and multi-knotoid diagrams in $\mathbb{R}^2$ is based on a closure operation that is defined in analogy with the closure of braid diagrams in handlebodies [13] (see the Chapter 7 for further details).

In this closure operation, each pair of the corresponding ends of a labeled braidoid diagram is connected with an arc that goes entirely over or under the rest of the diagram according to the label of ends, as illustrated in Figure 37.11 and in Figure 37.12.

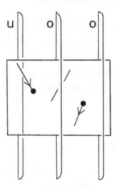

FIGURE 37.11: Closure of a labeled braidoid diagram.

Precisely, if a pair is labeled with o then the connecting arc goes entirely over the diagram and if a pair is labeled with u then the connecting arc goes entirely under the diagram. The connecting arcs lie on the right-hand side of the pairs of corresponding ends, in an arbitrarily close distance to the vertical lines determined by these pairs so that there is no endpoint of the braidoid diagram between the vertical lines and the connecting arcs. The resulting diagram is clearly a knotoid or a multi-knotoid diagram in the plane. Note that the closure of the trivial braidoid diagram shown in Figure 37.5 that has no pairs of corresponding braidoid ends, is assumed to be the trivial knotoid. Also, the isotopy class of the resulting multi-knotoid diagram depends on the labeling. A braidoid diagram equipped with two different labelings might yield to non-isotopic closures, as shown in Figure 37.12 (it is left as an exercise for the interested reader to verify the knotoid on the right-hand side of the figure is actually non-trivial).

It is shown in [6, 7] that the closure operation induces a well-defined map on the set of labeled braidoids to the set of multi-knotoids in $\mathbb{R}^2$.

Figure 37.13 illustrates an unrestricted swing move on an endpoint of a labeled braidoid diagram that corresponds to a knotoid forbidden move upon closure.

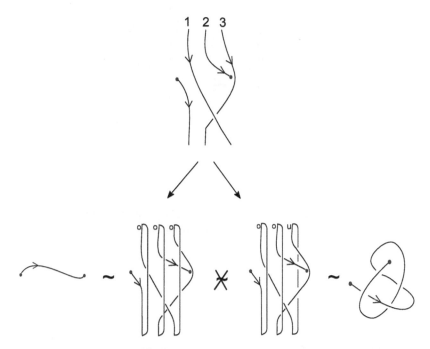

FIGURE 37.12: Two different labelings and nonequivalent knotoids upon closure.

37.3.3 How to Turn a Knotoid Diagram into a Braidoid Diagram

Let K be a knotoid or a multi-knotoid diagram in a plane equipped with the top-to-bottom direction. One way to obtain an inverse operation to the closure we define above, that is, to transform K to a labeled braidoid diagram whose closure is isotopic to K, is to manipulate K by eliminating its arcs oriented upward, namely the *up-arcs*. Before the elimination begins, the knotoid diagram K is subdivided (marked with dots) starting from its local maxima and minima until each of its up-arcs contains only one type of crossings (either over- or under- crossings) or contains no crossings at all. We label each up-arc with o or u according to the type of crossings it contains. If an up-arc does not contain any crossings, then we are free to label the up-arc either with o or u. Braidoiding moves are analogues of the braiding moves defined for classical braids [11, 12, 14].

Definition 4. A *braidoiding move* consists of cutting an up-arc at a point (we call a *cut-point*), and then pulling the resulting ends entirely over or under the rest of the diagram according to the label of the up-arc, to $I \times \{0\}$ and $I \times \{1\}$ preserving the alignment with the cut-point (see Figure 37.14). A cut-point is chosen so that it is not vertically aligned with another cut-point or with an endpoint of K. To obtain a pair of braidoid strands (that is, monotonically descending strands with a label), we complete the braidoiding move by Δ-moves applied to the upward-directed pieces of the resulting strands (see the second step shown in Figure 37.14). Finally, we label the resulting braidoid strands with o or u with respect to the label of the up-arc eliminated. Notice that in the last instance of Figure 37.14, the pair of the resulting corresponding ends is joined up together with a connecting arc that goes entirely over the diagram in accord with the labeling. By this, we obtain a closed strand which can be retracted back to the initial up-arc QP. Note that during this isotopy, a

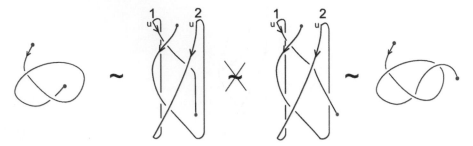

FIGURE 37.13: Swing moves may give rise to nonequivalent knotoids.

violation of the knotoid forbidden moves is avoided by choosing the connecting arc close enough to the vertical line determined by the pair of corresponding ends.

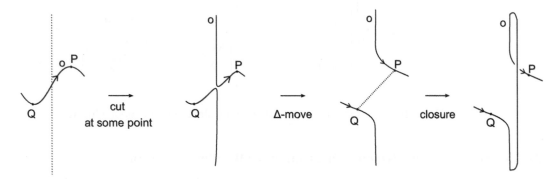

FIGURE 37.14: A braidoiding move and its closure.

In [6, 7] we present two *braidoiding algorithms* for turning a planar multi-knotoid diagram into a braidoid diagram. Both of these algorithms are based on the *braidoiding moves*. Figure 37.15 exhibits one of the braidoiding algorithms which consists of rotating a crossing of a given multi-knotoid diagram by 90 degrees if the crossing is contained in one up-arc and by 180 degrees if the crossing is contained in two up-arcs and then applying the braidoiding moves on the resulting multi-knotoid diagram. The reader is encouraged to see the closure of the labeled braidoid diagram obtained in Figure 37.15 is isotopic to the given knotoid diagram, and is referred to [6, 7] for the technical details of the algorithms.

The closure defined for labeled braidoid diagrams and the braidoiding algorithms provide us the following theorem [6, 7, 5] that is an analogue of Alexander's theorem [1].

Theorem 5. Any multi-knotoid diagram in $\mathbb{R}^2$ is isotopic to the closure of a labeled braidoid diagram.

The braidoiding algorithm presented in Figure 37.15 also provides the following sharpened version of Theorem 5 [6, 5].

Theorem 6. Any multi-knotoid diagram in $\mathbb{R}^2$ is isotopic to the closure of a labeled braidoid diagram whose corresponding ends labeled only with u.

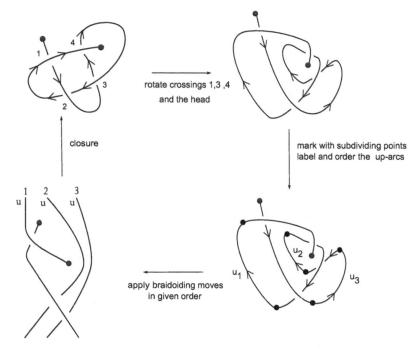

FIGURE 37.15: A knotoid diagram and the associated labeled braidoid diagram.

37.3.4 A Geometric Markov Theorem for Braidoids

It is possible that two non-isotopic labeled braidoid diagrams in a plane close to isotopic planar knotoids or multi-knotoids (consider the closures of the first three braidoid diagrams in Figure 37.5 with some labeling). A natural question arising is the following: *Is there a way to define a relation between labeled braidoids that have isotopic closures?*

In [7] we give a geometric relation between labeled braidoid diagrams closing to isotopic knotoids or multi-knotoids. This relation is induced by the *L*-moves [6, 7] and some special swing moves defined on labeled braidoid diagrams. The *L*-moves were originally defined for classical braids by Lambropoulou [11, 12, 14], and utilized to prove a one-move Markov theorem for classical braids [11].

There are two types of *L*-moves as shown in Figure 37.16. An L_o- move (respectively an L_u-move) consists of cutting a labeled braidoid strand at an interior point and pulling the ends of the resulting sub-strands to the top and bottom lines over the rest of the diagram (respectively under the rest of the diagram) so that a new pair of strands is obtained whose ends are vertically aligned with the cut-point. The resulting strands are labeled with o (respectively with u). Notice that when the corresponding ends of the resulting strands are connected with an overpassing arc (respectively with an underpassing arc), we obtain a closed strand which is isotopic to the original braidoid strand. This observation is sufficient to deduce that two labeled braidoid diagrams which are related to each other via *L*-moves have isotopic closures. We also observe that there are some type of unrestricted swing moves that yield isotopic knotoids or multi-knotoids through the closure. With this type of swing moves, the endpoint surpasses the vertical line determined by a pair of

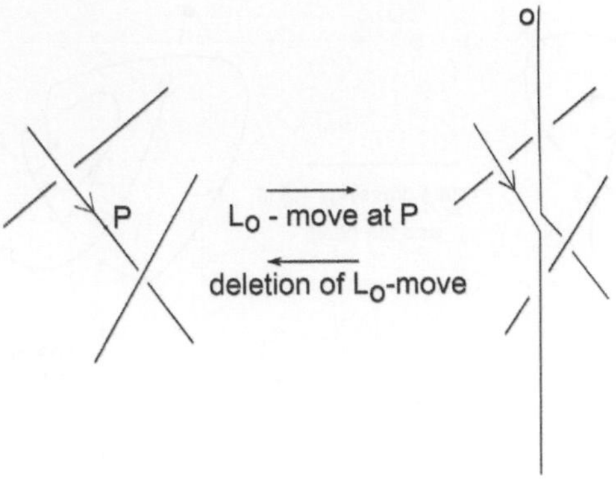

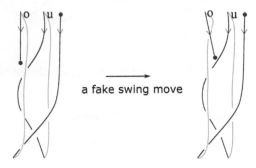

FIGURE 37.16: An L_o-move on a braidoid strand.

corresponding ends but this do not affect the isotopy class of the resulting (multi-)knotoid diagram. See Figure 37.17. We call these moves *fake swing moves*.

FIGURE 37.17: A fake swing move.

The *L*-moves together with the fake swing moves extend the labeled braidoid isotopy and provide us the following theorem that is an analogue of Markov's theorem [15] for labeled braidoids.

Theorem 7. The closures of two labeled braidoid diagrams are isotopic multi-knotoid diagrams in $\mathbb{R}^2$ if and only if the labeled braidoid diagrams relate to each other by a sequence of labeled braidoid isotopy moves, addition/deletion of *L*-moves, and fake swing moves.

For proving this theorem, we examine the effect of all possible algorithmic choices made in the braidoiding algorithms and the transformation of the knotoid isotopy moves under the braidoiding moves. The reader is referred to [6] for the details of the proof.

37.4 Discussion

Braidoids provide a new diagrammatic theory that is, in fact, a braid theory for the theory of knotoids. An underlying combinatorial structure for braidoids is discussed in [5, 7], In this combinatorial structure, braidoid diagrams can be partitioned into a finite number of *building blocks* extending the braiding generators. With the algebraic expressions corresponding to the building blocks and the Alexander and Markov-type theorems discussed throughout this paper, braidoids suggest an algebraic tabulation for polymers, specifically for proteins [7].

Acknowledgment

The author is grateful to Sofia Lambropoulou for fruitful discussions on braidoids.

37.5 Bibliography

[1] J.W. Alexander. A lemma on systems of knotted curves, *Proc. Nat. Acad. Sci. U.S.A.* (1923), 9, pp. 93–95.

[2] J. Dorier, D. Goundaroulis, F. Benedetti, A. Stasiak. Knoto-ID: A tool to study the entanglement of open protein chains using the concept of knotoids, *Bioinformatics*, Volume 34, Issue 19, 01 October 2018, pp. 3402–3404, https://doi.org/10.1093/bioinformatics/bty365.

[3] N. Gügümcü, L.H. Kauffman. New invariants of knotoids, *European Journal of Combinatorics*, 65C, 2017, pp. 186–229, https://doi.org/10.1016/j.ejc.2017.06.004.

[4] N. Gügümcü, L.H. Kauffman. On the height of knotoids, Algebraic Modeling of Topological and Computational Structures and Applications, THALES, Athens, Greece, July 1-3, 2015, *Springer Proceedings in Mathematics & Statistics (PROMS)* Volume 219; S. Lambropoulou et al. Eds; (2017); https://doi.org/10.1007/987-3-319-681030.

[5] N. Gügümcü. On knotoids, braidoids and their applications, PhD thesis, National Technical University of Athens (2017).

[6] N. Gügümcü, S. Lambropoulou. Braidoids, to appear in *Israel Journal of Mathematics*.

[7] N. Gügümcü, S. Lambropoulou. Knotoids, Braidoids and Applications, *Symmetry* (2017), 9(12), 315, https://doi.org/10.3390/sym9120315.

[8] D. Goundaroulis, J. Dorier, F. Benedetti, A. Stasiak. Studies of global and local entanglements of individual protein chains using the concept of knotoids. *Sci. Reports* (2017), 7, 6309.

[9] D. Goundaroulis, N. Gügümcü, S. Lambropoulou, J. Dorier, A. Stasiak, L.H. Kauffman. Topological models for open knotted protein chains using the concepts of knotoids and bonded knotoids, *Polymers*, Special issue: Knotted and Catenated Polymers, Dusan Racko and Andrzej Stasiak Eds. (2017), 9(9), 444, DOI: 10.3390/polym9090444.

[10] L.H. Kauffman. Virtual knot theory, *European Journal of Combinatorics*, Vol:20, Issue 7 (1999), pp. 663–691; https://doi.org/10.1006/eujc.1999.0314.

[11] S. Lambropoulou. Short proofs of Alexander's and Markov's theorems, Warwick preprint (1990).

[12] S. Lambropoulou. A study of braids in 3-manifolds, PhD thesis, University of Warwick (1993).

[13] R. H. Oldenburg, S. Lambropoulou. Knot theory in handlebodies, *Journal of Knot Theory and Its Ramifications*, 11(2002).

[14] S. Lambropoulou, C.P. Rourke. Markov's theorem in 3-manifolds, *Topology and Its Applications* (1997), 78, pp. 95–122.

[15] A.A. Markov. Über die freie Äquivalenz geschlossener Zöpfe, *Rec. Math. Moscou* (1936), 1,(43), pp. 73–78.

[16] V. Turaev. Knotoids, *Osaka Journal of Mathematics*, 49 (2012), no. 1, 195–223. See also arXiv:1002.4133v4.

Chapter 38

What Is a Singular Knot?

Zsuzsanna Dancso, University of Sydney

38.1 Introduction

A **singular knot** is an immersed circle in $\mathbb{R}^3$, whose singularities are limited to finitely many *transverse double points*. We will focus on *oriented* singular knots, which are equipped with a direction: Figure 38.1 shows an example. Other knotted objects, such as braids, tangles, or knotted graphs have similar singular versions, see Figure 38.1 for an example of a singular braid. A **singular knot diagram** is a planar projection of a singular knot, which has two types of *crossings*: regular crossings with over/under strand information, and double points. Reidemeister moves can be formulated to describe singular knots modulo ambient isotopy, see Figure 38.2. For minimal sets of oriented Reidemeister moves see [4].

38.2 The Vassiliev Filtration

Singular knots give rise to a decreasing filtration, called the *Vassiliev filtration*, on the infinite-dimensional vector space generated by isotopy classes of knots. Knot invariants which vanish on some step of this filtration are called *finite type* or *Vassiliev* invariants. The theory originates in Vassiliev's papers [20, 21], though his original approach was quite different from what we present here. Examples of Vassiliev invariants include many famous knot invariants, for example the coefficients of any knot polynomial – after an appropriate variable substitution – are of finite type

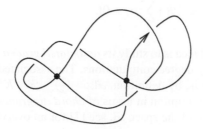 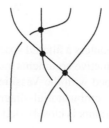

FIGURE 38.1: An example of a 2-singular knot – that is, a knot with two double points; and a 3-singular braid on four strands.

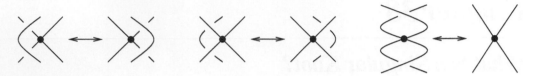

FIGURE 38.2: Reidemeister moves for singular knots, in addition to the usual Reidemeister 1, 2 and 3 moves.

FIGURE 38.3: Resolution of singularities.

[5, 1]. In this section we describe the Vassiliev filtration for knots and other knotted objects. For more on finite type invariants see Chapter 82 of this book.

Let $\mathcal{K}$ denote the set of (isotopy classes of) oriented knots, and let $\mathbb{Q}\mathcal{K}$ be the $\mathbb{Q}$-vector space[1] spanned by elements of $\mathcal{K}$. In other words, elements of $\mathbb{Q}\mathcal{K}$ are formal linear combinations of knots.

Let $\mathcal{K}^n$ denote the set of (isotopy classes of) n-singular oriented knots, and let $K \in \mathcal{K}^n$. Each singularity of K can be **resolved** in two ways: by replacing the double point with an over-crossing or with an under-crossing. Note that the notion of "over" and "under" crossings don't depend on the choice of knot projection – this is true not only in $\mathbb{R}^3$ but in any oriented manifold. We define a **resolution map** $\rho : \mathcal{K}^n \to \mathbb{Q}\mathcal{K}$ as follows: replace each singularity of K by the difference of its two resolutions, as shown in Figure 38.3. This produces a linear combination of 2^n knots with ± 1 coefficients, which we call the **resolution** of K. By an abuse of notation, we denote also by $\mathcal{K}^n$ the linear span of the image $\rho(\mathcal{K}^n)$ in $\mathbb{Q}\mathcal{K}$.

The subspaces $\mathcal{K}^n$ form a decreasing filtration on $\mathbb{Q}\mathcal{K}$, called the **Vassiliev filtration**:

$$\mathbb{Q}\mathcal{K} = \mathcal{K}^0 \supset \mathcal{K}^1 \supset \mathcal{K}^2 \supset \mathcal{K}^3 \supset \dots \qquad (38.1)$$

It's a worthwhile exercise to show that $\mathcal{K}^1$ is the set of elements of $\mathbb{Q}\mathcal{K}$ whose coefficients sum to zero:

$$\mathcal{K}^1 = \{ \sum_i \alpha_i K_i \in \mathbb{Q}\mathcal{K} \mid \sum_i \alpha_i = 0 \}.$$

When one encounters a filtered space, a natural idea is to study its *associated graded* space instead: this enables inductive arguments and degree-by-degree computations. The associated graded space of knots equipped with the Vassiliev filtration is, by definition, $\mathcal{A} := \bigoplus_{n=0}^{\infty} \mathcal{K}^n / \mathcal{K}^{n+1}$. The space $\mathcal{A}$ has a useful combinatorial–diagrammatic description in terms of *chord diagrams*. The study of finite type invariants is closely tied to the study of the space $\mathcal{A}$, see [1] for an overview and many further references.

We now examine the Vassiliev filtration from a more algebraic perspective through the example of braid groups. Recall that the **braid group** B_n consists of braids on n strands up to braid isotopy, and

---

[1] One may replace $\mathbb{Q}$ with one's favourite field of characteristic zero, at no cost.

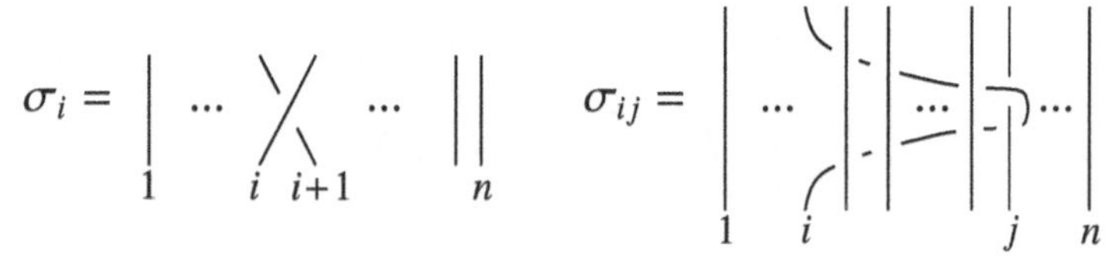

$$\sigma_i = \left| \quad \cdots \quad \times \quad \cdots \quad \right| \right| \qquad \sigma_{ij} = \left| \quad \cdots \quad \right|$$

FIGURE 38.4: The generators σ_i of B_n, and the generators σ_{ij} of PB_n.

the group operation is given by vertical stacking; see Chapter 7. The *Artin presentation* is a finite presentation for the group B_n in terms of generators and relations (see also Figure 38.4):

$$B_n = \langle \sigma_1, ..., \sigma_{n-1} \mid \sigma_i \sigma_{i+1} \sigma_i = \sigma_{i+1} \sigma_i \sigma_{i+1}, \text{and } \sigma_i \sigma_j = \sigma_j \sigma_i \text{ when } |i - j| \geq 2 \rangle.$$

Following the strands of a braid from bottom to top determines a permutation of n points; this gives rise to a group homomorphism $S : B_n \to S_n$ to the symmetric group. For example, (any resolution of) the braid of Figure 38.1 is mapped to the cycle (1324). The image $S(b) \in S_n$ of a braid b is called the **skeleton** of b; the kernel of this homomorphism is the **pure braid group** PB_n. The pure braid group also has a finite presentation due to Artin, in terms of the generators σ_{ij} $(i < j)$, shown in Figure 38.4, and somewhat more complicated relations. For more on pure braid presentations see [14].

The Vassiliev filtration can be defined for braid groups the same way as it is defined for knots; we first discuss it for the pure braid group PB_n. In this case $\mathbb{Q}PB_n$ is the *group algebra* of the pure braid group over the field $\mathbb{Q}$. Denote the linear subspace generated by the ρ-image of n-singular pure braids by $\mathcal{B}^n$. Just like in the case of knots, $\mathcal{B}^1 = \{\sum_i \alpha_i b_i \in \mathbb{Q}PB_n \mid \sum_i \alpha_i = 0\}$. This is a two-sided ideal in $\mathbb{Q}PB_n$, called the **augmentation ideal**.

However, in this case much more is true. Any n-singular braid can be written as a product of n 1-singular braids: one can cut the braid into n horizontal levels so that each level has exactly one singularity (one might need to "jiggle" the braid first, if there are multiple double points at the same horizontal level). In other words, $\mathcal{B}^n$ is simply the n-th power of the ideal $\mathcal{B}^1$.

A similar statement holds for the braid group B_n, with the modification that in $\mathbb{Q}B_n$ linear combinations are only allowed within each coset of PB_n; in other words, only braids of the same skeleton can be combined. Steps of the Vassiliev filtration once again correspond to powers of the augmentation ideal. In fact, this phenomenon is much more general. Many flavours of knotted objects can be finitely presented as an algebraic structure of some kind: *tangles* form a *planar algebra*, see Part III of this book; *knotted trivalent graphs* have their own special set of operations [19]; virtual and welded tangles form *circuit algebras* [3]. The Vassiliev filtration – or its appropriate generalizations – coincide with the powers of the augmentation ideal in all of these examples.

A notable exception is the case of knots. A "multiplication operation" does exist for knots: it is called a **connected sum** and denoted #; see Figure 38.5 for a pictorial definition. It is not true, however, that any n-singular knot is the connected sum of n 1-singular knots. Knots which cannot be expressed as a non-trivial connected sum are called **prime knots**, and there exist prime knots of arbitrary crossing number, in particular all *torus knots* are prime. In other words, knots are not *finitely generated* as an "algebraic structure."

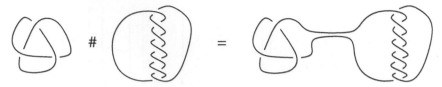

FIGURE 38.5: The connected sum operation of knots: the reader is encouraged to check that it is well defined.

The algebraic description of the Vassiliev filtration we outlined here gives rise to a deep connection between knot theory and quantum algebra. Let $\mathcal{K}$ denote some class of knotted objects (such as knots, braids, tangles, etc). A central question in the study of finite type invariants is to find a *universal finite type invariant*, which contains all of the information that any finite type invariant can retain about $\mathcal{K}$. More precisely, a universal finite type invariant is a filtered map $Z : \mathbb{Q}\mathcal{K} \to \hat{\mathcal{A}}$ which takes values in the *degree completed* associated graded space, and which satisfies a certain universality property. Perhaps the most famous universal finite type invariant is the *Kontsevich integral* of knots [13]. It is still an open problem whether the Kontsevich integral separates knots.

Assume that some class of knotted objects $\mathcal{K} = \langle g_1, ..., g_k | R_1, ..., R_l \rangle$ forms a finitely presented algebraic structure with generators g_i and relations R_j (e. g. $\mathcal{K}$ may be the braid group). Assume one looks for a universal finite type invariant Z which respects operations in the appropriate sense (e. g. Z is an algebra homomorphism). Then it is enough to find the values of the generators $Z(g_i) \in \mathcal{A}$, subject to equations arising from each R_j. In other words, one needs to solve a set of equations in a graded space. For several examples of $\mathcal{K}$, this set of equations turns out to be interesting in its own right: for knotted trivalent graphs or *parenthesized braids* they are the equations which define *Drinfeld associators* in quantum algebra [15, 2, 7]; for *welded foams* they are the *Kashiwara-Vergne* equations of Lie theory [3].

38.3   Invariants of Singular Knots

Vassiliev's idea in the early '90s was to extend number-valued knot invariants to singular knots using the resolution of singularities discussed above. This led to an explosion of activity in knot theory at the time. Since then, singular knots have been studied in their own right by many researchers, and various non-numerical invariants have been extended to singular knots. Here we summarise some of these results.

In [16], Murakami, Ohtsuli and Yamada developed a skein theory for the HOMFLY polynomial using a generalisation of singular knots called *abstract singular knots*. We present a simplified version of their result, which is used in several applications. The HOMFLY polynomial is determined by the skein relation shown on the left of Figure 38.7, and its value on the unknot; for more detail see Chapter 66.

To define abstract singular knots, one replaces double points with an oriented *thick edge* from the

FIGURE 38.6: Creating an oriented abstract singular knot from a singular knot, and an abstract singular knot which cannot be oriented.

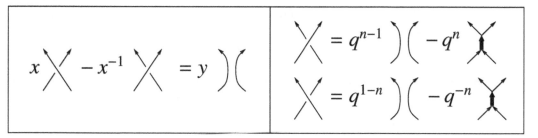

FIGURE 38.7: The original HOMFLY skein relation on the left; the HOMFLY skein relations for abstract singular knots [16] on the right.

incoming to the outgoing edges of the double point, creating an oriented trivalent graph embedded in $\mathbb{R}^3$, as in Figure 38.6. These trivalent graphs are called **oriented abstract singular knots** and they are characterised by two properties:

1. Each vertex is incident to one thick and two thin edges.

2. At each vertex the thin edges are oriented the same way, while the thick edge is oppositely oriented.

Embedded trivalent graphs which satisfy condition (1) are called **abstract singular knots**. Not every such abstract singular knot can be oriented to satisfy (2), and hence they don't all arise from ordinary singular knots with double points: an example is shown in Figure 38.6.

Let P_n be the one-variable specialization of the HOMFLY polynomial with the substitution $x = q^n$, $y = q - q^{-1}$. In particular, P_0 is the Alexander polynomial, and P_2 is the Jones polynomial. Then P_n satisfies the skein relations shown on the right of Figure 38.7, reducing to the case of planar trivalent graphs, that is, abstract singular knots with no crossings. For these graphs [16] provides a further set of skein relations which uniquely determine the value of P_n.

Khovanov and Rozansky [11] use the [16] calculus, and the homological algebra of *matrix factorizations*, in their categorification of the polynomials P_n. In [18], Ozsvath, Stipsitz and Szabo generalize knot Floer homology to abstract singular knots. For more on categorification and knot homologies, see Part XII.

In [12] Kauffman and Vogel extend the Kauffman polynomial to singular knots – in fact, to knotted four-valent graphs with rigid vertices. In [10] Juyumaya and Lambropoulou introduced a Jones-type invariant for singular knots, using the theory of singular braids and a Markov trace on Yokonuma–Hecke algebras. In [8], Fiedler extends the Kauffman state models of the Jones and Alexander polynomials to the context of singular knots. To read more about knot polynomials see Part XI.

Quandle-type invariants have been generalised and studied for singular knots and other types of knot-like objects (virtual knots, flat knots, pseudoknots). Authors who have contributed to this research include Churchill, Elhamdadi, Hajij, Henrich, Nelson, Oyamaguchi, and Sazdanovich [9, 17, 6]. For more on quandles see Part XIII.

This short summary cannot aim to be a comprehensive treatment of the rich body of research, spanning nearly three decades, on singular knots and related knotted objects. We hope that the reader will be inspired to explore some of the many pointers and references, or contribute to the future of the subject.

38.4   Bibliography

[1] D. Bar-Natan: On the Vassiliev knot invariants, *Topology* **34** (1995) 423–472.

[2] D. Bar-Natan: Non-associative tangles, *Geometric Topology, Proceedings of the Georgia international Topology Conference*, W. H. Kazez, ed., 139–183, Amer. Math. Soc. and International Press, Providence, 1997.

[3] D. Bar-Natan, Z. Dancso: Finite type invariants of w-knotted objects II: Tangles, foams and the Kashiwara–Vergne problem, *Math. Annalen* **367**(3-4) 1517–1586.

[4] K. Bataineh, M. Elhamdadi, M. Hajij, and W. Youmans: Generating sets of Reidemeister moves of oriented singular links and quandles, arXiv:1702.01150.

[5] J. S. Birman, X. S. Lin: Knot polynomials and Vassiliev's invariants *Invent. Math.* **111-1** (1993) 225–270.

[6] I. R. U. Churchill, M. Elhamdadi, M. Hajij, S. Nelson: Singular knots and involutive quandles, *J. Knot Theory Ramifications* **26-14** (2017).

[7] Z. Dancso: On the Kontsevich integral for knotted trivalent graphs, *Alg. Geom. Topol.* **10** (2010), 1317–1365.

[8] T. Fiedler: The Jones and Alexander polynomials for singular links, *J. Knot Theory Ramifications* **19** (2010), no. 7, 859–866.

[9] A. Henrich and S. Nelson: Semiquandles and flat virtual knots, *Pacific J. Math.* **248** (2010), no. 1, 155–170.

[10] J. Juyumaya, S. Lambropoulou: An invariant for singular knots, *J. Knot Theory Ramifications* **18** (2009), no. 6, 825–840.

[11] M. Khovanov and L. Rozansky: Matrix factorizations and link homology, *Fund. Math.* **199-1** (2008), 1–91.

[12] L. H. Kauffman and P. Vogel: Link polynomials and a graphical calculus, *J. Knot Theory Ramifications* **1** (1992), no. 1, 59–104.

[13] M. Kontsevich: Vassiliev's knot invariants, *Adv. Sov. Math.* **16**(2) (1993) 137–150.

[14] D. Margalit, J. McCammond: Geometric presentations of the pure braid group, *J. Knot Theory Ramifications* **18-1** (2009), 1–20.

[15] J. Murakami, T. Ohtsuki: Topological quantum field theory for the universal quantum invariant, *Commun. Math. Phys.* **188** (1997), 501–520.

[16] H. Murakami, T. Ohtsuki and S. Yamada: HOMFLY polynomial via an invariant of coloured plane graphs, *Enseign. Math.* **2** 44 (1998), no. 3-4, 325–360.

[17] S. Nelson, N. Oyamaguchi, R. Sazdanovic: Psyquandles, singular knots and pseudoknots, arXiv:1710.08481.

[18] P. Ozsváth, A. Stipsitz, Z. Szabó: Floer homology and singular knots, *Journal of Topology* **2-2** (2016), 380–404.

[19] D. P. Thurston: The algebra of knotted trivalent graphs and Turaev's shadow world, *Geom. Toplol. Monographs* **4** (2002), 337–362.

[20] V. A. Vassiliev: Cohomology of knot spaces. *Theory of Singularities and Its Applications* (V. I. Arnold, ed.), Amer. Math. Soc., Providence, 1990.

[21] V. A. Vassiliev: Complements and discriminants of smooth maps: Topology and applications *Trans. of Math. Mono.* **98**, Amer. Math. Soc., Providence, 1992.

Chapter 39

Pseudoknots and Singular Knots

Inga Johnson, Willamette University

In [2], Hanaki introduced the concept of a *pseudo diagram* as a kind of knot diagram where some crossing information may be missing. Ordinary knot diagrams and shadows, i.e., projections of knots where no crossing information is specified, are both special types of pseudo diagrams. The pseudo diagram definition was motivated by the study of DNA since, in many classical pictures of knotted DNA strands, some crossing information can be determined while other crossing information is elusive. In the terminology of Hanaki's pseudo diagram theory, crossings with unknown under/over information are called *precrossings*. We provide several examples of pseudo diagrams in Figure 39.1.

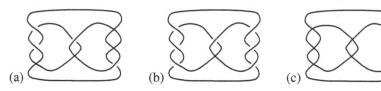

FIGURE 39.1: Pseudo diagram examples.

Before we discuss how pseudo diagrams can be used to define a theory of pseudoknots, we review a few key ideas from [2].

The *trivializing number* of a pseudo diagram P, denoted tr(P), is the smallest number of precrossings that must be *resolved*—that is, endowed with crossing information—to guarantee the diagram is the unknot, regardless of how the remaining precrossings are resolved. For example, in Figure 39.1 (a), a careful consideration of cases shows that one precrossing in each twist of the pretzel diagram must be resolved to ensure the resulting pseudo diagram is unknotted. Thus, the trivializing number of diagram (a) is 3. If, as in (b) of Figure 39.1, there is no complete resolution of the pseudo diagram that results in the unknot, then the trivializing number is infinite.

Similarly, the *knotting number*, denoted kn(P) of a pseudo diagram, is the minimum number of precrossings that must be resolved to guarantee that the resulting diagram is nontrivially knotted, regardless of how the remaining precrossings are resolved. Looking at the diagram shown in Figure 39.1 (c), we see that resolving the three precrossings in the leftmost twist to be alternating results in a pseudo diagram with infinite trivializing number; any complete resolution of the resulting pseudoknot must be at least as complex as a trefoil. Furthermore, no fewer than three crossing resolutions in the shadow does the job, so the knotting number of the pseudo diagram in Figure 39.1 (c) is 3.

Turning this diagrammatic theory into a knot theory as in [3], we say that *pseudoknots* are equivalence classes of pseudo diagrams where equivalence is defined by a sequence of ambient isotopies, the three classical Reidemeister moves, and the pseudo Reidemeister moves shown in Figure 39.2.

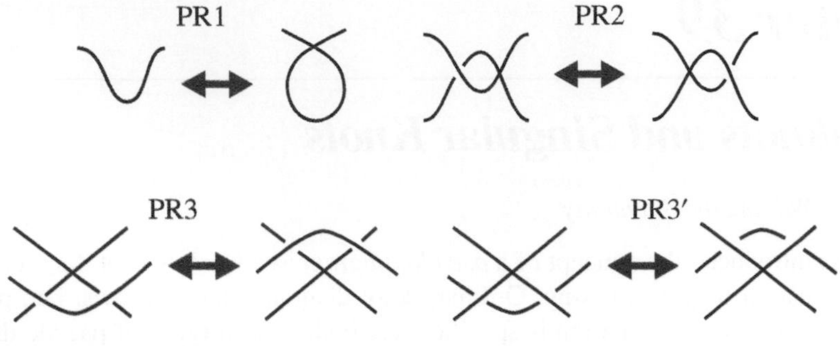

FIGURE 39.2: Pseudo Reidemeister moves.

Singular knots, similar to pseudoknots, are knots that contain a finite number of self-intersections. Singular knots are often depicted using the same diagrammatic conventions we use for pseudoknots, where precrossings are viewed as singularities. The difference between the two types of knotted objects can be found in the definition of equivalence. Singular knot equivalence does not include the PR1 move. For singular knots, the number of singularities is an invariant, while for pseudoknots, it is not. The reasoning behind the inclusion of the PR1 move in the theory of pseudoknots is that precrossings are viewed as crossings to be resolved. Taking the difficult-to-parse picture of knotted DNA as our inspiration, we might think of precrossings as having crossing information that we are simply unable to determine. In the loop that appears in the PR1 move, we see that regardless of which crossing information is assigned to the precrossing, the knot type remains the same. Hence, the move should philosophically be an allowable equivalence.

39.1　Pseudoknot Invariants

A natural question to ask when studying pseudoknots is: if we resolve the precrossings of a pseudoknot, choosing crossing information randomly, what is the probability that we will get a certain knot type? The weighted resolution set, first studied in [3], is an invariant of pseudoknots that provides an answer to just this question.

For a pseudoknot P, the *signed weighted resolution set* (or were-set) of P is the set of ordered pairs (K, p_K) where K is a knot obtained by a resolution of all the precrossing of P and p_K is the probability that K is obtained from P via a random resolution of the crossings, assuming that positive and negative crossings are equally likely. For example, the signed weighted resolution set of the pseudoknot in Figure 39.3a is $\{(0_1, \frac{3}{4}), (-3_1, \frac{1}{4})\}$ because, of the four possible resolutions of the two precrossings, three give the unknot and one results in the left-handed trefoil. Another curious example is shown in Figure 39.3b. For this pseudoknot, one resolution of the precrossing gives us 6_1 and the other resolution gives its mirror image, -6_1. So its were-set is $\{(6_1, \frac{1}{2}), (-6_1, \frac{1}{2})\}$.

The were-set of a pseudoknot is a powerful invariant and has been used to calculate a lower bound on the number of distinct equivalence classes of prime pseudoknots containing at least one pre-

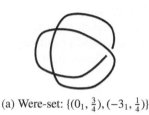

(a) Were-set: $\{(0_1, \frac{3}{4}), (-3_1, \frac{1}{4})\}$

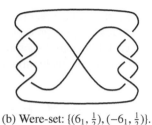

(b) Were-set: $\{(6_1, \frac{1}{2}), (-6_1, \frac{1}{2})\}$.

FIGURE 39.3: Two pseudoknots and their were-sets.

n	3	4	5	6	7	8	9
Number of pseudoknots	3	5	15	59	212	1344	7281

TABLE 39.1: Number of distinct prime pseudoknots with crossing plus precrossing number n as measured by the were-set [3].

crossing but fewer than a total of 10 crossings plus precrossings, as shown in Table 39.1. For example, there are 10 distinct pseudoknots derived from the knot 5_2 [3].

Unfortunately, the were-set is not a complete invariant of pseudoknots. Figure 39.4 illustrates two pseudoknots that differ by a local shadow flype move at the precrossing decorated by the grey disc. This flype relationship can be shown to give a one-to-one correspondence between equivalent knots in the resolution sets of these two pseudoknots [4]. In this example, the were-sets for both P_1 and P_2 are:

$$\{\left(0_1, \frac{72}{27}\right), \left(3_1, \frac{10}{27}\right), \left(-3_1, \frac{10}{27}\right), \left(4_1, \frac{20}{27}\right), \left(5_1, \frac{1}{27}\right), \left(-5_1, \frac{1}{27}\right), \left(5_2, \frac{2}{27}\right),$$
$$\left(-5_2, \frac{2}{27}\right), \left(6_1, \frac{2}{27}\right), \left(-6_1, \frac{2}{27}\right), \left(6_2, \frac{2}{27}\right), \left(-6_2, \frac{2}{27}\right), \left(-7_7, \frac{1}{27}\right), \left(7_7, \frac{1}{27}\right)\}.$$

Thus, the were-set invariant cannot distinguish between these two pseudoknots. However, using a pseudoknot invariant called the $\mathcal{I}$ invariant, we will see that the pseudoknots in Figure 39.4 are indeed distinct.

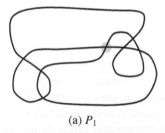

(a) P_1

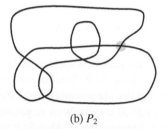

(b) P_2

FIGURE 39.4: These distinct pseudoknots have the same were-set.

The invariant $\mathcal{I}$, first defined in [5], is an invariant applied to the Gauss diagram of pseudoknot. Recall, for a classical knot K, a knot diagram of K is an immersion of a circle in the plane, $f_K : S^1 \rightarrow \mathbb{R}^2$, such that each double point is decorated with crossing information. The Gauss diagram of a classical knot starts with the domain circle along with chords that connect the double points of f_K. Each chord is decorated with an arrow pointing toward the understrand of the crossing and a sign, which records the sign (or local writhe) of the crossing. To extend Gauss diagrams to pseudoknots, the precrossings are represented by bold, or thicker, chords. A direction is assigned to each bold chord such that, if the precrossing were to be resolved positively, the arrow would point toward the understrand. An example pseudoknot and its Gauss diagram are given in Figure 39.5.

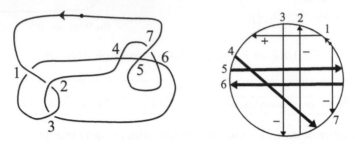

FIGURE 39.5: A pseudoknot and its Gauss diagram.

The invariant I is applied to the Gauss diagrams of a pseudoknot as follows: (1) remove the arrows on the bold chords, (2) decorate each bold chord with the integer value of the sum of the signs of the ordinary (unbold) chords that it intersects in the chord diagram, (3) delete all ordinary chords and their decorations, (4) delete any bold chord whose endpoints are adjacent along the circle if it is labeled with a value of 0. The result is a circle containing only bold chords, each assigned an integer value. This labeled chord diagram is the value of $I(P)$. For instance, the Gauss diagram from Figure 39.5 is the labeled chord diagram shown in Figure 39.6.

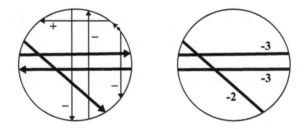

FIGURE 39.6: A pseudoknot Gauss diagram, P, and the value of $I(P)$.

When the invariant I is applied to the two pseudoknots of Figure 39.4, we obtain the distinct values shown in Figure 39.7. This measured difference between the pseudoknots P_1 and P_2 is encoding the fact that the precrossing where the flype is performed (those with grey shading in Figure 39.4) can be resolved positively or negatively. If the greyed precrossings in both P_1 and P_2 are resolved positively (denoted by P_1^+ and P_2^+), then the pseudoknot diagrams are equivalent by a flype move. Similarly, if they are both resolved negatively (denoted by P_1^- and P_2^-), the resulting pseudoknots are equivalent by a flype. On the other hand, if they are resolved with opposite sign, then the I invariant detects that the resulting pseudoknots are distinct, as shown in Figure 39.8.

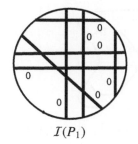
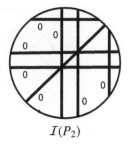

$$\mathcal{I}(P_1) \qquad\qquad \mathcal{I}(P_2)$$

FIGURE 39.7: The value of the $\mathcal{I}$ invariant for the two pseudoknots from Figure 39.4.

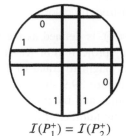
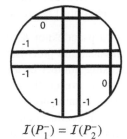

$$\mathcal{I}(P_1^+) = \mathcal{I}(P_2^+) \qquad\qquad \mathcal{I}(P_1^-) = \mathcal{I}(P_2^-)$$

FIGURE 39.8: The $\mathcal{I}$ invariant applied to equivalent pseudoknots P_1^+ and P_2^+ and applied to P_1^- and P_2^-. The invariant shows P_1^+ is not equivalent to P_1^-.

39.2 Classical Invariants Extended to Pseudoknots

There are several classical invariants that have been studied in the context of pseudoknots such as p-colorability, knot determinant, and crossing number, to name a few.

We begin with two seemingly distinct generalizations of p-colorability to pseudoknots first studied in [6] and extended in [7]. Similar to the classical setting, strands of a pseudoknot will begin and end only at a classical crossing. Therefore, all four pieces of the diagram emanating from a pre-crossing must be colored with the same value. A pseudoknot is *strong p-colorable* [6] if we can assign elements of $\mathbb{Z}/p\mathbb{Z}$ to the strands of the diagram so that at each classical crossing, twice the number on the overstrand is equal mod-p to the sum of the values assigned to the understrands, and all arcs emanating from a precrossing of a pseudoknot must be assigned the same value. Alternatively, a pseudoknot is said to be *p-colorable* if all of the resolutions of a pseudoknot are p-colorable. In [6] they show that strong p-colorable implies p-colorable. In [7], the authors prove that for p an odd prime, p-colorable implies strong p-colorable. Thus, for odd primes p, the two generalizations of p-colorability are, in fact, equivalent.

As is the case for classical knots, the concept of a knot determinant is used to determine for which p a given pseudoknot is p-colorable. The *pseudodeterminant* of a pseudoknot K is defined as the greatest common divisor of the determinants of all the resolutions of K and a pseudoknot is p-colorable for every value of p that divides its pseudodeterminant. The definition of p-colorability

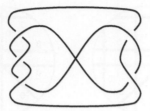

FIGURE 39.9: A pseudoknot with were-set $\{(5_1, \frac{1}{2}), (5_2, \frac{1}{2})\}$, but crossing number 6.

for pseudoknots is quite restrictive. In [6], the colorability of 8583 pseudoknots with 9 or fewer crossings is determined, and only 112 are actually non-trivially colorable.

The *crossing number* of a pseudoknot K, denoted $cr(K)$, is defined as the minimum number of total crossings (both classical and precrossings) in any projection of K. At first glance, it might seem as though the crossing number of a pseudoknot ought to equal to the maximum crossing number of its resolutions. However, the example in Figure 39.9 shows that this is not the case. However, if a pseudoknot contains no nugatory crossings and the precrossings are resolvable so that an alternating diagram is produced, then the crossing number of the pseudoknot is indeed the maximum crossing number of its resolutions [3].

In addition to these extensions of classical invariants, the bracket polynomial was extended to pseudoknots by Dye in [1], where she uses this pseudoknot invariant to determine a cosmetic crossing obstruction.

39.3 Bibliography

[1] H. Dye. Pseudo knots and an obstruction to cosmetic crossings. *J. of Knot Theory Ramifications* 26(04):1750022, 2017.

[2] R. Hanaki. Pseudo diagrams of knots, links and spatial graphs. *Osaka J. of Math.*, 47, 863–883, 2010.

[3] A. Henrich, R. Hoberg, S. Jablan, L. Johnson, E. Minten, L. Radović. The Theory of pseudo-knots. *J. of Knot Theory Ramifications* 22(7): 1350032, 12, 2013.

[4] A. Henrich, S. Jablan, I. Johnson, The signed Weighted resolution set is not a complete pseudoknot invariant. *J. of Knot Theory Ramifications* 25(9):1641007, 10, 2016.

[5] F. Dorais, A. Henrich, S. Jablan, I. Johnson. Isotopy and homotopy invariants of classical and virtual pseudoknots. *Osaka J. of Math.* 52, 409–423, 2015.

[6] A. Henrich, S. Jablan. On the coloring of pseudoknots. *J. of Knot Theory Ramifications* 23(12):1450061, 2014.

[7] S. Nelson, N. Oyamaguchi, R. Sazdanovic. Psyquandles, Singular Knots and Pseudoknots. *arXiv:1710.08481v1 [math.GT]* 2017.

Chapter 40

An Introduction to the World of Legendrian and Transverse Knots

Lisa Traynor, Bryn Mawr College

40.1 Introduction

Legendrian and transverse knots are smooth knots that naturally arise in contact geometry and topology. Roughly speaking, a contact structure on a 3-manifold consists of a field of twisting planes, and Legendrian and transverse knots are smooth knots that are always tangent or always transverse to these planes. Figures 40.1 and 40.2 show diagrams of some Legendrian and transverse knots.

Legendrian and transverse knot theory is an active field of research that has its roots in the broader fields of contact geometry and topology. Contact geometry has origins dating back to the 1600s in Huygens' work on geometric optics and was established as a subject of its own by Lie in the late 1800s [14]; more details on this history can be found in [10]. The topological flavor of contact geometry stems from Darboux's theorem, which says that all contact manifolds are locally modeled on $\mathbb{R}^3$ with its standard contact structure.

Legendrian and transverse knots are interesting on their own, but they also play an important role in the construction and classification of contact 3-manifolds. For example, every closed and connected contact 3-manifold can be obtained from S^3 with its standard contact structure by surgery on a Legendrian link or as a branched cover along a transverse link [5, 13, 12]; the existence of an "exotic" contact structure on $\mathbb{R}^3$ was discovered through understanding the behavior of Legendrian and transverse knots, [1]. Beyond contact geometry and topology, the study of Legendrian and

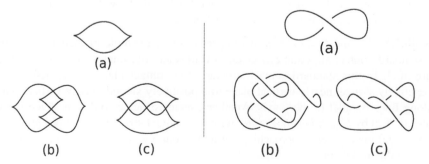

FIGURE 40.1: Front (on the left) and Lagrangian (on the right) diagrams of a Legendrian: (a) unknot, (b) negative trefoil, and (c) positive trefoil.

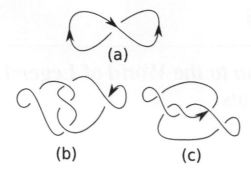

FIGURE 40.2: Front diagrams of a transversal: (a) unknot, (b) negative trefoil, and (c) positive trefoil.

transverse knots has a number of applications: for example, in geometric physics, Legendrian links play important roles in Lorentz geometry [4]; and in low-dimensional topology, Legendrian and transverse knot theory can be used to define new smooth knot invariants, such as the Thurston Bennequin number [2], and to shed light on known knot invariants, such as the slice genus [3].

In this chapter, a gentle introduction is given to Legendrian and transverse knots. Excellent sources for additional, more detailed background are [9] and [11].

40.2   Contact Manifolds

A **contact 3-manifold**, (M, ξ), is an orientable 3-manifold M together with a maximally noninte-grable plane field ξ. Here a *plane field* simply means that at each point $p \in M$, there is a specified 2-dimensional plane of the tangent space: $\xi_p \subset T_pM$. Near each point, ξ can be expressed as the kernel of a locally or globally defined 1-form α: $\xi_p = \ker \alpha_p$. The plane field is *maximally noninte-grable* if the plane field is never tangent to any surface, which happens precisely when $\alpha \wedge d\alpha \neq 0$ at all points. When ξ is globally defined by a 1-form, the contact manifold is *cooriented*, meaning that the quotient line bundle TM/ξ is trivial, which implies that there is a well-defined notion of vectors that are positively transverse to the contact planes.

Example 40.2.1 (ξ_{std} and ξ_{sym} on $\mathbb{R}^3$). An important basic example of a contact manifold is $\mathbb{R}^3$ with the standard contact structure ξ_{std} given in Cartesian coordinates (x, y, z) as $\ker(dz - ydx)$; see Figure 40.3. In this standard contact structure, the contact planes do a "left-handed" twist from a negatively sloped, nearly vertical plane to a positively sloped, nearly vertical plane as one moves along lines parallel to the y-axis. Another contact structure on $\mathbb{R}^3$ is given in cylindrical coordinates (r, θ, z) by $\xi_{sym} = \ker(dz + r^2 d\theta)$; with this contact structure there is a left-handed twist from a horizontal plane to one approaching vertical as one moves along any ray perpendicular to the z-axis; see the middle picture in Figure 40.3.

In fact, the contact manifolds $(\mathbb{R}^3, \xi_{sym})$ and $(\mathbb{R}^3, \xi_{std})$ are **equivalent** (also known as **contacto-morphic**): there is a diffeomorphism φ of $\mathbb{R}^3$ that takes ξ_{sym} to ξ_{std}. The previously mentioned

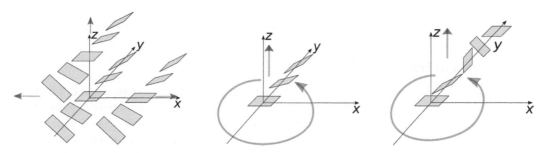

FIGURE 40.3: The contact structures ξ_{std}, ξ_{sym}, and ξ_{ot} on $\mathbb{R}^3$. The contact structure ξ_{std} does a half twist along the y-axis and is invariant as one moves in the x and z directions. Along rays perpendicular to the z-axis, ξ_{sym} does a quarter twist, and ξ_{ot} does an infinite number of full twists. Both ξ_{sym} and ξ_{ot} are invariant in the z-direction and have rotational symmetry.

Darboux's theorem says that every point in any contact manifold (M, ξ) has a neighborhood that is contactomorphic to $(\mathbb{R}^3, \xi_{std})$; see, for example, [11].

There are many other contact structures on $\mathbb{R}^3$. In particular, there is an "overtwisted" contact structure:

Example 40.2.2 (ξ_{ot} on $\mathbb{R}^3$). The contact structure ξ_{ot} is given in cylindrical coordinates as $\xi_{ot} = \ker(\cos r \, dz + \sin r \, d\theta)$; see Figure 40.3. Similar to ξ_{sym}, there is a rotational symmetry to the planes in ξ_{ot}, however as one moves out along rays perpendicular to the z-axis, the planes repeatedly do full twists.

In fact, the contact structure ξ_{ot} is not equivalent to ξ_{std}. Observe that for the disk $D = \{(r, \theta, z) : z = 0, r \leq \pi\}$, $T_pD = (\xi_{ot})_p$, for all $p \in \partial D$. Such a disk is called an *overtwisted disk*. A contact structure is **overtwisted** if there is an embedded overtwisted disk. When there does not exist such a disk, as is the case for ξ_{std}, the contact structure is called **tight**. This fundamental tight and overtwisted dichotomy was introduced by Eliashberg [7]. Overtwisted contact structures are well understood: two contact structures that are homotopic as plane fields are in fact isotopic through contact structures [6]. On $\mathbb{R}^3$, ξ_{std} is the unique tight contact structure, up to isotopy [8].

By a theorem of Martinet, all closed, orientable 3-manifolds admit contact structures; furthermore, by the Lutz twist construction, it is known that every cooriented tangent 2-plane field on a closed, orientable 3-manifold is homotopic to a contact structure. There are many proofs of the existence of contact structures on closed, orientable 3-manifolds using a variety of construction theorems for smooth 3-manifolds, including Dehn surgery, branched covering spaces, and open book decompositions; see [11]. In this chapter we will mainly focus on $(\mathbb{R}^3, \xi_{std})$.

Given a contact manifold (M, ξ), there are two natural types of embedded, smooth, oriented curves to consider: **Legendrian** curves are everywhere tangent to ξ, and **transverse** curves are always transverse to ξ. When these curves close to form knots or links, we get Legendrian and transverse knots and links. These special curves are explored more in the next two sections.

40.3   Legendrian Curves

Legendrian knots are smooth knots that are everywhere tangent to the contact structure ξ. Two Legendrian knots are **equivalent** if they can be connected through a 1-parameter family of Legendrian knots. Similar to the case of smooth knots in $\mathbb{R}^3$, it is convenient to represent Legendrian knots in $(\mathbb{R}^3, \xi_{std})$ through projections to a 2-dimensional plane. There are two special projections that are well adapted to the contact structure ξ_{std} known as the "front" and "Lagrangian" projections. Below we describe how to recover Legendrian curves from these projections and explain Reidemeister type theorems for the equivalence of Legendrians represented by these projections.

40.3.1   Front Diagrams of Legendrians

To become familiar with the front projection, first consider the graph of a smooth function $f(x)$ in the xz-plane, $\Gamma_f = \{(x, z = f(x))\}$. There are an infinite number of smooth curves in $\mathbb{R}^3$ that project to Γ_f, however Γ_f is the projection of precisely *one* Legendrian curve in $(\mathbb{R}^3, \xi_{std})$: as a Legendrian must be tangent to $\xi = \ker(dz - y\,dx)$, we lift the graph Γ_f to a Legendrian curve to $\mathbb{R}^3$ by letting $y = \frac{dz}{dx}$.

A more general (non-graphical) Legendrian curve Λ can be completely recovered from its xz-projection $D_{xz}(\Lambda)$, known as its **front diagram**, using a similar slope-lifting procedure along its "branches," which are the paths between cusps in the curve's projection. Observe that each branch is the graph of a function. The Legendrian condition guarantees that any parameterization $(x(t), y(t), z(t))$ of a Legendrian satisfies

$$z'(t) - y(t)x'(t) = 0.$$

In particular, the front diagram $D_{xz}(\Lambda)$ will never have any vertical tangencies: if $x'(t) = 0$, then $z'(t) = 0$. Generically, these projections can be parameterized by a map that is an immersion except at a finite number of points at which there are cusps with a well-defined tangent line. The smoothness of Λ requires that the cusps are "semi-cubic parabolas": after a change of coordinates, a cusp is parameterized as $z(t) = t^3$, $x(t) = t^2$, $t \in (-\varepsilon, \varepsilon)$. Along the entire front, the Legendrian is uniquely determined by the front diagram $D_{xz}(\Lambda)$ with y given as the slope of the tangent line at the cusps and along the branches. Embeddedness of the Legendrian curve guarantees that double points $(x_1, z_1) = (x_2, z_2)$ will have with distinct slopes, $y_1 \neq y_2$.

Any such cusped curve in the plane without vertical tangents and self-tangencies will represent an embedded Legendrian curve. The left-hand side of Figure 40.1 gives front diagrams of a Legendrian unknot, negative (left) trefoil, and positive (right) trefoil. In fact, *there are Legendrian representatives of every knot or link type*: given any continuous curve, there is a Legendrian curve that is C^0 close to it; see, for example, [9], [11]. Observe that in contrast to the setting for smooth knots where one needs to enhance a projection into a diagram by indicating the overstrand and understrand at each crossing, in the front projection the slopes at the crossing determine which strand is on top: to keep the standard orientation, the positive y-axis points into the page and thus the strand with lesser slope will always be the overstrand. Although it is common for smooth knot tables to only list one representative of a chiral knot, the "simplest" (in terms of the classical

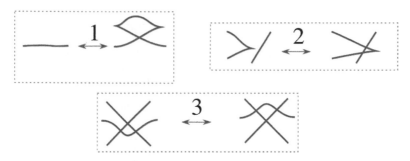

FIGURE 40.4: The three Legendrian Reidemeister moves for front diagrams. There is another type-1 move obtained by flipping the planar figure about a horizontal line, and there are three additional type-2 moves obtained by the planar figure about a vertical, a horizontal, and both a vertical and a horizontal line.

Thurston-Bennequin invariant defined later) Legendrian representatives of a smooth chiral knot and its mirror are often quite different.

Just as there are Reidemeister moves for topological knot diagrams, there is a set of "Reidemeister" moves for Legendrian front diagrams: two front diagrams represent equivalent Legendrian knots if and only if they are related by planar isotopies that do not introduce vertical tangent (or cusps) and a sequence of the moves shown in Figure 40.4; [16].

40.3.2 Lagrangian Projections of Legendrians

There is a second important projection that is often used to represent Legendrian curves $\Lambda \subset \mathbb{R}^3$; this is the projection to the xy-plane and is typically called the **Lagrangian diagram**, $D_{xy}(\Lambda)$. In contrast to the front diagram, the Lagrangian diagram of an embedded Legendrian curve is always parameterized by an immersion: $x'(t)$ and $y'(t)$ both vanishing forces $z'(t) \neq 0$, which violates the Legendrian condition. The Legendrian curve Λ can be recovered, up to translation in the z-direction, by integration: pick some $z_0 \in \mathbb{R}$ and define $z(0) = z_0$. Then define

$$z(t) = z_0 + \int_0^t y(s)x'(s)\,ds.$$

An arbitrary immersed curve in the xy-plane need not represent an embedded Legendrian: one needs to guarantee that double points $(x(t_1), y(t_1)) = (x(t_2), y(t_2))$ in the Lagrangian diagram lift to have different z-coordinates. This requires the non-vanishing of a corresponding integral: $\int_{t_1}^{t_2} y(s)x'(s)\,ds \neq 0$. To form a closed Legendrian, parameterized by $t \in [0, 2\pi]$, one needs $\int_0^{2\pi} y(t)x'(t)\,dt = 0$. These conditions are a bit more difficult to represent on the diagram; for this reason, it is the convention to use a broken line to indicate the crossing's understrand in a Lagrangian projection. The right-hand side of Figure 40.1 gives Lagrangian diagrams of the unknot and the trefoils.

There is a weak Reidemeister type theorem for equivalent Lagrangian projections: two equivalent Legendrian knots will have projections related by a sequence of moves; see, for example, [9]. However, these moves are not sufficient to guarantee Legendrian isotopy due to the integral conditions described above.

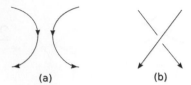

(a)   (b)

FIGURE 40.5: Forbidden vertical tangencies and crossings in the front projection of a transversal knot.

40.4   Transverse Curves

Recall that a transverse curve $\mathcal{T}$ is a smooth, oriented curve that is always transverse to the contact structure ξ. When ξ is cooriented, meaning defined as the kernel of a global 1-form α, it is a convention to give $\mathcal{T}$ a "positive" orientation, meaning that $\alpha(\mathcal{T}'(t)) > 0$. We will restrict attention to positively transverse knots in $(\mathbb{R}^3, \xi_{std})$.

40.4.1   Front Projections of Transverse Curves

In terms of a parameterization, $(x(t), y(t), z(t))$, the transversal condition means that

$$z'(t) - y(t)x'(t) > 0.$$

Observe that the transversal condition implies that $z'(t) > yx'(t)$; when $x'(t) > 0$, y is bounded above by the slope $\frac{dz}{dx}$, whereas when $x'(t) < 0$, $y(t)$ is bounded below by the slope.

Transverse knots are typically studied through their front projections to the xz-plane. Again starting with the graph of a smooth function $z = f(x)$, the parameterization $(t, f(t))$ can be lifted to many (positively) transverse curves by requiring that at each point (x, z) the associated y is less than the slope, $\frac{dz}{dx}$, of the graph at (x, z). By choosing a "right-to-left" parameterization such as $(-t, f(-t))$, we see that we now want to lift (x, z) to have a y value greater than the slope of the graph. In contrast to the Legendrian case, there are *many* transverse lifts of a parameterized graph, however all these transverse curves are **transversely equivalent**, meaning that they are isotopic through transverse curves. Observe that vertical tangents can occur in the front projections, however, *forbidden vertical tangencies* are those with the oriented tangent vector pointing down: $x'(t) = 0$ forces $z'(t) > 0$. There are also *forbidden crossings*, where the overstrand has greater slope than the understrand, and the overstrand is oriented right-to-left ($x'(t) < 0$) while the understrand is oriented left-to-right ($x'(t) > 0$). A crossing configuration as in Figure 40.5 contradicts the fact that the positive y-direction is into the page. In fact, any immersed curve without these forbidden vertical tangents and forbidden crossings will have a lift to a transverse curve $\mathcal{T}$ that is unique up to transversal isotopy. Figure 40.2 gives some examples of transverse representatives of some low crossing knot types. Just as with Legendrian knots, *all knots have transverse representatives*. For transverse knots, there is a Reidemeister theorem: two front diagrams will represent transverse knots in the same transverse isotopy class if and only if their diagrams are equivalent by the moves shown in Figure 40.6; [16].

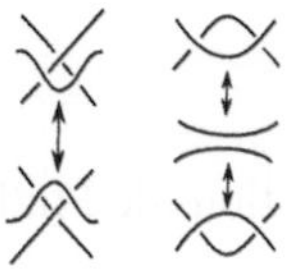

FIGURE 40.6: The Reidemeister moves for front diagrams of transversal knots; there are also the diagrams with the opposite crossings. Orientations on the strands are allowed in all ways that do not give the forbidden vertical tangencies or forbidden crossings of Figure 40.5.

40.4.2 Braid Representations of Transversal Curves

Another important way to study transversal knots is by using closed braids. To see this connection, it is convenient to work with the symmetric contact structure ξ_{sym} on $\mathbb{R}^3$ as described in Example 40.2.1. Any closed braid B in $\mathbb{R}^3$ can be parameterized as $(r(t), \theta(t), z(t))$ with $\theta'(t) > 0$. Furthermore, we can isotop B through closed braids so that all points on the braid are very far from the z-axis. We see that when r is very large, the planes of ξ_{sym} are almost "vertical," and thus the braid B is isotopic through braids to a transverse knot $\mathcal{T}$. Bennequin proved the converse: *any transverse knot in $(\mathbb{R}^3, \xi_{sym})$ is transversely isotopic to a closed braid* [1].

In the world of smooth knots, Markov's theorem says that two braids represent isotopic knots if and only if the braids are related by conjugation and positive and negative stabilization. There is a transverse version of Markov's theorem: now one is only allowed conjugation and *positive* stabilizations [17, 15].

40.5 Motivating Problems

Now that we have introduced the basic objects of Legendrian and transverse knots and have established the corresponding Reidemeister moves, we are naturally led to some concrete questions.

1. Are the Legendrian unknots Λ_1 and Λ_2 with front diagrams as follows equivalent?

       (a)   (b)

2. Are the transverse unknots $\mathcal{T}_1$ and $\mathcal{T}_2$ with front diagrams as follows equivalent?

       (a)   (b)

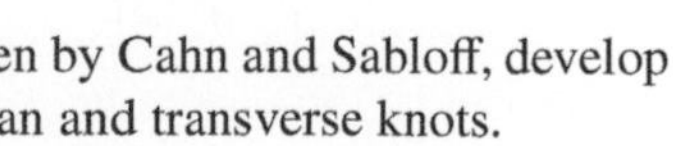

The following two chapters, written by Cahn and Sabloff, develop some classical and non-classical invariants to distinguish Legendrian and transverse knots.

40.6   Bibliography

[1] D. Bennequin. Entrelacements et equations de Pfaff. *Asterisque*, 107–108:87–161, 1983.

[2] J. C. Cha and C. Livingston. Knotinfo: Table of knot invariants.

[3] B. Chantraine. On Lagrangian concordance of Legendrian knots. *Algebr. Geom. Topol.*, 10:63–85, 2010.

[4] V. Chernov and S. Nemirovski. Legendrian links, causality, and the Low conjecture. *Geom. Funct. Anal.*, 19(5):1320–1333, 2008.

[5] F. Ding and H. Geiges. A Legendrian surgery presentation of contact 3-manifolds. *Math. Proc. Cambridge Philos. Soc.*, 136(3):583–598, 2004.

[6] Ya. Eliashberg. Classification of overtwisted contact structures on 3-manifolds. *Invent. Math.*, 98(3):623–637, 1989.

[7] Ya. Eliashberg. Contact 3-manifolds twenty years since J. Martinet's work. *Ann. Inst. Fourier (Grenoble)*, 42(1-2):165–192, 1992.

[8] Ya. Eliashberg. Classification of contact structures on $\mathbb{R}^3$. *Internat. Math. Res. Notices*, (3):87–91, 1993.

[9] J. Etnyre. Legendrian and transversal knots. In *Handbook of Knot Theory*, pages 105–185. Elsevier B. V., Amsterdam, 2005.

[10] H. Geiges. A brief history of contact geometry and topology. *Expo. Math.*, 19(1):25–53, 2001.

[11] H. Geiges. *An Introduction to Contact Topology*, volume 109 of *Cambridge Studies in Advanced Mathematics*. Cambridge University Press, Cambridge, 2008.

[12] E. Giroux. Géométrie de contact: de la dimension trois vers les dimensions supérieures. In *Proceedings of the International Congress of Mathematicians, Vol. II (Beijing, 2002)*, pages 405–414, Beijing, 2002. Higher Ed. Press.

[13] J. Gonzalo Perez. Branched covers and contact structures. *Proc. Amer. Math. Soc.*, 101(2):347–352, 1987.

[14] S. Lie. *Geometrie der Berhrungstransformationen*. B. J. Teubner, 1869.

[15] S. Orevkov and V. Shevchishin. Markov theorem for transverse links. *J. Knot Theory Ramifications*, 12(7):905–913, 2003.

[16] J. Świątkowski. On the isotopy of Legendrian knots. *Ann. Global Anal. Geom.*, 10(3):195–207, 1992.

[17] N. Wrinkle. The Markov Theorem for transverse knots. www.arxiv.org/abs/math/0202055, 2002.

Chapter 41

Classical Invariants of Legendrian and Transverse Knots

Patricia Cahn, Smith College

We approach the problem of classifying Legendrian knots by first studying their three *classical invariants*: the smooth knot type, the Thurston-Bennequin number and the Maslov (rotation) number. Here we define the latter two invariants, and discuss several ways of computing them in $\mathbb{R}^3$ with its standard contact structure. Then we survey key results which illustrate the role these invariants play in both the classification of Legendrian knots in a given contact manifold, and perhaps more surprisingly, in probing global information about the contact structure itself. We conclude with a brief discussion about classical invariants of transverse knots. For more thorough treatments of this material, we refer the reader to [7, 12].

41.1 Linking Number and Writhe

We begin by recalling two definitions. Each crossing of an oriented knot or link diagram can be assigned a sign, as shown in Figure 41.1. The *writhe* of a knot diagram D is defined by $w(D) = p(D) - n(D)$, where $p(D)$ is the number of positive crossings of D, and $n(D)$ is the number of negative crossings of D. Given a link diagram $D = D_1 \cup D_2$ of the 2-component link $K_1 \cup K_2$ the *linking number* of K_1 and K_2 is equal to $\mathrm{lk}(K_1, K_2) = \frac{1}{2}(P(D) - N(D))$, where $P(D)$ is the number of positive crossings between the two components D_1 and D_2, and $N(D)$ is the number of negative crossings between the two components. Note that the linking number is symmetric, and reversing the orientation on one of the components changes its sign.

It follows from these combinatorial formulas that the linking number of a knot diagram D with a push-off transverse to the plane containing it is equal to $w(D)$. This observation is used to derive several of the combinatorial formulas for the Thurston-Bennequin and Maslov numbers below.

While we focus on Legendrian knots in $\mathbb{R}^3$, we will briefly discuss definitions of classical invariants for Legendrian knots in other 3-manifolds. A knot K in a 3-manifold M is *null-homologous* if it represents the zero class in the first homology group $H_1(M; \mathbb{Z})$. Furthermore, a knot is null-homologous if and only if it bounds an orientable surface Σ in M, called a *Seifert surface* for K. The linking number is defined for any two null-homologous oriented knots K_1 and K_2 in an oriented 3-manifold M. Given a Seifert surface Σ for K_1, the linking number $\mathrm{lk}(K_1, K_2)$ is the algebraic intersection number of K_2 and Σ. We refer the reader to [2] for a proof that the linking number is well-defined and symmetric. For knots in $\mathbb{R}^3$, this linking number agrees with the combinatorial definition above.

FIGURE 41.1: A positive crossing (left) and a negative crossing (right).

41.2 The Thurston-Bennequin Number

Let Λ be a Legendrian knot in $\mathbb{R}^3$ with (coorientable) contact structure ξ. The Thurston-Bennequin number measures the number of full twists made by the contact planes along Λ, as follows. First, choose a vector field along Λ transverse to ξ. The *Legendrian push-off* Λ' of Λ is the push-off of Λ along this vector field. The *Thurston-Bennequin number* $tb(\Lambda)$ is the linking number of Λ and Λ'. Note that the linking number is defined for oriented links, so we choose an orientation for Λ and use the corresponding orientation for Λ'. Since reversing the orientation on Λ also reverses the orientation on Λ', the Thurston-Bennequin number is independent of our original choice of orientation.

When $\mathbb{R}^3$ is equipped with the standard contact structure $\xi_{std} = \ker(dz - y\,dx)$, there are simple formulas for $tb(\Lambda)$ involving combinatorial data; this data can be read off from either the front or Lagrangian projections of Λ.

41.2.1 Computing tb from the Lagrangian Projection

Let D_{xy} be the Lagrangian projection of Λ, which sits in the xy-plane. The vector field ∂_z is everywhere transverse to ξ_{std}. Let Λ' be a push-off of Λ in the z direction. By the discussion in Section 41.1, the linking number of Λ with Λ' is equal to the writhe $w(D_{xy})$ of D_{xy}. Hence $tb(\Lambda) = w(D_{xy})$.

41.2.2 Computing tb from the Front Projection

Let Λ be a Legendrian knot with front diagram D_{xz}. We do not need to show over-/under-crossing information in a front diagram; since the y-coordinate of a point on Λ is the slope at the corresponding point on its front diagram, and the positive y-axis points into the page, the strand with greater (positive) slope is always the understrand. Positive and negative crossings can now be viewed as having both arrows on the same side, or on opposite sides, of a vertical line, respectively; see the top row of Figure 41.2. The front diagram D'_{xz} of Λ' is obtained from D_{xz} by sliding D_{xz} off itself vertically in the xz-plane; see the bottom row of Figure 41.2. As shown in Figures 41.2 and 41.3, positive and negative crossings of D_{xz} contribute $+1$ and -1 to the linking number of Λ and Λ', respectively, and each cusp of D_{xz} contributes $-1/2$ to the linking number of Λ and Λ'. Let $c(D_{xz})$ denote the number of cusps in the diagram D_{xz}. Let $w(D_{xz})$ denote its writhe; the writhe of a cusped projection is computed as above, ignoring cusps. The Thurston-Bennequin number of Λ is given by the formula

$$tb(\Lambda) = w(D_{xz}) - \frac{1}{2}c(D_{xz}).$$

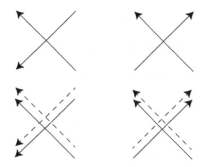

FIGURE 41.2: Positive and negative crossings contribute $+1$ and -1 to the Thurston-Bennequin number, respectively.

FIGURE 41.3: Right and left cusps each contribute $-\dfrac{1}{2}$ to the Thurston-Bennequin number, regardless of orientation.

41.2.3 Computing tb in Other Contact Manifolds

Let (M, ξ) be an oriented 3-manifold with a cooriented contact structure ξ. If Λ is null-homologous, then there exists an oriented surface $\Sigma \subset M$ such that $\partial\Sigma = \Lambda$. The Thurston-Bennequin number of Λ is the algebraic intersection number of Λ' with this Seifert surface Σ.

41.3 The Maslov Number

The *Maslov number*, or rotation number, $r(\Lambda)$, of an oriented Legendrian knot Λ in $(\mathbb{R}^3, \xi)$ is the number of full rotations of the tangent vector to Λ in ξ. More precisely, given a unit-speed parameterization $\gamma(t)$ of Λ, $r(\Lambda)$ is the degree of the map $S^1 \to S^1$ defined by $t \mapsto \gamma'(t)$. This degree is measured relative to a chosen *trivialization*, or smoothly varying basis, of ξ. When $\xi = \xi_{std}$, one choice of trivialization is $\{y\, \partial_z + \partial_x,\ \partial_y\}$.

41.3.1   Computing the Maslov Number from the Lagrangian Projection

The *index* of a unit-speed immersed plane curve C is the degree of the map $S^1 \to S^1$ defined by $t \mapsto C'(t)$; we denote this by $\operatorname{ind}(C)$. Intuitively, the index measures the number of full counterclockwise rotations made by the tangent vector C' as one travels around C. Since ∂_y is used to trivialize ξ_{std}, the Maslov number $r(\Lambda)$ of a Legendrian knot Λ with Lagrangian projection D_{xy} is equal to $\operatorname{ind}(D_{xy})$.

41.3.2   Computing the Maslov number from the Front Projection

Given an oriented front diagram D_{xz}, and a Legendrian knot Λ with front diagram D_{xz}, the tangent vector $\gamma'(t)$ to Λ passes ∂_y in the counterclockwise direction exactly when D_{xz} has a downward-pointing cusp, and in the clockwise direction exactly when D_{xz} has an upward-pointing cusp. This is most easily seen in the Lagrangian projection; see Figure 41.4. Hence up-cusps contribute $-\dfrac{1}{2}$ to the rotation number, while down-cusps contribute $\dfrac{1}{2}$, and

$$r(\Lambda) = \frac{1}{2}(d(D_{xz}) - u(D_{xz}))$$

where $u(D_{xz})$ is the number of up-cusps of D_{xz} and $d(D_{xz})$ is the number of down-cusps.

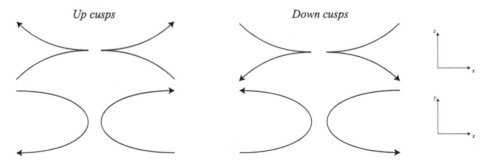

FIGURE 41.4: Up- and down-cusps in the front projection, and corresponding Lagrangian projections.

41.3.3   Maslov potentials

In the next chapter, we will define more sophisticated invariants of Legendrian knots, which make use of Maslov potentials, introduced in [3, 10]. A *branch* of a front diagram is a path between two of its cusps. Here we introduce Maslov potentials in their simplest form; namely, a *Maslov potential* is an assignment of integers to the branches of a front diagram such that the number of the upper branch at each cusp is one more than the number on the lower branch. A Maslov potential exists for D_{xz} if $r(D_{xz}) = 0$. For an example, see Figure 41.5. One can also define Maslov potentials for fronts with rotation number $r \neq 0$; the potential would be in $\mathbb{Z}/r\mathbb{Z}$-valued rather than $\mathbb{Z}$-valued.

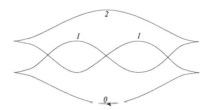

FIGURE 41.5: A Maslov potential for a Legendrian trefoil with $tb = 1$ and $r = 0$.

41.3.4   Computing the Maslov Number in Other Contact Manifolds

Let M be an oriented 3-manifold with cooriented contact structure ξ. The Maslov number of a Legendrian knot Λ in (M, ξ) can be defined in both of the following two settings:

1. The contact bundle ξ is a trivial 2-plane bundle.

2. The knot Λ is null-homologous.

In case (1), one can choose a trivialization of ξ and define $r(\Lambda)$ to be the degree of the map $t \mapsto \gamma'(t)$ as above. In case (2), one first chooses a Seifert surface Σ for Λ; as every 2-plane bundle over a surface with boundary is trivializable, we choose a trivialization of ξ over Σ. Then $r_\Sigma(\Lambda)$ is again the degree of the map $t \mapsto \gamma'(t)$, with respect to this trivialization. This rotation number does not depend on the choice of trivialization of ξ over Σ; it is also independent of the choice of Σ in a fixed relative homology class in $H_2(M, \Lambda; \mathbb{Z})$, but depends on the choice of such a relative class if the Euler class of ξ is nonzero [12].

41.4   Survey of Classification Results

We now restrict our attention to $(\mathbb{R}^3, \xi_{std})$, and ask:

Question 1. *For a fixed smooth, oriented, knot type $\mathcal{K}$, which ordered pairs of integers (tb, r) arise as the Thurston-Bennequin and Maslov numbers of a Legendrian representative of $\mathcal{K}$?*

Given an oriented Legendrian knot Λ, one can obtain two new Legendrian knots Λ^+ and Λ^- by inserting a pair of down- or up-cusps into its front diagram, as shown in Figure 41.6. These are called *positive* and *negative stabilizations* of Λ, respectively. By the formulas above, $tb(\Lambda^+) = tb(\Lambda^-) = tb(\Lambda) - 1$, and $r(\Lambda^\pm) = r(\Lambda) \pm 1$. By performing a sequence of stabilizations, one can construct infinitely many distinct Legendrian knots, of arbitrarily negative Thurston-Bennequin number, in a given smooth knot type.

We now turn to the question above. One common way to visualize the pairs (tb, r) which arise as

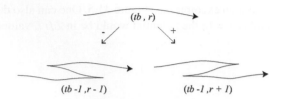

FIGURE 41.6: Negative and positive stabilization.

the Thurston-Bennequin and Maslov numbers of Legendrian knots in a given oriented knot type is to plot these pairs in the plane, with tb on the vertical axis and r on the horizontal axis. For reasons that will soon be clear, the corresponding picture is called a *mountain range*. An atlas of mountain ranges for small knots was developed by Chongchitmate and Ng [5]. The mountain range for the

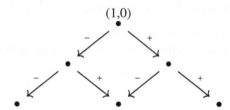

FIGURE 41.7: Mountain range for the right-handed (positive) trefoil [8].

right-handed trefoil is shown in Figure 41.7. Indeed, one can start with the $(tb, r) = (1, 0)$ trefoil in Figure 41.5 and perform positive and negative stabilizations to fill out the rest of the picture. However, it is not obvious that this is the complete mountain range for this knot.

While it is always possible to find Legendrian knots with arbitrarily negative Thurston-Bennequin number in a given smooth isotopy class, the Thurston-Bennequin number of a Legendrian knot in a given smooth isotopy class is bounded above. There are many such bounds; here we present the most well known, the Bennequin Inequality. In fact, this inequality holds for all null-homologous Legendrian knots in all tight contact 3-manifolds [6].

Theorem 1 (Bennequin [1]). *Let Λ be a Legendrian knot in $(\mathbb{R}^3, \xi_{std})$ with Seifert surface Σ. Then*

$$tb(\Lambda) + |r(\Lambda)| \leq -\chi(\Sigma),$$

where $\chi(\Sigma)$ denotes the Euler characteristic of Σ.

Because the trefoil bounds a surface of genus 1, the trefoil in Figure 41.5 has maximal Thurston-Bennequin number in its smooth isotopy class. This implies it cannot be *destabilized*, so it is indeed a peak in the mountain range. The fact that Figure 41.7 is the complete mountain range for the right-handed trefoil follows from Etnyre and Honda's classification of Legendrian torus knots in [8]. In contrast, the left-handed trefoil has two peaks in its mountain range; see Figure 41.8 [8].

The tb-coordinate of the highest peak in the mountain range of a smooth knot type $\mathcal{K}$ is called the *maximal Thurston-Bennequin number* of $\mathcal{K}$; it is a smooth knot invariant, denoted $TB(K)$ for

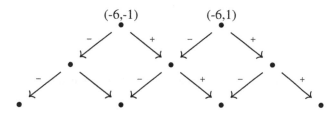

FIGURE 41.8: Mountain range for the left-handed (negative) trefoil [8].

a representative K of $\mathcal{K}$. In fact, there exist Legendrian knots that do not have maximal Thurston-Bennequin number and also cannot be destabilized. The first such example, which has the smooth knot type of the $(2, 3)$ cable of the $(2, 3)$ torus knot, was found by Etnyre and Honda [9]. A surprising application of the maximal Thurston-Bennequin number is its connection to the *slice genus* $g_s(K)$ of a smooth knot K, the minimum genus of a smoothly embedded, orientable surface in the four-ball with boundary $K \subset S^3$. Rudolf [13] proved the following bound on the slice genus:

$$g_s(K) \geq \frac{1}{2}(TB(K) + 1).$$

Another striking application of the Bennequin Inequality is its ability to distinguish between nonequivalent (non-contactomorphic) contact structures. For example, the Bennequin Inequality allows one to prove that ξ_{std} is tight, and therefore not equivalent to the structure ξ_{ot} introduced in the previous chapter, as follows: If D is an overtwisted disk, $\Lambda = \partial D$ must have $tb(\Lambda) = 0$; its Legendrian push-off can be taken to be disjoint from D. Hence the inequality in Theorem 1 is violated.

41.5 Classical Invariants of Transverse Knots

Transverse knots have just two classical invariants: the smooth knot type and the *self-linking number*. Let $\mathcal{T}$ be a transverse knot in $(\mathbb{R}^3, \xi)$. Since ξ must be a trivial bundle, we can choose a nowhere zero section X of ξ (a smoothly varying nowhere zero vector field on $\mathbb{R}^3$, whose vectors lie in ξ). Let $\mathcal{T}'$ be a push-off of $\mathcal{T}$ in the direction of X. The self-linking number sl of $\mathcal{T}$ is the linking number of $\mathcal{T}$ with $\mathcal{T}'$. When $\xi = \xi_{std}$, one can choose $X = \partial_y$. If D_{xz} is the front projection of $\mathcal{T}$ to the xz-plane, then $sl(\mathcal{T}) = w(D_{xz})$ by the discussion in Section 41.1. As in the case of the Maslov number for Legendrian knots, the self-linking number can be defined for any null-homologous transverse knot in any cooriented contact manifold (M, ξ), or for any transverse knot in (M, ξ) when ξ is trivializable.

One can define an analogue of a mountain range for transverse knots; in this case one just plots the values of sl realized by transverse representatives of $\mathcal{K}$ on a vertical axis. There is just one stabliization operation for transverse knots, which decreases sl by 2; see Figure 41.9.

The self-linking number is again bounded above, as in the Legendrian case. Eliashberg proved this inequality holds for all null-homologous knots in all tight contact 3-manifolds [6].

FIGURE 41.9: Stabilization of a transverse knot.

Theorem 2 (Bennequin [1]). *Let $\mathcal{T}$ be a transverse knot in $(\mathbb{R}^3, \xi_{std})$ with Seifert surface Σ. Then*

$$sl(\mathcal{T}) \leq -\chi(\Sigma).$$

41.6   Connections between Legendrian and Transverse Classification Problems

We now ask a second fundamental question concerning the classification of Legendrian knots in $(\mathbb{R}^3, \xi_{std})$.

Question 2. *Given a pair of smoothly isotopic Legendrian knots, with equal Thurston-Bennequin and Maslov numbers, must they be Legendrian isotopic?*

We say a smooth knot type $\mathcal{K}$ is *Legendrian simple*, or just *simple*, if any two Legendrian representatives Λ and Λ' of $\mathcal{K}$ with equal Thurston-Bennequin and Maslov numbers are Legendrian isotopic. The above question asks whether non-simple knot types exist. Similarly, we say $\mathcal{K}$ is *transversely simple* if any two transverse representatives $\mathcal{T}$ and $\mathcal{T}'$ of $\mathcal{K}$ with equal self-linking invariants are transversely isotopic.

We say two Legendrian knots Λ and Λ' are *stably isotopic* if Λ becomes Legendrian isotopic to Λ' after applying a sequence of positive and negative stabilizations. Fuchs and Tabachnikov [11] proved that Λ and Λ' are stably isotopic if and only if they are isotopic as smooth knots. This flexibility result is one step in the proof of the theorem below, which points to the difficulty of answering the question above. We refer the reader to [4] for an introduction to finite-type invariants.

Theorem 3 (Fuchs-Tabachnikov [11]). *Finite-type invariants cannot distinguish between smoothly isotopic Legendrian knots with the same Thurston-Bennequin and Maslov numbers.*

While allowing both positive *and* negative stabilizations introduces too much flexibility into the theory to be interesting, there is a surprising relationship between transverse simplicity and stabliziations of a fixed sign. Two Legendrian knots Λ and Λ' are said to be *negatively stably isotopic* if Λ becomes Legendrian isotopic to Λ' after applying a sequence of negative stabilizations. The *stable Thurston-Bennequin number* of a Legendrian knot Λ is defined to be $s(\Lambda) = tb(\Lambda) - r(\Lambda)$. The stable Thurston-Bennequin number is in fact the transverse self-linking invariant of the *transverse push-off* of Λ; we do not go into detail about the transverse push-off here, but refer the reader to [12] for more information. A smooth knot type $\mathcal{K}$ is *negatively stably simple* if any two Legendrian representatives Λ and Λ' of $\mathcal{K}$ with equal stable Thurston-Bennequin numbers are negatively stably isotopic.

Theorem 4 (Etnyre and Honda [8]). *A knot type is negatively stably simple if and only if it is transversely simple.*

Therefore, the transverse analogue of the question above reduces to a question about Legendrian simplicity. We leave it to the next chapter to address this question in the Legendrian case.

41.7 Bibliography

[1] Daniel Bennequin. *Entrelacements et équations de Pfaff.* PhD thesis, Université de Paris VII, 1982.

[2] Joan S Birman and Julian Eisner. *Seifert and Threlfall, A Textbook of Topology,* volume 89. Academic Press, 1980.

[3] Yu V Chekanov et al. Combinatorics of fronts of Legendrian links and the Arnol'd 4-conjectures. *Russian Mathematical Surveys,* 60(1):95, 2005.

[4] Sergei Chmutov, Sergei Duzhin, and Jacob Mostovoy. *Introduction to Vassiliev knot invariants.* Cambridge University Press, 2012.

[5] Wutichai Chongchitmate and Lenhard Ng. An atlas of Legendrian knots. *Experimental Mathematics,* 22(1):26–37, 2013.

[6] Yakov Eliashberg. Legendrian and transversal knots in tight contact 3-manifolds. *Topological methods in modern mathematics (Stony Brook, NY, 1991),* pages 171–193, 1993.

[7] John B Etnyre. Legendrian and transversal knots. In *Handbook of knot theory,* pages 105–185. Elsevier, 2005.

[8] John B Etnyre and Ko Honda. Knots and contact geometry i: torus knots and the figure eight knot. *Journal of Symplectic Geometry,* 1(1):63–120, 2001.

[9] John B Etnyre and Ko Honda. Cabling and transverse simplicity. *Annals of mathematics,* pages 1305–1333, 2005.

[10] Dmitry Fuchs. Chekanov–Eliashberg invariant of Legendrian knots: existence of augmentations. *Journal of Geometry and Physics,* 47(1):43–65, 2003.

[11] Dmitry Fuchs and Serge Tabachnikov. Invariants of Legendrian and transverse knots in the standard contact space. *Topology,* 36(5):1025–1054, 1997.

[12] Hansjörg Geiges. *An introduction to contact topology,* volume 109. Cambridge University Press, 2008.

[13] Lee Rudolph. The slice genus and the thurston-bennequin invariant of a knot. *Proceedings of the American Mathematical Society,* 125(10):3049–3050, 1997.

Chapter 42

Ruling and Augmentation Invariants of Legendrian Knots

Joshua M. Sabloff, Haverford College

42.1 Introduction

The classical invariants of Legendrian knots introduced in the previous chapter are quite powerful: they can be used to show that there are infinitely many distinct Legendrian knots realizing each smooth knot type, to prove that there exist exotic contact structures on $\mathbb{R}^3$ [1], and even to compute upper bounds on the slice genus of a smooth knot [31]. Further, as noted in the previous chapter, the classical invariants completely classify so-called Legendrian simple smooth knot types, including the unknot [6], torus knots [10], and the figure eight knot [10].

A reasonable question to ask is if any smooth knot type is *not* Legendrian simple, i.e. whether there exists a pair of Legendrian knots with the same classical invariants that are not Legendrian isotopic. To produce such a pair, we need more sophisticated invariants that go beyond the Thurston-Bennequin and rotation numbers. In this chapter, we shall introduce two of the simplest non-classical invariants, namely **rulings** and **augmentations**, apply them to the Legendrian mirror 5_2 knots Λ and Λ' that appear in Figure 42.1, and finally discuss connections between the two invariants.

While our definitions will be purely combinatorial, both of these invariants are derived from more sophisticated machinery. There are four main sources of non-classical invariants of Legendrian knots: pseudo-holomorphic curves and Legendrian contact homology (from which augmentations are derived) [2, 5, 7, 11], generating families (from which rulings are derived) [3, 15, 37], microlocal sheaves [36], and Heegaard-Floer homology [24, 30]. Invariants derived from Heegaard-Floer homology extend to be effective non-classical invariants of transverse knots; these invariants can be used to show that certain twist knots, among others, are not transversely simple [12]. Another source of non-classical transverse invariants comes from the theory of knot contact homology [4], a generalization of the Legendrian contact homology referred to above. Note that the references given here are but starting points for each topic. To keep this chapter elementary, however, we shall make no further mention of these larger theories until the very end.

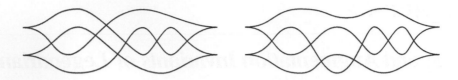

FIGURE 42.1: Front diagrams of two Legendrian mirror 5_2 knots Λ and Λ' that both have Thurston-Bennequin invariant 1 and rotation number 0, but are not Legendrian isotopic.

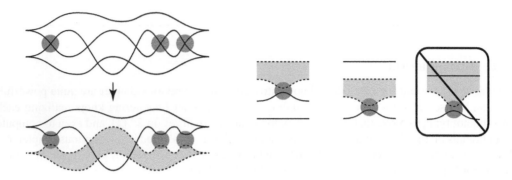

FIGURE 42.2: (a) Resolving a front diagram based on a ruling $\rho \subset X_0(D)$. (b) The normality condition for ruling disks, up to a reflection about the horizontal.

42.2   Rulings

Let us begin by examining **graded normal rulings** of a front diagram and the associated **graded ruling polynomial** of a Legendrian knot. The notion of a ruling was introduced independently by Fuchs [13] and Chekanov-Pushkar [3], who use the term "(Maslov) pseudo-involution" rather than "ruling," though the underlying idea appeared earlier in [8]. The definition we will introduce is closer to that of [32] .

To define a graded normal ruling of a front diagram D, we assume that the crossings and cusps of D all have distinct x coordinates; this is always possible after a small planar isotopy. We further assume, for simplicity of definition, that the rotation number of D is zero, and we choose a Maslov potential μ (see the previous chapter for a discussion of Maslov potentials). At each crossing of D, let o denote the overcrossing (upper left to lower right) strand, and let u denote the undercrossing strand. Let $X(D)$ denote the set of crossings and right cusps, and let $X_k(D) \subset X(D)$ consist of crossings for which $\mu(o) - \mu(u) = k$; right cusps belong to $X_1(D)$.

A ruling of a diagram D is, essentially, a decomposition of D into a set of "eyes" that interact with each other in controlled ways. More formally, a **graded ruling** of a front diagram D is a set of crossings $\rho \subset X_0(D)$ that satisfies several combinatorial conditions. Construct a new front diagram D^ρ from D by resolving each crossing in ρ into two horizontal line segments as in Figure 42.2(a).

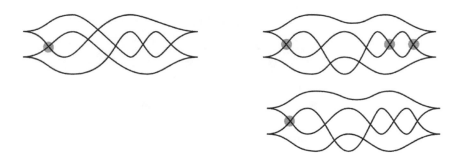

FIGURE 42.3: The rulings of the Legendrian knots from Figure 42.1. Note that the three rightmost crossings of the diagram at right are in $X_0(D)$, but they are in $X_{\pm 2}(D)$ for the diagram at left.

Note that D^ρ represents a different Legendrian link than does D. We say that the components of the link associated to D^ρ to which the new horizontal line segments belong are **incident** to the crossing. The components of the link represented by D^ρ must satisfy the following conditions:

1. Each component is planar isotopic to the standard "eye" diagram of a maximal $tb = -1$ unknot; in particular, each component bounds a **ruling disk** in the plane.

2. Exactly two components are incident to each crossing in ρ.

A graded ruling ρ is deemed **normal** if inside a small vertical strip around each crossing in ρ, the ruling disks incident to the crossing are either nested or disjoint; see Figure 42.2(b). Let us examine two examples. First, Figure 42.3 lists all of the graded normal rulings of the two Legendrian 5_2 knots in Figure 42.1; notice that Λ has two graded normal rulings while Λ' has only one. Second, notice that no stabilized front diagram admits a ruling, as Condition 1 cannot be satisfied no matter which crossings are resolved.

Following [3], we organize the count of rulings of a front diagram as follows. Denote the number of cusps of a front diagram D by $c(D)$ and define the **Euler number** of a ruling ρ to be

$$\chi(\rho) = c(D) - |\rho|,$$

where $|\rho|$ is the number of crossings of the ruling ρ. If we think of the ruling disks, connected by bands at the crossings in ρ, as an immersed surface, then $\chi(\rho)$ is the Euler characteristic of that surface [20]. We use the Euler number to define the **ruling polynomial** of a front diagram D:[1]

$$R_D(z) = \sum_{\text{Rulings } \rho} z^{1-\chi(\rho)}$$

Theorem 42.2.1 ([3]). *The ruling polynomial $R_D(z)$ associated to a front diagram D of a Legendrian link Λ is an invariant of Λ.*

The proof of the theorem follows from checking that $R_D(z)$ does not change under Legendrian Reidemeister moves. We see that the Legendrian mirror 5_2 knot Λ has ruling polynomial $1 + z^2$, while the knot Λ' has ruling polynomial 1. Thus, these two knots are not Legendrian isotopic

---

[1]There are several possible conventions for the ruling polynomial in the literature. Caveat lector!

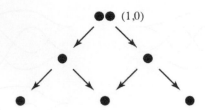

FIGURE 42.4: The mountain range for the mirror 5_2 knot, as described by [9, 12].

despite having the same classical invariants. In fact, the two Legendrian knots in Figure 42.1 are the only two maximal-tb mirror 5_2 knots [12]. The mountain range for this smooth knot type appears in Figure 42.4.

The material presented here provides only the combinatorial foundations of the ruling invariant of Legendrians in the standard contact $\mathbb{R}^3$. There are several avenues in which to generalize: to other ambient manifolds [21, 23, 33], to Morse Complex Sequences [16], or to generating family homology [15, 35, 37]. It is also interesting to explore connections to knot polynomials: the fundamental result is if we allow switches to lie in $\bigcup_{k \equiv 0(2)} X_k(D)$ (resp. any crossing), then the ruling polynomial of a Legendrian knot appears as part of the HOMFLY (resp. Kauffman) polynomial of the underlying smooth knot [32].

42.3   Augmentations

The second non-classical invariant we will examine is the set of **augmentations** and the associated **augmentation number**. The concept and original definition of an augmentation are due to Chekanov [2], though the formulation here owes much to [25]; the augmentation number first appeared in [29].

The setup for the definition of an augmentation consists of a front diagram D whose right cusps all have the same x coordinate; this is always possible after a sequence of Legendrian Reidemeister II moves. While not strictly required, this setup makes the definition simpler; see, for example, [2, 13, 22, 25]. We once again assume that the rotation number is zero for simplicity, and we choose a Maslov potential.

The central objects in the definition of an augmentation are **admissible disks**, which play a similar role to ruling disks. An admissible disk is a continuous embedding $\phi : (D^2, \partial D^2) \to (\mathbb{R}^2, D)$, smooth on the interior, that satisfies the following conditions:

1. The restriction of ϕ to ∂D^2 is an embedding, smooth away from the crossings and cusps.

2. Near a crossing or a cusp, the image of the map ϕ matches one of the pictures in Figure 42.5(a), which we call terminating singularities, negative corners, or originating singularities.

3. The map ϕ has exactly one originating and one terminating singularity.

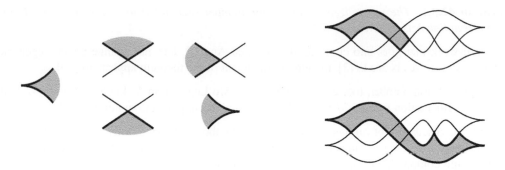

FIGURE 42.5: (a) Possible behavior of an admissible disk near a cusp or corner: (left) a terminating singularity, (center) negative corners, and (right) originating singularities. (b) Examples of admissible disks.

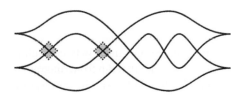

FIGURE 42.6: The leftmost Legendrian knot from Figure 42.1 has a unique augmentation.

See Figure 42.5(b) for examples of admissible disks.

An **augmentation** is a set $\varepsilon \subset X_0(D)$ of crossings such that:

- for every crossing of D, the number of admissible disks with originating singularity at the given crossing whose negative corners all lie in ε is even and

- for every right cusp of D, the number of admissible disks whose negative corners all lie in ε is odd.

For the two Legendrian mirror 5_2 knots in Figure 42.1, the leftmost diagram has but one augmentation, as shown in Figure 42.6, but the rightmost diagram has six. On the other hand, a stabilized diagram never has an augmentation.

To organize augmentations into an invariant, we must normalize their count using the shifted Euler characteristic of $X(D)$:

$$\chi^*(D) = \sum_{k \geq 0}(-1)^k |X_k(D)| + \sum_{k < 0}(-1)^{k+1}|X_k(D)|.$$

Let $N(D)$ denote the number of augmentations of D. We define the **normalized augmentation number** of the diagram D to be

$$\mathrm{Aug}(D) = 2^{-\chi^*(D)/2}N(D).$$

Theorem 42.3.1. *The normalized augmentation number associated to a front diagram D of a Legendrian link* Λ *is an invariant of* Λ.

The hard work that goes into checking directly that $\text{Aug}(D)$ does not change under Legendrian Reidemeister moves is due to [2], though the statement of the theorem appears in [29].

The augmentation number, too, can distinguish the two Legendrian 5_2 knots in Figure 42.1: the first has augmentation number $3/\sqrt{2}$, while the second has augmentation number $\sqrt{2}$.

As with the discussion of rulings above, the material about augmentations is just the beginning. For more sophisticated ways to count augmentations, see [19, 27]. Augmentations are the first step in extracting computable invariants from Legendrian contact homology, a topic which has a vast literature beginning with [2, 5, 7, 11].

42.4 Connections

We finish the chapter by outlining some connections between rulings and augmentations. Fuchs motivated his definition of a ruling by proving that if a front diagram has a ruling, then it also has an augmentation [13]. The converse is also true [14, 34]. The correspondence between augmentations and rulings can be pushed further, beginning with the following theorem:

Theorem 42.4.1 ([29]). *The ruling polynomial determines the augmentation number:* $\text{Aug}(D) = 2^{1/2} R_D(2^{-1/2})$.

These connections have been deepened in several ways, with direct generalizations of Theorem 42.4.1 appearing in [19, 22, 23, 27]. As mentioned before, the ruling and augmentation invariants are but combinatorial shadows of the larger generating family and contact homology theories. Connections between these larger theories have long been expected even though existing proofs are combinatorial rather than analytic [15, 17, 18, 26, 28].

42.5 Bibliography

[1] D. Bennequin. Entrelacements et equations de Pfaff. *Asterisque*, 107–108:87–161, 1983.

[2] Yu. Chekanov. Differential algebra of Legendrian links. *Invent. Math.*, 150:441–483, 2002.

[3] Yu. Chekanov and P. Pushkar. Combinatorics of Legendrian links and the Arnol'd 4-conjectures. *Russ. Math. Surv.*, 60(1):95–149, 2005.

[4] T. Ekholm, J. Etnyre, L. Ng, and M. Sullivan. Filtrations on the knot contact homology of transverse knots. *Math. Ann.*, 355(4):1561–1591, 2013.

[5] Ya. Eliashberg. Invariants in contact topology. In *Proceedings of the International Congress of Mathematicians, Vol. II (Berlin, 1998)*, number Extra Vol. II, pages 327–338 (electronic), 1998.

[6] Ya. Eliashberg and M. Fraser. Classification of topologically trivial Legendrian knots. In *Geometry, Topology, and Dynamics (Montreal, PQ, 1995)*, pages 17–51. Amer. Math. Soc., Providence, RI, 1998.

[7] Ya. Eliashberg, A. Givental, and H. Hofer. Introduction to symplectic field theory. *Geom. Funct. Anal.*, (Special Volume, Part II):560–673, 2000.

[8] Ya. M. Eliashberg. The structure of 1-dimensional wave fronts, nonstandard Legendrian loops and Bennequin's theorem. In *Topology and Geometry—Rohlin Seminar*, volume 1346 of *Lecture Notes in Math.*, pages 7–12. Springer, Berlin, 1988.

[9] J. Epstein, D. Fuchs, and M. Meyer. Chekanov-Eliashberg invariants and transverse approximations of Legendrian knots. *Pacific J. Math.*, 201(1):89–106, 2001.

[10] J. Etnyre and K. Honda. Knots and contact geometry I: Torus knots and the figure eight knot. *J. Symplectic Geom.*, 1(1):63–120, 2001.

[11] J. Etnyre, L. Ng, and J. Sabloff. Invariants of Legendrian knots and coherent orientations. *J. Symplectic Geom.*, 1(2):321–367, 2002.

[12] J. Etnyre, L. Ng, and V. Vértesi. Legendrian and transverse twist knots. 2013.

[13] D. Fuchs. Chekanov-Eliashberg invariant of Legendrian knots: Existence of augmentations. *J. Geom. Phys.*, 47(1):43–65, 2003.

[14] D. Fuchs and T. Ishkhanov. Invariants of Legendrian knots and decompositions of front diagrams. *Mosc. Math. J.*, 4(3):707–717, 2004.

[15] D. Fuchs and D. Rutherford. Generating families and Legendrian contact homology in the standard contact space. *J. Topol.*, 4(1):190–226, 2011.

[16] M. B. Henry. Connections between Floer-type invariants and Morse-type invariants of Legendrian knots. *Pacific J. Math.*, 249(1):77–133, 2011.

[17] M. B. Henry and D. Rutherford. A combinatorial DGA for Legendrian knots from generating families. *Commun. Contemp. Math.*, 15(2):1250059, 60, 2013.

[18] M. B. Henry and D. Rutherford. Equivalence classes of augmentations and Morse complex sequences of Legendrian knots. *Algebr. Geom. Topol.*, 15(6):3323–3353, 2015.

[19] M. B. Henry and D. Rutherford. Ruling polynomials and augmentations over finite fields. *J. Topol.*, 8(1):1–37, 2015.

[20] T. Kálmán. Rulings of Legendrian knots as spanning surfaces. *Pacific J. Math.*, 237(2):287–297, 2008.

[21] M. Lavrov and D. Rutherford. On the $S^1 \times S^2$ HOMFLY-PT invariant and Legendrian links. *J. Knot Theory Ramifications*, 22(8):1350040, 21, 2013.

[22] C. Leverson. Augmentations and rulings of Legendrian knots. Preprint available as arXiv:1403.4982, 2014.

[23] C. Leverson. Augmentations and rulings of Legendrian links in $\#^k(S^1 \times S^2)$. *Pacific J. Math.*, 288(2):381–423, 2017.

[24] P. Lisca, P. Ozsváth, A. I. Stipsicz, and Z. Szabó. Heegaard Floer invariants of Legendrian knots in contact three-manifolds. *J. Eur. Math. Soc. (JEMS)*, 11(6):1307–1363, 2009.

[25] L. Ng. Computable Legendrian invariants. *Topology*, 42(1):55–82, 2003.

[26] L. Ng and D. Rutherford. Satellites of Legendrian knots and representations of the Chekanov-Eliashberg algebra. *Algebr. Geom. Topol.*, 13(5):3047–3097, 2013.

[27] L. Ng, D. Rutherford, V. Shende, and S. Sivek. The cardinality of the augmentation category of a Legendrian link. Preprint available as arXiv:1511.06724, 2015.

[28] L. Ng, D. Rutherford, V. Shende, S. Sivek, and E. Zaslow. Augmentations are sheaves. Preprint available as arXiv:1502.04939, 2015.

[29] L. Ng and J. Sabloff. The correspondence between augmentations and rulings for Legendrian knots. *Pacific J. Math.*, 224(1):141–150, 2006.

[30] P. Ozsváth, Z. Szabó, and D. Thurston. Legendrian knots, transverse knots and combinatorial Floer homology. *Geom. Topol.*, 12(2):941–980, 2008.

[31] L. Rudolph. Quasipositivity as an obstruction to sliceness. *Bull. Amer. Math. Soc. (N.S.)*, 29(1):51–59, 1993.

[32] D. Rutherford. Thurston-Bennequin number, Kauffman polynomial, and ruling invariants of a Legendrian link: the Fuchs conjecture and beyond. *Int. Math. Res. Not.*, pages Art. ID 78591, 15, 2006.

[33] D. Rutherford. HOMFLY-PT polynomial and normal rulings of Legendrian solid torus links. *Quantum Topol.*, 2(2):183–215, 2011.

[34] J. Sabloff. Augmentations and rulings of Legendrian knots. *Int. Math. Res. Not.*, (19):1157–1180, 2005.

[35] J. Sabloff and L. Traynor. Obstructions to Lagrangian cobordisms between Legendrian submanifolds. *Algebr. Geom. Topol.*, 13:2733–2797, 2013.

[36] V. Shende, D. Treumann, and E. Zaslow. Legendrian knots and constructible sheaves. 2014.

[37] L. Traynor. Generating function polynomials for Legendrian links. *Geom. Topol.*, 5:719–760, 2001.

Part VIII

Higher Dimensional Knot Theory

Chapter 43

Broken Surface Diagrams and Roseman Moves

J. Scott Carter, University of South Alabama

Masahico Saito, University of South Florida

43.1   Introduction

Just as classical knots can be described and studied via their diagrams, so too can surfaces that are embedded in 4-dimensional space.

In the classical case, a knot diagram consists of a generic projection of a knot in 3-space into a plane in which over-/under-crossing information is given at each transverse double point of the image curve. The crossing information is depicted by a broken arc convention: the arc that is further from the plane of projection is cut into two. We imagine the plane of projection to be the retina of the observer. Thus the arc that is closer to the viewer is the unbroken one. Take a look at the cords that are connecting your electronic devices to the wall to envision the analogy.

The diagrams prove to be useful in defining, for example, the fundamental group and quandle, the Jones polynomial, and other quantum-type invariants. Such an algebraic entity that is associated to a particular diagram, but that remains unchanged when a different representative diagram is chosen, is called a *knot invariant*.

The first notice that surfaces, indeed spheres, could be knotted in 4-space was given in [1]. Artin's spinning construction indicated that a knotted object in n-space could be spun to create a knotted object in $(n+1)$-space. Thus knotted $(n-2)$-dimensional spheres exist in all dimensions $n \geq 3$. Fox's seminal article [2] laid the groundwork for the study of knotted surfaces. Additional geometric constructions (twist-spinning) were given by Zeeman [9]. In the same era (circa 1962-1979), great strides were made in the understanding of higher-dimensional knots via their infinite cyclic covers and homological invariants thereof (e.g., [6]).

Additional diagrammatic techniques were pioneered in [3, 4, 5], and [8]. Notably, Roseman's paper [7] informs the presentation given here.

43.2   Knotted Surfaces and Broken Surface Diagrams

A *knotted surface*, or a *surface-knot*, is a closed (compact without boundary) surface smoothly embedded in $\mathbb{R}^4$ or S^4 with the standard smooth structures.

An orientable knotted surface $F \subset \mathbb{R}^4$ is *unknotted* if it bounds a solid handlebody. This is equivalent to the condition that it is isotopic to a surface *standardly embedded* in $\mathbb{R}_0^3 = \{(x, y, z, 0) : x, y, z \in \mathbb{R}\}$. A surface $F \subset \mathbb{R}^4$ is *knotted* if it is not unknotted.

Since diagrammatic approaches in classical knot theory are successful, a diagrammatic method to study embedded surfaces is highly desirable. Thus we choose a 3-dimensional affine subset of 4-space that is disjoint from F, choose a generic projection of F into this hyperplane, and indicate the relative height of crossing points by means of an illustrative convention.

A *general position surface* in 3-dimensional space is one for which each point has a 3-ball neighborhood $N \approx (-1, 1)^3$ in which $F \cap N$ is of one of the forms below:

1. *non-singular point:* $(\{(x, y, z) \in (-1, 1)^3 : z = 0\}$;

2. *double point:* $\{(x, y, z) \in (-1, 1)^3 : yz = 0\}$;

3. *triple point:* $\{(x, y, z) \in (-1, 1)^3 : xyz = 0\}$;

4. *branch point:* Cone on a figure $8 \subset (-1, 1)^3$.

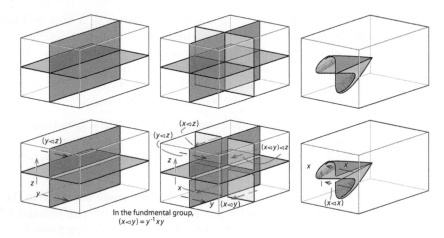

FIGURE 43.1: General position maps and their associated broken surfaces.

The crossings and branch point are depicted in Fig. 43.1 together with the associated *broken surface diagram*. The sheet(s) that are further from the hyperplane into which the image is projected are broken in order for the closer sheets to appear to be passing above them. The broken surface diagram, then, conveys the crossing information of the embedding of the surface F in 4-space. To clarify, the branch point is the cone point of the cone on a figure 8, and in any sufficiently small 3-ball neighborhood thereof, the surface F resembles the surface of the cone.

Definition 1. A *broken surface diagram of a knotted surface* $F \subset \mathbb{R}^4$ is a generic projection of a surface into a hyperplane for which crossing information is indicated as in Fig. 43.1. That is, along each arc of double curves in the projection, one of the two transverse sheets is broken to indicate that it is further from the hyperplane of projection than the other sheet.

In Fig. 43.1, a method to compute the fundamental group of the complement of the surface $(\mathbb{R}^4 \setminus F)$ is also indicated, that is an analogue of the Wirtinger presentation of classical knot groups. The short arrows that are transverse to the sheets of the local pictures of the diagram schematize meridional loops that pass around the sheets of the surface. Labels on these arcs represent generators for the fundamental group. Within the figure, relations are given. These are expressed both in quandle form and as conjugacy relations to serve the companion articles. See the discussion below in Section 43.4.

43.3 Roseman Moves

Reidemeister's theorem states that two knot diagrams represent the same knot if and only if they are related by a sequence of Reidemeister moves or a plane isotopy. Roseman [7] provided an analogous theorem for broken surface diagrams.

Theorem 1 (Roseman [7]). *Two broken surface diagrams represent the same knotted surface if and only if they are related by a finite sequence of Roseman moves, depicted in Fig. 43.3, or a spatial isotopy.*

We remark that the depiction of the Roseman moves follows the tradition given in the original paper [7] in which moves are drawn in projection for simplicity. In this case, it is up to the reader to interpolate the breaks in the surfaces so that the moves can be performed consistently. The movies that accompany the projections of the surfaces represent 3-dimensional cross sections and indicate one possible choice of over/under information. See Section 43.4. Furthermore, the moves are also encoded via information on the *chart*: this is a projection of the double point set and the fold set into an affine plane in 3-space, with exemplary choices of projection directions. The label $D(i, j)$ indicates that this is a double curve with i sheets behind it and j sheets in front of it. The label $F(i, j)$ indicates a fold in a similar location. If there are no hidden sheets, then $i = j = 0$.

The idea of proof is to classify generic singularities. This idea also applies to Reidemeister's theorem, although the original proof was combinatorial. In the classical case, the projection of the trace of the Reidemeister type III move forms an isolated triple point in space-time. Specifically, as the move occurs, the three arcs form three sheets as in the middle of Fig. 43.2. The whole of the continuous trace, then, forms a triple point of three sheets in general position, that is, the intersection among three planes. In the right of the figure, the resulting triple point is depicted in a projection of the double curves into a plane.

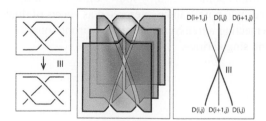

FIGURE 43.2: Reidemeister type III move and a triple point of broken surface diagram.

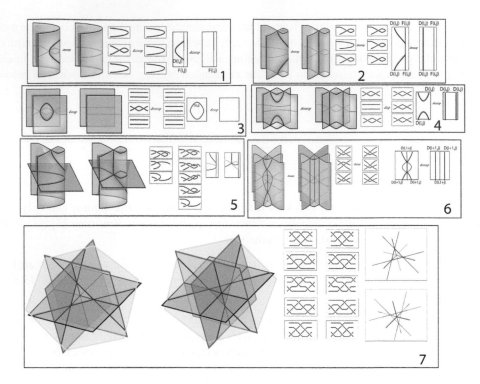

FIGURE 43.3: Roseman moves, movies, and charts.

Let us examine the Roseman moves that appear in Fig. 43.3. The double points of the projection of an embedded surface form an immersed 1-manifold. It has transverse self-intersection points at triple points and has branch points as its end points. During an isotopy, the 1-dimensional double point set can have a generic critical point. Such a critical point, then, would be a birth, death, or saddle. These correspond to the moves numbered 3 (birth/death) and 4 (saddle) in Fig. 43.3. A critical point for the branch point set is a birth or death of a pair of branch points; either is depicted in moves numbered 1 or 2. Next, observe that a branch point might pass through a transverse sheet (move 5) . A critical point for the triple point set appears as move number 6. The analogue of the Reidemeister type III move among Roseman moves is the *tetrahedral move* that involves moving four planes intersecting at the moment of the move in a quadruple point (move 7). Thus this corresponds to a quadruple point formed by four 3-dimensional hyperplanes in $\mathbb{R}^4$ that are in general position. This quadruple point can be interpreted as sheets of double points intersecting or a line of triple points intersecting a transverse sheet. The proof consists of showing that these are the complete list of generic singularities.

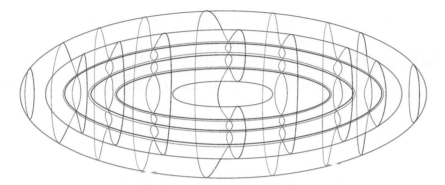

FIGURE 43.4: A broken surface diagram of a spun trefoil knot.

43.4 Fundamental Group and Quandle

The most important invariants that can be defined using a broken surface diagram are the fundamental group and the fundamental quandle. In general, a quandle is a set, Q, upon which a binary operation $\triangleleft$ is defined and which satisfies:

I. $\forall x \in Q, x \triangleleft x = x$;

II. $\forall x, y \in Q, \exists! z \in Q$ such that $z \triangleleft x = y$;

III. $\forall x, y, z \in Q, (x \triangleleft y) \triangleleft z = (x \triangleleft z) \triangleleft (y \triangleleft z)$.

Given a group G, there is an associated quandle $\mathrm{Conj}(G)$ which consists of the set of elements of G on which the operation $\triangleleft$ is given by $x \triangleleft y = y^{-1}xy$.

Within Fig. 43.1, representatives for homotopy classes of loops in 4-space are indicated by short arrows that are transverse to surfaces in the figures. These correspond to meridional generators of the fundamental group or quandle. When x is the label on the lower sheet away from which the normal to the upper sheet points, and y is the label on the sheet above, then the other lower sheet receives a label $x \triangleleft y$. In the fundamental group, we have $x \triangleleft y = y^{-1}xy$. In this way, labels are assigned to the sheets of the diagram, relations occur along the double curves, and a group or quandle presentation arises.

Using this methodology, one can compute that the fundamental quandle of the spun trefoil depicted in Fig. 43.4 is

$$\pi_Q = \langle a, b, c : a \triangleleft b = c, b \triangleleft c = a, c \triangleleft a = b \rangle.$$

A presentation for the fundamental group (using the operation $a \triangleleft b = b^{-1}ab$) is

$$\langle a, b : aba = bab \rangle.$$

This group is known to be isomorphic to the three-string braid group, and consequently is not cyclic. Thus the corresponding sphere is knotted.

43.5 Bibliography

[1] Emil Artin. Zur Isotopie zweidimensionaler Flächen im R_4. *Abh. Math. Sem. Univ. Hamburg*, 4(1):174–177, 1925.

[2] Ralph H. Fox. A quick trip through knot theory. In *Topology of 3-Manifolds and Related Topics (Proc. The Univ. of Georgia Institute, 1961)*, pages 120–167. Prentice-Hall, Englewood Cliffs, N.J., 1962.

[3] Cole A. Giller. Towards a classical knot theory for surfaces in $\mathbf{R}^4$. *Illinois J. Math.*, 26(4):591–631, 1982.

[4] Akio Kawauchi, Tetsuo Shibuya, and Shin'ichi Suzuki. Descriptions on surfaces in four-space. I. Normal forms. *Math. Sem. Notes Kobe Univ.*, 10(1):75–125, 1982.

[5] Akio Kawauchi, Tetsuo Shibuya, and Shin'ichi Suzuki. Descriptions on surfaces in four-space. II. Singularities and cross-sectional links. *Math. Sem. Notes Kobe Univ.*, 11(1):31–69, 1983.

[6] Jerome Levine. Knot modules. I. *Trans. Amer. Math. Soc.*, 229:1–50, 1977.

[7] Dennis Roseman. Reidemeister-type moves for surfaces in four-dimensional space. In *Knot theory (Warsaw, 1995)*, volume 42 of *Banach Center Publ.*, pages 347–380. Polish Acad. Sci. Inst. Math., Warsaw, 1998.

[8] Katsuyuki Yoshikawa. An enumeration of surfaces in four-space. *Osaka J. Math.*, 31(3):497–522, 1994.

[9] E. C. Zeeman. Twisting spun knots. *Trans. Amer. Math. Soc.*, 115:471–495, 1965.

Chapter 44

Movies and Movie Moves

J. Scott Carter, University of South Alabama

Masahico Saito, University of South Florida

44.1 Introduction

Movie descriptions of knotted surfaces date back to the 1960s [8, 7]. Artin's *spinning* construction [1] of non-trivial knotted spheres can be also visualized via a movie description (see Chapter 43, the companion chapter to this one). Since each still (except for critical levels) is a classical knot or link, a movie description has some advantages over the broken surface diagram description of knotted surfaces. In particular, Fox [7] used movie descriptions to present the fundamental group and the Alexander ideals for several key examples.

In the middle to late 1980s, several authors became interested in finding a movie move description for knotted surface isotopies since they might be useful in developing 2-categorical descriptions of knotted surfaces. In the classical case, [9] characterized the category of tangles as the braided monoidal category with duals on one generating object (see also [13]). Such a category theoretical description gave rise to invariants via functors from the category of tangles to categories of representations. These generalize the Jones polynomial [10].

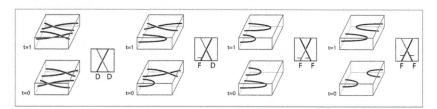

FIGURE 44.1: Tangles as 1-morphisms and critical exchanges as 2-morphisms.

The 2-category of 2-tangles is described as follows. The set of objects are the set of dots arranged along a line segment in a rectangular face in some 4-dimensional box. The set of 1-morphisms is the set of tangles in a 3-dimensional cube that connect dots to dots. By choosing projections and heights, arcs, crossings, and critical points are described by I, X, $\overline{X}$, U, and $\cap$. Here, I is the identity morphism on an object, X represents a positive braiding with inverse $\overline{X}$, U represents a minimal point, and $\cap$ represents a maximal point (Fig. 44.1). The 2-morphisms are given by birth, death, saddles, exchanges between crossings and/or critical points (Fig. 44.1), and the generalized Reidemeister moves that are depicted in Fig. 44.2. Relations among 2-morphisms include that generalized Reidemeister moves are natural 2-isomorphisms. For example, the Roseman moves encapsulate that the Reidemeister moves are natural 2-isomorphisms. More details will appear below. Analogues of the Freyd-Yetter [9] and Shum [13] theorems are given in [6, 11], and [2]. It states roughly that the category of 2-tangles is the free braided monoidal 2-category with duals that is generated by one self-dual object generator.

In Khovanov homology theory [12], a version of the movie moves was used to prove functoriality. See [3], for example.

44.2 Movie Descriptions

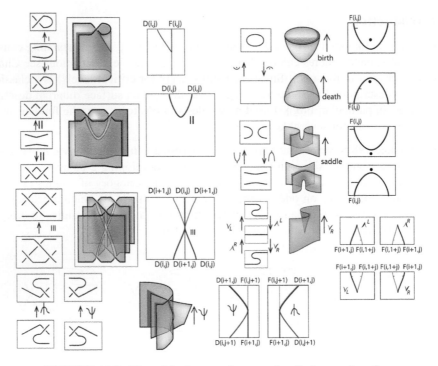

FIGURE 44.2: The critical events in a movie of a knotted surface.

Here we give an overview of the movie moves for surfaces $F \subset \mathbb{R}^4 = \{(x, y, z, t) : x, y, z, t \in \mathbb{R}\}$ that are embedded in 4-dimensional space. The 4th coordinate is regarded as time while a space-like slice $\mathbb{R}^3_t = \mathbb{R}^3 \times \{t\}$ is the hyperplane (*still, frame, slice*) at time t. We will describe the movie moves from a 2-categorical point of view and present some of those that are not among the seven Roseman moves that appear in Chapter 43.

We put the embedding $F \subset \mathbb{R}^4$ in a nice position with respect to several projections.

Step 1. Perturb the image of F so that the restriction of the projection p_4 onto the time direction has only finitely many non-degenerate critical points, and each of these lies at distinct critical values. The critical points are births (index 0), deaths (index 2), or saddle points (index 1). See the right-hand side of Fig. 44.2.

Step 2. The projection of F into the $z = 0$ hyperplane can be used to create a broken surface diagram as described in Chapter 43. In particular, for a non-critical level t_0, the intersection $F \cap \mathbb{R}^3(t_0)$ is a classical knot or link. Create a diagram by breaking the arc at a transverse double point that is further away from the $z = 0$ plane. These broken arc diagrams are consistent through time.

Step 3. Assume that the projection of $F \cap \mathbb{R}^3(t_0)$ to the y-axis is a generic height function such that crossings and critical points lie at distinct heights.

Now at successive non-critical levels $t_0 < t_1$ that are separated by a singular point, the diagrams in the (x, y) planes $\{(x, y, 0, t_j) : x, y \in \mathbb{R}\}$ differ at most by one of the relations:

1. a critical point of the projection p_4;

2. a generalized Reidemeister move;

3. an exchange of levels of distant crossings or critical points.

Figure 44.2 indicates local pictures of type 1 and 2 in movie form and via representative broken surface diagrams (cf. Chapter 43) for one choice of crossing. Figure 44.1 indicates several cases of the exchange of distant crossings (item 3) and the projection of these into the (y, t)-plane $(0, y, 0, t)$ with double curves and folds labeled D and F is also indicated. The indices in an expression such as $D(i, j)$ describe the number of possible sheets, i, that are behind, and j, that are in front of the indicated level.

44.3   Movie Moves

In the 2-categorical description of knotted surfaces, a classical knot or tangle is thought of as a 1-morphism while the critical points and generalized Reidemeister moves are thought of as 2-morphisms. In this way, the seven Roseman moves and a number of the remaining movie moves can be described precisely in terms of identities among 2-morphisms.

We establish notation in which the iconography suggests the nature of the singularities in a 2-dimensional projection. In this way the movie moves can be expressed as equalities among compositions of symbols. The Reidemeister type II moves are denoted as follows: reading from bottom to top, $\cup$ indicates adding a pair of crossings while $\cap$ indicates canceling them. Type I moves are denoted by $\measuredangle$, $\curlyvee$, $\measuredangle$, and $\curlywedge$, with $\curlyvee$ indicating the specific type I move illustrated in Fig. 44.2. The type III move illustrated in Fig. 44.2 is indicated as $\mathbf{X}$. Vertical juxtaposition in the time direction is indicated $a \circ b$ with a appearing above b. As isomorphisms, the successive application of a Reidemeister move and its undoing is isotopic to doing nothing to a knotted surface. Thus we have,

$$\cup \circ \cap = \mathrm{id}, \ \cap \circ \cup = \mathrm{id}, \ \measuredangle \circ \curlywedge = \mathrm{id}, \ \curlywedge \circ \measuredangle = \mathrm{id}, \ \mathbf{X} \circ (\mathbf{X}^{-1}) = \mathrm{id},$$

and so forth. Here and below, the inverse of a move is the same move with a time reversal. These are, of course, expressions of five of the Roseman moves.

There are four possible cusps denoted as $\vee_L$, $\wedge^L$, $\vee_R$, and $\wedge^R$. Use κ to denote a crossing passing over a minimum, and $\rtimes$ for it passing over a maximum (Fig. 44.2).

That the remaining generalized Reidemeister moves are isomorphisms is to assert, for $D = L$ or R, that $\wedge^D \circ \vee_D = $ id — lips, $\vee_D \circ \wedge^D = $ id — beak-to-beak. Furthermore, $\kappa \circ \kappa^{-1} = $ id, $\kappa^{-1} \circ \kappa = $ id, $\rtimes \circ \rtimes^{-1} = $ id, and $\rtimes^{-1} \circ \rtimes = $ id. These identities have been expressed as movie moves in [4] and [5].

The move in which a branch point is pushed through a transverse sheet and a triple point is added or subtracted is explained categorically by saying that the type III move is natural with respect to another (type I) 2-morphism. That the naturality of the type III move also implies the quadruple point move—the bottom move in the list of Roseman moves in Chapter 43—is demonstrated in [5] by decomposing the permutahedron into 3-cells that represent 2-categorical isomorphisms.

Denote the birth of a simple closed curve by $\smile$, the death of a simple closed curve by $\frown$, a saddle by $\underset{.}{\cup}$, and an upside-down saddle by $\overset{.}{\cap}$ (Fig. 44.2). Denote by $\dashv$ a fold in the projection onto the (y, t) plane ($0 < y$ is right and $0 < t$ is up) that is the time trace of a maximum of a knot diagram in (x, y), and let $\vdash$ denote a fold that is the time trace of a minimum. The short horizontal segments demonstrate where surface appears. Horizontal juxtaposition indicates that these events occur in the same time interval. Then

$$\left(\dashv \frown\right) \circ \left(\underset{.}{\cup} \dashv\right) \;=\; \dashv \;=\; \left(\overset{.}{\cap} \dashv\right) \circ \left(\dashv \smile\right),$$
$$\left(\frown \vdash\right) \circ \left(\vdash \underset{.}{\cup}\right) \;=\; \vdash \;=\; \left(\vdash \overset{.}{\cap}\right) \circ \left(\smile \vdash\right).$$

These equalities indicate the cancellation of critical points of successive indices (Fig. 44.3). Note that the corresponding movie move is found in Fig. 44.4 where the stills in all moves are rotated ninety degrees from those here and in the Roseman moves.

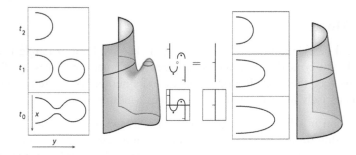

FIGURE 44.3: Canceling critical points and indicating some iconography.

The saddle/death or birth/saddle cancellation can be used to show that a branch point passing over an optimum follows from passing it over a saddle. A similar result holds for κ.

We describe the list of movie moves in the following form due to limited space. See [4] for details and more explicit descriptions.

Theorem 1 ([4]). *A pair of movies represent the same knotted surface if and only if they are related by the movie moves that describe the Roseman moves (see Chapter 43), the moves depicted in Fig. 44.4, naturality and invertibility of crossing and critical exchanges, and the invertibilities*

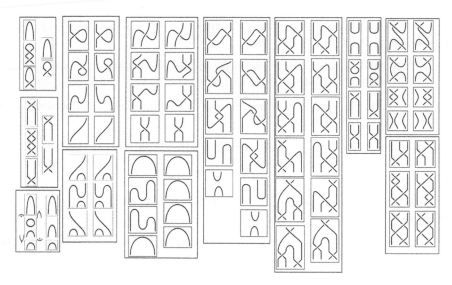

FIGURE 44.4: Several movie moves.

of κ, ⋊, and cusps ⋎$_D$, ⋏$_D$ for D = L or R. All variations of time reversal and consistent crossing information are also assumed to be included in the list.

44.4 Bibliography

[1] Emil Artin. Zur Isotopie zweidimensionaler Flächen im R_4. *Abh. Math. Sem. Univ. Hamburg*, 4(1):174–177, 1925.

[2] John C. Baez and Laurel Langford. Higher-dimensional algebra. IV. 2-tangles. *Adv. Math.*, 180(2):705–764, 2003.

[3] Dror Bar-Natan. Khovanov's homology for tangles and cobordisms. *Geom. Topol.*, 9:1443–1499, 2005.

[4] J. Scott Carter, Joachim H. Rieger, and Masahico Saito. A combinatorial description of knotted surfaces and their isotopies. *Adv. Math.*, 127(1):1–51, 1997.

[5] J. Scott Carter and Masahico Saito. *Knotted Surfaces and Their Diagrams*, volume 55 of *Mathematical Surveys and Monographs*. American Mathematical Society, Providence, RI, 1998.

[6] John E. Fischer, Jr. 2-categories and 2-knots. *Duke Math. J.*, 75(2):493–526, 1994.

[7] Ralph H. Fox. A quick trip through knot theory. In *Topology of 3-Manifolds and Related Topics (Proc. The Univ. of Georgia Institute, 1961)*, pages 120–167. Prentice-Hall, Englewood Cliffs, N.J., 1962.

[8] Ralph H. Fox and John W. Milnor. Singularities of 2-spheres in 4-space and cobordism of knots. *Osaka J. Math.*, 3:257–267, 1966.

[9] Peter J. Freyd and David N. Yetter. Braided compact closed categories with applications to low-dimensional topology. *Adv. Math.*, 77(2):156–182, 1989.

[10] V. F. R. Jones. Hecke algebra representations of braid groups and link polynomials. *Ann. of Math. (2)*, 126(2):335–388, 1987.

[11] V. M. Kharlamov and V. G. Turaev. On the definition of the 2-category of 2-knots. In *Mathematics in St. Petersburg*, volume 174 of *Amer. Math. Soc. Transl. Ser. 2*, pages 205–221. Amer. Math. Soc., Providence, RI, 1996.

[12] Mikhail Khovanov. A categorification of the Jones polynomial. *Duke Math. J.*, 101(3):359–426, 2000.

[13] Mei Chee Shum. Tortile tensor categories. *J. Pure Appl. Algebra*, 93(1):57–110, 1994.

Chapter 45

Surface Braids and Braid Charts

Seiichi Kamada, Osaka University

A surface braid is a higher-dimensional generalization of a braid. Surface braids can be presented by the motion picture method. However, it is often inconvenient to use only motion pictures when we deform surface braids globally up to equivalence. We introduce the chart description method, which is a method of presenting surface braids using graphs on a 2-disk. Using this method, one can easily deform surface braids up to equivalence. As classical links are represented by braids, surface links are also represented by surface braids. Thus we can use surface braids to study surface links, and vice versa. Refer to [1, 9, 10, 11] for details and related topics on surface braids.

Throughout this chapter, X_m stands for a fixed set of m interior points of B^2 such that they are on the same line in B^2.

45.1 Surface Braids

Recall that a classical braid b of degree m is the union of m arcs embedded in the cylinder $B^2 \times I$ such that the projection map $b \to I$ is a covering map of degree m and $b \cap (B^2 \times I) = X_m \times \partial I$. A surface braid is defined similarly.

Definition 1. A *surface braid* S of degree m is a compact surface embedded in $B^2 \times D^2$ such that (i) the projection map $S \to D^2$ of S to the D^2 factor is a branched covering map of degree m and (ii) $S \cap \partial(B^2 \times D^2) = X_m \times \partial D^2$.

A surface braid is called *simple* if the branched covering map $S \to D^2$ is a simple branched covering map, i.e., for every point $y \in D^2$ the inverse image consists of m or $m - 1$ points.

A classical braid $b \subset B^2 \times I$ is oriented with an orientation compatible with the orientation of I. A surface braid $S \subset B^2 \times D^2$ is oriented with an orientation compatible with the orientation of D^2.

Definition 2. Two surface braids S and S' are said to be *equivalent* if there is an isotopy $\{h_s\}_{s \in [0,1]}$ of $B^2 \times D^2$ satisfying the following conditions: (i) h_0 is the identity map and $h_1(S) = S'$, (ii) for any s in $[0, 1]$, h_s is a fiber-preserving homeomorphism of $B^2 \times D^2$, i.e., there is a homeomorphism $\underline{h}_s : D^2 \to D^2$ such that $pr_2 \circ h_s = \underline{h}_s \circ pr_2$, and (iii) for any s in $[0, 1]$, the restriction of h_s to $B^2 \times \partial D^2$ is the identity map.

Let D_1^2 and D_2^2 be copies of D^2 and we fix an identification of D^2 with the boundary connected sum $D_1^2 \natural D_2^2$. For surface braids $S_1 \subset B^2 \times D_1^2$ and $S_2 \subset B^2 \times D_2^2$ of the same degree, the union $S_1 \cup S_2 \subset B^2 \times (D_1^2 \natural D_2^2) \cong B^2 \times D^2$ is a surface braid. It is called the *product* of S_1 and S_2 and denoted by $S_1 \cdot S_2$. The equivalence class of $S_1 \cdot S_2$ does not depend on the identification $D_1^2 \natural D_2^2 \cong D^2$. The product is well defined on equivalence classes of surface braids of the same degree.

Let $\mathcal{B}_m$ (or $S\mathcal{B}_m$) be the set of equivalence classes of surface braids (or simple surface braids) of degree m. It forms a commutative monoid with the product. The trivial element of $\mathcal{B}_m$ (or $S\mathcal{B}_m$) is the equivalence class of the *trivial surface braid* $X_m \times D^2$. (We call a surface braid which is equivalent to the trivial surface braid $X_m \times D^2$ a *trivial surface braid*.)

A surface braid S is trivial if and only if the associated branched covering map $S \to D^2$ is a covering map.

45.2   Motion Pictures: Braid Movies

Let us consider motion pictures of surface braids. Identify D^2 with $I_3 \times I_4$, where I_3 and I_4 are the interval $[0, 1]$. Then $B^2 \times D^2 = (B^2 \times I_3) \times I_4$. (We often identify B^2 with $I_1 \times I_2$. However, it is not used here.)

Let S be a surface braid of degree m. For each $t \in I_4 = [0, 1]$, let b_t be a subset of $B^2 \times I_3$ such that $S \cap (B^2 \times I_3) \times \{t\} = b_t \times \{t\}$. From the definition of a surface braid, we see that b_t are m-braids in $B^2 \times I_3$ for all t but a finite number of exceptional values $t_1, \ldots, t_n$. For each exception value t_i, b_{t_i} is a singular m-braid. The *motion picture* or the *braid movie* of S is the 1-parameter family $\{b_t\}_{t \in [0,1]}$. (This depends on the identification $D^2 \cong I_3 \times I_4$.)

Let $\{b_t\}_{t \in [0,1]}$ be the motion picture of S. Then b_0 and b_1 are trivial m-braids $X_m \times I_3 \subset B^2 \times I_3$. When $t \in I_4$ moves continuously avoiding exceptional values, b_t moves continuously and the braid type does not change. When t moves passing an exceptional value t_i, the braid b_t changes through a singular braid b_{t_i}. If S is a simple surface braid, then the change is easy to understand.

The following is a characterization of the motion picture of a simple surface braid with one singular point in each singular braid.

Theorem 1. *(1) Let $\{b_t\}_{t \in [0,1]}$ be the motion picture of a simple surface braid S. Let $t_1, \ldots, t_n$ be exceptional values and assume that there exists only one singular point in the singular braid b_{t_i} for $i = 1, \ldots, n$. Let $b_{t_i}^-$ and $b_{t_i}^+$ be braids before and after passing a singular braid b_{t_i}. Then one of the following holds.*

(a) *$b_{t_i}^+$ is obtained from $b_{t_i}^-$ by inserting a standard generator.*

(b) *$b_{t_i}^+$ is obtained from $b_{t_i}^-$ by inserting the inverse of a standard generator.*

(c) *$b_{t_i}^+$ is obtained from $b_{t_i}^-$ by removing a standard generator.*

(d) *$b_{t_i}^+$ is obtained from $b_{t_i}^-$ by removing the inverse of a standard generator.*

For each $i = 1, \ldots, n$, $b_{t_i}^+$ and $b_{t_{i+1}}^-$ belong to the same braid type, and b_0 and b_1 are the trivial braid.

(2) If $\{b_t\}_{t\in[0,1]}$ is a 1-parameter family of braids and singular braids satisfying the conditions above then it is the motion picture of a simple surface braid.

The assumption that there exists only one singular point in each singular braid is inessential. By perturbing the identification $D^2 \cong I_3 \times I_4$ slightly, we can always assume this. Figure 45.1 is a motion picture of a surface braid of degree 3. There are two singular braids, one appears at $t = 1/3$ and the other at $t = 2/3$. Each singular braid has two singular points.

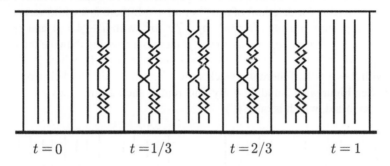

$$t=0 \qquad t=1/3 \qquad t=2/3 \qquad t=1$$

FIGURE 45.1: A motion picture.

45.3 Braid Charts

Definition 3. A *chart* or a *braid chart* of degree m is a finite graph Γ in the interior of D^2, possibly having *hoops* that are closed edges without vertices, such that every vertex has degree one (called a *black vertex*), degree six (called a *white vertex*) or degree four (called a *crossing*), and every edge is oriented and is labeled with integers from $\{1, \ldots, m-1\}$ as in Figure 45.2. Here we assume that $|i-j| = 1$ for each degree-6-vertex and $|i-j| > 1$ for each degree-4 vertex.

FIGURE 45.2: Vertices of degree 1, degree 6 and degree 4.

For a chart, we construct a surface braid $S(\Gamma)$. Assume that $D^2 = I_3 \times I_4$ such that I_3 is in the vertical direction and I_4 is in the horizontal direction. Deforming Γ slightly by an isotopy of D^2, we assume that the map $\Gamma \to I_4$ obtained from the projection $D^2 = I_3 \times I_4 \to I_4$ is a generic map. We assume that for each white vertex (or crossing), three (or two) edges are left and the other three (or two) edges are right of the vertex. Let $\text{Reg}(\Gamma \to I_4)$ be the set of values $t \in I_4$ such that there are no vertices and no critical points of Γ in $I_3 \times \{t\}$.

For $t \in \text{Reg}(\Gamma \to I_4)$, each intersection of Γ and the interval $I_3 \times \{t\}$ is a point on an edge of Γ. If the orientation of the edge is from left to right (or right to left) and if the label is i then we associate the intersection with σ_i (or σ_i^{-1}), where $\sigma_1, \ldots, \sigma_{m-1}$ are standard generators of the braid group.

Consider a braid movie $\{b_t\}_{t \in [0,1]}$ such that for each $t \in \text{Reg}(\Gamma \to I_4)$, b_t is an m-braid presented by the braid word obtained from the intersection of Γ and the interval $I_3 \times \{t\}$.

Note that the braid word does not change when t moves continuously on $\text{Reg}(\Gamma \to I_4)$. For $t \notin \text{Reg}(\Gamma \to I_4)$, if there is a local minima (or a local maxima) of an edge with label i then the braid word changes by insertion (or deletion) of $\sigma_i \sigma_i^{-1}$ or $\sigma_i^{-1} \sigma_i$. If there is a white vertex then the braid word changes by replacing a triple $\sigma_i^{\epsilon_1} \sigma_j^{\epsilon_2} \sigma_i^{\epsilon_3}$ by $\sigma_j^{\epsilon_3} \sigma_i^{\epsilon_2} \sigma_j^{\epsilon_1}$ with $|i - j| = 1$ and $\epsilon_1, \epsilon_2, \epsilon_3 \in \{\pm 1\}$, which is a consequence of $\sigma_i \sigma_{i+1} \sigma_i = \sigma_{i+1} \sigma_i \sigma_{i+1}$. If there is a crossing then the braid word changes by replacing a pair $\sigma_i^{\epsilon_1} \sigma_j^{\epsilon_2}$ by $\sigma_j^{\epsilon_2} \sigma_i^{\epsilon_1}$ with $|i - j| > 1$ and $\epsilon_1, \epsilon_2 \in \{\pm 1\}$, which is a consequence of $\sigma_i \sigma_j = \sigma_j \sigma_i$ with $|i - j| > 1$. If there is a black vertex then the braid word changes by insertion or deletion of σ_i^ϵ. See Figures 45.3 and 45.4. By Theorem 1, such a braid movie is a motion picture of a simple surface braid, which is our $S(\Gamma)$.

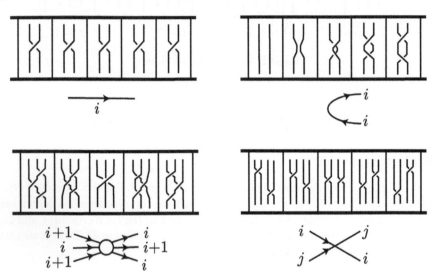

FIGURE 45.3: Motion pictures and charts, I.

Theorem 2 ([3]). *(1) For a chart Γ in D^2, the simple surface braid $S(\Gamma)$ is uniquely determined up to equivalence.*

(2) For any simple surface braid S, there is a chart Γ such that S is equivalent to $S(\Gamma)$.

In this theorem, we call Γ a *chart describing S* or a *chart description* of S. A chart description of S is not unique. We introduce moves on charts which do not change equivalence classes of surface braids.

(1) Let Γ be a chart. Let E be a region of D^2 homeomorphic to a 2-disk such that (i) there are no black vertices in E and (ii) $\Gamma \cap \partial E$ is empty or consists of transverse double points of some edges of Γ. Replace $\Gamma \cap E$ arbitrarily as long as there are no black vertices in E. Such a replacement is called a *CI-move*. All CI-moves are generated by CI-moves depicted in Figure 45.5 [2, 14].

(2) *CII-moves* are moves depicted in Figure 45.6, where $|i - j| > 1$.

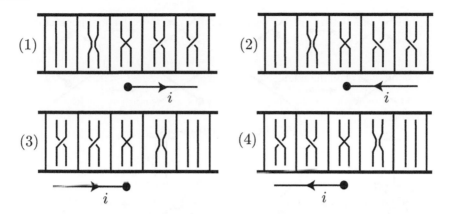

FIGURE 45.4: Motion pictures and charts, II.

(3) *CIII-moves* are moves depicted in Figure 45.7, where $|i - j| = 1$.

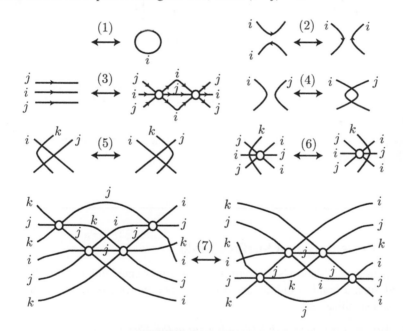

FIGURE 45.5: CI-moves.

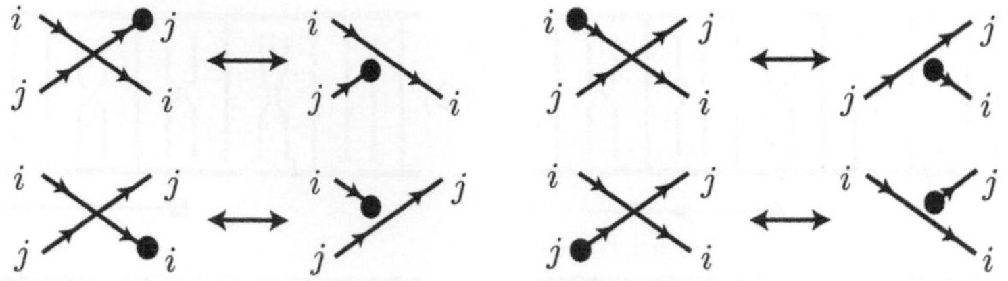

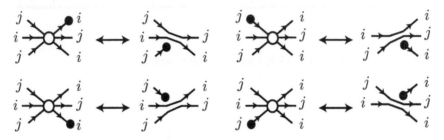

FIGURE 45.6: CII-moves.

FIGURE 45.7: CIII-moves.

Theorem 3 ([7]). *A chart description of a simple surface braid is unique up to C-move equivalence, which is the equivalence relation generated by chart moves (CI, CII and CIII) and ambient isotopies of D^2.*

A chart description is related to the singular set of a surface diagram [4]. For non-simple surface braids and their chart description, refer to [8, 12].

45.4   Braid Presentation of Surface Links

Alexander's theorem and Markov's theorem state that any oriented link is equivalent to the closure of a braid and such a braid is unique up to equivalence generated by braid ambient isotopies, conjugations and stabilizations.

Theorem 4. *Any oriented surface link is equivalent to the closure of a (simple) surface braid and such a surface braid is unique up to equivalence generated by braid ambient isotopies, conjugations and stabilizations.*

For the definitions of closures, braid ambient isotopies, conjugations and stabilizations, refer to [1, 9, 11].

On Theorem 4, at his lecture talk at Osaka City University in September, 1990, O. Viro stated that any oriented surface link is equivalent to the closure of a surface braid. It is proved in [5] that any surface link is equivalent to the closure of a simple surface braid. Uniqueness of braid presentation in Theorem 4 was announced in [6] and the full proof was given in [9].

For an oriented surface link F, the *braid index* Braid(F) is defined by the minimum among the degrees of all simple surface braids whose closures are equivalent to F. It is known ([3]) that Braid(F) = 1 if and only if F is a trivial 2-knot, Braid(F) = 2 if and only if F is an unknotted connected surface with positive genus or a trivial 2-component 2-link. Any surface link with Braid(F) = 3 is a ribbon surface link [3]. Thus we see that the braid index of the 2-twist-spun trefoil is 4. In [13] it is shown that the braid index is not additive for the connected sum of 2-knots.

45.5   Bibliography

[1] J.S. Carter, S. Kamada, and M. Saito. *Surfaces in 4-Space*. Springer-Verlag, Berlin, 2004.

[2] J.S. Carter and M. Saito. *Knotted Surfaces and Their Diagrams*. American Mathematical Society, Providence, RI, 1998.

[3] S. Kamada. Surfaces in R^4 of braid index three are ribbon. *J. Knot Theory Ramifications*, 1(2):137–160, 1992.

[4] S. Kamada. 2-dimensional braids and chart descriptions. In *Topics in Knot Theory*, pages 277–287. Kluwer Acad. Publ., 1993.

[5] S. Kamada. A characterization of groups of closed orientable surfaces in 4-space. *Topology*, 33(1):113–122, 1994.

[6] S. Kamada. Alexander's and Markov's theorems in dimension four. *Bull. Amer. Math. Soc. (N.S.)*, 31(1):64–67, 1994.

[7] S. Kamada. An observation of surface braids via chart description. *J. Knot Theory Ramifications*, 5(4):517–529, 1996.

[8] S. Kamada. On braid monodromies of non-simple braided surfaces. *Math. Proc. Cambridge Philos. Soc.*, 120(2):237–245, 1996.

[9] S. Kamada. *Braid and Knot Theory in Dimension Four*. American Mathematical Society, Providence, RI, 2002.

[10] S. Kamada. Theory of 2-dimensional braids and knots in dimension four. *Sugaku Expositions*, 25(1):1–18, 2012.

[11] S. Kamada. *Surface-Knots in 4-Space. An Introduction*. Springer, Singapore, 2017.

[12] S. Kamada and T. Matumoto. Chart descriptions of regular braided surfaces. *Topology Appl.*, 230:218–232, 2017.

[13] S. Kamada, S. Satoh, and M. Takabayashi. The braid index is not additive for the connected sum of 2-knots. *Trans. Amer. Math. Soc.*, 358(12):5425–5439, 2006.

[14] K. Tanaka. A note on ci-moves. In *Intelligence of Low Dimensional Topology 2006*, pages 307–314. World Sci. Publ., Hackensack, NJ, 2007.

Chapter 46

Marked Graph Diagrams and Yoshikawa Moves

Sang Youl Lee, Pusan National University

This chapter will introduce some of the basic notions of marked graphs and their diagrams and then discuss a method of presenting surface-links by marked graph diagrams which is due to S. J. Lomonaco and K. Yoshikawa as well as contributed to by F. J. Swenton and C. Kearton and K. Kurlin. We will see that a surface-link can be presented by many distinct marked graph diagrams and this complexity can be explained by some local moves on marked graph diagrams (called Yoshikawa moves) that entirely capture ambient isotopy types of surface-links presented by marked graph diagrams.

46.1   Marked Graphs and Diagrams

We begin with the definition of a marked graph.

Definition 1. *A **marked graph** or **ch-link** is a graph G embedded in $\mathbb{R}^3$ (piecewise linearly) satisfying the following conditions:*

(1) G has a finite number of 4-valent vertices and edges (possibly loops).

(2) Every vertex v is rigid, that is, it has a small rectangular neighborhood $N(v)$ homeomorphic to $\{(x, y)| - 1 \leq x, y \leq 1\}$, where v corresponds to the origin and the edges incident to v are represented by $x^2 = y^2$.

(3) Every vertex v has a marker represented by the thickened line segment on $N(v)$ whose core corresponds to the line segment $\{(x, 0)| - \frac{1}{2} \leq x \leq \frac{1}{2}\}$ as shown in Figure 46.1 (left).

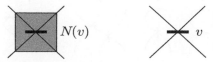

FIGURE 46.1: A neighborhood of a marked vertex v.

*Two marked graphs G and G' are **equivalent** if there is an ambient isotopy of $\mathbb{R}^3$ that sends G onto G' and takes rectangular neighborhoods and markers of G to those of G'. The equivalence class of a marked graph is called a **marked graph type**.*

433

As usual in the case of link diagrams, a marked graph in $\mathbb{R}^3$ can be represented by its generic projection on the plane $\mathbb{R}^2$ such that all multiple points are double points and the restriction of the projection to the rectangular neighborhood $N(v)$ of each 4-valent vertex v is a homeomorphism. Such a projection of G on $\mathbb{R}^2$ with additional over/under information (called over-/under-crossing) at every double point is called a **marked graph diagram** or **ch-link diagram** (simply, **ch-diagram**) of G, where the homeomorphic images of the rectangular neighborhoods $N(v)$ are omitted for the sake of simplicity as shown in Figure 46.1 (right). In fact, a marked graph diagram is a link diagram on $\mathbb{R}^2$ possibly with some marked 4-valent vertices. See Figure 46.2 for examples.

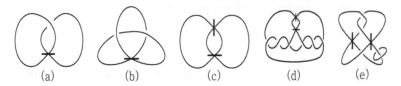

(a)   (b)   (c)   (d)   (e)

FIGURE 46.2: Marked graph diagrams.

In [9], L. H. Kauffman proved that piecewise linear ambient isotopy of rigid vertex graphs in $\mathbb{R}^3$ is generated diagrammatically by a set of local moves on rigid vertex graph diagrams that generalizes the Reidemeister moves on link diagrams. This leads to the following:

Theorem 1. *Two marked graphs are equivalent if and only if their marked graph diagrams are transformed into each other by a finite sequence of ambient isotopies of $\mathbb{R}^2$, the moves $\Omega_1, \Omega_2, \Omega_3, \Omega_4, \Omega_4', \Omega_5$ in Figure 46.3 and their mirror image moves.*

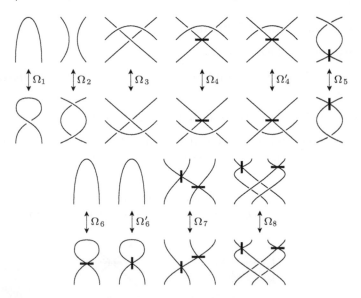

FIGURE 46.3: Yoshikawa moves.

For a marked graph G, let $L_-(G)$ and $L_+(G)$ be the links obtained from G by smoothing every marked vertex $\times$ with $)($ and $\smile$, which are called the **negative resolution** and the **positive**

resolution of G, respectively. If G has no marked vertices, then we define $L_-(G) = L_+(G) = G$.

Definition 2. *A marked graph G is* **admissible** *if both resolutions $L_-(G)$ and $L_+(G)$ are all trivial links.*

For a marked graph diagram D, we similarly define the negative resolution $L_-(D)$ and the positive resolution $L_+(D)$ of D (see Figure 46.4). A marked graph diagram D is said to be **admissible** if both resolutions $L_-(D)$ and $L_+(D)$ are all trivial link diagrams. The marked graph diagrams (a), (c) and (d) in Figure 46.2 are all admissible but not (b) and (e). It is straightforward to check that two resolutions on the moves $\Omega_1, \ldots, \Omega_5$ in Figure 46.3 and their mirror image moves do not change the resulting link type. Hence we can see from Theorem 1 that if G is an admissible marked graph, then any diagram of G is admissible and, conversely, if D is an admissible marked graph diagram, then the marked graph represented by D is also admissible.

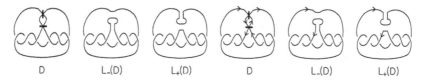

FIGURE 46.4: Resolutions of a marked graph diagram.

Definition 3. *An* **orientation** *of a marked graph (diagram) G is a choice of an orientation for every edge of G in such a way that two edges incident to v in a diagonal position are oriented toward v and the other two incident edges are oriented outward. Such a marked vertex,* $\times$ *or* $\times$ *, is called an* **oriented marked vertex**. *A marked graph (diagram) is said to be* **orientable** *if it admits an orientation. Otherwise, it is said to be* **non-orientable**. *An* **oriented marked graph (diagram)** *is an orientable marked graph (diagram) with a fixed orientation.*

An oriented marked graph (diagram) is indeed an oriented link (diagram) with possibly oriented marked vertices. We should note that not every marked graph diagram admits an orientation. For example, the marked graph diagrams (b), (c), (d) and (e) in Figure 46.2 are all orientable, while (a) is non-orientable. Every diagram of an oriented marked graph G inherits the orientation of G. For oriented marked graphs, it follows from Theorem 1 that two oriented marked graphs are equivalent if and only if their diagrams can be transformed into each other by a finite sequence of ambient isotopies of $\mathbb{R}^2$ and the moves $\Omega_1, \ldots, \Omega_4, \Omega_4', \Omega_5$ in Figure 46.3 and their mirror-images move with all possible orientations.

46.2 Marked Graph Diagram Presentations of Surface-Links

In this subsection, we will explore the correspondence between marked graph diagrams and surface-links. Let D be an admissible marked graph diagram with marked vertices $v_1, \ldots, v_n$ and let $\mu_+(D)$ and $\mu_-(D)$ denote the numbers of components of the resolutions $L_+(D)$ and $L_-(D)$, respectively. Then we can construct a surface-link in $\mathbb{R}^4$ as follows: Let $B_i (1 \leq i \leq n)$ be a band attached to $L_-(D)$ at the marked vertex v_i as shown in Figure 46.5. Let $\Delta_1, \ldots, \Delta_a \subset \mathbb{R}^3$ be mutually disjoint 2-disks with $\partial(\cup_{j=1}^a \Delta_j) = L_+(D)$ and let $\Delta'_1, \ldots, \Delta'_b \subset \mathbb{R}^3$ be mutually disjoint 2-disks with $\partial(\cup_{k=1}^b \Delta'_k) = L_-(D)$, where $a = \mu_+(D)$ and $b = \mu_-(D)$. We define a closed surface $\mathcal{S}(D) \subset \mathbb{R}^4$ by

$$(\mathbb{R}^3 \times \{t\}, \mathcal{S}(D) \cap \mathbb{R}^3 \times \{t\}) = \begin{cases} (\mathbb{R}^3, \phi) & \text{for } t > 1, \\ (\mathbb{R}^3, L_+(D) \cup (\cup_{j=1}^a \Delta_j)) & \text{for } t = 1, \\ (\mathbb{R}^3, L_+(D)) & \text{for } 0 < t < 1, \\ (\mathbb{R}^3, L_-(D) \cup (\cup_{i=1}^n B_i)) & \text{for } t = 0, \\ (\mathbb{R}^3, L_-(D)) & \text{for } -1 < t < 0, \\ (\mathbb{R}^3, L_-(D) \cup (\cup_{k=1}^b \Delta'_k)) & \text{for } t = -1, \\ (\mathbb{R}^3, \phi) & \text{for } t < -1. \end{cases}$$

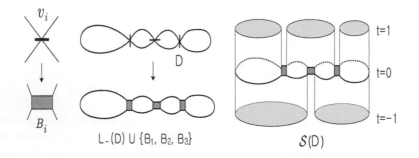

$L_-(D) \cup \{B_1, B_2, B_3\}$ $\mathcal{S}(D)$

FIGURE 46.5: A surface-link $\mathcal{S}(D)$ associated to D.

It is known that the ambient isotopy type of $\mathcal{S}(D)$ does not depend on the choices of trivial disks $\{\Delta_i | 1 \le i \le a\}$ and $\{\Delta'_j | 1 \le j \le b\}$ by Horibe and Yanagawa's lemma. Hence the surface-link $\mathcal{S}(D)$ is uniquely determined by D up to ambient isotopy of $\mathbb{R}^4$ (cf. [13], [7]), which is called the **surface-link associated to** D (see Figure 46.5).

Definition 4. *Let $\mathcal{L}$ be a surface-link and let D be an admissible marked graph diagram. We say that $\mathcal{L}$ is **presented** by D or D is a **marked graph diagram** of $\mathcal{L}$ if $\mathcal{L}$ is ambient isotopic to the surface-link $\mathcal{S}(D)$.*

By definition, D is a marked graph diagram of the surface-link $\mathcal{S}(D)$.

Theorem 2. *Every surface-link is presented by an admissible marked graph diagram.*

Proof. Let $\mathcal{L}$ be a surface-link and let $p : \mathbb{R}^4 \to \mathbb{R}$ denote the projection given by $p(x_1, x_2, x_3, x_4) = x_4$. It turns out that any surface-link $\mathcal{L}$ can be perturbed into a surface-link $\mathcal{L}'$ by an ambient isotopy of $\mathbb{R}^4$ such that the restriction $p : \mathcal{L}' \to \mathbb{R}$ satisfies that p has only finitely many elementary critical points; all the index 0 critical points (minimal points) are in $\mathbb{R}^3 \times \{-1\}$; all the index 1 critical points (saddle points) are in $\mathbb{R}^3 \times \{0\}$; all the index 2 critical points (maximal points) are in $\mathbb{R}^3 \times \{1\}$; the cross-section $\mathcal{L}'_t = \mathcal{L}' \cap (\mathbb{R}^3 \times \{t\})$ in any level t except $0, \pm 1$ is empty or a trivial link in $\mathbb{R}^3 \times \{t\}$ (cf. [13], [7]). Then the cross-section $\mathcal{L}'_0$ at the level $t = 0$ is a 4-valent graph in $\mathbb{R}^3 \times \{0\}$. Giving a marker at every 4-valent vertex (saddle point) so that it indicates how the saddle point opens up

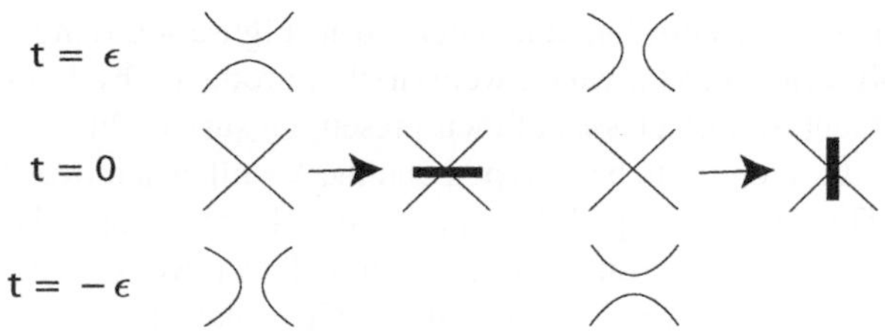

$$t = \epsilon$$

$$t = 0$$

$$t = -\epsilon$$

FIGURE 46.6: Marking of a 4-valent vertex.

above as illustrated in Figure 46.6, we obtain a marked graph $G(\mathcal{L})$ in $\mathbb{R}^3$ that is clearly admissible. It follows by construction that $S(G(\mathcal{L}))$ is ambient isotopic to $\mathcal{L}'$ and thus $\mathcal{L}$, i.e., $\mathcal{L}$ is presented by $G(\mathcal{L})$. This completes the proof. □

The surface-link $\mathcal{L}'$ is called a **normal form** or **hyperbolic splitting** of $\mathcal{L}$. A schematic picture of a hyperbolic splitting of the trivial S^2-knot is shown in Figure 46.7 (cf. Figure 46.5).

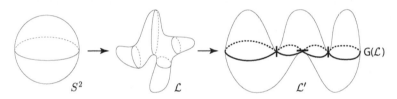

FIGURE 46.7: A hyperbolic splitting of an S^2-knot.

From the construction $S(D)$ and Theorem 2, we get the following corollary.

Corollary 1. *Let D be a marked graph diagram. Then D presents a surface-link if and only if D is admissible.*

Theorem 3. *Let D be a marked graph diagram of a surface-link $\mathcal{L}$.*

(1) D is orientable if and only if $\mathcal{L}$ is orientable.

(2) Let $g(\mathcal{L})$ denote the genus of the surface-link $\mathcal{L}$. Then

$$g(\mathcal{L}) = \begin{cases} \frac{1}{2}(|V(D)| - \mu_+(D) - \mu_-(D) + 2) & \text{if } D \text{ is orientable,} \\ |V(D)| - \mu_+(D) - \mu_-(D) + 2 & \text{if } D \text{ is non-orientable.} \end{cases}$$

Proof. (1) If D is orientable, then the resolutions $L_-(D)$ and $L_+(D)$ have orientations induced from the orientation of D as shown in Figure 46.4. Then the surface-link $S(D)$ associated to D can be oriented so that the induced orientation on the cross-section $S(D) \cap (\mathbb{R}^3 \times \{1\})$ matches the orientation of $L_+(D)$ and so does $\mathcal{L}$. Conversely, if $\mathcal{L}$ is orientable, then any hyperbolic splitting $\mathcal{L}'$ of $\mathcal{L}$ is also orientable. Hence the marked graph $G(\mathcal{L}) = \mathcal{L}' \cap \mathbb{R}^3 \times \{0\}$ presenting $\mathcal{L}$ can be oriented so that it coincides with the induced orientation on the boundary of $\mathcal{L}' \cap (\mathbb{R}^3 \times (-\infty, 0])$.

(2) Let $\chi(D)$ denote the Euler characteristic of $S(D)$. Then $\chi(D) = \mu_-(D) - |V(D)| + \mu_+(D)$. Hence $g(\mathcal{L}) = g(S(D)) = 1 - \frac{\chi(D)}{2}$ if D is orientable. Otherwise, $g(\mathcal{L}) = 2 - \chi(D)$. This gives the desired assertion. □

The local moves on marked graph diagrams depicted in Figure 46.3 and all their mirror image moves are called **Yoshikawa moves**, which were firstly introduced by K. Yoshikawa in [22] and do not change the ambient isotopy classes of their presenting surface-links. The Yoshikawa moves $\Omega_1, \Omega_2, \Omega_3, \Omega_4, \Omega'_4$ and Ω_5 are said to be of **type I** and the Yoshikawa moves $\Omega_6, \Omega'_6, \Omega_7$ and Ω_8 are said to be of **type II**. The moves of type I do not change ambient isotopy classes of marked graphs in $\mathbb{R}^3$ (see Theorem 1) and hence presenting surface-links. The moves of type II are transformations of surface-links in $\mathbb{R}^4$ not in $\mathbb{R}^3$. We also notice that all Yoshikawa moves preserve admissibility. The following theorem 4 shows that a surface link is completely presented by its marked graph diagram modulo Yoshikawa moves.

Theorem 4. *Let D and D' be marked graph diagrams and let $\mathcal{L}$ and $\mathcal{L}'$ be the surface-links presented by D and D', respectively. Then $\mathcal{L}$ and $\mathcal{L}'$ are equivalent if and only if D can be transformed into D' by a finite sequence of ambient isotopies of $\mathbb{R}^2$ and Yoshikawa moves.*

For a proof of this theorem, we refer to F. J. Swenton [21] and C. Kearton and V. Kurlin [10]. On many occasions it is necessary to minimize the number of Yoshikawa moves when one checks that a certain function from marked graph diagrams defines a surface-link invariant. A collection S of Yoshikawa moves is called a **generating set** of Yoshikawa moves if any Yoshikawa move is obtained by a finite sequence of plane isotopies and the moves in the set S. In [12], J. Kim, Y. Joung and S. Y. Lee derived some generating sets of Yoshikawa moves determining surface-link types, and also oriented surface-link types.

From now on, we present a numerical invariant, the ch-index, of surface-links and Yoshikawa's table of all weakly prime surface-links with up to 10 ch-indices.

Definition 5. *Let D be a marked graph diagram and let $c(D)$ and $h(D)$ denote the number of all crossings and marked vertices of D, respectively. The sum $ch(D) = c(D) + h(D)$ is called the **ch-index** of D. Then the **ch-index** of a surface-link $\mathcal{L}$ is the minimum number of all ch-indices $ch(D)$ of marked graph diagrams D presenting $\mathcal{L}$ and denoted by $ch(\mathcal{L})$, i.e., $ch(\mathcal{L}) = \min_D\{ch(D)\}$.*

Definition 6. *A surface-link $\mathcal{L}$ is said to be **prime** if $\mathcal{L}$ is not the connected sum of any two surface-links that are not trivial 2-knots. A surface-link $\mathcal{L}$ is said to be **weakly prime** if $\mathcal{L}$ is not the connected sum of any two surface-links $\mathcal{L}_1$ and $\mathcal{L}_2$ such that $ch(\mathcal{L}_i) \leq ch(\mathcal{L})$ for $i = 1, 2$.*

By definition, the ch-index $ch(\mathcal{L})$ of a surface-link $\mathcal{L}$ is clearly an invariant of $\mathcal{L}$. It has a property that for each $n \geq 0$, the number of the non-splittable surface-link types with ch-index n is finite. Any prime surface-link is weakly prime. In [22], K. Yoshikawa gave a table of all weakly prime surface-links, denoted by $I_k^{g_1, g_2, \ldots, g_n}$, whose ch-indices are less than or equal to ten as shown in Table 46.1, where $I_k^{g_1, g_2, \ldots, g_n}$ means the k-th surface-link with ch-index I and n components whose genera are $g_1, g_2, \ldots, g_n$. If $g_i < 0$, then it means non-orientable genus. For a 2-knot, I_k^0 is abbreviated as I_k. Among all 23 surface-links in Yoshikawa's table, there are 15 orientable surface-links and 8 non-orientable surface-links.

If a surface-link $\mathcal{L}$ is the split union of two surface-links $\mathcal{L}_1$ and $\mathcal{L}_2$, then $ch(\mathcal{L}) = ch(\mathcal{L}_1) + ch(\mathcal{L}_2)$ and, furthermore, for arbitrary orientable surface-links $\mathcal{L}_1$ and $\mathcal{L}_2$, we see that $ch(\mathcal{L}_1 \sharp \mathcal{L}_2) \leq ch(\mathcal{L}_1) + ch(\mathcal{L}_2)$, where $\mathcal{L}_1 \sharp \mathcal{L}_2$ is the connected sum of $\mathcal{L}_1$ and $\mathcal{L}_2$. But it is still an open question whether or not $ch(\mathcal{L}_1 \sharp \mathcal{L}_2) = ch(\mathcal{L}_1) + ch(\mathcal{L}_2)$. For the standard positive projective plane 2_1^{-1} and a non-trivial 2-knot K such that $K \sharp 2_1^{-1} = 2_1^{-1}$ (e.g., the 3-twist spun 2-knot 10_3 of the trefoil), it follows from $ch(K) > 0$ that $ch(K \sharp 2_1^{-1}) = ch(2_1^{-1}) = 2 < ch(K) + ch(2_1^{-1})$. Thus, in general, the additivity of the ch-index does not hold for non-orientable surface-links. For more, we refer to [22] and [7].

We will conclude this chapter by remarking that some other invariants of surface-links can be defined, reinterpreted and computed by using marked graph diagrams. For example, the surface-link groups ([16], [2]), fundamental biquandles [1] and, especially, Alexander biquandles [11], ribbon biquandles [17], (shadow) quandle cocycle invariants [5], bikei counting invariants ([18], [19]), biquasile counting invariants [20], and (shadow) biquandle cocycle invariants [8]. Besides these invariants, some invariants involving skein relations are discussed in [14], [3], [4], [15], and [6].

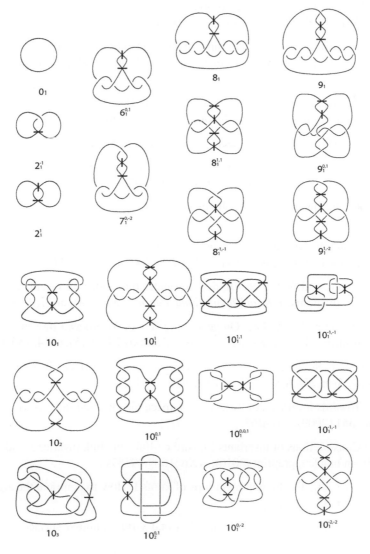

TABLE 46.1: Yoshikawa's table.

46.3   Bibliography

[1] S. Ashihara, Calculating the fundamental biquandles of surface-links from their ch-diagrams. *J. Knot Theory Ramifications* **21** (2012), no. 10, 1250102 (23 pages).

[2] J. S. Carter, S. Kamada and M. Saito, Surfaces in 4-Space, Springer, 2004.

[3] Y. Joung, J. Kim and S. Y. Lee, Ideal coset invariants for surface-links in $\mathbb{R}^4$, *J. Knot Theory Ramifications* **22** (2013), no. 9, 1350052 (25 pages).

[4] Y. Joung, S. Kamada and S. Y. Lee, Applying Lipson's state models to marked graph diagrams of surface-links, *J. Knot Theory Ramifications* **24** (2015), no. 10, 1540003 (18 pages).

[5] S. Kamada, J. Kim and S. Y. Lee, Computations of quandle cocycle invariants of surface-links using marked graph diagrams, *J. Knot Theory Ramifications* **24** (2015), no. 10, 1540010 (35 pages).

[6] Y. Joung, S. Kamada, A. Kawauchi, and S. Y. Lee, Polynomial of an oriented surface-link diagram via quantum A_2 invariant, *Topology Appl.* **231** (2017), 159–185.

[7] S. Kamada, Surface-Knots in 4-Space. An Introduction, Springer, 2017.

[8] S. Kamada, A. Kawauchi, J. Kim, S. Y. Lee, Biquandle cohomology and state-sum invariants of links and surface-links, preprint (2018), arXiv:1803.03137.

[9] L. H. Kauffman, Invariants of graphs in three-space, *Trans. Amer. Math. Soc.* **311** (1989), 697–710.

[10] C. Kearton and V. Kurlin, All 2-dimensional links in 4-space live inside a universal 3-dimensional polyhedron, *Algebr. Geom. Topol.* **8** (2008), 1223–1247.

[11] J. Kim, Y. Joung and S. Y. Lee, On the Alexander biquandles of oriented surface-links via marked graph diagrams, *J. Knot Theory Ramifications* **23** (2014), no. 7, 1460007 (26 pages).

[12] J. Kim, Y. Joung and S. Y. Lee, On generating sets of Yoshikawa moves for marked graph diagrams of surface-links, *J. Knot Theory Ramifications* **24** (2015), no. 4, 1550018 (21 pages).

[13] A. Kawauchi, T. Shibuya and S. Suzuki, Descriptions on surfaces in four-space, I Normal forms, *Math. Sem. Notes Kobe Univ.* **10** (1982), 75–125.

[14] S. Y. Lee, Towards invariants of surfaces in 4-space via classical link invariants, *Trans. Amer. Math. Soc.* **361** (2009), 237–265.

[15] S. Y. Lee, Constructions of invariants for surface-links via link invariants and applications to the Kauffman bracket, preprint (2016), arXiv:1609.09587v1.

[16] S. J. Lomonaco, Jr., The homotopy groups of knots I. How to compute the algebraic 2-type, *Pacific J. Math.* **95** (1981), 349–390.

[17] S. Nelson and P. Rivera, Ribbon biquandles and virtual knotted surfaces, preprint (2014), arXiv:1409.7756v1.

[18] S. Nelson and P. Rivera, Bikei invariants and Gauss diagrams for virtual knotted surfaces. *J. Knot Theory Ramifications* **25** (2016), no. 3, 1640008 (14 pp).

[19] S. Nelson and J. Rosenfield, Bikei homology, *Homotopy & Applications* **19** (2017), 23–35.

[20] J. Kim and S. Nelson, Biquasile colorings of oriented surface-links. *Topology Appl.* **236** (2018), 64–76.

[21] F. J. Swenton, On a calculus for 2-knots and surfaces in 4-space, *J. Knot Theory Ramifications* **10** (2001), 1133–1141.

[22] K. Yoshikawa, An enumeration of surfaces in four-space, *Osaka J. Math.* **31** (1994), 497–522.

Chapter 47

Knot Groups

Alexander Zupan, University of Nebraska, Lincoln

47.1 Knot and Link Groups

One of the foundational structures in algebraic topology is the fundamental group of a topological space. Roughly speaking, the fundamental group of X records the collection of "lassos" thrown by a stationary observer living at a fixed point $x_0 \in X$, called a basepoint, where these lassos are considered to be equivalent up to continuous deformation. More rigorously, a *loop* in X at x_0 is a (continuous) map $f : I \to X$ such that $f(0) = f(1) = x_0$. Two loops f and g are *homotopic* if there exists a continuous family of maps $h_t : I \to X$ such that $h_0 = f$, $h_1 = g$, and h_t is a loop at x_0 for each fixed value of t (the family h_t is called a *homotopy*). The group elements of the fundamental group of X based at x_0, denoted $\pi_1(X, x_0)$, are homotopy classes of loops at x_0, and the group operation is a natural concatenation $f \cdot g$; loosely, $f \cdot g$ represents the loop obtained by first traversing the loop f and then traversing the loop g. The identity element of $\pi_1(X, x_0)$ is the constant loop $f : I \to X$ given by $f(s) = x_0$.

As an example, consider $X = \mathbb{R}^n$ with x_0 the origin. Any loop in $\mathbb{R}^n$ can be continuously contracted to the constant loop, and thus $\pi_1(\mathbb{R}^n, x_0)$ is the trivial group. For a more interesting example, $\pi_1(S^1, x_0) \cong \mathbb{Z}$ for any basepoint $x_0 \in S^1$. Intuitively, this is apparent, since $\pi_1(S^1, x_0)$ should be generated by the loop that goes around S^1 once. Verifying this fact, however, takes considerable development, and the interested reader is encouraged to reference an introductory algebraic topology textbook; for instance, see [6] or [13].

In this chapter, the term *n-knot* will be used to refer to an embedded S^n in S^{n+2} or $\mathbb{R}^{n+2}$, the term *n-link* will refer to an embedded disjoint union of copies of S^n in S^{n+2} or $\mathbb{R}^{n+2}$, and a *surface knot* will refer to an embedded surface in S^4 or $\mathbb{R}^4$. A 1-knot or 1-link will also be called a *classical knot* or *classical link*. The term *link* will refer to any of these embedded objects.

An incredible amount of information about links can be gleaned from studying the topology of the space around a given link K, called the exterior of K. The *exterior* $E(K)$ is defined to be $E(K) = S^{n+2} - K$ or $E(K) = \mathbb{R}^{n+2} - K$. In turn, the topology of the space $E(K)$ can be better understood by looking at its fundamental group. The *link group* of K (or *knot group*, if K is connected), denoted πK, is the fundamental group of $E(K)$. There are several observations to keep in mind: First, the choice of basepoint is suppressed, as any choice of basepoint yields the same group, up to isomorphism. Second, the choice between the ambient spaces S^{n+2} and $\mathbb{R}^{n+2}$ does not matter from the perspective of πK, since $\mathbb{R}^{n+2}$ is homeomorphic to S^{n+2} with a point removed, and removing a point from an n-dimensional space, $n \geq 3$, does not change its fundamental group. Finally, if two links L_1 and L_2 are isotopic, then $E(L_1)$ and $E(L_2)$ are homeomorphic and thus $\pi_1 L_1 = \pi_1 L_2$. The reverse implication is not true in general—not even for classical knots—but it is true for prime classical knots, a fact which was proved by Gordon-Luecke [5] and Whitten [19]. Together, these two results imply that two prime classical knots with distinct knot groups must be inequivalent knots.

Despite the fact that groups do not distinguish all pairs of knots, they do a very good job of telling the difference between trivial and non-trivial knots, as shown by the next theorem.

Theorem 1. *For $n = 1$ or $n \geq 3$, an n-knot K is equivalent to the unknot if and only if $\pi K \cong \mathbb{Z}$. For $n = 2$, an n-knot K is topologically equivalent to the unknot if and only if $\pi K \cong \mathbb{Z}$.*

For $n = 1$, the theorem follows from the proof of Dehn's lemma [14]; for $n \geq 3$, the theorem is implied by [11, 17, 18] . For $n = 2$, the theorem was proved in the topological category by Freedman [4]. It is not currently known whether the theorem holds in the smooth category when $n = 2$. This is the famous Unknotting Conjecture.

Conjecture 1. *A 2-knot K is smoothly equivalent to the unknot if and only if $\pi K \cong \mathbb{Z}$.*

47.1.1 Computing Fundamental Groups

Some classical knot groups are straightforward to compute, and perhaps the easiest calculation is the knot group of the torus knot $K = T(p, q)$. By definition, a torus knot is contained in an unknotted torus T in S^3, which means that T cuts S^3 into two solid tori, homeomorphic to $S^1 \times D^2$. In addition, K cuts T into an annulus A, and thus A is an annulus in $E(K)$ that cuts $E(K)$ into two solid tori as well, say V_1 and V_2. In this case, the Seifert-Van Kampen theorem can be applied to give a presentation for πK (by slightly enlarging V_1 and V_2 so that they overlap in an open neighborhood of A). Each solid torus is homeomorphic to $S^1 \times D^2$, where $\pi_1(S^1 \times D^2) \cong \mathbb{Z}$. The annulus A also has fundamental group $\mathbb{Z}$, and the inclusion of A into V_1 sends a generator of $\pi_1(A)$ to pth power of a generator of $\pi_1(V_1)$, while the inclusion of A into V_2 sends the same generator of $\pi_1(A)$ to the qth power of a generator of $\pi_1(V_2)$. This setup is shown in Figure 47.1. It follows from the Seifert-Van Kampen Theorem that

$$\pi K = \langle x, y \,|\, x^p = y^q \rangle.$$

Note that if $(p, q) \neq (p', q')$ for $p, q, p', q' > 0$, then πK distinguishes $T(p, q)$ from $T(p', q')$ [16]; however, it is not a powerful enough tool to show that, for instance, $T(3, 2)$ is not isotopic to $T(-3, 2)$.

Although this technique will not work in general—not every classical knot exterior admits such a convenient decomposition—there are other algorithms to find the presentation of a classical knot group. The most famous is the *Wirtinger method*, which produces a presentation for πK using a diagram D of K. Let D be an oriented link diagram, viewed as an embedding in $\mathbb{R}^3$ such that $K \subset \{(x, y, z) : -\epsilon < z < \epsilon\}$, and choose a basepoint x_0 along the z-axis, significantly above the diagram; $(0, 0, 100\epsilon)$ will work. As in the case of colorability, the diagram D can be viewed as containing n connected planar arcs $A_1, \ldots, A_n$, numbered and oriented using the orientation of D. For each arc, let a_i be the loop in $E(K)$ based at x_0 consisting of three linear paths, one segment from x_0 to a point x_1^i directly to the "right" of the arc A_i (where "right" is determined by the orientation of A_i) at height $z = -2\epsilon$, the second segment from x_1^i to a point x_2^i directly to the "left" of A_i at height $z = -2\epsilon$, and the third segment from x_2^i back to x_0, completing the loop. The loops $a_1, \ldots, a_n$ determine a set of generators for πK. The relators $r_1, \ldots, r_n$ arise from the n crossings in

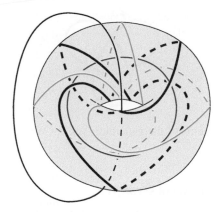

FIGURE 47.1: The exterior of the $(2, 3)$-torus knot decomposes as two solid tori V_1 and V_2 that meet along annulus A. A generator of $\pi_1(A)$ (green) is homotopic to x^2 and y^3, where x is red and y is blue.

the diagram. A positive crossing of strand A_i over strands A_j and A_{j+1} gives rise to the equivalence $a_j a_i = a_i a_{j+1}$, so that $r_i = a_j^{-1} a_i a_{j+1} a_i^{-1}$ is a relator for πK. The homotopy from $a_j a_i$ to $a_i a_{j+1}$ is shown in Figure 47.2.

FIGURE 47.2: The loops $a_j a_i$ (left) and $a_i a_{j+1}$ (right) are homotopic.

A similar argument shows that a negative crossing of A_i over A_j and A_{j+1} yields the equivalence $a_i a_j = a_{j+1} a_i$ and relator $r_i = a_{j+1} a_i a_j a_i^{-1}$. The Wirtinger method then gives a presentation for πK,

$$\pi K = \langle a_1, \ldots, a_n \mid r_1, \ldots, r_{n-1} \rangle.$$

Note that as indicated in the formula above, it is always that case that one of the relators is redundant. A very readable proof of this fact using the Seifert-Van Kampen theorem appears in [15], for instance. As an example, consider the diagram for the trefoil K shown in Figure 47.3. Using the diagram D, the Wirtinger presentation for πK coming from D has generators a_1, a_2, a_3 and relators $r_1 = a_2^{-1} a_1 a_3 a_1^{-1}$, $r_2 = a_3^{-1} a_2 a_1 a_2^{-1}$, $r_3 = a_1^{-1} a_3 a_2 a_3^{-1} = r_2^{-1} r_1^{-1}$. The relators arise from the equivalences $a_2 a_1 = a_1 a_3$, $a_3 a_2 = a_2 a_1$, and $a_1 a_3 = a_3 a_2$; by inspection, a relator is redundant (as predicted).

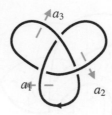

FIGURE 47.3: The lower segments of the generators for the Wirtinger presentation for the trefoil group.

47.1.2   Characterizations of Knot Groups

A natural and well-studied question about knot and link groups is the following: Which groups G occur as the group πK for an knot or link G? First, observe that every classical knot has a presentation coming from the Wirtinger method described above. More generally, a group presentation $P = \langle a_1, \ldots, a_n \mid r_1, \ldots, r_m \rangle$ is called a *Wirtinger presentation* if every relator is of the form $r_i = a_j^{-1} w a_k w^{-1}$ for a pair of generators a_j, a_k and a word w in $a_1, \ldots, a_n$. The groups of classical links are completely characterized by the Alexander-Artin theorem (see [2] for a proof).

Theorem 2. *A group G is the group πL for a classical link L if and only if G has a Wirtinger presentation $\langle a_1, \ldots, a_n \mid r_1, \ldots, r_n \rangle$ and a collection $w_1, \ldots, w_n$ of words in $a_1, \ldots, a_n$ such that*

1. *there is a permutation σ of $1, \ldots, n$ such that $r_i = a_i^{-1} w_i a_{\sigma(i)} w_i^{-1}$; and*

2. *in the free group generated by $a_1, \ldots, a_n$, it holds that $a_1 a_2 \cdots a_n = w_1 a_{\sigma(1)} w_1^{-1} w_2 a_{\sigma(2)} w_2^{-1} \cdots w_n a_{\sigma(n)} w_n^{-1}$.*

More generally, let $\mathcal{K}^n$ denote the collection of all n-knot groups G. In order to compare the collections $\mathcal{K}^n$, it is helpful to invoke Artin's spinning construction [1], which turns a classical knot into a 2-knot (or an n-knot into an $(n+1)$-knot). Consider an arc τ embedded in the upper half-plane $\mathbb{R}_+^2$, and observe that revolving $\mathbb{R}_+^2$ about its boundary axis sweeps out all of $\mathbb{R}^3$. Under this revolution, the arc τ sweeps out an embedded sphere $S^2 \subset \mathbb{R}^3$. Now, given a classical knot K in $\mathbb{R}^3$, let τ be a 1-strand tangle in upper half-space $\mathbb{R}_+^3$ such that connecting the endpoints of τ in the planar boundary of $\mathbb{R}_+^3$ yields a knot isotopic to K. Analogous to what happens one dimension lower, there is a revolution of $\mathbb{R}_+^3$ in $\mathbb{R}^4$ that fixes the planar boundary of $\mathbb{R}_+^3$ and sweeps out all of $\mathbb{R}^4$. The image of the arc τ under this revolution is an embedded 2-sphere denoted $S(K)$, called a *spun knot*. This construction is depicted in Figure 47.4.

Most importantly to the discussion at hand, it can be shown that $\pi K \cong \pi S(K)$. Furthermore, the spinning construction generalizes to higher dimensions in a straightforward way, and if $\pi K \cong G$ for an n-knot K, then $\pi S(K) \cong G$ for the $(n + 1)$-knot $S(K)$. Therefore,

$$\mathcal{K}^1 \subset \mathcal{K}^2 \subset \mathcal{K}^3 \subset \ldots$$

Theorem 2 gives conditions for a group G to be a 1-knot group. For higher-dimensional knots, the situation is less restrictive. Kervaire proved necessary and sufficient conditions for a group G to be an n-knot for an $n \geq 3$ [9]. Let G' denote the commutator subgroup of G, normally generated by elements of the form $g_1 g_2 g_1^{-1} g_2^{-1}$ for $g_1, g_2 \in G$. The second homology group $H_2(G)$ is defined to be the second homology group of any aspherical, locally path-connected topological space X

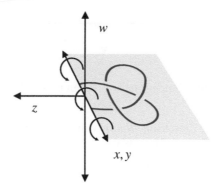

FIGURE 47.4: Spinning a trefoil in $\mathbb{R}^4$.

such that $\pi_1(X) = G$. Finally, a group is said to have *weight one* if it is normally generated by a single element—the presentations produced using the Wirtinger method, for example, have this property.

Theorem 3. *[9] A finitely presented group G is an n-knot group for all $n \geq 3$ if and only if $G/G' \cong \mathbb{Z}$, $H_2(G) = 0$, and G has weight one.*

Kervaire extended this result to show that G is the group of a c-component n-link ($n \geq 3$) precisely when $G/G' \cong \mathbb{Z}^c$, $H_2(G) = 0$, and G has weight c (i.e., G is normally generated by c elements) [10]. It follows from Kervaire's theorem that $\mathcal{K}^3 = \mathcal{K}^n$ for all $n \geq 3$. On the other hand, it can be shown using the Alexander polynomial that $\mathcal{K}^1 \subsetneq \mathcal{K}^2$, and in addition, it is known that $K^2 \subsetneq K^3$ (see [8] for relevant details and additional references).

In dimension four, further distinctions can be made between the groups of surface knots according to their genera. Let $\mathcal{F}_g$ denote the collection of all surface knot groups corresponding to genus g surfaces, so that $\mathcal{K}^2 = \mathcal{F}_0$. Since taking the connected sum of a surface knot K with an unknotted torus does not change πK, it follows that

$$\mathcal{F}_0 \subset \mathcal{F}_1 \subset \mathcal{F}_2 \subset \ldots$$

Litherland proved that each of these inclusions is strict; for every $n > 0$ there exists a group $G \in \mathcal{F}_n$ such that $G \notin \mathcal{F}_{n-1}$ [12]. Finally, in [8] it is proved that

$$\mathcal{K}^3 \subset \bigcup_{g=0}^{\infty} \mathcal{F}_g.$$

For related results and a more comprehensive discussion of knot groups, the interested reader is encouraged to start with the excellent treatments in [3], [7], and [8].

47.2 Bibliography

[1] Emil Artin. Zur Isotopie zweidimensionaler Flächen im R_4. *Abh. Math. Sem. Univ. Hamburg*, 4(1):174–177, 1925.

[2] Gerhard Burde and Heiner Zieschang. *Knots*, volume 5 of *De Gruyter Studies in Mathematics*. Walter de Gruyter & Co., Berlin, 1985.

[3] Scott Carter, Seiichi Kamada, and Masahico Saito. *Surfaces in 4-Space*, volume 142 of *Encyclopaedia of Mathematical Sciences*. Springer-Verlag, Berlin, 2004. Low-Dimensional Topology, III.

[4] Michael H. Freedman. The disk theorem for four-dimensional manifolds. In *Proceedings of the International Congress of Mathematicians, Vol. 1, 2 (Warsaw, 1983)*, pages 647–663. PWN, Warsaw, 1984.

[5] C. McA. Gordon and J. Luecke. Knots are determined by their complements. *Bull. Amer. Math. Soc. (N.S.)*, 20(1):83–87, 1989.

[6] Allen Hatcher. *Algebraic Topology*. Cambridge University Press, Cambridge, 2002.

[7] Seiichi Kamada. *Braid and Knot Theory in Dimension Four*, volume 95 of *Mathematical Surveys and Monographs*. American Mathematical Society, Providence, RI, 2002.

[8] Akio Kawauchi. *A Survey of Knot Theory*. Birkhäuser Verlag, Basel, 1996. Translated and revised from the 1990 Japanese original by the author.

[9] Michel A. Kervaire. Les nœuds de dimensions supérieures. *Bull. Soc. Math. France*, 93:225–271, 1965.

[10] Michel A. Kervaire. On higher dimensional knots. In *Differential and Combinatorial Topology (A Symposium in Honor of Marston Morse)*, pages 105–119. Princeton Univ. Press, Princeton, N.J., 1965.

[11] J. Levine. Unknotting spheres in codimension two. *Topology*, 4:9–16, 1965.

[12] R. A. Litherland. The second homology of the group of a knotted surface. *Quart. J. Math. Oxford Ser. (2)*, 32(128):425–434, 1981.

[13] William S. Massey. *A Basic Course in Algebraic Topology*, volume 127 of *Graduate Texts in Mathematics*. Springer-Verlag, New York, 1991.

[14] C. D. Papakyriakopoulos. On Dehn's lemma and the asphericity of knots. *Ann. of Math. (2)*, 66:1–26, 1957.

[15] Dale Rolfsen. *Knots and Links*. Publish or Perish, Inc., Berkeley, Calif., 1976. Mathematics Lecture Series, No. 7.

[16] Otto Schreier. Über die gruppen $A^a B^b = 1$. *Abh. Math. Sem. Univ. Hamburg*, 3(1):167–169, 1924.

[17] Julius L. Shaneson. Embeddings with codimension two of spheres in spheres and H-cobordisms of $S^1 \times S^3$. *Bull. Amer. Math. Soc.*, 74:972–974, 1968.

[18] C. T. C. Wall. Unknotting tori in codimension one and spheres in codimension two. *Proc. Cambridge Philos. Soc.*, 61:659–664, 1965.

[19] Wilbur Whitten. Knot complements and groups. *Topology*, 26(1):41–44, 1987.

Chapter 48

Concordance Groups

Kate Kearney, Gonzaga University

48.1 Concordance Groups, Slice Knots and Ribbon Knots

Recall that from two knots K and J we can create a new knot, the connected sum, $K\#J$. For those with a mind for algebra, this raises the question of whether knots form a group. We have a natural candidate for an identity element (the unknot), and it is easy to see that the property of additive identities is satisfied. That is, for any knot K, we can see that $K\#U = K$.

As we are feeling encouraged, we can explore the other requirements for a group. Knots are closed under connected sum (if K and J are knots, $K\#J$ is also a knot). We also have associativity for connected sum. Unfortunately, we run into trouble when we try to define an additive inverse. A good guess for an inverse is $-K$ (the reverse of the mirror image of K), but we have no reason to expect that $K\# - K$ might be the unknot. We'll see in this section that a solution lies in taking equivalence classes of knots under the relation of knot concordance. We can then create a group, C, the knot concordance group.

Concordance was originally considered by Murasugi [27] and Fox and Milnor [11, 12] to answer questions about slices of knotted 2-spheres in $\mathbb{R}^4$ that were raised by Artin [1] in the 1920s. This is why we use the word *slice* to refer to knots concordant to the unknot. We'll discuss both the algebraic and topological implications of concordance. This chapter gives a brief overview of the study of concordance. For a more in-depth survey, we refer the reader to [23].

48.1.1 Definitions

We'll begin with several important definitions. We have been discussing knots embedded in the three–sphere, S^3. In the following definitions we will think of S^3 either as the boundary of a four–ball, B^4, or as part of the boundary of an annulus, $S^3 \times [0, 1]$.

Definition 1. *A knot, $K \subset S^3$, is slice if it bounds a disk embedded in B^4.*

Example 1. *The unknot is slice.*

Example 2. *The trefoil is not slice. This follows from Murasugi's work [27]. We will discuss the signature of a knot as an obstruction to sliceness in Section 48.1.2. Since the signature of the trefoil is -2, it cannot be slice.*

Considering prime knots up to 9 crossings, the nontrivial slice knots are $6_1, 8_8, 8_9, 8_{20}, 9_{27}, 9_{41}, 9_{46}$. Amongst the prime knots, there are 14 slice knots of 10 crossings and 27 slice knots of 11 crossings.

Definition 2. *Two knots K and J are concordant if K# − J is slice.*

It is also useful to consider an equivalent definition.

Definition 3. *Two knots K and J are concordant if they cobound an annulus embedded in $S^3 \times [0, 1]$ where $K \subset S^3 \times \{0\}$ and $J \subset S^3 \times \{1\}$.*

It is much easier to see from Definition 3 that concordance is an equivalence relation. Examples in Section 48.1.3 will show knots that are concordant and, in Section 48.1.2, knots that are not concordant. First we observe that we now have the means to define the concordance group.

Definition 4. *The concordance group, C, is the group formed by concordance classes of knots. The addition in the group is the connected sum of knots. The identity element is the class of slice knots (which in particular includes the unknot). The additive inverse of a knot K is −K, the reverse of the mirror image of K.*

Note: In the definitions above we have not made reference to category. We can specify the definition of concordance to either the smooth or topological category, and it is important to note that we will have different results. There are knots that are topologically slice, but not smoothly slice [14]. With some exceptions, in the discussion that follows we will focus on the smooth category.

We can head from here in two directions, studying the concordance group as an algebraic object, and studying concordance as an equivalence relation on knots. We'll first look at the concordance group and some of its structure and properties in Section 48.1.2. Beginning here, we'll discuss obstructions to concordance and the knot invariants connected to the study of concordance. Then we'll continue to explore concordance by way of discussing how to find concordances in Section 48.1.3. We'll wrap up our concordance discussion with a related notion: Section 48.1.4 addresses ribbon knots.

48.1.2   The Concordance Group and Obstructions to Concordance

We observed in Section 48.1.1 that the equivalence classes of concordant knots form the concordance group, C. We'd like to see some of the algebraic properties of this group.

There is a surjection from C to $\mathbb{Z}^\infty \oplus \mathbb{Z}_2^\infty \oplus \mathbb{Z}_4^\infty$ via the algebraic concordance group [20]. The algebraic concordance group arises from considering the Seifert form of a knot and defining a notion of algebraic concordance. Rather than go into detail about the definition of algebraic concordance, we'll focus on several algebraic concordance invariants and observe how they show us both elements of infinite order and elements of finite order in the algebraic concordance group.

Levine-Tristram signature: The signature, $\sigma(K)$, of a knot is the signature of $V + V^T$, where V is a Seifert matrix for the knot K. The Levine-Tristram signature [19, 30] is a generalized version of this: $\sigma_t(K) = \lim_{\epsilon \to 0} \frac{1}{2}(\sigma'_{t-\epsilon}(K) + \sigma'_{t+\epsilon}(K))$ where $\sigma'_t(K)$ is the signature of $(1 - e^{2\pi i t})V + (1 - e^{-2\pi i t})V^T$ for $t \in [0, 1]$. Each of the Levine-Tristram signatures is a concordance invariant. That is, if two knots are concordant they have the same values of Levine-Tristram signatures. Since there are knots with non-trivial signatures, we can find families of non-slice knots via signatures. Since signature is additive under connected sum, this produces infinite families of non-concordant knots. Here we see the surjection to $\mathbb{Z}^\infty$, and can find infinite order elements of C.

Alexander polynomial: The Alexander polynomial is not a concordance invariant, but Fox and Milnor [12] showed that the Alexander polynomial of a slice knot must be of the form $\Delta_K(t) = \pm t^k f(t) f(t^{-1})$ for some polynomial $f(t) \in \mathbb{Z}[t, t^{-1}]$. This can be used to find a surjection from C to $\mathbb{Z}_2^{\infty}$.

Example 3. *In particular, 4_1 is an example of a knot which is order 2 in the concordance group. That is, 4_1 is not slice, but $4_1 \# 4_1$ is slice. Comparing to our statement above, the Alexander polynomial of 4_1 is $\Delta_{4_1}(t) = 1 - 3 * t + t^2$ so 4_1 cannot be slice, but $4_1 \# 4_1$ can be. In fact, since 4_1 is amphichiral, 4_1 is equivalent to -4_1, so we see that $4_1 \# 4_1$ is slice.*

Further consideration of the Alexander polynomial illuminates the distinction between the smooth and topological category. All knots with Alexander polynomial one ($\Delta_K(t) = 1$) are topologically slice [13, 14], but many knots with Alexander polynomial one are not smoothly slice [9]. On the other hand, in the topological category we know that if $\Delta_K(t)$ is a nontrivial Alexander polynomial, then there are two nonconcordant knots having that Alexander polynomial [18, 21].

Witt groups: Without going into detail, we include Witt groups in this discussion as a source of $\mathbb{Z}_4$ torsion in the algebraic concordance group.

We have observed that 4_1 has finite order in the concordance group. There are significant differences in concordance order between the algebraic concordance group, the topological concordance group, and the smooth concordance group. The algebraic concordance group has finite elements of orders 1, 2 and 4, and infinite order elements. The topological and smooth concordance groups have elements of infinite order as well as elements of order 1 and 2. There is a wealth of examples of knots that represent elements of order 4 in the algebraic concordance group, but have infinite smooth concordance order and topological concordance order. Tools such as the homology of branched covers and twisted Alexander polynomials can help us identify finite or infinite order elements of the concordance group [8]. On the other hand, it is unknown whether there might be elements of order 4 in the smooth or topological concordance groups [25, 26].

Open question: Are there elements of the concordance group with finite orders other than 1, 2 and 4?

Open question: Are there elements of the smooth concordance group with order 4?

Let's get back to the algebraic structure of the concordance group. We have focused up to this point on properties of the algebraic concordance group. Casson-Gordon invariants are used to find knots that are algebraically slice but are not slice [2]. This opens a wider discussion of studying the concordance group at a deeper level than algebraic concordance.

Cochran-Orr-Teichner defined the (n)-solvable filtration of the concordance group:

$$\cdots \mathcal{F}_2 \subset \mathcal{F}_{1.5} \subset \mathcal{F}_1 \subset \mathcal{F}_{0.5} \subset \mathcal{F}_0 \subset C.$$

This filtration, defined in [4, 5], uses topological invariants, such as Casson-Gordon invariants and the Arf invariant, to algebraically decompose the concordance group. Subsequent work has shown that each quotient is infinite [3, 6, 7] and continues to explore the structures of the successive quotients.

Much of the research in concordance has been in studying the interaction between various knot invariants and concordance. Here are several invariants that were not yet discussed above and their connection to the study of concordance.

Genus: The genus of a knot was discussed in Chapter 5. We can extend this notion in two ways to connect to the study of concordance. The four-ball genus, $g_4(K)$, is the minimal genus of a surface embedded in B^4 with boundary K. This is a concordance invariant and is bounded above by the 3-genus, $g(K)$. If K is slice, then $g_4(K) = 0$. The concordance genus, $g_c(K)$, is the minimal genus of all knots concordant to K. This invariant, defined by Livingston [22], is bounded below by the four-ball genus and bounded above by the 3-genus.

Epsilon and Upsilon: Recent work has also developed concordance invariants from Knot Floer Homology (Chapter 12). The epsilon invariant [15] and upsilon invariant [28] are both concordance invariants developed from Floer homology. These two invariants are independent and provide bounds on the concordance genus [16, 24].

48.1.3   Slicing Knots and Finding Concordances

We have discussed many of the invariants that provide obstructions to concordances. On the other hand, there are many slice knots, and many concordances. In this section we'll look at two examples to illustrate the construction of concordances.

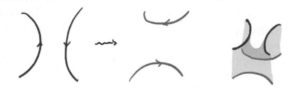

FIGURE 48.1: Band move.

A *band move* connects two points on an oriented knot by an oriented band (see Figure 48.1). We may use band moves to build surfaces in $S^3 \times I$ (or B^4) to find concordances. Isotopy of our knot can be thought of as tracing out a cylinder in $S^3 \times I$. Each oriented band move creates a saddle point. We can cap off unlinked unknotted circles (or unlinked slice-knotted circles) with disks to create an embedded annulus in $S^3 \times I$ relating two knots.

Example 4. *The knot 6_1 is slice. Figure 48.2a illustrates a band move on 6_1 that splits the knot into an unknot of two components. Capping off these two unknots gives us a slice disk for 6_1.*

Example 5. *The knot 8_{10} is concordant to 3_1. In considering possible concordances, we may examine known invariants, such as the Alexander polynomial for clues. In this case, $\Delta_{8_{10}}(t) = 1 - 3t + 6t^2 - 7t^3 + 6t^4 - 3t^5 + t^6 = (t^2 - t + 1)^3$. Since $\Delta_{3_1}(t) = t^2 - t + 1$, we wonder if the trefoil is concordant to 8_{10}. Figure 48.2b confirms this.*

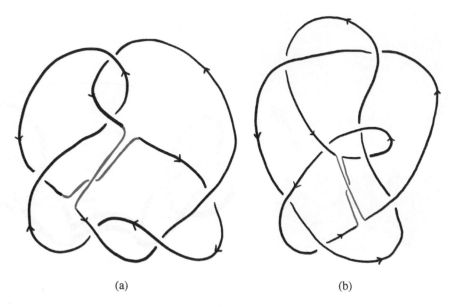

FIGURE 48.2: Band moves applied to (a) 6_1 and (b) 8_{10}.

The band move illustrated splits 8_{10} into a trefoil and an unlinked unknot. Capping off the unknot gives a concordance from 8_{10} to 3_1.

Many more examples and a census of concordances of prime knots up to 11 crossings can be found in [17, 22].

48.1.4 Ribbon Knots

In considering concordance as an equivalence relation, we make the observation that any knot K is concordant to itself. In particular, taking an immersed annulus cobounded by K and $-K$ in S^3, we can carefully cut out a neighborhood of a curve on the side of an annulus to get a disk bounded by $K\# - K$. Figure 48.3b shows such a disk for $3_1\# - 3_1$. We can perturb the intersection points in B^4 to get a slice disk for $K\# - K$, but we also notice that by construction we had fairly "nice" intersections. Each self-intersection consists of two arcs (the green arcs in Figure 48.3b), one of which is interior to the disk. A disk with this property is called a ribbon disk.

Definition 5. *A knot K is called ribbon if it bounds a singular disk in S^3 which has the property that each self-intersection is an arc whose pre-image is exactly two arcs in the disk, one of which is interior. [29]*

This definition is easy to assess diagrammatically, although it can be trickier to work with mathematically. An alternative definition considers slice disks for the knot and insists that the knot have a slice disk with no index 2 critical points with respect to a certain Morse function. It is clear from this second definition that every ribbon knot is slice, since ribbon disks are a special case of slice disks. A long-standing open question is whether the converse is true.

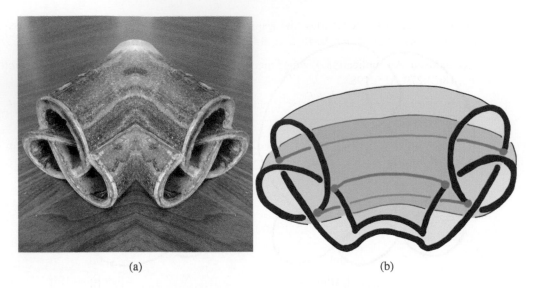

(a)   (b)

FIGURE 48.3: A ribbon disk for $3_1 \# - 3_1$.

Open question: Is every smoothly slice knot a ribbon knot? [10]

48.2 Bibliography

[1] Emil Artin. Zur Isotopie zweidimensionaler Flächen im R_4. *Abh. Math. Sem. Univ. Hamburg*, 4(1):174–177, 1925.

[2] A. J. Casson and C. McA. Gordon. Cobordism of classical knots. In *À la recherche de la topologie perdue*, volume 62 of *Progr. Math.*, pages 181–199. Birkhäuser Boston, Boston, MA, 1986. With an appendix by P. M. Gilmer.

[3] Tim D. Cochran, Shelly Harvey, and Constance Leidy. Knot concordance and higher-order Blanchfield duality. *Geom. Topol.*, 13(3):1419–1482, 2009.

[4] Tim D. Cochran, Kent E. Orr, and Peter Teichner. Knot concordance, Whitney towers and L^2-signatures. *Ann. of Math. (2)*, 157(2):433–519, 2003.

[5] Tim D. Cochran, Kent E. Orr, and Peter Teichner. Structure in the classical knot concordance group. *Comment. Math. Helv.*, 79(1):105–123, 2004.

[6] Tim D. Cochran, Kent E. Orr, and Peter Teichner. Structure in the classical knot concordance group. *Comment. Math. Helv.*, 79(1):105–123, 2004.

[7] Tim D. Cochran and Peter Teichner. Knot concordance and von Neumann ρ-invariants. *Duke Math. J.*, 137(2):337–379, 2007.

[8] Julia Collins, Paul Kirk, and Charles Livingston. The concordance classification of low cross-ing number knots. *Proc. Amer. Math. Soc.*, 143(10):4525–4536, 2015.

[9] S. K. Donaldson. An application of gauge theory to four-dimensional topology. *J. Differential Geom.*, 18(2):279–315, 1983.

[10] Rob Kirby (Ed.) and Edited Rob Kirby. Problems in low-dimensional topology. In *Proceedings of Georgia Topology Conference, Part 2*, pages 35–473. Press, 1995.

[11] R. H. Fox. A quick trip through knot theory. In *Topology of 3-manifolds and related topics (Proc. The Univ. of Georgia Institute, 1961)*, pages 120–167. Prentice-Hall, Englewood Cliffs, N.J., 1962.

[12] Ralph H. Fox and John W. Milnor. Singularities of 2-spheres in 4-space and cobordism of knots. *Osaka J. Math.*, 3:257–267, 1966.

[13] Michael H. Freedman and Frank Quinn. *Topology of 4-Manifolds*, volume 39 of *Princeton Mathematical Series*. Princeton University Press, Princeton, NJ, 1990.

[14] Michael Hartley Freedman. The topology of four-dimensional manifolds. *J. Differential Geom.*, 17(3):357–453, 1982.

[15] Jennifer Hom. An infinite-rank summand of topologically slice knots. *Geom. Topol.*, 19(2):1063–1110, 2015.

[16] Jennifer Hom. A note on the concordance invariants epsilon and upsilon. *Proc. Amer. Math. Soc.*, 144(2):897–902, 2016.

[17] M. Kate Kearney. The concordance genus of 11-crossing knots. *J. Knot Theory Ramifications*, 22(13):1350077, 17, 2013.

[18] Taehee Kim. Filtration of the classical knot concordance group and Casson-Gordon invariants. *Math. Proc. Cambridge Philos. Soc.*, 137(2):293–306, 2004.

[19] J. Levine. Invariants of knot cobordism. *Invent. Math. 8 (1969), 98–110; addendum, ibid.*, 8:355, 1969.

[20] J. Levine. Knot cobordism groups in codimension two. *Comment. Math. Helv.*, 44:229–244, 1969.

[21] Charles Livingston. Seifert forms and concordance. *Geom. Topol.*, 6:403–408, 2002.

[22] Charles Livingston. The concordance genus of knots. *Algebr. Geom. Topol.*, 4:1–22, 2004.

[23] Charles Livingston. A survey of classical knot concordance. In *Handbook of Knot Theory*, pages 319–347. Elsevier B. V., Amsterdam, 2005.

[24] Charles Livingston. Notes on the knot concordance invariant upsilon. *Algebr. Geom. Topol.*, 17(1):111–130, 2017.

[25] Charles Livingston and Swatee Naik. Obstructing four-torsion in the classical knot concordance group. *J. Differential Geom.*, 51(1):1–12, 1999.

[26] Charles Livingston and Swatee Naik. Stabilizing four-torsion in classical knot concordance. *Topology Proc.*, 37:119–128, 2011.

[27] Kunio Murasugi. On a certain numerical invariant of link types. *Trans. Amer. Math. Soc.*, 117:387–422, 1965.

[28] Peter S. Ozsváth, András I. Stipsicz, and Zoltán Szabó. Concordance homomorphisms from knot Floer homology. *Adv. Math.*, 315:366–426, 2017.

[29] Dale Rolfsen. *Knots and links*, volume 7 of *Mathematics Lecture Series*. Publish or Perish, Inc., Houston, TX, 1990. Corrected reprint of the 1976 original.

[30] A. G. Tristram. Some cobordism invariants for links. *Proc. Cambridge Philos. Soc.*, 66:251–264, 1969.

Part IX

Spatial Graph Theory

Chapter 49

Spatial Graphs

Stefan Friedl, Universität Regensburg, Germany

Gerrit Herrmann, AXA AG Deutschland

Before we talk about spatial graphs we need to settle what we mean by a graph.

Definition 1. An *abstract graph* G is a triple (V, E, φ) where V is a finite non-empty set, E is a finite set and φ is a map

$$\varphi \colon E \ \to \ \{\text{subsets of } V \text{ with one or two elements}\}.$$

The elements of V are called *vertices of G* and the elements of E are called the *edges of G*. Furthermore, given $e \in E$, the points in $\varphi(e)$ are called the *endpoints of e*.

Our goal is to introduce the notion of a "spatial graph." The idea is that a spatial graph should be a graph in S^3, the same way that a link is a union of circles in S^3. The problem is to make precise what "in S^3" is supposed to mean.

In the following we first recall the definition of a link and the notion of a tubular neighborhood of a link. We then proceed with the definition of a spatial graph and a discussion of graphical neighborhoods which are a generalization of tubular neighborhoods.

To us it seems like the most natural definition of a link is to define it as a closed 1-dimensional submanifold of S^3. (It is worth pointing out that the definitions of a link in the various textbooks [1, p. 1], [5, p. 1], [6, p. 2 and 100] and [4, p. 4] are all somewhat different, but they lead to equivalent theories.) One advantage of the above definition is that it allows us to dip into the well-developed theory of smooth manifolds.

One of the key technical tools in studying links are tubular neighborhoods. In this context there are again different definitions that are being used. Therefore we fix our conventions. Let M be a 3-manifold. (Here and throughout by a 3-manifold we mean an orientable, connected 3-dimensional manifold, possibly with boundary). Furthermore let C be a compact proper 1-dimensional submanifold of M. (Here, proper means that $\partial C = C \cap \partial M$ and that C intersects ∂M transversally.) A *tubular embedding of C* is an embedding $F \colon C \times \overline{B}^2 \to M$ with $F(P, 0) = P$ for all $P \in C$ such that $F(\partial C \times \overline{B}^2) = \partial M \cap F(C \times \overline{B}^2)$ and such that $F(C \times \overline{B}^2)$ is a submanifold with corners of M. A *tubular neighborhood of C* is the image N of a tubular embedding.

Remark 1. Note that if C has non-empty boundary, then N is a submanifold with non-empty corners, thus strictly speaking it is not a smooth submanifold. Fortunately, in practice this is not a problem. For example, we are mostly interested in considering the exterior $E_C := M \setminus \overset{\circ}{N}$ where $\overset{\circ}{N}$ is the interior of N. The exterior E_C is a smooth manifold with corners, but by "straightening of corners," see [8, Proposition 2.6.2] we can view $E_C = M \setminus \overset{\circ}{N}$ as a smooth manifold in a canonical way.

The following theorem shows that tubular neighborhoods always exist.

Theorem 49.0.1. *[8, Theorem 2.3.3] Every compact proper 1-dimensional manifold of every compact 3-manifold admits a tubular neighborhood.*

The following theorem, see [8, Chapter 2.5], shows that tubular neighborhoods are unique in an appropriate sense.

Theorem 49.0.2. *Let M be a 3-manifold and let C be a proper 1-dimensional submanifold of M. If N and N′ are two tubular neighborhoods of C, then there exists a map $\Phi \colon M \times [0, 1] \to M$ with the following properties:*

1. *if given $t \in [0, 1]$ we denote by $\Phi_t \colon M \to M$ the map defined by $\Phi_t(P) := \Phi(P, t)$, then each $\Phi_t \colon M \to M$ is a diffeomorphism,*

2. *each Φ_t restricts to the identity on C,*

3. *$\Phi_0 = \mathrm{id}_M$,*

4. *the restriction of Φ_1 to N defines a diffeomorphism from N to N′.*

Given a closed 1-dimensional manifold C of a 3-manifold M we pick a tubular neighborhood N. We denote by $\overset{\circ}{N} = N \setminus \partial N$ the interior of N. We refer to $M_C := M \setminus \overset{\circ}{N}$ as the *exterior of C*. The tubular neighborhood N and the exterior of C have the following four useful properties:

(a) by Theorem 49.0.2 the diffeomorphism type of the exterior is in fact well-defined, i.e. it is independent of the choice of the tubular embedding,

(b) M_C is a compact 3-manifold,

(c) M_C is a deformation retract of $M \setminus C$,

(d) C is a deformation retract of the tubular neighborhood N.

After this long discussion of links and 1-dimensional submanifolds of 3-manifolds we turn to the notion of a spatial graph. Our goal is to give a definition of spatial graphs that has the following two features:

1. the definition is flexible enough to incorporate all "reasonable examples," for example we want the pictures shown in Figure 49.1 to be examples of spatial graphs (here we follow the usual topological convention of viewing S^3 as $\mathbb{R}^3 \cup \{\infty\}$),

2. the definition ensures that spatial graphs have good technical properties, which we interpret as asking for an analogue of tubular neighborhoods, in the sense that we want an analogue of the above statements (a), (b), (c) and (d).

In preparation for our definition of a spatial graph we need to introduce the notion of an arc.

Definition 2. An *arc* in S^3 is a subset E of S^3 for which there exists a map $\varphi \colon [0, 1] \to S^3$ with the following properties:

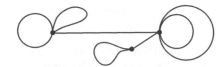 

FIGURE 49.1: Examples of spatial graphs.

1. the map φ is smooth, i.e. all derivatives are defined on the open interval $(0, 1)$ and they extend to continuous maps on the closed interval $[0, 1]$ that we also call derivatives,

2. the first derivative $\varphi'(t)$ is non-zero for all $t \in [0, 1]$,

3. the restriction of φ to $(0, 1)$ is injective,

4. $\varphi((0, 1)) \cap \varphi(\{0, 1\}) = \emptyset$, and

5. $\varphi((0, 1)) = E$.

Given an arc E as above, we refer to $\varphi(0)$ and $\varphi(1)$ as the *endpoints of E*. (Note that the endpoints of E are *not* points on E.)

one endpoint   two endpoints

FIGURE 49.2: Illustration of arcs with one or two endpoints.

Remark 2. If an arc has two different endpoints, then an arc is precisely the same as the image of an embedding of the interval $[0, 1]$ in S^3. The more technical features of the definition are necessary to get the right degree of control in the case that an arc has only one endpoint.

Now we can give our definition of a spatial graph.

Definition 3. A *spatial graph G* is a pair (V, E) with the following properties:

1. V is a finite non-empty subset of S^3.

2. E is a subset of S^3 with the following properties:

   (a) E is disjoint from V,

   (b) E has finitely many components,

   (c) each component of E is an arc and the endpoints of each arc lie on V.

We refer to the points in V as the *vertices of G* and we refer to the components of E as the *edges of G*. Furthermore, given a spatial graph $G = (V, E)$ we write $|G| = V \cup E \subset S^3$. A spatial graph defines an abstract graph in obvious way.

It should be clear that Figure 49.1 shows three examples of spatial graphs. In the figure the red dots correspond to the vertices and the blue segments correspond to the edges.

Next we want to introduce a generalization of the tubular neighborhood for a link. We start out with the following definition.

Definition 4. Let $G = (V, E)$ be a spatial graph with vertex set $V = \{v_1, \ldots, v_m\}$. A *small neighborhood of* V is a compact 3-dimensional manifold X of S^3 with components $X_1, \ldots, X_m$ such that for each $i \in \{1, \ldots, m\}$ there exists a diffeomorphism $\Phi_i \colon \overline{B}^3 \to X_i$ with $\Phi_i = v_i$ and such that for each $r \in (0, 1]$ the image $\Phi_i(S^2_r)$ is a submanifold of $S^3 \setminus V$ that is transverse to the submanifold E of $S^3 \setminus V$. (Here S^2_r is the sphere of radius r in $\mathbb{R}^3$ around the origin).

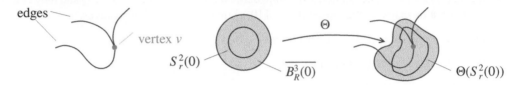

FIGURE 49.3: Illustration of a small neighborhood.

We point out that $E \cap (S^3 \setminus \mathring{X})$ is a proper submanifold of $S^3 \setminus \mathring{X}$. Now we can define a graphical neighborhood of a spatial graph.

Definition 5. Let $G = (V, E)$ be a spatial graph. A *graphical neighborhood of* (V, E) is a subset N of S^3 that can be written as a union $N = X \cup Y$ where X is a small neighborhood of V and Y is a tubular neighborhood of $E \cap (S^3 \setminus \mathring{X})$ in $S^3 \setminus \mathring{X}$.

Remark 3. As remarked above, after "straightening of corners," we can view $S^3 \setminus \mathring{N}$ as a smooth manifold in a canonical way.

FIGURE 49.4: Illustration of a graphical neighborhood.

In the following we will see that every spatial graph admits a graphical neighborhood and that graphical neighborhoods have properties that are very similar to the properties of tubular neighborhoods.

We start out with the following theorem that plays the role of Theorem 49.0.1.

Theorem 49.0.3. *Every spatial graph admits a graphical neighborhood.*

Sketch of proof. Let $G = (V, E)$ be a spatial graph. It is fairly elementary to see that V admits a small neighborhood X (Here one needs to use properties (1) and (2) of arcs.) By the Tubular Neighborhood Theorem 49.0.1, there exists a tubular neighborhood Y of the proper submanifold $E \cap (S^3 \setminus \mathring{X})$ of $S^3 \setminus \mathring{X}$. Then $X \cup Y$ is a graphical neighborhood of G. $\square$

The most difficult aspect is to show that graphical neighborhoods are unique in an appropriate sense. The following theorem plays the role of Theorem 49.0.2.

Theorem 49.0.4. *Let G be a spatial graph. If N and N' are two graphical neighborhoods of G, then there exists a map $\Phi: S^3 \times [0, 1] \to S^3$ with the following properties:*

1. *if given $t \in [0, 1]$ we denote by $\Phi_t: S^3 \to S^3$ the map defined by $\Phi_t(P) := \Phi(P, t)$, then each $\Phi_t: S^3 \to S^3$ is a diffeomorphism,*

2. *each Φ_t is the identity on the vertex set and it preserves each edge setwise,*

3. *$\Phi_0 = \mathrm{id}$,*

4. *we have $\Phi_1(N) = N'$.*

The proof of Theorem 49.0.4 is somewhat involved. We refer to [3] for more details. The next proposition summarizes some of the key properties of graphical neighborhoods.

Proposition 49.0.5. *Let G be a spatial graph and let N be a graphical neighborhood of G.*

1. *N contains $|G|$ in the interior $\mathring{N} = N \setminus \partial N$ of N,*

2. *$|G|$ is a deformation retract of N,*

3. *∂N is a deformation retract of $N \setminus |G|$,*

4. *the exterior E_G is a compact 3-manifold that is a deformation retract of $S^3 \setminus |G|$, and*

5. *the diffeomorphism type of the exterior $E_G := S^3 \setminus \mathring{N}$ does not depend on the choice of N.*

Sketch of proof. Let $N = X \cup Y$ be a graphical neighborhood of G. Statement (1) is immediate. Statement (4) is a consequence of statement (3). Statements (2) and (3) are a fairly immediate consequence of the fact that Y is a product and the observation that for every component X' of X the pair $(X', E \cap X')$ is homeomorphic to a pair $(\overline{B}^3, R)$ where R consists of straight segments emanating from the origin. Finally, (5) is an immediate consequence of Theorem 49.0.4 and the above discussion of smoothening corners. More details will be provided in [3]. □

Remark 4. 1. One can easily generalize the definition of a spatial graph in S^3 to a spatial graph in any closed 3-manifold. The above results can also easily be generalized to this more general context. We will consider this more general case in [3].

2. In the literature, sometimes a spatial graph is defined as a PL-embedding of a graph in S^3 and as a neighborhood one uses a regular neighborhood as defined in [7, p. 33]. By [7, Theorem 3.8] these are unique in an appropriate sense and by [7, Corollary 3.30] the PL-spatial graph is a deformation retract of each regular neighborhood. It is not so clear, though, whether there are analogues of statements (3) and (4) of Proposition 49.0.5.

49.1 Bibliography

[1] G. Burde, H. Zieschang and M. Heusener. *Knots*, 3rd fully revised and extended edition. De Gruyter Studies in Mathematics 5 (2014).

[2] P. Cromwell. *Knots and links*, Cambridge University Press (2004).

[3] S. Friedl and G. Herrmann. *Graphical neighborhoods of spatial graphs*, Preprint (2019). Available at `https://www.uni-regensburg.de/Fakultaeten/nat_Fak_I/friedl/papers/graphical-nhd.pdf`.

[4] A. Kawauchi. *A Survey of Knot Theory*, Basel: Birkhäuser (1996).

[5] W. B. R. Lickorish. *An Introduction to Knot Theory*, Graduate Texts in Mathematics 175, Springer Verlag (1997).

[6] D. Rolfsen. *Knots and Links*, Mathematics Lecture Series. 7. Houston, TX: Publish or Perish. (1990).

[7] C. P. Rourke and B. J. Sanderson. *Introduction to Piecewise-Linear Topology*, Ergebnisse der Mathematik und ihrer Grenzgebiete 69 (1972).

[8] C. T. C. Wall. *Differential Topology*, Cambridge Studies in Advanced Mathematics 156. Cambridge University Press (2016).

Chapter 50

A Brief Survey on Intrinsically Knotted and Linked Graphs

Ramin Naimi, Occidental College

50.1   Introduction

In the early 1980s, Sachs [36, 37] showed that if G is one of the seven graphs in Figure 50.1, known as the Petersen family graphs, then every spatial embedding of G, i.e., embedding of G in S^3 or $\mathbb{R}^3$, contains a nontrivial link — specifically, two cycles that have odd linking number. Henceforth, spatial embedding will be shortened to embedding; and we will not distinguish between an embedding and its image. A graph is **intrinsically linked** (IL) if every embedding of it contains a nontrivial link. For example, Figure 50.1 shows a specific embedding of the first graph in the Petersen family, K_6, the complete graph on six vertices, with a nontrivial 2-component link highlighted. At about the same time, Conway and Gordon [4] also showed that K_6 is IL. They further showed that K_7 in **intrinsically knotted** (IK), i.e., every spatial embedding of it contains a nontrivial knot.

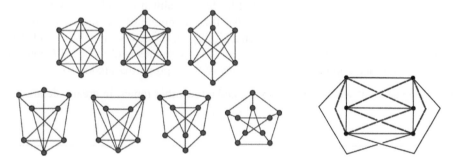

FIGURE 50.1: Left: The Petersen family graphs [42]. Right: An embedding of K_6, with a nontrivial link highlighted.

A graph H is a **minor** of another graph G if H can be obtained from a subgraph of G by contracting zero or more edges. It's not difficult to see that if G has a linkless (resp. knotless) embedding, i.e., G is not IL (IK), then every minor of G has a linkless (knotless) embedding [6, 30]. So we say the property of having a linkless (knotless) embedding is **minor closed** (also called *hereditary*).

A graph G with a given property is said to be **minor minimal** with respect to that property if no minor of G, other than G itself, has that property. Kuratowski [26] and Wagner [41] showed that a graph is nonplanar, i.e., cannot be embedded in $\mathbb{R}^2$, iff it contains K_5 or $K_{3,3}$ (the complete bipartite graph on $3 + 3$ vertices) as a minor. Equivalently, these two graphs are the only minor minimal nonplanar graphs. It's easy to check that each of the seven graphs in the Petersen family is minor minimal IL (MMIL), i.e., if any edge is deleted or contracted, the resulting graph has a linkless embedding. Sachs conjectured that these are the only MMIL graphs, which was later proved by Robertson, Seymour, and Thomas:

Theorem 1. [35] A graph is IL iff it contains a Petersen family graph as a minor.

This gives us an algorithm for deciding whether any given graph G is IL: check whether one of the Petersen family graphs is a minor of G.

In contrast, finding all minor minimal IK (MMIK) graphs has turned out to be more difficult. Robertson and Seymour's Graph Minor Theorem [34] says that in any infinite set of (finite) graphs, at least one is a minor of another. It follows that for any property whatsoever (minor closed or not), there are only finitely many graphs that are minor minimal with respect to that property. In particular, there are only finitely many MMIK graphs. If we knew the finite set of all MMIK graphs, we would be able to decide whether or not any given graph is IK. So far there are at least 264 known MMIK graphs [9], and, for all we know, this could be just the tip of the iceberg — we don't even have an upper bound on the number of MMIK graphs.

For $n \geq 3$, a graph is **intrinsically n-linked** ($I n L$) if every spatial embedding of it contains a nonsplit n-link (a link with n components). It was shown in [12] that K_{10} is I3L; and it was shown in [5] that removing from K_{10} four edges that share one vertex, or two nonadjacent edges, yields I3L graphs; but it's not known if they are MMI3L. Examples of MMInL graphs were given for every $n \geq 3$ in [8].

Other "measures of complexity" have also been studied. For example: given any pair of positive integers λ and $n \geq 2$, every embedding of a sufficiently large complete graph contains a 2-link with linking number at least λ in magnitude [7, 38]; contains a nonsplit n-link with all linking numbers even [13]; contains a knot K such that the magnitude of the second coefficient of its Conway polynomial, i.e., $|a_2(K)|$, is larger than λ [7, 38]; and contains a nonsplit n-link L such that for every component C of L, $|a_2(C)| > \lambda$ and for any two components C and C' of L, $|lk(C, C')| > \lambda$ [10]. No minor minimal graphs with respect to any of these properties are known.

In the following sections we discuss the above, and a few other topics, in greater detail.

50.2 IL Graphs

The proof that K_6 is IL is short and beautiful: There are 20 triangles (3-cycles) in K_6. For each triangle, there is exactly one triangle disjoint from it. Thus there are exactly 10 pairs of disjoint triangles, i.e., 2-links, in K_6. Any pair of disjoint edges is contained in exactly two such links. So, given any embedding of K_6, any crossing change between any two disjoint edges affects the linking number of exactly two links, and the magnitude of each of their linking numbers changes by 1. Thus, the sum of all linking numbers does not change parity under any crossing change. Now, in the embedding of K_6 shown in Figure 50.1, the sum of all ten linking numbers is odd. Therefore the same is true in every embedding of K_6, since any embedding can be obtained from any other embedding by isotopy and crossing changes. Hence every embedding contains at least one link with odd linking number.

Sachs observed that a similar argument can be used to show all seven graphs in the Petersen family are IL. He also observed that each of these graphs can be obtained from any other by one or more ∇Y and $Y\nabla$ moves, as defined in Figure 50.2. Furthermore, this family is closed under ∇Y and $Y\nabla$ moves. It follows, by Theorem 1, that ∇Y and $Y\nabla$ moves preserve the property of being MMIL.

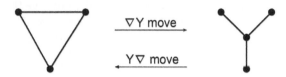

FIGURE 50.2: ∇Y and $Y\nabla$ moves.

In fact, Sachs observed that a ∇Y move on any IL graph yields an IL graph; or, equivalently, that a $Y\nabla$ move on a "linklessly embeddable" graph yields a linklessly embeddable graph. The proof of this is elementary and straightforward, which we outline here. Suppose G' is obtained from a linklessly embeddable graph G by a $Y\nabla$ move. Take a linkless embedding Γ of G, and replace the Y involved in the $Y\nabla$ move by a triangle ∇ whose edges are "close and parallel" to the edges of the Y. This gives an embedding Γ' of G'. It's easy to show that any link in Γ' that doesn't have ∇ as a component is isotopic to a link in Γ, and hence is trivial. And any link in Γ' that does have ∇ as a component is also trivial since ∇ bounds a disk whose interior is disjoint from Γ'.

It is also true that a $Y\nabla$ move on any IL graph yields an IL graph, but the only known proofs of it rely on Theorem 1 or the following result of [35]: If G has a linkless embedding, then it has a paneled embedding, i.e., an embedding Γ such that every cycle in Γ bounds a disk whose interior is disjoint from Γ.

Let's say a graph is $\mathbb{Z}_2$-IL if every embedding of it contains a 2-link with linking number nonzero mod 2. Thus, each of the Petersen family graphs is $\mathbb{Z}_2$-IL. This, together with Theorem 1, implies that G is IL iff it is $\mathbb{Z}_2$-IL.

It is possible to determine if a graph G is $\mathbb{Z}_2$-IL by simply solving a system of linear equations, without even using Theorem 1. We give an outline here. First, pick an arbitrary embedding Γ of G, and compute the linking numbers mod 2 for all 2-links (pairs of disjoint cycles) in Γ. An arbitrary embedding Γ' of G can be obtained from Γ by adding some number of full twists between each pair of disjoint edges, plus isotopy (adding twists is equivalent to letting edges "pass through" each other). Say there are d pairs of disjoint edges in G. Let $x_1, \cdots, x_d$ be variables representing the number of full twists to be added to the d disjoint pairs of edges to obtain Γ' from Γ. Then the linking number of any 2-link in Γ' can be written in terms of $x_1, \cdots, x_d$ and the linking number of that 2-link in Γ. Setting each of these expressions equal to zero gives us a system of linear equations in d variables. This system of equations has a solution in $\mathbb{Z}_2$ iff G is not $\mathbb{Z}_2$-IL. Note that the number of cycles in a graph can grow exponentially with the graph's size, so this algorithm is exponential in time and space. In [25, 40], polynomial time algorithms are given for finding linkless embeddings of graphs.

50.3 IK Graphs

Essentially the same argument that shows the ∇Y move preserves ILness also shows the ∇Y move preserves IKness. However, the $Y\nabla$ move does not necessarily preserve IKness [11]. For example, there are twenty graphs that can be obtained from K_7 by zero or more ∇Y and $Y\nabla$ moves. Six of these graphs cannot be obtained from K_7 by ∇Y moves only — they require $Y\nabla$ moves also. And it turns out all these six graphs have knotless embeddings [11, 19, 20].

Given two disjoint graphs G_1 and G_2, let $G_1 * G_2$, the **cone** of G_1 with G_2, be the graph obtained by adding all edges from vertices of G_1 to vertices of G_2, i.e., $G_1 * G_2 = G_1 \cup G_2 \cup \{v_1 v_2 \mid v_i \in V(G_i)\}$.

For about twenty years, the only known IK graphs were K_7 and its **descendants**, i.e., graphs obtained from K_7 by ∇Y moves only. It was suspected that $K_{3,3,1,1}$ (the complete 4-partite graph on $3 + 3 + 1 + 1$ vertices) is also IK. Recall that K_5 and $K_{3,3}$ are minor minimal nonplanar. Coning with one vertex on each of these graphs gives K_6 and $K_{3,3,1}$, both of which are in the Petersen family and hence MMIL. Coning again gives K_7 and $K_{3,3,1,1}$; and K_7 was shown to be MMIK; so it was natural to ask if $K_{3,3,1,1}$ is too. Foisy [15] proved that $K_{3,3,1,1}$ indeed is IK. His technique, partially outlined below, led to finding many more MMIK graphs later on [16, 17, 19].

Figure 50.3 shows a multi-graph (i.e., double edges and loops are allowed) commonly called D_4, with four of its cycles labeled $C_1, \cdots, C_4$. Let's say an embedding of D_4 is **double linked** mod 2 if $lk(C_1, C_3)$ and $lk(C_2, C_4)$ are both nonzero mod 2. To show $K_{3,3,1,1}$ is IK, Foisy proved the following key lemma. A more general version of the lemma was proved, independently, by Taniyama and Yasuhara [39].

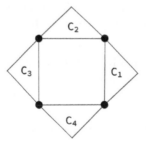

FIGURE 50.3: The D_4 graph.

Lemma 2 (D_4 Lemma). [15, 39] Every embedding of D_4 that is double linked mod 2 contains a knot K with $a_2(K) \neq 0$ mod 2.

Foisy proved that $K_{3,3,1,1}$ is IK by showing that every embedding of it contains as a minor a D_4 that is double linked mod 2.

Let's say a graph G is ID_4 mod 2 if every embedding of G contains as a minor a D_4 that is double linked mod 2; and G is Ia_2 mod 2 if every embedding of G contains a knot K such that $a_2(K) \neq 0$ mod 2. Thus, the D_4 Lemma says if G is ID_4 mod 2 then it's Ia_2 mod 2. We also know that if G is

Ia_2 mod 2 then it's IK. Let's abbreviate these two implications as ID_4 mod 2 $\implies$ Ia_2 mod 2 $\implies$ IK. It is natural to ask if the converse of each of these implications is also true.

Question 1. (a) IK $\implies$ Ia_2 mod 2? (b) Ia_2 $\implies$ ID_4 mod 2? (c) IK $\implies$ ID_4 mod 2?

The question "InL $\implies$ InL mod 2?" is also open.

It turns out that every known MMIK graph[1] is Ia_2 mod 2 and ID_4 mod 2. But this is not necessarily evidence that the answer to either part of Question 1 is yes, because most of the known MMIK graphs were found by looking for graphs that are ID_4 mod 2.

Determining if a graph is ID_4 mod 2 can be done by solving systems of linear equations [29]. If it is true that IK $\iff$ ID_4 mod 2, then the algorithm of [29] can be used to decide whether an arbitrary graph is IK.

There may be (a lot) more MMIK graphs than have been found so far; and trying to find some of them might not be too hard. For example, one can start with a non-MMIK graph G obtained by a Y∇ move from a known MMIK graph, and keep adding new edges to G or expanding vertices of G into edges (the reverse of contracting edges) until one obtains an IK graph. But finding more and more MMIK graphs doesn't seem to have advanced our understanding of IK graphs very much. In trying to understand IK graphs better, another approach has been to try to classify all IK graphs with a given number of edges. For example, it has been shown that there are no IK graphs with 20 or fewer edges, and the only IK graphs with exactly 21 edges are K_7 and its descendants [1, 23, 24, 28]; IK graphs with 22 edges have also been partially classified [21, 22]. However, this approach doesn't seem to have led to significant insights or advances in the theory either.

50.4 Miscellaneous Facts and Open Problems

In [31] it was shown that if G is IK and e is an edge of a 3-cycle in G, then $G \setminus e$ is IL. The following related questions might be useful in trying to answer Question 1.

Question 2. Suppose G is IK. (a) Is $G \setminus e$, or G/e, IL for every edge, or for some edge, e of G? (b) Does G have at least two distinct nonsplit links?

Sachs observed that a graph G is non-planar iff the graph $G * v$, the cone of G with one vertex v, is IL. This can be seen as follows. If G is nonplanar, then it contains K_5 or $K_{3,3}$ as a minor. So $G * v$ contains $K_5 * v = K_6$ or $K_{3,3} * v = K_{3,3,1}$ as a minor, and hence $G * v$ is IL. Conversely, if G is planar, it is easy to construct a linkless embedding of $G * v$: embed G in the plane, put v above the plane, and connect v with straight edges to all vertices of G.

A graph G is said to be n-**apex** if there exist vertices $v_1, \cdots, v_n$ in G such that $G - \{v_1, \cdots, v_n\}$, i.e., the graph obtained by removing $\{v_1, \cdots, v_n\}$ and all edges incident to them, is planar. A 1-apex graph is called **apex**. Thus, by the above, apex graphs are not IL. It can similarly be shown that 2-apex graphs are not IK. In fact, G is planar iff the graph $G * v * w$ (i.e., $G * K_2$) is not IK [3, 32]. The reason is similar to the one given above: If G is nonplanar, then $G * v * w$ contains $K_5 * v * w = K_7$

---

[1]That K_7 and its descendants are ID_4 is not in the literature, but is believed to be true if the computer program of [29] is correct.

or $K_{3,3} * v * w = K_{3,3,1,1}$ as a minor; and since both of these graphs are IK, $G * v * w$ is IK too. If G is planar, we can construct a knotless embedding of $G * v * w$ as follows: embed G in the plane, put v above the plane, put w below the plane, and connect v and w to all vertices of G and to each other with straight edges. The list of all minor minimal non-n-apex graphs is not known for any n, even $n = 1$. A detailed survey of results on apex and 2-apex graphs can be found in [9].

The crossing number $C(K)$ of a knot K is the fewest number of crossings among all regular projections of K. It's easy to see that for every n, the set $\{K : C(K) \leq n\}$ is finite; so $A(n) = \max\{|a_2(K)| : C(K) \leq n\}$ is well-defined and finite. As mentioned before, given a fixed n, every embedding of a sufficiently large complete graph contains a knot K with $|a_2(K)| > A(n)$; hence, every embedding of a sufficiently large complete graph contains a knot with crossing number larger than n. The **bridge number** of a knot K is the minimum number of local maxima with respect to height (z-coordinate in $\mathbb{R}^3$) among all isotopic embeddings of K. Given any integer $n \geq 2$, there are infinitely many n-bridge knots. So the above argument for the crossing number doesn't work for the bridge number. This leads to the question: Does there exist, for each n, a graph G such that every embedding of G contains a knot with bridge number at least n?

Suppose G' is obtained by a ∇Y move from G. It turns out that if G has any of the following properties, then G' has that property too: IL; IK; Ia_2; ID_4; InL; nonplanar; non-n-apex. The proofs for all of these are elementary and short, and most of them are similar to the one we saw for IL. But, curiously, IL is the only property from the above list known to be preserved by $Y\nabla$ moves.

The **complement** of a graph G is a graph G^c with the same vertices as G and with exactly those edges not in G. In [2] it was shown that if G has 9 or more vertices, then G or G^c is nonplanar. This result is sharp: there exists a graph G on 8 vertices such that both G and G^c are planar. We can ask a similar question of IL graphs: What is the smallest integer v such that for every graph G with v vertices, G or G^c is IL? Here is a partial answer. In [27] it was shown that for all $n \leq 5$, if G has v vertices, e edges, and $e > nv - \binom{n+1}{2}$, then G contains K_{n+2} as a minor. Now, K_{15} has 105 edges, so if G has 15 vertices, then G or G^c has at least $\lceil 105/2 \rceil = 53$ edges. Letting $n = 4$, we have $nv - \binom{n+1}{2} = 4(15) - \binom{5}{2} = 50$; since $53 > 50$, G or G^c contains K_6 as minor, and hence is IL. But this is not sharp. In [33], it is shown that: (i) if G has 13 vertices, then G or G^c is IL, and (ii) there is a graph G with 10 vertices such that neither G nor G^c is IL. The question for graphs with 11 or 12 vertices remains open.

One can similarly show that for every graph G with 18 or more vertices, G or G^c contains a K_7 minor and hence is IK. It is unknown what the minimum number of vertices is that would guarantee that G or G^c is IK.

Let's say a graph is **strongly intrinsically linked** (SIL) if every embedding of it contains a 2-link with linking number at least 2 in magnitude. Then K_{10} is SIL, since, by [12], K_{10} contains a 3-link, two of whose linking numbers are nonzero, and by [7], any embedded complete graph that contains such a 3-link contains a "strong" 2-link. What about K_9? By [12], K_9 does not contain such a 3-link; but it's not known whether or not K_9 is SIL.

A **digraph** (directed graph) is a graph each of whose edges is oriented. A **consistently oriented** cycle in a digraph is a cycle $x_0, x_1, \cdots, x_n$, where $x_n = x_0$, such that each edge $x_i x_{i+1}$ is oriented from x_i to x_{i+1}. A digraph is said to be InL (resp. IK) if every spatial embedding of it contains a nonsplit n-link (resp. nontrivial knot) consisting of consistently oriented cycles. In [14], an IK digraph and an I4L digraph are constructed. It is not known whether there exists an InL digraph with $n \geq 5$. In [18] it was shown that (unlike all the other graph properties we have discussed) the property of having a linkless embedding is not minor closed for digraphs.

50.5 Bibliography

[1] J. Barsotti, T. W. Mattman. Graphs on 21 edges that are not 2-apex. To appear, *Involve* Vol. 9, no. 4 (2016), 591–621. .

[2] J. Battle, F. Harary, Y. Kodama. Every planar graph with nine points has a nonplanar complement. *Bulletin of the American Mathematical Society*, 68 (1962), 569–571. doi10.1090/S0002-9904-1962-10850-7

[3] P. Blain, G. Bowlin, T. Fleming, J. Foisy, J. Hendricks, J. Lacombe. Some results on intrinsically knotted graphs. *J. Knot Theory Ramifications* 16 (2007), no. 6, 749–760.

[4] J. Conway and C. Gordon. Knots and links in spatial graphs. *J. Graph Theory*, 7 (1983) 445–453.

[5] G. Bowlin, J. Foisy. Some new intrinsically 3-linked graphs. *J. Knot Theory Ramifications* 13 (2004), no. 8, 1021–1027.

[6] M. R. Fellows, M. A. Langston. Nonconstructive tools for proving polynomial-time decidability. *J. Assoc. Comput. Mach.* 35 (1988), no. 3, 727–739.

[7] E. Flapan. Intrinsic Knotting and Linking of Complete Graphs. *Algebraic and Geometric Topology*, Vol 2, (2002) 371–380.

[8] E. Flapan, J. Foisy, R. Naimi, J. Pommersheim. Intrinsically n-linked graphs. *Journal of Knot Theory and Its Ramifications*, Vol. 10, (2001) 1143–1154.

[9] E. Flapan, T. W. Mattman, B. Mellor, R. Naimi, R. Nikkuni. Recent developments in spatial graph theory. *Contemporary Mathematics*, Vol 689 (2017), American Mathematical Society.

[10] E. Flapan, B. Mellor, and R. Naimi. Intrinsic linking and knotting are arbitrarily complex. *Fundamenta Mathematicae*, Vol 201 (2008), 131–148.

[11] E. Flapan and R. Naimi. The Y∇ move does not preserve intrinsic knottedness. *Osaka J. Math.*, 45 (2008), 107–111.

[12] E. Flapan, R. Naimi, J. Pommersheim. Intrinsically triple linked complete graphs. *Topology and Its Applications*, Vol. 115, (2001) 239–246.

[13] T. Fleming, A. Diesl Intrinsically linked graphs and even linking number. *Algebr. Geom. Topol.* 5 (2005), 1419–1432.

[14] T. Fleming, J. Foisy. Intrinsically knotted and 4-linked directed graphs. *J. Knot Theory Ramifications* (2018). doi:10.1142/S0218216518500372.

[15] J. Foisy. Intrinsically knotted graphs. *J. Graph Theory*, 39 (2002), 178–187.

[16] J. Foisy. A newly recognized intrinsically knotted graph. *J. Graph Theory*, 43 (2003) 199–209.

[17] J. Foisy. More intrinsically knotted graphs. *J. Graph Theory* 54 (2007), no. 2, 115–124.

[18] J. Foisy, H. N. Howards, N. R. Rich. Intrinsic linking in directed graphs. *Osaka J. Math.*, Volume 52, no. 3, (2015), 817–833.

[19] N. Goldberg, T. W. Mattman, R. Naimi. Many, many more intrinsically knotted graphs: Appendix. arXiv:1109.1632.

[20] R. Hanaki, R. Nikkuni, K. Taniyama, A. Yamazaki. On intrinsically knotted or completely 3-linked graphs. *Pacific J. Math.*, Vol. 252, (2011), 407–425.

[21] H. Kim, T. W. Mattman, S. Oh. More intrinsically knotted graphs with 22 edges and the restoring method. arXiv:1708.03925.

[22] H. Kim, H. J. Lee, M. Lee, T. W. Mattman, S. Oh. Triangle-free intrinsically knotted graphs with 22 edges. arXiv:1407.3460v1 (2014).

[23] B. Johnson, M. E. Kidwell, T. S. Michael. Intrinsically knotted graphs have at least 21 edges. *J. Knot Theory Ramifications*, Vol. 19, (2010), 1423–1429.

[24] M. J. Lee, H. J. Kim, H. J. Lee, S. Oh. Exactly fourteen intrinsically knotted graphs have 21 edges. *Algebr. Geom. Topol.* Vol. 15, (2015), 3305–3322.

[25] K. Kawarabayashi, S. Kreutzer, B. Mohar. Linkless and flat embeddings in 3-space. *Discrete Comput. Geom.* 47 (2012) 731–755.

[26] K. Kuratowski. Sur le probléme des courbes gauches en topologie. *Fund. Math.* (in French), 15 (1930), 271–283.

[27] W. Mader. Existenz n-fach zusammenhangender Teilgraphen in Graphen genugend großen Kantendichte. *Abh. Math. Sem. Univ. Hamburg.* 37 (1972), 86–97.

[28] T. W. Mattman. Graphs of 20 edges are 2–apex, hence unknotted. *Alg. Geom. Top.*, 11 (2011) 691–718, arXiv:0910.1575.

[29] J. Miller and R. Naimi. An algorithm for detecting intrinsically knotted graphs. *Experimental Mathematics*, 23:1 (2014), 6–12, doi: 10.1080/10586458.2014.852033 arXiv:1109.1030.

[30] J. Nesetril, R. Thomas. A note on spatial representations of graphs. *Commentat. Math. Univ. Carolinae* 26 (1985), 655–659.

[31] R. Naimi, E. Pavelescu, H. Schwartz. Deleting an edge of a 3-cycle in an intrinsically knotted graph gives an intrinsically linked graph. *J. Knot Theory and Its Ramifications*, 23 (2014), no. 14, 1450075 (6 pages).

[32] M. Ozawa. Y. Tsutsumi. Primitive spatial graphs and graph minors. *Rev. Mat. Complut.* 20 (2007), no. 2, 391–406.

[33] A. Pavelescu, E. Pavelescu. The complement of a nIL graph with thirteen vertices is IL. *Alg. Geom. Top.* 20 (2020) 395–402. doi:10.2140/agt.2020.20.395.

[34] N. Robertson, P. Seymour. Graph minors. XX. Wagner's conjecture. *J. Combin. Theory Ser. B*, 92, (2004), 325–357.

[35] N. Robertson, P. Seymour, R. Thomas. Sachs' linkless embedding conjecture. *J. Combin. Theory Ser. B*, 64 (1995) 185–227.

[36] H. Sachs. On spatial representations of finite graphs, *Colloq. Math. Soc. János Bolyai* (A. Hajnal, L. Lovasz, V.T. Sós, eds.), 37, North Holland, Amsterdam, New York, 1984, 649–662.

[37] H. Sachs. On a spatial analogue of Kuratowski's theorem on planar graphs: An open problem. Borowiecki M., Kennedy J.W., Syso M.M. (eds) *Graph Theory*. Lecture Notes in Mathematics, Vol 1018. Springer, Berlin, Heidelberg, 1983.

[38] M. Shirai, K. Taniyama. A large complete graph in a space contains a link with large link invariant. *J. Knot Theory Ramifications* 12 (2003), 915–919.

[39] K. Taniyama, A. Yasuhara. Realization of knots and links in a spatial graph. *Topology and Its Applications*, 112 (2001), 87–109.

[40] H. van der Holst. A polynomial-time algorithm to find a linkless embedding of a graph. *J. Combin. Theory, Series B*, 99 (2009), 512–530.

[41] K. Wagner. Über eine Eigenschaft der ebenen Komplexe. *Math. Ann.*, 114 (1937), 570–590. doi:10.1007/BF01594196.

[42] Petersen Family. *Wikipedia*. https://en.wikipedia.org/wiki/Petersen_family.

Chapter 51

Chirality in Graphs

Hugh Howards, Wake Forest University

51.1 Introduction

Questions about the chirality of spatial graphs are interesting not only for understanding molecules, but also for their own sake. Chirality was first studied in order to better understand molecular behavior. Chemists say that a molecule is *chiral* if it can chemically change over to its mirror image at room temperature, and if not it is called *achiral*. Small molecules tend to be rigid so for these molecules the question is geometric, but a molecule which is more flexible and can rotate around certain bonds may be geometrically achiral even though a rigid model might be distinct from its mirror image. This leads to a more topological notion of equivalence. Chirality of graphs questions are interesting for understanding molecules, but also for their own sake.

We now recall some basic definitions from graph theory and knot theory. A graph γ is made up of a set of *vertices* and a collection of *edges* which connect pairs of vertices. A graph with two edges between the same pair of vertices is said to have *multiple edges*. An edge from a vertex back to itself is called a *loop*. A graph is called a *simple graph* if it has no loops or multiple edges. In this chapter, the term graph will refer to simple graphs unless otherwise specified. A graph γ is *3 connected*, if at least three vertices together with the edges containing them must be removed to disconnect γ or reduce γ to a single vertex.

Graphs exist as abstract objects, but we can also think about how to place a graph in space. A *spatial embedding* Γ of an abstract graph γ is a specific mapping of the vertices and edges into $\mathbb{R}^3$ or some other manifold. The vertices go to points and the edges go to a collection of tame arcs which are disjoint except at their endpoints (the vertices).

Many results describe the chirality of classes of graphs so we now define a few common classes of graphs. The *complete graph on n vertices*, K_n, consists of n vertices and one edge between each pair of vertices. Figure 51.1 shows one spatial embedding of the complete graph on seven vertices in $\mathbb{R}^3$. A connected graph with n vertices where each vertex is of degree 2 is called a *cycle graph* denoted C_n (Figure 51.2 shows a knotted spatial embedding of C_{10} on the right). A *Möbius ladder with n rungs*, denoted M_n, is formed from C_{2n} by adding edges joining each pair of antipodal vertices (See Figure 51.3).

We say that an embedding of a knot or graph Γ in $M = \mathbb{R}^3$ (or S^3 or some other 3 manifold) is *achiral (in M)*, if there is an orientation-reversing homeomorphism h of M leaving Γ setwise invariant. If it is not achiral, it is said to be *chiral*. Chiral is derived from the Greek word "kheir" meaning hand. Hands are not the same as their mirror images and this in turn makes it easier to keep track of chiral vs. achiral (Figure 51.4). It is well known in knot theory that some knots, such as the unknot and figure-eight knot (see Figure 51.5), are isotopic to their mirror images and thus

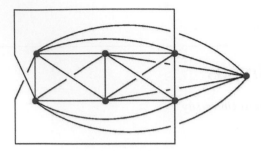

FIGURE 51.1: Here we see a spatial embedding of K_7, the complete graph on 7 vertices.

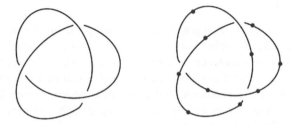

FIGURE 51.2: The trefoil is chiral since it is not isotopic to its reflection. On the left is a traditional picture of the trefoil. On the right we see the C_{10} graph embedded as a trefoil.

are achiral and others, such as the trefoil (see Figure 51.2), are not isotopic to their mirror images and thus are chiral.

Although most of the results on chirality of spatial graphs have been about simple graphs one can certainly ask questions and prove theorems about graphs with loops and multiple edges, too.

As in knot theory, for spatial graphs we can picture S^3 as the one-point compactification of $\mathbb{R}^3$ and most results that can be proven in one setting can be proven in the other setting using essentially an identical proof.

From a spatial graphs perspective, we can think of the trefoil as Γ, a chiral spatial embedding of an abstract graph γ in S^3. A specific example of this is shown in Figure 51.2, where the abstract graph γ is a cyclic graph with 10 vertices and 10 edges (C_{10}). On the other hand in Figure 51.5 we have a new embedding of the same abstract graph $\gamma = C_{10}$, but this time the embedding is achiral.

51.2 Chirality in S^3

As stated above, we see one embedding of C_{10} in Figure 51.2 which is chiral, but another embedding of the same abstract graph in Figure 51.5 which is achiral showing chirality of the C_{10} depends on the specific chosen spatial embedding. This leads to the question of if there are graphs which

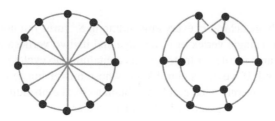

FIGURE 51.3: The Möbius ladder with 6 rungs pictured in two ways (in the picture on the left the 6 edges that represent the rungs cross over each other in the center, but if thought of as a spatial embedding, those edges do not intersect at that point and do not form a vertex).

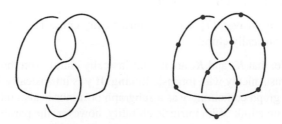

FIGURE 51.4: Hands are chiral.

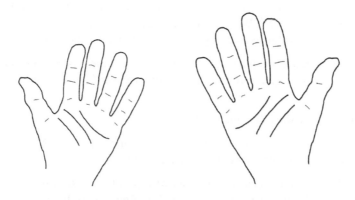

FIGURE 51.5: On the left we see a figure-eight knot. On the right we see an embedding of C_{10} which is achiral since it is essentially a figure-eight knot, which unlike the trefoil is isotopic to its reflection.

are achiral in every embedding in S^3 or alternatively ones which are chiral in every embedding in S^3.

Looking at graphs which are achiral in every embedding in S^3 turns out not to be very interesting. A graph that consists of a single vertex or of an edge and its two vertices is, of course, achiral in every embedding. The same is true for trees (graphs containing no cycles). So we might call these graphs intrinsically achiral. If we add a chiral knot (whose mirror image is not already contained in the graph) to a cycle of an embedded graph we obtain a chiral embedding of the graph, thus it is easy to find chiral embeddings of graphs which contain cycles.

Instead we focus on the more interesting direction which is the question of if there are graphs which have only chiral embeddings in S^3.

Figure 51.1 depicts a spatial embedding of the complete graph on 7 vertices, K_7. We can count the total number of cycles in the graph and then check manually which of those cycles are knotted. The process shows that this embedding only contains one knotted cycle, and that knot is a trefoil (this was one of the steps Conway and Gordon used to prove K_7 is intrinsically knotted in their seminal paper on the subject [1]). Since the trefoil is chiral, this implies that this embedding of K_7 is chiral. Conway and Gordon showed in that paper that K_7 contains a knot, not just in this embedding, but in every possible embedding in S^3. We will soon see a theorem that says K_7 is chiral not just in this embedding, but in every possible embedding in S^3. This motivates the definition below.

We say that an abstract graph γ is *intrinsically chiral in* S^3, if there is no embedding Γ of γ in S^3 such that there exists an orientation-reversing homeomorphism h of S^3 leaving Γ setwise invariant.

For Möbius ladders we see the following result.

Theorem 51.2.1. Flapan, [3]: For every odd $n > 3$, every embedding of a Möbius ladder M_n in S^3 is chiral. For every even n, M_n has an achiral embedding in S^3.

Thus we see that M_{2n+1} is intrinsically chiral, but M_{2n} is not for $n \in \mathbb{Z}, n > 1$! The spirit of the proof that M_{2n} has an achiral embedding is captured in Figure 51.6. In each of those pictures we see a Möbius ladder, that we can rotate 90 degrees around the z-axis and then reflect through the horizontal circle to see the embedding of the graph is achiral. Although the figure gets harder to draw and visualize as n goes up, the same type of construction continues to work.

For complete graphs we see the following result.

Theorem 51.2.2. Flapan, Weaver, [7]: K_{4n+3} is intrinsically chiral for $n \in \mathbb{Z}, n > 0$, but no other complete graphs are intrinsically chiral.

This shows, for example, that K_6 and K_8 are not intrinsically chiral even though K_7 is intrinsically chiral. This result contrasts nicely with intrinsic linking. If γ is intrinsically linked then it is nearly trivial to show that any graph containing γ as a subgraph or minor is also intrinsically linked since that subgraph will contain a link. With intrinsic chirality, however, the parallel result does not hold since K_8 is not intrinsically chiral even though it contains K_7 which is.

Theorem 51.2.3. Flapan, [2] Every non-planar graph γ with no order 2 automorphism is intrinsically chiral in S^3.

Recall that an automorphism of a graph is an isomorphism of the graph with itself so vertices are

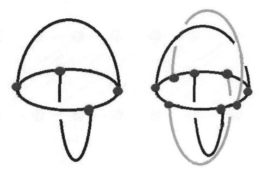

FIGURE 51.6: Rotate either of the pictured Möbius graphs 90 degrees and then reflect through the horizontal circle to see the embedding of the graph is achiral. The same rotation and reflection trick works for higher even numbers, but it is slightly harder to visualize.

mapped to vertices and edges to edges (but, of course, the vertices need not be fixed and may be permuted by the isomorphism).

51.3   Chirality in Other 3-Manifolds

Since chirality can be defined for graphs embedded in a 3-manifold other than S^3 or $\mathbb{R}^3$, it is natural to ask whether a graph which is intrinsically chiral in S^3 would necessarily be intrinsically chiral in other 3-manifolds.

Given a 3-manifold M, we say that an abstract graph γ is *intrinsically chiral in M*, if there is no embedding Γ of γ in M such that there exists an orientation-reversing homeomorphism h of M leaving Γ setwise invariant. We observe that, of course, if there is no orientation-reversing homeomorphism of M, then every graph is trivially intrinsically chiral, so for the rest of the chapter we consider only manifolds which do have an orientation-reversing homeomorphism.

We note that in [5] Flapan, Howards, Lawrence, and Mellor show that if a graph is intrinsically linked in S^3, then it is intrinsically linked in every closed 3-manifold. Again the parallel to intrinsic chirality is not perfect as we see below.

For chirality we do see the following parallel.

Theorem 51.3.1. Flapan, Yu [8]: Let M be an orientable 3-manfold with an orientation-reversing homeomorphism. If a graph γ has an achiral embedding in S^3, then γ has an achiral embedding in M.

However, Flapan and Howards show that no graph can be intrinsically chiral in every 3-manifold, or even in every "nice" 3-manifold.

Theorem 51.3.2. Flapan, Howards [6]: For every graph γ, there are infinitely many closed, con-

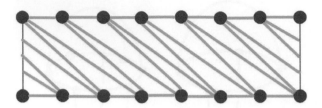

FIGURE 51.7: The rectangle above represents a torus with the opposite sides identified so the four vertices in the corner represent the same point and each of the other six vertices on the top edge is identified with the vertex directly below it. The green line segments are edges showing yielding a complete graph on seven vertices embedded on a torus.

nected, orientable, irreducible 3-manifolds M such that some embedding of γ in M is pointwise fixed by an orientation-reversing involution of M.

This, of course, implies that every graph fails to be intrinsically chiral in some manifold even the ones that are intrinsically chiral in S^3 (so intrinsic chirality isn't completely intrinsic). The proof is by construction and we can sketch it relatively quickly by skipping some technical details.

Recall that the *genus* of a graph is the least genus surface a graph can be embedded on. K_7, for example is genus 1. This is shown since Figure 51.7 gives K_7 embedded on a torus (viewed as a rectangle with opposite edges identified), but K_7 is not planar and thus cannot be embedded on a sphere confirming it must be genus exactly 1.

Let γ be a graph of genus g and embed it on the boundary of a handlebody H of genus g. Doubling H gives the graph in a closed 3-manifold M such that reflection across the boundary of the handlebody fixes the graph pointwise. Unfortunately, M is not irreducible but one can use a simple cut-and-paste argument on the interior of H to get a new manifold H' with a copy of γ in its boundary and doubling H' gives the graph embedded in a new closed, irreducible manifold M'. There are an infinite number of ways to do the cut and paste (simply corresponding to pasting in different knot complements) yielding an infinite number of distinct manifolds and proving the theorem.

We now turn to one final theorem that contrasts with Theorem 51.3.2. Given any manifold M, a constant n_M is defined strictly based on M and this constant is used in the statement of the theorem below.

Theorem 51.3.3. (Flapan, Howards) For every closed, connected, orientable, irreducible 3-manifold M, any graph with no order 2 automorphism which has a 3-connected minor λ with genus(λ) > n_M is intrinsically chiral in M.

Tools used to prove this as well as many of the other results in this chapter include the work of Jaco, Shalen, [9] and Johannson, [11], Thurston [13], Mostow [12], and Waldhausen [15], [14].

51.4 Bibliography

[1] Conway, John; Gordon, Cameron: Knots and links in spatial graphs. *J. Graph Theory* 7 (1983), no. 4, 445–453.

[2] E. Flapan, Rigidity of graph symmetries in the 3-sphere, *Journal of Knot Theory and Its Ramifications* **4**, (1995), 373–388.

[3] E. Flapan, Symmetries of Möbius ladders, *Mathematische Annalen* **283** (1989), 271–283.

[4] E. Flapan, Intrinsic chirality, *Journal of Molecular Structure (Theochem)* **336** (1995), 157–164.

[5] E. Flapan, H. Howards, D. Lawrence and B. Mellor, Intrinsic linking and knotting in arbitrary 3-manifolds, *Algebr. Geom. Topol.* 6 (2006), 1025-1035.

[6] E. Flapan and H. Howards Intrinsic chirality of graphs in 3-manifolds,*Communications in Analysis and Geometry* 26 no. 6 (2018), 1223–1250.

[7] E. Flapan and N. Weaver, Intrinsic chirality of complete graphs, *Proc. AMS* **115** (1992), 233–236.

[8] E. Flapan and S. Yu, Symmetries of spatial graphs in homology spheres, Preprint

[9] W. Jaco and P. Shalen, *Seifert fibred spaces in 3-manifolds*, *Memoirs Amer. Math. Soc.* **220**, Amer. Math. Soc., Providence (1979).

[10] B. Jiang and S. Wang, Achirality and planarity, Communications in Contemporary Mathematics **2**, (2000), 299–305.

[11] K. Johannson, *Homotopy equivalences of 3–manifolds with boundaries*, *Lecture Notes in Mathematics* **761**, Springer-Verlag, New York, Berlin, Heidelberg (1979).

[12] G. Mostow, Strong rigidity of locally symmetric spaces, *Annals of Mathematics Studies* **78**, Princeton University Press, Princeton, NJ, (1973).

[13] W. Thurston, *Three-dimensional manifolds, Kleinian groups and hyperbolic geometry*, *Bull. Amer. Math. Soc.* **6**, (1982), 357–381.

[14] F. Waldhausen, Eine Klasse von 3-dimensionaler Mannigfaltigkeiten I, *Inventiones* **3**, (1967), 308–333.

[15] F. Waldhausen, On irreducible 3-manifolds which are sufficiently large, *Ann. Math.* **87**, (1968), 56–88.

Chapter 52

Symmetries of Graphs Embedded in S^3 and Other 3-Manifolds

Erica Flapan, Pomona Collage

52.1   Molecular Symmetries

The study of symmetries of graphs embedded in $\mathbb{R}^3$ was developed in response to questions from chemists about symmetries of non-rigid molecules. Molecular symmetries are chemically important for many reasons, including their role in identifying molecular structures that are too small to see with electron microscopy, their role in predicting reactions, and their role in classifying molecules.

Chemists normally focus on the group of rigid symmetries of a molecule. In particular, the *point group* of a molecule is its group of rotations, reflections, and reflections composed with rotations. It is called the *point group* because the group fixes a point of $\mathbb{R}^3$. But not all molecules are rigid. For example, Figure 52.1 illustrates a molecule which has three symmetrically positioned chlorines which spin while the rest of the molecule is fixed. This spinning at the top of the molecule cannot be achieved by a rigid motion of the entire molecule. In fact, a reflection in the plane of the paper is the only rigid symmetry of this molecule, and hence its point group is $\mathbb{Z}_2$. Thus in this case, the point group does not represent all of the symmetries of this molecule.

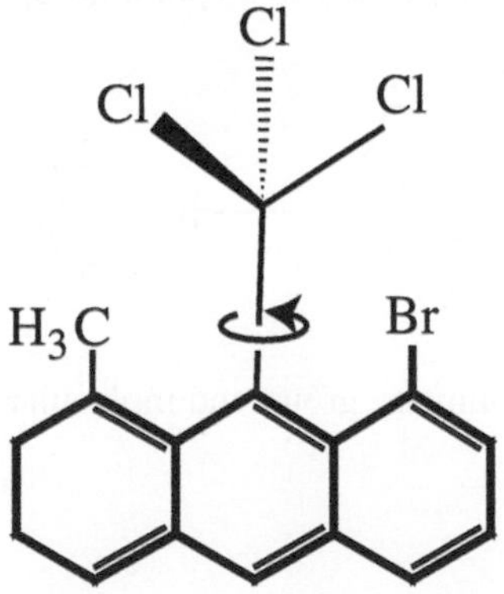

FIGURE 52.1: This molecule has three chlorines at the top that spin while the rest of the molecule is fixed.

In order to define the molecular symmetry group so that it includes both rigid and non-rigid symmetries, we focus on the automorphisms induced by molecular motions rather than the motions themselves. In particular, we have the following definition.

Definition 52.1.1. An *automorphism* of an abstract graph γ is a permutation of the vertices preserving adjacency. For molecular graphs, automorphisms must also preserve types of atoms (e.g., oxygens go to oxygens, carbons go to carbons). The *molecular symmetry group* of a molecule Γ is the subgroup of the automorphism group $\text{Aut}(\Gamma)$ induced by chemically achievable motions together with reflections taking Γ to itself.

For example, if we label the chlorines in Figure 52.1 with the numbers 1, 2, and 3, then the molecular symmetry group of this molecule is given by $\langle(123),(12)\rangle$. This group is isomorphic to the dihedral group D_3 with 6 elements.

Most small molecules are rigid, but some (like the one in Figure 52.1) have parts which rotate around certain bonds, and large molecules can be rather flexible. On the extreme end of this, DNA is quite flexible. However, even molecules which are much smaller than DNA can be somewhat flexible. We cannot give a purely mathematical definition of the molecular symmetry group, since which motions are chemically achievable depends on the chemical properties of a molecule, not just on its geometry or topology. While the point group treats molecules as rigid, we now consider the *topological symmetry group* which treats molecules as flexible.

Definition 52.1.2. The *topological symmetry group* $\text{TSG}(\Gamma)$ of a molecular graph Γ is the subgroup of $\text{Aut}(\Gamma)$ induced by homeomorphisms of $\mathbb{R}^3$ leaving Γ setwise invariant and taking atoms of a given type to atoms of the same type.

For example, the molecular Möbius ladder illustrated on the left in Figure 52.2 is rather flexible. In order to define its topological symmetry group we label the vertices with the numbers 1 through 6, as illustrated in the right. The homeomorphism which turns the graph over induces the automorphism $(23)(56)(14)$. Rotating the graph around a vertical axis by $120°$ and then twisting the ladder so that the crossing will go back to being in the front, induces the automorphism (123456). These two automorphisms together generate the topological symmetry group of the Möbius ladder. Since both of these actions are chemically possible, the topological symmetry group and molecular symmetry group of the molecular Möbius ladder are $\langle(23)(56)(14),(123456)\rangle = D_6$. However, $(23)(56)(14)$ is its only rigid symmetry in $\mathbb{R}^3$, so its point group is $\mathbb{Z}_2$.

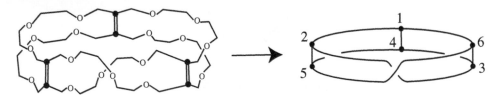

FIGURE 52.2: The topological symmetry group and molecular symmetry group of this molecule are $\langle(23)(56)(14),(123456)\rangle = D_6$.

52.2 Topological Symmetry Groups

While the original motivation for studying the topological symmetry group came from chemistry, in fact, we can consider the topological symmetry group of any graph embedded in S^3 or in any

3-manifold. In this sense, topological symmetry groups are a generalization of the study of symmetries of knots and links.

Definition 52.2.1. The *topological symmetry group* TSG(Γ, M) of a graph Γ embedded in a 3-manifold M is the subgroup of Aut(Γ) induced by homeomorphisms of (M, Γ). If we only consider orientation-preserving homeomorphisms then we obtain the *orientation-preserving topological symmetry group* TSG$_+(\Gamma, M)$.

In 1939, Robert Frucht [7] proved that every finite group occurs as the automorphism group of some connected graph. It is natural to ask what finite groups can occur as topological symmetry groups of graphs embedded in S^3. In Figure 52.3, we give examples to show that every finite abelian group and every symmetric group can occur as the topological symmetry group of a graph embedded in S^3. In particular, the embedded graph Γ_1 on the left has three wheels which can each rotate independently, and no two of the wheels can be interchanged. Also because the trefoil knots on the spokes of one wheel are chiral, no orientation-reversing homeomorphism of S^3 can take Γ_1 to itself. Thus TSG$(\Gamma_1, S^3) = $ TSG$_+(\Gamma_1, S^3) = \mathbb{Z}_2 \times \mathbb{Z}_2 \times \mathbb{Z}_4$. For the embedded graph Γ_2 on the right, every transposition (ij) can be achieved by interchanging two strands going between v and w. Each of the strands with numbered vertices contain the non-invertible knot 8_{17} and hence vertices v and w cannot be interchanged. Finally, because of the trefoil knot on the edge on the left, no orientation-reversing homeomorphism of S^3 can take the graph to itself. Thus TSG$(\Gamma_2, S^3) = $ TSG$_+(\Gamma_2, S^3) = S_n$, the symmetric group on n elements.

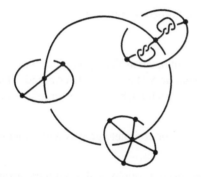 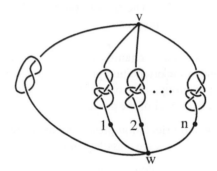

FIGURE 52.3: The graph on the left has TSG$(\Gamma_1, S^3) = \mathbb{Z}_2 \times \mathbb{Z}_2 \times \mathbb{Z}_4$ while the one on the right has TSG$(\Gamma_2, S^3) = S_n$.

In contrast with the above examples, the following theorem shows that not all finite groups can occur as topological symmetry groups.

Theorem 52.2.2 ([5])**.** TSG(Γ) can be the alternating group A_n for some graph Γ embedded in S^3 if and only if $n \leq 5$.

The following theorem classifies exactly which finite groups can occur as orientation-preserving topological symmetry groups for 3-connected graphs (i.e., those where at least three vertices together with their incident edges need to be removed to disconnect the graph or reduce it to a single vertex) in S^3. This theorem follows from [5] together with the proof of the Geometrization Conjecture [11].

Theorem 52.2.3. Let Γ be a 3-connected graph embedded in $\mathbb{R}^3$, and let $G = $ TSG$_+(\Gamma)$. Then Γ can be re-embedded in $\mathbb{R}^3$ so that G is induced by a finite group of orientation-preserving isometries of S^3.

This result is somewhat surprising given that, in general, the homeomorphisms that induce $TSG_+(\Gamma)$ need not have finite order. We can see this in Figure 52.3 as well as for the 3-connected embedded graph illustrated in Figure 52.4. In particular, in Figure 52.4, the automorphism (153426) is induced by slithering the graph along itself then interchanging the grey and black figure-eight knots. However, no finite-order homeomorphism of S^3 can induce (153426) because no order 3 homeomorphism of S^3 takes a figure-eight knot to itself [9, 13], and by the proof of the Smith Conjecture [12] a knot cannot be the fixed point set of a finite-order homeomorphism.

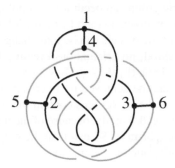

FIGURE 52.4: The automorphism (153426) is induced by a homeomorphism but not by a finite-order homeomorphism.

For some specific families of graphs, all possible groups that can occur as the topological symmetry group or the orientation-preserving topological symmetry group in S^3 have been determined. These lists of groups are complicated, so we give the references for each such family of graphs, rather than listing all of the groups.

- All orientation-preserving topological symmetry groups for complete graphs with at least 7 vertices are classified in [4].

- All topological symmetry groups and orientation-preserving topological symmetry groups for complete graphs with less than 7 vertices are classified in [1].

- All orientation-preserving topological symmetry groups for Möbius ladders with $n \geq 3$ rungs are classified in [3].

- For the bipartite graphs $K_{n,n}$, all n are determined such that some embedding of $K_{n,n}$ has A_4, S_4, or A_5 as its orientation-preserving topological symmetry group [10]. In [8], all n are determined such that some embedding of $K_{n,n}$ has an orientation-preserving topological symmetry group which is cyclic, dihedral, the product of two cyclic groups, or $(\mathbb{Z}_r \times \mathbb{Z}_s) \ltimes \mathbb{Z}_2$.

Next we consider graphs embedded in other 3-manifolds. We see from the following theorem that no closed, connected, orientable, irreducible 3-manifold has the property that every finite group occurs as the topological symmetry group of some graph embedded in the manifold.

Theorem 52.2.4 ([6]). *For every closed, connected, orientable, irreducible 3-manifold M, there exists a finite simple group G which is not isomorphic to $TSG(\Gamma, M)$ for any graph Γ embedded in M.*

On the other hand, if we allow the manifold M to vary then we can obtain every finite group. In fact, we can even put restrictions on both the manifold and the graph and still get every finite group, as we see in the following result.

Theorem 52.2.5 ([6])**.** For every finite group G, there is a hyperbolic rational homology sphere M and a 3-connected graph Γ embedded in M such that $\mathrm{TSG}(\Gamma, M) \cong G$.

52.3 Mapping Class Groups

Topological symmetry groups are not the only way to classify the symmetries of embedded graphs. We define the mapping class group of a graph embedded in a 3-manifold as follows.

Definition 52.3.1. The *mapping class group* $\mathrm{MCG}(\Gamma, M)$ of a graph Γ embedded in a 3-manifold M is the group of isotopy classes of (M, Γ). If we only consider orientation-preserving homeomorphisms, then we obtain the *orientation-preserving mapping class group* $\mathrm{MCG}_+(\Gamma, M)$.

Cho and Koda [2] have explored the relationship between the mapping class group and the topological symmetry group for graphs in S^3. In particular, they proved the following theorem showing that for many graphs Γ embedded in S^3, we have $\mathrm{MCG}_+(\Gamma, S^3) \cong \mathrm{TSG}_+(\Gamma, S^3)$ and $\mathrm{MCG}(\Gamma, S^3) \cong \mathrm{TSG}(\Gamma, S^3)$.

Theorem 52.3.2. [2] Let Γ be a graph embedded in S^3. Then the following hold.

- $\mathrm{MCG}_+(\Gamma, S^3) \cong \mathrm{TSG}_+(\Gamma, S^3)$ if and only if $\mathrm{MCG}_+(\Gamma, S^3)$ is a finite group.

- If Γ is 3-connected, then $\mathrm{MCG}_+(\Gamma, S^3) \cong \mathrm{TSG}_+(\Gamma, S^3)$ if and only if $S^3 - \Gamma$ contains no incompressible tori.

- If Γ is not a knot and $\mathrm{MCG}_+(\Gamma, S^3) \cong \mathrm{TSG}_+(\Gamma, S^3)$, then $\mathrm{MCG}(\Gamma, S^3) \cong \mathrm{TSG}(\Gamma, S^3)$ if and only if Γ is not planar.

For a graph embedded in an arbitrary 3-manifold M, the relationship between the mapping class group and the topological symmetry group has not yet been explored.

Acknowledgments

The author was supported in part by NSF Grant DMS-1607744.

52.4 Bibliography

[1] Dwayne Chambers and Erica Flapan. Topological symmetry groups of small complete graphs. *Symmetry*, 6(2):189–209, 2014.

[2] Sangbum Cho and Yuya Koda. Topological symmetry groups and mapping class groups for spatial graphs. *Michigan Math. J.*, 62(1):131–142, 2013.

[3] Erica Flapan and Emille Davie Lawrence. Topological symmetry groups of Möbius ladders. *J. Knot Theory Ramifications*, 23(14):1450077, 13, 2014.

[4] Erica Flapan, Blake Mellor, Ramin Naimi, and Michael Yoshizawa. Classification of topological symmetry groups of K_n. *Topology Proc.*, 43:209–233, 2014.

[5] Erica Flapan, Ramin Naimi, James Pommersheim, and Harry Tamvakis. Topological symmetry groups of graphs embedded in the 3-sphere. *Comment. Math. Helv.*, 80(2):317–354, 2005.

[6] Erica Flapan and Harry Tamvakis. Topological symmetry groups of graphs in 3-manifolds. *Proc. Amer. Math. Soc.*, 141(4):1423–1436, 2013.

[7] R. Frucht. Herstellung von Graphen mit vorgegebener abstrakter Gruppe. *Compositio Math.*, 6:239–250, 1939.

[8] Kathleen Hake, Blake Mellor, and Matt Pittluck. Topological symmetry groups of complete bipartite graphs. *Tokyo J. Math.*, 39(1):133–156, 2016.

[9] Richard Hartley. Knots with free period. *Canad. J. Math.*, 33(1):91–102, 1981.

[10] Blake Mellor. Complete bipartite graphs whose topological symmetry groups are polyhedral. *Tokyo J. Math.*, 37(1):135–158, 2014.

[11] John Morgan and Gang Tian. *The Geometrization Conjecture*, volume 5 of *Clay Mathematics Monographs*. American Mathematical Society, Providence, RI; Clay Mathematics Institute, Cambridge, MA, 2014.

[12] John W. Morgan and Hyman Bass, editors. *The Smith Conjecture*, volume 112 of *Pure and Applied Mathematics*. Academic Press, Inc., Orlando, FL, 1984. Papers presented at the symposium held at Columbia University, New York, 1979.

[13] H. F. Trotter. Periodic automorphisms of groups and knots. *Duke Math. J.*, 28:553–557, 1961.

Chapter 53

Invariants of Spatial Graphs

Blake Mellor, Loyola Marymount University

53.1 Introduction

The fundamental problem in knot theory is the classification of knots; in other words, the classification up to isotopy of all embeddings of S^1 in S^3. Similarly, the fundamental problem in the theory of spatial graphs is, for any given graph G, the classification up to isotopy of the embeddings of G in S^3. In this sense, knot theory is simply a special case of the theory of spatial graphs, since a knot (resp. link) can be viewed as a cycle graph (resp. disjoint union of cycle graphs).

The two tasks in such a classification are determining when two embeddings are isotopic, and when they are not. Both tasks are generally difficult. As with knots, two spatial graphs are isotopic if and only if their diagrams are equivalent modulo a set of *Reidemeister moves*, shown in Figure 53.1; however, finding the specific sequence of moves between two equivalent diagrams can be extremely challenging. In fact, there are two slightly different notions of isotopic that are commonly considered. Moves (I)–(V) in Figure 53.1 generate *rigid (or flat) vertex isotopy*, in which the cyclic order of the edges at each vertex is fixed. Including move (VI) generates *pliable vertex isotopy*, or simply *isotopy*, in which the order of the edges around a vertex may be changed.

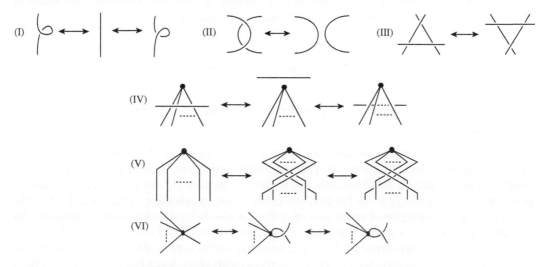

FIGURE 53.1: Reidemeister moves for spatial graphs.

Conversely, to determine that two embeddings are *not* isotopic requires an *invariant*—a function of the embeddings whose output is not changed by isotopies, and which takes different values on the two embeddings. Over the last century, many such invariants have been defined and studied for knots; in this chapter, we will look at some ways in which those ideas have been extended to other spatial graphs, with a particular emphasis on combinatorial and polynomial invariants. In this chapter, we will mostly discuss invariants of pliable vertex isotopy, though we will discuss one important invariant of rigid vertex isotopy, the Yamada polynomial.

53.2 Knots in Graphs

If a graph has no cycles (i.e. the graph is a tree), then all of its embeddings in S^3 are isotopic, so in any case of interest, the graph will have cycles. In the embedded graph, each of these cycles becomes a knot (possibly a trivial knot), and any collection of pairwise disjoint cycles becomes a link (possibly a trivial link). So to each embedding G of a graph we can associate a collection $T(G)$ of knots and links (with multiplicity, as a particular knot type may appear many times). More precisely, at each vertex, choose two edges to connect, and delete the others; making this choice at every vertex leaves some number (possibly 0) of closed loops that form a link. The set of all such links, for all choices of two edges at each vertex, is $T(G)$. Kauffman [15] first observed that $T(G)$ is an isotopy invariant of G, and that many other (more easily comparable) invariants can be defined by applying our favorite knot and link invariants to the elements of $T(G)$. Conway and Gordon's famous result that every embedding of the complete graph on seven vertices contains a knot uses this perspective—they consider the sum of the Arf invariants of all Hamiltonian cycles in the embedded graph (mod 2) and show that this invariant is always nonzero [3].

As an example, consider the θ-graph in Figure 53.2 (this is θ-graph 5_4 in Moriuchi's table [22]). In this case $T(G)$ is the collection of *constituent knots* formed by selecting two of the three edges of the graph; these are a (2,5) torus knot, a trefoil knot, and an unknot.

53.3 The Fundamental Group and the Alexander Polynomial of a Spatial Graph

Perhaps the most important knot invariant is the *fundamental group* of the knot (i.e. the fundamental group of the knot complement). The fundamental group can be used to define many other invariants, most famously the *Alexander polynomial* of the knot [1]. Similarly, the *fundamental group of a spatial graph G* is the fundamental group of the complement of the embedded graph, which is invariant under pliable vertex isotopy, and it can also be used to define other invariants. An Alexander polynomial for spatial graphs was first defined by Kinoshita [16]. As with the Alexander polynomial of knots, the polynomial for spatial graphs arises from the first homology of the infinite cyclic cover of the graph complement. However, also as with knots, it can be easily computed from the coefficient matrix for a system of linear relations that we can read off from the diagram of the spatial graph.

As with knots, the computation of the fundamental group for a spatial graph depends on an *orien-*

$G =$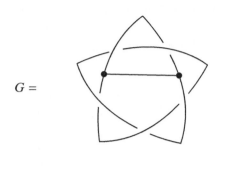

$\longrightarrow \quad T(G) =$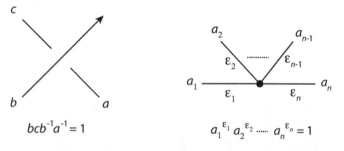

FIGURE 53.2: $T(G)$ for a θ-graph.

tation of the graph—i.e., an orientation of each edge of the graph. For an oriented spatial graph G, the fundamental group $\pi_1(S^3 - G)$ has a Wirtinger presentation constructed from a diagram for the graph, where the generators correspond to the arcs in the diagram. The presentation has a relation at each crossing and vertex in the diagram, as shown in Figure 53.3. At a vertex, the *local sign* ε_i of arc a_i is $+1$ if the arc is directed into the vertex, and -1 if the arc is directed out from the vertex.

$$bcb^{-1}a^{-1} = 1 \qquad\qquad a_1^{\varepsilon_1} a_2^{\varepsilon_2} \cdots a_n^{\varepsilon_n} = 1$$

FIGURE 53.3: Wirtinger relations for $\pi_1(S^3 - G)$.

The Alexander polynomial is defined for a *balanced* oriented spatial graph. This means each edge is given an integral *weight* so that the oriented sum of weights at each vertex is zero. Then we get a linear relation at each crossing and vertex of a diagram of the graph, where the variables are the arcs of the diagram as shown in Figure 53.3 (note that the orientation of the undercrossing arc on the left is not specified, since either orientation leads to the same relation). In the crossing relation we will assume arc b has weight w_1, and arcs a and c (on the same edge) have weight w_2. In the vertex relation, arc a_i has weight w_i. Let $m_i = \varepsilon_1 w_1 + \varepsilon_2 w_2 + \cdots + \varepsilon_{i-1} w_{i-1} + \min\{\varepsilon_i, 0\} w_i$. Then the *Alexander relations* are:

Crossing relation: $(1 - t^{w_2})b + t^{w_1} c - a = 0$ and

Vertex relation: $\sum_{i=1}^{n} \varepsilon_i t^{m_i} a_i = 0$.

The Alexander matrix is the coefficient matrix for this system of linear relations. For knots, there are only crossing relations, and the Alexander matrix is a square $r \times r$ matrix. The Alexander polynomial is then any $(r - 1) \times (r - 1)$ minor of the matrix. For graphs, however, the matrix is generally *not* square; rather, it is an $r \times s$ matrix with $s \geq r$. In this case the Alexander polynomial is the greatest common divisor of the $(r-1) \times (r-1)$ minors of the matrix. As for knots, the Alexander polynomial is well defined only up to multiplication by t^k.

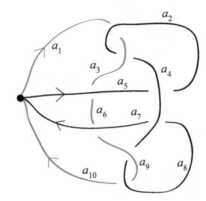

FIGURE 53.4: A spatial graph diagram D.

As an example, consider the spatial graph G with diagram D shown in Figure 53.4, with the edges given weights x and y (the edge containing arc a_1 has weight x). Then the Alexander matrix is shown below (where the columns correspond to the arcs $a_1, \ldots, a_{10}$ and the last row is the vertex relation):

$$
\begin{bmatrix}
-1 & 1 - t^x & t^y & 0 & 0 & 0 & 0 & 0 & 0 & 0 \\
0 & -1 & 1 - t^y & t^x & 0 & 0 & 0 & 0 & 0 & 0 \\
0 & 0 & t^y & 0 & 1 - t^x & -1 & 0 & 0 & 0 & 0 \\
0 & t^y & 0 & 1 - t^y & -1 & 0 & 0 & 0 & 0 & 0 \\
0 & 0 & 0 & 0 & 0 & -1 & 1 - t^x & 0 & t^y & 0 \\
0 & 0 & 0 & 1 - t^y & 0 & 0 & -1 & t^y & 0 & 0 \\
0 & 0 & 0 & t^x & 0 & 0 & 0 & -1 & 1 - t^y & 0 \\
0 & 0 & 0 & 0 & 0 & 0 & 0 & 1 - t^x & t^y & -1 \\
-t^{-x} & 0 & 0 & 0 & -t^{-x-y} & 0 & t^{-x-y} & 0 & 0 & t^{-x}
\end{bmatrix}
$$

If both edges are given weight 1, then the Alexander polynomial is $\Delta(G)(t) = t^2 - 2t + 2$ (normalized so that the lowest term is the constant term).

Since the fundamental group of a spatial graph is a topological invariant of the exterior, we can contract a spanning tree of the graph without changing its fundamental group. For example, the three graphs in Figure 53.5 all have isomorphic fundamental groups, since G_1 and G_2 are each the result of contracting one edge of the θ-graph G. The Alexander polynomial, however, also depends on the weights on the edges. For example, if the two arcs in G_1 are given weight 1, the Alexander polynomial is $\Delta(G_1)(t) = t^2 - t + 1$. But if we give both arcs of G_2 weight 1, the Alexander polynomial is $\Delta(G_2)(t) = t^4 - t^2 + 1$. The reason for the difference is that these weightings result from different weightings on G; if we try to give all three arcs of G a weight of 1, then the weighting is not balanced. On the other hand, if we give arcs a and b weight 1 and arc c weight 2, then G, G_1 and G_2 will all have an Alexander polynomial of $t^4 - t^2 + 1$.

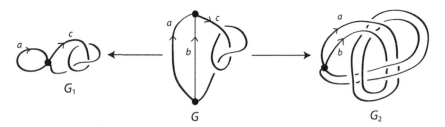

FIGURE 53.5: Contracting edges of a θ-graph.

Further invariants can be derived from the Alexander matrix, such as the *determinant* of the spatial graph (which is calculated by substituting -1 for t in the Alexander matrix, and then computing the greatest common divisor of the $(r - 1) \times (r - 1)$ minors), or whether the graph is *p-colorable* [20, 21]. Litherland [18] defined a more refined, and more powerful, Alexander polynomial for θ_n-graphs (graphs with two vertices, and n edges connecting them). In his treatment, the edges of the graph were all oriented the same way, and there were no weights on the edges (though the edges are ordered).

While the Alexander polynomial of a spatial graph is constructed similarly to the Alexander polynomial of a knot, it is not as well understood. For example, it is well known which polynomials can be realized as the Alexander polynomial of a knot; for spatial graphs this question is still open. Also, the Alexander polynomial satisfies a nice *skein relation* (particularly when normalized to give the *Alexander-Conway polynomial*) which provides an alternative method of computation. We do not know of a similar skein relation for the Alexander polynomial of a spatial graph.

There are other algebraic invariants for spatial graphs as well. For example, Ghuman [9] defined a *longitude* of a cycle in a graph, and used them to define invariants analogous to Milnor's $\bar{\mu}$-invariants for links. However, these are only isotopy invariants if the longitude is unique, which occurs when all vertices on the cycle have valence 3; where they are defined, however, they can contain more information than the fundamental group. Fleming [7] generalized Milnor's invariants in another way, looking at "links" of spatial graphs with multiple connected components.

53.4 The Fundamental Quandle of a Spatial Graph

Another important algebraic object we can associate to a knot or spatial graph is the *fundamental quandle*. The *fundamental quandle* of a knot was introduced in the early 1980s by Joyce [14].

Like the fundamental group, the fundamental quandle is defined by Wirtinger-type relations at the crossings. However, Joyce proved that the fundamental quandle, unlike the fundamental group, is a complete knot invariant (of course, since nothing is free, quandles are much harder to compare than groups!). Niebrzydowski [25] extended the fundamental quandle to spatial graphs. Though it seems unlikely that the fundamental quandle of a spatial graph is a complete invariant, as it is for knots, it can be used to define many useful invariants of a graph [11, 13, 25, 27].

A *quandle* is a set Q equipped with two binary operations $\rhd$ and $\rhd^{-1}$ that satisfy the following three axioms:

1. $x \rhd x = x$ for all $x \in Q$

2. $(x \rhd y) \rhd^{-1} y = x = (x \rhd^{-1} y) \rhd y$ for all $x, y \in Q$

3. $(x \rhd y) \rhd z = (x \rhd z) \rhd (y \rhd z)$ for all $x, y, z \in Q$

The operation $\rhd$ is, in general, not associative. It is useful to adopt the exponential notation introduced by Fenn and Rourke in [5] and denote $x \rhd y$ as x^y and $x \rhd^{-1} y$ as $x^{\bar{y}}$. With this notation, x^{yz} will be taken to mean $(x^y)^z = (x \rhd y) \rhd z$ whereas x^{y^z} will mean $x \rhd (y \rhd z)$. If n is an integer, we will also let $x^{y^n} = x^{yy\cdots y}$, where the y is repeated n times.

Given a diagram for a spatial graph G, the fundamental quandle $Q(G)$ is the quandle whose generators are the arcs of the diagram, with additional Wirtinger-type relations at each crossing and vertex. Using the labels on the arcs in Figure 53.3, the additional relations are:

4. $a^b = c$, and

5. For *every* generator x of the quandle, $x^{a_1^{\varepsilon_1} \cdots a_n^{\varepsilon_n}} = x$, where $a_i^{\varepsilon_i} = a_i$ if $\varepsilon_i = +1$ and $a_i^{\varepsilon_i} = \overline{a_i}$ if $\varepsilon_i = -1$.

Relation (4) is the same as for the fundamental quandle of a knot; relation (5) ensures invariance under Reidemeister moves (IV)–(VI), so the quandle is an invariant of pliable vertex isotopy.

Given a quandle Q, there is an *associated group* $As(Q)$, obtained by interpreting the quandle operation as conjugation (i.e. $x^y = y^{-1}xy$). For a knot, the associated group of the fundamental quandle is isomorphic to the fundamental group. However, for spatial graphs, these groups are generally not isomorphic (though there is always an epimorphism from the associated group of the fundamental quandle to the fundamental group). In particular, while the abelianization of the fundamental group of a knot is always isomorphic to $\mathbb{Z}$, the abelianization of $As(Q(G))$ is equal to $\mathbb{Z}^E$, where E is the number of edges in the graph. So two cycles with different numbers of edges will be distinguished, even if their fundamental groups are isomorphic.

A particularly rich source of invariants derived from the fundamental quandle are *quandle colorings*. A coloring of a diagram D of a spatial graph G by a quandle X is an assignment of elements of X to the arcs of D so that relations (4) and (5) are satisfied at each crossing and vertex. Essentially, this is a quandle homomorphism from $Q(G)$ to X. For a fixed X, the number of quandle colorings is an invariant [25], and other coloring invariants of spatial graphs (such as those in [12, 20, 21]) can be interpreted in terms of quandles. These also lead to quandle homology and cohomology invariants [11]. There is still much room to extend the work that has been done on knot quandles to spatial graphs.

53.5 The Yamada Polynomial

In the 1980's, knot theorists discovered a family of new knot invariants, including the Jones, Kauffman and HOMFLY polynomials. Like the Alexander polynomial, these invariants satisfied nice skein relations which make them relatively easy to compute—several of them, however, are defined entirely in terms of this skein relation, rather than derived from a deeper topological invariant such as the fundamental group. The proof of invariance then relies on using the skein relation to show the value of the invariant is unchanged by Reidemeister moves.

This inspired attempts to construct similar, combinatorially defined, polynomial invariants for spatial graphs. The best-known of these is the Yamada polynomial [35], which can be defined as the unique polynomial $R(G)(A)$ which satisfies the following axioms:

1. $R\left(\times\right) = AR\left(\right)\left(\right) + A^{-1}R\left(\asymp\right) + R\left(\times\right).$

2. $R(G) = R(G - e) + R(G/e)$, where e is a nonloop edge in G, $G - e$ is the result of deleting e, and G/e is the result of contracting e.

3. $R(G_1 \sqcup G_2) = R(G_1)R(G_2)$, where $\sqcup$ denotes disjoint union.

4. $R(G_1 \vee G_2) = -R(G_1)R(G_2)$, where $G_1 \vee G_2$ is the graph obtained by joining G_1 and G_2 at any single vertex.

5. $R(B_n) = -(-\sigma)^n$, where B_n is the n-leafed bouquet of circles and $\sigma = A + 1 + A^{-1}$. In particular, if G is a single vertex, $R(G) = R(B_0) = -1$.

6. $R(\emptyset) = 1$.

Using these skein relations, $R(G)$ can be computed by reducing G to combinations of bouquets of circles. $R(G)$ is an invariant of spatial graphs up to *regular rigid vertex isotopy*, meaning that it is invariant under moves (II), (III) and (IV) in Figure 53.1, but not moves (I), (V) or (VI). The behavior of R under these moves is given by the following formulas:

1. $R\left(\rho\right) = A^2 R\,(\vert)$ and $R\left(\rho\right) = A^{-2}R\,(\vert)$

2. $R\left(\bigotimes\right) = (-A)^n R\left(\bigcap\right)$ and $R\left(\bigotimes\right) = (-A)^{-n} R\left(\bigcap\right)$

3. $R\left(\times\right) = -AR\left(\times\right) - (A^2 + A)R\left(\right)\left(\right)$

4. $R\left(\times\right) = -A^{-1}R\left(\times\right) - (A^{-2} + A^{-1})R\left(\right)\left(\right)$

We can obtain an invariant of rigid vertex isotopy (invariance under moves (I)–(V)) by defining $\overline{R}(G) = (-A)^{-m}R(G)$, where m is the smallest power of A in $R(G)$. This will still not be invariant under move (VI), however, so it is not an invariant of pliable vertex isotopy. The exception is when the maximum degree of the vertices of the graph is 3 or less, since a move of type (VI) on a vertex of degree 3 is equivalent to a move of type (V) followed by moves of types (IV) and (I). So if the maximum degree of the vertices is 3 or less, $\overline{R}(G)$ is an invariant of pliable vertex isotopy. If the graph G is simply a link, then $R(G)$ is a specialization of the Dubrovnik version of the Kauffman polynomial; if G is a knot, then $R(G)$ is the Jones polynomial of the $(2, 0)$-cabling of the knot [35].

The skein relations make the Yamada polynomial convenient to work with and relatively easy to compute (with the help of a computer); it is probably the most popular invariant of spatial graphs. It has been used to provide necessary conditions for a spatial graph to be *p-periodic* (i.e. symmetric under an action of $\mathbb{Z}_p$ on S^3 by homeomorphisms) [2, 19]. The Yamada polynomial has also been used to provide a lower bound on the number of crossings in a spatial graph [24]. The Yamada polynomial is often used to help develop "knot tables" for particular graphs, such as θ-graphs, handcuff graphs and bouquet graphs [4, 22, 23, 28]. For example, Figure 53.6 shows the θ-graphs 5_3 and 5_4 from Moriuchi's table [22], along with their Yamada polynomials $\overline{R}(G)$. (These graphs are also distinguished by their constituent knots, since $T(5_3)$ contains only one knot, while $T(5_4)$ contains two.) The Yamada polynomial is important enough that is has become an object of study in its own right; for example, Li, Lei, Li and Vesnin [17] have studied the distribution of zeros of the Yamada polynomials of certain classes of spatial graphs.

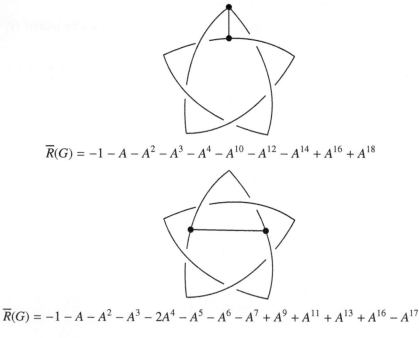

$$\overline{R}(G) = -1 - A - A^2 - A^3 - A^4 - A^{10} - A^{12} - A^{14} + A^{16} + A^{18}$$

$$\overline{R}(G) = -1 - A - A^2 - A^3 - 2A^4 - A^5 - A^6 - A^7 + A^9 + A^{11} + A^{13} + A^{16} - A^{17}$$

FIGURE 53.6: Yamada polynomials for θ-graphs 5_3 and 5_4.

The fact that the Yamada polynomial is, in general, only an invariant of rigid vertex isotopy is its primary drawback. Other polynomials are invariants of pliable vertex isotopy, but these are generally harder to compute. In addition to the Alexander polynomial discussed in the last section, the Yokota polynomial [36] is an isotopy invariant which agrees with the Yamada polynomial for graphs of constant valence 3; however, its computation is far more complex, involving transforming the graph into a large linear combination of link diagrams, and summing invariants of all the terms. The Thompson polynomial [34] is also an isotopy invariant, which can be computed recursively, but each stage of the recursion requires determining topological facts about the boundary of a three-manifold, such as whether it is compressible.

53.6   Other Invariants of Spatial Graphs

We have only scratched the surface of spatial graph invariants. We have discussed some invariants that are defined for all (or at least most) graphs, but there are many others defined for smaller classes of graphs that are still useful. For example, the Simon invariants were initially defined only for embeddings of the complete graphs K_5 and $K_{3,3}$, and yet they have turned out to be surprisingly useful [6, 26, 29]. Or, we can look at invariants for equivalence relations other than rigid vertex isotopy and pliable vertex isotopy. Many such relations have been studied for knots and links, and nearly all of these have an analogue (or even several analogues) for spatial graphs. For instance, Taniyama introduced homotopy and homology equivalence relations on spatial graphs [30, 31, 32], and many others have studied these ideas as well [8, 29]. Other ideas from knot theory, such as *finite-type invariants*, can also be extended to spatial graphs [10, 33]. We could go on, but it is enough to point out that invariants of spatial graphs have been studied for only 30 years, while invariants of knots and links have been studied for over a century; there is much to be done!

53.7   Bibliography

[1] J.W. Alexander, Topological invariants of knots and links, *Trans. Amer. Math. Soc.* (1928), pp. 275–306.

[2] N. Chbili, Skein algebras of the solid torus and symmetric spatial graphs, *Fundamenta Math.* v. 190 (2006).

[3] J. Conway and C. Gordon, Knots and links in spatial graphs, *J. of Graph Theory* v. 7 (1983), pp. 445–453.

[4] A. Dobrynin and A. Vesnin, The Yamada polynomial for graphs, embedded knot-wise into three-dimensional space, *Vychisl. Sistemy* v.155 (1996), pp. 37–86 (in Russian). An English translation is available at https://www.researchgate.net/publication/266336562.

[5] R. Fenn and C. Rourke, Racks and links in codimension two, *J. of Knot Theory and Its Ramif.* v. 1 (1992), pp. 343–406.

[6] E. Flapan, W. Fletcher and R. Nikkuni, Reduced Wu and generalized Simon invariants for spatial graphs, *Math. Proc. Camb. Phil. Soc.* v. 156 (2014), pp. 521–544.

[7] T. Fleming, Milnor invariants for spatial graphs, *Topology Appl.*, v. 155 (2008), pp. 1297–1305.

[8] T. Fleming and R. Nikkuni, Homotopy on spatial graphs and the Sato-Levine invariant, *Trans. Amer. Math. Soc.* v. 361 (2009), pp. 1885–1902.

[9] S. Ghuman, Invariants of graphs, *J. Knot Theory Ramif.* v. 9 (2000), pp. 31–92.

[10] Y. Huh and G.T. Jin, θ-curve polynomials and finite-type invariants, *J. Knot Theory Ramif.* v. 11 (2002), pp. 555–564.

[11] A. Ishii and M. Iwakiri, Quandle cocycle invariants for spatial graphs and knotted handlebodies, *Canad. J. Math.* v. 64 (2012), pp. 102–122.

[12] Y. Ishii and A. Yasuhara, Color invariant for spatial graphs, *J. Knot Theory Ramif.*, v. 6 (1997), pp. 319–325.

[13] Y. Jang and K. Oshiro, Symmetric quandle coloring for spatial graphs and handlebody-links, *J. Knot Theory Ramif.* v. 21 (2012).

[14] D. Joyce, A classifying invariant of knots, the knot quandle, *J. of Pure and Applied Algebra* v. 23 (1982), pp. 37–65.

[15] L.H. Kauffman, Invariants of graphs in three-space, *Trans. Amer. Math. Soc.* (1989), pp. 697–710.

[16] S. Kinoshita, Alexander Polynomials as Isotopy Invariants I, *Osaka Math. J.* v. 10 (1958), pp. 263–271.

[17] M. Li, F. Lei, F. Li and A. Vesnin, On Yamada polynomial of spatial graphs obtained by edge replacements, *J. Knot Theory Ramif.* v.27 (2018).

[18] R. Litherland, The Alexander module of a knotted theta-curve, *Math. Proc. Camb. Phil. Soc.* v. 106 (1989), pp. 95–106.

[19] Y. Marui, The Yamada polynomial of spatial graphs with $\mathbb{Z}_n$-symmetry, *Kobe J. Math.* v. 18 (2001), pp. 23–49.

[20] J. McAtee, D. Silver and S. Williams, Coloring spatial graphs, *J. Knot Theory Ramif.* v. 10 (2001), pp. 109–120.

[21] B. Mellor, T. Kong, A. Lewald and V. Pigrish, Colorings, determinants and Alexander polynomials for spatial graphs, *J. Knot Theory Ramif.* v. 25 (2016).

[22] H. Moriuchi, An enumeration of theta-curves with up to seven crossings, *J. Knot Theory Ramif.* v. 18 (2009), pp. 167–197.

[23] H. Moriuchi, A table of θ-curves and handcuff graphs with up to seven crossings, in *Noncommutativity and Singularities*, Adv. Stud. Pure Math., v. 55, Math. Soc. Japan, Tokyo, 2009, pp. 281–290.

[24] T. Motohashi, Y. Ohyama and K. Taniyama, Yamada polynomial and crossing number of spatial graphs, *Rev. Mat. Univ. Complut. Madrid* v. 7 (1994), pp. 247–277.

[25] M. Niebrzydowski, Coloring invariants of spatial graphs, *J. Knot Theory Ramif.* v. 19 (2010), pp. 829–841.

[26] R. Nikkuni and K. Taniyama, Symmetries of spatial graphs and Simon invariants, *Fund. Math.* v. 205 (2009), pp. 219–236.

[27] K. Oshiro, On pallets for Fox colorings of spatial graphs, *Topology and Its Applications* v. 159 (2012), pp. 1092–1105.

[28] N. Oyamaguchi, Enumeration of spatial 2-bouquet graphs up to flat vertex isotopy, *Topology and Its Appl.* v. 196 (2015), pp. 805–814.

[29] R. Shinjo and K. Taniyama, Homology classification of spatial graphs by linking numbers and Simon invariants, *Topology Appl.* v. 134 (2003), pp. 53–67.

[30] K. Taniyama, Link homotopy invariants of graphs in $\mathbb{R}^3$, *Rev. Mat. Univ. Complut. Madrid* v. 7 (1994), pp. 129–144.

[31] K. Taniyama, Cobordism, homotopy and homology of graphs in $\mathbb{R}^3$, *Topology* v. 33 (1994), pp. 509–523.

[32] K. Taniyama, Homology classification of spatial embeddings of a graph, *Topology Appl.* v. 65 (1995), pp. 205–228.

[33] K. Taniyama and A. Yasuhara, Local moves on spatial graphs and finite type invariants, *Pacific J. Math.* v. 211 (2003), pp. 183–200.

[34] A. Thompson, A polynomial invariant of graphs in 3-manifolds, *Topology* v. 31 (1992), pp. 657–665.

[35] S. Yamada, An invariant of spatial graphs, *J. Graph Theory* v. 13 (1989), pp. 537–551.

[36] Y. Yokota, Topological invariants of graphs in 3-space, *Topology* v. 35 (1996), pp. 77–87.

Chapter 54

Legendrian Spatial Graphs

Danielle O'Donnol, Marymount University

54.1 Legendrian Spatial Graphs

In this section, we will consider spatial graphs in a three-space with more structure. We will look at spatial graphs in contact manifolds. A contact manifold is three-space together with a plane field that follows certain rules. At every single point in the space there is a plane, and as you go from one point to the next, the position of the plane smoothly changes. For it to be a contact manifold there is the additional rule: There is no way to put a surface in the space so that it is everywhere tangent to the planes. For more information, see Chapter 40. We call these planes *contact planes*. While a surface cannot be everywhere tangent to the contact planes, a knot or spatial graph can. Spatial graphs that are everywhere tangent to the contact planes in a contact manifold are *Legendrian spatial graphs*, or just *Legendrian graphs*.

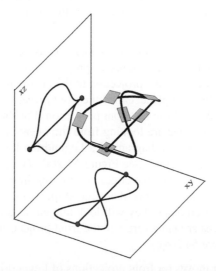

FIGURE 54.1: Legendrian θ-curve in $(\mathbb{R}^3, \xi_{std})$ with front projection (on (xz)-plane) and Lagrangian projection (on (xy)-plane). Here, the center edge of the θ-curve is along the x-axis and the negative side of the y-axis is on the right.

Legendrian graphs were first used as a tool to understand the structure of contact manifolds. The two articles 'Géométrie de contact: de la dimension trois vers les dimensions supérieures" [7], and "Topologically trivial Legendrian knots" [4] are examples of foundational work in the field of contact topology that both use Legendrian graphs. Studying Legendrian graphs in their own right started later with the article "On Legendrian graphs" [11]. In addition to exploring Legendrian graphs, there has been recent interest in Legendrian singular links [2]. These can be viewed as the special case of four-valent Legendrian graphs.

54.1.1   How We Work with Legendrian Graphs

Legendrian graphs are not rigid objects: they can move as long as the spatial graph is always Legendrian. Two Legendrian graphs are Legendrian isotopic if you can move from one to the other through Legendrian graphs. At the vertices, the edges all come from different directions; they are never tangent to each other. The cyclic order of the edges around a vertex in a contact plane is not changed as the graph moves. In other words, an edge cannot move passed another edge next to a vertex.

It is tempting to think that the cycles of a Legendrian graph are Legendrian knots, but it is a bit more complicated. Legendrian knots are smooth, but at the vertices there can be a kink as we go from one edge to the next. So a cycle is a piecewise smooth Legendrian knot, because at each vertex it is allowed to kink as it moves. It is always possible to move the graph so that a pair of edges match up to go straight through the vertex and have no kink at the vertex. Any Legendrian graph can be moved through Legendrian graphs, such that a chosen cycle is made smooth. This is called the *smoothing* of the cycle. The smoothing is unique up to Legendrian isotopy.

Much of the work on Legendrian graphs has been working with the graphs in $\mathbb{R}^3$ with what is called the *standard contact structure*, denoted ξ_{std}. The pair $\mathbb{R}^3$ with the standard contact structure will be denoted by $(\mathbb{R}^3, \xi_{std})$. For more information on the standard contact structure see Chapter 40.

Just as with knots, spatial graphs, and Legendrian knots, it is useful to work with diagrams of Legendrian graphs, that are a particularly nice choice of projection of the original object. In $(\mathbb{R}^3, \xi_{std})$, there are two projections that are used: the projection to the (xy)-plane, called the *Lagrangian projection*, and the projection to the (xz)-plane, called the *front projection*. See Figure 54.1. For more information see Chapter 40. We will use the front projection. For the front projection, we project to the (xz)-plane and imagine that we are looking from the negative side of the y-axis. The front projection is very nice because from a diagram of a Legendrian knot in the front projection, we know all of the coordinates of each point. See Figure 54.2(a). The front projection can be a little confusing at the vertices. Because all of the edges are coming into the same point, the vertex, in the front projection all the edges must have the same slope at the vertex. So while none of the edges are actually tangent to each other, they will look tangent as a result of the front projection. Additionally, the order of edges around a vertex in the front projection is not the same as the order in the contact plane. See Figure 54.2(b,c).

There is a set of Reidemeister moves for front projections of Legendrian graphs. Two generic front projections of a Legendrian graph are related by Reidemeister moves I, II and III together with two moves given by the mutual position of vertices and edges [3]. See Figure 54.3.

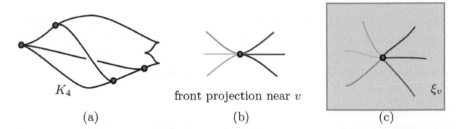

FIGURE 54.2: (a) An example of a front projection of a Legendrian K_4. (b) The front projection of a single vertex with the edges highlighted in different colors. Remember that the edges with the more negative slope are closer. (c) On the right is the same vertex with a view looking down at the contact plane. From the view of the contact plane we can see that none of the edges are tangent to one another.

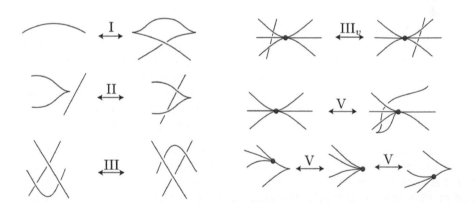

FIGURE 54.3: Legendrian graph Reidemeister moves. Reidemeister moves I, II, and III, an edge passing under or over a vertex (III_v), an edge adjacent to a vertex rotates to the other side of the vertex (V). Move V is shown with edges on both sides and on one side of the vertex. Reflections of these moves that are Legendrian front projections are also allowed.

54.1.2   Invariants

Because Legendrian graphs can move it can be hard to tell if two diagrams are of the same Legendrian graph or different Legendrian graphs. To help with this we use invariants. For more information on invariants in general see Part VI, Chapter 53, or Part XI.

For each abstract graph, there are infinitely many spatial graphs that are embeddings of that abstract graph. In a similar way, for each spatial graph there are infinitely many Legendrian graphs that have the topological type of that spatial graph. These Legendrian graphs are called *Legendrian realizations* of the spatial graph, and the spatial graph is called the *topological type* or *topological class* of the Legendrian graph. The same is true for knots.

One natural invariant for Legendrian graphs is the spatial graph type of the Legendrian graph. As a Legendrian graph moves it will not change the spatial graph type of the Legendrian graph. Another invariant that naturally arises is the cyclic order of the edges around each vertex. If two Legendrian embeddings of the same spatial graph have different cyclic ordering of the edges around any vertex, then they must be different, because we cannot move from one to the other.

A number of different invariants for Legendrian graphs have been developed. Many of these invariants are generalizations of invariants for Legendrian knots. The two classical invariants of Legendrian knots, the Thurston–Bennequin number (denoted tb) and the rotation number (denoted rot) have been generalized to Legendrian graphs [11]. The Thurston–Bennequin number measures the amount of twisting of the contact planes when you go around the knot once. The rotation number (also called the Maslov number) is approximately the winding number of the tangent vectors as they traverse the knot once. For more information see Chapter 40.

To extend tb and rot to Legendrian graphs, they are first defined for the cycles of Legendrian graphs, that is, piecewise smooth Legendrian knots. As discussed in the last section the cycles can be moved to be smooth. Let a cycle be called K and the smooth Legendrian knot that it can be moved to K_{st}. Then we can define $\text{tb}(K) = \text{tb}(K_{st})$ and $\text{rot}(K) = \text{rot}(K_{st})$ [11]. This is possible because for any piecewise smooth Legendrian knot when it is smoothed, there is only one Legendrian knot that can result. Then, given a fixed order on the cycles of a Legendrian graph G, the *Thurston–Bennequin number of* G, denoted by $\text{tb}(G)$, is the ordered list of the Thurston–Bennequin numbers of the cycles of G. In the same way, the *rotation number of* G, $\text{rot}(G)$, is the ordered list of the rotation numbers of the cycles of G. Additionally, there is the *total Thurston–Bennequin number* (TB) which is the sum of the Thurston–Bennequin numbers for each cycle in the graph [13].

Various more complex invariants have also been developed and explored. Here I will highlight these with minimal explanation. The Legendrian ribbon of the graph has been explored in a few different ways. It has been worked with as the invariant itself [8], and used to determine the transverse push-off of a Legendrian graph [12]. There has also been a DGA developed for Legendrian graphs [1].

54.1.3   The Geography Question, the Botany Question, and Stabilizations

For knots there is a relationship between all of the Legendrian realizations of a given knot type. There are two special moves that can be done which change the Legendrian type of the knot (but not the topological type) called stabilization and destabilization. A *stabilization* means replacing an arc of K in the front projection of K by one of the zig-zags as in Figure 54.4. So a straight arc is replaced with a spiral that goes around once. The stabilization is said to be positive if down cusps are introduced and negative if up cusps are introduced. If two Legendrian knots are of the same topological type, then there is a finite sequence of stabilizations and destabilizations between the two Legendrian knots [6]. Given a Legendrian knot K, stabilizing the Legendrian knots always results with a different Legendrian knot with the same topological type.

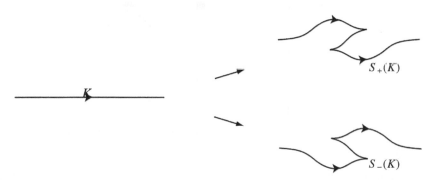

FIGURE 54.4: Positive and negative stabilizations in the front projection of a knot or edge.

We are going to focus on some questions that have to do with tb and rot. For a given knot type there are two big questions to ask. The Geography Question: Which pairs of (tb, rot) can occur for this knot type? The Botany Question: For each pair of (tb, rot) that can occur for this knot type, how many Legendrian realizations are there of that knot with the given (tb, rot)? There are examples like the unknot [4], trefoil, and figure-8 knot [5] where there is exactly one Legendrian realization of the knot for each (tb, rot) that can occur. When this happens we say that the knot is determined by its (tb, rot), because once you know the (tb, rot), then you know which Legendrian realization it is. This is also called *Legendrian simple*.

As an example, we will look at the unknot. The answer to both the Geography and Botany questions can be nicely summed up in one image for each knot type, which is its mountain range. For the mountain range, a dot is placed for each Legendrian realization of the knot on a grid where the vertical axis is the tb, and the horizontal axis is the rot. The mountain range for the unknot is shown in Figure 54.5. Each dot represents a different Legendrian realization of the knot. In our example the (tb, rot) = $(-1, 0)$ and (tb, rot) = $(-2, -1)$ unknots are shown. Since the unknot is determined by its (tb, rot), there is a single point at each (tb, rot) that can occur. However, if the knot is not Legendrian simple, there will be one or more points on the grid that have multiple dots. The lines represent stabilizations as we go down (positive stabilizations to the right and negative to the left), and destabilizations as you go up. We are only looking at the top of the mountain and it continues indefinitely downward.

There is a special knot at the top of the mountain range, with (tb, rot) = $(-1, 0)$, often called the *trivial unknot*. This is the realization of the unknot with maximal Thurston–Bennequin number, and it is the one non-destabilizable realization of the unknot. Since all of the Legendrian realizations of a given knot are related by de/stabilizations, we would have the answer to the Geography question for a knot if we knew its full set of non-destabilizable realizations. In all known examples, there is a finite set of non-destabilizable realizations for each knot type.

The Geography Question and the Botany Question are questions that you ask about a knot type.

$$\text{rot} = \qquad -3 \quad -2 \quad -1 \quad 0 \quad 1 \quad 2 \quad 3$$

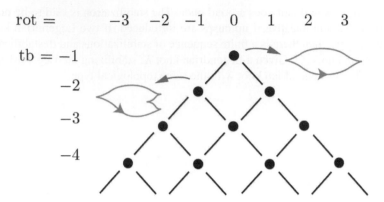

FIGURE 54.5: The mountain range for the Legendrian unknot.

So when we think about these questions we must choose a spatial graph to ask the question about, not an abstract graph. These questions have been answered for a number of planar graphs with topologically planar embeddings [11, 12, 8, 9]. A *topologically planar* embedding is a spatial graph that can be moved to a planar embedding. Of course, this is only possible for a graph that can be embedded in the plane. Here we will look at the example of the θ-graph. This is a nice example because it highlights some of the additional complexity that occurs with Legendrian graphs.

First the Geography Question: Which tb and rot can occur for a Legendrian θ-graph where all of the cycles are unknots? There are two constraints on triples of integers (tb_1, tb_2, tb_3) and (rot_1, rot_2, rot_3) to be the tb and rot of a topologically planar Legendrian θ-graph. First, there is a constraint on each pair of integers (tb_i, rot_i), for $i = 1, 2, 3$, because they are the Thurston-Bennequin number and the rotation number of a Legendrian unknot. So each pair of integers (tb_i, rot_i), for $i = 1, 2, 3$ have different parities and $tb_i + |rot_i| \leq -1$. The second constraint is a result of the structure of the graph. Since the rotation number depends on the orientation of the cycles, we must first give the cycles orientations, before we can work with the rotation number. Let the vertices of our graph be labeled v_1 and v_2 and the edges be labeled e_1, e_2, and e_3. We will represent the cycle made up of edges e_i and e_j which is oriented from v_1 to v_2 on e_i, and v_2 to v_1 on e_j by (e_i, e_j). Let rot_1, rot_2 and rot_3 be integers representing rotation numbers for the oriented cycles (e_1, e_2), (e_2, e_3) and (e_3, e_1), then $rot_1 + rot_2 + rot_3 \in \{1, 0, -1\}$. Because the cycles together traverse each of the edges in both directions, this sum is capturing information about the cyclic ordering of the edges at the vertices. These are the only constraints on the tb and rot. In other words, any two triples of integers (tb_1, tb_2, tb_3) and (rot_1, rot_2, rot_3) which satisfy the above described constraints can be realized as the Thurston-Bennequin number and the rotation number of a Legendrian θ-graph with all cycles unknotted [12].

Next the Botany question: How many different topologically planar Legendrian θ-graphs are there for each tb and rot that can occur? For two triples of integers $tb = (tb_1, tb_2, tb_3)$ and $rot = (rot_1, rot_2, rot_3)$ that can be realized as the Thurston-Bennequin number and the rotation number of a Legendrian θ-graph, there are two different things that can happen and again it depends on the sum of the rotation numbers [8, 9]. If $rot_1 + rot_2 + rot_3 = \pm 1$ there exists one topologically planar embedding θ with the given tb_θ and rot_θ. If $rot_1 + rot_2 + rot_3 = 0$ there exist two distinct topologically planar embeddings θ_+ and θ_- with the given tb and rot. These results use convex surface theory and a combinatorial argument.

We answered our two big questions for topologically planar Legendrian θ-graphs, but there is more to think about to understand the relationship between the different θ-graphs. With the mountain range for a knot we see how each Legendrian realization of the knot is related by stabilizations. So for our Legendrian θ-graphs (and all Legendrian graphs) we should think about these additional questions: Which topologically planar Legendrian θ-graphs are non-destabilizable? Are all topologically planar Legendrian θ-graphs related by stabilization and destabilization? Will we need to consider new moves between different Legendrian graphs to be able to relate all of the topologically planar Legendrian θ-graphs?

What does it mean for a Legendrian graph to be non-destabilizable and what is the set of non-destabilizable Legendrian θ-graphs? For an edge to be non-destabilizable it must be part of a cycle that is a non-destabilizable piecewise smooth Legendrian knot. If each edge of the graph is in at least one cycle that is a non-destabilizable knot, then all of the edges are non-destabilizable, so the Legendrian graph is non-destabilizable. For topologically planar Legendrian θ-graphs there is an infinite family of different non-destabilizable Legendrian graphs, shown in Figure 54.6 [9]. This is surprising because there is no known knot that has infinitely many non-destabilizable Legendrian realizations. Label these graphs G_l, where l denotes the number of full twists between the lower two edges as shown in the figure. Notice $l \in \frac{1}{2}\mathbb{Z}$ and $l \geq 0$. This infinite family is the complete list of non-destabilizable Legendrian realizations of the θ-graph.

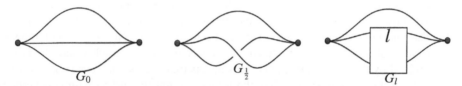

FIGURE 54.6: The non-destabilizable topologically planar Legendrian θ-graphs. There is a curve for each $l \in \frac{1}{2}\mathbb{Z}$ where $l \geq 0$, where the number in the box indicates the number of full twists of the two edges around each other.

Are all of the Legendrian realizations of the θ-graph related by stabilizations and destabilization? All Legendrian realizations of a particular knot type are related by stabilizations and destabilization [6]. If we turn our attention to the two graphs G_0 and $G_{\frac{1}{2}}$, there is a simple way to see that they will not be related by stabilizations and destabilization. If we color the edges of these two graphs so the cyclic order is the same around the first vertex, then the cyclic order of the edges around the second vertex is different. Since stabilization and destabilizations happen on an edge, they can never change the cyclic order of edges around a vertex. Thus there is no way to go from G_0 to $G_{\frac{1}{2}}$ using only (edge) stabilization and destabilization. We need another move, something that will change the cyclic order of the edges at the vertex. The new move is called *vertex stabilization*; it flips the vertex over. See Figure 54.7. If reversed, the move is called *vertex destabilization*. To avoid confusion, the usual stabilization and destabilizations are called *edge stabilization* and *edge destabilizations*, respectively. With this addition to the moves we can now get from G_0 to $G_{\frac{1}{2}}$. In

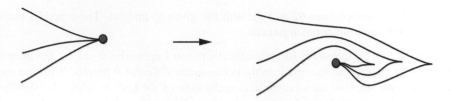

FIGURE 54.7: Vertex stabilization.

fact, using both edge stabilization and destabilization, and vertex stabilization and destabilization, we can get between any pair of topologically planar Legendrian θ-graphs [9]. It is conjectured that these moves are sufficient to move between any two Legendrian realizations of the same spatial graph, for any trivalent graph [9]. However, it is known that it will take more moves or a different set of moves for higher valence (pliable vertex) spatial graphs.

54.1.4 Legendrian Simplicity, Trivial Unknots, and the Total Thurston-Bennequin Number

In this section, we will highlight a few other questions and results that have been explored.

As discussed earlier, some knots are Legendrian simple, i.e. the Legendrian class is determined by their tb and rot. We can explore similar questions for graphs. What types of spatial graphs have their Legendrian class determined by the pair (tb, rot)? What types do not? In the negative, there are easy to find uninteresting examples; just pick a knot that is not Legendrian simple and have that knot in one of the cycles of your spatial graph. To say something meaningful about these questions we need to either consider spatial graphs where all of the cycles are Legendrian simple, or answer the more general question of how the structure of the abstract graph plays a role in the Legendrian simplicity. The question of the abstract graph has been addressed: No graph containing at least one cycle and at least one cut edge, or one cut vertex has any spatial graph where the Legendrian class is determined by the pair (tb, rot) [11]. This means that even some very simple graphs have no chance of being Legendrian simple, like the lollipop graph (the graph with two vertices, one edge between them and one loop). Another question that has been explored is whether there is a different set of invariants that would determine the Legendrian type, and what would be a natural choice [8].

It is natural to also consider if there are obstructions to having Legendrian realizations of graphs with all trivial unknots, (or other particular Legendrian types). Recall that the trivial unknot is the Legendrian unknot that is non-destabilizable. Additionally, it is the one unknot that attains the maximal Thurston-Bennequin number of its topological type. In [10], Mohnke proved that the Borromean rings and the Whitehead link cannot be represented by Legendrian links of trivial unknots. When looking at these questions it is always important to keep in mind the intrinsic properties of graphs, like being intrinsically knotted. (For more information see Chapter 50.) If an abstract graph is intrinsically knotted, it will always contain a nontrivial knot in its embedding. Thus it cannot have a Legendrian realization with all cycles as trivial unknots, because it cannot have all cycles unknots at all. Another uninteresting example of this is to choose a spatial graph with the Borromean rings or the Whitehead link as subgraphs. Since these knots have this property so will the spatial graph. But this does not give us any more insight into how the graph structure impacts whether it can be realized with all cycles as trivial unknots. How the abstract graph structure impacts this has been fully characterized. A graph G has a Legendrian realization in $(\mathbb{R}^3, \xi_{std})$ with all its cycles trivial unknots if and only if G does not contain K_4 as a minor [11]. The proof of this theorem relies partly on the fact that the trivial unknot has an odd Thurston-Bennequin number, that is tb $= -1$. So the reverse implication holds in more generality. Let G be a graph that contains K_4 as a minor. Let L_{odd}, which represents the set of topological knot classes with odd maximal Thurston-Bennequin number. There does not exist a Legendrian realization of G such that all its cycles are knots in L_{odd} realizing their maximal Thurston-Bennequin number. However, there are examples of Legendrian realizations of K_4 where all of the cycles have maximal Thurston-Bennequin number, when one of the knot types has an even maximal Thurston-Bennequin number [14].

Recall that the total Thurston-Bennequin number of a Legendrian graph is the sum of the tbs of all of its cycles, denoted as TB. While in general this invariant has much less information than the Thurston-Bennequin number, in some cases it is enough to tell us both the tb and |rot|. This is the case for unknotted Legendrian realizations of K_4 with all its 3-cycles trivial unknots, and for unknotted Legendrian realizations of $K_{3,3}$ with all its 4-cycles trivial unknots. In general the total Thurston-Bennequin number of a complete graph is determined by the Thurston-Bennequin number of the 3-cycles and the total Thurston-Bennequin number of a complete bipartite graph is determined by the Thurston-Bennequin number of the 4-cycles. For a Legendrian embedding of the complete graph on n vertices, K_n, in $(\mathbb{R}^3, \xi_{std})$,

$$TB(K_n) = TB_3(K_n) \cdot \sum_{r=3}^{n} \frac{(n-3)!}{(n-r)!}$$

where $TB_3(K_n)$ is the sum of *tb*s over all 3-cycles of K_n. For a Legendrian embedding of the complete bipartite graph, $K_{n,m}$ with $m \leq n$ in $(\mathbb{R}^3, \xi_{std})$,

$$TB(K_{n,m}) = TB_4(K_{n,m}) \cdot \sum_{r=2}^{m} \frac{(m-2)!(n-2)!}{(m-r)!(n-r)!},$$

where $TB_4(K_{n,m})$ is the sum of *tb*s over all 4-cycles of $K_{n,m}$. The proofs of these facts are strongly combinatorial; for more details see [13].

54.2 Bibliography

[1] Byung Hee An and Youngjin Bae. A Chekanov-Eliashberg algebra for Legendrian graphs. *Journal of Topology*, 13, 2018.

[2] Byung Hee An, Youngjin Bae, and Seonhwa Kim. Legendrian singular links and singular connected sums. *Journal of Symplectic Geometry*, 16(4), 2018.

[3] Sebastian Baader and Masaharu Ishikawa. Legendrian graphs and quasipositive diagrams. *Ann. Fac. Sci. Toulouse Math. (6)*, 18(2):285–305, 2009.

[4] Yakov Eliashberg and Maia Fraser. Topologically trivial Legendrian knots. *J. Symplectic Geom.*, 7(2):77–127, 2009.

[5] John B. Etnyre and Ko Honda. Knots and contact geometry. I. Torus knots and the figure eight knot. *J. Symplectic Geom.*, 1(1):63–120, 2001.

[6] Dmitry Fuchs and Serge Tabachnikov. Invariants of Legendrian and transverse knots in the standard contact space. *Topology*, 36(5):1025–1053, 1997.

[7] Emmanuel Giroux. Géométrie de contact: de la dimension trois vers les dimensions supérieures. In *Proceedings of the International Congress of Mathematicians, Vol. II (Beijing, 2002)*, pages 405–414. Higher Ed. Press, Beijing, 2002.

[8] Peter Lambert-Cole and Danielle O'Donnol. Planar Legendrian graphs, in process.

[9] Peter Lambert-Cole and Danielle O'Donnol. Planar Legendrian Θ-graphs, in process.

[10] Klaus Mohnke. Legendrian links of topological unknots. In *Topology, Geometry, and Algebra: Interactions and New Directions*, volume 279 of *Contemp. Math.*, pages 209–211. Amer. Math. Soc., Providence, RI, 2001.

[11] Danielle O'Donnol and Elena Pavelescu. On Legendrian graphs. *Algebr. Geom. Topol.*, 12(3):1273–1299, 2012.

[12] Danielle O'Donnol and Elena Pavelescu. Legendrian θ-graphs. *Pacific J. Math.*, 270(1):191–210, 2014.

[13] Danielle O'Donnol and Elena Pavelescu. The total Thurston-Bennequin number of complete and complete bipartite Legendrian graphs. In *Advances in the Mathematical Sciences*, volume 6 of *Assoc. Women Math. Ser.*, pages 117–137. Springer, [Cham], 2016.

[14] Toshifumi Tanaka. On the maximal Thurston-Bennequin number of knots and links in spatial graphs. *Topology Appl.*, 180:132–141, 2015.

Chapter 55

Linear Embeddings of Spatial Graphs

Elena Pavelescu, University of South Alabama

A *spatial graph* is an embedding of a graph in $\mathbb{R}^3$. In what follows, we consider graphs without multiple edges or loops and a special class of embeddings, *linear embeddings*, in which each edge is a straight line segment. Chemists are interested in this type of spatial graph since the vertices can be thought of as atoms, while edges are bonds, which behave more like rigid sticks than flexible rope. As a subclass of a larger class, the linear embeddings of a graph share any property pertaining to all embeddings. By contrast, linear embbedings have properties not shared by all embeddings. The study of linear embeddings has seen approaches using combinatorial and geometric arguments, Conway-Gordon-type results, and oriented matroid theory.

Taking a step back, we look at *planar graphs*, graphs which can be drawn in the plane without any crossings. Fáry [5], Wagner [29] and Stein [28] independently proved that any planar graph can be drawn without crossings so that its edges are straight line segments. That is, the constraint of drawing edges as straight line segments instead of curves does not restrict to a smaller class of graphs. As seen in this chapter, restricting to straight edges may result in a smaller class of embeddings for non-planar graphs.

The cornerstone of the theory of spatial graphs is marked by results of Conway and Gordon [3] and Sachs [27], who showed that K_6, the complete graph on 6 vertices, is an *intrinsically linked* graph, a graph such that every embedding in $\mathbb{R}^3$ contains a non-trivial link. Conway and Gordon [3] also showed that K_7 is an *intrinsically knotted* graph, a graph such that every embedding in $\mathbb{R}^3$ contains a cycle which is a non-trivial knot. These results have generated a significant amount of work, including the classification of intrinsically linked graphs by Robertson, Seymour and Thomas [25]: a graph is intrinsically linked if and only if it contains one of the graphs in the Petersen family of graphs as a minor. The classification of intrinsically knotted graphs is not yet complete. Sachs [26, 27] conjectured that if a graph has an embedding in $\mathbb{R}^3$ with no non-trivial links, then it has a linear embedding with no non-trivial links. This conjecture is still open.

One of the most significant results concerning linear embeddings of graphs was obtained by Negami [21]. He showed that given any knot or link, for all sufficiently large n, every linear embedding of K_n contains that knot or link. Miyauchi [18] extended this result to complete bipartite graphs. In addition, several authors have characterized what links and knots can occur in linear embeddings of specific graphs.

Given their relatively small size and their intrinsic topological properties, the graphs K_6 and K_7 are natural candidates for the study of their linear embeddings. The *stick number* of a knot or link K is the minimal number of line segments required to realize K as a polygonal knot or link. This means that a linear embedding of a graph with n vertices can contain knots or links with stick number no more than n. An (i, j)–link in a spatial graph is a 2-component link whose two components contain i and j vertices, respectively. Six is the minimal possible stick number for non-trivial knots and links, and the trefoil and the $(3, 3)$–Hopf link are the only curves with stick number six [24]. As such, these two are the only knots or links which can appear in a linear embedding of K_6. There are three knots and links with stick number 7: the figure-8 knot, the $(3, 4)$–Hopf link and the $(3, 4)$–torus link. These three, together with the trefoil and the $(3, 3)$–Hopf link, can appear in a linear embedding of K_7. Figure 55.1 presents embeddings of knots and links realizing the stick numbers of their knot or link type.

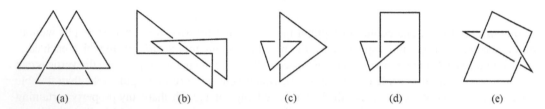

FIGURE 55.1: (a) Polygonal trefoil with 6 edges; (b) polygonal figure-8 knot with 7 edges; (c) $(3, 3)$–Hopf-link; (d) $(3, 4)$–Hopf-link ; (e) $(3, 4)$–torus link.

The graph K_6 has ten distinct pairs of disjoint triangles. If flexible edges are allowed, for each $1 \leq k \leq 10$, one can build a spatial embedding of K_6 where exactly k pairs of triangles are non-trivially linked. By contrast, as shown by Hughes [12], any linear embedding of K_6 contains either 1 or 3 nontrivial links. She showed that any linear embedding of K_6 is ambient isotopic with one of the two embeddings in Figure 55.2. The embedding in Figure 55.2(a) has exactly one non-trivial link and the embedding in Figure 55.2(b) has three distinct non-trivial links. Huh and Jeon independently showed these same results, adding that the embedding in Figure 55.2(b) is the only linear embedding of K_6 that contains a trefoil [14]. An alternate proof using an integral lift of the Conway-Gordon theorem [3] was given by Nikkuni [22]. We present the ideas of this proof at the end of the chapter.

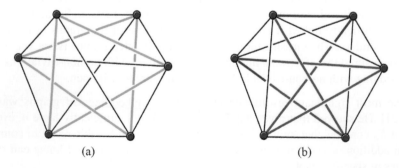

FIGURE 55.2: (a) A linear embedding of K_6 containing one non-trivial link (highlighted in green); (b) a linear embedding of K_6 containing three non-trivial links and one trefoil (highlighted in orange).

Ludwig and Arbisi [17] extended the work of Hughes, Huh and Jeon to characterize the linear embeddings of K_7 which form convex polyhedra with seven vertices. They showed the number of non-trivial links in any such linear embedding of K_7 is at least twenty-one and at most forty-eight. They also determined the number of $(3, 3)$–links and $(3, 4)$–links in such embeddings. Independently, Brown [2], Foisy et. al. [9], Ramirez Alfonsin [23], and Nikkuni [22] showed that every linear embedding of K_7 contains a trefoil knot. And Huh [13] showed no linear embedding of K_7 contains more than three figure-eight knots. It's interesting to note that the three proofs for the existence of a trefoil use different methods: combinatorial analysis, matroid theory and Conway-Gordon-type identities. These results are in contrast to those for topological embeddings of K_7, where one can vary the number of non-trivial knots by simply allowing knotted edges. Jeon et. al. [15] used oriented matroids to completely characterize linking and knotting in linear embeddings of K_7. They gave the type and number of knots and links in each one of 462 linear embeddings of K_7 (representing all possible embeddings up to isotopy through linear graphs).

A characterization of linear embeddings of K_8 has not been done yet. Finding links in a linear embedding of K_8 requires the inclusion of $(3, 5)$–links and $(4, 4)$–links, which adds much complexity to the analysis. Work has been done on linear embeddings of $K_{3,3,1}$ and $K_{3,3,1,1}$, the complete 3–partite graph and the complete 4–partite graph, respectively. Naimi and Pavelescu [20] characterized the number and type of links in a linear embedding of $K_{3,3,1}$. They showed that a linear embedding of $K_{3,3,1}$ contains between one and five non-trivial links. They also found how many of these links are Hopf links and how many are torus links. Hashimoto and Nikkuni studied linear embeddings of $K_{3,3,1,1}$ [11]. They showed that every linear embedding of $K_{3,3,1,1}$ contains a non-trivial Hamiltonian knot, a knotted cycle which contains all vertices of the graph. By contrast, it is still unknown whether every spatial embedding of $K_{3,3,1,1}$ contains a Hamiltonian non-trivial knot. The graphs K_6 and $K_{3,3,1}$ together are necessary and sufficient for generating the entire Petersen graph family using delta-Y moves [27]. The graphs K_7 and $K_{3,3,1,1}$ are minor minimal intrinsically knotted and by performing repeated delta-Y moves starting from these two graphs, one finds more minor minimal intrinsically knotted graphs [16], [10]. If the effect of delta-Y moves on the number of knots or links in a linearly embedded graph has a nice characterization, then this information can be used to determine linking and knotting patterns in linear embeddings of the Petersen family of graphs, the K_7 family and the $K_{3,3,1,1}$ family.

Going back to "not-necessarily-linear embeddings," a graph is called *intrinsically n-linked* if every embedding (not necessarily linear) of it in $\mathbb{R}^3$ contains a *non-split n*-component link, i.e., for every embedded 2-sphere S^2 disjoint from the link, all n components of the link are contained in the same component of $\mathbb{R}^3 \setminus S^2$. The smallest complete graph that could possibly be intrinsically 3–linked is K_9, since each component must contain at least three vertices. Flapan, Naimi, and Pommersheim [8] showed that K_9 is not intrinsically 3–linked, but K_{10} is. They illustrated an embedding of K_9 that contains no non-split 3–link. This embedding, however, was not a linear embedding. Naimi and Pavelescu [19] used oriented matroid theory to show that every linear embedding of K_9 does contain a non-split 3–link. This means that whether true or not, Sachs's conjecture [26, 27] cannot be extended to non-split links with three components. The result in [8] was generalized by Drummond-Cole and O'Donnol [4] who proved that for $n > 1$, every embedding of $K_{\lfloor \frac{7n}{2} \rfloor}$ contains a non-split link with n–components. Within the class of linear embeddings of complete graphs, intrinsic n–linkedness is yet to be characterized. While every linear embedding of $K_{\lfloor \frac{7n}{2} \rfloor}$ does contain a non-split n–component link, the K_9 result in [19] suggests the bound might be lower than $\lfloor \frac{7n}{2} \rfloor$ for linear embeddings.

As a way to model entanglements of polymers in a confined region, Flapan and Kozai [7] considered random linear embeddings of K_n in the cube $C = [0, 1]^3$. These are linear embeddings with vertices given by a random uniform distribution of n points in C. They proved that for such embeddings, the mean sum of squared linking numbers of links and the mean sum of squared writhes of cycles are both of order $O(n(n!))$. Remember that a linear K_6 contains either one or three non-trivial links, while a linear $K_{3,3,1}$ contains between one and five links. As a consequence, Flapan and Kozai show that the probability that a random linear embedding of K_6 contains exactly one non-trivial link is $p = 0.7380 \pm 0.0003$. They also show that the probability that a random linear embedding of $K_{3,3,1}$ contains exactly one non-trivial link is $p \geq 0.5856 \pm 0.0004$.

We have previously mentioned oriented matroids and Conway-Gordon-type arguments as tools for the analysis of linear embeddings. We end by describing these two approaches.

Oriented matroids and linear embeddings. An *oriented matroid* of rank k on n elements, $k \leq n$, is a pair $\Omega = (V, \mathcal{B})$, where V is a set with n elements, and $\mathcal{B}$ is a collection of signed subsets of V with k elements, satisfying certain compatibility conditions. A thorough treatment of oriented matroids can be found in [1]. Here we present how oriented matroids are associated to linear embeddings of graphs. A spatial embedding of K_n is completely determined by the position of its vertices, $v_1, v_2, \ldots, v_n$. Using a fixed order on the set of vertices, each subset of four vertices $v_1 < v_2 < v_3 < v_4$ is labeled as $+$ or $-$, according to whether the ordered basis $\{\overrightarrow{v_1 v_2}, \overrightarrow{v_1 v_3}, \overrightarrow{v_1 v_4}\}$ is a positively or a negatively oriented basis of $\mathbb{R}^3$. In this way, the embedding generates a uniform oriented matroid of rank 4 on n elements, $\Omega = (V, \mathcal{B})$. *Uniform* means that the set $\mathcal{B}$ contains *all* subsets of V of size 4. The set of all uniform matroids of rank 4 on n elements is denoted by $OM(4, n)$. Each linear embedding of K_n generates an element of $OM(4, n)$. On the other hand, a (realizable) matroid specifies an isotopy class of spatial embeddings of K_n. The key observation which helps in determining linking and knotting patterns in linear embeddings of a K_n is that any five vertices in an embedding are in one of two positions, as in Figure 55.3. With the procedure described above, the ordered list of signed bases for Figure 55.3(a) is $- + + - +$ and the ordered list of signed bases for Figure 55.3(b) is $- - + - +$. The list of signed bases in the matroid captures this arrangement of vertices and so the oriented matroid tells which edge pierces which triangle. This is sufficient for computing the linking number between any two triangles, and between any two cycles, by decomposing each cycle into triangles. Uniform matroids with various ranks and up to

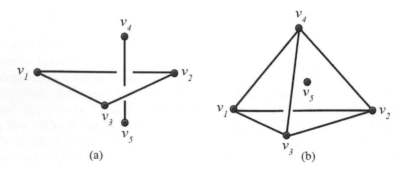

(a)   (b)

FIGURE 55.3: (a) Edge $v_4 v_5$ pierces the triangle $v_1 v_2 v_3$; (b) vertex v_5 lies inside the tetrahedron $v_1 v_2 v_3 v_4$.

10 elements have been found using computers and are available in [6]. In [19], Naimi and Pavelescu determined that every linear embeddings of K_9 contains a non-split link with three components. A computer program checked that each matroid of $OM(4, 9)$ contains three triangles such that at least one of them has non-zero linking number with each of the other two. In [23], Ramirez Alfonsin listed three geometric conditions for the oriented matroid under which a linear embedding of K_7 contains a trefoil. Then, a computer verified that every oriented matroid in $OM(4, 7)$ satisfies at least one of the three conditions. Since any linear embedding of K_7 generates an element of $OM(4, 7)$, it contains a trefoil. Using oriented matroids, Jeon et. al. [15] completely characterized linking and knotting in linear embeddings of K_7.

Conway-Gordon-type theorems. For a spatial graph, let Γ_i represent the set of cycles of length i and $\Gamma_{i,j}$ the set of (i, j)–links. Conway and Gordon [3] showed that for any spatial embedding of K_6,

$$\sum_{\gamma \in \Gamma_{3,3}} \mathrm{lk}(\gamma) \equiv 1 (\mathrm{mod}\, 2), \qquad (1)$$

and that for any spatial embedding of K_7,

$$\sum_{\gamma \in \Gamma_7} \mathrm{Arf}(\gamma) \equiv 1 (\mathrm{mod}\, 2). \qquad (2)$$

Nikkuni [22] gave integer refinements of these identities. In particular, he proved that for any spatial embedding of K_6

$$2\{\sum_{\gamma \in \Gamma_6} a_2(\gamma) - \sum_{\gamma \in \Gamma_5} a_2(\gamma)\} = \sum_{\gamma \in \Gamma_{3,3}} (\mathrm{lk}(\gamma))^2 - 1, \qquad (*)$$

where a_2 represents the second coefficient of the Conway polynomial. Nikkuni used identity $(*)$ to characterize linking and knotting in linear embeddings of K_6 as follows. Since the smallest possible stick number is 6 and is realized by the trefoil only, in a linear K_6 no 5-cycle can be knotted and every non-trivially knotted 6-cycle is a trefoil. Since $a_2(\text{trefoil}) = 1$, the left-hand side of equation $(*)$ represents twice the number of trefoils in the embedding, $2n_T$. Moreover, non-trivial links in a linear K_6 are $(3, 3)$–links with linking number 1. This means the right hand side of equation $(*)$ represents $n_L - 1$, one less than the number of $(3, 3)$–links in the embedding. A parity argument, plus the fact that n_L is at most 10, restrict the pair (n_T, n_L) to the set $\{(k, 2k + 1), k = 0, 1, 2, 3, 4\}$. A geometric argument shows that the only two possible (n_T, n_L) are $(0, 1)$ and $(1, 3)$. A refinement of identity (2) given by Nikkuni [22] was used by Hashimoto and Nikkuni [11] to prove that any linear embedding of $K_{3,3,1,1}$ contains a non-trivial Hamiltonian knot.

55.1 Bibliography

[1] B. Björner, M. Las Vergnas, B. Sturmfels, N. White, G. Ziegler. Oriented matroids. *Ecyclopedia of Mathematics and its Applications*. Addison-Wesley, Reading, MA. 1993.

[2] A. F. Brown. Embeddings of graphs in E^3. Ph.D. dissertation, Kent State University, (1977).

[3] J. Conway, C. Gordon. Knots and links in spatial graphs. *J. Graph Theory* 7 (1983), 445–453.

[4] C. Drummond-Cole, D. O'Donnol. Intrinsically n–linked complete graphs. *Tokyo J. of Math.* **32**, Number 1 (2009), 113–125.

[5] I. Fáry. On straight-line representation of planar graphs. *Acta Sci. Math.* (Szeged) **11** (1948), 229–233.

[6] L. Finschi. *Homepage of oriented matroids.* http://www.om.math.ethz.ch/.

[7] E. Flapan, K. Kozai. Linking numbers and writhe in random linear embeddings of graphs *J. Math Chem.* **54** (2016), 1117–1133.

[8] E. Flapan, R. Naimi, J. Pommersheim. Intrinsically triple linked complete graphs. *Topol. Appl.* **115** (2001), 239–246.

[9] J. Bustamante, J. Ferderman, J. Foisy, K. Kozai, K. Matthews, K. McNamara, E. Stark, K. Tricky. Intrinsically linked graphs in projective space. *Alg. Geom. Topology* **9** (2009), 1255–1274

[10] N. Goldberg, T. Mattman, R. Naimi. Many, many more minor minimal intrinsically knotted graphs. *Alg. Geom. Topology* **14** (2014), 1801–1823.

[11] H. Hashimoto, R. Nikkuni. Conway-Gordon type theorem for the complete four-partite graph $K_{3,3,1,1}$. *New York J. Math* **20** (2014), 471–495.

[12] C. Hughes. Linked triangle pairs in a straight edge embedding of K_6. $\Pi M E$ *Journal* **12** (2006), No. 4, 213–218.

[13] Y. Huh. Knotted Hamiltonian cycles in linear embeddings of K_7 into $\mathbb{R}^3$. *J. Knot Theory Ramifications* **21** (2012), 1250132

[14] Y. Huh, C. B. Jeon. Knots and links in linear embeddings of K_6. *J. Korean Math. Soc.* **44**, no.3 (2007), 661–671.

[15] C. B. Jeon, G. T. Jin, H. J. Lee, S. J. Park, H. J. Huh, J. W. Jung, W. S. Nam, M. S. Sim. Number of knots and links in linear K_7. Slides from the International Workshop on Spatial Graphs (2010), http://www.f.waseda.jp/taniyama/SG2010/talks/19-7Jeon.pdf.

[16] T. Kohara, S. Suzuki. Some remarks on knots and links in spatial graphs. *Knots* 90, Osaka, 1990, de Gruyter (1992), 435–445.

[17] L. Ludwing, P. Arbisi. Linking in straight-edge embeddings of K_7. *J. Knot Theory Ramifications* **19**, No. 11 (2010), 14431–1447.

[18] M. S. Miyauchi. Topological Ramsey theorem for complete bipartite graphs. *J. of Combinatorial Theory*, Series B **62**, Number 1 (1994) ,164–179.

[19] R. Naimi, E. Pavelescu. Linear embeddings of K_9 are triple linked. *J. Knot Theory Ramifications* **24**, No. 8 (2015), 1550041

[20] R. Naimi, E. Pavelescu. On the number of links is a linearly embedded $K_{3,3,1}$. *J. Knot Theory Ramifications* **23**, No. 3 (2014), 1420001

[21] S. Negami. Ramsey theorems for knots, links, and spatial graphs. *Trans. Amer. Math. Soc.* (1991), 527–541.

[22] R. Nikkuni. A refinement of the Conway-Gordon theorems. *Topol. Appl.* **156** (2009), 2782–2794.

[23] J. L. Ramirez Alfonsin. Spatial graphs and oriented matroids: The trefoil. *Discrete and Computational Geometry* **22** (1999), 149–158.

[24] K. Reideimeister. Homotopieringe und Linsenräume. *Sem. Hamburg* **11** (1936), 102–109.

[25] N. Robertson, Neil, P. Seymour, R. Thomas Linkless embeddings of graphs in 3-space. *Bull. Amer. Math. Soc.* **28** (1993), No. 1, 84–89.

[26] H. Sachs. On a spatial analogue of Kuratowski's theorem on planar graphs—an open problem. *Graph theory* (Łagow, 1981), 230–241, Lecture Notes in Math. **1018**, Springer, Berlin, (1983).

[27] H. Sachs. On spatial representations of finite graphs. Finite and infinite sets, Vol. I, II (Eger, 1981), 649–662, *Colloq. Math. Soc. János Bolyai* **37**, North-Holland, Amsterdam (1984).

[28] S. K. Stein. Convex maps. *Proc. Amer. Math. Soc.* **2** (3) (1951), 464–466.

[29] K. Wagner. Bemerkungen zum Vierfarbenproblem. *Jahresbericht der Deutschen Mathematiker-Vereinigung*, **46** (1936), 26–32.

Chapter 56

Abstractly Planar Spatial Graphs

Scott A. Taylor, Colby College

Beginning courses in graph theory prove many wonderful theorems about planar graphs. An even more wonderful theory arises when we put planar graphs (which we'll henceforth refer to as **abstractly planar** graphs) into 3-dimensional space. One way of doing this is to choose an embedding of an abstractly planar graph G in the sphere S^2 and then include S^2 in the 3-sphere $S^3 = \mathbb{R}^3 \cup \{\infty\}$ (tamely). As an abstractly planar graph may have several planar embeddings, we may wonder if we can end up with different embeddings in S^3. It turns out that in S^3, all planar embeddings give rise to **equivalent** (that is, ambient isotopic) spatial graphs [16]. Such spatial graphs are called **trivial**. A nontrivial spatial graph is **knotted**. Are there knotted embeddings? Yes there are!

Some of the most important spatial graphs have very few vertices and edges, and are thus abstractly planar. Important classes include spatial θ-graphs, handcuff graphs, the tetrahedral graph, θ_n-graphs, and bouquets. We depict the abstract graph type for these graphs in Figure 56.1. Of these families, spatial θ and θ_n graphs have received the most attention in the literature. One reason spatial θ-graphs are so prevalent is that we can create one by attaching an arc to a knot so that the endpoints of the arc are distinct points on the knot. This construction arises naturally in knot theory, where the arc may record some information about the knot K. Typical examples include knot tunnels or an arc recording the location of some crossing change, as in the first two diagrams of Figure 56.2. The knot K becomes a cycle, or **constituent knot**, in the resulting spatial θ-graph. What can we say about the other constituent knots? Perhaps surprisingly, Kinoshita [13] showed that, given three knots, there is a spatial θ-graph whose three constituent knots are precisely the three given knots. An example of a θ-graph whose three constituent knots are all the figure-8 knot is shown in Figure 56.2. Kinoshita's construction can be applied recursively to construct θ_n-graphs whose constituent knots are specified beforehand.

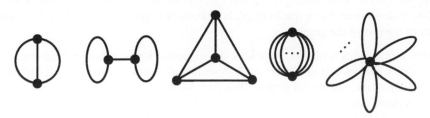

FIGURE 56.1: From left to right, we have the abstract graph type of θ-graphs, handcuff graphs, the tetrahedral graph, θ_n-graphs, and bouquets.

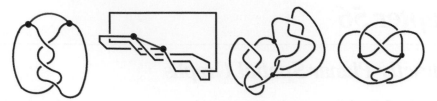

FIGURE 56.2: Four examples of spatial θ-graphs. From left to right: a trefoil knot with tunnel (in red); a knot with an arc (in red) marking the location of a crossing change; a θ-graph whose every constituent knot is a figure 8 knot; the Kinoshita graph.

The trivial θ-graph has every cycle an unknot. Are there other θ-graphs with this property? A spatial graph having the property that every collection of disjoint cycles is an unlink is a **ravel**. Similarly, a spatial graph that has the property that every proper subgraph is trivial has the **Brunnian property**. A knotted graph with the Brunnian property is **Brunnian, almost unknotted**, or **minimally knotted**. A θ-graph is Brunnian if and only if it is a ravel; but the same is not necessarily true for other graphs.

Kinoshita [12] also provided the first example of a Brunnian θ-graph, now named after him. It is the rightmost diagram in Figure 56.2. Wolcott [29] later generalized this construction to a family now known as the Kinoshita-Wolcott graphs. More examples of Brunnian θ-graphs are given in [15] and [9]. Suzuki [24] generalized Kinoshita's construction to θ_n graphs. Every abstractly planar graph without degree zero and degree one vertices has a Brunnian spatial embedding [11, 31]. Ravels are of interest to chemists [3]; Flapan and Miller [5] have constructed many examples.

How can we be sure that Kinoshita's graph really is knotted, or indeed that any given spatial embedding of an abstractly planar graph really is knotted?

An equivalence between spatial graphs takes the constituent knots of one to the constituent knots of the other (see [10]). Thus, if one spatial graph has a constituent knot K and another has no constituent knot of the same knot type, the graphs can't be equivalent. This doesn't help us show minimally knotted spatial graphs are knotted, though; we need other tools. As always in knot theory, we might ask for an invariant and there are some very nice invariants available. In general, Brunnian graphs and ravels provide good tests for the strength of invariants of spatial graphs. The three most popular are the Yamada polynomial [32], Litherland's version of the Alexander polynomial [14], and Thompson's polynomial invariant [28]. This last polynomial is defined recursively, but is zero if and only if the graph is trivial. It is based on an earlier algorithm of Scharlemann and Thompson [22] for determining if a spatial graph is unknotted. Their results were also adapted by Wu [30], who showed that a spatial graph is unknotted if and only if each cycle bounds a disc disjoint from the rest of the graph.

We can also turn to other tools from topology and algebra. How accessible these are, depends, of course, on one's background and interests. Kinoshita and Suzuki used Alexander ideals to prove the nontriviality of their Brunnian graphs. However, McAtee, Silver, and Williams [18] point out that Suzuki's proof contains an error. The first complete proof of their nontriviality is likely given by Scharlemann [21], using topological techniques stemming from the braid groups. In [19, Example 22], the topology of surfaces containing the spatial graph is used to prove the Kinoshita graph is knotted, and in [18], quandle colorings are used. Perhaps the simplest proofs that the Kinoshita graph is knotted rely on classical knot invariants. The article [20] provides two. In [9], a combination of handlebody theory and rational tangles are applied to an infinite family of θ-graphs. One popular and beautifully simple approach for θ_n-graphs is to use branched covers. Livingston [15] uses these to prove Suzuki's graphs are nontrivial and Calcut and Metcalf-Burton [2] use them to show Kinoshita's graph is prime, in a sense which we now explore.

Whenever mathematicians are introduced to some new mathematical object, we want to be able to create more of them and to understand how the object fits into the larger context of known mathematics. For the remainder, we take up the question of creating new spatial graphs from old ones and understanding how spatial graphs are related to knot theory and 3-manifold theory.

Throughout the study of manifolds, the connected sum is one of the most important methods of combining two manifolds. Recall that if M_1 and M_2 are manifolds of the same dimension, we form their connected sum $M_1 \# M_2$ by removing an open ball from each of them and gluing the resulting manifolds with boundary together along the new spherical boundary components. The centers of the balls are called the **summing points**. Classical results show that if the summands are connected and we pay attention to orientations, the sum is unique up to homeomorphism. In particular $S^3 \# S^3$ is homeomorphic to S^3. When we consider 3-manifolds containing spatial graphs, on the other hand, there are potentially many more options. For starters, we have a choice of where to perform the sum. Given two spatial graphs G_1 and G_2 in distinct copies M_1 and M_2 of S^3, we make the choice by picking summing points $p_1 \in M_1$ and $p_2 \in M_2$. We can pick the points to be both disjoint from the graphs, or we can pick them to be contained in the interiors of edges in the graphs, or we can pick them both to be vertices of the graphs having the same degree. We'll denote the result by $G_1 \#_k G_2$ where $k = 0$ if the points are disjoint from the graphs; $k = 2$ if the points are interior to edges; and, otherwise, k is the degree of the vertices.[1] Figure 56.3 depicts the case when $k = 0$ (the **distant sum**), $k = 2$ (the **connected sum**), and $k = 3$ (**the trivalent vertex sum**). Even for a fixed k, in general, $G_1 \#_k G_2$ is not uniquely defined. Summing operations are associative. If G is equivalent to $G_1 \#_k G_2$ and neither is a trivial θ_k-graph, then we say that G is k-**composite**. If G is neither trivial nor k-composite, it is k-**prime**.

For simplicity, let's consider only spatial θ-graphs. We also assume our θ-graphs are oriented. To orient a θ-graph, choose one vertex as source, one vertex as sink, and color the edges red, blue, and green. We then restrict $\#_3$ so that, when forming $M_1 \#_3 M_2$, the summing point p_1 is the sink vertex of G_1, the summing point p_2 is the source vertex of G_2, and the gluing map takes red, blue, and green endpoints to red, blue and green endpoints respectively. Under the operation $\#_3$, the set $\mathbb{G}$ of oriented θ-graphs in S^3 is particularly rich. The operation $\#_3$ is well defined [29] and it makes the set $\mathbb{G}$ into a semigroup with the trivial graph as the identity. In other words, the operation $\#_3$ is associative and if $G \in \mathbb{G}$ and if T is the trivial θ-graph, then $G \#_3 T = T \#_3 G = G$. In general, elements of $\mathbb{G}$ do not have inverses. If they did, our semigroup would be a group. The center of the semigroup is the subset of elements commuting with all other elements. In our case, it consists of the θ-graphs that are a connected sum of a trivial θ-graph and a knot. Elements of $\mathbb{G}$ have 3-prime factorizations. This means that for each nontrivial $G \in \mathbb{G}$, there are 3-prime elements $G_1, \ldots, G_n$ such that $G = G_1 \#_3 \cdots \#_3 G_n$. Furthermore, except for the fact that elements of the center commute with all other elements, this factorization is unique [17]. Conjecturally, (3-manifold, graph) pairs more generally have unique prime factorizations [8].

For θ-graphs, the property of being Brunnian also persists under $\#_k$. Indeed, for $G_1, G_2 \in \mathbb{G}$, the

---

[1]This creates conflicting notation when $k = 2$, but we will ignore this.

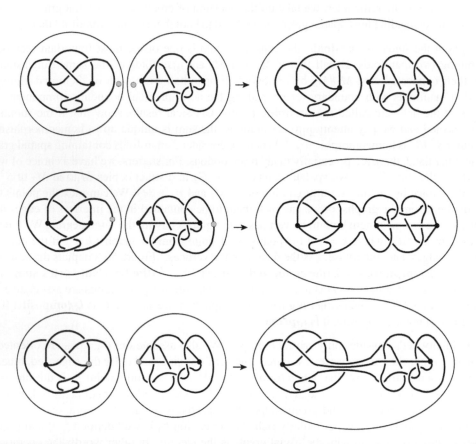

FIGURE 56.3: From top to bottom we have the distant sum, the connected sum, and the trivalent vertex sum of two spatial θ-graphs. In each case, the large circles and ellipses denote distinct copies of S^3 and the gray dots indicate the points where the summing occurs.

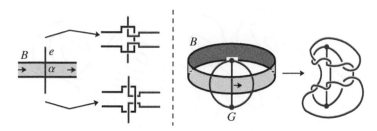

FIGURE 56.4: The left side shows how to form $G(B)$. We take the union of the graph G and the buckle ∂B, but wherever an edge e of G intersects B as on the left, we replace it with either of the pictured two "belt buckles." On the right, we see that using the indicated belt B to buckle the trivial theta-graph G produces the trivalent vertex sum of the Kinoshita graph with its mirror image.

sum, $G_1\#_3G_2$ is Brunnian if and only if G_1 and G_2 are both Brunnian. (Exercise!) For general spatial graphs, the property of being Brunnian may not persist under trivalent vertex sum. (Another exercise!) The property of being a ravel does persist under trivalent vertex sum. However, for $k \geq 4$, the property of being a ravel need not persist under $\#_k$. (Briefly: there exist knots with essential tangle decompositions and such knots result from summing bouquets.) Even for θ_k graphs, if we choose the gluing map for the connected sum to be very complicated, we may end up with knotted cycles after performing the sum.

How can we construct infinitely many *3-prime* Brunnian θ-graphs? The Kinoshita and Kinoshita-Wolcott graphs are 3-prime [14, 2], as are the Brunnian θ-graphs found in [9] (see [27] for an indication of how this might be proved). Here is a very general method (essentially found in [23]) that likely produces arbitrarily complicated Brunnian θ-graphs, most of which are probably 3-prime. For a spatial graph $G \subset S^3$, a new spatial graph $G(B)$, called a **buckling** of G, is determined by a choice of oriented annulus $B = S^1 \times [0, 1]$, called a **belt**, intersecting G in intervals of the form $\{\text{point}\} \times [0, 1]$. We create $G(B)$ as follows. At each intersection arc α between B and G we replace a neighborhood of α in S^3 with a **belt buckle** as on the left of Figure 56.4 and include the remaining portions of ∂B as part of $G(B)$. The right of Figure 56.4 shows how a certain buckling of the trivial θ-graph produces the trivalent vertex sum of the Kinoshita graph with its mirror image.

Not very much is known about how buckling affects a spatial graph. It is not difficult to see, however, that G and $G(B)$ are abstractly isomorphic and that if e is an edge of G intersecting the band B, then the subgraphs $G(B) - e$ and $G - e$ are equivalent. In particular, if G has the Brunnian property, and if B intersects every edge of G, then $G(B)$ also has the Brunnian property. It seems difficult to determine whether or not $G(B)$ is trivial. Nevertheless, we conjecture:

Conjecture 1. There is no θ-graph G in S^3 such that there is a belt B intersecting all the edges of G with $G(B)$ either the trivial θ-graph or the Kinoshita graph.

The exteriors of $G \in \mathbb{G}$ provide fertile ground for exploring topological and geometric questions. By work of Thurston (see Chapter 19), most knot complements admit a geometric structure based on 3-dimensional hyperbolic space. Likewise, the exteriors of most spatial graphs similarly admit hyperbolic structures. If our spatial graph is hyperbolic, we have access to a number of useful invariants. In particular, we may consider the *volume* of the exterior. The simplest type of hyperbolicity for a spatial graph is **hyperbolicity with parabolic meridians** [7]. It follows from work of Thurston (see [7, Corollary 2.5]) that $G \in \mathbb{G}$ is hyperbolic with parabolic meridians if and only

FIGURE 56.5: Four different spines for a genus 2 handlebody.

if it is 2-prime and the exterior of G does not contain an essential torus. In particular, if $G \subset S^3$ is a Brunnian θ-graph, then G is hyperbolic with parabolic meridians whenever its exterior does not contain an essential torus. In [9], there is an example of a buckling producing a Brunnian θ-graph having an essential torus in its exterior. That graph is, therefore, non-hyperbolic. Thurston's work can also be used to show that if $G = G_1 \#_3 G_2$ is some trivalent vertex sum of nontrivial elements of $\mathbb{G}$, then G is hyperbolic with parabolic meridians if and only if both G_1 and G_2 are. This suggests that volume vol(G) is a particularly interesting invariant for $G \in \mathbb{G}$. We ask (based on [1]):

Question 1. Suppose that $G_i \in \mathbb{G}$ for $i = 1, 2$ are both hyperbolic with parabolic meridians. How different can vol($G_1 \#_3 G_2$) and vol(G_1) + vol(G_2) be?

Finally, returning to topology, we consider the uniqueness up to homeomorphism of the exterior of connected spatial graphs. The relevant issues are illuminated by considering not just a spatial graph G but also a regular neighborhood of it $N(G)$. The neighborhood $N(G)$ is a 3-manifold with boundary. It is called a **handlebody** and the graph G is called a **spine** for the handlebody. The boundary of $N(G)$ is a connected, orientable surface. Its genus is called the **genus** of the handlebody $N(G)$. Handlebodies may have many different spines. That is, if $N(G)$ is a handlebody with a spine G, there may exist many other graphs $G' \subset N(G)$ such that $N(G) = N(G')$. Indeed, if the genus of the handlebody is at least 2, it will have infinitely many spines. Figure 56.5 depicts four different spines for a genus 2 handlebody. We say that two spatial graphs G and G' are **neighborhood-equivalent** if $N(G)$ and $N(G')$ are ambient isotopic. Equivalently, G and G' are neighborhood-equivalent if there exists a handlebody in S^3 having two spines, one of which is equivalent to G and the other to G'. If two graphs are equivalent then they are neighborhood-equivalent, but the converse is not necessarily true. Since the exterior of a spatial graph G is identical to the exterior of $N(G)$, neighborhood-equivalent spatial graphs have homeomorphic exteriors. In particular, spatial graphs are not determined by their complements, unlike knots in S^3 [6].

If two spatial graphs are neighborhood-equivalent, we can also ask how their constituent knots are related. For θ-graphs, this question was studied extensively in [25, 26], where it was connected to an operation on knots and 2-component links called **boring**. Many operations in knot theory, such as rational tangle replacement on knots (an important operation in studying DNA, see e.g. [4]), are examples of boring. One result (see [26, Theorem 6.5]) arising from that work is that two Brunnian θ-graphs are equivalent if and only if they are neighborhood-equivalent. This suggests that Brunnian θ-graphs may be determined by their complements.

Conjecture 2. If two Brunnian θ-graphs have homeomorphic exteriors, then they are equivalent.

In general, the topology of Brunnian graphs (including, but not limited to, Brunnian θ-graphs) is an area ripe for further study.

I would like to thank the referee and my students Tara Brownson and Qidong He for helpful comments.

56.1 Bibliography

[1] Colin C. Adams. Thrice-punctured spheres in hyperbolic 3-manifolds. *Trans. Amer. Math. Soc.*, 287(2):645–656, 1985.

[2] Jack S. Calcut and Jules R. Metcalf-Burton. Double branched covers of theta-curves. *J. Knot Theory Ramifications*, 25(8):1650046, 9, 2016.

[3] Toen Castle, Myfanwy E. Evans, and S.T. Hyde. Ravels: Knot-free but not free. novel entanglements of graphs in 3-space. *New J. Chem.*, 32:1484–1492, 2008.

[4] C. Ernst and D. W. Sumners. A calculus for rational tangles: Applications to DNA recombination. *Math. Proc. Cambridge Philos. Soc.*, 108(3):489–515, 1990.

[5] Erica Flapan and Allison N. Miller. Ravels arising from Montesinos tangles. *Tokyo J. Math.*, 40(2):393–420, 2017.

[6] C. McA. Gordon and J. Luecke. Knots are determined by their complements. *J. Amer. Math. Soc.*, 2(2):371–415, 1989.

[7] Damian Heard, Craig Hodgson, Bruno Martelli, and Carlo Petronio. Hyperbolic graphs of small complexity. *Experiment. Math.*, 19(2):211–236, 2010.

[8] C. Hog-Angeloni and S. Matveev. Roots in 3-manifold topology. In *The Zieschang Gedenkschrift*, volume 14 of *Geom. Topol. Monogr.*, pages 295–319. Geom. Topol. Publ., Coventry, 2008.

[9] Byoungwook Jang, Anna Kronaeur, Pratap Luitel, Daniel Medici, Scott A. Taylor, and Alexander Zupan. New examples of Brunnian theta graphs. *Involve*, 9(5):857–875, 2016.

[10] Louis H. Kauffman. Invariants of graphs in three-space. *Trans. Amer. Math. Soc.*, 311(2):697–710, 1989.

[11] Akio Kawauchi. Almost identical imitations of (3, 1)-dimensional manifold pairs. *Osaka J. Math.*, 26(4):743–758, 1989.

[12] Shin'ichi Kinoshita. On elementary ideals of polyhedra in the 3-sphere. *Pacific J. Math.*, 42:89–98, 1972.

[13] Shin'ichi Kinoshita. On θ_n-curves in $\mathbf{R}^3$ and their constituent knots. In *Topology and Computer Science (Atami, 1986)*, pages 211–216. Kinokuniya, Tokyo, 1987.

[14] Rick Litherland. The Alexander module of a knotted theta-curve. *Math. Proc. Cambridge Philos. Soc.*, 106(1):95–106, 1989.

[15] Charles Livingston. Knotted symmetric graphs. *Proc. Amer. Math. Soc.*, 123(3):963–967, 1995.

[16] W. K. Mason. Homeomorphic continuous curves in 2-space are isotopic in 3-space. *Trans. Amer. Math. Soc.*, 142:269–290, 1969.

[17] Sergei Matveev and Vladimir Turaev. A semigroup of theta-curves in 3-manifolds. *Mosc. Math. J.*, 11(4):805–814, 822, 2011.

[18] Jenelle McAtee, Daniel S. Silver, and Susan G. Williams. Coloring spatial graphs. *J. Knot Theory Ramifications*, 10(1):109–120, 2001.

[19] Makoto Ozawa. Bridge position and the representativity of spatial graphs. *Topology Appl.*, 159(4):936–947, 2012.

[20] Makoto Ozawa and Scott A. Taylor. Two More Proofs that the Kinoshita Graph is Knotted. *Amer. Math. Monthly*, 126(4):352–357, 2019.

[21] Martin Scharlemann. Some pictorial remarks on Suzuki's Brunnian graph. In *Topology '90 (Columbus, OH, 1990)*, volume 1 of *Ohio State Univ. Math. Res. Inst. Publ.*, pages 351–354. de Gruyter, Berlin, 1992.

[22] Martin Scharlemann and Abigail Thompson. Detecting unknotted graphs in 3-space. *J. Differential Geom.*, 34(2):539–560, 1991.

[23] Jonathan K. Simon and Keith Wolcott. Minimally knotted graphs in S^3. *Topology Appl.*, 37(2):163–180, 1990.

[24] Shin'ichi Suzuki. Almost unknotted θ_n-curves in the 3-sphere. *Kobe J. Math.*, 1(1):19–22, 1984.

[25] Scott A. Taylor. Boring split links. *Pacific J. Math.*, 241(1):127–167, 2009.

[26] Scott A. Taylor. Comparing 2-handle additions to a genus 2 boundary component. *Trans. Amer. Math. Soc.*, 366(7):3747–3769, 2014.

[27] Scott A. Taylor and Maggy Tomova. Additive invariants for knots, links and graphs in 3-manifolds. *Accepted by Geom. Top.*, 2018.

[28] Abigail Thompson. A polynomial invariant of graphs in 3-manifolds. *Topology*, 31(3):657–665, 1992.

[29] Keith Wolcott. The knotting of theta curves and other graphs in S^3. In *Geometry and Topology (Athens, Ga., 1985)*, volume 105 of *Lecture Notes in Pure and Appl. Math.*, pages 325–346. Dekker, New York, 1987.

[30] Ying Qing Wu. On planarity of graphs in 3-manifolds. *Comment. Math. Helv.*, 67(4):635–647, 1992.

[31] Ying Qing Wu. Minimally knotted embeddings of planar graphs. *Math. Z.*, 214(4):653–658, 1993.

[32] Shuji Yamada. An invariant of spatial graphs. *J. Graph Theory*, 13(5):537–551, 1989.

Part X

Quantum Link Invariants

Chapter 57

Quantum Link Invariants

D. N. Yetter, Kansas State University

57.1   Introduction

What are now called "quantum link invariants" began in the mid-1980s as "knot polynomials" determined by skein relations (see chapters in Part XI of this volume).

Jones's construction of his polynomial [13], as well as the proofs of Freyd and Yetter and of Ocneanu of the existence of the HOMFLY-PT polynomial [10, 23], all fit comfortably into the framework provided by Markov's theorem [2], that all classical links can be represented by elements of the Artin braid group [1] as "closed braids," and that two elements of braid groups represent the same link if and only if one can be reached from the other by a sequence of conjugation (or equivalently binary factorization and exchange of the factors) within a braid group, and "stabilization" or "destabilization" moves: applying the inclusion from the braid group B_n into the braid group B_{n+1} then multiplying by the added generator or its inverse, or the reverse.

The next developments, however did not. Both Kauffman's reconstruction of the Jones polynomial from his "bracket polynomial" [14], and the construction of Brandt, Lickorish and Millet [5] and independently, Ho [11], which led to the construction of the Kauffman polynomial (see Chapter 67) involved "turnarounds" in their skein relations shown in Figure 57.1 (in the context where braids are written with strands descending the page) – the summands in the skein relations designated K_∞ in [5].

FIGURE 57.1: K_∞.

57.2   Functorial Knot Theory: Categories of Tangles

Motivated by this, Yetter [29] and, independently Turaev [26], gave expositions of an algebraic structure in which both braids and the turnarounds lived quite naturally: (monoidal) categories of tangles.

Definition 57.2.1. A *geometric tangle* is a tamely embedded 1-submanifold with boundary T of $[0, 1]^3$ with $\partial T = T \cap (0, 1) \times (0, 1) \times \{0, 1\}$.

A typical geometric tangle is shown in Figure 57.2.

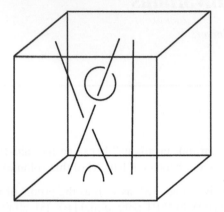

FIGURE 57.2: A tangle.

Note that "tangle" as used here is a generalization of Conway's tangles (see Chapter 12) in that, rather than two incoming and two outgoing strands, any finite number of strands (including 0) may be incoming and any (possibly different) number outgoing.

Definition 57.2.2. The *category of tangles* (resp. *oriented tangles, framed tangles*), **Tang** (resp. **OTang**, **FrTang**), has as objects finite sets of points (resp. signed points, signed points with a framing of their normal bundle) in $(0, 1) \otimes (0, 1)$, and as arrows isotopy rel boundary classes of geometric tangles (resp. geometric tangles equipped with an orientation, geometric tangles equipped with an orientation and framing).

The source of an arrow $[T]$ is $T \cap (0, 1) \times (0, 1) \times \{0\}$, the target is $T \cap (0, 1) \times (0, 1) \times \{1\}$ (resp. $T \cap (0, 1) \times (0, 1) \times \{1\}$ with its orientation reversed, $T \cap (0, 1) \times (0, 1) \times \{1\}$ with its orientation reversed). Composition is induced by the map $[0, 1]^3 \coprod [0, 1]^3 \to [0, 1]^3$ given on the first summand by $(x, y, z) \mapsto (x, y, z/2)$ and on the second by $(x, y, z) \mapsto (x, y, (z + 1)/2)$.

The identity arrow for an object $X \subset (0, 1) \times (0, 1)$ in any category of tangles, is represented by the a geometric tangle (with appropriate auxiliary structures) given by $X \times [0, 1] \subset [0, 1]^3$.

This category is equipped with a natural monoidal structure with the monoidal product $\otimes$ induced on arrows by the map $[0, 1]^3 \coprod [0, 1]^3 \to [0, 1]^3$ given on the first summand by $(x, y, z) \mapsto (x/2, y, z)$ and on the second by $(x, y, z) \mapsto ((x + 1)/2, y, z)$, and monoidal identity given by $I = \emptyset \subset (0, 1) \times (0, 1)$. These satisfy the usual pentagon and triangle coherence relations in the definition of monoidal categories shown in Figure 57.3.

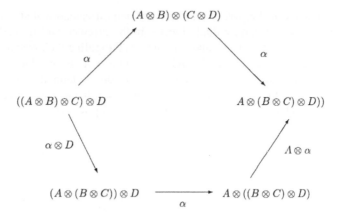

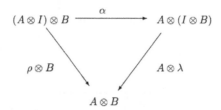

FIGURE 57.3: Coherence conditions for monoidal categories.

Crossings equip categories of tangles with a structure that generalizes the more classical notion of a symmetric monoidal category – a braiding $\sigma_{A,B} : A \otimes B \to B \otimes A$. In the case of a symmetry in which the inverse natural transformation has the same component maps as the symmetry, a single hexagon coherence condition suffices. For a braiding the condition must hold both for σ and σ^{-1} as given in Figure 57.4.

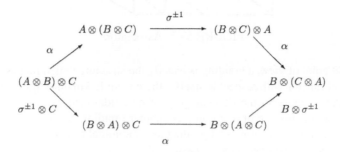

FIGURE 57.4: The hexagons.

The "turnarounds" turn out to be part of another very natural categorical structure: they are the structure maps for an object to have a dual object – in the oriented and framed cases, an object given by mirror-imaging the set of point (and framings) and toggling their orientations, and simply the mirror image in the unoriented case. A (right) dual object X^* to an object X (once one does not have a symmetry, even with a braiding, duality has a side) is defined by being equipped with structure maps generalizing the evaluation and projective coordinate system maps for dual vector spaces, $\epsilon : X \otimes X^* \to I$ and $\eta : I \to X^* \otimes X$, for which the compositions

$$X \xrightarrow{\rho^{-1}} X \otimes I \xrightarrow{X \otimes \eta} X \otimes (X^* \otimes X) \xrightarrow{\alpha^{-1}} (X \otimes X^*) \otimes X \xrightarrow{\epsilon \otimes X} I \otimes X \xrightarrow{\lambda} X$$

and

$$X^* \xrightarrow{\lambda^{-1}} I \otimes X^* \xrightarrow{\eta \otimes X^*} (X^* \otimes X) \otimes X^* \xrightarrow{\alpha} X^* \otimes (X \otimes X^*) \xrightarrow{X^* \otimes \epsilon} X^* \otimes I \xrightarrow{\rho} X^*$$

are both identity arrows.

In each case the structure maps for the additional categorical structure – the associator and unitor for the monoidal structure, the braiding and its inverse, and the evaluation and coevaluation maps for the dual objects, respectively – are given by the image of the geometric tangle inducing an identity arrow under an appropriate inclusion of $(0, 1)^2 \times [0, 1]$ (or for the associator and the braiding a disjoint union of three or two copies of this, respectively) into $[0, 1]^3$, shown schematically in Figures 57.5 to 57.8

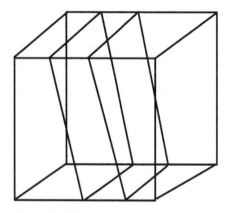

FIGURE 57.5: Associator.

From a categorical point of view, a braiding is exactly the structure which makes a monoidal product into a monoidal functor with respect to itself – there is an Eckmann-Hilton-type argument that shows "a monoidal category in the (2-) category of monoidal categories is a braided monoidal category." Of course, the inverse of a braiding is another braiding, so any braided monoidal category has two ways to regard its monoidal product as a monoidal functor. A balancing is exactly a monoidal natural isomorphism between the two.

Viewed in this way the natural coherence equation satisfied by a balancing θ_A is

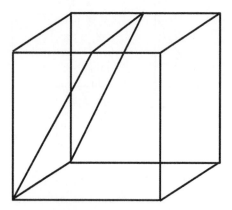

FIGURE 57.6: Left unitor.

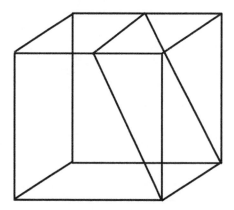

FIGURE 57.7: Right unitor.

$$[\sigma_{B,A}]^{-1}(\theta_{A\otimes B}) = \sigma_{A,B}(\theta_A \otimes \theta_B),$$

though it is more usual to write the equivalent condition

$$\theta_{A\otimes B} = \sigma_{B,A}(\sigma_{A,B}(\theta_A \times \theta_B)).$$

Categories of tangles, are in fact balanced, and, as with all the other structures, the balancing is induced by an inclusion of $[0, 1]^3$ into $[0, 1]^3$, shown schematically in Figure 57.9.

That is, the balancing in a category of tangles is obtained by applying a full twist to the tangle giving an identity arrow, from which the second formulation of the coherence condition is immediate! For this reason, even in more general ribbon categories, the balancing is often called the *ribbon twist*.

The balancing in categories of tangles also interacts nicely with the duality structure – the braiding

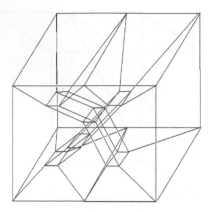

FIGURE 57.8: Braiding.

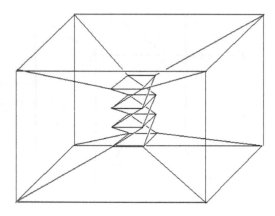

FIGURE 57.9: Balancing.

makes right dual objects into left dual objects and the balancing in turn makes them into two-sided duals (cf. [32]), though the usual condition stated in terms of right duals only is simply that

$$\theta_{X^*} = (\theta_X)^*$$

using the fact that a choice of dual objects for each object extends to a contravariant functor.

Collecting all of this structure together, we make:

Definition 57.2.3. A *ribbon category* is a monoidal category $(C, \otimes, I, \alpha, \rho, \lambda)$ moreover equipped with a braiding σ and a balancing θ in which every object X admits a right dual X^* and in which $\theta_{X^*} = (\theta_X)^*$.

Theorem 57.2.4. [Shum's coherence theorem][1] *The ribbon category freely generated by a single object is equivalent as a ribbon category to* **FrTang**, *the category of framed tangles.*

---

[1] Shum [25], who called ribbon categories "tortile categories," actually proved a more general theorem, characterizing up to equivalence the ribbon category freely generated by any (small) category. It is, however, the special case given which is most relevant to classical knot theory and low-dimensional topology in general.

This results in a reasonable definition of a quantum link invariant:

Definition 57.2.5. A *quantum link invariant* is an invariant of framed links (resp. links, oriented links) which arises from a monoidal functor from **FrTang** to a monoidal category linear over some ring or field in which the monoidal product is bilinear on arrows (resp. such a functor factoring through the forgetful functor to **Tang**, such a functor factoring through the forgetful functor to **OTang**), taking values in the ring of endomorphism of the monoidal identity object.

Or rather a definition which is reasonable once it is observed that the knot polynomials (or at least Zariski dense sets of their values) arise from precisely this sort of construction with the target category for the monoidal functor being the representations of a quantum group, in the sense of a quantized universal enveloping algebra, $U_q(\mathfrak{g})$ for some Lie algebra $\mathfrak{g}$ (for the HOMFLY-PT polynomial a Lie algebra of type A_n, for the Kauffman polynomial, a Lie algebra of type D_n), the functor being induced by sending the downward-oriented strand to the fundamental representation of the QUEA.

Most often the monoidal product in the target category is (or is induced by) the tensor product over the ring or field, but the slightly more generous view taken in Definition 57.2.5 allows the whole two-variable HOMFLY-PT and Kauffman polynomials to be seen as quantum link invariants if one is willing to settle for a target category constructed explicitly from the skein relations: linearize **FrTang** with respect to the ring of Laurent polynomials and quotient the resulting category by the congruence of monoidal categories generated by the skein relations.

Some other link invariants usually not constructed in terms of functors from a category of tangles also turn out to be quantum link invariants in this sense: for instance, the invariants which count the number of colorings of a link by a given quandle or biquandle (or in the framed case rack or birack) (see Chapter 79): a braiding can be constructed on the category whose objects are rational vector spaces, which are tensor products of the vector-space X_Q with the given quandle, biquandle, rack, or birack Q as a basis and its dual space in such a way that it fits with the actual projective coordinate system and evaluation maps to produce a ribbon category. The counting invariant is then the quantum link invariant arising from the functor that sends the downward-oriented strand to X_Q.

Shum's Coherence Theorem assures us that there are plenty of quantum link invariants: given any linear ribbon category, and a choice of object in it, there is an induced functor from **FrTang** and a resulting quantum link invariant valued in the ring of endomorphisms of the monoidal identity object.

57.3 Why "Quantum"?

The simple answer sometimes given, "These are quantum invariants because they come from quantum groups," is either perfectly defensible if one takes a very broad notion "quantum groups" or rather inadequate if one does not. In either case, though, it begs the question by leaving unsaid why quantum groups, which are most assuredly not groups, are, in fact quantum.

Actually, the most satisfying answer as to why link invariants arising from the functors described in the last section are properly called "quantum link invariants" makes no reference to quantum groups: the functors are, in fact, toy models of quantum field theory (possibly generalized by allowing scalars other than the real or complex numbers) – the ground or vacuum state is the empty set of points, the state spaces assigned to sets of (possibly signed and framed) points tensor when disjoint unions are taken, the structure maps for the duality are creation and annihilation operators for particle-antiparticle pairs, and the values of the link invariant are vacuum-to-vacuum amplitudes – at least formally, everything has the same structure as occurs in actual quantum physics (leaving aside the fact that the pairing defining dual spaces is bilinear, rather than sequilinear).

There is, however, another answer, more closely tied to quantum groups.

If one applies Mac Lane's coherence theorem and moves to a strict monoidal category equivalent to the one started with so that all instances of the associator α and its inverse are now identity arrows, the hexagon coherence condition becomes

$$\sigma_{X \otimes Y, Z} = \sigma_{X,Z} \otimes Y(X \otimes \sigma_{Y,Z}).$$

The naturality square for the map $\sigma_{A,B} \otimes C$, written as an equation

$$\sigma_{B \otimes A, C}(\sigma_{A,B} \otimes C) = C \otimes \sigma_{A,B}(\sigma_{A \otimes B, C})$$

can thus be rewritten as

$$\sigma_{B,C} \otimes A(B \otimes \sigma_{A,C}(\sigma_{A,B} \otimes C)) = C \otimes \sigma_{A,B}(\sigma_{A,C} \otimes B(A \otimes \sigma_{B,C}))$$

or more briefly as

$$\sigma \otimes 1(1 \otimes \sigma(\sigma \otimes 1)) \;\; = \;\; 1 \otimes \sigma(\sigma \otimes 1(1 \otimes \sigma)) \tag{57.1}$$

that is, it is precisely the version of the quantum Yang-Baxter equation (qYBE) without the spectral parameter. Thought of in that way, σ would traditionally be denoted R, solutions to the qYBE being called (quantum) R-matrices.

The construction of solutions to the qYBE in the guise in which it appears in quantum inverse problems was Drinfel'd's original motivation for studying quantum groups in the narrow sense of Hopf algebra deformations of universal enveloping algebras (or dually of function algebras on Lie groups) [8]. Thus a second reason for calling these invariants "quantum link invariants" is that they are the link invariants which arise from (spectral parameter-free) solutions to the quantum Yang-Baxter equation.

However, with a generous notion of "quantum group," the view that the class of invariants "comes from quantum groups" is perfectly sound:

Linear ribbon categories with a forgetful functor to a category of vector-spaces over a field, all arise as categories of (co)modules over the broadest reasonable notion of "quantum group" – a (co)ribbon quasi-Hopf algebra, via Tannaka-Krein reconstruction theory (see, for example Majid [21]). If the forgetful functor is monoidal, the "quasi-" can be omitted.[2]

---

[2]For technical reasons beyond the scope of this chapter, reconstruction theorems always produce a given category as a

57.4 Ribbon Hopf Algebras

The vast bulk of ribbon categories from which link invariants have been constructed either explicitly or implicitly via Shum's theorem are categories of modules over ribbon Hopf algebras:

Definition 57.4.1. A *ribbon Hopf algebra A* over a field K is an associative K-algebra, whose multiplication we write either as the null infix or as m as a prefix, and whose unit, whether as an element or as a map from K, we denote by 1, moreover equipped with the following:

1. K-algebra homomorphisms $\Delta : A \to A \otimes_K A$ and $\epsilon : A \to K$ called the *comultiplication* and *counit*, respectively, satisfying the formal duals of associativity and unitalness:

$$(A \otimes_K \Delta)(\Delta) = (\Delta \otimes_K A)(\Delta) \quad (A \otimes_K \epsilon)(\Delta) = Id_A = (\epsilon \otimes_K A)(\Delta)$$

   where as is customary we have used Mac Lane's coherence theorem to suppress mention of the associator and unitors in the category of K-vector spaces.

2. A K-linear map $S : A \to A$, called the *antipode*, satisfying

$$m(S \otimes A(\Delta)) = 1(\epsilon) = m(A \otimes S(\Delta)).$$

3. A unit R in the algebra $A \otimes A$, called the *universal R-matrix*, satisfying

   - $R\Delta = \sigma(\Delta)R$, where σ denotes the symmetry in K-**vect** and the null infix denotes multiplication in $A \otimes_K A$,

   - $(\Delta \otimes A)(R) = R_{13}R_{23}$,

   - $(A \otimes \Delta)(R) = R_{13}R_{12}$,

   where R_{ij} denotes the image of R under the inclusion of the tensor square of A into its tensor cube with the unit 1 in the other tensorand, and the first and second tensorand sent to the ith and j^{th}, respectively.

   And, finally,

4. An element v, called the *ribbon element*, satisfying

   - $\Delta(v) = (\sigma(R)R)^{-1}(v \otimes_K v)$, and

   - $v^2 = uS(u)$,

   - $S(v) = v$,

   - $\epsilon(v) = 1$.

(subcategory of) a category of comodules, with additional structure on the coalgebra corresponding to additional structure on the category and forgetful functor – e.g., bialgebra corresponding to monoidal structures on both. In the discussion in the next section, we will instead treat the more familiar categories of modules. The reader may, of course, dualize everything by reversing arrows, thereby swapping elements with functionals, actions with coactions, multiplication with comultiplication, and so forth.

Here, again, σ denotes the symmetry in K-**vect**.

An algebra equipped with (1) only is called a bialgebra; with (1) and (2), a Hopf algebra; with (1), (2) and (3) a quasi-triangular Hopf algebra.

Examples are plentiful: besides the quantized universal enveloping algebras, any cocommutative Hopf algebra is a ribbon Hopf algebra (though these are not very interesting from the point of view of knot theory), as are the Drinfel'd doubles [7] of finite-dimensional Hopf algebras.

If some of the axioms and names seemed to parallel the discussion of ribbon categories above, this is not accidental, in fact we have:

Theorem 57.4.2. *The category of modules* **Rep**(A) *over a ribbon Hopf algebra A over K, equipped with $\otimes_K$, $I = K$ and the associator and unitors inherited from K – **vect** is a ribbon category.*

The bits of structure in the ribbon Hopf algebra with names suggestive of names in the structure of ribbon categories are those which induce the corresponding structures, for instance, the braiding on $\mathbb{R}ep(A)$ is given by $\sigma(R\cdot)$ with $\cdot$ denoting the obvious action of $A \otimes_K A$ on a tensor product of two A modules, and σ the symmetry in K – **vect**, while the balancing is given by multiplication by the ribbon element.

A more general algebraic structure, called ribbon quasi-Hopf algebras (cf. [9, 21]), also have categories of modules with a ribbon category structure. In these, there is a unit in $A \otimes_K A \otimes_K A$, which induces a new associator in the same way the universal R-matrix replaces the symmetry with a braiding. And, all of this can be done not only with algebras, but with "super algebras" living in a category of graded vector spaces with the symmetry following Koszul's sign convention (see, for example [28]) .

57.5   *R*-Matrices and Markov Traces

Having reconceived the braiding as a quantum R-matrix with its defining property being Equation 57.1 suggests an alternative point of view of what should constitute quantum link invariants that remains closer to constructions of Jones and Ocneanu, and is favored in, for instance, Birman [3]. In this view the "quantum-ness" of the invariants is precisely that they come from a solution to the quantum Yang-Baxter equation.

Any R-matrix, in the sense of a single linear transformation from $R : V \otimes V \rightarrow V \otimes V$, satisfying Equation 57.1 for V a vector space over K produces representation of all of the Artin braid groups

$$\rho_{R,n} : B_n \rightarrow Gl(V^{\otimes n})$$

mapping σ_i, the generator of B_n which crosses the $i + 1^{st}$ strand over the ith strand, to $Id_V^{\otimes i-1} \otimes R \otimes$

$Id_V^{\otimes n-i-1}$. In fact, these representations assemble into a monoidal functor from the category **Braids** to K-**vect**. [3]

Markov's theorem makes it clear that if one has traces on the images of these representations (automatically giving invariance under the move $\beta\gamma \mapsto \gamma\beta$), that can somehow be normalized to give invariance under including one braid group into the next and multiplying by the new generator or its inverse, the traces will give rise to an invariant of oriented links.

The usual way of doing this is to require that the traces collectively form what is called a *Markov trace*. Following Birman [3] we make:

Definition 57.5.1. Given a sequence of representation $\rho_n : B_n \to Gl(d_n, K)$, a *Markov trace* is a sequence of functionals $tr_n : \rho_n(B_n) \to K$ such that

- $tr_n(1) = 1$,

- $tr_n(\rho_n(\beta\gamma)) = tr_n(\rho_n(\gamma\beta))$ whenever $\beta, \gamma \in B_n$, and

- $\exists z \in K$ such that $tr_{n+1}(\rho_{n+1}(\iota_n(\beta)\sigma_n^{\pm 1})) = z tr_n(\rho_n(\beta))$ whenever $\beta \in B_n$

The one-variable specializations of the HOMFLY-PT and Kauffman polynomials corresponding to the functor sending the downward-oriented strand to the fundamental representation of a QUEA of type A or D can all be constructed from Markov traces.

57.6 Deformations, Series Expansions and Vassiliev Invariants

It turns out that Vassiliev or finite-type invariants (see Chapter 82) are intimately related to quantum link invariants. In fact, in the absence of torsion all finite-type invariants are linear combinations of those arising from perturbative expansions of quantum link invariants.

Almost immediately upon the introduction of finite-type invariants, Birman and Lin [4] showed that the coefficients of x in the invariant obtained by substituting $\exp(x)$ for t in any of the usual univariate versions of the HOMFLY-PT or Kauffman polynomials (those arising by mapping the generating object of **FrTang** to the fundamental representation of a quantized universal enveloping algebra of type A or D), were finite-type invariants. In fact their proof showed more – substituting any formal power series with leading coefficient 1 for the variable would give coefficients which are a sequence of Vassiliev invariants.

The same relation can be found directly in the functorial setting: if a linear rigid symmetric monoidal category is formally deformed into a braided monoidal category, the resulting category is ribbon [30] and the now-formal-series valued quantum link invariants resulting from mapping the generating object of **FrTang** to any object have Vassiliev invariants as coefficients [31]. In fact, in the torsion-free setting, all Vassiliev invariants arise in this way: the framed Kontsevich integral as applied to links is the universal Vassiliev invariant (see Chapter 82), but as applied to tangles, it is

---

[3]The objects of **Braids** are the natural numbers, the only arrows being the braid groups B_n as endomorphisms of n and monoidal structure given by the well-known inclusion of groups $B_n \times B_m \to B_{n+m}$.

the universal (extrinsic) braided deformation of the rational linearization of the free rigid symmetric monoidal category [31, 32], which category (unlinearized) Kelly and Laplaza [18] described via earlier work of Kelly [16, 17] in terms of 'flat tangles' (Indeed, despite having semantics from symmetric monoidal categories, and thus corresponding to 'flat tangles,' the pictures in the 1972 papers [16, 17] are exactly the now familiar diagrams for oriented tangles!)

57.7   Not Quite Functorial Invariants and Quantum Invariants of 3-Manifolds

Quantum link invariants have a fairly direct application to 3-manifold topology by way of the Kirby calculus.

Recall that any compact oriented 3-manifold arises as the boundary of a 4-dimensional 2-handlebody, and thus can be presented by the framed link in S^3 describing the attaching maps or 2-handles to the 4-ball (as a 0-handle). Kirby [19] showed that two such presentations (each defined up to ambient isotopy of framed links) give the same 3-manifold if and only if there are related by a sequence of moves of two types:

1. Forming the separated union of the framed link with a ±1-framed unknot, or removing a separated ±1-framed unknotted component from the framed link.

2. Replacing one component of the framed link with its band-sum (along a band disjoint from the link) with a parallel copy of another component, with the parallel copy framed and twisted around the other component according to the framing of the other component.

As was first discovered by Reshetikhin and Turaev [24], certain quantum link invariants can be modified and normalized to produce framed link invariants, which are, moreover, invariant under the two Kirby moves, thus giving invariants of the 3-manifold described by the framed link. The target categories for the original construction of [24], a certain subcategory of the representations of the quantized universal enveloping algebra, $U_q(sl_2)$, when the deformation parameter q is a root of unity, are the first examples of a class of ribbon categories for which their construction works: modular tensor categories.

Definition 57.7.1. An **Ab**-*category* is a category all of whose hom-sets are abelian groups, with composition being bilinear, and in which every pair of objects admits a biproduct (usually called the direct sum and denoted $\oplus$). It is k-linear if the hom-sets are, moreover, k-vector spaces and composition is bilinear.

A *tensor category* over k is a k-linear **Ab**-category equipped with a monoidal structure, for which $\otimes$ is k-bilinear on the hom-vector spaces.

An object in a k-linear category is *simple* if its endomorphism vector space is 1-dimensional (that is, every endomorphism is a scalar multiple of its identity map).

A k-linear **Ab**-category is *semisimple* if every object is a direct sum of finitely many simple objects. A *fusion category* over k is a semisimple tensor category over k in which there are only finitely many isomorphism classes of simple objects, among which is the monoidal identity object I.

And finally, a *modular tensor category over k* is a ribbon fusion category over k in which an object X being *transparent*, that is

$$\forall Y \in Ob(C)\ \sigma_{Y,X}(\sigma_{X,Y}) = Id_{X \otimes Y},$$

implies that X is isomorphic to a direct sum of copies of I.[4]

In a modular tensor category, the value of the 0-framed unknot under the quantum link invariant obtained by mapping the downward-oriented strand in **FrTang** to an object, X, is called the *quantum dimension of X*, and we denote it by $\dim(X)$.

The quantum link invariant which, is modified to produce invariants of 3-manifolds by the Reshtikhin-Turaev, is that obtained from the functor from **FrTang** that sends the downward strand to the object $\oplus_{i=1}^{n} X_i$, where the X_i range over representatives of the n isomorphism classes of simple objects.

The first modification is to insert one copy of the endomorphism of $\oplus_{i=1}^{n} X_i$, given by multiplying each direct summand by its quantum dimension, on each component of the framed link.[5] The resulting scalar (the multiple of Id_I given by the image of the link with the endomorphisms inserted) is an invariant of framed links since an isotopy invariant of framed links with a basepoint on each component is an invariant of framed links, but one whose behavior under the Kirby moves is tractable. If we denote the value assigned to a framed link L in this way by $\{L\}$, and let $D = \sum_{i=1}^{n} \dim^2(X_i)$, $\Delta = \{U_{-1}\}$ for U_{-1} the -1-framed unknot, then we have:

Theorem 57.7.2. *For any framed link L*

$$\Delta^{\sigma(L)} D^{-\sigma(L)-m-1}\{L\},$$

where m is the number of components of L and $\sigma(L)$ is the signature of the framed linking matrix of L, is invariant under the Kirby moves, and thus is an invariant of the 3-manifold presented by L.

Chapter II of [27] is devoted to a detailed proof of this theorem.

57.8 Bibliography

[1] Artin, E., Theory of braids, *Ann. Math.* **48** (1) (1947) 101–126.

[2] Birman, J.S., Braids, links and mapping class groups, *Ann. Math. Studies* **82**, Princeton U. Press, Princeton, NJ (1975).

[3] Birman, J.S., New points of view in knot theory, *Bull. AMS* **28** (2) (1993) 253–287.

---

[4]The original "modular" condition on a ribbon fusion category was given not in terms of a paucity of transparent objects, but as the condition that a matrix obtained as the traces of the squares of the braiding ranging over pairs of representatives of the isomorphism classes of simple objects be invertible. Bruguières [6] and Müger [22] showed this condition which is completely arcane to those with no familiarity with conformal field theory is equivalent to the more categorically obvious condition given here.

[5]This modification is usually described as coloring the components of the link with the formal linear combination of the simple objects with their quantum dimensions as coefficients.

[4] Birman, J.S. & Lin, XS., Knot Polynomials and Vassiliev's invariants, *Invent. Math.* (1993) 111: 225. https://doi.org/10.1007/BF01231287.

[5] Brandt, R.D., Lickorish, W.B.R. and Millet, K.C., A polynomial invariant for unoriented knots and links, *Invent. Math.*, **84** (1986) 563–573.

[6] Bruguières A., Catégories prémodulaires, modularisations et invariants des variétés de dimension 3, *Math. Ann.* **316** (2) (2000), 215–236.

[7] Drinfel'd, V.G., Quantum groups, in *Proceedings of the ICM* (A. Gleason, ed.) AMS, Providence, RI (1987) 789–820.

[8] Drinfel'd, V.G., Quantum groups, *J. Sov. Math* **41** (2) (1988) 898–915.

[9] Drinfel'd, V.G., Quasi-Hopf algebras, *Leningrad J. of Math.* **1** (1990) 1419–1457.

[10] Freyd, P.J., Yetter, D.N.; Hoste, J.; Lickorish, W.B.R., Millet, K.; and Ocneanu, A., A new polynomial invariant of knots and links, *Bull. Amer. Math. Soc.*, **12** (1985) 239–249

[11] Ho, C.F., A new polynomial for knots and links; preliminary report, *Abstracts Amer. Math. Soc.*, **6** (4) (1985) 300.

[12] Hoste, J., A polynomial invariant of knots and links, *Pacific J. Math.*, **124** (1986) 295–320.

[13] Jones, V.F.R., A new polynomial invariant of knots and links, *Bull. AMS* **12** (1985) 103–111.

[14] Kauffman, L.H., State models and the Jones polynomial, *Topology* **26** (1987) 395–407.

[15] Kauffman, L.H., *Knots and Physics* World Scientific, Singapore (1994).

[16] Kelly, G.M., Many variable functorial calculus, I, in *Coherence in Categories* SLNM 281, Springer, Berlin (1972) 66–105.

[17] Kelly, G.M., An abstract approach to coherence, in *Coherence in Categories* SLNM 281, Springer, Berlin (1972) 106–147.

[18] Kelly, G.M. and Laplaza, M.L., Coherence for compact closed categories, *J. Pure A Alg.* **19** (1980) 193–213.

[19] Kirby, R., A calculus for framed links in S^3, *Invent. Math.* **45** (1978) 35–56.

[20] Lickorish, W.B.R. and Millet, K., A polynomial invariant of oriented links, *Topology*, **26** (1987) 107–141.

[21] Majid, S., Tannaka-Krein theorems for quasi-Hopf algebras and other results, in *Deformation Theory and Quantum Groups with Applications to Mathematical Physics* (M. Gerstenhaber and J.D. Stasheff, eds.) AMS Contemp. Math. vol. 134 (1992), 219–219.

[22] Müger M., On the structure of modular categories, *Proc. London Math. Soc.* (3) **87** (2) (2003), 291–308.

[23] Przytyck, J. and Traczyk, P., Invariants of links of Conway type, *Kobe J. Math.*, **4** (1987) 115–139

[24] Reshetikhin, N. and Turaev, V.G., Invariants of 3-manifolds via link polynomials and quantum groups, *Invent. Math.* **103** (3) (1991), 547–597.

[25] Shum, M.-C., *Tortile Tensor Categories*, doctoral dissertation, Macquarie University, 1989.

[26] Turaev, V.G., Operator invariants of tangles, and R-matrices, *Mathematics of the USSR – Izvestiya*, **35** (2) (1990) 411–444.

[27] Turaev, V.G., *Quantum Invariants of Knots and 3-Manifolds*, Walter deGruyter, New York, 1994.

[28] Yamane, H., Quantized enveloping algebras associated to simple Lie superalgebras and their universal R-matrices, *Publ. RIMS, Kyoto Univ.*, **30** (1994), 15–87.

[29] Yetter, D.N., Markov algebras, in *Braids* (Birman, J.S. and Libgober, A., eds) AMS Contemp. Math. Vol. 78, AMS Providence, RI (1988) 705–730.

[30] Yetter, D.N., Framed tangles and a theorem of deligne on braided deformations of Tannakian categories, in *Deformation Theory and Quantum Groups with Applications to Mathematical Physics* (M. Gerstenhaber and J.D. Stasheff, eds.) AMS Contemp. Math. vol. 134 (1992) 325–350.

[31] Yetter, D.N., Braided deformations of monoidal categories and Vassiliev invariants, in *Higher Category Theory: Workshop on Higher Category Theory and Physics, March 28-30, 1997, Northwestern University, Evanston, IL* (E. Getzler and M. Kapranov, eds.), AMS Contemp. Math Series vol. 230, AMS, Providence, RI (1998) 117–134, also q-alg e-print #9710010.

[32] Yetter, D.N., *Functorial Knot Theory: Categories of Tangles, Coherence, Categorical Deformations and Topological Invariants*, World Scientific, Singapore (2001).

Chapter 58

Satellite and Quantum Invariants

H. R. Morton, University of Liverpool

58.1 Algebras and Knot Polynomials

One of the most exciting and fruitful features of the burgeoning of ideas released by the discovery of the invariant known as the Jones polynomial is the extent of the different strands which have come to be drawn together in the course of explorations. These include algebraic, combinatoric and geometric aspects, and the involvement of quantum groups originating from theoretical physics.

A key strand from early times has been the appearance of the Hecke algebras (initially those of type A), and their role in both defining the Jones, and subsequently the Homfly polynomials [7], while at the same time their algebraic involvement with the quantum groups of the A series, and their very satisfactory modelling by pieces of knot diagrams [18].

From this standpoint the Hecke algebras H_n of type A are best defined as linear combinations of n-braids, in the sense of Artin, in which the elementary Artin braids $\{\sigma_i\}$ with

$$\sigma_i \quad = \quad$$

each satisfy a quadratic relation in addition to the braid relations.

Up to isomorphism, the quadratic relation can take the form

$$\sigma^2 - z\sigma - 1 = 0$$

for a parameter z, or equally

$$(\sigma - s)(\sigma + s^{-1}) = 0,$$

where $z = s - s^{-1}$.

Replacing σ by $x^{-1}\sigma$ gives an isomorphic algebra with the relation $x^{-1}\sigma - x\sigma^{-1} = z$, so that any other quadratic relation can be used instead. The 2-parameter version lies at the heart of the Homfly polynomial.

In algebraic literature the letters g_i or T_i often replace σ_i, and the quadratic relation $(g_i - q)(g_i + 1) = 0$ translates to $x = s, q = s^2$ above, while the notation $x = \alpha$ or $x = v$ occurs at an early stage in the work of Kauffman or Morton [11].

As an aside, the cases $v = s^N, z = s - s^{-1}$ showed up very early in the Hecke algebra explorations of

547

Jones, [7], and relate closely to the basic invariants from the quantum groups $sl_q(N)$, with $q = s^2$. The original Jones polynomial fits readily here with the case $N = 2$.

From the knot-theoretic viewpoint the most useful approach is to work with linear combinations of knot diagrams, using a ground ring Λ which is often $\mathbb{Z}[v^{\pm 1}, s^{\pm 1}]$ or some variant with Laurent polynomials in two independent parameters, and possibly some allowed denominators, such as $s^r - s^{-r}$.

At a crossing in the diagram, impose the linear relation

between diagrams which only differ as shown. This is a direct counterpart to the quadratic relation $\sigma_i - \sigma_i^{-1} = z$ Id and gives the basic version of the Hecke algebra H_n, when using braids on n strings as diagrams.

The more general case with

$$ x^{-1} \qquad - x \qquad = z $$

corresponds to using the quadratic with roots $xs, -xs^{-1}$ and $z = s - s^{-1}$. This approach is developed in [12, 1] and [10].

The relation

$$ l \qquad + l^{-1} \qquad + m \qquad = 0, $$

which is the form adopted by Lickorish and Millett for their initial diagram-based construction of the Homfly polynomial, corresponds to $l = ix^{-1}, m = iz$.

Remark 58.1.1. The version using x and s adapts more readily to quantum group invariants, and also to finite-type invariants, and is now generally adopted, up to the letters chosen for the parameters, in favour of the equivalent l, m form.

The parameter x can be sidelined quite a bit in calculations by working with framed (or 'banded') diagrams, which carry a specified ribbon neighbourhood for every curve. This is often given implicitly as the 'blackboard framing' in which the band lies parallel to the curve as drawn in the diagram. Then, use of Reidemeister moves R_{II} and R_{III} on diagrams preserve the implicit bands, and the basic relation

can be used, along with a parameter v to handle the effect of R_I on a diagram, which introduces a twist in the implicit band. This shows up as the additional relation

Banded diagrams can be adapted to incorporate a parameter x by simply multiplying any diagram

D by $x^{\mathrm{wr}(D)}$, where $\mathrm{wr}(D)$ is the 'writhe' of the diagram, in other words the number of positive crossings minus the number of negative crossings. In this adaptation the relations become

$$x^{-1}\ \overset{\nearrow\ \nwarrow}{\underset{}{\bigtimes}}\ -\ x\ \overset{\nearrow\ \nwarrow}{\underset{}{\bigtimes}}\ =\ z\ \Big)\Big(\ \ \ )($$

and

$$\overset{\uparrow}{\mathbb{b}}\ =\ xv^{-1}\ \overset{\uparrow}{\int}\ ,\quad \overset{\uparrow}{\mathbb{d}}\ =\ x^{-1}v\ \overset{\uparrow}{\int}\ .$$

This allows for any adjustment of the quadratic relation that may be wanted when comparing invariants.

Early Hecke algebra calculations of the Homfly polynomial [16] made use of the case $x = 1$ to simplify the work with braids, so that only the parameter z was needed for a great part of the calculations, and the second parameter v was only incorporated in the closing stages. The general method of working with framed diagrams followed Kauffman's similar approach to the second variable in his construction of the 2-variable Kauffman polynomial.

58.2 Satellite Invariants

Framed (banded) knots and links form the natural setting for the use of satellites in developing extra invariants based on the Homfly polynomial.

Calculations in [16] showed at an early stage that, unlike the special case of the Alexander polynomial, there were potential further invariants available for a knot K by the simple device of calculating the Homfly polynomial of chosen satellites of K. For example, to compare K and K' we could look at the 2-cable of each (with the same number of twists). If K and K' are equivalent knots then so are the corresponding 2-cables, and hence the cables will have the same Homfly polynomials. However, it can happen that cables have different polynomials even when the original knots K, K' have the same polynomial, showing that the polynomial of the chosen cable is an invariant which can carry information that is independent from the polynomial of the knot itself [16, 19].

To give a systematic account of satellites we start with a banded knot K, and 'decorate' K with a 'pattern' Q.

For example, when

$$K\quad =\quad$$

and Q is the simple 2-cable pattern

$$Q \quad = \quad$$

then the resulting satellite is

$$K * Q \quad = $$

In general, a pattern is a framed diagram $Q \subset A \times I$ lying in the standard thickened annulus $A \times I$. Placing the annulus around the chosen band neighbourhood of K carries Q to the satellite $K * Q$.

For each choice of Q we can then regard the Homfly polynomial $P(K * Q)$ as an invariant of the banded knot K, written $P(K : Q)$.

While very many choices of Q are available, giving potentially a large number of different invariants for K, it is possible to spot relations between these, where the choices for a pattern are closely related as diagrams within $A \times I$. If, for example, we restrict attention to patterns formed from closed m-string braids in $A \times I$, then these will result in at most $p(m)$ linearly independent resulting invariants, where $p(m)$ is the number of partitions of the positive integer m.

A partition of m is often displayed as a Young diagram with m cells. For example, the partition of 8 into three parts $8 = 4 + 2 + 2$ can be visualised as a diagram

with 3 rows.

58.2.1   Organising Satellite Invariants

There are a number of ways to organise satellite invariants in terms of linear combinations of framed diagrams in the annulus, modulo skein relations. Among these are some particularly useful combinations Q_λ, one for each partition of m.

The resulting invariants $P(K : Q_\lambda)$ span all the $p(m)$ invariants arising from closed m-braid patterns. They have good integrality properties in the variables v, s, and change by a scalar multiple depending only on the partition λ when the twist of the band around K is changed.

The elements Q_λ can be constructed in a number of equivalent ways. The original method was as the closure of suitably chosen idempotents in the Hecke algebra H_m, [4, 1, 2, 22]. Later they appeared as the analogue of the Schur functions s_λ when the possible decorating elements in the annulus are interpreted as symmetric polynomials in a large number of variables, [9]. More recently they have been identified with eigenvectors of a simple operation on linear combinations of diagrams in the annulus, [10]. An extension of this approach can be found in [5] and the consequent integrality results in [13], while an account of all the different approaches can be found in [15, 14].

58.3   Quantum Group Invariants

One unexpected development, which emerged shortly after the knot polynomial discoveries, was the use of quantum groups in constructing 1-parameter invariants of framed knots and links. The initial work by Kirillov and Reshetikhin [8] on invariants derived from the quantum group $sl_q(2)$ was followed by a more systematic general approach by Reshetikhin and Turaev [20] based on the properties of the universal R-matrix in the algebraic formulations of quantum groups arising from work of Jimbo [6] and Drinfeld [3].

Connections between these invariants and knot polynomials were established at a very early stage [17, 12], although the exact details and general consequences were the subject of much subsequent work.

58.3.1   Quantum Groups

Quantum groups are algebras $\mathcal{G}$ over the formal power series ring $\mathbf{Q}[[h]] = \Lambda$. They are typically a deformed version of a classical semi-simple Lie algebra, which shows up in the limit when $h \to 0$.

Knot invariants are constructed from finite-dimensional $\mathcal{G}$-modules, making use of crucial features of the algebra $\mathcal{G}$. The major properties are that the tensor product $V \otimes W$ of two $\mathcal{G}$-modules is again a $\mathcal{G}$-module, and that there is an invertible $\mathcal{G}$-module homomorphism $R_{V,W} : V \otimes W \to W \otimes V$, determined by the universal R-matrix in $\mathcal{G}$. In addition there is a dual module V^* for each V, and homomorphisms $V \otimes V^* \to \Lambda$ and $\Lambda \to V \otimes V^*$, where the ground ring Λ acts as the trivial $\mathcal{G}$-module.

58.3.2   Colouring Tangles

Any oriented diagram consisting of closed curves and arcs connecting points at the top and bottom, as shown,

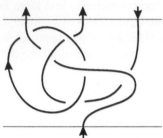

can be *coloured* by making a choice of G-module for each component.

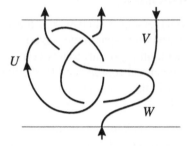

The resulting *coloured tangle* T determines a module homomorphism

$$J(T)\colon T_{\text{bottom}} \to T_{\text{top}},$$

where T_{bottom} is the tensor product of the outgoing string colours at the bottom of T, using the dual colour for any incoming string, and T_{top} is the tensor product of the incoming strings at the top of T, with again the dual colour for any outgoing strings.

So the example T above yields $J(T)\colon W \to W \otimes V \otimes V^*$, while the coloured tangle

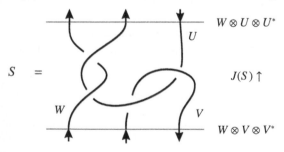

gives $J(S)\colon W \otimes V \otimes V^* \to W \otimes U \otimes U^*$.

Set $T_{\text{top}} = \Lambda$ when no strings of T meet the top, and equally $T_{\text{bottom}} = \Lambda$ when there are no strings at the bottom.

Placing consistently coloured tangles S and T one above the other results in the composite $J(S)J(T)$ of the homomorphisms $J(S)$ and $J(T)$,

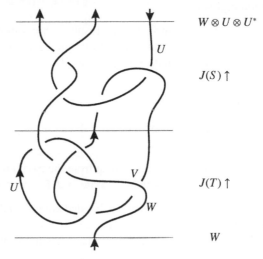

while placing them alongside each other

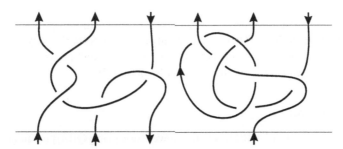

represents their tensor product $J(S) \otimes J(T)$.

58.3.3 Construction of Homomorphisms

To construct $J(T)$ for a general coloured tangle T, dissect the tangle into elementary pieces

and build up the homomorphism $J(T)$ from a definition of these pieces. Combine the homomorphisms for the elementary pieces, by taking their tensor product when they lie side by side, and composing them when consistently coloured pieces are placed one on top of the other.

Use $R_{V,W}$ or its inverse on a simple crossing of strands coloured V and W, alongside the homomorphisms $\Lambda \to V \otimes V^*$ and $V \otimes V^* \to \Lambda$ for the cup and cap and the identity 1_V for a single strand $\Big\{$ coloured by V. Where a string orientation is reversed, use the dual module in the homomorphism.

The key feature of this construction is that the resulting homomorphism $J(T)$ can be shown, using algebraic properties of a quantum group, to be unaltered when the tangle T is changed by Reidemeister moves R_{II} and R_{III} on the strings inside it.

58.3.4   Knot Invariants

Any knot diagram K when coloured by a module W can be regarded as a coloured tangle with no strings at the bottom or top.

The resulting homomorphism $J(K)$ is then a map from the trivial module $\Lambda = \mathbf{Q}[[h]]$ to itself. This is simply multiplication by a scalar, which we write as $J(K\colon W) \in \mathbf{Q}[[h]]$.

The scalar $J(K\colon W)$ depends only on W and the banded knot K, and gives the 1-parameter quantum group invariant of K for the $\mathcal{G}$-module W. The construction similarly provides an invariant of a banded link depending on a choice of module for each oriented component.

The invariant $J(K)$ is additive under a direct sum of modules,

$$J(K\colon W) = J(K\colon W_1) + J(K\colon W_2)$$

when $W = W_1 \oplus W_2$. It is usual to look at the case where $\mathcal{G}$ is semi-simple, so that W decomposes as the sum of irreducible modules. Then we need only consider colourings by *irreducible* modules. These have the added property that introducing a twist in the band around K multiplies $J(K\colon W)$ by a scalar f_W which depends only on the irreducible module W.

58.3.5   Dependence on h

While the quantum invariant $J(K\colon W)$ lies in the power series ring $\mathbf{Q}[[h]]$ it can generally be written as a multiple, depending only on W and the amount of twisting in the band around K, of a Laurent polynomial in $\mathbb{Z}[q^{\pm 1}]$, where $q = e^h$.

The Lie algebras $\{sl(N)\}$ of the A series have corresponding quantum groups $\{sl(N)_q\}$. The knot invariants derived from these have a close relationship with the Homfly satellite invariants discussed above.

For each N there is a 'fundamental' N-dimensional irreducible $sl(N)_q$-module $V_\square^{(N)}$. Formulae derived from the universal R-matrix show [9] that the homomorphism

$$R = R_{V_\square^{(N)}, V_\square^{(N)}} \colon V_\square^{(N)} \otimes V_\square^{(N)} \to V_\square^{(N)} \otimes V_\square^{(N)}$$

satisfies the quadratic equation

$$e^{\frac{h}{2N}} R - e^{-\frac{h}{2N}} R^{-1} = (e^{\frac{h}{2}} - e^{-\frac{h}{2}})\mathrm{Id}.$$

The invariant where all components of a link are coloured by $V_\square^{(N)}$ then satisfies the equation

$$x^{-1} \diagup\!\!\!\!\diagdown - x \diagdown\!\!\!\!\diagup = (s - s^{-1}) \Big) \Big(,$$

with $s = e^{\frac{h}{2}}$ and $x = e^{-\frac{h}{2N}} = s^{-\frac{1}{N}}$.

Calculations for the module endomorphism of $V_\square^{(N)}$ given by ρ show [9] that $J\left(\rho\right) = xv^{-1}$

with $v = s^{-N}, x = s^{-\frac{1}{N}}$. The consequence is that for a banded knot K we have

$$J(K \colon V_\square^{(N)}) = x^{wr(K)}P(K)$$

where $P(K)$ is the framed Homfly polynomial of K in v and s, normalised to take the value 1 on the empty knot, and v, s, x are given as above in terms of h.

Remark 58.3.1. The classical Jones polynomial of a link is given from the Homfly polynomial by setting $v = t = s^2$. With $v = s^{-2}$ we get the Jones polynomial up to a sign depending on the number of components of the link.

In the quantum group case with $N = 2$, the modules are all self-dual, and so string orientation in the tangles can be omitted.

For the fundamental irreducible 2-dimensional module $V_\square^{(2)}$ we then have the relation

$$x^{-1} \diagdown\!\!\!\diagup - x \diagup\!\!\!\diagdown = (s - s^{-1}) \,)\,($$

and also

$$x^{-1} \diagup\!\!\!\diagdown - x \diagdown\!\!\!\diagup = (s - s^{-1}) \smile\!\!\frown,$$

with $s = e^{\frac{h}{2}}, x = e^{-\frac{h}{4}}$. These give

$$(x^{-2} - x^2) \diagdown\!\!\!\diagup = (s - s^{-1})\left(x^{-1}\,)\,\right)\left(+x \smile\!\!\frown\right)$$

so, since $x^2 = s^{-1}$ we get

$$\diagdown\!\!\!\diagup = x^{-1}\,)\,\left(+x \smile\!\!\frown\right).$$

The relations here are very close to the classical Kauffman bracket relations with $A = x^{-\frac{1}{2}}$. Sikora [21] looks at the skein theory $J(K \colon V_\square^{(N)})$ and its extension to links, and notes that when $N = 2$ it coincides with the normalised Kauffman bracket up to a sign depending on the number of crossings and link components.

58.3.6 Dependence Among Invariants

In light of the huge array of available knot invariants arising from the use of satellites and quantum groups, the questions of their interrelations and independence become important. There is, for example, no point in trying to distinguish pairs of knots by use of an invariant which depends on invariants that are already known to agree on the knots in question.

We have noted above that the 2-variable Homfly polynomial $P(K)$ of a knot specialises to the family of 1-variable quantum invariants $\{J(K: V_{\square}^{(N)}\}$. Conversely, if we know enough of these quantum invariants then we can recover the whole Homfly polynomial.

There is a much wider result about the whole collection of invariants arising from the A-series of quantum groups $\{sl(N)_q\}$, showing that these quantum invariants of a knot are collectively equivalent to its Homfly satellite invariants.

The relations between the invariants come in a very attractive form, linking up the irreducible quantum group modules in a nice way with the natural set of satellite invariants $\{P(K: Q_\lambda)\}$ discussed earlier.

58.3.7　Quantum and Satellite Invariants

The most striking correspondence comes when we use the irreducible $\{sl(N)_q\}$-modules to colour a knot or link. As for the classical case of $sl(N)$ there is an irreducible $sl(N)_q$-module $V_\lambda^{(N)}$ for every partition λ of an integer into at most $N-1$ parts. The 1-variable quantum invariants $J(K: V_\lambda^{(N)})$ are all special cases of the 2-variable Homfly satellite invariant $P(K: Q_\lambda)$ corresponding to the same partition λ. The following explicit result, due essentially to Wenzl [22], is carefully discussed in Chapter 11 of [9] .

Theorem 58.3.2. For a partition λ of m into at most $N-1$ parts

$$J(K: V_\lambda^{(N)}) = x^{m^2 \mathrm{wr}(K)} P(K: Q_\lambda)$$

with $s = e^{\frac{h}{2}}, v = s^{-N} = e^{-\frac{Nh}{2}}, x = s^{-\frac{1}{N}} = e^{-\frac{h}{2N}}$ replacing the variables x, s and v on the right-hand side, and $\mathrm{wr}(K)$ is the sum of the crossings in K, counted with sign.

From this theorem, and its natural extension to links, we can recover any of the quantum invariants once we know sufficiently many satellite invariants. Conversely, we can find the satellite invariants from a knowledge of the quantum invariants for sufficiently many N. More detailed comments about the exact requirements can be found in [14].

58.3.8　Other Quantum Invariants

Wenzl goes on in [23] to study the quantum invariants arising from the quantum groups of the B, C and D series of classical Lie algebras, and relates them to satellite invariants based on the Kauffman 2-variable polynomial for unoriented banded knots and links. The connection in these cases comes from the Birman-Wenzl-Murakami algebras, in place of the Hecke algebras.

58.4　Bibliography

[1] Aiston, A. K.: Skein theoretic idempotents of Hecke algebras and quantum group invariants. PhD. thesis, University of Liverpool, (1996), http://livrepository.liverpool.ac.uk/7155.

[2] Aiston, A. K. and Morton, H. R.: Idempotents of Hecke algebras of type *A*, *J. Knot Theory Ramifications* **7** (1998), 463–487.

[3] Drinfeld, V.G.: Quantum groups, in *Proceedings of the International Congress of Mathematicians, Berkeley 1986*, Amer. Math. Society (1987), 798–820.

[4] Gyoja, A.: A q-analogue of Young symmetriser. *Osaka J. Math.*, **23** (1986), 841–852.

[5] Hadji, R. J. and Morton, H. R.: A basis for the full Homfly skein of the annulus. *Math. Proc. Camb. Phil. Soc.* **141** (2006), 81–100.

[6] Jimbo, M.: A q-analogue of $U(gl(n + 1))$, Hecke algebra and Yang Baxter equation. *Lett. Math. Phys.*, **11** (1986), 247–252.

[7] Jones, V.F.R.: Hecke algebra representations of braid groups and link polynomials, *Annals of Math.* **126** (1987), 335–388.

[8] Kirillov, A.N. and Reshetikhin, N.Y. : Representations of the algebra $U_q(Sl(2))$, q-orthogonal polynomials and invariants of links, in *Infinite dimensional Lie algebras and groups*, ed. V.G. Kac, World Scientific (1989), 285–342.

[9] Lukac, S. G.: Homfly skeins and the Hopf link. PhD. thesis, University of Liverpool, (2001), http: //livrepository.liverpool.ac.uk/7273.

[10] Lukac, S. G.: Idempotents of the Hecke algebra become Schur functions in the skein of the annulus. *Math. Proc. Camb. Phil. Soc.* **138** (2005), 79–96.

[11] Morton, H.R.: Polynomials from braids. In *Braids*, ed. Joan S. Birman and Anatoly Libgober, *Contemporary Mathematics* **78**, Amer. Math. Soc. (1988), 375–385.

[12] Morton, H.R.: Invariants of links and 3-manifolds from skein theory and from quantum groups. In *Topics in Knot Theory; Proceedings of the NATO Summer Institute in Erzurum 1992*, NATO ASI Series C 399, ed. M. Bozhüyük. Kluwer (1993), 107–156.

[13] Morton, H.R.: Integrality of Homfly 1-tangle invariants. *Algebraic and Geometric Topology*, **7** (2007), 327–338.

[14] Morton, H.R.: Knots, satellites and quantum groups. In *Introductory Lectures on Knot Theory (Trieste 2009)*, ed. L.H. Kauffman et al., World Scientific Press, Singapore (2011), 379–406.

[15] Morton, H. R. and Manchon, P. M. G.: Geometrical relations and plethysms in the Homfly skein of the annulus. *J. London Math. Soc.* **78** (2008), 305–328.

[16] Morton, H.R. and Short, H.B.: The 2-variable polynomial of cable knots. *Math. Proc. Camb. Philos. Soc.* **101** (1987), 267–278.

[17] Morton, H.R. and Strickland, P.M.: Jones polynomial invariants for knots and satellites, *Math. Proc. Camb. Philos. Soc.* **109** (1991), 83–103.

[18] Morton, H.R. and Traczyk, P.: Knots and algebras, in *Contribuciones Matematicas en homenaje al profesor D. Antonio Plans Sanz de Bremond*, ed. E. Martin-Peinador and A. Rodez Usan, University of Zaragoza, (1990), 201–220.

[19] Jun Murakami: The parallel version of polynomial invariants of links. *Osaka J. Math.* **26** (1989), 1–55.

[20] Reshetikhin, N. Y. and Turaev, V. G.: Ribbon graphs and their invariants derived from quantum groups, *Commun. Math. Phys.* **127** (1990), 1–26.

[21] Sikora, A.S.: Skein theory for $SU(n)$-quantum invariants *Algebr. Geom. Topol.* **5** (2005) 865–897.

[22] Wenzl H.: Representations of braid groups and the quantum Yang-Baxter equation. *Pacific J. Math.* **145** (1990), 153–180.

[23] Wenzl, H.: Quantum groups and subfactors of Lie type B, C and D, *Comm. Math. Phys.* **133** (1990), 383–433.

Chapter 59

Quantum Link Invariants: From QYBE and Braided Tensor Categories

Ruth Lawrence, Hebrew University of Jerusalem

59.1 Introduction

Link invariants can be constructed from any of the many different presentations of a link. Their construction is very much like fitting a plug into a socket — a geometric plug of the link in its chosen presentation must be fitted into a matching type of algebraic socket. When put together, a link invariant results. Typically there are many possible algebraic sockets (procedures for calculating an invariant) of each type, and quantum groups are a source of them, for which reason those invariants are known as *quantum* link invariants.

Historically, the sorts of link invariants that are called quantum link invariants began with the discovery of the Jones polynomial [26] and its relationships with statistical mechanics [27, 69, 32, 33]. From this point of view, new knot invariants were constructed by forming state summations related to link diagrams that were analogous to partition functions in statistical mechanics. As will be seen in this exposition, such structures can be given a remarkable number of different formulations that are here summarized in terms of tensor categories.

Some possible presentations (geometric plugs) of links are as

(a) the braid closure of a braid;

(b) a special case of a tangle;

(c) a special case of a colored framed tangle;

(d) a link embedded in three-dimensional space with a chosen direction in which the link is in general position, thus considering the link in a space identified with $\mathbb{C} \times \mathbb{R}$;

(e) a Gauss diagram of a link;

(f) a link diagram in the plane, with a chosen direction in the plane (height function);

(g) a link diagram in the plane or on a thickened surface.

Some matching algebraic sockets are

(a) a Markov trace on braid groups;

(b) a representation of the category of tangles;

(c) a ribbon category;

(d) the Kontsevich invariant, which is rather a 'converter' plug (to chord diagrams) which must be plugged into another socket (weight system, again which can be generated from quantum groups and their representations) to give a scalar-valued invariant — see Vasiliev invariants;

(e) arrow diagrams [58];

(f) quantum group to give the 'universal link invariant';

(g) skein algebra [59], Turaev shadow world [70].

Quantum link invariants, when suitably normalised, are functorial invariants, which implies that independently of how we cut up the link, the result obtained will be the same, and so almost any quantum invariant can be obtained through any of the presentation techniques just mentioned, though there are some subtle differences in some of the invariants obtained (such as coloring). In this chapter we discuss only link presentations by braids, tangles and link diagrams.

59.2 Invariants from Braids: Markov Traces

Let B_n denote the n-strand Artin braid group [3], with standard presentation [8]

$$B_n = \langle \sigma_1, \ldots, \sigma_{n-1} \, | \, \sigma_i \sigma_j = \sigma_j \sigma_i \ (|i - j| > 1), \ \sigma_i \sigma_{i+1} \sigma_i = \sigma_{i+1} \sigma_i \sigma_{i+1} \rangle$$

Since the relations preserve the exponent sum, there is a well-defined group homomorphism $e : B_n \longrightarrow (\mathbb{Z}, +)$ for which $e(\sigma_i) = 1$ and for a braid $\beta \in B_n$, $e(\beta)$ is known as the *exponent sum* of β.

By Alexander's theorem [2], any link can be presented as the braid closure of a braid on a sufficiently large number of strands, and moreover, by Markov's theorem [54], the braid closures of two braids are isotopy equivalent if, and only if, the braids can be linked by a sequence of Markov moves, so that

$$\{\text{links}\} \simeq \left(\coprod_{n \in \mathbb{N}} B_n \right) \Big/ \Big\langle \{\beta \sim \theta \beta \theta^{-1} \, | \, \beta, \theta \in B_n\}, \{(\beta \in B_n) \sim ((\beta \otimes 1)\sigma_n^{\pm 1} \in B_{n+1})\} \Big\rangle$$

where $\beta \otimes 1 \in B_{n+1}$ denotes the $(n + 1)$-strand braid in which the first n strands braid according to β while the last strand is untouched.

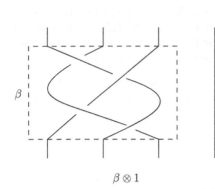

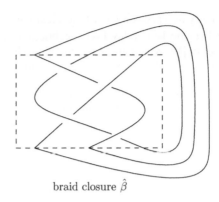

$$\beta \otimes 1 \qquad\qquad\qquad \text{braid closure } \hat{\beta}$$

This shifts the problem of construction of, say K-valued, link invariants to the algebraic problem of constructing maps $t_n : B_n \longrightarrow K$ for $n \in \mathbb{N}$, for which

$$t_n(\beta) = t_n(\theta\beta\theta^{-1}), \qquad t_n(\beta) = t_{n+1}((\beta \otimes 1)\sigma_n^{\pm 1})$$

since now $t_n(\beta)$ will be an invariant of the link presented as the braid closure $\hat{\beta}$ for $\beta \in B_n$. The first condition is precisely that t_n be a *class function* on B_n, in the usual group-theoretical sense, and a collection of functions $\{t_n\}$ satisfying the pair of conditions is known as a *Markov trace*. Commonly, a collection of maps tr_n satisfying the weaker conditions

$$tr_n(\beta) = tr_n(\theta\beta\theta^{-1}), \qquad tr_{n+1}((\beta \otimes 1)\sigma_n^{\pm 1}) = a.b^{\pm 1} tr_n(\beta)$$

for some invertible constants $a, b \in K$ is also known as a Markov trace, although then $t_n(\beta) = a^{-n}b^{-e(\beta)}tr_n(\beta)$ defines a renormalisation which satisfies the strict conditions.

The two main techniques for constructing Markov traces on the braid groups are

- as linear combinations of characters of (finite-dimensional) braid group representations; for this a source of braid group representations is needed, the main families being

    - R-matrix representations;

    - monodromy representation of the Knizhnik-Zamolodchikov equation, an integrable system of n first-order partial differential equations for a vector-valued function of complex numbers $z_1, \ldots, z_n$ with singularities on the diagonals $z_i = z_j$ ([68], [66]);

    - representations based on topology of configuration spaces, using the fact that the n-strand braid group is the fundamental group of the configuration space, $\widetilde{X_n}$, of n unordered points in the plane, so that any vector bundle with base space $\widetilde{X_n}$ and a given flat connection, will generate as its monodromy representation, a representation of the braid group; a typical source of vector bundles with flat connections is twisted homology of fibrations of configuration spaces with the Gauss-Manin connection ([47], [6]);

    - representations which factor through 'small' enough quotients of $\mathbb{C}B_n$ that their representation theory can be completely understood, such as that of the Iwahori-Hecke algebra which is generically a deformation of the completely understood representation theory of $\mathbb{C}S_n$;

- directly as traces on 'small' quotient algebras of the group algebra, $\mathbb{C}B_n$, of the braid group, such as the Temperley-Lieb algebra ([67], [57]), Iwahori-Hecke algebra [26], Birman-Wenzl algebra [9], which are small enough to have enumerable linear bases for direct construction of a trace in the algebraic sense, meaning a linear map with $tr(ab) = tr(ba)$, $tr(1) = 1$.

The first three subcases of the first approach can be implemented with essentially equivalent initial data, and give rise to quantum link invariants. This data is that of a quantum group and representation (in the first case), a Lie group and representation (in the second case), and a set of twisting data (coming from a Cartan matrix and weights) in the third case. The fact that the resulting braid group representations are essentially the same (with some *caveats*) is non-trivial; the equivalence of the first two is known as the Kohno-Drinfeld theorem (see [41], [15]) while Schechtman-Varchenko (see [66], [49]) investigated the correspondence between the second and third. Being that they give rise to the same braid group representations, they of course extend to the same (quantum) link invariants, but allow them to be looked at from different perspectives: the first of quantum groups and their (rigid braided tensor) category of representations; the second of conformal field theory, topological quantum field theory and Kontsevich's universal invariant; and the third via some sort of more classical configuration space pictures (see [48], [7]). It is the first which we follow in this chapter.

The rest are a collection of *ad hoc* constructions, although they usually are also equivalent to specific quantum group-generated representations. The following are examples of the last approach.

Example 59.2.1: The *Temperley-Lieb algebra TL_n* is generated as a linear space by unoriented planar tangles without isolated loops (that is, embeddings of n disjoint unoriented intervals in $\mathbb{R} \times [0, 1]$ whose endpoints are $\{1, 2, \ldots, n\} \times \{0, 1\}$, up to ambient isotopy). It has linear dimension given by the Catalan number $C_n = \frac{1}{n+1}\binom{2n}{n}$, which is the same as the number of ways of bracketing an ordered product of $(n + 1)$ symbols.

Any such planar tangle, x, can be closed using the same connection of upper and lower end-points as for usual braid closure; the result will be a collection of non-intersecting loops whose number will be denoted $a(x)$. The operation of vertical juxtaposition, in which isolated loops are counted with a factor d, provides a (non-commutative) multiplication which makes TL_n into a unital algebra $TL_n(d)$, with generators e_i, $i = 1, \ldots, n-1$ in which there are $n - 2$ vertical strands in positions $j \neq i, i+1$ and a cup/cap joining the points in the i-th, $(i+1)$-th positions. This gives presentation

$$TL_n(d) = \langle e_1, \ldots, e_{n-1} \mid e_i^2 = de_i, \ e_i e_j = e_j e_i \ (|i - j| > 1), \ e_i e_{i \pm 1} e_i = e_i \rangle$$

Define a linear map $tr_n : TL_n(d) \longrightarrow \mathbb{C}$ by linearly extending its value on the above linear basis, $tr_n(x) = d^{a(x)-n}$. This satisfies the properties

$$tr_{n+1}(x \otimes 1) = tr_n(x), \quad tr_{n+1}((x \otimes 1)e_n) = d^{-1}tr_n(x)$$

and is an algebra trace in the sense that $tr_n(ab) = tr_n(ba)$ and $tr_n(1) = 1$.

The Temperley-Lieb algebra can be considered as a quotient of the group algebra of the braid group, as demonstrated by the map

$$\iota : \mathbb{C}B_n \longrightarrow TL_n(d)$$
$$\sigma_i \longmapsto A.e_i + A^{-1}.1$$
$$\sigma_i^{-1} \longmapsto A^{-1}.e_i + A.1$$

where $d = -A^2 - A^{-2}$. One easily checks that this is well-defined, namely that the images of the relations in B_n hold in $TL_n(d)$, for example $(Ae_i+A^{-1})(Ae_{i+1}+A^{-1})(Ae_i+A^{-1}) = (Ae_{i+1}+A^{-1})(Ae_i+A^{-1})(Ae_{i+1}+A^{-1})$. By composing with ι we obtain a trace on the braid groups $t_n(\beta) = tr_n(\iota(\beta))$ which satisfies

$$t_{n+1}((\beta \otimes 1)\sigma_n) = tr_{n+1}((\iota(\beta) \otimes 1)(Ae_i + A^{-1})) = (Ad^{-1} + A^{-1})tr_n(\iota(\beta))$$

But $Ad^{-1} + A^{-1} = -d^{-1}A^{-3}$ and similarly $A^{-1}d^{-1} + A = -d^{-1}A^3$ giving

$$t_{n+1}((\beta \otimes 1)\sigma_n^{\pm 1}) = -d^{-1}A^{\mp 3}t_n(\beta)$$

which makes t_n into a Markov trace on the braid groups. Consequently $d^n(-A)^{3e(\beta)}t_n(\beta)$ is a link invariant, in fact (d times) the one-variable Jones polynomial, evaluated very closely to the original definition of Jones. In his case, Jones [25] was working on type II_1-subfactors in von Neumann algebras, and the algebra of projectors coming from a generated tower satisfied the Temperley-Lieb algebra relations (in our notation, $d^{-1}e_i$ are projectors). Note that the trace without the exponent sum normalisation, that is, $d^n t_n(\beta)$ is (d times) the *Kauffman bracket* [32] of the link diagram $\hat{\beta}$.

Variants of this approach, realising the trace as the matrix trace of a unitary representation of the Temperley-Lieb algebra, have been used to provide quantum algorithms for good approximations to the (one-variable) Jones polynomial at roots of unity; see [18], [1] and the survey [43].

Example 59.2.2: Using the Iwahori-Hecke algebra quotient of $\mathbb{C}B_n$ given by adding a quadratic relation

$$H_n(q) = \mathbb{C}B_n/\langle(\sigma_i - 1)(\sigma_i + q) = 0, \ 1 \le i \le n-1\rangle$$

Ocneanu defined a trace on $H_n(q)$ which becomes a Markov trace on B_n, dependent on an additional parameter λ. For generic q (away from roots of unity), the representation theory of $H_n(q)$ is a deformation of that of the symmetric groups, irreducible representations are labelled by Young diagrams and have the same dimensions as for the symmetric groups. This enabled the construction of the two-variable Jones polynomial, often known as the HOMFLY-PT polynomial, see [20], [60] for several methods of construction, including Ocneanu's trace, and [26] for the explicit decomposition according to Hecke algebra representations. Note that the fact that σ_i satisfies a quadratic relation in $H_n(q)$ (so that there is a linear relation between σ_i, 1 and σ_i^{-1}) means that any invariant obtained from its representations (or traces on it) will satisfy a three-term skein relation.

Similar work was carried on with the Birman-Wenzl algebra yielding the Kauffman polynomial ([9], [35]), as well as invariants coming from other related algebras, for example [17]. Even for invariants first defined via quantum groups, to actually compute them there are various combinatorial techniques based on the structure of the relevant representation theory, which reduces the calculation to one which is essentially diagrammatic and combinatorial. See for example *spin networks* [34], *recoupling theory* for the Temperley-Lieb algebra [36], *spiders* [44], *webs* and *ladder diagrams* [11].

59.3 *R*-Matrix Representations and the QYBE

Definition By an *R-matrix* is meant a vector space V, together with a map $R : V \otimes V \longrightarrow V \otimes V$ which satisfies the braid relation $R_{12} \circ R_{23} \circ R_{12} = R_{23} \circ R_{12} \circ R_{23}$ as maps $V \otimes V \otimes V \longrightarrow V \otimes V \otimes V$, where R_{ij} indicates the action of R on the ith and j^{th} factors of $V \otimes V \otimes V$, and the identity on the third.

By the associated *R-matrix representation* is meant the map $\rho : B_n \longrightarrow \text{End}(V^{\otimes n})$ defined on generators by $\rho(\sigma_i) = R_{i,i+1}$, which acts according to R on the ith and $(i + 1)^{\text{th}}$ factors and as the identity on the rest. The *R*-matrix condition is precisely what is required to guarantee that ρ is a representation, since $R_{i,i+1}, R_{j,j+1}$ automatically commute when $|i - j| > 1$, as $\{i, i + 1\} \cap \{j, j + 1\} = \emptyset$.

Example 59.3.1: For a 2-dimensional vector space V, say with basis $\{e_1, e_2\}$, the following 4×4-matrix gives an *R*-matrix, with respect to the basis $\{e_1 \otimes e_1, e_1 \otimes e_2, e_2 \otimes e_1, e_2 \otimes e_2\}$:

$$R = \begin{pmatrix} A^{-1} & 0 & 0 & 0 \\ 0 & A^{-1} - A^3 & A & 0 \\ 0 & A & 0 & 0 \\ 0 & 0 & 0 & A^{-1} \end{pmatrix}$$

Then $R' = A \cdot R$ is also an *R*-matrix, with eigenvalues 1 and $-A^4$ and so the associated *R*-matrix representation of B_n has dimension 2^n and factors through $H_n(q)$ where $A^4 = q$.

In order to make a link invariant from an *R*-matrix, a little more data is required, what Turaev [69] called an *enhanced Yang-Baxter operator*, which in addition to an *R*-matrix (V, R), contains a map $\mu : V \longrightarrow V$ and non-zero scalars a, b for which

$$(\mu \otimes \mu)R = R(\mu \otimes \mu), \quad \text{Tr}_2((\text{id}_V \otimes \mu) \circ R^{\pm 1}) = ab^{\pm 1}.\text{id}_V$$

where Tr_2 denotes the contraction (trace) $\text{End}(V \otimes V) \longrightarrow \text{End}(V)$ on the second factor. Then $tr_n(\beta) \equiv \text{Tr}(\mu^{\otimes n} \circ \rho(\beta))$ defines a Markov trace on the braid groups, giving the link invariant $V_R(L) \equiv a^{-n}b^{-e(\beta)}tr_n(\beta)$ to the link L represented as the braid closure $\hat{\beta}$ of $\beta \in B_n$. Here Tr is the usual matrix trace (sum of all diagonal entries).

Example 59.3.2: The above *R*-matrix on two-dimensional V can be completed to an enhanced

Yang-Baxter operator by $\mu = \begin{pmatrix} A^2 & 0 \\ 0 & A^{-2} \end{pmatrix}$, $a = -1$, $b = -A^{-3}$. Since $AR^{-1} - A^{-1}R = (A^2 - A^{-2})I$, the associated link invariant V_R will be such that for any braid $\beta \in B_n$,

$$-A^4 V_R(\widehat{\beta\sigma_i^{-1}}) + A^{-4} V_R(\widehat{\beta\sigma_i}) = (A^2 - A^{-2})V_R(\hat{\beta})$$

giving the skein relation which identifies it with the one-variable Jones polynomial in the variable $q = A^4$.

QYBE: Let $P_{V,V} : V \otimes V \longrightarrow V \otimes V$ denote the transposition $P_{V,V}(v \otimes w) = w \otimes v$. Then (V, R) is an R-matrix if, and only if, $\tilde{R} \equiv P_{V,V} \circ R$ satisfies the *Quantum Yang-Baxter Equation* (QYBE)[1]

$$\tilde{R}_{12} \circ \tilde{R}_{13} \circ \tilde{R}_{23} = \tilde{R}_{23} \circ \tilde{R}_{13} \circ \tilde{R}_{12}.$$

The geometric picture of the QYBE is as for the braid relation, only instead of labelling crossings by the two strands which cross in their position counted from the left (first/second strands, or second/third strands), we label the strands 1,2,3 and then describe a crossing by the labels on the two strands which have crossed.

braid relation   QYBE

The reason for the importance of the QYBE as opposed to the braid relation for R-matrices, is that the natural quantum group source of R-matrices, which are known as *universal R-matrices*, come as solutions *at the algebra level* to the QYBE, and *not* to the braid relation.

59.4 Universal R-Matrices and Braided Hopf Algebras

Bialgebras: Usually an *associative algebra with unit* is thought of as a vector space A together with a multiplication operation $(a, b) \longmapsto a.b$ and a unit $1 \in A$ for which $(ab)c = a(bc)$ and $1.a = a = a.1$. To emphasize the operations, we will consider it as a triple (A, m, η) of a vector space A over a field k, a linear map $m : A \otimes A \longrightarrow A$ and a linear map $\eta : k \longrightarrow A$ (which takes a

---

[1] The QYBE first arose in connection with exactly solvable models in statistical mechanics [5], and factorizability of S-matrices in scattering theory [74]. See [56] for a brief history. See also [45].

scalar $x \in k$ to the element $x.1 \in A$), making the following diagrams commute

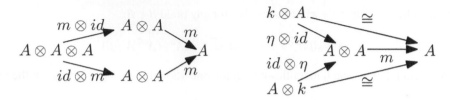

A *bialgebra* is a collection $(A, m, 1, \Delta, \epsilon)$ of a vector space A over a field k, with multiplication $m : A \otimes A \longrightarrow A$, unit $1 \in A$ (or equivalently multiplication of scalars by the unit, $\eta : k \longrightarrow A$), comultiplication $\Delta : A \longrightarrow A$ and counit $\epsilon : A \longrightarrow k$, for which (A, m, η) forms an associative algebra with unit, (A, Δ, ϵ) forms a coassociative coalgebra with counit (the properties are given by the identical commutative diagrams as for associative unital algebras with all the arrows reversed and m, η replaced everywhere by Δ, ϵ respectively), while the two structures are consistent in the sense that Δ and ϵ are algebra morphisms, η is a coalgebra morphism (that is, $\Delta(1) = 1 \otimes 1$) and $\epsilon \circ \eta = $ id. (Saying that Δ is an algebra morphism is equivalent to saying that m is a coalgebra morphism, so this definition of bialgebra is symmetric.)

Let $P : A \otimes A \longrightarrow A \otimes A$ denote the map which interchanges factors, $P(v \otimes w) = w \otimes v$. A bialgebra is called *cocommutative* if $P \circ \Delta = \Delta$. It is called *quasi-cocommutative* if there exists an invertible $R \in A \otimes A$ such that $P(\Delta(x)) = R \cdot \Delta(x) \cdot R^{-1}$ for all $x \in A$, and then R is called its *universal R-matrix*. If in addition R satisfies

$$(\Delta \otimes \text{id})R = R_{13}R_{23}, \quad (\text{id} \otimes \Delta)R = R_{13}R_{12}$$

then the bialgebra is said to be *braided* (see [30]), or in Drinfeld's original terminology, *quasi-triangular* (see [12]). Here R_{ij} is the element of $A \otimes A \otimes A$ given by R placed in the ith and j^{th} factors with $1 \in A$ in the third factor (in particular, $R_{12} = R \otimes 1$ and $R_{23} = 1 \otimes R$). As a consequence of quasi-triangularity, it follows that R satisfies the QYBE,

$$R_{12} \cdot R_{13} \cdot R_{23} = R_{12} \cdot ((\Delta \otimes \text{id})R = ((P \circ \Delta) \otimes \text{id})R \cdot R_{12}$$
$$= P_{12}(R_{13} \cdot R_{23}) \cdot R_{12} = R_{23} \cdot R_{13} \cdot R_{12}$$

where now the two sides of the equation are products (using the multiplication in the algebra) of elements of $A \otimes A \otimes A$. This means that for any A-module, V, which gives a map $\pi_V : A \longrightarrow \text{End } V$, we will now obtain an R-matrix (V, R_V) in which $R_V = P_{V,V} \circ ((\pi_V \otimes \pi_V)(R)) \in \text{End}(V \otimes V)$. So a braided bialgebra A will generate from its universal R-matrix, a whole collection of R-matrices, one from each representation of A (A-module).

Hopf algebras and ribbon algebras: In order to have the data to construct quantum link invariants, as we know from the previous sections, a little more than an R-matrix is required. A *Hopf algebra* is a bialgebra with an additional map $S : A \longrightarrow A$, called the *antipode*, which is such that (as maps $A \longrightarrow A$),

$$m \circ (\text{id} \otimes S) \circ \Delta = m \circ (S \otimes \text{id}) \circ \Delta = \eta \circ \epsilon$$

A braided bialgebra with an antipode is called a *braided* ($\equiv$ *quasi-triangular*) *Hopf algebra*. If in addition S is invertible, then it can be shown [14] (see also [61] without the invertibility assumption) that $(S \otimes S)R = R$, $(S \otimes \text{id})R = R^{-1} = (\text{id} \otimes S^{-1})R$ and moreover the square of the antipode is given by conjugation,

$$S^2(x) = uxu^{-1}, \quad \text{for all } x \in A$$

by the special element $u \equiv (m \circ (S \otimes \text{id}) \circ P)R \in A$.

A braided Hopf algebra, A, is called a *ribbon algebra* if there is an element v in the centre $Z(A)$ of A, known as the *ribbon element*, which interacts with the comultiplication, counit and antipode, according to

$$\Delta(v) = (R_{21}R)^{-1} \cdot (v \otimes v), \ \epsilon(v) = 1, \ S(v) = v.$$

When A is finite-dimensional, it can be shown that $v^2 = uS(u)$.

Example 59.4.1 For any finite group G, turn $A = \mathbb{C}G$ into a bialgebra in which m is linearly extended from the group, 1 is the unit in the group, while $\Delta(g) = g \otimes g$ for any $g \in G$ and $\epsilon(g) = 0$ when $g \in G$, $g \neq 1$ while $\epsilon(1) = 1$. This is a cocommutative Hopf algebra with antipode $S(g) = g^{-1}$.

Example 59.4.2 For any Lie algebra $\mathfrak{g}$, the universal enveloping algebra $A = U\mathfrak{g}$ is defined as the quotient of the tensor algebra on $\mathfrak{g}$ by the two-sided ideal generated by all elements of the form $xy - yx - [x, y]$ for $x, y \in \mathfrak{g}$. It has the structure of a cocommutative Hopf algebra in which $\Delta(x) = x \otimes 1 + 1 \otimes x$, $\epsilon(x) = 0$, $S(x) = -x$ for any $x \in \mathfrak{g}$ with Δ, ϵ extended to be homomorphisms and S extended to be an anti-homomorphism.

Example 59.4.3 Arising out of investigation of the quantum inverse scattering method in statistical mechanics, Kulish and Reshetikhin [42] defined and investigated the algebra $U_q\mathfrak{sl}(y)$, a one-parameter Hopf algebra deformation of the universal enveloping algebra $U\mathfrak{sl}(2)$. It has algebra generators $E, F, K^{\pm 1}$ with relations

$$KE = qEK, \ KF = q^{-1}FK, \ [E, F] = \frac{K - K^{-1}}{v - v^{-1}}$$

where $q = v^2$. This definition has many variants, for example

- it can be considered for generic q;

- when q is a root of unity, say of order n, a finite-dimensional quotient $\bar{U}_q\mathfrak{sl}(2)$ can be constructed by adding the relations $E^n = F^n = 0$, $K^n = 1$ (see [52]);

- over the (field of fractions of the algebra of) formal power series $\mathbb{C}[[h]]$, see [12], it can be embedded in $U_h\mathfrak{sl}(2)$, generated as an algebra by X, Y, H with

$$[H, X] = 2X, \ [H, Y] = -2Y, \ [X, Y] = \frac{\sinh(hH/2)}{\sinh(h/2)}$$

by $E = Xe^{\frac{hH}{4}}$, $F = e^{-\frac{hH}{4}}Y$, $K = e^{\frac{hH}{2}}$ and $v = e^{h/2}$.

In the latter approach, the data (Δ, ϵ, S) of the Hopf algebra is given by

$$\Delta(H) = H \otimes 1 + 1 \otimes H, \ \Delta(a) = e^{-\frac{hH}{4}} \otimes a + a \otimes e^{\frac{hH}{4}} \text{ for } a = X, Y$$

$\epsilon(H) = \epsilon(X) = \epsilon(Y) = 0$ and $S(X) = -vX$, $S(Y) = -v^{-1}Y$, $S(H) = -H$.

Drinfeld double: From *any* finite-dimensional Hopf algebra, A (with invertible antipode), Drinfeld [13] showed how to construct a braided Hopf algebra, called the *Drinfeld double $D(A)$*, which as a vector space is $A^* \otimes A$. This will already generate R-matrices (and thus braid group representations) from its finite-dimensional modules, but not necessarily link invariants.

Drinfeld-Jimbo quantum group: For any complex semisimple Lie algebra $\mathfrak{g}$, Drinfeld [12] and Jimbo [24], showed how to construct a deformation $U_q\mathfrak{g}$ or $U_h\mathfrak{g}$ of the universal enveloping algebra $U\mathfrak{g}$. Drinfeld's double construction applied to a Hopf subalgebra, B_q, coming from the Borel sub-algebra, produces a braided Hopf algebra structure on $D(B_q)$, which is bigger than $U_q\mathfrak{g}$, roughly speaking the piece coming from the Cartan subalgebra has been doubled. Taking a suitable quotient recovers $U_q\mathfrak{g}$, making $U_q\mathfrak{g}$ into a braided (quasi-triangular) quasi-cocommutative Hopf algebra. Furthermore, $U_q\mathfrak{g}$ is a ribbon algebra with ribbon element v which is the quantum Casimir element. There are some subtleties involved: when q is a root of unity, this can be done exactly on a suitable finite-dimensional version; when q is generic (or in the h-adic setting), there are completion issues, and only a quasi-R matrix is obtained (since it is an infinite sum), but they do not affect the construction of quantum link invariants which operate on specific finite-dimensional representations.

When q is not a root of unity, any finite-dimensional irreducible representation of $\mathfrak{g}$ can be deformed to a corresponding finite-dimensional $U_h\mathfrak{g}$-module, see [51], [62], thus generating solutions of QYBE and R-matrix representations of the braid groups. At roots of unity, this can also be done for a certain finite family of finite-dimensional irreducible representations.

Example 59.4.4 In the case of $\mathfrak{g} = \mathfrak{sl}(2)$, the universal R-matrix [12] is

$$R = \sum_{n=0}^{\infty} \frac{(1 - q^{-1})^n}{[n]!} e^{\frac{h}{4}(H \otimes H + n(H \otimes 1 - 1 \otimes H))}(X^n \otimes Y^n)$$

where $[n]! \equiv \prod_{k=1}^{n} \frac{q^k - 1}{q - 1}$. (See [63] for a similar formula for $\mathfrak{g} = \mathfrak{sl}(\mathfrak{n})$. See [16], Ex 8.3.7 for an R-matrix for the finite-dimensional Sweedler Hopf algebra and Ex 7.14.11 for the small quantum group $\bar{U}_q\mathfrak{sl}_2$) at a root of unity.) The ribbon element [65] is given by $v = uK^{-1}$; it is central and thus constant on irreducible representations.

Finite-dimensional irreducible representations of $\mathfrak{sl}(2)$ are indexed by their weight (twice the spin, one less than the dimension), a non-negative integer n. The same is true of finite-dimensional (topologically free) $U_h\mathfrak{sl}(2)$-modules. The smallest is two-dimensional, V, with

$$\pi : H \longmapsto \begin{pmatrix} 1 & 0 \\ 0 & -1 \end{pmatrix}, X \longmapsto \begin{pmatrix} 0 & 1 \\ 0 & 0 \end{pmatrix}, Y \longmapsto \begin{pmatrix} 0 & 0 \\ 1 & 0 \end{pmatrix}$$

and the above universal R-matrix evaluates in this case to the 4×4 matrix $R_V = \begin{pmatrix} A & 0 & 0 & 0 \\ 0 & 0 & A^{-1} & 0 \\ 0 & A^{-1} & A - A^{-3} & 0 \\ 0 & 0 & 0 & A \end{pmatrix}$ which is the inverse of the example given in section 59.2 above. The ribbon element evaluates to $(-1)^n A^{-n(n+2)}$ on the $(n+1)$-dimensional irreducible representation.

59.5 Braided Tensor Categories and Tangles

Rather than considering a braided (quasi-triangular) Hopf algebra, A, as a source of R-matrices, one from each A-module, we can turn this on its head and consider the collection of *all* A-modules. The comultiplication in a bialgebra, A, allows the tensor product $V \otimes W$ of two A-modules V, W to be constructed as an A-module, in which $\pi_{V \otimes W}(a) \equiv (\pi_V \otimes \pi_W)(\Delta(a)) \in \text{End}(V \otimes W)$. Similarly the antipode allows (left) duals by $(\pi_{V^*}(a).\phi)(v) \equiv \phi(\pi_V(S(a)).v)$ where $\phi \in V^*$, $v \in V$; note that generally $S^2 \neq$ id and so $(V^*)^* \neq V$. The braiding gives, for any pair of A-modules, U and V, a canonical map $R_{U,V} : U \otimes V \longrightarrow V \otimes U$ and the QYBE guarantees commutativity of the hexagon appearing from the possible reorderings of a tensor product of three modules. In fact, the quasi-triangularity condition gives commutativity of two triangles appearing from the interaction of the tensor product and interchanging factors,

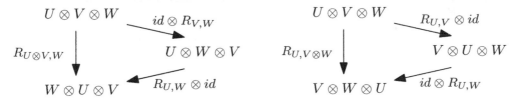

The counit makes k into a one-dimensional A-module, I, with $\pi_I(a).v = \epsilon(a)v$. The result is that for a braided Hopf algebra A, the category of A-modules becomes a *strict* (meaning with trivial associativity constraint), *braided* (meaning with morphisms in $\text{Hom}(U \otimes V, V \otimes U)$ making the above triangles commute), *(left) rigid* (meaning with (left) duals), *monoidal* (meaning with a tensor product and 1) *category*. Here the notion of (left) dual assigns to every object V, a dual object V^*, as well as special morphisms $b_V : I \longrightarrow V \otimes V^*$ (coevaluation) and $d_V : V^* \otimes V \longrightarrow I$ (contraction/evaluation). The little extra data of the ribbon element ν provides a morphism known as the *twist* $\theta_V : V \longrightarrow V$ (by the action of ν^{-1}) for every object V, compatible with 1, duals, tensor product and braiding, in the sense that

$$\theta_1 = id_1, \quad \theta_{V^*} = (\theta_V)^*, \quad \theta_{U \otimes V} = R_{V,U} \circ R_{U,V} \circ (\theta_U \otimes \theta_V)$$

This makes the category of (finite-dimensional) representations Mod_A of a ribbon algebra A into a *ribbon category* and this is exactly the correct data from which invariants of framed links (ribbons) can be defined.

The name *ribbon category* was given by Turaev [71]; the notion had been previously introduced under the name *tortile category* by Joyal and Street [28], see also [19]. The twist is also known as a *balancing*, and a ribbon category is also known as a strict balanced braided rigid tensor category.

The ribbon category of framed tangles: We now define the category of framed tangles, $\mathcal{T}$. Objects are finite sequences of signs, that is, an object is a pair of a non-negative integer n and an element of $\{+, -\}^n$. A morphism from an object $(m, (\epsilon_1, \ldots, \epsilon_m))$ to an object $(n, (\eta_1, \ldots, \eta_n))$ is a framed tangle, that is, a PL embedding (up to ambient isotopy equivalence) of a finite disjoint union of oriented circles and intervals in $\mathbb{C} \times [0, 1]$ supplied with a framing (choice of normal vector field), whose boundary is the same as its intersection with $\mathbb{C} \times \{0, 1\}$ and consists of the m points on the bottom plane $(j, 0)$, $1 \le j \le m$, and the n points on the top plane $(j, 1)$, $1 \le j \le n$, with orientations matching ϵ_j on the bottom and η_j on the top, and such that the framing at boundary points is in the direction of the positive real axis. Here we choose to call a downward orientation +, the source and target of a tangle are on the bottom and top, respectively, and similarly composition of tangles $g \circ f$ requires juxtaposition with g on top of f. Choosing a realisation in general position, any tangle can be depicted by a two-dimensional diagram of projected ribbons and annuli between a pair of horizontal lines; if we fix blackboard framing, then a tangle diagram (one side of the ribbon) is sufficient. Tangles contain both braids and links as examples, and in particular a framed link is a framed tangle from $(0, \emptyset)$ to itself.

This category becomes a ribbon category $\mathcal{T}$ in which (see [64])

- the unit object is $I = (0, \emptyset)$,

- the tensor product of objects $U \otimes V$ concatenates the strings of signs, and the tensor product of tangles places them side-by-side;

- the dual object to $(n, (\epsilon_1, \ldots, \epsilon_n))$ is $(n, (-\epsilon_n, \ldots, -\epsilon_1))$ and the duality morphisms are

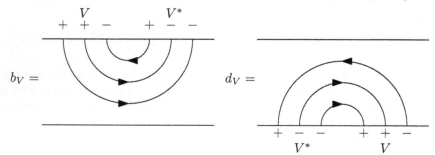

- the braiding morphism $R_{V,W}$ is

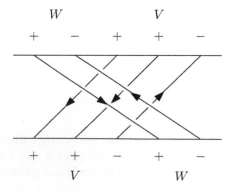

- the twist morphism θ_V is

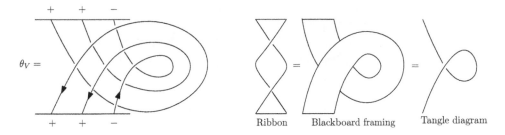

Ribbon   Blackboard framing   Tangle diagram

Quantum link invariants: The ribbon category $\mathcal{T}$ is universal in the sense that for any ribbon category $\mathcal{R}$ and object V, there is a unique functor between ribbon categories $\mathcal{T} \longrightarrow \mathcal{R}$ for which $(1, (+)) \longmapsto V$. In particular, for any finite-dimensional representation V of $U_h\mathfrak{g}$ there is a unique functor preserving the structures of a ribbon category, $\mathcal{T} \longrightarrow \mathrm{Mod}_{U_q\mathfrak{g}}$ and taking $(1, (+))$ to V, and the image of a framed link L is the quantum link invariant J_L associated to the pair $(U_h\mathfrak{g}, V)$.

More generally, given any ribbon category $\mathcal{V}$, one can define the category of colored framed tangles, colored by $\mathcal{V}$, in which objects are non-negative integers n, along with an ordered string of n objects in (V). Morphisms are oriented framed tangles with components colored by the objects in $\mathcal{V}$, which match the source and target objects in the sense that a strand labelled V oriented downwards at an end counts as V, while a strand labelled V and oriented upwards at an end counts as V^*. This category becomes a ribbon category, $\mathcal{T}_\mathcal{V}$ in which the duality, braiding and twist morphisms are as above for the usual category of framed tangles, except with the addition of coloring on components and endpoints by objects in $\mathcal{V}$. Now there is a unique ribbon functor $\mathcal{T}_\mathcal{V} \longrightarrow \mathcal{V}$ for which $(1, (V)) \longmapsto V$ for every object V in $\mathcal{V}$. The image of a colored framed link under this functor is known its *quantum link invariant* (see [64], [31], [72], [75]).

Note that to evaluate the invariant on a given framed link, the link should be considered as a framed tangle (given by a tangle diagram) and then broken into elementary parts. These elementary tangles are crossings and cups, caps with all the possible orientations. It is easy to see that the differently oriented crossings can all be 'turned' so as to be written in terms of one crossing (say with both strands downwards) by the addition of suitable cups and caps. Thus the tangle category has generators

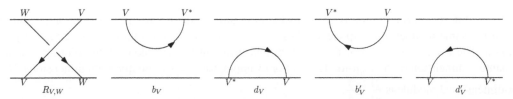

To compute the image of a link under the functor $\mathcal{T}_\mathcal{V} \longrightarrow \mathcal{V}$, it remains to know the images of these elementary tangles, namely $R_{V,W}$ for the crossing, coevaluation b_V, evaluation d_V for the right-going cup, cap and then 'twisted coevaluation/evaluation' b'_V, d'_V for the left-going cup, cap (which can be expressed in terms of the right-going cup and cap by using the twist):

$$b'_V = (\mathrm{id}_{V^*} \otimes \theta_V)R_{V,V}\cdot b_V, \quad d'_V = d_V c_{V,V}\cdot(\theta_V \otimes \mathrm{id}_{V^*})$$

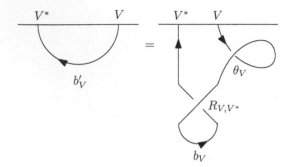

The conclusion is that the correct data (algebraic socket) for computing quantum link invariants of framed links is a ribbon category, that is a balanced, rigid, braided, strict monoidal (tensor) category and then the geometric plug we use to supply the link is a framed tangle.

59.6 Some Properties of Quantum Link Invariants

Quantum link invariants possess an enormous number of special properties, which indeed makes the subject all the more interesting. We list a few of them here.

Values. For a simple complex Lie algebra $\mathfrak{g}$, using $\mathcal{V} = \mathrm{Mod}_{U_q\mathfrak{g}}$ gives an invariant, $J_L(M_1, \ldots, M_c)$, of a framed link L whose c components are colored by finite-dimensional representations M_i, of $\mathfrak{g}$. It takes values in $\mathbb{Q}[[h]]$, or in $\mathbb{Z}[q^{\pm \frac{1}{2D}}]$ where D is an integer dependent on $\mathfrak{g}$ ($D = 1$ in the simply-laced case). See [50].

Disjoint union and normalisation. By their functorial definition in the previous section, quantum link invariants are multiplicative under disjoint union,

$$J_{L \sqcup L'}(M_1, \ldots, M_c, N_1, \ldots, N_{c'}) = J_L(M_1, \ldots, M_c).J_{L'}(N_1, \ldots, N_{c'}).$$

Often a different normalisation is used in which its value on the unknot is 1 or even on the unlink with the same number of components; these normalisations are not multiplicative under disjoint union. The evaluation $J_U(V)$ of the quantum link invariant of the unknot with coloring V is the quantum dimension of the $U_q\mathfrak{g}$-module V. For example, for $\mathfrak{g} = \mathfrak{sl}_2$, the quantum dimension of the n-dimensional module is $\frac{q^{\frac{n}{2}} - q^{-\frac{n}{2}}}{q^{\frac{1}{2}} - q^{-\frac{1}{2}}}$.

Integrality. Using results of Lusztig [53] on special bases for quantum groups at roots of unity and associated R-matrices and others, Le [50] showed the stronger integrality result, that for any simple complex $\mathfrak{g}$, link L colored by simple $\mathfrak{g}$-modules, there exists a rational $p \in \frac{1}{2D}\mathbb{Z}$ (dependent on L and the coloring) for which $J_L \in q^p \mathbb{Z}[q^{\pm 1}]$. In particular, for knots with framing zero (the knot parallel has zero linking number with the knot), the quantum link invariant of the knot colored by any simple module, when normalised to 1 on the unknot, lies in $\mathbb{Z}[q^{\pm 1}]$. The integrality of coefficients and powers means that quantum link invariants lend themselves to categorification.

Direct sum, tensor product and dual. From the functorial nature of the definition of quantum link invariants it follows that

- when a component of a link is colored by the trivial (one-dimensional) representation, the value of the link invariant is the same as that of the link with that component removed;

- when a component of a link is colored by the direct sum of two modules M_1 and M_2, the value of the link invariant is the sum of the values on the link with the two colorings given by coloring that one component by M_1 or by M_2;

- when a component of a link is colored by the tensor product $M_1 \otimes M_2$ of two modules, the value of the quantum link invariant is the same as that of the link obtained by doubling (cabling) that component, according to its framing, and coloring the components by M_1 and M_2;

- the quantum link invariant of a link is unchanged if the orientation of one component is reversed, and simultaneously the color V is replaced by the dual V^*.

Jones and HOMFLYPT polynomials. In the special case of $\mathfrak{g} = \mathfrak{sl}_2$ and where all components are colored by the two-dimensional representation, V_L normalised to one on the unknot, is the one-variable Jones polynomial. In the case of $\mathfrak{g} = \mathfrak{sl}_n$ and where all components are colored by the n-dimensional (vector) representation, V_L normalised to one on the unknot is the slice $X_L(q, q^{n-1})$ of the two-variable HOMFLYPT ([20], [60]) polynomial $X_l(q, \lambda)$ (see [26]).

The general properties of quantum link invariants under direct sum and tensor product imply that all quantum invariants based on $\mathfrak{sl}_2$ can be obtained as linear combinations of the one-variable Jones polynomial of suitable cablings of the link. Investigation [29] of the values of the $\mathfrak{sl}_2$ quantum link invariants of a knot colored by the n-dimensional representation (n-colored Jones polynomial) at the nth root of unity as $n \longrightarrow \infty$ led to the formulation of the volume conjecture.

The dependence of the colored Jones polynomial on the color is known to be q-holonomic [22], meaning that there exists a recurrence relation for each knot, producing a quick calculational method for computing the colored Jones polynomial of a fixed knot, for large values of the color. Furthermore, as a power series in $q - 1$, the coefficients are polynomials in the color and this allows sense to be made of the Jones polynomial even at non-integer colors, where there is no obvious representation-theoretic meaning; this is called the *Jones function*. The Melvin-Morton-Rozansky conjecture [4] gives even stronger information on the color dependence in a power series expansion, and is related to the origin of these invariants in topological quantum field theory.

Similarly, $\mathfrak{sl}_n$ quantum link invariants at more general representations, leads to a colored HOMPLYPT invariant, where the color is described by a partition and they also possess q-holonomic structure [21].

Symmetry properties of quantum link invariants at roots of unity. Le [50] showed that when q is a primitive r^{th} root of unity, quantum link invariants evaluated with colors which are simple modules, change by at most a sign under the action of the affine Weyl group on any one color; he also gave an extension to an action of an extended Weyl group which is important in understanding integrality properties of quantum 3-manifold invariants.

This is a generalisation of a symmetry property found for $\mathfrak{g} = \mathfrak{sl}_2$ in [40]. At an rth root of unity, they found that the colored $\mathfrak{sl}_2$ link invariant (in the functorial normalisation) changes by a simple factor (a sign and a power of A) when the color on a single component is changed from the k-dimensional representation to the $(r - k)$-dimensional representation.

Reconstructing ribbon categories from link invariants: It is possible from a 'suitably good' link invariant to construct a ribbon category. The basic idea is first to extend the link invariant to a 3-manifold invariant, in fact to the structure of a TQFT (with corners) and then the category associated to a circle, which is a codimension-two object, will be the desired ribbon category. From the Jones polynomial, this reconstructs (see [10], [72]) the category of representations of $U_q\mathfrak{sl}_2$ while from the Kauffman polynomial [73] it is the category of representations of $U_q\mathfrak{sp}_n$. This shows that there is indeed an intimate relation between quantum link invariants and their generating ribbon category, and even the quantum group from which they were generated [39].

$(1, 1)$-tangle representation of a link and universal link invariant. A useful tool for calculating the quantum link invariant of a link for a given quantum group $A = U_q\mathfrak{g}$, is given by expressing the link as the closure of a $(1, 1)$-tangle, namely one opens *one* strand of the link. The functor $\mathcal{T}_\mathcal{V} \longrightarrow \mathcal{V}$, where $\mathcal{V} = \text{Mod}_A$, associates to the tangle a morphism $V \longrightarrow V$ where V is the color of the strand opened (a finite-dimensional $U_q\mathfrak{g}$-module). If V is irreducible, then this map is multiplication by a constant, namely the quantum link invariant, normalised to 1 on the unknot. It is the evaluation in V of a certain element of A, known as the *universal link invariant* (see [61], [46]) which can be read off a tangle-diagram as follows. Let the universal R-matrix in $A \otimes A$ be denoted as an infinite sum $R = \sum_i \alpha_i \otimes \beta_i$, where i runs over a suitable indexing set. We assume that the tangle diagram has been drawn with only downward oriented crossings. Number crossings by an index j. We decorate the tangle diagram by placing elements of the algebra at certain points on the diagram,

- at the jth crossing, if it is a positively oriented crossing (slash), place elements α_{i_j}, β_{i_j} on the two involved strands, just below the crossing and if it is a negatively-oriented crossing (backslash), place elements β_{i_j}, $S\alpha_{i_j}$;

- decorate left-oriented caps with the element $K = v^{-1}u$ and left-oriented cups with the element K^{-1}.

Note that right-oriented cups and caps receive no decoration. Now read off an element of A by tracing the $(1, 1)$ tangle in the direction of its orientation and picking up all decorations in order, leaving an algebra product which is summed over all indices i_j, each of which appears twice. The resulting element of A is the universal link invariant. See [38], [37], [46].

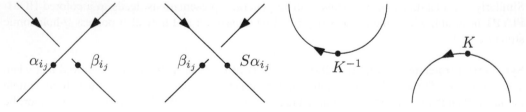

This method is particularly useful when quantum link invariants are used to compute quantum 3-manifold invariants, e.g. [23], [55].

59.7 Bibliography

[1] D. Aharonov, V.F.R. Jones, Z. Landau, A polynomial quantum algorithm for approximating the Jones polynomial, *Algorithmica* 55 (2009) no. 3, 395–421; `arXiv:quant-ph/0511096`.

[2] J. W. Alexander, A lemma on a system of knotted curves, *Proc. Nat. Acad. Sci. USA* **9** (1923) 93–95.

[3] E. Artin, Theorie der Zöpfe, *Abh. Math. Sem. Univ. Hamb.* **4** (1926) 47–72.

[4] D. Bar-Natan, S. Garoufalidis, On the Melvin-Morton-Rozansky conjecture, *Invent. Math.* **125** (1996) 103–133.

[5] R. Baxter, *Exactly Solved Models in Statistical Mechanics*, Academic Press, London (1982).

[6] S. Bigelow, Homological representations of braid groups, Ph.D. Thesis, University of California at Berkeley (2000).

[7] S. Bigelow, A homological definition of the HOMFLY polynomial, *Algebr. Geom. Topol.* 7 (2007) 1409–1440.

[8] J. Birman, Braids, links and mapping class groups, *Ann. Math. Stud.*, Princeton Univ. Press (1974) 37–69.

[9] J. Birman, H. Wenzl, Braids, link polynomials and a new algebra, *Trans. AMS* 313 (1989) 249–273.

[10] C. Blanchet, N. Habegger, G. Masbaum, P. Vogel, Topological quantum field theories derived from the Kauffman bracket, *Topology* **34** (1995) no. 4 883–927.

[11] S. Cautis, J. Kamnitzer, S. Morrison, Webs and quantum skew Howe duality, *Math. Ann.* 360 (2014) 351–390.

[12] V. Drinfeld, Hopf algebras and the quantum Yang-Baxter equation, *Dokl. Akad. Nauk USSR* 283:5 (1985) 1060–1064, English translation *Soviet Math. Dokl.* 32 (1985) 254–258.

[13] V. Drinfeld, Quantum groups, *Proc. Int. Cong. Math. Berkeley,* 1986, Amer. Math. Soc. (1987) 798–820.

[14] V. Drinfeld, On almost cocommutative Hopf algebras, *Algebra i Analiz,* **1** (1989) no. 2, 30–46.

[15] V. Drinfeld, Quasi-Hopf algebras and Knizhnik-Zamolodchikov equations, in *Problems of modern quantum field theory* (Alushta, 1989) Res. Rep. Phys., Springer Verlag (1989) 1–13.

[16] P. Etingof, S. Gelaki, D. Nikshych, V. Ostrik, *Tensor Categories*, Mathematical Surveys and Monographs 205 AMS (2015).

[17] M. Flores, J. Juyumaya, S. Lambropoulou, A framization of the Hecke algebra of type B, *J. Pure Appl. Algebra* 222 (2018) no. 4, 778–806.

[18] M. H. Freedman, A. Kitaev, Z. Wang, Simulation of topological field theories by quantum computers, *Commun. Math. Phys.* 227 (2002) 587–603; `arXiv:quant-ph/0001071`.

576 Encyclopedia of Knot Theory

[19] P. Freyd, D. Yetter, Braided compact closed categories with applications to low-dimensional topology, *Adv. Math.* 77 (1989) 156–182.

[20] P. Freyd, J. Hoste, W.B.R. Lickorish, K. Millett, A. Ocneanu, D. Yetter, A new polynomial invariant of knots and links, Bull. Amer. Math. Soc. (N.S.) 12 (1985), no. 2, 239–246.

[21] S. Garoufalidis, A. Lauda, T. Q. T. Le, The colored HOMFLYPT function is q-holonomic, *Duke Math. J.* 167 (2018) no. 3, 397–447.

[22] S. Garoufalidis, T. Q. T. Le, The colored Jones function is q-holonomic, *Geom. Topol.* 9 (2005) 1253–1293.

[23] K. Habiro, T. Q. T. Le, Unified quantum invariants for homology spheres associated with simple Lie algebras, *Geom. Topol.* 20 (2016) no. 5, 2687–2835.

[24] M. Jimbo, A q-analogue of $U(gl(N + 1))$, Hecke algebra, and the Yang-Baxter equation, *Letters in Mathematical Physics II* (1986) 247–252.

[25] V. F. R. Jones, Index for subfactors, *Invent. Math.* 72 (1983) 1–25.

[26] V. F. R. Jones, Hecke algebra representations of braid groups and link polynomials, *Amm. Math.* 126 (1987) 335–388.

[27] V. F. R. Jones, On knot invariants related to some statistical mechanical models, *Pacific J. Math.* 137 (1989) no. 2, 311–334.

[28] A. Joyal, R. Street, Tortile Yang-Baxter operators in tensor categories, *J. Pure Appl. Algebra* 71 (1991) 43–51.

[29] R. Kashaev, The hyperbolic volume of knots from the quantum dilogarithm, *Lett. Math. Phys.* 39 (1997) no. 3, 269–275.

[30] C. Kassel, *Quantum Groups*, Graduate Texts in Mathematics 155, Springer-Verlag (1995).

[31] C. Kassel, M. Rosso, V. Turaev, Quantum Groups and Knot Invariants, *Panoramas et syntheses* 5, Soc. Math. de France (1997).

[32] L. H. Kauffman, State models and the Jones polynomial, *Topology* 26 (1987) no. 3, 395–407.

[33] L. H. Kauffman, Statistical mechanics and the Jones polynomial, in *Braids (Santa Cruz, 1986), Contemp. Math. Pub.* 78 Amer. Math. Soc. (1988) 263–297.

[34] L. H. Kauffman, Spin networks and knot polynomials, *Internat. J. Modern Phys. A* 5 (1990) no. 1, 93–115.

[35] L. H. Kauffman, An invariant of regular isotopy, Trans. *Amer. Math. Soc.* 318 (1990) 417–471.

[36] L. H. Kauffman, S. Lins, Temperley-Lieb recoupling theory and invariants of 3-manifolds, *Ann. Math. Studies* 134 Princeton UP, Princeton NJ (1994).

[37] L. H. Kauffman, D. Radford, *Invariants of 3-manifolds derived from finite dimensional Hopf algebras*, *J. Knot Th. Ramif.* 4. (1995) no. 1, 131–162.

[38] L. H. Kauffman, D. Radford, S. Sawin, Centrality and the KRH invariant, *J. Knot Th. Ramif.* 7 (1998) no. 5, 571–624.

[39] D. Kazhdan, H. Wenzl, Reconstructing monoidal categories, *Adv. Soviet Math.* **16** Part 2, Amer. Math. Soc. Providence RI (1993) 111–136.

[40] R. Kirby, P. Melvin, The 3-manifold invariants of Witten and Reshetikhin-Turaev for $\mathfrak{sl}(2, \mathbb{C})$, *Invent. Math.* 105 (1991) 473–545.

[41] T. Kohno, Monodromy representations of braid groups and Yang-Baxter equations, *Ann. Inst. Fourier* (Grenoble) 37 (1987) 139–160.

[42] P. P. Kulish, N. Yu. Reshetikhin, Quantum linear problem for the sine-Gordon equation and higher representations, *Zap. Nauchn. Sem. Leningrad Otdel. Mat. Inst. Steklov (LOMI)* 101 (1981) 101–110, *J. Soviet Math.* 23 (1983) 2435–2441.

[43] G. Kuperberg, How hard is it to approximate the Jones polynomial?, *Theory Comput.* 11 (2015) 183–219, arXiv:0908.0512v2 [quant-ph].

[44] G. Kuperberg, Spiders for rank 2 Lie algebras, *Comm. Math. Phys.* 180 (1996) no. 1, 109–151; arXiv:q-alg/9712003.

[45] L. Lambe, D. Radford, Introduction to the quantum Yang Baxter equations and quantum groups: an algebraic approach, *Mathematics and Applications* **423** (1997), Kluwer Academic Publishers, Dordrecht.

[46] R. Lawrence, A universal link invariant, in *The Interface of Mathematics and Particle Physics*, Oxford (1988) 151–156, Inst. Math. Appl/ Conf. Ser. New Ser. **24** (1990) Oxford University Press, New York.

[47] R. Lawrence, Homological representations of the Hecke algebra, *Commun. Math. Phys.* **135** (1990) 141–191.

[48] R. Lawrence, A functorial approach to the one-variable Jones polynomial, *J. Differential Geom.* 37 (1993) no. 3, 689–710.

[49] R. Lawrence, Connections between CFT and topology via knot theory, in *Differential Geometric Methods in Theoretical Physics* (Rapallo, 1990)", Lecture Notes in Phys. 375 Springer (1991) 245–254.

[50] T. Q. T. Le, Integrality and symmetry of quantum link invariants, *Duke Math. J.* 102 (2000) no. 2, 273–306.

[51] G. Lusztig, Quantum deformations of certain simple modules over enveloping algebras, *Adv. Math.* 70 (1988) 237–249.

[52] G. Lusztig, Finite-dimensional Hopf algebras arising from quantised universal enveloping algebra, *J. Amer. Math. Soc.* 3 (1990) no. 1, 257–296.

[53] G. Lusztig, Introduction to quantum groups, *Progr. Math.* 110 (1994) Birkhäuser.

[54] A. A. Markov, Über die freie Äquivalenz geschlossner Zöpfe, *Recueil Mathematique* [Mat. Sb.] N.S 1(43) no. 1 (1936) 73–78.

[55] T. Ohtsuki, Invariants of 3-manifolds derived from universal invariants of framed links, Math. Proc. Cambridge Philos. Soc. **117** (1995) no. 2, 259–273.

[56] J. Park, H. Au-Yang, Yang-Baxter equations, in *Encyclopedia of Math. Physics*, eds J.-P. Françoise, G. L. Naber, Tsou S. T., Oxford, Elsevier **5** (2006) 465–475; arXiv:math-ph/0606053.

[57] M. Pimsner, S. Popa, Entropy and index for subfactors, *Ann. Sci. École Norm. Sup.* **19** (1986) 57–106.

[58] M. Polyak, O. Viro, Gauss diagram formulas for Vassiliev invariants, *Int. Math. Res. Not.* **11** (1994) 445–453.

[59] J. Przytycki, Skein modules of 3-manifolds, *Bull. Polish Acad. Sci. Math.* **39** (1991) 91–100.

[60] J. H. Przytycki, P. Traczyk, Invariants of Conway type, *Kobe J. Math.* 4 (1988) no. 2, 115–139.

[61] D. Radford, On the antipode of a quasitriangular Hopf algebra, *J. Algebra* **151** (1992) no. 1, 1–11.

[62] M. Rosso, Finite-dimensional representations of the quantum analog of the enveloping algebra of a complex simple Lie algebra, *Comm. Math. Phys.* 117 (1988) 581–593.

[63] M. Rosso, An analogue of PBW theorem and the universal R-matrix for $U_h\mathfrak{sl}(N + 1)$, *Comm. Math. Phys.* 124 (1989), 307–318.

[64] N. Yu. Reshetikhin, V. Turaev, Ribbon graphs and their invariants derived from quantum groups, *Comm. Math. Phys.* 127 (1990) no. 1, 1–26.

[65] N. Yu. Reshetikhin, V. Turaev, Invariants of 3-manifolds via link polynomials and quantum groups, *Invent. Math.* **103** (1991) 547–597.

[66] V. Schechtman, A.N. Varchenko, Arrangements of hyperplanes and Lie algebra homology, *Invent. Math.* 106 (1991) no. 1, 139–194.

[67] H. Temperley, E. Lieb, Relations between the "percolation" and "colouring" problem and other graph-theoretical problems associated with regular planar lattices: some exact results for the "percolation" problem, *Proc. Roy. Soc. London Ser. A* **322** (1971) 251–280.

[68] A. Tsuchiya, Y. Kanie, Vertex operators in conformal field theory on $\mathbb{P}^1$ and monodromy representations of braid groups, in *Conformal Field Theory and Solvable Lattice Models* (Kyoto, 1986)', *Adv. Stud. Pure Math.* **16** Academic Press, Boston (1988) 297–376.

[69] V. Turaev, The Yang-Baxter equation and invariants of links, *Invent. Math.* 92 (1988) no. 3, 527–553.

[70] V. Turaev, Shadow links and face models of statistical mechanics, *J. Diff. Geom.* **36** (1992) no. 1, 35–74.

[71] V. Turaev, Modular categories and 3-manifold invariants, *Intern. J. Modern Phys.* B **6** (1992) 1807–1824.

[72] V. Turaev, Quantum invariants of knots and 3-manifolds, *Studies in Mathematics* 18 Walter de Gruyter (1994).

[73] V. Turaev, H. Wenzl, Semisimple and modular categories from link invariants, *Math. Ann.* **309** (1997) no. 3, 411–461.

[74] C. N. Yang, Some exact results for the many-body problem in one dimension with repulsive delta-function interaction, *Phys. Rev. Lett.* **19** (1967) 1312–1314.

[75] D. N. Yetter, *Functional Knot Theory*, World Scientific (2001).

Chapter 60

Knot Theory and Statistical Mechanics

Louis H. Kauffman, University of Illinois at Chicago

60.1 Introduction

Here we discuss the relationship of knot theory with statistical mechanics, and how quantum link invariants arise in this connection.

Partition functions in statistical mechanics take the form

$$Z_G = \sum_{\sigma} e^{(-1/kT)E(\sigma)}$$

where σ runs over the different physical states of a system G and $E(\sigma)$ is the energy of the state σ. The probability for the system to be in the state σ is taken to be

$$prob(\sigma) = e^{(-1/kT)E(\sigma)}/Z_G.$$

Since Onsager's work, showing that the partition function of the Ising model has a phase transition, it has been a significant subject in mathematical physics to study the properties of partition functions for simply defined models based on graphs G. The underlying physical system is modeled by a graph G and the states σ are certain discrete labellings of G. The reader can consult Baxter's book [5] for many beautiful examples. The Potts model, discussed below, is a generalization of the Ising model and it is an example of a statistical mechanics model that is intimately related to knot theory and to the Jones polynomial [9, 10].

The partition function for the Potts model is given by the formula

$$P_G(Q, T) = \sum_{\sigma} e^{(J/kT)E(\sigma)} = \sum_{\sigma} e^{KE(\sigma)}$$

where σ is an assignment of one element of the set $\{1, 2, \cdots, Q\}$ to each node of the graph G, and $E(\sigma)$ denotes the number of edges of a graph G whose end-nodes receive the same assignment from σ. In this model, σ is regarded as a *physical state* of the Potts system and $E(\sigma)$ is the *energy* of this state. Here $K = J\frac{1}{kT}$ where J is plus or minus one (ferromagnetic and anti-ferromagnetic cases), k is Boltzmann's constant and T is the temperature. The Potts partition function can be expressed in terms of the dichromatic polynomial of the graph G. Letting

$$v = e^K - 1,$$

it is shown in [13, 16] that the dichromatic polynomial $Z[G](v, Q)$ for a plane graph can be expressed in terms of a bracket state summation of the form

$$\{\times\} = \{\asymp\} + Q^{-\frac{1}{2}}v\{\rangle\langle\}\{\rangle\langle\}$$

with

$$\{\bigcirc\} = Q^{\frac{1}{2}}.$$

Then the Potts partition function is given by the formula

$$P_G = Q^{N/2}\{K(G)\}$$

where $K(G)$ is an alternating link diagram associated with the plane graph G so that the projection of $K(G)$ to the plane is a medial diagram for the graph. This translation of the Potts model in terms of a bracket expansion makes it possible to examine how the states of the bracket model are related to the evaluation of the partition function.

In Section 60.2 we recall the definition of the dichromatic polynomial and the fact [13] that the dichromatic polynomial for a planar graph G can be expressed as a special bracket state summation on an associated alternating knot $K(G)$. We recall that, for special values of its two parameters, the Potts model can be expressed as a dichromatic polynomial.

We analyze those cases of the Potts model where $\rho = 1$ and we find that at criticality, this requires $Q = 4$ and $e^K = -1$ ($K = J/kT$ as in the second paragraph of this introduction), hence an imaginary value of the temperature. When we simply require that $\rho = 1$, not necessarily at criticality, then we find that for $Q = 2$ we have $e^K = \pm i$. For $Q = 3$, we have $e^K = \frac{-1 \pm \sqrt{3}i}{2}$. For $Q = 4$ we have $e^K = -1$. For $Q > 4$ it is easy to verify that e^K is real and negative. Thus in all cases of $\rho = 1$ we find that the Potts model has complex temperature values.

Section 60.3 discusses quantum link invariants from the point of view of amplitudes for space-time processes in two dimensions of space and one dimension of time. We show how these amplitudes correspond to partition functions on the link diagram, and explain the role of the Yang-Baxter-Equation in constructing link invariants. The models in this section lend themselves to category theory formulations since the diagrams are arranged as temporal processes and hence can be viewed as compositions of elementary processes corresponding to diagrammatic cups, caps and crossings. In this way of thinking we use the Yang-Baxter equation in the same conceptual framework as did C. N. Yang in his approach related quantum field theory [36].

Section 60.4 returns to the spatial statistical mechanics point of view for oriented knots and links in piecewise linear form. We formulate partition functions whose vertex weights occur at the corners of the piecewise linear directed graph of the knot, and at the crossings just as in the previous section. It is of interest that the geometry of the piecewise linear graph underlies the structure of these invariants.

It should be remarked that all the state summations treated in this chapter that involve vertex weights are formulated in terms of tensor nets or abstract tensors in the sense of Roger Penrose [32, 15]. That is, certain graphical loci correspond to matrices and the indices of these matrices correspond to edges emanating from these loci. The partition function is evaluated by contracting the tensor corresponding to a given network (summing over the products of matrix entries for each choice of indices for the edges of the graph as a whole).

Remark. In [35] Edward Witten proposed a formulation of a class of 3-manifold invariants and associated invariants of links in 3-manifolds via quantum field theory. He used a generalized Feynman integral in the form $Z(M, K)$ where

$$Z(M) = \int dA e^{(ik/4\pi)S(M,A)} W_K(A).$$

Here M denotes a 3-manifold without boundary and A is a gauge field (also called a gauge potential or gauge connection) defined on M. The gauge field is a one-form on a trivial G-bundle over M with values in a representation of the Lie algebra of G. The group G corresponding to this Lie algebra is said to be the gauge group. In this integral the "action" $S(M, A)$ is taken to be the integral over M of the trace of the Chern-Simons three-form $CS = AdA + (2/3)AAA$. (The product is the wedge product of differential forms.) The term measuring the knot or link is $W_K(A)$, the trace of the holonomy of the gauge connection along the knot (product of such traces for links).

$Z(M, K)$ integrates over all gauge fields modulo gauge equivalence. (See [1] for a discussion of the definition and meaning of gauge equivalence.)

Witten's functional integral model of link invariants places them in the context of quantum field theory and quantum statistical mechanics. In the form of this integral, this is the first time that we see the invariants expressed directly in terms of the embedding of the knot or link into three-dimensional space. All models described up to the point of Witten's work used diagrammatic representations for the topology. Witten's approach was a breakthrough into three dimensions and into new relationships between topology and quantum field theory. The formalism and internal logic of Witten's integral supports the existence of a large class of topological invariants of 3-manifolds and associated invariants of knots and links in these manifolds.

The invariants associated with this integral have been given rigorous descriptions [29], [22], [26], [17], [2]. The upshot of these descriptions is that the three-dimensional character of the invariants can be seen via differential geometric expressions that arise in the perturbation expansion of the functional integral [2], but the original three-dimensional vision of the integral remains problematic. Questions and conjectures arising from the functional integral formulation are still outstanding. Specific conjectures about this integral take the form of just how it involves invariants of links and 3-manifolds, and how these invariants behave in certain limits of the coupling constant k in the integral. Many conjectures of this sort can be verified through the combinatorial and algebraic models. Some of the most perspicuous of these models use the work of Drinfeld [6, 28] on Hopf algebras to capture just the right context for the Yang-Baxter equation to reappear in the models in relation to the structure of the gauge groups. Drinfeld showed how solutions to the Yang-Baxter equation appear naturally in new algebras (the Drinfeld Double Construction) that are directly related to the classical Lie algebras and to Hopf algebras more generally. In this way these algebraic results are deeply connected with the quantum field theory.

The really outstanding conjecture about the Witten functional integral is that it exists. At the present time there is no measure theory or generalization of measure theory that supports it. Here is a formal structure of great beauty. It is also a structure whose consequences can be verified by a remarkable variety of alternative means. Perhaps in the course of exploration the true nature of this form of integration will come to light.

This remark is a short indication of a large topic in the relationship of knot theory and statistical mechanics that we do not cover in this chapter. In this article we concentrate on the connections with statistical mechanics and the use of solutions to the Yang-Baxter equation (see Sections 60.3 and 60.4) as a source of vertex weights. Other articles on quantum invariants in this volume deal with the relationships with category theory and with quantum groups.

60.2 The Dichromatic Polynomial and the Potts Model

We define the *dichromatic polynomial* as follows:

$$Z[G](v, Q) = Z[G'](v, Q) + vZ[G''](v, Q)$$

$$Z[\bullet \sqcup G] = QZ[G]$$

where G' is the result of deleting an edge from G, while G'' is the result of contracting that same edge so that its end-nodes have been collapsed to a single node. In the second equation, $\bullet$ represents a graph with one node and no edges, and $\bullet \sqcup G$ represents the disjoint union of the single-node graph with the graph G.

In [13, 16] it is shown that the dichromatic polynomial $Z[G](v, Q)$ for a plane graph can be expressed in terms of a bracket state summation of the form

$$\{ \times \} = \{ \asymp \} + Q^{-\frac{1}{2}}v\{ \}\langle \rangle$$

with

$$\{ \bigcirc \} = Q^{\frac{1}{2}}.$$

Here

$$Z[G](v, Q) = Q^{N/2}\{K(G)\}$$

where $K(G)$ is an alternating link diagram associated with the plane graph G so that the projection of $K(G)$ to the plane is a medial diagram for the graph. Here we use the opposite convention from [16] in associating crossings to edges in the graph. We set $K(G)$ so that smoothing $K(G)$ along edges of the graph give rise to B-smoothings of $K(G)$. See Figure 4. The formula above, in bracket

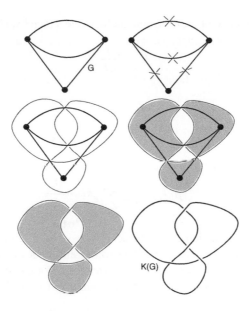

FIGURE 60.1: Medial graph, Tait checkerboard graph and K(G).

expansion form, is derived from the graphical contraction-deletion formula by translating first to the medial graph as indicated in the formulas below:

$$Z[\text{⧓}] = Z[\text{⧓}] + vZ[\text{)(}].$$

$$Z[R \sqcup K] = QZ[K].$$

Here the shaded medial graph is indicated by the shaded glyphs in these formulas. The medial graph is obtained by placing a crossing at each edge of G and then connecting all these crossings around each face of G as shown in Figure 4. The medial can be checkerboard shaded in relation to the original graph G (this is usually called the Tait Checkerboard Graph after Peter Guthrie Tait who introduced these ideas into graph theory), and encoded with a crossing structure so that it represents a link diagram. R denotes a connected shaded region in the shaded medial graph. Such a region corresponds to a collection of nodes in the original graph, all labeled with the same color. The proof of the formula $Z[G] = Q^{N/2}\{K(G)\}$ then involves recounting boundaries of regions in correspondence with the loops in the link diagram. The advantage of the bracket expansion of the dichromatic polynomial is that it shows that this graph invariant is part of a family of polynomials that includes the Jones polynomial and it shows how the dichromatic polynomial for a graph whose medial is a braid closure can be expressed in terms of the Temperley-Lieb algebra. This in turn reflects on the structure of the Potts model for planar graphs, as we remark below.

It is well known that the partition function $P_G(Q, T)$ for the Q-state Potts model in statistical mechanics on a graph G is equal to the dichromatic polynomial when

$$v = e^{J\frac{1}{kT}} - 1$$

where T is the temperature for the model and k is Boltzmann's constant. Here $J = \pm 1$ according as we work with the ferromagnetic or anti-ferromagnetic models (see [5] Chapter 12). For simplicity we denote

$$K = J\frac{1}{kT}$$

so that

$$v = e^K - 1.$$

We have the identity

$$P_G(Q, T) = Z[G](e^K - 1, Q).$$

The partition function is given by the formula

$$P_G(Q, T) = \sum_\sigma e^{KE(\sigma)}$$

where σ is an assignment of one element of the set $\{1, 2, \cdots, Q\}$ to each node of the graph G, and $E(\sigma)$ denotes the number of edges of the graph G whose end-nodes receive the same assignment from σ. In this model, σ is regarded as a *physical state* of the Potts system and $E(\sigma)$ is the *energy* of this state. Thus we have a link diagrammatic formulation for the Potts partition function for planar graphs G.

$$P_G(Q, T) = Q^{N/2}\{K(G)\}(Q, v = e^K - 1)$$

where N is the number of nodes in the graph G.

This bracket expansion for the Potts model is very useful in thinking about the physical structure of the model. For example, since the bracket expansion can be expressed in terms of the diagrammatic Temperley-Lieb algebra, one can use this formalism to express the expansion of the Potts model in terms of the Temperley-Lieb algebra. This method clarifies the fundamental relationship of the Potts model and the algebra of Temperley and Lieb. Furthermore, the conjectured critical temperature for the Potts model occurs for T when $Q^{-\frac{1}{2}}v = 1$. We see clearly in the bracket expansion that this value of T corresponds to a point of symmetry of the model where the value of the partition function does not depend upon the designation of over- and undercrossings in the associated knot or link. This corresponds to a symmetry between the plane graph G and its dual.

We now adopt yet another bracket expansion as indicated below. We call this two-variable bracket expansion [20] the ρ-*bracket*. It reduces to a topological version of the bracket as a function of q when ρ is equal to one.

$$[\asymp] = [\asymp] - q\rho[\,)(\,]$$

with

$$[\bigcirc] = q + q^{-1}.$$

We can regard this expansion as an intermediary between the Potts model (dichromatic polynomial) and the topological bracket. When $\rho = 1$ we have the topological bracket expansion in Khovanov form [24, 3, 4].

$$[\asymp] = [\asymp] - q[\,)(\,]$$

with

$$[\bigcirc] = q + q^{-1}.$$

For this special value we can define $J_K(q) = (-1)^{n_-} q^{n_+ - 2n_-} [K]$ and retrieve the Jones polynomial in the variable q, invariant under all three Reidemeister moves. With this formulation we can see what value of ρ corresponds to the Potts model, and thus how the Potts model and the Jones polynomial partake of a 1-parameter family of state sum models.

When

$$-q\rho = Q^{-\frac{1}{2}} v$$

and

$$q + q^{-1} = Q^{\frac{1}{2}},$$

we have the Potts model.

We now look more closely at the Potts model by writing a translation between the variables q, ρ and Q, v. We have

$$-q\rho = Q^{-\frac{1}{2}} v$$

and

$$q + q^{-1} = Q^{\frac{1}{2}},$$

and from this we conclude that

$$q^2 - \sqrt{Q} q + 1 = 0.$$

Whence

$$q = \frac{\sqrt{Q} \pm \sqrt{Q - 4}}{2}$$

and

$$\frac{1}{q} = \frac{\sqrt{Q} \mp \sqrt{Q - 4}}{2}.$$

Thus

$$\rho = -\frac{v}{\sqrt{Q} q} = v(\frac{-1 \pm \sqrt{1 - 4/Q}}{2})$$

For physical applications, Q is a positive integer greater than or equal to 2. Let us begin by analyzing the Potts model at criticality (see discussion above) where $-\rho q = 1$. Then

$$\rho = -\frac{1}{q} = \frac{-\sqrt{Q} \pm \sqrt{Q - 4}}{2}.$$

For the Jones polynomial to be identical with the partition function we want

$$\rho = 1.$$

Thus we want

$$2 = -\sqrt{Q} \pm \sqrt{Q - 4}.$$

Squaring both sides and collecting terms, we find that $4 - Q = \mp \sqrt{Q} \sqrt{Q - 4}$. Squaring once more, and collecting terms, we find that the only possibility for $\rho = 1$ is $Q = 4$. Returning to the equation for ρ, we see that this will be satisfied when we take $\sqrt{4} = -2$. Then the partition function will have topological terms. However, note that with this choice, $q = -1$ and so $v/\sqrt{Q} = -\rho q = 1$ implies that $v = -2$. Thus $e^K - 1 = -2$, and so

$$e^K = -1.$$

From this we see that in order to have $\rho = 1$ at criticality, we need a 4-state Potts model with imaginary temperature variable $K = (2n + 1)i\pi$. It is worthwhile considering the Potts models at imaginary temperature values. For example, the Lee-Yang Theorem [25] shows that under certain circumstances the zeros on the partition function are on the unit circle in the complex plane. We take the present calculation as an indication of the need for further investigation of the Potts model with real and complex values for its parameters.

Now we go back and consider $\rho = 1$ without insisting on criticality. Then we have $1 = -v/(q\sqrt{Q})$ so that

$$v = -q\sqrt{Q} = \frac{-Q \mp \sqrt{Q}\sqrt{Q-4}}{2}.$$

From this we see that

$$e^K = 1 + v = \frac{2 - Q \mp \sqrt{Q}\sqrt{Q-4}}{2}.$$

From this we get the following formulas for e^K : For $Q = 2$, we have $e^K = \pm i$. For $Q = 3$, we have $e^K = \frac{-1 \pm \sqrt{3}i}{2}$. For $Q = 4$, we have $e^K = -1$. For $Q > 4$ it is easy to verify that e^K is real and negative. Thus in all cases of $\rho = 1$ we find that the Potts model has complex temperature values. It is a research problem to analyze the influence of the Jones polynomial at these complex values on the behaviour of the model for real temperatures.

It should be mentioned that these explorations of relationship among the bracket state sum, statistical mechanics and the Jones polynomial can be extended to explorations of Khovanov homology [24, 20, 21]. This, in turn, may lead to new understanding of statistical mechanics in terms of the categorical framework initiated in the Khovanov homological generalization of the bracket polynomial.

60.3   Quantum Link Invariants

In this section we describe the construction of quantum link invariants from knot and link diagrams that are arranged with respect to a given direction in the plane. This special direction will be called "time." Arrangement with respect to the special direction means that perpendiculars to this direction meet the diagram transversely (at edges or at crossings) or tangentially (at maxima and minima). The designation of the special direction as time allows the interpretation of the consequent evaluation of the diagram as a generalized scattering amplitude.

In the course of this discussion we find the need to reformulate the Reidemeister moves for knot and

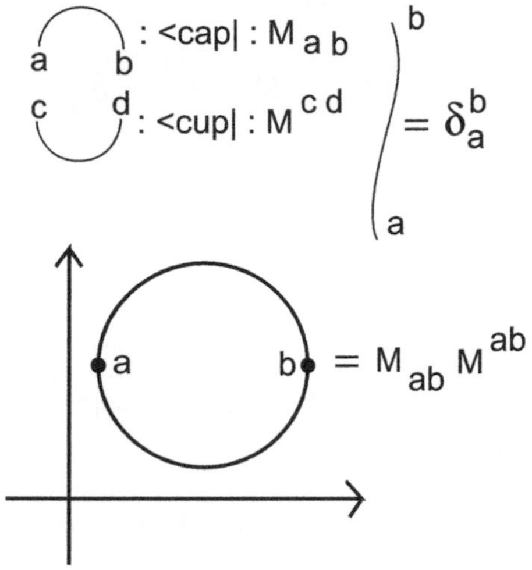

FIGURE 60.2: Spacetime circle.

link diagrams that are arranged to be transverse (except for a finite collection of standard critical points) to the specific special direction introduced in the previous paragraph. This brings us back to our theme of diagrams and related structures. This particular reformulation of the Reidemeister moves is quite far-reaching. It encompasses the relationship of link invariants with solutions to the Yang-Baxter equation and the relationship with Hopf algebras (to be dealt with in Section 60.5).

60.3.1 Knot Amplitudes

Consider first a circle in a spacetime plane with time represented vertically and space horizontally as in Figure 60.2.

The circle represents a vacuum to vacuum process that includes the creation of two "particles" and their subsequent annihilation. We could divide the circle into these two parts (creation "cup" and annihilation "cap") and consider the amplitude $< cap|cup >$. Since the diagram for the creation of the two particles ends in two separate points, it is natural to take a vector space of the form $V \otimes V$ as the target for the bra and as the domain of the ket. We imagine at least one particle property being catalogued by each factor of the tensor product. For example, a basis of V could enumerate the spins of the created particles.

Any non-self-intersecting differentiable curve can be rigidly rotated until it is in general position with respect to the vertical. It will then be seen to be decomposed into an interconnection of minima and maxima. We can evaluate an amplitude for any curve in general position with respect to a vertical direction. Any simple closed curve in the plane is isotopic to a circle, by the Jordan Curve Theorem. If these are topological amplitudes, then the value for any simple closed curve should be equal to the original amplitude for the circle. What condition on creation (cup) and annihilation (cap) will insure topological amplitudes? The answer derives from the fact that isotopies of the sim-

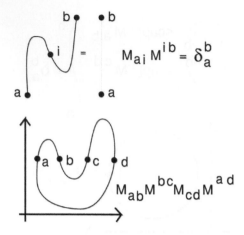

FIGURE 60.3: Spacetime Jordan curve.

ple closed curves are generated by the cancellation of adjacent maxima and minima as illustrated in Figure 60.3.

This condition is articulated by taking a matrix representation for the corresponding operators. Specifically, let $\{e_1, e_2, ..., e_n\}$ be a basis for V. Let $e_{ab} = e_a \otimes e_b$ denote the elements of the tensor basis for $V \otimes V$. Then there are matrices M_{ab} and M^{ab} such that

$$|cup> (1) = \sum M^{ab} e_{ab}$$

with the summation taken over all values of a and b from 1 to n. Similarly, $< cap|$ is described by

$$< cap|(e_{ab}) = M_{ab}.$$

Thus the amplitude for the circle is

$$< cap|cup > (1) =< cap| \sum M^{ab} e_{ab}$$

$$= \sum M^{ab} < cap|(e_{ab}) = \sum M^{ab} M_{ab}.$$

In general, the value of the amplitude on a simple closed curve is obtained by translating it into an "abstract tensor expression" in the M^{ab} and M_{ab}, and then summing over these products for all cases of repeated indices. Note that here the value "1" corresponds to the vacuum.

Returning to the topological conditions we see that they are just that the matrices M^{ab} and M_{cd} are inverses in the sense that

$$\sum_i M^{ai} M_{ib} = \delta_b^a$$

where δ_b^a denotes the identity matrix. See Figure 60.3.

One of the simplest choices is to take a 2 x 2 matrix M such that $M^2 = I$ where I is the identity

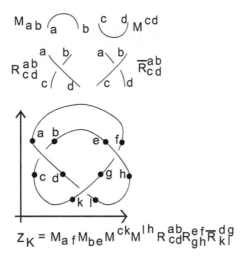

FIGURE 60.4: Cups, caps and crossings.

matrix. Then the entries of M can be used for both the cup and the cap. The value for a loop is then equal to the sum of the squares of the entries of M:

$$< cap|cup > (1) = \sum M^{ab} M_{ab} = \sum M_{ab} M_{ab} = \sum M_{ab}^2.$$

In particular, consider the following choice for M. Its square is equal to the identity matrix and yields a loop value of $d = -A^2 - A^{-2}$, just the right loop value for the bracket polynomial model for the Jones polynomial [13], [12].

$$M = \begin{bmatrix} 0 & iA \\ -iA^{-1} & 0 \end{bmatrix}$$

Any knot or link can be represented by a picture that is configured with respect to a vertical direction in the plane. The picture will decompose into minima (creations), maxima (annihilations) and crossings of the two types shown in Figure 60.4. Here the knots and links are unoriented. These models generalize easily to include orientation.

Next to each of the crossings we have indicated mappings of $V \otimes V$ to itself, called R and R^{-1} respectively. These mappings represent the transitions corresponding to elementary braiding. We now have the vocabulary of *cup*, *cap*, R and R^{-1}. Any knot or link can be written as a composition of these fragments, and consequently a choice of such mappings determines an amplitude for knots and links. In order for such an amplitude to be topological (i.e. an invariant of regular isotopy the equivalence relation generated by the second and third of the classical Reidemeister moves), we want it to be invariant under a list of local moves on the diagrams as shown in Figure 60.5. These moves are an augmented list of Reidemeister moves, adjusted to take care of the fact that the diagrams are arranged with respect to a given direction in the plane. The proof that these moves generate regular isotopy is composed in exact parallel to the proof that we gave for the classical

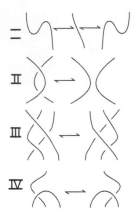

FIGURE 60.5: Regular isotopy with respect to a vertical direction.

Reidemeister moves in Section 60.2. In the piecewise linear setting, maxima and minima are replaced by upward and downward pointing angles. The fact that the triangle, in the Reidemeister piecewise linear triangle move, must be projected so that it is generically transverse to the vertical direction in the plane introduces the extra restriction that expands the move set.

In this context, the algebraic translation of Move *III* is the Yang-Baxter equation that occurred for the first time in problems of exactly solved models in statistical mechanics [5].

All the moves taken together are directly related to the axioms for a quasi-triangular Hopf algebra (aka quantum group). Many seeds of the structure of Hopf algebras are prefigured in the patterns of link diagrams and the structure of the category of tangles. The interested reader can consult [29], [37], [18] and [16], [19] for more information on this point.

Here is the list of the algebraic versions of the topological moves. Move 0 is the cancellation of maxima and minima. Move *II* corresponds to the second Reidemeister move. Move *III* is the Yang-Baxter equation. Move *IV* expresses the relationship of switching a line across a maximum. (There is a corresponding version of *IV* where the line is switched across a minimum.)

$$0. \quad M^{ai} M_{ib} = \delta^a_b$$

$$II. \quad R^{ab}_{ij} \overline{R^{ij}_{cd}} = \delta^a_c \delta^b_d$$

$$III. \quad R^{ab}_{ij} R^{jc}_{kf} R^{ik}_{de} = R^{bc}_{ij} R^{ai}_{dk} R^{kj}_{ef}$$

$$IV. \quad R^{ai}_{bc} M_{id} = M_{bi} \overline{R^{ia}_{cd}}$$

In the case of the Jones polynomial we have all the algebra present to make the model. It is easiest to indicate the model for the bracket polynomial: Let *cup* and *cap* be given by the 2 x 2 matrix M, described above so that $M_{ij} = M^{ij}$. Let R and R^{-1} be given by the equations

$$R^{ab}_{cd} = AM^{ab}M_{cd} + A^{-1}\delta^a_c \delta^b_d,$$

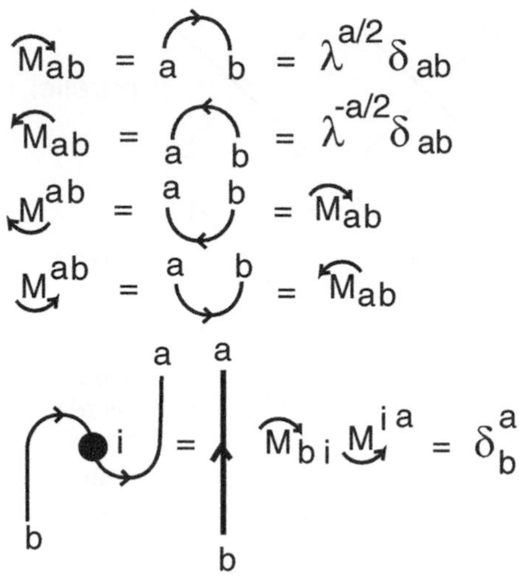

FIGURE 60.6: Right and left cups and caps.

$$(R^{-1})^{ab}_{cd} = A^{-1} M^{ab} M_{cd} + A \delta^a_c \delta^b_d.$$

This definition of the R-matrices exactly parallels the diagrammatic expansion of the bracket, and it is not hard to see, either by algebra or diagrams, that all the conditions of the model are met.

60.3.2 Oriented Amplitudes

Slight but significant modifications are needed to write the oriented version of the models we have discussed in the previous section. See [16], [34, 30], [28], [23]. In this section we sketch the construction of oriented topological amplitudes.

The generalization to oriented link diagrams naturally involves the introduction of right and left oriented caps and cups. These are drawn as shown in Figure 60.6.

A right cup cancels with a right cap to produce an upward pointing identity line. A left cup cancels with a left cap to produce a downward pointing identity line.

Just as we considered the simplifications that occur in the unoriented model by taking the cup and cap matrices to be identical, let's assume here that right caps are identical with left cups and that consequently left caps are identical with right cups. In fact, let us assume that the right cap and left cup are given by the matrix

$$M_{ab} = \lambda^{a/2} \delta_{ab}$$

where λ is a constant to be determined by the situation, and δ_{ab} denotes the Kronecker delta. Then the left cap and right cup are given by the inverse of M:

$$M^{-1}_{ab} = \lambda^{-a/2} \delta_{ab}.$$

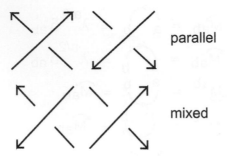

FIGURE 60.7: Oriented crossings.

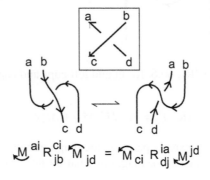

$$\underset{\leftharpoonup}{M}{}^{ai} R^{ci}_{jb} \overset{\frown}{M}{}_{jd} = \overset{\frown}{M}{}_{ci} R^{ia}_{dj} \underset{\leftharpoonup}{M}{}^{jd}$$

FIGURE 60.8: Conversion.

We assume that along with M we are given a solution R to the Yang-Baxter equation, and that in an oriented diagram the specific choice of R^{ab}_{cd} is governed by the local orientation of the crossing in the diagram. Thus a and b are the labels on the lines going into the crossing and c and d are the labels on the lines emanating from the crossing.

Note that with respect to the vertical direction for the amplitude, the crossings can assume these aspects: both lines pointing upward, both lines pointing downward, one line up and one line down (two cases). See Figure 60.7.

Call the cases of one line up and one line down the *mixed* cases and the upward and downward cases the *parallel* cases. A given mixed crossing can be converted, in two ways, into a combination of a parallel crossing of the same sign plus a cup and a cap. See Figure 60.8.

This leads to an equation that must be satisfied by the R matrix in relation to powers of λ (again we use the Einstein summation convention):

$$\lambda^{a/2}\delta^{ai} R^{ci}_{jb}\lambda^{-d/2}\delta_{jd} = \lambda^{-c/2}\delta_{ic}R^{ia}_{dj}\lambda^{b/2}\delta^{jb}.$$

This simplifies to the equation

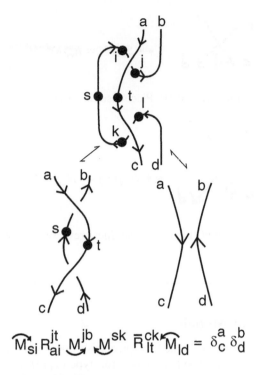

$$\widehat{M_{si}} R_{ai}^{jt} \underset{\searrow}{M^{jb}} \underset{\nwarrow}{M^{sk}} \overline{R}_{lt}^{ck} \overline{\widehat{M}_{ld}} = \delta_c^a \delta_d^b$$

FIGURE 60.9: Antiparallel second move.

$$\lambda^{a/2} R_{db}^{ca} \lambda^{-d/2} = \lambda^{-c/2} R_{db}^{ca} \lambda^{b/2},$$

from which we see that R_{db}^{ca} is necessarily equal to zero unless $b+d = a+c$. We say that the R matrix is *spin preserving* when it satisfies this condition. Assuming that the R matrix is spin preserving, the model will be invariant under all orientations of the second and third Reidemeister moves just so long as it is invariant under the anti-parallel version of the second Reidemeister move as shown in Figure 60.9.

This antiparallel version of the second Reidemeister move places the following demand on the relation between λ and R:

$$\sum_{st} \lambda^{(s-b)/2} \lambda^{(t-c)/2} R_{as}^{bt} \overline{R_{dt}^{cs}} = \delta_c^a \delta_d^b.$$

Call this the $R - \lambda$ *equation*. The reader familiar with [9] or with the piecewise linear version as described in [16] will recognise this equation as the requirement for regular homotopy invariance in these models.

$$\leftrightarrow S_{cd}^{ab}(\theta) = R_{cd}^{ab}\lambda^{(d-a)\theta/2\pi}.$$

FIGURE 60.10: Piecewise linear crossing.

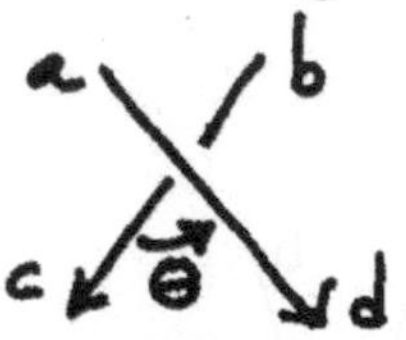

$$\leftrightarrow \overline{S}_{cd}^{ab}(\theta) = \overline{R}_{cd}^{ab}\lambda^{(d-a)\theta/2\pi}.$$

FIGURE 60.11: Piecewise linear reverse crossing.

60.4 Piecewise Linear Models

The models we now describe are designed for piecewise linear knot diagrams. In the piecewise linear category Kauffman first described, models of this type in [13] were designed to give a model in common for both the bracket polynomial and a generalization of the Potts model in statistical mechanics. Vaughan Jones [11] gave a more general version of vertex models for link invariants using the smooth category. The smooth models involve integrating an angular parameter along the link diagram. Here we give a piecewise linear version of Jones models (see also [14, 15, 16]) where the angular information is concentrated at the vertices of the diagram. In statistical mechanics these sorts of models appear in the work of Perk and Wu [33] and Zamolodchikov [38]. A piecewise linear link diagram is composed of straight line segments so that the vertices are either crossings (locally 4-valent) or corners (2-valent). The diagram is oriented, and we assume the usual two types of local crossing. A matrix $S_b^a(\theta)$ is associated with each crossing as indicated in Figure 60.10.

Here θ is the angle in radians between the two crossing segments measured in the counter-clockwise direction. It is assumed that R_{cd}^{ab} is a solution to the Yang-Baxter Equation for $a, b, c, d \in I$ (I is a specified ordered index set). Furthermore, we assume that R_{cd}^{ab} is spin-preserving in the sense that $R_{cd}^{ab} = 0$ unless $a + b = c + d$. To the reverse crossing we associate $\overline{S}_b^a(\theta)$ as shown in Figure 60.11.

Here $\overline{R}$ is the inverse matrix for R in the sense that $R_{in}^{ab}\overline{R}_{cd}^{ij} = \delta_c^a \delta_d^b$. Note that the angular part of the contribution is independent of the type of crossing. Since the model is to be piecewise linear, we also have vertices of valence two. These acquire vertex weights according to the angle between the segments incident to the vertex, as shown in Figure 60.12.

The partition function (or amplitude) for a given piecewise linear link diagram K is then defined by the summation $\langle K \rangle = \sum_\sigma [K|\sigma]$ where $[K|\sigma]$ denotes the product of these vertex weights and a state (configuration) σ is an assignment of spins to the edges of the link diagram (each edge

FIGURE 60.12: Piecewise linear corner.

extends from one vertex to another) such that spins are constant at the corners (two-valent vertices) and preserved (a+b = c +d) at the crossings.

We can rewrite this product of vertex weights as $[KIa] = \langle Kl\sigma\rangle\lambda^{\|\sigma\|}$ where $\langle Kl\sigma\rangle$ is the product of R values from the crossings, and $\|\sigma\|$ is the sum of angle exponents from all the crossings and corners. Thus

$$\langle K\rangle = \sum_\sigma \langle Kl\sigma\rangle\lambda^{\|\sigma\|}.$$

A basic theorem [11, 34, 16] gives a sufficient condition for $\langle K\rangle$ to be an invariant of regular isotopy:

Theorem. If the matrix R^{ab}_{cd} satisfies the Yang-Baxter Equation and if R^{ab}_{cd} also satisfies the cross-channel inversion

$$\sum_{ij} \bar{R}^{ia}_{jc} R^{jb}_{id} \lambda^{(c-i)/2+(d-j)/2)} = \delta^a_d \delta^b_c$$

then the model $\langle K\rangle$ as described above, is an invariant of regular isotopy.

Remark. The cross-channel condition corresponds to invariance under the re-versed type II move. Figure 60.13 shows the pattern of index contractions corresponding to $R^{ia}_{jc}R^{jb}_{id}$.

Remark. Note that the result obtained with the piecewise linear oriented model in this section is equivalent to the result obtained with the oriented amplitude model at the end of the previous section. Both are fundamental ways to view the quantum invariants of knots and links.

60.5   Bibliography

[1] M.F. Atiyah, *The Geometry and Physics of Knots*, Cambridge University Press, (1990).

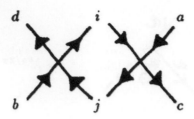

FIGURE 60.13: Cross channel contraction.

[2] D. Bar-Natan, Perturbative Chern-Simons theory. *J. Knot Theory Ramifications* 4 (1995), no. 4, 503547.

[3] D. Bar–Natan, On Khovanov's categorification of the Jones polynomial, *Algebraic and Geometric Topology*, **2**(16), (2002), pp. 337–370.

[4] D. Bar-Natan, Khovanov's homology for tangles and cobordisms, *Geometry and Topology*, **9–33**, (2005), pp. 1465–1499. arXiv:mat.GT/0410495

[5] R.J. Baxter. *Exactly Solved Models in Statistical Mechanics.* Acad. Press (1982).

[6] V. G. Drinfeld, Quantum groups. *Proceedings of the International Congress of Mathematicians*, Vol. 1, 2 (Berkeley, Calif., 1986), 798820, Amer. Math. Soc., Providence, RI, (1987).

[7] D.S. Freed and R.E. Gompf, Computer calculation of Witten's three-manifold invariant, *Commun. Math. Phys.*, No. 141, (1991), pp. 79–117.

[8] L.C. Jeffrey, Chern-Simons-Witten invariants of lens spaces and torus bundles, and the semi-classical approximation, *Commun. Math. Phys.*, No. 147, (1992), pp. 563–604.

[9] V.F.R. Jones, A polynomial invariant for links via von Neumann algebras, *Bull. Amer. Math. Soc.* **129** (1985), 103–112.

[10] V.F.R. Jones. Hecke algebra representations of braid groups and link polynomials. *Ann. of Math.* 126 (1987), pp. 335–388.

[11] V.F.R. Jones. On knot invariants related to some statistical mechanics models. *Pacific J. Math.*, vol. 137, no. 2 (1989), pp. 311–334.

[12] L.H. Kauffman, State models and the Jones polynomial, *Topology* **26** (1987), 395–407.

[13] L.H. Kauffman, Statistical mechanics and the Jones polynomial, *AMS Contemp. Math. Series* **78** (1989), 263–297.

[14] L. H. Kauffman. Knot polynomials and Yang-Baxter models. *IXth International Congress on Mathematical Physics - 17–27 July 1988 - Swansea, Wales*, Ed. Simon, Truman, Davies, Adam Hilger Pub. (1989), pp. 438–441.

[15] L. H. Kauffman. Knots, abstract tensors, and the Yang-Baxter Equation. In *Knots, Topology and Quantum Field Theories - Proceedings of the Johns Hopkins Workshop on Current*

Problems in Particle Theory 13. Florence (1989). Ed. by L. Lussana. World Scientific Pub. (1989), pp. 179–334.

[16] L.H. Kauffman, *Knots and Physics*, World Scientific Publishers (1991), Second Edition (1993), Third Edition (2002), Fourth Edition (2012).

[17] L.H. Kauffman and S. L. Lins, *Temperley-Lieb Recoupling Theory and Invariants of 3- Manifolds*, Annals of Mathematics Study 114, Princeton Univ. Press, (1994).

[18] L.H. Kauffman, Gauss Codes, quantum groups and ribbon Hopf algebras, *Reviews in Mathematical Physics* **5** (1993), 735–773. (Reprinted in L. H. Kauffman, *Knots and Physics*, World Scientific Pub. (2012), 551–596.)

[19] L.H. Kauffman and D. Radford, Invariants of 3-Manifolds derived from finite dimensional Hopf algebras, *Journal of Knot Theory and Its Ramifications.* Vol. 4, No. 1 (1995), 131–162.

[20] L. H. Kauffman. Remarks on Khovanov homology and the Potts model. In *Perspectives in analysis, geometry, and topology: on the occasion of the 60th birthday of Oleg Viro*, (Progress in Mathematics), pages 237–262. Birkh user/Springer, New York, (2012).

[21] H. A. Dye, A. Kaestner, L. H. Kauffman, Khovanov homology, Lee homology and a Rasmussen invariant for virtual knots, *J. Knot Theory Ramifcations* 26 (2017), no. 3, 1741001, 57 pp.

[22] R. Kirby and P. Melvin, On the 3-manifold invariants of Reshetikhin-Turaev for sl(2,C), *Invent. Math.* 105, 473–545, (1991).

[23] M.A. Hennings, Hopf algebras and regular isotopy invariants for link diagrams, *Math. Proc. Camb. Phil. Soc.*, Vol. 109 (1991), pp. 59–77

[24] Khovanov, M., A categorification of the Jones polynomial, *Duke Math. J*,**101** (3), pp.359–426. (1997).

[25] Lee, T. D., Yang, C. N., Statistical theory of equations of state and phase transitions II., lattice gas and Ising model, *Physical Review Letters*, **87**, 410–419. (1952).

[26] W.B.R. Lickorish, The Temperley-Lieb Algebra and 3-manifold invariants, *Journal of Knot Theory and Its Ramifications*, Vol. 2, (1993), pp. 171–194.

[27] L. Rozansky, Witten's invariant of 3-dimensional manifolds: loop expansion and surgery calculus, In *Knots and Applications*, edited by L. Kauffman, World Scientific Pub. Co. (1995).

[28] N.Y. Reshetikhin, Quantized universal enveloping algebras, the Yang-Baxter equation and invariants of links, I and II, *LOMI reprints E-4–87 and E-17–87*, Steklov Institute, Leningrad, USSR, (1986).

[29] N.Y. Reshetikhin and V. Turaev, Invariants of Three-Manifolds via link polynomials and quantum groups, *Invent. Math.*,Vol.103, (1991), pp. 547–597.

[30] V.G. Turaev and O.Y. Viro, State sum invariants of 3-manifolds and quantum 6j symbols, *Topology 31*, (1992), pp. 865–902.

[31] V.G. Turaev and H. Wenzl, Quantum invariants of 3-manifolds associated with classical simple Lie algebras, *International J. of Math.*, Vol. 4, No. 2, (1993), pp. 323–358.

[32] R. Penrose. Applications of negative dimensional tensors. In *Combinatorial Mathematics and its Applications*. Edited by D. J. A. Welsh, Academic Press (1971) .

[33] J. H. H. Perk and F. Y. Wu. Nonintersecting string model and graphical approach: equivalence with a Potts model. *J. Stat. Phys.*, vol. 42, nos. 5/6, (1986), pp. 727–742.

[34] Turaev, V. G. The Yang-Baxter equation and invariants of links. *Invent. Math.* 92 (1988), no. 3, 527553.

[35] E. Witten. Quantum field theory and the Jones polynomial. Comm. in *Math. Phys.* Vol. 121 (1989), 351–399.

[36] C. N. Yang. Some exact results for the many-body problem in one dimension with repulsive delta-function interaction *Phys. Rev. Lett.* 19, 1312, (1967).

[37] D. Yetter, Quantum groups and representations on monoidal categories. *Math. Proc. Camb. Phil. Soc.*, Vol. 108 (1990), pp. 197–229.

[38] Zamolodchikov, Alexander B.; Zamolodchikov, Alexey B. Factorized S-matrices in two dimensions as the exact solutions of certain relativistic quantum field theory models. *Ann. Physics* 120 (1979), no. 2, 253 - 291.

Part XI

Polynomial Invariants

Chapter 61

What Is the Kauffman Bracket?

Charles Frohman, University of Iowa

61.1 Introduction

The **Kauffman bracket** is a Laurent polynomial-valued invariant of framed links, that was introduced by Louis Kauffman [5] in order to make the computation and invariance of the **Jones polynomial** [2] more obvious. We begin by reviewing the diagrammatic presentation of links and framed links. Next we describe the Kauffman bracket. Finally, we relate the Kauffman bracket to the Jones polynomial.

Kauffman [5] used a slightly different partition function than we use here. This is because he wanted to relate the Kauffman bracket to the Jones polynomial. The normalization given here is the one used in Temperley-Lieb recoupling theory [4, 11].

The definition of the Kauffman bracket was a revolutionary event in knot theory that spilled over into category theory [6, 7] and representation theory [9]. It is now common to use diagrammatic methods in both these fields. The Jones normalization is important in quantum topology as it is the normalization used in the **volume conjecture** [12, 8]. Two of the Tate conjectures were settled using the Jones polynomial [5, 17, 11, 15].

A skein relation is a local relation that occurs inside any ball you can place in space. Locality is a fundamental property of Feynman integrals. Witten realized this and built a connection between the Jones polynomial and quantum field theory [18].

The connection with representation theory was codified by Kuperberg [9], with the introduction of the concept of a **spider**. Up to a scalar factor and change of variable, the Kauffman bracket is the unique locally defined topological invariant based on the combinatorics of the representation theory of sl_2. The uniqueness of a local invariant given combinatorial data from the representation theory of a Lie algebra was used implicitly in [9, 13] and [16] to determine link polynomials associated with $SU(n)$, A_2 and G_2. The idea that the dimension of the space of invariants determines skein relations was explored further in [14].

61.2 Flavors of Links

A link in $\mathbb{R}^3$ is encoded by a four-valent planar graph, with data at the vertices indicating overcrossings. The link can be reconstructed up to isotopy from such a diagram. Two diagrams represent the same link if you can pass from one to the other by isotopies that preserve the combinatorial data (sometimes called R0 moves) and three different moves that change the combinatorics of the diagram [1].

These are the Reidemeister moves. The first Reidemeister move, or RI is

$$\text{———} \longleftrightarrow \text{⌒⌒}. \tag{61.1}$$

The second Reidemeister move or RII is,

$$)(\longleftrightarrow)(. \tag{61.2}$$

Finally, the third Reidemeister move or RIII is

$$\text{⤬} \longleftrightarrow \qquad . \tag{61.3}$$

There are two common ways of decorating links. An **oriented link** is a link in which each component of the link has been assigned an orientation. This is indicated in diagrams by putting a directional arrow on each component. Two projections represent the same oriented link if you can pass from one to another by the same moves as regular links, but so that the arrows coincide.

The second is framed links. A **framed link** in $\mathbb{R}^3$ is an embedding of a disjoint union of annuli into M. Diagrammatically framed links are depicted by showing the core of the annuli. You should imagine the annuli lying parallel to the plane of the paper; this is sometimes called **blackboard framing**. Two framed links in $\mathbb{R}^3$ are equivalent if they are isotopic.

Every framed link can be isotoped so that it has a projection into the plane that is blackboard framed. Two diagrams represent the same framed link with blackboard framing if you can pass from one to the other by isotopies that don't change the combinatorics of intersection, RII, or RIII.

The Kauffman bracket is an invariant of framed links; the Jones polynomial is an invariant of oriented links. To pass from the Kauffman bracket of the framed link coming from a diagram for an oriented link we need to use the **total writhe** of the diagram. This the sum of the signs of each crossing of the oriented link, where the sign of a crossing is given by the convention below.

FIGURE 61.1: The sign convention.

61.3 The Kauffman Bracket

Given a diagram of a link there are two ways of smoothing a crossing.

FIGURE 61.2: Smoothing a crossing.

The middle diagram shows the positive smoothing of the crossing, the right diagram shows the negative smoothing. Here is how to tell. Put your hand on the overcrossing arc in the diagram, and sweep counterclockwise. The pair of opposite corners that you sweep out, are the ones that are joined in the positive smoothing.

If the diagram has n crossings, then there are 2^n ways of smoothing all the crossings to get a collection of simple closed curves. These are the **states** of the diagram. Given a state s, there are three obvious numbers. The first is $p(s)$, the number of positive smoothings that were made to arrive at the state. The second is $n(s)$, the number of negative smoothings. Notice that $p(s) + n(s)$ is the number of crossings of the diagram. The third number, $c(s)$, is the number of connected components of the state.

Suppose that D is a diagram and S is the set of states of the diagram. We define a **partition function** of the diagram that takes on values in the ring Laurent polynomials with integer coefficients in the variable A, $\mathbb{Z}[A, A^{-1}]$. The **Kauffman bracket** of the diagram D is given by

$$< D >= \sum_{s \in S} A^{p(s)-n(s)}(-A^2 - A^{-2})^{c(s)}. \tag{61.4}$$

The Kauffman bracket is in fact an invariant of the framed link that D represents. It is most usually understood via the **skein relations** that the Kauffman bracket satisfies. A skein relation is a linear relation between diagrams with coefficients in some ring. In this case the ring is $\mathbb{Z}[A, A^{-1}]$. Skein relations are **local**. This means that the diagrams appearing in the relation are identical outside of some disk. We depict the relation by only showing the part of the disk where the diagrams are different. In linear algebra, a function from the cartesian product of vector spaces to the base field is called **multilinear** if, when you fix all the variables except one, you get a linear functional. Skein relations are a topological analog of multilinearity. If you fix the link away from a ball, you get a linear relation between the values of invariants of links.

Theorem 61.3.1. [5] The Kauffman bracket satisfies the skein relations

$$\left\langle \quad \right\rangle - A \left\langle \quad \right\rangle - A^{-1} \left\langle \quad \right\rangle = 0$$

and

$$\langle \bigcirc \cup L \rangle + (A^2 + A^{-2}) \langle L \rangle = 0.$$

Proof. Let S be the states of the diagram represented at the far left of the first skein relation. The set S can be partitioned into the states S_p where the smoothing at the pictured crossing is positive, and S_n where the smoothing at the pictured crossing is negative. We break the sum for $\langle \times \rangle$ into sums over S_p and S_n.

$$\left\langle \times \right\rangle = \sum_{s \in S_p} A^{p(s)-n(s)}(-A^2 - A^{-2})^{c(s)} + \sum_{s \in S_n} A^{p(s)-n(s)}(-A^2 - A^{-2})^{c(s)}. \tag{61.5}$$

Notice that the quantity summed over S_p is the partition function applied to states of the positively smoothed diagram times A. Also the sum over S_n is A^{-1} times the partition function applied to states of the negatively smoothed diagram. This yields the first skein relation,

$$\left\langle \times \right\rangle = A \left\langle \asymp \right\rangle + A^{-1} \left\langle \,)(\, \right\rangle. \tag{61.6}$$

The second skein relation is even more obvious. If you delete a trivial circle you are lowering the number of components of the corresponding state by 1, hence the answer is $-A^2 - A^{-2}$ times the partition function of the diagram with the circle deleted. □

Theorem 61.3.2. [5] The Kauffman bracket is an invariant of framed links.

Proof. We only need to show that it is unchanged by isotopies that don't change the combinatorics of crossings, flipping monogons, and the Reidemeister II and III moves.

The partition function only depends on the combinatorics of the crossings so isotopies that preserve the combinatorics of intersection do not change the bracket. We need to explore the two sides of the second and third Reidemeister moves.

FIGURE 61.3: The second Reidemeister mover.

Resolving the crossings using the Kauffman bracket skein relations, everything cancels except the term that coincides with the right side.

$$\left\langle \asymp\!\!\asymp \right\rangle = \tag{61.7}$$

$$A^2 \left\langle \,)(\, \right\rangle \prec + \left\langle \,)(\, \right\rangle \circ \left\langle \,)(\, \right\rangle + \left\langle \asymp \right\rangle + A^{-2} \left\langle \succ \right\rangle \left\langle \,)(\, \right\rangle = \tag{61.8}$$

$$A^2 \left\langle \qquad \right\rangle + (-A^2 - A^{-2}) \left\langle \,)(\, \right\rangle \quad \left\langle \,)(\, \right\rangle + \left\langle \qquad \right\rangle + A^{-2} \left\langle \qquad \right\rangle = \tag{61.9}$$

$$\left\langle \qquad \right\rangle. \tag{61.10}$$

To understand invariance under Reidemeister III, resolve the lowest crossing on the left, the highest crossing on the right, and apply a Reidemeister II move to both diagrams.

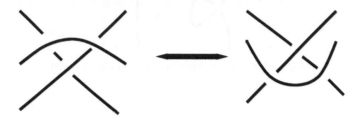

FIGURE 61.4: The third Reidemeister move.

First we expand the left-hand side.

$$\text{(diagram)} = A\,\text{(diagram)} + A^{-1}\,\text{(diagram)} \tag{61.11}$$

Apply the Reidemeister II move to get,

$$= A \qquad\quad + A^{-1}\,\text{(diagram)}. \tag{61.12}$$

Apply the Reidemeister II move to the right-hand side.

$$\text{(diagram)} = A\,\text{(diagram)} + A^{-1}\,\text{(diagram)} \tag{61.13}$$

Finally we apply the Reidemeister II move to the term on the right to see we have the same expression as in Equation 61.12.

□

Using the skein relations it is easy to see

$$\text{(diagram)} = -A^3\,\text{(diagram)}, \tag{61.14}$$

and

$$\text{(kink diagram)} = -A^{-3} \; \Big| \; . \tag{61.15}$$

The top monogon is often called a positive kink and the bottom a negative kink, because of the exponents of A.

From the state sum formula for the Kauffman bracket, and the fact that a state always has at least one circle, you can see that the Kauffman bracket is always divisible by $(-A^2 - A^{-2})$. Let

$$<L>' = \frac{<L>}{-A^2 - A^{-2}}. \tag{61.16}$$

This is the version of the bracket polynomial that is directly related to the Jones polynomial.

61.4 Connection with the Jones Polynomial

Suppose that the link L is oriented. The **Jones polynomial** $V(L)$ is a Laurent polynomial in the variable $t^{\frac{1}{2}}$, that satisfies the skein relations

$$t^{-1} V\left(\text{crossing}\right) - t V\left(\text{crossing}\right) = (t^{\frac{1}{2}} - t^{-\frac{1}{2}}) V\left(\text{smoothing}\right) \tag{61.17}$$

and $V(\bigcirc) = 1$.

If D is a diagram for L and $w(D)$ is its total writhe, then

$$(-A^3)^{-w(D)} < D >' \tag{61.18}$$

is an invariant of the oriented link. Making the substitution $A = t^{-\frac{1}{4}}$ in this formula yields the Jones polynomial.

61.5 Bibliography

[1] Crowell, Richard H.; Fox, Ralph H. *Introduction to Knot Theory*, Reprint of the 1963 original. Graduate Texts in Mathematics, No. 57. Springer-Verlag, New York-Heidelberg, 1977. x+182 pp.

[2] Jones, V. F. R. Hecke algebra representations of braid groups and link polynomials, *Ann. of Math.* (2) **126** (1987), no. 2, 335-388.

[3] Hoffman, Kenneth; Kunze, Ray. *Linear Algebra*, Second edition Prentice-Hall, Inc., Englewood Cliffs, N.J. 1971 viii+407 pp.

[4] Kauffman, Louis H.; Lins, Sóstenes L. Temperley-Lieb Recoupling Theory and Invariants of 3-Manifolds, *Annals of Mathematics Studies*, **134** Princeton University Press, Princeton, NJ, 1994. x+296 pp. ISBN: 0-691-03640-3 .

[5] Kauffman, Louis H. State models and the Jones polynomial, *Topology* **26** (1987), no. 3, 395-407.

[6] Khovanov, Mikhail *A categorification of the Jones polynomial*, *Duke Math. J.* **101** (2000), no. 3, 359-426.

[7] Khovanov, Mikhail; Lauda, Aaron D. *A diagrammatic approach to categorification of quantum groups II*, *Trans. Amer. Math. Soc.* **363** (2011), no. 5, 2685-2700.

[8] Kashaev, R. M. An invariant of triangulated links from a quantum dilogarithm, (Russian) *Zap. Nauchn. Sem. S.-Peterburg. Otdel. Mat. Inst. Steklov. (POMI)* **224** (1995), *Voprosy Kvant. Teor. Polya i Statist. Fiz.* 13, 208-214, 339; translation in *J. Math. Sci.* (New York) **88** (1998), no. 2, 244-248

[9] Kuperberg, Greg. Spiders for rank 2 Lie algebras, *Comm. Math. Phys.* **180** (1996), no. 1, 109-151.

[10] Lickorish, W. B. Raymond. What is a skein module?,

[11] Lickorish, W. B. Raymond *An Introduction to Knot Theory*, Graduate Texts in Mathematics, **175** Springer-Verlag, New York, 1997. x+201 pp. ISBN: 0-387-98254-X. *Notices Amer. Math. Soc.* **56** (2009), no. 2, 240-242.

[12] Murakami, Hitoshi; Murakami, Jun. The colored Jones polynomials and the simplicial volume of a knot, *Acta Math.* **186** (2001), no. 1, 85-104.

[13] Murakami, Hitoshi; Ohtsuki, Tomotada; Yamada, Shuji. Homfly polynomial via an invariant of colored plane graphs, *Enseign. Math.* (2) **44** (1998), no. 3-4, 325-360.

[14] Morrison, Scott; Peters, Emily; Snyder, Noah. Categories generated by a trivalent vertex., *Selecta Math.* (N.S.) **23** (2017), no. 2, 817-868.

[15] Menasco, William W.; Thistlethwaite, Morwen B. The Tait flyping conjecture, *Bull. Amer. Math. Soc.* (N.S.)**25** (1991), no. 2, 403-412.

[16] Ohtsuki, Tomotada; Yamada, Shuji. Quantum $SU(3)$ invariant of 3-manifolds via linear skein theory, *J. Knot Theory Ramifications* **6** (1997), no. 3, 373-404.

[17] Turaev, V. G. A simple proof of the Murasugi and Kauffman theorems on alternating links, *Enseign. Math.* (2) **33** (1987), no. 3-4, 203-225.

[18] Witten, Edward. Quantum field theory and the Jones polynomial, *Comm. Math. Phys.* **121** (1989), no. 3, 351-399.

Chapter 62

Span of the Kauffman Bracket and the Tait Conjectures

Neal Stoltzfus, Louisiana State University

62.1 Key Idea

With the discovery of the state sum formulation for the Jones polynomial by Louis Kauffman [15], three different publications ([15],[22],[26]) soon appeared featuring the application of the state sum formulation to proving several long-open conjectures of Paul Guthrie Tait [27, 14] concerning diagrams of knots. The latter reference discusses the Tait conjectures using current mathematical usage in knot theory. Dan Silver ([25]) gives an excellent overview of Tait's early contributions to knot theory.

62.2 Basic Definitions & Properties

62.2.1 Span of a (Laurent) Polynomial

When knot invariants take their values in a ring of polynomials, e.g. $\mathbb{Z}[X]$, or Laurent polynomials, e.g. $\mathbb{Z}[t, t^{-1}]$, we can define the (t-)span of a (Laurent) polynomial, $p(t)$, as the non-negative integer: maximal degree of $p(t)$ in the variable t - minimal degree of $p(t)$. For example, the span of $p(t) = t^{-2} + 3t + 7t^3$ is $3 - (-2) = 5$.

Span is invariant under scalar multiplication. (Hence the span of Kauffman bracket, $< D >$, of a link diagram D is a link invariant even though the bracket is not (for a scalar multiple of the Kauffman bracket is the Jones polynomial link invariant (compatibly parameterized)).

62.2.2 Diagrams & State Surfaces

A link diagram is a four-valent graph embedded in the two-sphere (i.e. 4-valent genus zero combinatorial map [16, 11]) equipped with a choice of a pair of opposite edges at each vertex (the *over crossing*). Note that the embedding is fixed and is equivalent to the combinatorial map (or rotation graph) structure: $(E, \sigma_1 : E \to E)$ where σ_1 is a permutation of the edges, E, providing the choice of rotation of edges at each vertex. (Hence the orbits of σ_1 have cardinality four and are in bijection with the vertex set). Two edges at a vertex are opposite when they are **not** consecutive in the rotation order at the vertex. In general, we will assume the link diagram is connected; if not, we will make the following constructions and definitions for each component of the diagram separately.

Recalling the state diagrammatic definition of the Kauffman bracket [15], we will introduce a surface for each state, $S : V \rightarrow \{A, B\}$ of the link diagram. The dual state, $\hat{S} : V \rightarrow \{A, B\}$ is defined by choosing the opposite smoothing from S at each vertex. The number of circles resulting from the smoothing of the link diagram described by the state S is denoted $|S|$. Denote by S_A the state with constant value A, the all-A state. Thus, the dual to the all-A state is the all-B state, $S_B = \hat{S_A}$.

For each state S, we construct a *state surface*, denoted $Y[S]$. $Y[S]$ is a closed orientable surface whose genus is denoted $g[S] := g(Y[S])$.

The construction consists of a cobordism, denoted $W[S]$, interpolating from the circles of the $\hat{S}$-smoothing to the circles of the S-smoothing. $W[S]$ can be viewed as embedded in the cylinder over the two-sphere, $S^2 \times [-1, 1]$. In fact, there is a Morse function (in this setting, real-valued functions with only index 1 (saddles) critical points) on this cobordism given by the obvious height function, projection to the interval factor. Outside a small neighborhood of the vertex, the cobordism is a product, $E \times [-1, 1]$, along the edge E. In the neighbor of each vertex, the cobordism consists of index-one saddles centered at each crossing (vertex) interpolating from the $\hat{S}$-smoothing to the S-smoothing. Note that W is orientable but the normal direction at the vertices (index one critical saddle points) can be downwards (and changes direction when the crossings at either end of the edge do not alternate between over and under!) An excellent graphic of this cobordism for a non-alternating pretzel knot can be found in the work of Adam Lowrance [18] as well as on a T-shirt for the 2016 LSU Math Contest, https://www.math.lsu.edu/~contest/contest_logo_2016.html.

Attaching disks along the circles at the top and bottom of the cobordism we obtain a closed (orientable) surface to which the Morse function can be extended with maxima at the centers of the S-smoothing disks and minima at the centers of the $\hat{S}$-smoothing disks. The cobordism between the smoothing circles together with the attached disks form the state surface $Y[S]$. The process of taking duals, $S \rightarrow |^S$, is an involution on the set of states, $\hat{\hat{S}} = S$. The state surface $Y[\hat{S}]$ is isomorphic to $Y[S]$ (but with the Morse function replaced by $-f$).

We also define a combinatorial map (graph with a fixed full (complementary regions are disks) embedding in a surface [16]) associated to the state S as follows:

- vertex set:= centers of S-smoothing circles (maxima of Morse function)

- edge set:= ascending sheet for each saddle point (connecting two maxima)

- faces := the complementary regions in the associated surface: $Y[S]$

62.2.3 The Turaev Surface and Its Genus

For S_A, the all-A state, the surface, $Y[s_A]$, is dubbed the Turaev surface in [10], and its genus, the *Turaev genus*. If the link projection is alternating, the Turaev genus is zero, as the Turaev surface embeds in the two-sphere, in this case $Y[S]$ is equivalent to the checkerboard coloring of the link diagram. Defining the Turaev genus of a (non-split) link as the minimal genus of the all-A smoothing over all (non-split) diagrams for the given link, we obtain a link invariant. In fact, this link invariant characterizes the class of (non-split)-alternating links as the class of Turaev genus zero links ([10]). Hence the genus provides a measure of *deviation* of a link from being alternating. The Turaev genus is tabulated in Chuck Livingston's knotinfo database [17].

Moreover, the original link can be viewed as a *virtual* link in the cylinder over each state surface. For the Turaev surface, the link is alternating on that surface. (Recall that the normal vector at the saddle point located at a vertex changes to the opposite direction if the crossing does **not** alternate.)

62.2.4 Dual State Lemma & Genus of a State

The key technical result is the following lemma due to Lou Kauffman ([15]). Our exposition follows that of Turaev ([28]) but refines the analysis to incorporate the connection with the genus of the state surface, $g[S] := g(Y[S])$.

Let $|V|$ denote the number of vertices of the 4-valent graph (equivalently crossings of the link) and denote by $r(\Gamma)$, the number of components of the link diagram.

Lemma 1. Dual State Lemma: For every state S of a link diagram Γ, $|S| + |\hat{S}| \leq |V| + 2r$. Furthermore, the difference is twice the genus of $Y[S]$, $2g[S] = |V| + 2r - (|S| + |\hat{S}|)$.

Outline of verification. We will use a long exact sequence in the homology of the associated state cobordism pair, $(W[S], \partial W)$. The cobordism is a subset of a closed surface that embeds in the three-sphere, hence orientable and homology groups are all free. The relevant part is:

$$\to H_1(\partial W) \to H_1(W, \partial W) \to H_0(\partial W) \to H_0(W) \to H_0(\partial W) \to \mathbb{Z} \to 0.$$

Recalling that the boundary is a collection of circles resulting from the S and $\hat{S}$ smoothing, the first group in the sequence is free of rank $|S| + |\hat{S}|$. As the cobordism is orientable, the image of an orientation class is rank one in $H_1(\partial W)$.

The relative dimension one homology group of a surface of genus $g(W) = g(Y)$ with $|S| + |\hat{S}|$ boundary circles has rank $2g(Y) + |S| + |\hat{S}| - 1$.

The image of $H_1(W)$ in the relative group has rank $2g(W)$ (use exactness and the rank of zero-homology of $|S| + |\hat{S}|$ circles). The map is given by the adjoint map of the intersection pairing induced by Lefschetz duality on this surface with boundary.

Let $b_i(X)$ denote the i-th betti number (or rank) of the space X. Noting that the surface W deformation retracts onto the four-valent graph Γ, we have that $b_1(W) = b_0(W) - \chi(W) = r - (|V| - 2|V|)$, where χ is the Euler characteristic and, in a 4-valent graph, $|E| = 2|V|$.

Next recall that $b_1(W) = b_1)W, \partial W)$ by Lefschetz duality. From the homology exact sequence above, we have that $|S|+|\hat{S}| = b_0(\partial W) = b_0(W)+b_1(W, \partial W)-\text{rank}(\text{Image} \quad p_*) = r+(r+|V|)-2g(W)$.

As the genus is non-negative, the inequality follows.

□

62.3   Refinements & Applications

62.3.1   Alternating Diagrams & Turaev Genus Zero

In his simplification of the proof of the Tait conjectures, Turaev [28] introduced an auxiliary surface, viewed in the optic of [10] as the state surface of the all-A state (also, the Turaev surface). Turaev also realized that the corresponding diagram was alternating if and only if the genus of this state was zero [28][Cor. 1].

62.3.2   Adequate Diagrams

A link diagram is S-adequate if the underlying graph of the associated combinatorial map, $\mathbb{D}[S]$, has no loops. In [10] it is shown that if the all-A and the dual all-B states are adequate, then the degree contribution from the all-A state is maximal (all other states contribute lower degree terms and therefore the contribution from the all-A state cannot be cancelled), and the contribution from the dual all-B state persists and is of minimal degree. Given the usual parameterization of the Jones polynomial $J_K(q)$, in the parameter q, the relation to the Kauffman bracket is given by:

$$J_K(A^4) = -A^{-3\ \text{writhe(K)}} < D(K) >,$$

hence the q-span $J_K(q) = A$–span-$< D(K) >$.

Denoting by $|S|$ the number of circles in the S-smoothing, $a(S)$ the number of A-smoothings of the state S, and $b(S)$, the number of B-smoothings, the state sum formula for the bracket polynomial is:

$$< D(K) >= \sum_S (A^2 - A^{-2})^{|S|-1} A^{a(S)-b(S)}$$

Therefore the span of the Kauffman bracket is A-span $< K >=$ max $\deg_A < D > $ -min $\deg_A < D >\le 4|V| - 4g(S_A)$.

62.3.3   Crossing Number & Tait's Conjecture

There are several Tait conjectures (see the Wikipedia article on the Tait conjectures [29] and the survey by de la Harpe [14]). We will discuss the most striking.

Tait Crossing Number Conjecture: An alternating link always has an alternating diagram that has the minimal crossing number among all diagrams for that link.

Equality holds if D is adequate (e.g. if $g = 0$ and K is alternating). In fact, if D is a prime non-alternating diagram, then the Turaev genus is positive and there is strict inequality. Thus the q-span of the Jones, $V_L = |V| = c(L)$, the crossing number and Tait's conjecture follows.

62.3.4 Genus Bounds in Khovanov & Heegaard Floer Homology

In [6] the Turaev genus is demonstrated to bound the width of the Khovanov and, by the work of Adam Lowrance [19], a similar result holds for the Heegaard Floer homology.

62.4 Bibliography

[1] Tetsuya Abe, *The Turaev genus of an adequate knot*, Topology Appl. **156** (2009), no. 17, 2704–2712.

[2] Tetsuya Abe and Kengo Kishimoto, *The dealternating number and the alternation number of a closed 3-braid*, J. Knot Theory Ramifications **19** (2010), no. 9, 1157–1181.

[3] B. Bollobás and O. Riordan, *A polynomial invariant of graphs on orientable surfaces*, Proc. London Math. Soc. (3) **83** (2001), no. 3, 513–531.

[4] _____, *A polynomial of graphs on surfaces*, Math. Ann. **323** (2002), no. 1, 81–96.

[5] Abhijit Champanerkar and Ilya Kofman, *Spanning trees and Khovanov homology*, Proc. Amer. Math. Soc. **137** (2009), no. 6, 2157–2167.

[6] Abhijit Champanerkar, Ilya Kofman, and Neal Stoltzfus, *Graphs on surfaces and Khovanov homology*, Algebr. Geom. Topol. **7** (2007), 1531–1540.

[7] _____, *Quasi-tree expansion for the Bollobás-Riordan-Tutte polynomial*, Bull. Lond. Math. Soc. **43** (2011), no. 5, 972–984.

[8] Sergei Chmutov, *Generalized duality for graphs on surfaces and the signed Bollobás-Riordan polynomial*, J. Combin. Theory Ser. B **99** (2009), no. 3, 617–638.

[9] Peter R. Cromwell, *Knots and Links*, Cambridge University Press, Cambridge, 2004.

[10] Oliver T. Dasbach, David Futer, Efstratia Kalfagianni, Xiao-Song Lin, and Neal W. Stoltzfus, *The Jones polynomial and graphs on surfaces*, J. Combin. Theory Ser. B **98** (2008), no. 2, 384–399.

[11] Joanna A. Ellis-Monaghan and Iain Moffatt, *Graphs on Surfaces: Dualities, Polynomials, and Knots*, Springer Briefs in Mathematics, Springer, New York, 2013.

[12] P. Freyd, D. Yetter, J. Hoste, W. B. R. Lickorish, K. Millett, and A. Ocneanu. A new polynomial invariant of knots and links. *Bull. Amer. Math. Soc. (N.S.)*, 12(2):239–246, 1985.

[13] David Futer, Efstratia Kalfagianni, and Jessica Purcell, *Guts of Surfaces and the Colored Jones Polynomial*, Lecture Notes in Mathematics, vol. 2069, Springer, Heidelberg, 2013.

[14] Pierre de la Harpe, Michel Kervaire, and Claude Weber. On the Jones polynomial. *Enseign. Math. (2)*, 32(3-4):271–335, 1986.

[15] L. Kauffman, *State models and the Jones polynomial*, Topology **26** (1987), no. 3, 395–407.

[16] S. Lando and A. Zvonkin, *Graphs on Surfaces and Their Applications*, Encyclopaedia of Mathematical Sciences, vol. 141, Springer-Verlag, 2004.

[17] Charles Livingston, *Knotinfo* database. `http://www.indiana.edu/~knotinfo/descriptions/turaev_genus.html`.

[18] Adam M. Lawrence. *Homological Width and Turaev Genus*, `https://digitalcommons.lsu.edu/gradschool_dissertations/3797/`.

[19] Adam M. Lowrance, *On knot Floer width and Turaev genus*, Algebr. Geom. Topol. **8** (2008), no. 2, 1141–1162.

[20] Ciprian Manolescu, *An introduction to knot Floer homology*, (arXiv:1401.7107 [math.GT] (2014)).

[21] Ciprian Manolescu and Peter Ozsváth, *On the Khovanov and knot Floer homologies of quasialternating links*, Proceedings of Gökova Geometry-Topology Conference 2007, Gökova Geometry/Topology Conference (GGT), Gökova, 2008, pp. 60–81.

[22] Kunio Murasugi, *Jones polynomials and classical conjectures in knot theory*, Topology **26** (1987), no. 2, 187–194.

[23] Makoto Ozawa, *Essential state surfaces for knots and links*, J. Aust. Math. Soc. **91** (2011), no. 3, 391–404.

[24] Peter Ozsváth and Zoltán Szabó, *Heegaard Floer homology and alternating knots*, Geom. Topol. **7** (2003), 225–254 (electronic).

[25] Daniel S. Silver, *Knot Theory's Odd Origins*, American Scientist **94(2)** (2006), 158–165.

[26] M. Thistlethwaite, *A spanning tree expansion of the Jones polynomial*, Topology **26** (1987), no. 3, 297–309.

[27] P. G. Tait, *On Knots I, II, III. Scientific Papers Vol I* (1898), 273-347.

[28] V. G. Turaev, *A simple proof of the Murasugi and Kauffman theorems on alternating links*, Enseign. Math. (2) **33** (1987), no. 3-4, 203–225.

[29] Wikipedia, *Tait Conjectures*, `https://en.wikipedia.org/wiki/Tait~conjectures`.

Chapter 63

Skein Modules of 3-Manifolds

Rhea Palak Bakshi, George Washington University

Józef H. Przytycki, George Washington University and University of Gdańsk

Helen Wong, Claremont McKenna College

63.1 Introduction to Skein modules

A *skein module* is an algebraic object associated to a manifold, usually constructed as a formal linear combination of embedded (or immersed) submanifolds, modulo locally defined relations. In a more restricted setting, a skein module is a module associated to a 3-dimensional manifold, by considering linear combinations of links in the manifold, modulo properly chosen (skein) relations. In the choice of relations one takes into account several factors:

(i) Is the module we obtain accessible (computable)?

(ii) How good are our modules in distinguishing 3-manifolds and links in them?

(iii) Does the module reflect topology/geometry of a 3-manifold (e.g. surfaces in a manifold, geometric decomposition of a manifold)?

(iv) Does the module admit some additional structure (e.g. filtration, gradation, multiplication, Hopf algebra structure, categorification)?

We give here an overview of skein modules studied in the literature with a goal of giving an elementary introduction to the subject. Notice that in the case of a product $M = F \times [0, 1]$ where F is a surface, links in M have a structure of a monoid with a product $L_1 \cdot L_2$ obtained by placing L_1 above L_2 in $F \times [0, 1]$ and with the empty link as an identity element. This multiplicative structure extends to most of the skein modules we consider.

A skein module of a 3-manifold is a module that is defined, at least initially, by trying to generalize the skein relation of an Alexander or HOMFLYPT polynomial (see Chapter 66). In order to replace S^3 with an arbitrary 3-manifold, one must move away from projections and diagrams into a more general setting. This is easily done by considering links which are identical outside some ball B^3 contained in M^3 and which appear inside the ball as various prescribed tangles. However, one often wants to distinguish the links from one another based on their local appearance in B^3, and in order to do this it will usually be necessary to assume that M^3 is an oriented manifold.

We can now give a fairly general definition of a skein module of a 3-manifold M^3. But the reader will soon see that there seems to be no end to the amount of generality that can be introduced. For the present it seems clear that there can be no ultimate definition. At this point in the development of the theory there is simply too big a gap between which modules can be defined and which special cases of these modules can actually be computed and understood for reasonably interesting manifolds.[1]

We can start from the following fairly general definition based on oriented links up to ambient isotopy:

Definition 63.1.1. Let M^3 be an oriented 3-manifold, R a commutative ring with unit, $\mathcal{L}$ the set of all oriented links in M^3 considered up to isotopy, and $R\mathcal{L}$ the free R-module generated by $\mathcal{L}$. Let $T_0, T_1, ..., T_{k-1}$ be a sequence of k oriented tangles in a 3-ball B^3. (Recall that an oriented tangle is a properly embedded oriented 1-manifold in B^3 having a prechosen, and henceforth fixed, set of inputs and outputs on ∂D^3. Thus $\partial T_i = \partial B^3$ is the same oriented 0-manifold for each i.) Let $r_0, r_1, ..., r_{k-1}$ be a chosen sequence of elements of R. Now let S be the submodule of $R\mathcal{L}$ generated by all expressions of the form $r_0 L_{T_0} + ... + r_{k-1} L_{T_{k-1}}$ where $L_{T_0}, ..., L_{T_{k-1}}$ are k oriented links in M^3 which are identical outside some ball B^3 contained in M^3 but inside of B^3 they appear as the tangles $T_0, ..., T_{k-1}$. We will often abbreviate this "skein relation" as $r_0 T_0 + ... + r_{k-1} T_{k-1}$. Finally, the quotient module $\vec{S} = R\mathcal{L}/S$ is called a skein module of M^3. It should be denoted by something like $\vec{S}(M^3; R; r_0 T_0 + ... + r_{k-1} T_{k-1}; isotopy)$ to fully incorporate all the ingredients of its definition, but this is obviously unwieldy, and we shall always drop at least the last qualifier, and sometimes the last two or even three qualifiers, from the notation. The arrow over S is used to indicate that oriented links are used to define the module.

Notice that this definition is topological and 3-dimensional, not dependent on the existence of any particular projection of the 3-manifold to a 2-dimensional submanifold.

The most straightforward way to alter this definition, while still retaining a certain degree of generality, is to consider unoriented links, or to consider links up to link homotopy or some other relation. Another important variant is to consider framed links; in this case, the tangles must also be framed. It is now possible and usually desirable to introduce additional relations between any link and another which differ only by their framing. Many manifolds, for example homology spheres or products, possess natural framings. But for an arbitrary 3-manifold it seems very natural to consider framed links [16, 21, 35].[2]

We describe now several concrete examples of skein modules which are studied by researchers.

(1) (Framing skein module). Let M be an oriented 3-dimensional manifold and $\mathcal{L}^{fr}$ the set of framed unoriented links in M up to ambient isotopy. Consider the ring of Laurent polynomials $\mathbf{Z}[q^{\pm 1}]$ and the free $\mathbf{Z}[q^{\pm 1}]$-module with basis $\mathcal{L}^{fr}$. The skein relation is just related to framing change and q, that is, we consider the submodule of $\mathbf{Z}[q^{\pm 1}]\mathcal{L}^{fr}$ generated by expressions $L^{(1)} - qL$ where $L^{(1)}$ is obtained from L by performing a positive twist on its framing.

---

[1] In [28] the theory based on skein modules is called "Algebraic Topology based on knots." With the advent of Khovanov homology the name is more than justified. Categorification of skein modules should be the central object of algebraic topology based on knots.

[2] This part is taken from the survey [16] . Here is the abstract to the paper:

"This paper is a short survey and guide to the literature of the rapidly developing theory of skein modules. These are modules which can be associated to any 3-manifold M^3 and which strive to capture information about M^3 based on the knot theory that M^3 supports. There are many variations on the basic definition of a skein module and hence many different skein modules that can be defined for a given manifold. What all the various definitions have in common, however, is their attempt to capitalize on, and generalize, the so-called skein theory of the Alexander/Conway polynomial that was first promoted by Conway and later made popular by the emergence of the Jones polynomial. In some sense the theory of skein modules is the natural extension of the HOMFLY, Kauffman and Jones polynomials to links in arbitrary 3-manifolds."

We call this quotient the first skein module of M and denote by $S_0(M, q)$. If a link L cuts some 2-sphere in M exactly once, we see that L is a torsion element: $(q^2 - 1)L = 0$. The reason can be explained by the fact that $\pi_1(SO(3)) = \mathbf{Z}_2$. In fact, this is the only way of creating relations in $S_0(M)$ as we have:

Theorem 63.1.2. [15, 7, 2]

$S_0(M) = \mathbf{Z}[q^{\pm 1}]\mathcal{L}' \oplus \mathbf{Z}[q^{\pm 1}]/(q^2 - 1)\mathcal{L}''$ where $\mathcal{L}'$ is the set of ambient isotopy classes of links which do not cut any S^2 in exactly one point and $\mathcal{L}''$ are ambient isotopy classes of links which cut some S^2 in exactly one point. In particular, if M is an irreducible manifold then $S_0(M)$ is a free $\mathbf{Z}[q^{\pm 1}]$-module.

The torsion obtained from framing relations are present in many other skein modules.

(2) q-deformation of the first homology group of a 3-manifold M, denoted by $S_2(M; q)$. It is based on the skein relation (between oriented framed links in M): $L_+ = qL_0$. Already this simply defined skein module "sees" nonseparating surfaces in M. These surfaces are responsible for torsion part of our skein module. We have the following result [30].

Theorem 63.1.3.

$$S_2(M, q) = \mathbf{Z}[q^{\pm 1}]T(H_1(M, \mathbf{Z})) \oplus \bigoplus_{\alpha \in H_1(M, \mathbf{Z}) - T(H_1(M, \mathbf{Z}))} \mathbf{Z}[q^{\pm 1}]/(q^{2mul(\alpha)} - 1),$$

where $T(H_1(M, \mathbf{Z}))$ and the multiplicity of α, $mul(\alpha)$, are defined as follows: Let $\phi : H_1(M, \mathbf{Z}) \times H_2(M, \mathbf{Z}) \to \mathbf{Z}$ be the bilinear form of the intersection of 1-cycles with 2-cycles on M. Then $\alpha \in T(H_1(M, \mathbf{Z}))$ if and only if $\phi_\alpha(\beta) = \phi(\alpha, \beta) = 0$ for any β. Otherwise $mul(\alpha)$ is defined as the positive generator of $im\phi_\alpha(H_2(M, \mathbf{Z}))$.

One should notice that for a closed 3-manifold M, $T(H_1(M, \mathbf{Z}))$ is the torsion part of $H_1(M, \mathbf{Z})$.

(3) q-deformation of the fundamental group of a 3-manifold M, denoted by $S_{\pm 1}(M; q)$ [28, 19].

(4) The Kauffman bracket skein module, KBSM.
This is the skein module based on the *Kauffman bracket skein relation*, $L_+ = AL_0 + A^{-1}L_\infty$, and denoted by $S_{2,\infty}(M)$. It is best understood among Jones-type skein modules and we devote a separate entry to it (Chapter 69).

(5) HOMFLYPT Skein Module of a 3-manifold. Just as the Kauffman bracket skein module is based on the Kauffman bracket of unoriented links in S^3, one can define a skein module based on the 2-variable HOMFLYPT polynomial for oriented links in S^3. More precisely, the *HOMFLYPT skein module* of an oriented 3-manifold M is the free $\mathbf{Z}[v^{\pm}, z^{\pm}]$-module consisting of finite linear combinations of isotopy classes of oriented links in M, quotiented by the submodule generated by skein relations $v^{-1}L_+ - vL_- - zL_0$ where the oriented links are identical everywhere except in the small ball, as illustrated in the following figure:

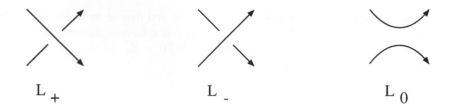

$$L_+ \qquad\qquad L_- \qquad\qquad L_0$$

The HOMFLYPT skein module of M is denoted by $S_3(M)$. The HOMFLYPT skein module is computed for a few manifolds [11, 14, 24, 25, 12, 13]. For $M = F \times I$, a product of a surface and the interval, $S_3(F \times I)$ is a Hopf algebra (usually neither commutative nor co-commutative) [36, 26] . $S_3(F \times [0, 1])$ is a free module and can be interpreted as a quantization [14, 36, 25]. There is also growing evidence that the HOMFLYPT skein module is related to elliptic Hall algebras from quantum algebra [22] and to $SL_n(\mathbb{C})$ character varieties [6, 31, 32].

(6) Skein modules based on the *Kauffman polynomial* relation.
This is denoted by $S_{3,\infty}$ and is known to be free for $M = F \times [0, 1]$. In fact there are two versions of the skein module: one based on the original Kauffman relation and one on its Dubrovnik version. They already differ on the projective space $\mathbb{R}P^3$ [23] .

(7) Homotopy skein modules. In these skein modules, $L_+ = L_-$ for self-crossings. The best studied example is the q-homotopy skein module with the skein relation $q^{-1}L_+ - qL_- = zL_0$ for mixed crossings. For $M = F \times [0, 1]$ it is a quantization, [15, 36, 29], and as noted by Kaiser they can be almost completely understood using the singular tori technique of Lin.

(8) Skein modules based on relations deforming n-moves. The only studied skein module of this type is the fourth skein module $S_4(M) = R\mathcal{L}/b_0L_0 + b_1L_1 + b_2L_2 + b_3L_3$, with a possible additional framing relation. It is conjectured that in S^3 this module is generated by trivial links. Motivation for this is the *Montesinos-Nakanishi three-move conjecture*. The fact that the conjecture does not hold (i.e. $(\sigma_1\sigma_2\sigma_3\sigma_4)^{10}$ is a counterexample [8]) implies that even in S^3 the cubic skein module is not generated by trivial links [33].

(9) Skein modules based on Vassiliev-Gusarov filtration.
We extend the family of knots, $\mathcal{K}$, by singular knots, and resolve a singular crossing by $K_{cr} = K_+ - K_-$. This allows us to define the Vassiliev-Gusarov filtration: $\dots \subset C_3\dots \subset C_2\dots \subset C_1\dots \subset C_0 = R\mathcal{K}$, where C_k is generated by knots with k singular points. The k'th Vassiliev-Gusarov skein module is defined to be a quotient: $W_k(M) = R\mathcal{K}/C_{k+1}$. The completion of the space of knots with respect to the Vassiliev-Gusarov filtration, $\hat{R\mathcal{K}}$, is a *Hopf algebra* (for $M = S^3$). Functions dual to Vassiliev-Gusarov skein modules are called *finite type* or *Vassiliev invariants* of knots [27].

(10) Bar-Natan skein modules [3, 18]. Incompressible surfaces in 3-manifolds are important in this theory.

(11) Relative skein modules. If we consider an oriented 3-dimensional manifold with a boundary and a set of relative links (that is properly embedded one manifolds modulo proper relations) we deal with relative skein modules. There is extensive literature about them starting from [24, 16, 28].

(12) Topological quantum field theory (TQFT) and Witten-Reshetikhin-Turaev invariants of 3-manifolds come from Kauffman bracket skein modules [37, 34, 4].

(13) Homology built on skein modules.

The homology built on skein modules was proposed in [5, 20]. Later, motivated by Khovanov homology (see Chapters 70 and 73), the categorification of the Kauffman bracket skein modules is described in [1, 10] for interval bundles over surfaces.

63.2 Bibliography

[1] M. Asaeda, J. H. Przytycki, A. S. Sikora, Categorification of the Kauffman bracket skein module of I-bundles over surfaces, *Algebraic & Geometric Topology (AGT)*, 4, 2004, 1177-1210.

[2] R. P. Bakshi, D. Ibarra, G. Montoya-Vega, J. H. Przytycki, D. Weeks, On framings of links in 3-manifolds, *Canad. Math. Bull.* DOI: `https://doi.org/10.4153/S000843952000079X`10.4153/S000843952000079X, e-print: `https://arxiv.org/abs/2001.07782`arXiv:2001.07782 [math.GT] .

[3] M. Asaeda, C. Frohman, A note on the Bar-Natan skein module. Internat. J. Math. 18,(2007, no. 10, 1225–1243

[4] C. Blanchet, N. Habegger, G. Masbaum, P. Vogel, Topological quantum field theories derived from the Kauffman bracket. Topology 34 (1995), no. 4, 883-927.

[5] D. Bullock, C. Frohman, J. Kania-Bartoszyńska, Skein homology, Canad. Math. Bull. 41(2), 1998, 140-144.

[6] S. Cautis, J. Kamnitzer and S. Morrison, Webs and quantum skew Howe duality, *Mathematische Annalen* 360(1-2), 2014, 351–390.

[7] V. Chernov, Framed knots in 3-manifolds and affine self-linking numbers. *J. Knot Theory Ramifications* 14 (2005), no. 6, 791818. `https://arxiv.org/abs/math/0105139`arXiv:math/0105139 [math.GT].

[8] M. K. Dabkowski, J. H. Przytycki, Burnside obstructions to the Montesinos-Nakanishi 3-move conjecture, (with M.K. Dąbkowski), *Geometry and Topology*, 6, June, 2002, 355-360;
e-print: `http://front.math.ucdavis.edu/math.GT/0205040`

[9] M. K. Dabkowski, C. Li, J. H. Przytycki, Catalan states of lattice crossing, *Topology and its Applications*, 182, March, 2015, 1-15;
e-print: `arXiv:1409.4065 [math.GT]`

[10] B. Gabrovšek, The categorification of the Kauffman bracket Skein module of $\mathbb{R}P^3$. *Bull. Aust. Math. Soc.* 88 (2013), no. 3, 407422. `https://arxiv.org/abs/1809.03540`arXiv:1809.03540 [math.GT].

[11] B. Gabrovšek, M. Mroczkowski, The HOMFLYPT skein module of the lens spaces $L_{p,1}$, Topology Appl. 175, 2014, 72–80.

[12] Gilmer, Patrick M. and Zhong, Jianyuan K., On the HOMFLYPT skein module of $S^1 \times S^2$, *Mathematische Zeitschrift* 237(4), 2001, 769–814.

[13] Gilmer, Patrick M and Zhong, Jianyuan K and others, The HOMFLYPT skein module of a connected sum of 3–manifolds, *Algebraic & Geometric Topology* 1(1), 2001, 605–625.

[14] J. Hoste, M. Kidwell, Dichromatic link invariants, *Trans. Amer. Math. Soc.*, 321(1), 1990, 197-229.

[15] J. Hoste, J. H. Przytycki, Homotopy skein modules of oriented 3-manifolds, *Math. Proc. Cambridge Phil. Soc.*, (1990) 108, 475-488.

[16] J. Hoste, J. H. Przytycki, A survey of skein modules of 3-manifolds. in Knots 90, Proceedings of the International Conference on Knot Theory and Related Topics, Osaka (Japan), August 15-19, 1990, Editor A. Kawauchi, Walter de Gruyter 1992, 363-379.

[17] J.Hoste, J. H. Przytycki, The Kauffman bracket skein module of $S^1 \times S^2$, *Math. Z.*, 220(1), 1995, 63-73.

[18] U. Kaiser, Frobenius algebras and skein modules of surfaces in 3-manifolds, in Algebraic Topology–Old and New, *Banach Center Publications*, Vol. 85 (Polish Acad. Sci. Inst. Math., Warsaw, 2009), pp. 59-81.

[19] U. Kaiser, Quantum deformations of fundamental groups of oriented 3-manifolds. *Trans. Amer. Math. Soc.* 356(1), 2003.

[20] Kania-Bartoszyńska, J. H. Przytycki, A. S. Sikora, Estimating the Size of Skein Homologies, Knots in Hellas 98; The Proceedings of the International Conference on Knot Theory and its Ramifications; Volume 1. In the Series on Knots and Everything, Vol. 24, September 2000, pp. 138-142.

[21] L. H. Kauffman, New invariants in the theory of knots, *Amer. Math. Monthly*, 95, 1988, 195-242.

[22] Morton, Hugh and Samuelson, Peter and others, The HOMFLYPT skein algebra of the torus and the elliptic Hall algebra, *Duke Mathematical Journal* 166(5), 2017, 801–854.

[23] M. Mroczkowski, The Dubrovnik and Kauffman skein modules of lens spaces $L_{p,1}$, *J. Knot Theory Ramifications* 27 (2018), no. 3. 1840004, 15pp.

[24] J.H. Przytycki, Skein modules of 3-manifolds, *Bull. Polish Acad. Science*, 39(1-2), 1991, 91-100.

[25] J.H. Przytycki, Skein module of links in a handlebody, Topology 90, Proc. of the Research Semester in Low Dimensional Topology at OSU, Editors: B.Apanasov, W.D.Neumann, A.W.Reid, L.Siebenmann, De Gruyter Verlag, 1992; 315-342.

[26] J.H. Przytycki, Quantum group of links in a handlebody *Contemporary Math: Deformation Theory and Quantum Groups with Applications to Mathematical Physics*, M.Gerstenhaber and J.D.Stasheff, Editors, Volume 134, 1992, 235-245.

[27] J.H. Przytycki, Vassiliev-Gusarov skein modules of 3-manifolds and criteria for periodicity of knots, Low-Dimensional Topology, Knoxville, 1992 ed.: Klaus Johannson, International Press Co., Cambridge, MA 02238, 1994, 157-176.

[28] J.H. Przytycki, Algebraic topology based on knots: an introduction, *Knots 96*, Proceedings of the Fifth International Research Institute of MSJ, edited by Shin'ichi Suzuki, 1997 World Scientific Publishing Co., 279-297.

[29] J.H. Przytycki, Homotopy and q-homotopy skein modules of 3-manifolds: an example in Algebra Situs; In: *Knots, Braids, and Mapping Class Groups: Papers dedicated to Professor Joan Birman*, Ed. J. Gilman, W. Menasco, and X.-S. Lin, International Press., AMS/IP Series on Advanced Mathematics, Vol 24, Co., Cambridge, MA, 2001, 143-170. e-print: http://front.math.ucdavis.edu/math.GT/0402304

[30] J.H. Przytycki, A q-analogue of the first homology group of a 3-manifold, *Contemporary Mathematics* 214, Perspectives on Quantization (Proceedings of the joint AMS-IMS-SIAM conference on Quantization, Mount Holyoke College, 1996); Ed. L.A.Coburn, M.A.Rieffel, AMS 1998, 135-144.

[31] A. S. Sikora, and others, Skein theory for $SU(n)$–quantum invariants, *Algebraic & Geometric Topology* 5(3), 2005, 865–897.

[32] A. S. Sikora, SL_n-character varieties as spaces of graphs, *Transactions of the American Mathematical Society* 353(7), 2001, 2773–2804.

[33] J.H. Przytycki, T. Tsukamoto, The fourth skein module and the Montesinos-Nakanishi conjecture for 3-algebraic links, *J. Knot Theory Ramifications*, 10(7), 2001, 959–982; e-print: http://front.math.ucdavis.edu/math.GT/0010282

[34] N. Reshetikhin, V. G. Turaev, Invariants of 3-manifolds via link polynomials and quantum groups. *Invent. Math.* 103 (1991), no. 3, 547597.

[35] V.G. Turaev, The Conway and Kauffman modules of the solid torus, *Zap. Nauchn. Sem. Lomi* 167 (1988), 79-89. English translation: *J. Soviet Math.*, 52, 1990, 2799-2805.

[36] V.G. Turaev, Skein quantization of Poisson algebras of loops on surfaces, *Ann. Scient. Éc. Norm. Sup.*, 4(24), 1991, 635-704.

[37] E. Witten, Quantum field theory and the Jones polynomial. *Comm. Math. Phys.* 121 (1989), no. 3, 351399.

Chapter 64

The Conway Polynomial

Sergei Chmutov, Ohio State University at Mansfield

The classical Alexander polynomial $\Delta_K(t)$ of a knot K is well-defined up to a sign and multiplication by a power of the variable t. In 1969, J. Conway [4] observed that the *skein relation* for the Alexander polynomial together with the initial value on the unknot allows us to fix this ambiguity. Thus he introduced the CONWAY polynomial which is sometimes called the ALEXANDER-CONWAY polynomial.

Definition 64.0.1. The CONWAY polynomial $\nabla_K(z)$ is a link invariant determined by the (skein) relation and the initial value

$$\nabla \underset{}{\bigtimes}(z) - \nabla \underset{}{\bigtimes}(z) = z\nabla \underset{}{)(} , \qquad \nabla \bigcirc = 1. \qquad (64.1)$$

Here the three diagrams in the skein relation represent the diagrams identical everywhere except for the small discs shown.

With such a definition one needs to prove that these two equations give a unique well-defined link invariant. The relation to the Alexander polynomial is given by equation $\nabla_L(x - x^{-1}) = \Delta_L(x^2)$ for an appropriate normalization of the Alexander polynomial.

64.1 Examples of Calculation

(i) $\nabla \underset{}{\bigcirc\bigcirc} = \frac{1}{z}\nabla \underset{}{\bigcirc\!\!\bigcirc} - \frac{1}{z}\nabla \underset{}{\bigcirc\!\!\bigcirc} = 0,$

because both knots in the expression in the middle are trivial.

(ii) $\nabla \underset{}{\bigcirc\!\!\bigcirc} = \nabla \underset{}{\bigcirc\!\!\bigcirc} - z\nabla \underset{}{\bigcirc\!\!\bigcirc} = \nabla \qquad - z\nabla \qquad = -z.$

(iii) ∇ ⬡ $= \nabla$ ⬡ $- z\nabla$ ⬡ $= \nabla$ ◯ $- z\nabla$ ⬭ $= 1 + z^2$.

Similarly, one may check the following skein-type local relations for the Conway polynomial which we borrow from [1, Chapter 2, Exercise 8].

(iv) ∇ ⬡ $+ \nabla$ ⬡ $= (2 + z^2)\nabla$ ⬭ .

(v) ∇ ⬡ $+ \nabla$ ⬡ $= 2\nabla$ ⬭ .

(vi) ∇ ⬡ $+ \nabla$ ⬡ $= \nabla$ ⬡ $+ \nabla$ ⬡ .

64.2 Properties of the Conway Polynomial

1. The Conway polynomial does not distinguish between a link and its mirror image, as well as between a link and one obtained from it by the simultaneous change of orientations of all the components.

2. The Conway polynomial of a *split link* is equal to zero, $\nabla_L = 0$. A link is called *split* if it is isotopic to a link which has some of its components located inside a ball while all the other components are located outside of the ball.

3. The Conway polynomial is multiplicative with respect to the connected sum of knots, $\nabla_{K_1 \# K_2} = \nabla_{K_1} \cdot \nabla_{K_2}$.

4. The Conway polynomial of the *Kinoshita-Terasaka knot* (the mirror image of the $11n42$ knot) on the right is equal to 1, $\nabla_{KT}(z) = 1$. $KT =$ (diagram)

5. For any knot K, the Conway polynomial ∇_K is an even polynomial in z with constant term 1:

$$\nabla_K(z) = 1 + c_2(K)z^2 + c_4(K)z^4 + \dots$$

   The coefficient $c_2(K)$ is called the Casson *invariant* of K. Its mod 2 reduction is called the Arf *invariant* of K. L. Kauffman [11] described the moves between two knots with the same Arf invariant.

6. The Conway polynomial of a link L with two components K_1 and K_2 is an odd polynomial in z whose linear coefficient is equal to the linking number $lk(K_1, K_2)$:

$$\nabla_L(z) = lk(K_1, K_2)z + c_3(L)z^3 + c_5(L)z^5 + \dots$$

7. The Conway polynomial of a link L with μ components is divisible by $z^{\mu-1}$ and is odd or

even depending on the parity of μ:

$$\nabla_L(z) = c_{\mu-1}(L)z^{\mu-1} + c_{\mu+1}(L)z^{\mu+1} + c_{\mu+3}(L)z^{\mu+3} + \dots$$

8. For a knot K, $\nabla_K(2i) \equiv 1$ or 5 (mod 8) depending on the parity of the Casson invariant $c_2(K)$.

9. For any even polynomial $g(z)$ with integer coefficients satisfying $g(0) = 1$, there is a knot K such that $\nabla_K(z) = g(z)$.

10. The *genus* of a non-split link L with μ components is bounded below in terms of the degree of the CONWAY polynomial:

$$g(L) \geq \frac{1}{2}(\deg \nabla_L(z) - \mu + 1).$$

64.3   Jaeger's State Model

F. Jaeger formulated [8] a state model description of the HOMFLY polynomial. Here we specialize his description to the CONWAY polynomial of knots.

A state S is a subset of the crossings of a knot diagram K. It is said to be *one-component* if the curve obtained from K by smoothing all the crossings of S according to the orientation has one component.

Assume that the diagram K has a base point and S is a one-component state. Let us travel along the smoothened curve starting at the base point. In this journey we pass a neighborhood of every smoothing twice. We call a one-component state S *ascending* if, for every smoothing, the first time we approach its neighborhood on an overpass of K (so we need to jump down to perform the smoothing) and upon returning to the neighborhood we approach it on the underpass (jumping up).

Theorem 64.3.1 ([8]). For a knot K,

$$\nabla_K(z) \quad := \quad \sum_{\substack{S \text{ ascending} \\ \text{one-component}}} \left(\prod_{x \in S} \text{wr}(x)\right) z^{|S|},$$

where $\text{wr}(x)$ is the usual local writhe of the crossing x. If S is the empty set, then we set the product to be equal to 1 by definition. Therefore the free term of $\nabla_K(z)$ always equals 1.

Example 64.3.2. For the knot 6_2,

with local writhes indicated, there are eleven one-component states with two crossings, $\{1, 2\}$, $\{1, 4\}$,

$\{1, 5\}, \{1, 6\}, \{2, 4\}, \{2, 5\}, \{2, 6\}, \{3, 4\}, \{3, 5\}, \{4, 6\}, \{5, 6\}$. However, only three of these states, $\{2, 5\}$, $\{2, 6\}$, and $\{4, 6\}$, are ascending:

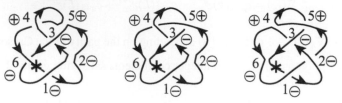

The products of the writhes for the states $\{2, 5\}$, $\{2, 6\}$, and $\{4, 6\}$ are equal to -1, $+1$, and -1, respectively. Hence the coefficient at z^2 in the polynomial $\nabla_{6_2}(z)$ equals $-1 + 1 - 1 = -1$.

There is only one ascending one-component state with four crossings, $\{2, 3, 5, 6\}$:

with the product of their writhes equal to -1. So the CONWAY polynomial of the knot 6_2 is equal to $CP_{6_2}(z) = 1 - z^2 - z^4$.

It turns out that the state sum does not depend on the choice of the base point. One may define one-component *descending* states and make the state sum over them. Both of them will be equal to the CONWAY polynomial. The state model can be formulated for links and virtual knots. However, for virtual knots, the polynomial z is given by this state model depends on the choice of the base point. Also ascending and descending versions of the polynomial may be different for virtual knots. This gives a method for proving that a virtual knot is not classical.

This state model was used in [2] to derive the Polyak-Viro formulas mentioned in Section 82.

64.4   Gaps in the Sequence of Coefficients

According to property 7 of Section 64.2, for a link L with μ components, the first $\mu - 2$ coefficients of the CONWAY polynomial vanish. The next theorem of F. Hosokawa, R. Hartley, and J. Hoste gives a formula la for the first non-trivial coefficient $c_{\mu-1}(L)$.

Theorem 64.4.1 ([6, 5, 7])**.** For a link $L = K_1 \sqcup K_2 \sqcup \cdots \sqcup K_\mu$ with μ components, $c_{\mu-1}(L) = \det \Lambda^{(p)}$, where $\Lambda = (\lambda_{ij})$ is a matrix of linking numbers: $\lambda_{ij} := -lk(K_i, K_j)$ for $i \neq j$, $\lambda_{ii} := \sum_{k \neq i} lk(K_i, K_k)$, and $\Lambda^{(p)}$ is obtained from Λ by deleting the p-th row and p-th column. The determinant in the right-hand side does not depend on p.

The determinant of a matrix from this theorem can be computed by the classical *Matrix-Tree Theorem* from combinatorics.

A link is called *algebraically split* if any pair of its components has linking number 0. For algebraically split links, of course, $c_{\mu-1}(L) = 0$. The remarkable fact is that for algebraically split links the next $\mu-2$ coefficients also vanish, and the next theorem expresses the first non-trivial coefficient in this situation in terms of triple Milnor numbers.

Theorem 64.4.2 ([17, 18, 12]). For an algebraically split link $L = K_1 \sqcup K_2 \sqcup \cdots \sqcup K_\mu$ with μ components,

$$c_0(L) = c_1(L) = \cdots = c_{2\mu-3}(L) = 0, \qquad \text{and} \qquad c_{2\mu-2}(L) = \det M^{(p)},$$

where $M = (m_{ij})$ is a matrix formed by the triple Milnor numbers: $m_{ij} := \sum_k \mu_{ijk}(L)$, and $M^{(p)}$ is obtained from M by deleting the p-th row and p-th column.

On the way to interpret this theorem from the point of view of the Matrix-Tree Theorem, G. Masbaum and A. Vaintrob discovered the *Pfaffian Matrix-Tree Theorem* [14, 15].

In general, if all the higher Milnor numbers vanish, $\mu_{i_1,\dots,i_p} = 0$ for $p \le n$, then the CONWAY polynomial starts with the term of degree $n(\mu - 1)$ and there is a similar formula for the non-trivial coefficient, see [18, 13, 15].

A geometric construction of the CONWAY polynomial in terms of a *Seifert matrix* was given by L. Kauffman in [10]. It was generalized to the multivariable CONWAY potential function of links in [3] using the skein relations for the multivariable ALEXANDER polynomial from [16]. The latter were recently simplified in [9].

64.5 Bibliography

[1] S. Chmutov, S. Duzhin, and Y. Mostovoy. *Introduction to Vassiliev Knot Invariants*. Cambridge University Press, 2012.

[2] S. Chmutov, M. Khoury, and A. Rossi. Polyak-Viro formulas for coefficients of the Conway polynomial. *Jour. of Knot Theory and Its Ramifications*, 18(6):773–783, 2009.

[3] D. Cimasoni. A geometric construction of the Conway potential function. *Comment. Math. Helv.*, 79:124–146, 2004.

[4] J.H. Conway. An enumeration of knots and links and some of their algebraic properties. In *Computational Problems in Abstract Algebra*, pages 329–358. Pergamon Press, NY, 1970.

[5] R. Hartley. The Conway potential function for links. *Comment. Math. Helv.*, 58:365–378, 1983.

[6] F. Hosokawa. On ∇-polynomials of links. *Osaka Math. J.*, 10:273–282, 1958.

[7] J. Hoste. The first coefficient of the Conway polynomial. *Proc. AMS*, 95:299–302, 1985.

[8] F. Jaeger. A combinatorial model for the Homfly polynomial. *European Jour. of Combinatorics*, 11:549–558, 1990.

[9] B.J. Jiang. On Conway's potential function for colored links. *Acta Mathematica Sinica*, 32(1):25–39, 2016.

[10] L. Kauffman. The Conway polynomial. *Topology*, 20(1):101108, 1981.

[11] L. Kauffman. The Arf invariant of classical knots. In *Combinatorial Methods in Topology and Algebraic Geometry. Contemporary Mathematics*, volume 44, pages 101–106. AMS, RI, 1985.

[12] J. Levine. The Conway polynomial of an algebraically split link. In *Proceedings of Knots'96*, page 2329. World Scientific, Singapore, 1997.

[13] J. Levine. A factorization of the Conway polynomial. *Comment. Math. Helv.*, 74:27–53, 1999.

[14] G. Masbaum and A. Vaintrob. A new matrix-tree theorem. *International Mathematics Research Notices*, 27:1397–1426, 2002.

[15] G. Masbaum and A. Vaintrob. Milnor numbers, spanning trees, and the Alexander-Conway polynomial. *Advances in Mathematics*, 180:765–797, 2003.

[16] J. Murakami. On local relations to determine the multi-variable Alexander polynomial of colored links. In *Proceedings of Knots'90*, pages 455–464. de Gruyter, Berlin, 1992.

[17] L. Traldi. Milnors invariants and the completions of link modules. *Trans. AMS*, 284:401–424, 1984.

[18] L. Traldi. Conway's potential function and its Taylor series. *Kobe J. Math.*, 5:233–264, 1988.

Chapter 65

Twisted Alexander Polynomials

Stefano Vidussi, University of California, Riverside

65.1 Introduction

The Alexander polynomial $\Delta_K(t) \in \mathbb{Z}[t^{\pm 1}]$ of a classical knot $K \subset S^3$, well defined up to units of $\mathbb{Z}[t^{\pm 1}]$, is one of the first and still one of the most useful tools in knot theory (and, via its various generalizations, 3–manifold topology). Its definition depends ultimately only on the knot group $\pi := \pi_1(S^3 \setminus \nu K)$. More precisely, it is determined by its maximal metabelian quotient: denote $\pi' = [\pi, \pi]$ as the commutator subgroup, and $\pi'' = [\pi', \pi']$; the abelianization sequence

$$0 \to \pi' \to \pi \to \mathbb{Z} \to 0$$

induces, taking the quotients of the first two terms by $\pi'' \leq \pi' \leq \pi$, the short exact sequence

$$0 \to \pi'/\pi'' \to \pi/\pi'' \to \pi/\pi' \cong \mathbb{Z} \to 0$$

where π/π'' is the maximal metabelian quotient of π and $\pi'/\pi'' = H_1(\pi'; \mathbb{Z})$ is the Alexander module of K, whose $\mathbb{Z}$–module structure is as usual induced by conjugation in π.

The knot groups of inequivalent (even prime) knots may have isomorphic metabelian quotients. To extract information which is lost by the metabelian quotient of knot groups, we can try to supplement the algebraic information encoded in the Alexander module with a finite-dimensional representation of the knot group. This is the path initiated (albeit with a fairly different point of view) by Xiao–Song Lin in the 90s (appeared in [15]), followed by seminal work by Bo Jung Jiang and Shi Cheng Wang ([9]), Masaaki Wada ([17]), Teruaki Kitano ([12]) and (notably in the development of applications) by Paul Kirk and Charles Livingston ([10, 11]). Many mathematicians contributed to further developments on the subject; due to the scope and length of this presentation (as well as our own biases), we don't even hope to do justice to these contributions, and we apologize for this neglect to the authors. The reader interested in detailed reference to the literature will find [5, 13] as more extensive starting points.

65.2 Definitions

In what follows, we will give a definition of twisted Alexander polynomials that mirrors the homological definition of the ordinary Alexander polynomial, as outlined above. We will do so in the realm of 3–manifolds with vanishing Euler characteristic, namely those whose boundary, if nonempty, is a disjoint union of tori.

Notation 1. *Let N be a smooth, orientable, connected 3–manifold with $\chi(N) = 0$; denote by A a $\mathbb{Z}$–module, and assume that the fundamental group $\pi := \pi_1(N)$ of N acts by automorphisms on A via a representation ("twist") $\alpha : \pi \to \mathrm{Aut}(A)$. Next, let F be a free abelian group, and denote by $\phi : \pi \to F$ a homomorphism; $\mathbb{Z}[F]$ becomes a (left) $\mathbb{Z}[\pi]$–module. We can thus endow $A[F] = A \otimes_{\mathbb{Z}} \mathbb{Z}[F]$ with the structure of a (left) $\mathbb{Z}[\pi]$–module, which is compatible with the natural structure of a (right) $\mathbb{Z}[F]$–module given by multiplication.*

(It is at times convenient, for both computational and theoretical purposes, to replace the ring $\mathbb{Z}$ with another integral domain or a field, usually the complex numbers or a finite field; this may require suitable adjustments in the definition.)

Definition 1. *The twisted Alexander module of the pair (N, ϕ) associated with the twist α is $H_1(\pi; A[F])$, the first homology group of π with coefficients in $A[F]$. This group has the structure of a (right) $\mathbb{Z}[F]$–module.*

The twisted Alexander polynomial $\Delta^\alpha_{N,\phi} \in \mathbb{Z}[F]/ \pm F$ is then defined as the order of the twisted Alexander module

$$\Delta^\alpha_{N,\phi} = \mathrm{ord}_{\mathbb{Z}[F]} H_1(\pi; A[F]) \in \mathbb{Z}[F]/ \pm F.$$

The order of the $\mathbb{Z}[F]$–module is defined by choosing a free resolution

$$\mathbb{Z}[F]^m \xrightarrow{A} \mathbb{Z}[F]^n \to H_1(\pi; A[F]) \to 0$$

(where we can always assume $m \geq n$) and then taking the g.c.d. of the $n \times n$ minors of the presentation matrix A (which has $\mathbb{Z}[F]$-entries) . This is an element of $\mathbb{Z}[F]$, well defined up to units of that ring, which entails the indeterminacy up to multiplication by $\pm F$. In what follows, we will often overlook this indeterminacy, as it usually does not affect our discussion. Note that when $H_1(\pi; A[F])$ has a positive rank as a $\mathbb{Z}[F]$–module, $\Delta^\alpha_{N,\phi} = 0$.

When the group π is provided in terms of generators and relations, it is possible to compute explicitly the twisted Alexander polynomials using Fox calculus, in a vein similar to that of the ordinary case.

In many applications, the homomorphism $\phi : \pi \to F$ arises as one of the following two cases: In the first case, $\phi \in \mathrm{Hom}(\pi, \mathbb{Z})$ is identified with a primitive element of $H^1(N; \mathbb{Z})$, so that after an identification, $\mathbb{Z} = \langle t \rangle$, $\Delta^\alpha_{N,\phi} \in \mathbb{Z}[t^{\pm 1}]$ is a single–variable Laurent polynomial, defined up to multiplication by $\pm t^m$, $m \in \mathbb{Z}$. In the second case, ϕ is the maximal free abelian quotient map $\phi : \pi \to H := H_1(\pi)/\mathrm{Tor}$, and after an identification, $H = \langle t_1, ..., t_n \rangle$ with $n = b_1(N)$, $\Delta^\alpha_{N,\phi} \in \mathbb{Z}[t_1^{\pm 1}, ..., t_n^{\pm 1}]$ is a multivariable Laurent polynomial, defined up to multiplication by $\pm t_1^{m_1} t_2^{m_2} ... t_n^{m_n}$, with $m_1, m_2 ..., m_n \in \mathbb{Z}$. When $b_1(\pi) = 1$ (in particular, for knots) we can omit reference to ϕ, to get the more streamlined notation $\Delta^\alpha_K(t) \in \mathbb{Z}[t^{\pm 1}]$.

If we choose $\alpha : \pi \to \mathrm{Aut}(A)$ to be the trivial representation on $A = \mathbb{Z}$, and $\phi : \pi \to H$ to be the maximal free abelian quotient map, we can use Eckmann–Shapiro's Lemma to rewrite the (un)twisted Alexander module as $H_1(\pi; \mathbb{Z}[H]) = H_1(\ker \phi; \mathbb{Z})$. In the case of a knot or a link $L \subset S^3$, $H_1(\pi, \mathbb{Z}) = H_1(S^3 \setminus L; \mathbb{Z})$ has no torsion, hence $\ker \phi = \pi'$. Therefore, we recast the homological definition of the ordinary single- or multi-variable Alexander polynomial.

The role of the representation α (the twist) is to obviate the dependence of the ordinary Alexander polynomial solely on the metabelian quotient of π and, at least in principle, retain further information on π. It is not difficult to verify that conjugate representations yield the same invariants.

It is oftentimes convenient to define and use the *twisted Reidemeister torsion*, which (roughly speaking) relates to the twisted Alexander polynomial in the same way the Reidemeister–Franz–Milnor torsion relates to the Alexander polynomial.

65.3 Applications

The first line of application of twisted Alexander polynomials is in distinguishing and relating knots. An immediate instance of this approach (that *per se* justifies the rationale of their definition) is the fact that twisted Alexander polynomials detect the unknot. Namely, we have the following ([16]):

Theorem 1. *Let $K \subset S^3$ be a nontrivial knot; there exists a representation of the knot group $\alpha : \pi \to Aut(A)$ such that $\Delta_K^\alpha(t) \neq 1$.*

In contrast, it is easy to show that all twisted Alexander polynomials of the unknot equal 1. (The statement holds even accounting for the usual indeterminacy arising from multiplication by $\pm t^m$.)

To further distinguish pairs of knots (or links) we need to take into account that the definition of the invariants requires the choice of a twist, that in general will not be canonical. Namely, if we were to choose *different* representations $\alpha_i : \pi \to Aut(A)$ on the same module for two knots K_i which actually have the *same* knot type (in particular isomorphic knot group) we may end up with the "false positive" of different polynomials. To circumvent this problem, we can compare *characteristic* collections of twisted invariants, associated to all representations $\alpha : \pi \to Aut(A)$ of a given type. Following this approach, twisted Alexander invariants can be effective in distinguishing pairs of mutant knots: for instance, in [17], the Kinoshita–Terasaka knot and the Conway knot (which have trivial a Alexander polynomial, as well as identical HOMFLY and Jones polynomials) are distinguished by computing the twisted Alexander polynomials arising from the *two* parabolic representations of their respective groups into $SL(2, \mathbb{F}_7)$.

A further method to obtain characteristic collections of twisted invariants comes by considering representations arising from *all* homomorphisms $Hom(\pi, G)$ to a finite group G, composed with a given representation of G on the module A. (See below for an illustration of this template.)

The second line of application of twisted Alexander polynomials comes from the generalization to the twisted case of constraints arising from topological conditions on knots and 3–manifolds. We will discuss several instances of this.

To start, recall that a knot is *topologically slice* if $K \subset S^3 = \partial D^4$ bounds a locally flat disk $D^2 \subset D^4$. Furthermore we say that two knots K_1, K_2 are *concordant* if the sum $K_1 \# - K_2$ is topologically slice. The Alexander polynomial of a topologically slice knot factorizes as $\Delta_K(t) = \pm t^m f(t) f(t^{-1})$. By analyzing representations $\alpha : \pi \to Aut(A)$ that suitably extend to $\pi_1(D^4 \setminus D^2)$, it is possible to show that a similar factorization occurs for $\Delta_K^\alpha(t)$. This is the starting point to find obstruction to sliceness and, consequently, obstruction to the concordance of two knots, see [10, 11].

A second application is to the study of a partial order on knots which can be defined as follows: given two knots K_i, we say $K_1 \geq K_2$ if there exists an epimorphism $\varphi : \pi_1(S^3 \setminus K_1) \to \pi_1(S^3 \setminus K_2)$ which preserves the respective meridians. It can be verified that as knot groups are Hopfian, this is actually a partial order. Generalizing a result holding for the Alexander polynomials, [14] proved that if $K_1 \geq K_2$, for an appropriate class of twists $\alpha : \pi_1(S^3 \setminus K_2) \to \mathrm{Aut}(A)$, the twisted Alexander polynomial $\Delta^\alpha_{K_2}(t)$ is a factor of $\Delta^{\varphi \circ \alpha}_{K_1}(t)$. This allows one to narrow down, via explicit calculations, possible ordered pairs.

To discuss further applications of the twisted polynomial, we will describe now an interesting class of twists that comes by considering finite quotients of π. Given a finite group G, we consider the permutation representation of G on its group ring $\mathbb{Z}[G]$ induced by left multiplication of G on itself. Given an epimorphism $\alpha : \pi \to G$ to a finite group, this induces a representation $\alpha : \pi \to \mathrm{Aut}(\mathbb{Z}[G])$. (The conflation of notation, hopefully, does not cause ambiguity.) As a $\mathbb{Z}$–module, $\mathbb{Z}[G]$ has rank $|G|$, hence the representation takes value in the general linear group $GL(|G|, \mathbb{Z})$. In this case, we can find a fruitful identification of the twisted Alexander module: In fact, the module of coefficients is

$$(\mathbb{Z}[G])[F] = \mathbb{Z}[G] \otimes_{\mathbb{Z}} \mathbb{Z}[F] = (\mathbb{Z}[F])[G],$$

so using Eckmann–Shapiro's Lemma we have

$$H_1(\pi; \mathbb{Z}[G][F]) = H_1(\pi; \mathbb{Z}[F][G]) = H_1(\tilde{\pi}; \mathbb{Z}[F])$$

where $\tilde{\pi} = \ker \alpha \leq_f \pi$ is the fundamental group of the regular cover $\tilde{N} \to N$ determined by the epimorphism $\alpha : \pi \to G$. This way, the Alexander polynomial of N twisted by the representation α can be interpreted in terms of an ordinary Alexander polynomial of a finite cover of N. This observation allows one to repurpose to twisted Alexander polynomials the recent progress, initiated by Agol, Wise, and their collaborators, on *virtual* properties of 3–manifolds, i.e. properties that emerge for finite covers of a 3–manifold. We cite two applications, respectively, to the study of fiberability and genus of a knot.

A classical result asserts that when a knot is *fibered* (i.e. its exterior is a locally trivial surface bundle over S^1), then its Alexander polynomial is *monic*, i.e. the leading coefficient equals ± 1. Successive generalizations of this result to the twisted case appear in [2, 8, 4] and others. Plenty of nonfibered knots have a monic Alexander polynomial. In contrast, separability properties of 3–manifold groups entail (see [6]) that a suitable twist will detect nonfiberability:

Theorem 2. *Let K be a knot that is not fibered; then there exists an epimorphism $\alpha : \pi \to G$ to a finite group such that the corresponding twisted Alexander polynomial $\Delta^\alpha_K(t) = 0$.*

Next, we remind the reader of the classical fact that the degree of the Alexander polynomial gives the lower bound $\deg \Delta_K(t) \leq 2g(K)$ for the genus of a knot. A similar bound holds in terms of the degree of twisted Alexander polynomials (and, in all cases, gives an equality when K is fibered), see [4]. Again, we can easily find (nonfibered) knots where the bound coming from $\deg \Delta_K(t)$ does not give much information. Using virtual fiberability of (most) 3–manifolds, which guarantees, roughly speaking, that those manifold admit a finite cover which fibers, it is possible to show that some twisted Alexander polynomial of K will determine the genus of K: the twist is obtained by modifying the one determined by the deck transformation group G of the cover. We refer to [7] for the complete statement.

We point out that analogous statements exist for any pair (N, ϕ), where $\phi \in H^1(N; \mathbb{Z}) = [N, S^1]$, to determine fiberability of the homotopy class determined by ϕ, and to determine the *Thurston norm* $\|\phi\|_T$, which generalizes the notion of the genus of a knot.

While these results are of use (in particular they have some fruitful application in the study of symplectic 4–manifolds, see [6]) they face the obvious computational limit that, to the best of our understanding, we don't have a way to determine (or even bound in any sense) the twist requires to determine genus, or exclude fiberability, of a given knot. From this viewpoint, a conjecture of Dunfield–Friedl–Jackson provides a way forward, at least in the case of hyperbolic knots. In fact, given one such knot, there exists a canonical representation $\alpha : \pi \to SL(2, \mathbb{C})$ that arises as a lift of the holonomy representation associated to the hyperbolic structure. Dunfield, Friedl, and Jackson posited in [3] the following conjecture, and they numerically verified it for knots having up to 15 crossings:

Conjecture 1. *Let $K \subset S^3$ be an hyperbolic knot, and let $\alpha : \pi \to SL(2, \mathbb{C})$ be the representation lifting the holonomy representation. Then the twisted Alexander polynomial Δ_K^α satisfies $\deg(\Delta_K^\alpha(t)) = 4g(K) - 2$.*

The authors further conjecture that the same representation allows one to decide fiberedness of a knot, using a suitably defined twisted invariant: we refer to [3] for details. The conjecture has been further validated for an infinite class of knots in [1].

Acknowledgments

It is a pleasure to thank Stefan Friedl for many years of collaboration and discussions about this subject.

65.4   Bibliography

[1] I. Agol, N. M. Dunfield, *Certifying the Thurston norm via $SL(2, \mathbb{C})$-twisted homology*, to appear in the Thurston memorial conference proceedings.

[2] J. Cha, *Fibred knots and twisted Alexander invariants*,. Transactions of the AMS 355: 4187–4200 (2003).

[3] N. M. Dunfield, S. Friedl, and N. Jackson, *Twisted Alexander polynomials of hyperbolic knots*, Exp. Math. 21, 329352 (2012).

[4] S. Friedl and T. Kim, *Thurston norm, fibered manifolds and twisted Alexander polynomials*, Topology, Vol. 45: 929–953 (2006).

[5] S. Friedl and S. Vidussi, *A survey of twisted Alexander polynomials*, The Mathematics of Knots, 45–94, Contrib. Math. Comput. Sci., 1, Springer (2011).

[6] S. Friedl and S. Vidussi, *A vanishing theorem for twisted Alexander polynomials with applications to symplectic 4-manifolds* J. Eur. Math. Soc. 15 no. 6, 2027–2041 (2015).

[7] S. Friedl and S. Vidussi, *The Thurston norm and twisted Alexander polynomials*, J. Reine Angew. Math. 707, 87–102 (2015).

[8] H. Goda, T. Kitano and T. Morifuji, *Reidemeister torsion, twisted Alexander polynomial and fibred knots*, Comment. Math. Helv. 80, no. 1: 51–61 (2005).

[9] B. Jiang and S. Wang, *Twisted topological invariants associated with representations*, Topics in knot theory (Erzurum, 1992), 211–227, NATO Adv. Sci. Inst. Ser. C Math. Phys. Sci., 399, Kluwer Acad. Publ., Dordrecht, 1993.

[10] P. Kirk and C. Livingston, *Twisted Alexander invariants, Reidemeister torsion and Casson–Gordon invariants*, Topology 38, no. 3: 635–661 (1999).

[11] P. Kirk and C. Livingston, *Twisted knot polynomials: Inversion, mutation and concordan*, Topology **38** , no. 3, 663–671, (1999).

[12] T. Kitano, *Twisted Alexander polynomials and Reidemeister torsion*, Pacific J. Math. **174**, no. 2, 431–442, (1996).

[13] T. Kitano, *Introduction to twisted Alexander polynomials and related topics*. Winter Braids Lect. Notes 2 (2015), Winter Braids V (Pau, 2015).

[14] T. Kitano, M. Suzuki and M. Wada, *Twisted Alexander polynomials and surjectivity of a group homomorphism*, Algebr. Geom. Topol. 5, 1315–1324 (2005), Erratum: Algebr. Geom. Topol. 11, 2937–2939 (2011).

[15] X.-S. Lin, *Representations of knot groups and twisted Alexander polynomials*, Acta Math. Sin. (Engl. Ser.) 17, no. 3: 361–380 (2001).

[16] D. Silver and S. Williams, *Twisted Alexander polynomials detect the unknot*, Algebraic and Geometric Topology, 6 (2006), 1893–1907.

[17] M. Wada, *Twisted Alexander polynomial for finitely presentable groups*, Topology 33, no. 2: 241–256 (1994).

Chapter 66

The HOMFLYPT Polynomial

Jim Hoste, Pitzer College

Suppose that L_+, L_-, and L_0 are three oriented link diagrams that are nearly identical but differ only by *changing* a right-handed crossing in L_+ to produce L_- and by *smoothing* that crossing to obtain L_0. Near the crossing that has been changed and smoothed, the three links appear as shown in Figure 66.1. Away from this crossing the diagrams are identical.

FIGURE 66.1: The skein triple L_-, L_0, L_+.

In his 1928 paper, J. Alexander defined what is now known as the *Alexander polynomial* $\Delta(t)$ and showed how the polynomials of the three links are related. The relationship is most beautifully described by changing to the Conway polynomial $\nabla(z)$, a normalized version of $\Delta(t)$ introduced by Conway. Conway called L_+, L_-, and L_0 a *skein* triple and showed that $\nabla(z)$ satisfies the following *skein* relation

$$\nabla_{L_+}(z) - \nabla_{L_-}(z) = z\nabla_{L_0}(z). \tag{66.1}$$

This relation, together with the fact that the Conway polynomial of the unknot is 1, suffices to compute the Conway polynomial for any oriented diagram.

In 1984, Vaughan Jones introduced a new polynomial invariant of oriented links, $V(t)$, which was shown to satisfy the skein relation

$$t^{-1}V_{L_+}(t) - tV_{L_-}(t) = (t^{\frac{1}{2}} - t^{-\frac{1}{2}})V_{L_0}(t). \tag{66.2}$$

Again, this relation coupled with the fact that the Jones polynomial of the unknot is 1, suffices to compute $V(t)$.

The obvious similarity between the two skein relations was immediately recognized by the community of knot theorists and shortly after Jones's discovery, five independent sets of researchers,

P. Freyd and D. Yetter, J. Hoste, R. Lickorish and K. Millett, A. Ocneanu, and J. Przytycki and P. Traczyk announced a 2-variable generalization, $P(v, z)$, that satisfies the skein relation

$$v^{-1}P_{L_+}(v, z) - vP_{L_-}(v, z) = zP_{L_0}(v, z) \tag{66.3}$$

and which takes the value 1 on the unknot. Because Poland was then under martial law, making it difficult for information to go in and out of the country, it was not until some time later that the contribution of Przytycki and Traczyk was known. Thus $P(v, z)$ was originally dubbed the "HOM-FLY" polynomial by Peter Freyd, using the initials of its discoverers. Later, various attempts to add "P" and "T" gave rise to HOMFLYPT, Flypmoth, P.T. Homfly and others. A joint announcement (still ignorant of Przytycki and Traczyk's work [18]) appeared in the *Bulletin of the American Mathematical Society* [5].

Interestingly, the proofs given by the various authors varied considerably. Hoste, Lickorish and Millett, and Przytycki and Traczyk, all gave similar arguments based on a diagrammatic approach, showing that the skein relation, together with the value of P on the unknot, can be used to define P. Ocneanu generalized Jones's definition of the Jones polynomial, defining a function on the braid group that is then shown to be preserved by the Markov moves on braids. Finally, Freyd and Yetter used Markov's Theorem and conditions given by Knuth and Bendix for equivalence classes defined by a rewrite system to contain a canonical reduced element.

The polynomial $P(v, z)$ specializes to both the Conway and Jones polynomials by $\nabla(z) = P(1, z)$ and $V(t) = P(t, t^{\frac{1}{2}} - t^{-\frac{1}{2}})$ but is truly more general than both. That is, there exist pairs of knots that are distinguished by P but not by either ∇ or V. However, it is not a complete invariant of knots. For example, links related by mutation have the same P polynomial but are typically not the same link. Nevertheless, it does a "good" job of distinguishing knots. Of the 313,230 prime knots with 15 or less crossings, 167,579 have distinct values of P while various sets of knots ranging in size from two to 23 share the same polynomial. In contrast, only 83,150 have a unique Jones polynomial with sets as large as 85 sharing the same polynomial. The situation is even worse for the Conway polynomial—35,111 have a unique Conway polynomial with the largest set with the same polynomial having 275 knots. While mutant pairs of knots have the same P polynomial, comparing hyperbolic volume (also preserved by mutation) shows that mutation cannot account for all duplications. For whatever reason, there are simply "not enough polynomials to go around," to quote Morwen Thistlethwaite.

Many nontrivial prime knots have a trivial Alexander polynomial, but to date, it is still unknown if such examples exist for either the Jones or HOMFLYPT polynomials. There do exist nontrivial links with two or more components having the same Jones polynomial as the unlink, but such examples have not been found for HOMFLYPT.

It is not hard to prove, using the skein relation (66.3), that if L is an oriented link with $|L|$ components, then $P_L(v, z)$ has the form

$$P_L(v, z) = z^{1-|L|}(q_0(v) + q_2(v)z^2 + \cdots + q_{2m}(v)z^{2m}), \tag{66.4}$$

where each $q_{2i}(v)$ is a Laurent polynomial in v with integer coefficients, $q_0(v) \neq 0$, and every v-exponent has parity opposite to $|L|$. It is not known if every polynomial of this form occurs for some link. If L is of a special form, then a number of results have been proven that further restrict the form of $P_L(v, z)$. For example, a sample of typical results of this kind are:

- Equation 66.3 implies that $P_{L^*}(v, z) = P_L(v^{-1}, -z)$, where L^* is the obverse of L. Thus, if L is amphichiral, then $q_i(v^{-1}) = (-1)^{|L|-1} q_i(v)$, for all i.

- If L is the (m, n)-torus knot, then Jones gives an explicit, but rather complicated, formula for $P_L(v, z)$ in [7].

- (Cromwell and Morton [3]) If L is a positive link, then $P(v, z)$ is a polynomial in z with positive coefficients for all $0 < v < 1$.

- (Morton [13]) If L is the closure of a braid β on n strings, then

  1. $M \leq \text{length}(\beta) - (n - 1)$, where M is the highest z-exponent in $P_L(v, z)$,

  2. $\tilde{c}(\beta) - (n - 1) \leq e \leq E \leq \tilde{c}(\beta) + (n - 1)$, where $\tilde{c}(\beta)$ is the total exponent sum of β and e and E are the lowest and highest v-exponents, respectively, in $P_L(v, z)$.

- (Przytycki [16]) Let $\mathcal{R}$ be the subring of $\mathbb{Z}[v^{\pm 1}, z^{\pm 1}]$ generated by $v^{\pm 1}$, $\frac{v + v^{-1}}{z}$ and z. If L is periodic with prime period p, then

$$P_L(v, z) \equiv P_L(v^{-1}, z) \bmod (p, z^p)$$

where (p, z^p) is the ideal in $\mathcal{R}$ generated by p and z^p.

Along similar lines, some evaluations of $P_L(v, z)$ are known to have topological significance. To give just two examples, Lickorish and Millett show in [11] that if D_L and T_L are the 2- and 3-fold cyclic covers, respectively, of S^3 branched along L, then

- $P_L(i, i) = (-2)^{\frac{1}{2} \text{Dim} H_1(T_L; \mathbb{Z}_2)}$

- $P_L(ie^{-i\pi/6}, i)) = \pm i^{|L|-1} (i\sqrt{3})^{\text{Dim} H_1(D_L; \mathbb{Z}_3)}$.

In addition to using the HOMFLYPT polynomial to tell knots apart, or to study various properties of knots as described above, a significant application was in the proof of the so-called "Tait flyping conjecture." This result states that any two oriented, reduced, alternating link diagrams represent the same link if and only if they are related by a sequence of flypes. The proof, due to Menasco and Thistlethwaite [12], used the HOMFLYPT polynomial in an essential way. The discovery of the HOMFLYPT polynomial almost certainly contributed to the discovery of other skein polynomials, in particular the Brandt-Lickorish-Millett and Ho polynomial [1], [4], as well as the Kauffman 2-variable polynomial [8]. These skein polynomials in turn, all led to the development of *skein modules* of 3-manifolds (see Chapter 63).

Given a knot K, one can replace K by an n-string cable and compute P for the cable knot instead. These "n-colored" polynomials give, in general, stronger invariants. Additionally, evaluating the colored polynomials at certain complex numbers, one can obtain invariants of manifolds built by surgery on the knot or link [19]. This work, involving representations to the quantum groups $SU(N)_q$, takes one into the realm of topological quantum field theory, which provides a unifying point of view for all the various "skein" invariants. Entré to this area of research can be found in the papers of Morton [14], Lickorish [10], and Yokota [20], to name just a few.

It is known that both the Jones polynomial and the HOMFLYPT polynomial are, in general, #P-hard to compute [6]. However, if a link is presented as the closure of a braid, then for a fixed number of strings, P can be computed in polynomial time with respect to the crossing number [15]. Alternatively, if one wishes only to compute the polynomial $q_{2i}(v)$ in (66.4) from an n-crossing diagram of the link, then Vertigan has shown that this can be done in time complexity of $O(n^{2+3i})$ [17]. Recently, Burton [2] has shown that P is fixed-parameter tractable with respect to treewidth. Using this result, he showed that for any diagram of an oriented link with n crossings, P can be computed in $e^{O(\sqrt{n} \cdot \log n)}$ time. In other words, the HOMFLYPT polynomial can be computed in subexponential time.

Different knot homology theories have been defined that *categorify* various knot invariants. That is to say, there exists some doubly-graded homology theory $H_{i,j}(K)$, depending on the knot K, whose graded Euler characteristic with respect to one of the gradings gives the knot invariant. For example, Khovanov homology categorifies the Jones polynomial and Floer homology provides a categorification of the Alexander polynomial. In [9], Khovanov and Rozansky provide a categorification of the HOMFLYPT polynomial. Starting from a closed braid representation of a given oriented link L, they define a complex of bigraded vector spaces. The Euler characteristic of this complex, and of its triply-graded cohomology groups, is (after a suitable change of variables) the polynomial $P_L(v, z)$.

66.1   Bibliography

[1] R.D. Brandt, W.B.R. Lickorish and K.C. Millett, *A polynomial invariant for unoriented knots and links*, Invent. Mth. **84**, 1986, 563–573.

[2] B. Burton, *The HOMFLY-PT polynomial is fixed parameter tractable*, arXiv:1712.05776 (2017).

[3] P.R. Cromwell and H.R. Morton, *Positivity of knot polynomials on positive links*, J. Knot Theory and Its Ramifications, **1** (1992), 203-206.

[4] C.F. Ho, *On polynomial invariants for knots and links*, Ph.D. thesis, California Institute of Technology, 1986.

[5] P. Freyd, J. Hoste, W.B.R. Lickorish, K. Millett, A. Ocneanu, and D.Yetter, *A new polynomial invariant of knots and links*, Bull. Amer. Math. Soc. **12** No. 2 (1985), 239–246.

[6] F. Jaeger, D.L. Vertigan, and D.J.A. Welsh, *On the computational complexity of the Jones and Tutte polynomials*, Math. Proc. Cambridge Philos. Soc. **108** no. 1 (1990), 35–53.

[7] V.F.R. Jones, *Hecke Algebra Representations of Braid Groups and Link Polynomials*, Annals of Mathematics, Second Series, **126**, no. 2 (1987), 335–388.

[8] L.H. Kauffman, *An invariant of regular isotopy*, Trans. Am. Math. Soc., **318**, no. 2, (1990), 417–471.

[9] M. Khovanov, and L. Rozansky, *Matrix factorizations and link homology II*, Geometry and Topology, **12**, (2008), 1387–1425.

[10] W.B.R. Lickorish, *Skeins, $SU(N)$ three-manifold invariants and TQFT* Comment. Math. Helv. **75**, (2000), 45–64.

[11] W.B.R. Lickorish and K.C. Millett, *Some evaluations of link polynomials*, Commentarii Mathematici Helvetici, **61**, (1986), 349–359.

[12] W. Menasco and M.B. Thistlethwaite, *The classification of alternating links*, Ann. Math. **138**, (1993), 113–171.

[13] H.R. Morton, *Seifert circles and knot polynomials*, Math. Proc. Camb. Phil. Soc. **99**, (1986), 107–109.

[14] H.R. Morton, *Invariants of links and 3-manifolds from skein theory and from quantum groups*, Topics in Knot Theory, Proceedings of the NATO Advanced Study Institute held at Atatürk Üniversitesi, Erzurum, September 1–12, 1992. Edited by M. E. Bozhüyük.

[15] H.R. Morton and H.B. Short, *Calculating the 2-variable polynomial for knots presented as closed braids*, J. Algorithms, **11**, (1990), 117-131.

[16] J.H. Przytycki, *On Murasugi's and Traczyk's criteria for periodic links*, Math. Ann., **283** (1989), 465–478.

[17] J.H. Przytycki, *The first coefficient of Homflypt and Kauffman polynomials: Vertigan proof of polynomial complexity using dynamic programming*, Knots, Links, Spatial Graphs, and Algebraic Invariants, Contemp. Math., 689, Amer. Math. Soc., Providence, RI, 2017, 1–6.

[18] J.H. Przytycki, and P. Traczyk, *Invariants of links of Conway type*, Kobe J. Math., **4**, (1987), 115–139.

[19] N.Y. Reshetikhin and V.G. Turaev, *Invariants of 3-manifolds via link polynomials and quantum groups*, Invent. Math. **103** (1991), 547–597.

[20] Y. Yokota, *Skeins and quantum $SU(N)$ invariants of 3-manifolds*, Math. Ann., **307**, (1997), 109–138.

Chapter 67

The Kauffman Polynomials

Jianyuan K. Zhong, California State University, Sacramento

67.1   The Kauffman Polynomials

The original Kauffman polynomial was a two-variable Laurent polynomial invariant for classical unoriented links introduced by Kauffman [2]. There are various specializations of the variables in *the Kauffman polynomial* in the literature. We adopt the version by Beliakava and Blanchet in [1].

Definition 1. *The Kauffman polynomial of a knot or link L is the unique two-variable rational function < L > in variables α, s that satisfies the following relations:*

$$(i) < \times > - < \times > \; = \; (s - s^{-1})\left(< \;) (\; > - < \times > \right),$$

$$(ii) < \mathcal{P} > \; = \; \alpha \; < / >,$$

$$(iii) < L \sqcup \bigcirc > \; = \; \delta < L >,$$

where $\delta = \left(\dfrac{\alpha - \alpha^{-1}}{s - s^{-1}} + 1 \right)$. Relations (i) and (ii) are local relations; diagrams in each relation are identical except where shown. Relation (iii) follows from the first two when L is nonempty. A trivial closed curve in (iii) is a curve which is null-homotopic and contains no crossing. We use the normalization that the Kauffman polynomial of the empty link $\emptyset$ is 1. Kauffman [2] showed $< L >$ is a regular isotopy invariant of unoriented links in three-dimensional space. It follows that $< L >$ is an ambient isotopy invariant of framed (banded) unoriented links in three-dimensional space.

Example: We demonstrate how to calculate the Kauffman polynomial of the Hopf link. Consider the unoriented Hopf link with linking number 1 as below:

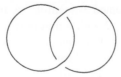

We first apply relation (i) as highlighted in dotted squares, then we apply relations (ii) and (iii).

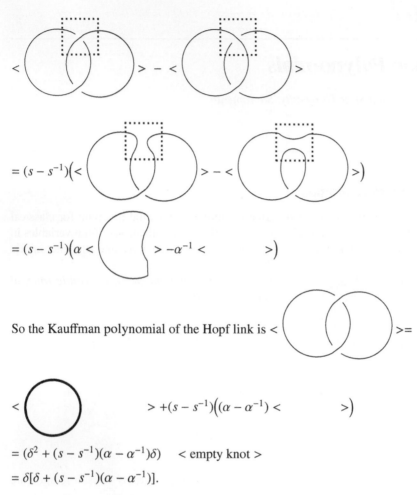

So the Kauffman polynomial of the Hopf link is < ( ) >=

$$= (\delta^2 + (s - s^{-1})(\alpha - \alpha^{-1})\delta) \quad < \text{empty knot} >$$
$$= \delta[\delta + (s - s^{-1})(\alpha - \alpha^{-1})].$$

The knot diagrams in the Kauffman polynomial relations allow us to study knots and links in three-dimensional spaces in terms of knot diagrams. This leads to the Kauffman skein theory below.

67.2   The Kauffman Skein Modules

Let M be a smooth, compact and oriented 3-manifold. A framed link in M is an unoriented link where each link component is endowed with a unit normal vector field which is also called the blackboard framing. Equivalently, a framed link can be understood as replacing the embedding of each component by an embedded band whose core is the component and bands for different components do not intersect each other. Let $\mathbb{Q}(\alpha, s)$ be the field of rational functions in α, s. Let k be an integral domain containing the invertible elements α and s. We assume that $s - s^{-1}$ is invertible in k.

Definition 2. **The Kauffman skein module** *of M, denoted by K(M), is the k-module freely gener-*
ated by isotopy classes of framed links in M including the empty link modulo the Kauffman skein
relations:

$$ \diagup\!\!\!\!\diagdown - \diagdown\!\!\!\!\diagup \;=\; (s - s^{-1})\left(\;)(\; \right)\left(- \;\asymp\; \right) , $$

$$ \text{(curl)} = \alpha \; | \; , $$

$$ L \sqcup \bigcirc \;=\; \left(\frac{\alpha - \alpha^{-1}}{s - s^{-1}} + 1 \right) \; L \; . $$

The last relation follows from the first two when L is nonempty.

Remark: (1) Since every link in S^3 can be simplified to its Kauffman polynomial multiple of the
empty link, it follows that the Kauffman Skein module $K(S^3)$ is generated by the empty link. (2)
One can define the corresponding relative skein modules in a similar fashion.

Definition 3. **The relative Kauffman skein module.** *Let $X = \{x_1, x_2, \cdots, x_n\}$ be a finite set of*
framed points (called input points) in ∂M, and let $Y = \{y_1, y_2, \cdots, y_n\}$ be a finite set of framed
points (called output points) in the boundary ∂M. Define the relative skein module $K(M, X, Y)$ to
be the k-module generated by relative framed links in $(M, \partial M)$ such that $L \cap \partial M = \partial L = \{x_i, y_i\}$
with the induced framing, considered up to an ambient isotopy fixing ∂M modulo the Kauffman
skein relations.

As a special case, in the cylinder $D^2 \times I$, let X_n be a set of n distinct framed points on a diameter
$D^2 \times \{1\}$ and Y_n be a set of n distinct framed points on a diameter $D^2 \times \{0\}$, then the relative
Kauffman skein module $K(D^2 \times I, X_n, Y_n)$ gives a geometric realization of the Birman-Murakami-
Wenzl algebra K_n, which is the quotient of the braid group algebra $k[B_n]$ by the Kauffman skein
relations.

The Birman-Murakami-Wenzl algebra K_n is generated by the identity $\mathbf{1}_n$, positive transpositions
$e_1, e_2, \cdots, e_{n-1}$ and hooks $h_1, h_2, \cdots, h_{n-1}$ as the following:

$$ e_i = \;\Big|\; \cdots \;\Big|\; \overset{i \quad i+1}{\diagdown\!\!\!\!\diagup} \;\Big|\; \cdots \;\Big| $$

$$ h_i = \;\Big|\; \cdots \;\Big|\; \overset{i \quad i+1}{\smile\!\!\!\frown} \;\Big|\; \cdots \;\Big| $$

for $1 \le i \le n - 1$. Then K_n is the braid group algebra $k[B_n]$ quotient by the following relations:

$$ (B_1)\; e_i e_{i+1} e_i = e_{i+1} e_i e_{i+1} $$

$$ (B_2)\; e_i e_j = e_j e_i, \; |i - j| \ge 2 $$

$$ (R_1)\; h_i e_i = \alpha^{-1} h_i $$

$$(R_2)\ h_i e_{i-1}{}^{\pm 1} h_i = \alpha^{\pm 1} h_i$$

$$(K)\ e_i - e_i^{-1} = (s - s^{-1})(\mathbf{1}_n - h_i).$$

The quotient of K_n by the ideal I_n generated by h_{n-1} is isomorphic to the nth Hecke algebra H_n of type A. Assume $s^{2n} - 1 \neq 0$ and $1 \pm \alpha^2 s^{2n} \neq 0$ for all integers $n > 0$ in k, Beliakava and Blanchet gave an alternative basis for the Birman-Murakami-Wenzl algebra K_n in [1].

The natural wiring of the cylinder $D^2 \times I$ into the solid torus $S^1 \times D^2$ induces a homomorphism: $K(D^2 \times I, X_n, Y_n) \to K(S^1 \times D^2)$.

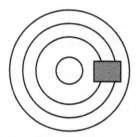

We denote the image of K_n under the wiring by $\widehat{K_n}$. Let L be a framed link in $S^1 \times D^2$, then L is the closure of an n-strand braid for some integer $n \geq 0$; the braid modulo the Kauffman skein relations, is an element in K_n. Therefore in $K(S^1 \times D^2)$, $L \in \widehat{K_n}$. This shows $K(S^1 \times D^2) = \bigcup_{n \geq 0} \widehat{K_n}$. Over $k = \mathbb{Q}(\alpha, s)$, it was shown in [3] that the Kauffman Skein module $K(S^1 \times D^2)$ has a countable basis indexed by the set of Young diagrams.

The Kauffman skein module of $S^1 \times S^2$ was studied in [3]. The space $S^1 \times S^2$ can be obtained from the solid torus $S^1 \times D^2$ by first attaching a 2-handle along the meridian γ and then attaching a 3-handle. Adding a 3-handle induces isomorphism between skein modules; while adding a 2-handle to $S^1 \times D^2$ adds relations to the module $K(S^1 \times D^2)$. The natural inclusion $i \colon S^1 \times D^2 \to S^1 \times S^2$ induces an epimorphism $i_* \colon K(S^1 \times D^2) \to K(S^1 \times S^2)$. Over $k = \mathbb{Q}(\alpha, s)$, it was proved in [3] that $K(S^1 \times S^2)$ is one-dimensional and generated by the empty link ϕ.

67.3   Calculating the Kauffman Polynomials Using Linear Skein Method

The Kauffman polynomial leads to the study of the Kauffman skein modules. On the other hand, we can also use linear skein techniques to calculate the Kauffman polynomials of knots and links with special patterns, such as the 2-bridge knots (rational knots) and pretzel knots [4] [5], etc.

The 2-bridge knots (links) are a family of knots with bridge number 2. A 2-bridge knot (link) has at most 2 components. The regular diagram D of a 2-bridge knot can be drawn as follows.

In the diagram, $d_1, d_2, \cdots, d_n, b_1, b_2, \cdots, b_{n+1}$ are integers whose absolute values indicate the number of crossings. We allow the possibility that some d_i, b_i are zero. By an isotopy, a 2-bridge knot can also be drawn as

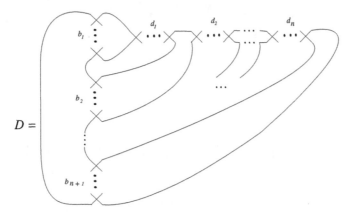

The regular diagram D of a pretzel knot can be drawn as follows,

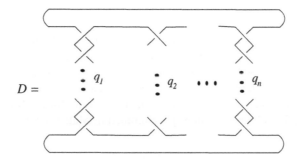

where $q_1, q_2, \cdots, q_n$ are nonzero integers whose absolute values indicate crossing numbers.

The 2-bridge knots and the pretzel knots can be studied locally as connecting diagrams in 3-balls with four boundary points through concatenation, juxtaposition, and closure. We place a distinguished set of four coplanar points $\{N, E, S, W\}$ on the sphere S^2, the boundary of the 3-ball B^3. A link in $(B^3, NESW)$ is a collection of closed curves and arcs up to isotopy joining the distinguished boundary points N, E, S, W. We define the Kauffman skein space $K(B^3, NESW)$ to be the $\mathbb{Q}(\alpha, s)$-space freely generated by framed links L in (B^3, S^2) such that $L \cap S^2 = \partial L = \{N, E, S, W\}$, considered up to an ambient isotopy fixing S^2, quotient by the subspace generated by the Kauffman skein relations. There are two natural multilinear multiplication operations in $K(B^3, NESW)$:

1. **Concatenation.** By stacking the first on top of the second through gluing points W, S in the first with N, E in the second,

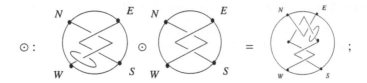

2. **Juxtaposition.** By putting two skein elements next to each other through gluing points E, S in the first with N, W in the second,

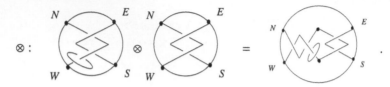

The Kauffman skein space $K(B^3, NES\,W)$ has an orthogonal basis $\{e_1, e_2, e_3\}$ with respect to the $\odot$ operation [1] given by

$$e_1 = \frac{1}{s + s^{-1}}\left(s^{-1}\;\asymp\;+\;\diagdown\!\!\!\diagup\;-(\delta^{-1}s^{-1} + \delta^{-1}\alpha^{-1})\;\smile\!\!\!\frown\right);$$

$$e_2 = \frac{1}{s + s^{-1}}\left(s\;\asymp\;-\;\diagdown\!\!\!\diagup\;+(-\delta^{-1}s + \delta^{-1}\alpha^{-1})\;\smile\!\!\!\frown\right);$$

$$e_3 = \delta^{-1}\;\smile\!\!\!\frown\;.$$

If we rotate the basis elements e_1, e_2, e_3 in the plane by $90°$, we obtain an orthogonal basis $\{e_{1h}, e_{2h}, e_{3h}\}$ with respect to the $\otimes$ operation. There is a base change formula between these base. Then we can use the linear skein method to write each section of horizontal (vertical) crossings in a 2-bridge knot or pretzel knot in terms of the orthogonal base, calculate the matrices for concatenation, juxtaposition, and closure maps. Therefore, we obtain a formula for the Kauffman polynomials as a product of matrices. The algorithm can be programmed in Mathematica, etc.

67.4 Bibliography

[1] A. Beliakava and C. Blanchet, *Skein construction of Idempotents in Birman-Murakami-Wenzl algebras*, Math Ann 321 (2001) 2, 347-373.

[2] L. Kauffman, *An invariant of regular isotopy*, Transactions of the American Mathematical Society, 318 (2) (1990): 417-471.

[3] B. Lu and J. K. Zhong, *On the Kauffman skein modules*, Manuscripta Math. 109 (2002) 1, 29-47.

[4] B. Lu and J. K. Zhong, *An algorithm to compute the Kauffman polynomial of 2-bridge knots*, the Rocky Mountain Journal of Mathematics, Vol. 40, Number 3, (2010) 977-993.

[5] B. Lu and J. K. Zhong, *The Kauffman polynomials of pretzel knots*, Journal of Knot Theory and its Ramifications, Vol. 17, No. 2 (2008) 157-169.

Chapter 68

Kauffman Polynomial on Graphs

Carmen Caprau, California State University, Fresno

68.1   Introduction

In the late 80s, François Jaeger found a relationship between the two-variable Kauffman polynomial [5] and the regular isotopy version of the HOMFLYPT polynomial [4, 9]. He showed that the Kauffman polynomial of an unoriented link L can be obtained as a weighted sum of HOMFLYPT polynomials of certain oriented links associated with L. A brief description of Jaeger's model can be found in [6, Part 1, Section 14]. We illustrate here Jaeger's model for the $SO(2n)$ Kauffman polynomial by considering the $sl(n)$-link invariant instead of the HOMFLYPT polynomial. The $SO(2n)$ Kauffman polynomial and the $sl(n)$ polynomial are one-variable specializations of the Kauffman (two-variable) polynomial, and respectively, the HOMFLYPT polynomial.

The $SO(2n)$ *Kauffman polynomial*, $[\![L]\!]$, of an unoriented link L is a Laurent polynomial valued in $\mathbb{Z}[q, q^{-1}]$ that is uniquely defined by the following skein relations:

$$\left[\!\!\left[\,\times\,\right]\!\!\right] - \left[\!\!\left[\quad\right]\!\!\right] = (q - q^{-1})\left(\left[\!\!\left[\,\underset{\frown}{\smile}\,\right]\!\!\right] - \left[\!\!\left[\,\supset\subset\,\right]\!\!\right]\right),$$

$$\left[\!\!\left[\,\bigcirc\,\right]\!\!\right] = [2n - 1] + 1 \,, \quad \left[\!\!\left[\,\underset{\bigcirc}{\varphi}\,\right]\!\!\right] = q^{2n-1}\left[\!\!\left[\,\cap\,\right]\!\!\right], \quad \left[\!\!\left[\quad\right]\!\!\right] = q^{1-2n}\left[\!\!\left[\quad\right]\!\!\right],$$

and such that $[\![L_1]\!] = [\![L_2]\!]$, whenever L_1 and L_2 are regular isotopic links.

We are using the notation $[k] = \dfrac{q^k - q^{-k}}{q - q^{-1}}$ to denote the (q-symmetric) *quantum integers* for every $k \in \mathbb{N}$. We remark that the $SO(2n)$ Kauffman polynomial is related to Chern-Simons gauge theory for $SO(2n)$, in the sense that the expectation value of Wilson loop operators in 3-dimensional $SO(2n)$ Chern-Simons gauge theory gives the $SO(2n)$ Kauffman polynomial.

We work with a state-sum formula for the $sl(n)$-invariant derived from the results in [8] applied to oriented 4-valent planar graphs, and we extend Jaeger's construction to incorporate these types of graphs. In doing this, we arrive at a state model for the $SO(2n)$ Kauffman polynomial of unoriented links, where the states are unoriented 4-valent planar graphs. For details on this construction, we refer the reader to [1].

68.2   A State Model for the $sl(n)$ Polynomial Invariant

For our purpose, we work with the Murakami-Ohtsuki-Yamada [8] state model for the regular isotopy version of the $sl(n)$ polynomial (for $n \geq 2$), $R(L)$, of an oriented link L. We describe now this model. Let L be an oriented link with a generic diagram D containing c crossings. We resolve each crossing of D in the two ways shown below:

$$)(\leftarrow \times \rightarrow \times \qquad\qquad \leftarrow \times \rightarrow$$

This process yields 2^c resolutions (states) corresponding to the link diagram D. A resolution Γ of D is a 4-valent oriented planar graph, possibly with loops with no vertices, such that each vertex is crossing-type oriented: . We seek to have a consistent way to evaluate these types of graphs.

Indeed, there is a well-defined Laurent polynomial $R(\Gamma) \in \mathbb{Z}[q, q^{-1}]$ associated to a resolution Γ, such that it satisfies the skein relations depicted in Figure 68.1. We will refer to $R(\Gamma)$ as the *MOY graph polynomial*. For more details on this polynomial, we refer the reader to [8].

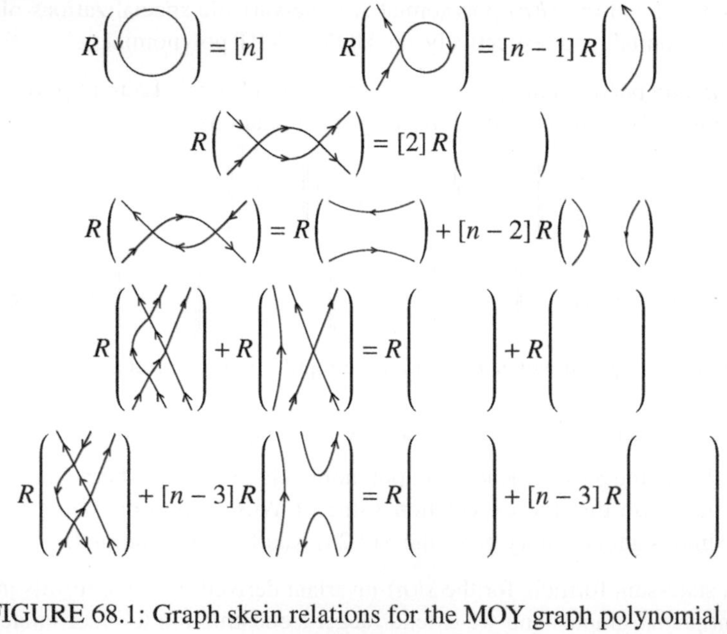

$$R\left(\bigcirc\right) = [n] \qquad R\left(\bigotimes\right) = [n-1]R\left(\Big\uparrow\right)$$

$$R\left(\bowtie\right) = [2]R\left(\quad\right)$$

$$R\left(\bowtie\right) = R\left(\asymp\right) + [n-2]R\left(\big)\big(\right)\quad\big(\big)$$

$$R\left(\bigotimes\right) + R\left(\big|\big|\right) = R\left(\quad\right) + R\left(\quad\right)$$

$$R\left(\bigotimes\right) + [n-3]R\left(\big)\big(\right) = R\left(\quad\right) + [n-3]R\left(\quad\right)$$

FIGURE 68.1: Graph skein relations for the MOY graph polynomial R.

These graph skein relations provide recursive formulas for evaluating any 4-valent planar graph with crossing-type oriented vertices in terms of evaluations of similar graphs with fewer vertices. In particular, this graphical calculus associates to a resolution Γ of the link diagram D a unique polynomial $R(\Gamma)$.

It remains to know how to piece these evaluations together, to obtain a polynomial evaluation for D. For that, we decompose each crossing in D using the skein relations below, where the diagrams in each relation are parts of larger diagrams that are identical except in a small neighborhood where they differ as shown:

$$R\left(\times\right) = qR\left(\right)\left(\right) - R\left(\times\right)$$

$$R\left(\times\right) = q^{-1}R\left(\quad\right) - R\left(\quad\right).$$

This yields $R(D)$ as a $\mathbb{Z}[q, q^{-1}]$-linear combination of MOY graph polynomials associated to all of the 2^c resolutions Γ of D. That is,

$$R(D) = \sum_{\Gamma} a_{\Gamma} R(\Gamma),$$

where the coefficients $a_{\Gamma} \in \mathbb{Z}[q, q^{-1}]$ are given by the above rules for decomposition of crossings. Note that these decomposition rules of crossings yield Conway's skein relation:

$$R\left(\quad\right) - R\left(\quad\right) = (q - q^{-1})R\left(\quad\right). \qquad (68.1)$$

We leave to the reader to verify that $R(D_1) = R(D_2)$, whenever diagrams D_1 and D_2 differ by a Reidemeister II or III move, and that the following holds:

$$R\left(\gtrdot\right) = q^n R\left(\right)\right) \quad \text{and} \quad R\left(\quad\right) = q^{-n}R\left(\quad\right).$$

By defining $R(L) := R(D)$, we arrive at the regular isotopy version of the $sl(n)$ polynomial of the link L.

68.3 $SO(2n)$ Kauffman Polynomial via Jaeger's Construction

We introduce now the relationship between the $SO(2n)$ Kauffman polynomial and the $sl(n)$ polynomial, due to Jaeger. Given an unoriented link diagram K, splice some of the crossings of K and orient the resulting link. This results in a state for the expansion $[\![K]\!]$, where $[\![K]\!]$ is the $SO(2n)$ Kauffman polynomial of K.

Specifically, each crossing in K is written as a linear combination of weighted evaluations of oriented link diagrams, according to the following skein relation:

$$\left[\!\!\left[\times\right]\!\!\right] := (q - q^{-1})\left(\left[\!\!\left[\begin{smallmatrix}\curlyvee\\\curlywedge\end{smallmatrix}\right]\!\!\right] - \left[\!\!\left[\succ\,\prec\right]\!\!\right]\right) + \left[\quad\right] + \left[\quad\right] + \left[\quad\right] + \left[\quad\right] \qquad (*)$$

By applying this skein relation to all crossings in the unoriented diagram, K results in 6^c states associated with K, with arrow configurations, where c is the number of crossings in K. Many of these states have arrow configurations that are not globally compatible, and therefore they are discarded. That is, the formula $(*)$ requires states that are oriented in a globally compatible way as oriented link diagrams. Then, each of the surviving states σ receives a weight b_{σ}, where $b_{\sigma} \in$

$\mathbb{Z}[q, q^{-1}]$ and is obtained by taking the product of the weights $\pm(q - q^{-1})$ or 1 according to the skein relation (∗).

We remark that Jaeger's model requires the following conventions:

$$\left[\!\!\left[\, \cap \, \right]\!\!\right] = \left[\, \cap \, \right] + \left[\quad\right] \quad \text{and} \quad \left[\!\!\left[\, \times \, \right]\!\!\right] = \left[\, \times \, \right] + \left[\, \times \, \right] + \left[\, \times \, \right] + \left[\, \times \, \right]$$

Using this formalism, the $SO(2n)$ Kauffman polynomial $[\![K]\!]$ of the unoriented link diagram K is given by:

$$[\![K]\!] = \sum_{\sigma} b_\sigma [\sigma], \quad \text{where} \quad [\sigma] = (q^{1-n})^{\mathrm{rot}(\sigma)} R(\sigma), \tag{68.2}$$

and where the sum is over all states σ associated with K that have globally compatible orientations, $\mathrm{rot}(\sigma)$ is the rotation number of the oriented link diagram σ, and $R(\sigma)$ is the $sl(n)$ polynomial of σ.

Recall that the *rotation number*, also called the Whitney degree, of an oriented link diagram is the sum of the rotation numbers of the Seifert circles obtained by splicing every crossing according to its orientation. A counterclockwise oriented circle contributes a $+1$, and a clockwise oriented circle contributes a -1. The rotation number is a regular isotopy invariant for oriented links.

Note that the Conway identity holds for $[\,\cdot\,]$:

$$\left[\, \times \,\right] - \left[\, \times \,\right] \stackrel{(68.2)}{=} q^{(1-n)\mathrm{rot}\left(\;\right)} R\left(\;\right) - q^{(1-n)\mathrm{rot}\left(\;\right)} R\left(\;\right)$$

$$= q^{(1-n)\mathrm{rot}\left(\;\right)} \left(R\left(\;\right) - R\left(\;\right) \right)$$

$$\stackrel{(68.1)}{=} (q - q^{-1}) q^{(1-n)\mathrm{rot}\left(\;\right)} R\left(\;\right)$$

$$\stackrel{(68.2)}{=} (q - q^{-1}) \left[\quad\right].$$

Then, the following holds:

$$\left[\!\!\left[\, \times \,\right]\!\!\right] - \left[\quad\right] \stackrel{(\star)}{=} (q - q^{-1})\left\{\left(\left[\quad\right] - \left[\, \supset \subset \,\right]\right) - \left(\left[\quad\right] - \left[\quad\right]\right)\right\}$$

$$+ \left[\quad\right] + \left[\quad\right] + \left[\quad\right] + \left[\quad\right]$$

$$- \left[\quad\right] - \left[\quad\right] - \left[\quad\right] - \left[\quad\right]$$

$$= (q - q^{-1})\left(\left[\quad\right] + \left[\quad\right] + \left[\quad\right] + \left[\quad\right]\right)$$

$$- (q - q^{-1})\left(\left[\quad\right] + \left[\quad\right] + \left[\quad\right] + \left[\quad\right]\right)$$

$$= (q - q^{-1})\left(\left[\!\!\left[\quad\right]\!\!\right] - \left[\, \supset \subset \,\right]\right).$$

Moreover,

$$\left[\!\left[\bigcirc\right]\!\right] \;=\; \left[\!\left[\bigcirc\right]\!\right] + \left[\!\left[\quad\right]\!\right]$$

$$\overset{(68.2)}{=}\; q^{(1-n)\mathrm{rot}\left(\quad\right)} R\left(\quad\right) + q^{(1-n)\mathrm{rot}\left(\quad\right)} R\left(\quad\right)$$

$$=\; (q^{1-n} + q^{n-1})[n] = \frac{q^{2n-1} - q^{1-2n}}{q - q^{-1}} + 1 = [2n-1] + 1.$$

Similarly, it can be verified that

$$\left[\!\left[\,\mathcal{Q}\,\right]\!\right] = q^{2n-1}\left[\!\left[\;\frown\;\right]\!\right] \quad \text{and} \quad \left[\!\left[\,\mathcal{Q}\,\right]\!\right] = q^{1-2n}\left[\!\left[\qquad\right]\!\right],$$

and that $[\![\,\cdot\,]\!]$ is invariant under the Reidemeister II and III moves. (For detailed proofs, see [1].) Therefore, $[\![\,\cdot\,]\!]$ is the $SO(2n)$ Kauffman polynomial for unoriented links.

68.4 A State Model for the $SO(2n)$ Kauffman Polynomial

We seek to construct a state-sum model for the $SO(2n)$ Kauffman polynomial, similar in spirit to the MOY model for the $sl(n)$ polynomial. Moreover, we want to derive such a state model by implementing the MOY model into Jaeger's construction. The states corresponding to an unoriented link diagram D will be unoriented 4-valent planar graphs and the desired state model requires a consistent method to evaluate these types of graphs.

To this end, we note that implementing the MOY state summation into Jaeger's model requires the bracket evaluation $[\Gamma]$, where Γ is an oriented 4-valent planar graph whose vertices are crossing-type oriented. We define

$$[\Gamma] := (q^{1-n})^{\mathrm{rot}(\Gamma)} R(\Gamma), \tag{68.3}$$

where $\mathrm{rot}(\Gamma)$, the *rotation number of the graph* Γ, is the sum of the rotation numbers of the disjoint oriented circles obtained by splicing each vertex of Γ according to the orientation of its edges. In particular, we have the following evaluation:

$$\left[\,\bigtimes\,\right] = (q^{1-n})^{\mathrm{rot}\left(\quad\right)} R\left(\quad\right) = (q^{1-n})^{\mathrm{rot}\,\bigcirc\,\bigcirc} R\left(\quad\right).$$

It was shown in [1] that we need to impose the following skein relation as the method to evaluate an unoriented vertex:

$$\left[\,\bigtimes\,\right] :\;=\; \left[\,\bigtimes\,\right] + \left[\,\bigtimes\,\right] + \left[\,\bigtimes\,\right] + \left[\,\bigtimes\,\right]$$

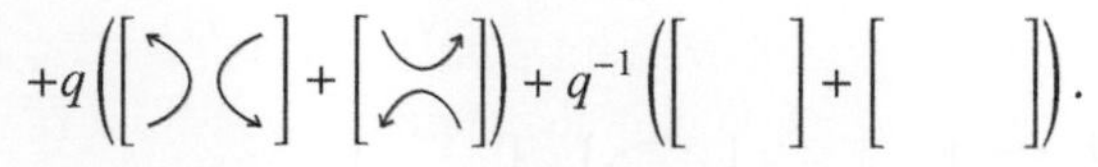

$$+q\left(\left[\,\right)\,\left(\,\right]+\left[\,\overset{\smile}{\underset{\frown}{\times}}\,\right]\right)+q^{-1}\left(\left[\quad\right]+\left[\quad\right]\right).$$

Implementing the above definitions and conventions into Jaeger's formula, we obtain that the $SO(2n)$ Kauffman polynomial satisfies the following skein relation:

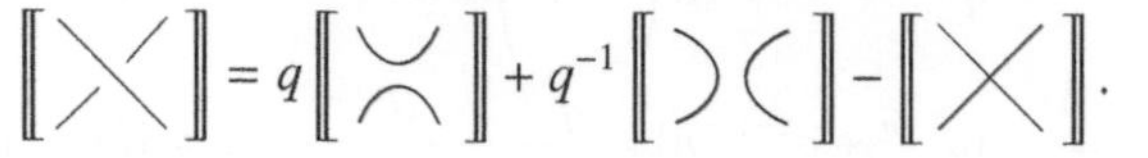

$$\left[\!\!\left[\,\overset{\times}{}\,\right]\!\!\right]=q\left[\!\!\left[\,\overset{\smile}{\frown}\,\right]\!\!\right]+q^{-1}\left[\!\!\left[\,\right)\,\left(\,\right]\!\!\right]-\left[\!\!\left[\,\times\,\right]\!\!\right]. \tag{68.4}$$

In addition, it was shown in [1] that the polynomial $[\![\,\cdot\,]\!]$ satisfies the graph skein relations depicted in Figure 68.2.

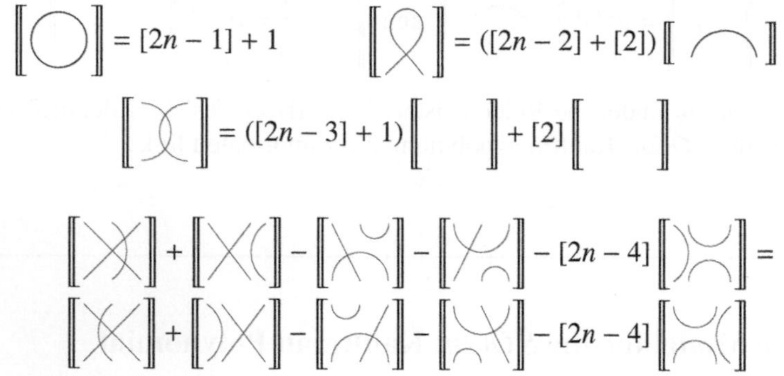

$$\left[\!\!\left[\,\bigcirc\,\right]\!\!\right]=[2n-1]+1 \qquad \left[\!\!\left[\,\overset{\bigcirc}{\times}\,\right]\!\!\right]=([2n-2]+[2])\left[\!\!\left[\,\frown\,\right]\!\!\right]$$

$$\left[\!\!\left[\,\overset{\times}{\times}\,\right]\!\!\right]=([2n-3]+1)\left[\!\!\left[\quad\right]\!\!\right]+[2]\left[\!\!\left[\quad\right]\!\!\right]$$

$$\left[\!\!\left[\,\times\,\right]\!\!\right]+\left[\!\!\left[\,\times(\,\right]\!\!\right]-\left[\!\!\left[\,\overset{\smile}{\times}\,\right]\!\!\right]-\left[\!\!\left[\,\overset{\smile}{\times}\,\right]\!\!\right]-[2n-4]\left[\!\!\left[\,\right)\overset{\smile}{\frown}\,\right]\!\!\right]=$$

$$\left[\!\!\left[\,\times\,\right]\!\!\right]+\left[\!\!\left[\,\right)\times\,\right]\!\!\right]-\left[\!\!\left[\,\overset{\smile}{\times}\,\right]\!\!\right]-\left[\!\!\left[\,\overset{\smile}{\times}\,\right]\!\!\right]-[2n-4]\left[\!\!\left[\,\overset{\smile}{\frown}(\,\right]\!\!\right]$$

FIGURE 68.2: Graph skein relations for the $SO(2n)$ Kauffman polynomial.

Therefore, given an unoriented link diagram D, the Kauffman $SO(2n)$ polynomial of D can be computed as follows: We resolve each of the crossings in D using the skein relation (68.4), to write $[\![D]\!]$ as a $\mathbb{Z}[q, q^{-1}]$-linear combination of evaluations of all states associated with D. These states are unoriented 4-valent planar graphs which are recurrently evaluated using the graph skein relations from Figure 68.2, which assign well-defined Laurent polynomials in $\mathbb{Z}[q, q^{-1}]$ to such graphs.

Final comments. Kauffman and Vogel [7] extended the Dubrovnik polynomial (a version of the two-variable Kauffman polynomial) to a three-variable rational function for rigid-vertex embeddings of unoriented 4-valent graphs. The results in [7] implied that there is a state model for the Kauffman two-variable polynomial via planar 4-valent graphs. This model can also be deduced from Carpentier's work [3] on the Kauffman-Vogel polynomial. A somewhat similar approach was used in [2] to construct a rational function in three variables which is an invariant of regular isotopy of unoriented links, and provides a state summation model for the Dubrovnik version of the two-variable Kauffman polynomial. The corresponding state model in [2] makes use of a special type of planar trivalent graphs.

Comparing the graph skein relations in Figure 68.2 with the graphical relations derived in [7], we see that the state model for the $SO(2n)$ Kauffman polynomial constructed here is essentially the same as that implied by the work in [7] (up to a negative sign for the weight received by the vertex-resolution of a crossing), and that given in [2, Section 5.1] (up to a change of variables). We would also like to point out that Hao Wu [10] used a different approach to write the Kauffman-Vogel graph polynomial [7] as a state sum of the Murakami-Ohtsuki-Yamada graph polynomial.

68.5 Bibliography

[1] C. Caprau, D. Heywood, D. Ibarra, *On a state model for the SO(2n) Kauffman polynomial*, Involve-A Journal of Mathematics, **7**, No. 4 (2014), 547–563.

[2] C. Caprau, J. Tipton, *The Kauffman polynomial and trivalent graphs*, Kyungpook Math. Journal **55**, No. 4 (2015), 779–806.

[3] R.P. Carpentier, *From planar graphs to embedded graphs - a new approach to Kauffman and Vogel's polynomial*, J. Knot Theory Ramifications **9**, Issue 8 (2000), 975–986.

[4] P. Freyd, D. Yetter, J. Hoste, W.B.R. Lickorish, K. Millett and A. Ocneanu, *A new polynomial invariant of knots and links*, Bull. Am. Math. Soc. **12** (1985), 239–246.

[5] L.H. Kauffman, *An invariant of regular isotopy*, Trans. Amer. Math. Soc. **318** No. 2 (1990), 417–471.

[6] L.H. Kauffman, **Knots and Physics**, Third edition. Series on Knots and Everything, Vol. 1, World Sci. Pub. (2001).

[7] L.H. Kauffman, P. Vogel, *Link polynomials and a graphical calculus*, J. Knot Theory Ramifications **1** (1992), 59–104.

[8] H. Murakami, T. Ohtsuki, S. Yamada, *Homfly polynomial via an invariant of colored plane graphs*, L'Enseignement Mathematique, **44** (1998), 325–360.

[9] J.H. Przytycki, P. Traczyk, *Invariants of links of Conway type*, Kobe J. Math. **2** (1987), 115–139.

[10] H. Wu, *On the Kauffman-Vogel and the Murakami-Ohtsuki-Yamada graph polynomials*, J. Knot Theory Ramifications **21**, No. 10 (2012) 1250098 (40 pages).

Chapter 69

Kauffman Bracket Skein Modules of 3-Manifolds

Rhea Palak Bakshi, George Washington University

Józef H. Przytycki, George Washington University and University of Gdańsk

Helen Wong, Claremont McKenna College

The Kauffman bracket skein module was originally introduced as a generalization of the Jones polynomial and Kauffman bracket for links in S^3, and as an integral part of a movement to create a theory for knots similar in spirit to constructions from homological algebra. This movement led to many definitions of other skein modules, some of which are covered in separate entries in this encyclopedia, and we refer the reader to [40, 42] for a historical discussion of the skein theoretic movement. Eventually, the movement led to the construction of the Witten-Reshetikhin-Turaev topological quantum field theory, which can be described combinatorially using the language of skein modules ([3], see also [33]). Because of its central role in these constructions and its deep ties to knot theory and geometric topology, the Kauffman bracket skein module has become a fundamental object of study in quantum topology.

69.1 Kauffman Bracket Skein Module of a 3-Manifold

Let M be an oriented 3-manifold, and let R be a commutative ring with unit and an invertible element, which we call A. For example, a good choice for R is $\mathbb{Z}[A^{\pm 1}]$.

Recall that we define a *framed knot* to be a piecewise-linear embedding of an annulus $S^1 \times [0, 1]$ into the interior of M. One boundary component is called the *underlying knot*, and the other is the *framing curve*. Equivalently, one can think of a framed knot as a knot with framing; that is, a piecewise linear, simple closed curve together with a continuous choice of non-zero, unit normal vector at every point on the curve. A *framed link* is the union of disjoint framed knots. The framed knot is said to be *trivial* or is the *unknot U* if there exists an embedding of the thickened disk $D^2 \times [0, 1]$ into M so that U is the image of $\partial D^2 \times [0, 1]$. We also allow the empty link to be a framed link, and denote it by $\emptyset$.

Throughout, we will consider framed links up to ambient isotopy. We choose a 3-ball in M which contains a portion of the link. We project the ball onto the 2-disk and work with link diagrams with respect to this projection. For simplicity in diagrams, we will only draw the underlying link with over- and under-crossing information, and assume that the framing curve is parallel to the link on the plane as drawn. In particular, we assume *blackboard framing*, where for every $x \in S^1$, the image of $\{x\} \times [0, 1]$ is parallel to the page of the paper. With this convention, any isotopy of a framed link corresponds to a sequence of local Reidemeister II and III moves on the diagrams.

We say that three framed links L_+, L_0, and L_∞ form a *Kauffman bracket skein triple* if they are identical everywhere except in a small 3-ball in M as follows:

To make sense of the above pictures, we rely on the orientation on the manifold M. In particular, the orientation of the 3-ball is inherited from M, and the diagrams of the framed links are the result of applying the natural projection of the ball onto a disk, with additional over- and under-crossing information and a blackboard framing.

The *Kauffman bracket skein module* $S(M)$ is defined in several steps. Recall that R is a commutative ring with unit and an invertible element A.

- First define $R\mathcal{L}$ to be the free $R[A, A^{-1}]$-module spanned by all (isotopy classes of) framed links in M. In other words, $R\mathcal{L}$ consists of all finite linear combinations of framed links, with coefficients from R.

- Then define S to be the submodule of $R\mathcal{L}$ spanned by elements of the form

  1. $[L_+] - A[L_0] - A^{-1}[L_\infty]$, where L_+, L_0, and L_∞ form a Kauffman bracket triple,

  2. $[L \sqcup U] - (-A^2 - A^{-2})[L]$, where L is any framed link and $L \sqcup U$ denotes the disjoint-union of L with the unknot U.

- Finally, $S(M)$ is the quotient $R\mathcal{L}/S$.

The Kauffman bracket skein module $S(M)$ is a particular example of a skein module. In the Chapter 63, it is denoted as $S_{2,\infty}(M)$, and other examples of skein modules are discussed there. Unless otherwise stated, $R = \mathbb{Z}[A^{\pm 1}]$.

An element of $S(M)$ is the equivalence class of some linear combination of (the isotopy classes of) framed links in M, and is referred to as a *skein* in M. The *skein relation* refers to the identities $[L_+] = A[L_0] + A^{-1}[L_\infty]$ for Kauffman triples L_+, L_0, L_∞, while sometimes the phrase *the unknot relation* is used for $[L \sqcup U] = (-A^2 - A^{-2})[L]$, where L is any framed link and U is the unknot. Notice that the skein relation and the unknot relation are exactly the defining relations for the Kauffman bracket for links in S^3. Because skeins in M are invariant under local Reidemeister II and III moves, they are invariant under isotopies of framed links, and hence $S(M)$ is well defined. We remark that a Reidemeister I move changes the isotopy type of the framed link, and changes the Kauffman bracket by a factor of $-A^{\pm 3}$.

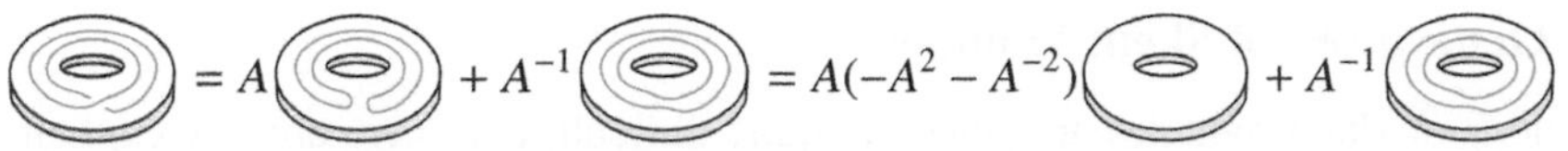

FIGURE 69.1: An element of the skein module of the thickened annulus $S(\mathcal{A} \times [0, 1])$.

69.1.1 Examples

69.1.1.1 The 3-Sphere

The skein relation allows any framed link in the three-sphere S^3 to be rewritten as an R-linear combination of disjoint unknots. The unknot relation replaces each trivial component with the scalar $-A^2 - A^{-2}$. Thus, every framed link is equivalent to some scalar times the empty link, and $S(S^3) = R[\emptyset] \cong R$. Moreover, the same reasoning shows that the skein module of S^3 with finitely many balls or punctures removed is also R.

69.1.1.2 The Thickened Annulus

Let $\mathcal{A}$ denote an annulus. As before, the skein relation allows any framed link in the thickened annulus $\mathcal{A} \times [0, 1]$ to be rewritten as an R-linear combination of disjoint loops without crossings, and then the unknot relation applies for the trivial components. Thus any framed link in the thickened annulus is an R-linear combination in $[z^n]$ for $n \in \mathbb{N}$, where z^n is the union of n disjoint, parallel copies of the core $S^1 \times \{(\frac{1}{2}, \frac{1}{2})\}$ with vertical framing on each component. For example, Figure 69.1 illustrates a skein which can be rewritten using the skein relations as a linear combination of $[z^0]$ and $[z^2]$. In [23] it is proved that $\{[z^n]\}_{n \in \mathbb{N}}$ is a linearly independent set, and hence forms a free basis for the skein module of the thickened annulus. Because a thickened annulus is homeomorphic to a solid torus $S^1 \times D^2$, the skein modules $S(\mathcal{A} \times [0, 1])$ and $S(S^1 \times D^2)$ are isomorphic.

Notice also in Figure 69.1 that the "wrapping number" of the link on the left is two and is detected by the highest power $[z^2]$ in the sum on the right (see [24] for a precise definition of wrapping number). In [24], it is conjectured that this phenomenon always holds. In particular, every link in $S(\mathcal{A} \times [0, 1])$ is a linear combination of $[z^n]$, and they conjecture that the highest degree is equal to the wrapping number of the link.

69.1.1.3 Lens Spaces

A lens space $L(p, q)$ can be described as the union of two solid tori, and any framed link in $L(p, q)$ can be isotoped into one of the two solid tori. Thus, the inclusion of one of the solid tori $i : S^1 \times D^2 \to L(p, q)$ induces a surjective module homomorphism $i_* : S(S^1 \times D^2) \to S(L(p, q))$, given by $i_*[L] = [i(L)]$ for a framed link L in $S^1 \times D^2$. It follows that any spanning set for the solid torus, say $\{i_*[z^n]\}_{n \in \mathbb{N}}$, is a spanning set for $S(L(p, q))$. However, i_* is not usually injective. In particular, in $L(p, q)$, a (p, q)-curve on the boundary of one solid torus bounds a disk in the other solid torus, and by sliding framed links across the disk, one finds many relations between the elements in the spanning set $\{i_*[z^n]\}_{n \in \mathbb{N}}$. Thus, $L(p, q)$ is potentially generated by fewer elements. In fact, in [25] it is proved that for $p \geq 1$ with $d = \lfloor \frac{p}{2} \rfloor$, the $S(L(p, q))$ is a free R-module with a finite basis $\{i_*[z^n]\}_{n=0}^{d}$. The case of $L(0, 1) = S^1 \times S^2$ was described in [26], and shows that its skein module has infinitely generated torsion.

69.1.2   Generators of Skein Modules

Computations of skein modules are generally quite difficult, and there are few explicit computations. Often, generators describing the free part of the skein module are known, but not the torsion.

We begin with the following theorem, which reduces the problem of finding generators for the skein module of any 3-manifold to prime 3-manifolds. In the following theorem, the ring is $\mathbb{Q}(A)$, the field of rational functions in the invertible variable A.

Proposition 1 ([41]). *If M and N are compact oriented 3-manifolds and M#N denotes there connected sum, then over the ring $R = \mathbb{Q}(A)$, $S(M\#N) = S(M) \otimes S(N)$.*

For compact prime 3-manifolds without boundary, E. Witten had conjectured that the Kauffman bracket skein module for a closed manifold is always finitely generated over $\mathbb{Q}(A)$ [21]. This has been verified by the computations of the skein module for the lens spaces $L(p, q)$ and $S^1 \times S^2$ [26, 25], all except two integral surgeries on a trefoil knot [8], the quaternionic manifold [20], a family of prism manifolds [35], and products of a closed oriented surface with S^1 [13, 19, 21]. Recently, a preprint [22] has appeared that shows the skein module of a closed 3-manifold is finite dimensional over the field $\mathbb{C}(A)$.

Of the compact prime 3-manifolds with boundary, the skein module is fully computed for I-bundles over a surface [40], the product of a 2-holed disk with S^1 [34], the exteriors of 2-bridged knots and links [11, 32, 7, 28].

In many cases, the skein modules were computed in order to establish a connection with the geometry and topology of the 3-manifold, e.g. whether incompressible surfaces exist inside the 3-manifold. In the next section, we will examine in detail the case of the thickened surface, where this connection is explicit.

69.2   Kauffman Bracket Skein Algebra of a Surface

Consider a thickened surface $\Sigma \times [0, 1]$, where Σ is an oriented surface. We say that a framed link in $\Sigma \times [0, 1]$ is a *multicurve* if its underlying link is a disjoint union of simple closed curves that lies entirely on a slice $\Sigma \times \{\frac{1}{2}\}$, the framing curve lies vertically above the underlying link at every point, and no component is trivial. For example, when Σ is the 2-sphere, the only multicurve is the empty link $\emptyset$. When Σ is an annulus $S^1 \times [0, 1]$, a thickened annulus is a solid torus, and then every multicurve is isotopic to some element of the set $\{z^n\}_{n\in\mathbb{N}}$ from Section 69.1.1.2.

Every element of $S(\Sigma \times [0, 1])$ can be rewritten as a linear combination of multicurves. It turns out that the multicurves are linearly independent.

Proposition 2 ([38, 40]). *Let Σ be an oriented surface. Then $S(\Sigma \times [0, 1])$ is a free R-module spanned by the (isotopy classes of) multicurves on Σ, including the empty link .*

Furthermore, the skein modules of thickened surfaces have a natural algebra structure from its

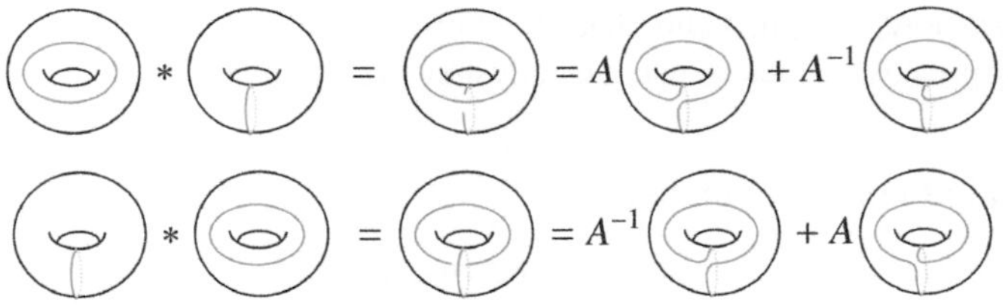

FIGURE 69.2: Examples of superposition in the skein algebra of a torus $\mathcal{S}(S^1 \times S^1)$.

product structure. More precisely, given two framed links L and L' in $\Sigma \times [0, 1]$ define their *superposition* $L \cdot L'$ to be the framed link obtained by putting L' on top of L. More precisely, $L \cdot L'$ is the disjoint union of natural embeddings $L \subset \Sigma \times [0, 1] \hookrightarrow \Sigma \times [0, \frac{1}{2}]$ and $L' \subset \Sigma \times [0, 1] \hookrightarrow \Sigma \times [\frac{1}{2}, 1]$. The superposition operation for skein is then defined by $[L] \cdot [L'] = [L \cdot L']$ for framed links L, L' and extending linearly. Figure 69.2 describes the superposition of skeins for a thickened torus. The superposition operation is compatible with the module structure, making $\mathcal{S}(\Sigma \times [0, 1])$ an algebra over R. This is the *Kauffman bracket skein algebra of* Σ, and we denote it by $\mathcal{S}(\Sigma)$.[1]

69.2.1 Examples

69.2.1.1 2-Sphere

As a module, $\mathcal{S}(S^2)$ is spanned by the empty link and is isomorphic to R. The superposition operation is commutative, since for any two framed links L and L' in a thickened 2-sphere, $L \cdot L'$ is isotopic to $L' \cdot L$. The superposition operation corresponds exactly to the usual multiplication of R. Thus $\mathcal{S}(S^2)$ is isomorphic to R.

69.2.1.2 Annulus

The multicurve basis for the thickened annulus is $\{[z^n]\}_{n \in \mathbb{N}}$, where z^n is the union of n disjoint, parallel copies of the core of an annular slice with vertical framing on each component. Notice that the n components of z^n can be isotoped to be parallel, so that $[z^n]$ is exactly the n-fold superposition $[z]^n$. Thus the skein algebra of the annulus is the commutative algebra $R[z]$.

Before moving to a non-commutative example, we remark that besides the 2-sphere and annulus, the only other surfaces with commutative skein algebras are the disk and the 2-holed disk (pair of pants) [12]. The skein algebra of a disk is R, and the skein algebra of a pair of pants is $R[x, y, z]$, where x, y, z correspond to the three distinct boundary parallel loops.

69.2.1.3 Torus

The multicurve basis for a thickened torus is infinite and consists of the (m, n)-torus links, with vertical framing. However, from Figure 69.2 we see that the superposition operation is not commutative for the torus. There are relations between the products of basis elements, and the skein algebra can be generated by exactly three elements, corresponding to the $(1, 0)$-, $(0, 1)$-, and $(1, 1)$-

---

[1]The new notation is intended to emphasize the dependence of the product structure of the thickened surface. Indeed, there exist surfaces (e.g. the 2-holed disk and 1-holed torus) whose skein algebras are isomorphic as modules, but not as algebras.

curves [12]. Furthermore, Frohman and Gelca [15] give a product-to-sum formula that completely describes the multiplicative structure using Chebyshev polynomials.

69.2.2   Algebraic Properties of $S(\Sigma)$

For every oriented surface, its skein algebra is known to be finitely generated [10, 43] and Noetherian with no zero divisors [44]. The skein algebra is commutative if Σ is a 2-sphere, disk, annulus, or pair of pants [38]. For the non-commutative cases, a finite presentation of the skein algebra is known for the torus with 0, 1, and 2 holes, and a sphere with 4 holes, [12]. In [43, 44] it was shown that the center of the skein algebra over $\mathbb{Z}[A^{\pm 1}]$ or a compact surface is generated by the boundary components.

In general, there are few explicit formulas describing the multiplicative structure of the skein algebra. One notable exception is the closed torus, for which [15] provides an explicit Product-to-Sum formula for multiplying curves involving Chebyshev polynomials of the first kind. For the four-holed sphere, [1] provides an algorithm to compute the product of any two elements of the algebra, and give an explicit formula for some families of curves.

The algebraic properties of $S(\Sigma)$ can change if we localize, replacing the variable A by a number. If A is replaced by 1 or -1, then the skein algebra is always commutative. In most other cases, for example excepting small surfaces like the 2-disk, 2-sphere, annulus, and pair of pants discussed earlier, the skein algebra is non-commutative. When A is replaced by a complex root of unity, in addition to the boundary parallel curves and curves bounding a single puncture, the center of the skein algebra can also contain other skeins, obtained from a threading procedure that uses Chebyshev polynomials of the first kind [4, 29, 14, 31].

Knowledge of the algebraic properties of $S(\Sigma)$ are useful, especially when relating them to different algebraic constructions from other fields of study. In the next section, we discuss one known relationship between $S(\Sigma)$ for A, a complex root of unity, and constructions from hyperbolic geometry and algebraic geometry.

69.2.3   Connections with Hyperbolic Geometry at Complex Roots of Unity

The skein algebra $S(\Sigma)$ is intimately tied to the hyperbolic geometry of the surface Σ. To see this, we introduce the $SL_2\mathbb{C}$-*character variety* $\mathcal{R}_{SL_2\mathbb{C}}(\Sigma)$, which consists of group homomorphisms from $\pi_1(\Sigma) \to SL_2\mathbb{C}$, up to conjugation in the sense of geometric invariant theory. Recall that the isotopy class of a complete, hyperbolic, finite-area metric on Σ corresponds uniquely, up to conjugation, to a monodromy representation $r : \pi_1(\Sigma) \to PSL_2\mathbb{R}$. One can lift the monodromy to $SL_2\mathbb{C}$ to obtain a point in $\mathcal{R}_{SL_2\mathbb{C}}(\Sigma)$. In this way, the $SL_2\mathbb{C}$-character variety contains information about all the complete, hyperbolic, finite-area metrics on Σ, and is thus an important object of study in hyperbolic geometry.

Proposition 3 ([9, 43, 2]). *The* $SL_2\mathbb{C}$-*character variety* $\mathcal{R}_{SL_2\mathbb{C}}(\Sigma)$ *is isomorphic as an algebra to the skein algebra* $S(\Sigma)$ *evaluated at* $A = 1$ *or* $A = -1$.

Notice that the algebras in the above proposition are commutative. Turaev [46] showed that the skein algebra $S(\Sigma)$ (without replacing A by a number) can be interpreted as a non-commutative deformation of $S(\Sigma)$ when A is replaced by 1 or -1, and that this interpretation is compatible with a quantization of $\mathcal{R}_{SL_2\mathbb{C}}(\Sigma)$ using the Poisson structure. This relationship was recently made more explicit, using representations of $S(\Sigma)$ [4, 5, 6, 14, 17, 18].

Along a similar vein, for an oriented 3-manifold M with a boundary which is a torus, e.g. M is a knot complement, one can think of the skein algebra of a thickened boundary torus as the ring, and the Kauffman bracket skein module $S(M)$ as a module over that ring. Such a set-up leads, for example, to a non-commutative generalization of an A-polynomial [16] . Thus skein modules and algebras may play an important role in tackling outstanding conjectures [37, 27], relating quantum topology and the geometry and topology of knot complements.

69.3 Relative Skein Modules

Suppose that M is an oriented 3-manifold with non-empty boundary, and that $v_1, v_2, \ldots, v_n$ are finitely many disjoint intervals embedded into ∂M. Then $(M; v_1, \ldots, v_n)$ is a *3-manifold with marked boundary*. A *framed arc* in it is an oriented embedding of a strip $[0, 1] \times [0, 1]$, so that $(0, 1) \times [0, 1]$ lies in the interior of M and $\{0, 1\} \times [0, 1] \subseteq \{v_1, \ldots, v_n\}$, and a framed link L is the union of disjoint framed arcs and framed knots, with the extra condition that $L \cap \partial = \{v_1, \ldots v_n\}$. Then, the *Kauffman bracket skein module of a 3-manifold with marked boundary* or *relative Kauffman bracket skein module* $S(M; v_1, \ldots, v_n)$ is defined analogously, as the R-module consisting of finite linear combinations of isotopy classes of framed links, quotiented by the submodule generated by elements of the form $[L] - A[L_\infty] - A^{-1}[L_0]$, where L, L_∞, L_0 are framed links in a Kauffman triple, and $[L \sqcup U] - (-A^2 - A^{-2})[L]$, where U is the unknot.

A particularly useful example of a relative Kauffman bracket skein module arises from considering a thickened $[0, 1] \times [0, 1]$ decorated with n intervals decorating two opposite sides of the disk. One can endow an algebra structure on the relative skein module by juxtaposition, and that algebraic structure leads to the Temperley-Lieb Algebra discussed in the entry on planar algebras. Relative Kauffman bracket skein modules have well-understood algebraic structure for other manifolds with marked boundary, e.g. annuli or handlebodies with marked boundary [38, 39, 29], and such knowledge is especially useful for computing the skein modules of larger 3-manifolds with or without marked boundary.

Finally, we mention two relative skein modules with additional relations between framed arcs, besides the usual Kauffman bracket skein relations. The arc module defined by Muller [36, 30] applies to 3-manifolds with marked boundary. If N is a 3-manifold, let T be a *trivial arc in N* if there exists an embedding of the thickened disk $D^2 \times [0, 1]$ into N so that the image of $\partial D^2 \times [0, 1]$ is $T \cup v_i$ for some i. Then the *Kauffman bracket arc module of a 3-manifold with marked boundary* is defined to be the Kauffman bracket skein module $S(M; v_1, \ldots, v_n)$, further quotiented by the submodule generated by $L \cup T = 0$, where T is a trivial arc in $M - L$. In the case of the thickened surface, Muller's arc module acquires the structure of an algebra, and is closely related to quantum cluster algebras [36] and to quantum Teichmüller spaces [30, 31]. Similarly, Roger and Yang [45] define an arc algebra for a closed surface with finitely many punctures, that uses a different relation between arcs and that is related to a quantization of the decorated Teichmüller space of the surface.

69.4 Bibliography

[1] Rhea Palak Bakshi, Sujoy Mukherjee, Józef H Przytycki, Marithania Silvero, and Xiao Wang. On multiplying curves in the Kauffman bracket skein algebra of the thickened four-holed sphere. *arXiv preprint arXiv:1805.06062, To appear in J. Knot Theory Ramifications*, 2020.

[2] John W. Barrett. Skein spaces and spin structures. *Math. Proc. Cambridge Philos. Soc.*, 126(2):267–275, 1999.

[3] C. Blanchet, N. Habegger, G. Masbaum, and P. Vogel. Topological quantum field theories derived from the Kauffman bracket. *Topology*, 34(4):883–927, 1995.

[4] Francis Bonahon and Helen Wong. Representations of the Kauffman bracket skein algebra I: Invariants and miraculous cancellations. *Invent. Math.*, 204(1):195–243, 2016.

[5] Francis Bonahon and Helen Wong. Representations of the Kauffman bracket skein algebra II: Punctured surfaces. *Algebr. Geom. Topol.*, 17(6):3399–3434, 2017.

[6] Francis Bonahon and Helen Wong. Representations of the Kauffman bracket skein algebra III: closed surfaces and naturality. *Quantum Topol.*, 10(2):325–398, 2019.

[7] Doug Bullock. The $(2, \infty)$-skein module of the complement of a $(2, 2p + 1)$ torus knot. *J. Knot Theory Ramifications*, 4(4):619–632, 1995.

[8] Doug Bullock. On the Kauffman bracket skein module of surgery on a trefoil. *Pacific J. Math.*, 178(1):37–51, 1997.

[9] Doug Bullock. Rings of $SL_2(\mathbf{C})$-characters and the Kauffman bracket skein module. *Comment. Math. Helv.*, 72(4):521–542, 1997.

[10] Doug Bullock. A finite set of generators for the Kauffman bracket skein algebra. *Math. Z.*, 231(1):91–101, 1999.

[11] Doug Bullock and Walter Lo Faro. The Kauffman bracket skein module of a twist knot exterior. *Algebr. Geom. Topol.*, 5:107–118, 2005.

[12] Doug Bullock and Józef H. Przytycki. Multiplicative structure of Kauffman bracket skein module quantizations. *Proc. Amer. Math. Soc.*, 128(3):923–931, 2000.

[13] Alessio Carrega. Nine generators of the skein space of the 3-torus. *Algebr. Geom. Topol.*, 17(6):3449–3460, 2017.

[14] C. Frohman, J. Kania-Bartoszyńska, and T. T. Q. Lê. Unicity for representations of the Kauffman bracket skein algebra. *Invent. Math.*, 215(2):609–650, 2019.

[15] Charles Frohman and Răzvan Gelca. Skein modules and the noncommutative torus. *Trans. Amer. Math. Soc.*, 352(10):4877–4888, 2000.

[16] Charles Frohman, Răzvan Gelca, and Walter Lofaro. The A-polynomial from the noncommutative viewpoint. *Trans. Amer. Math. Soc.*, 354(2):735–747, 2002.

[17] Charles Frohman and Joanna Kania-Bartoszyńska. The structure of the Kauffman bracket skein algebra at roots of unity. *Mathematische Zeitschrift*, pages 1–32, 2016.

[18] Iordan Ganev, David Jordan, and Pavel Safronov. The quantum Frobenius for character varieties and multiplicative quiver varieties. arXiv preprint arXiv:math/1901.11450[math.QA].

[19] Patrick M. Gilmer. On the Kauffman bracket skein module of the 3-torus. *Indiana Univ. Math. J.*, 67(3):993–998, 2018.

[20] Patrick M. Gilmer and John M. Harris. On the Kauffman bracket skein module of the quaternionic manifold. *J. Knot Theory Ramifications*, 16(1):103–125, 2007.

[21] Patrick M. Gilmer and Gregor Masbaum. On the skein module of the product of a surface and a circle. *Proc. Amer. Math. Soc.*, 147(9):4091–4106, 2019.

[22] Sam Gunningham, David Jordan, and Pavel Safronov. The finiteness conjecture for skein modules. arXiv preprint arXiv:math/1908.05233.

[23] Jim Hoste and Józef H. Przytycki. An invariant of dichromatic links. *Proc. Amer. Math. Soc.*, 105(4):1003–1007, 1989.

[24] Jim Hoste and Józef H. Przytycki. Homotopy skein modules of orientable 3-manifolds. *Math. Proc. Cambridge Philos. Soc.*, 108(3):475–488, 1990.

[25] Jim Hoste and Józef H. Przytycki. The $(2, \infty)$-skein module of lens spaces; a generalization of the Jones polynomial. *J. Knot Theory Ramifications*, 2(3):321–333, 1993.

[26] Jim Hoste and Józef H. Przytycki. The Kauffman bracket skein module of $S^1 \times S^2$. *Math. Z.*, 220(1):65–73, 1995.

[27] R. M. Kashaev. The hyperbolic volume of knots from the quantum dilogarithm. *Lett. Math. Phys.*, 39(3):269–275, 1997.

[28] Thang T. Q. Lê. The colored Jones polynomial and the A-polynomial of knots. *Adv. Math.*, 207(2):782–804, 2006.

[29] Thang T. Q. Lê. On Kauffman bracket skein modules at roots of unity. *Algebraic & Geometric Topology*, 15(2):1093–1117, 2015.

[30] Thang T. Q. Lê. Quantum Teichmüller spaces and quantum trace map. *Journal of the Institute of Mathematics of Jussieu*, pages 1–43, 2017.

[31] Thang T. Q. Lê and Jonathan Paprocki. On Kauffman bracket skein modules of marked 3–manifolds and the Chebyshev–Frobenius homomorphism. *Algebr. Geom. Topol.*, 19(7):3453–3509, 2019.

[32] Thang T. Q. Lê and Anh T. Tran. The Kauffman bracket skein module of two-bridge links. *Proc. Amer. Math. Soc.*, 142(3):1045–1056, 2014.

[33] W. B. Raymond Lickorish. Quantum invariants of 3-manifolds. In *Handbook of Geometric Topology*, pages 707–734. Elsevier, 2001.

[34] M. Mroczkowski and M. K. Dabkowski. KBSM of the product of a disk with two holes and S^1. *Topology Appl.*, 156(10):1831–1849, 2009.

[35] Maciej Mroczkowski. Kauffman bracket skein module of a family of prism manifolds. *Journal of Knot Theory and Its Ramifications*, 20(01):159–170, 2011.

[36] Greg Muller. Skein and cluster algebras of marked surfaces. *Quantum Topology*, 7(3):435–503, 2016.

[37] Hitoshi Murakami and Jun Murakami. The colored Jones polynomials and the simplicial volume of a knot. *Acta Math.*, 186(1):85–104, 2001.

[38] Józef Przytycki. Skein modules of 3-manifolds. *Bull. Polish Acad. Sci. Math.*, 39(1-2):91–100, 1991.

[39] Józef H. Przytycki. Skein module of links in a handlebody. In *Topology '90 (Columbus, OH, 1990)*, volume 1 of *Ohio State Univ. Math. Res. Inst. Publ.*, pages 315–342. de Gruyter, Berlin, 1992.

[40] Józef H. Przytycki. Fundamentals of Kauffman bracket skein modules. *Kobe J. Math.*, 16(1):45–66, 1999.

[41] Józef H. Przytycki. Kauffman bracket skein module of a connected sum of 3-manifolds. *Manuscripta Math.*, 101(2):199–207, 2000.

[42] Józef H. Przytycki. Chapter IX Skein Modules. In *KNOTS: From Combinatorics of Knot Diagrams to Combinatorial Topology Based on Knots.* arXiv preprint arXiv:math/0602264, 2006.

[43] Józef H. Przytycki and Adam S. Sikora. On skein algebras and $Sl_2(\mathbf{C})$-character varieties. *Topology*, 39(1):115–148, 2000.

[44] Józef H. Przytycki and Adam S. Sikora. Skein algebras of surfaces. *Trans. Amer. Math. Soc.*, 371(2):1309–1332, 2019.

[45] Julien Roger and Tian Yang. The skein algebra of arcs and links and the decorated Teichmüller space. *J. Differential Geom.*, 96(1):95–140, 2014.

[46] Vladimir G. Turaev. Skein quantization of Poisson algebras of loops on surfaces. *Ann. Sci. École Norm. Sup. (4)*, 24(6):635–704, 1991.

Part XII

Homological Invariants

Chapter 70

Khovanov Link Homology

Radmila Sazdanovic, North Carolina State University

70.1 Introduction

Skein relations are omnipresent and crucial in the world of polynomial knot invariants as in Chapter 61. At the end of 20th century Mikhail Khovanov [24] discovered a way to use skein relations to construct new and powerful knot invariants in the form of homology theories. This process is referred to as categorification: Khovanov link homology is a categorification of the Jones polynomial. Knot polynomials such as the HOMFLYPT, Kauffman 2-variable, and $\mathfrak{sl}(N)$ link polynomial [51, 29, 30, 17, 18], as well as the Kauffman bracket skein module [4, 11, 48] have also been categorified. These categorifications utilize results from different areas of mathematics, including representation and gauge theory. For example, Ozsváth, Szabó and Rasmussen constructed knot Floer homology [41, 49, 50, 38] that categorifies the Alexander polynomial.

Categorification offers new insights and relations: while Jones and Alexander polynomials only have a few evaluations in common, there is a plethora of relations in the form of spectral sequences between their categorifications [42, 16]. These were essential in proving that Khovanov homology distinguishes the unknot [32]. Khovanov's construction led to many significant advances in 3- and 4- dimensional topology [51], as well as many other areas of mathematics.

A connection between Khovanov homology and gauge theory was originally found by Gukov, Schwarz and Vafa [20]. Later Witten provided a more general connection by counting solutions to BPS equations. [67].

70.2 Construction of Khovanov Homology

This section contains the construction of Khovanov link homology using Bar Natan's approach [9] with a brief comment on Viro's enhanced state categorification of the Kauffman bracket [65] as an inspiration. A comprehensive account of the construction, development and history of Khovanov homology can be found in [62, 61], and the reader can further consult [24, 5, 23, 56]. We begin with a construction that follows the state sum formula for the Jones polynomial, organized in a so-called *cube of resolutions*.

Consider a directed graph determined by the 1-dimensional skeleton of the n-dimensional cube. The vertex set $\mathcal{V}(n) = \{0, 1\}^n$ of this graph comprises n-tuples of 0's and 1's. The edge set $\mathcal{E}(n)$ consists of directed edges between vertices $I, J \in \mathcal{V}(n)$ such that they only differ in exactly one

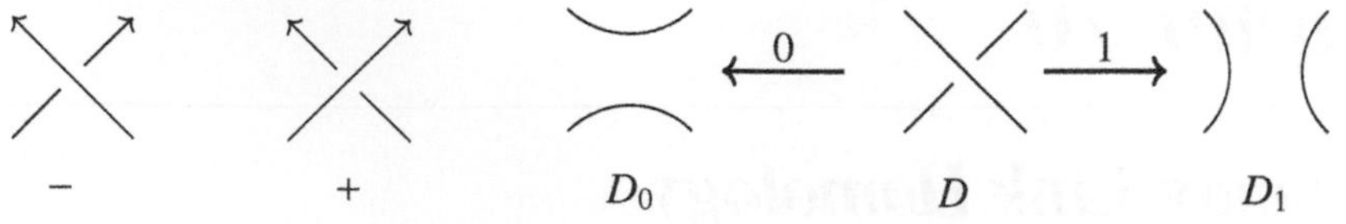

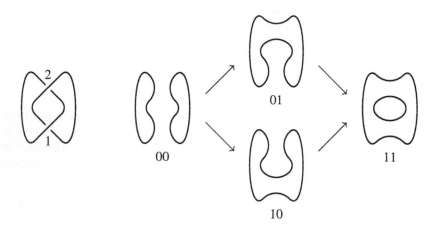

FIGURE 70.1: Conventions for crossings and resolutions in link diagrams.

entry which is 0 in I and 1 in J. An example of such a cube for $n = 2$ is shown in Figure 70.1, where there is no edge between $(0, 0)$ and $(1, 1)$ because they differ in more than one entry. The *height of a vertex* $I = (k_1, k_2, \ldots, k_n)$ is the number of entries in I equal one, $h(I) = k_1 + k_2 + \cdots + k_n$. If ε is an edge from vertex I to vertex J, affecting the r-th entry, the *height* of ε is $|\varepsilon| := h(I)$. The *sign* of an edge ε is -1 if the number of 1's before the r-th entry is odd, and is $+1$ if even, and can be expressed as $(-1)^\varepsilon := (-1)^{\sum_{i=1}^{r-1} k_i}$.

FIGURE 70.2: Hopf link and its two-dimensional cube of resolutions.

For a given link L consider a diagram D with n crossings $c_1, c_2, \ldots, c_n$. For positive and negative crossings, and for zero- and one-resolutions we use conventions as in Figure 70.1. Next, to each vertex $I \in \mathcal{V}(n)$ we associate a collection of circles $D(I)$, called a Kauffman state, obtained by 0-resolution those crossings c_i for which $k_i = 0$ and 1-resolutions of those crossings c_j for which $k_j = 1$, see Figure 70.2 as an illustration.

Since Khovanov homology is an invariant of oriented links $L \subset S^3$ with values in bigraded modules over a commutative ring R with identity, we proceed by introducing necessary algebra. A bigraded R-module is an R-module M with a direct sum decomposition of the form $M = \bigoplus_{i,j \in \mathbb{Z}} M^{i,j}$, with each submodule $M^{i,j}$ with *bigrading* (i, j). We will refer to i as the *homological grading* and to j as the *polynomial grading* or *quantum grading*. Given two bigraded R-modules $M = \bigoplus_{i,j \in \mathbb{Z}} M^{i,j}$ and $N = \bigoplus_{i,j \in \mathbb{Z}} N^{i,j}$, we define the direct sum $M \oplus N$ and tensor product $M \otimes_R N$ to be the bigraded R-modules with components $(M \oplus N)^{i,j} = M^{i,j} \oplus N^{i,j}$ and $(M \otimes N)^{i,j} = \bigoplus_{k+m=i, l+n=j} M^{k,l} \otimes N^{m,n}$. In order to have a grading-preserving differential in Khovanov homology we need homological and polynomial shift operators, denoted $[\cdot]$ and $\{\cdot\}$, respectively, by $(M[r])^{i,j} = M^{i-r,j}$ and $(M\{s\})^{i,j} = M^{i,j-s}$.

To each Kauffman state $D(I)$ we associate a bigraded R-module $C(D(I))$ as follows. Let $\mathcal{A}_2 = R[X]/X^2 = R1 \oplus RX$ be the bigraded module with generator 1 in bigrading $(0, 1)$ and generator X in bigrading $(0, -1)$. Then

$$C(D(I)) := \mathcal{A}_2^{\otimes |D(I)|}[h(I)]\{h(I)\},$$

where $|D(I)|$ denotes the number of circles in Kauffman state $D(I)$. Here, each tensor factor of $\mathcal{A}_2$ corresponds to a particular circle of the Kauffman state $D(I)$. In fact, $C(D(I))$ can be thought of as a bigraded module freely generated by *enhanced Kauffman states* – Kauffman states where each circle is labeled with 1 or X. Now that each vertex of the cube corresponds to an R-module, we can associate homomorphisms to the directed edges. These homomorphisms will be used to define differential on Khovanov homology. Define the following R-module homomorphisms

- multiplication $m : \mathcal{A}_2 \otimes \mathcal{A}_2 \to \mathcal{A}_2$ by

$$m(1 \otimes 1) = 1, \quad m(1 \otimes X) = X, \quad m(X \otimes 1) = X, \quad m(X \otimes X) = 0,$$

- comultiplication $\Delta : \mathcal{A}_2 \to \mathcal{A}_2 \otimes \mathcal{A}_2$

$$\Delta(1) = 1 \otimes X + X \otimes 1, \quad \Delta(X) = X \otimes X.$$

To an edge ε from I to J we associate the map $d_\varepsilon : C(D(I)) \to C(D(J))$ defined as follows.

1. If the number of circles in $D(J)$ is one less than in $D(I)$, then $D(J)$ is obtained by merging two circles of $D(I)$ into one. Algebraic analogue of this operation is the map d_ε which acts as multiplication m on the tensor factors associated to the circles being merged, and equals the identity on the remaining tensor factors.

2. If the number of circles $D(J)$ is one less than $D(I)$, then $D(J)$ is obtained by splitting one circle of $D(I)$ into two. In this case d_ε acts as comultiplication Δ on the tensor factor associated to the circle being split, and as the identity map on the remaining tensor factors.

Next, suppose that diagram D has n_+ positive crossings and n_- negative crossings.

Lemma 70.2.1 ([24]). *The bigraded module*

$$C^{*,*}(D) = \bigoplus_{I \in V(n)} C(D(I))[-n_-]\{n_+ - 2n_-\} \tag{70.1}$$

together with homomorphisms $d^i : C^{i,*}(D) \to C^{i+1,*}(D)$ *defined by* $d^i = \sum_{|\varepsilon|=i}(-1)^\varepsilon d_\varepsilon$ *form a (co)chain complex* $(C(D), d)$.

Due to this clever choice of polynomial shifts, the maps d_i turn out to preserve the polynomial degree. Therefore we get a bigraded homology R-module $H(L; R) = \bigoplus_{i,j\in\mathbb{Z}} H^{i,j}(L; R)$ called the *Khovanov homology of the link L with coefficients in R*. Khovanov homology is most often considered with rational, integer or $\mathbb{Z}_2$ coefficients. For $R = \mathbb{Z}$ we omit the coefficient ring $H(L) = H(L; \mathbb{Z})$.

Theorem 70.2.2 ([24, 65, 10])**.** *For any link L and commutative ring with identity R, the Khovanov homology $H(L; R)$ is a link invariant, i.e. independent of the choice of link diagram and the ordering of crossings within.*

Complete proofs of Lemma 70.2.1 and Theorem 70.2.2 can be found in [24, 65, 10]. We will provide the proof of invariance of Khovanov homology under the first Reidemeister move to illustrate the power, beauty, and elegance of categorification.

Proof. We want to show that $H\left(\,\rotatebox{0}{$\mathcal{R}$}\,\right)$ is the same as $H\left(\,\rotatebox{0}{$\mathcal{R}$}\,\right)$. Proving this requires a bit

of homological algebra in the form of the cancellation principle for chain complexes: if C' is an acyclic subcomplex of C then the homology of C is the same as that of C/C'; if the quotient C/C' is acyclic, then the homology of C is the same as that of C'.

Note that the Khovanov module $C = C\left(\,\rotatebox{0}{$\mathcal{R}$}\,\right)$ can be built out of modules for the same link with

the crossing of the first Reidemeister move replaced by resolutions according to the skein relation: $C\left(\,\rotatebox{0}{$\asymp$}\,\right)$ and $C\left(\,\rotatebox{0}{$\mathcal{R}$}\,\right)$ and connected by the multiplication map m. The subchain complex C'

of C obtained by only considering Kauffman states of $C\left(\,\rotatebox{0}{$\asymp$}\,\right)$ with the special circle on top

labeled 1 and all the states in the second term. C' is acyclic since the restricting multiplication to these states is an isomorphism. According to the above statement, C has the same homology as C/C', which can be thought of as only keeping label X on the special cycle, i.e. tensoring by R on the chain complex level, which does not change anything. Therefore, $C\left(\,\rotatebox{0}{$\mathcal{R}$}\,\right)$ has the same

homology as $C\left(\,\rotatebox{0}{$\wedge$}\,\right)$.

$\square$

Example 70.2.3. *Let us explicitly compute Khovanov homology of the unknot using the following diagram with one crossing $Kh\left(\,\rotatebox{0}{$\bowtie$}\,\right)$. Since the diagram has one crossing we get the 1-dimensional cube of resolutions. Next, $H^0\left(\,\rotatebox{0}{$\bowtie$}\,\right) = ker(\Delta) = 0$ and*

$$
\begin{aligned}
H^1\left(\,\rotatebox{0}{$\bowtie$}\,\right) &= \frac{\mathcal{A}_2 \otimes \mathcal{A}_2\{1\}}{im(\Delta)} \\
&= \frac{\langle 1 \otimes 1, 1 \otimes x, x \otimes x, x \otimes x \rangle}{\langle x \otimes x, 1 \otimes x + x \otimes 1 \rangle}\{1\} \\
&= \langle 1 \otimes 1, 1 \otimes x \rangle\{1\} = \mathbb{Z}_{(1)} \oplus \mathbb{Z}_{(3)}
\end{aligned}
$$

$$
\begin{array}{cc}
0 & 1 \\[4pt]
\mathcal{A}_2 & \xrightarrow{\Delta} \mathcal{A}_2 \otimes \mathcal{A}_2\{1\}
\end{array}
$$

After an overall shift by $[-n_-]\{n_+ - 2n_-\} = [-1]\{-2\}$ (where $[*]$ denotes homological shift and $\{*\}$

denotes quantum shift) and, as desired, we get:

$$H\left(\vcenter{\hbox{\includegraphics{kauffman}}} \right) = \mathbb{Z}\{-1\} \oplus \mathbb{Z}\{1\} = \mathcal{A}_2.$$

70.2.1 Functoriality of Khovanov Homology

The choice of algebra $\mathcal{A}_2$ is not arbitrary [10, 23]. There is a tight relationship between the properties of the algebra $\mathcal{A}_2$ and cobordisms that relate the Kauffman states. Each cobordism induces an algebra map, and the relations in the algebra, such as associativity, correspond to equivalences of cobordisms, see Figure 70.3. Disjoint circles correspond to a tensor product of the algebra $\mathcal{A}_2$; merging them corresponds to the multiplication operation in the algebra; and splitting one circle into two corresponds to the comultiplication. The same maps $\mathcal{A}_2^{\otimes 3} \to \mathcal{A}_2$ obtained by composing either $m \otimes id$ or $id \otimes m$ with m correspond to equivalent cobordisms; algebraically, this is associativity in $\mathcal{A}_2$.

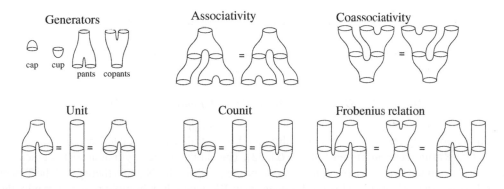

FIGURE 70.3: 2D Topological Quantum Field Theory: Generators of 2D cobordisms (top left) and the topological representations of the Frobenius algebra corresponding to 2D TQFT.

The bijection between commutative Frobenius algebras and $(1 + 1)$-dimensional TQFTs warrants that Khovanov homology, or the analogous construction with $\mathcal{A}_2$ replaced by any Frobenius algebra [31], is a functor on the category of trivial links and cobordisms. The choice of $\mathcal{A}_2$ specifically allows us to extend this functoriality to all knots and links [24, 25, 22, 15]. The existence of Khovanov-type theories for other Frobenius algebras, i.e. finding the necessary and sufficient conditions for existence of link homology, is discussed in [10, 28].

70.3 Khovanov Homology and the Jones Polynomial

That Khovanov homology categorifies the Jones polynomial carries many implications. The most immediate is that the Jones polynomial can be recovered from the free part of Khovanov homology.

Theorem 70.3.1. *[24] The (unnormalized) Jones polynomial $\hat{J}_L(q)$ is recovered as the graded Euler characteristic $\chi(H(L))$ of Khovanov homology:*

$$\hat{J}_L(q) = \chi(H(L)) = \sum_{i,j \in \mathbb{Z}} (-1)^i q^j \cdot \mathrm{rk}(H^{i,j}(L)).$$

Moreover, Khovanov homology is a stronger knot invariant than the Jones polynomial, e.g. it distinguishes 5_1 and 10_{132}. Khovanov homology of an alternating knot is determined by the Jones polynomial and the signature of the knot [34], yet when it comes to non-alternating knots, Khovanov homology contains more information than the Jones polynomial and signature.

The skein relation which defines the Jones polynomial is categorified by a long exact sequence in Khovanov homology [24, 65], which is the main computational tool. Given an oriented link diagram D, there is a long exact sequence relating the Khovanov homology of D to that of the 0- and 1- resolutions D_0 and D_1 at a given crossing of D. If the crossing to be resolved is negative, set $c = n_-(D_0) - n_-(D)$; there is a long exact sequence:

$$\overset{\delta_*}{\to} H^{i,j+1}(D_1) \to H^{i,j}(D) \to H^{i-c,j-3c-1}(D_0) \overset{\delta_*}{\to} H^{i+1,j+1}(D_1) \to .$$

If the crossing to be resolved is positive, we set $c = n_-(D_1) - n_-(D)$ and the long exact sequence takes the following form

$$\overset{\delta_*}{\to} H^{i-c-1,j-3c-2}(D_1) \to H^{i,j}(D) \to H^{i,j-1}(D_0) \overset{\delta_*}{\to} H^{i-c,j-3c-2}(D_1) \to .$$

Khovanov homology features a range of additional differentials and associated spectral sequences such as Lee-Rasmussen and Turner [34, 64, 63, 52]. The choice of the coefficient ring is essential: Lee theory is defined over $\mathbb{Q}$, $\mathbb{Z}$, or $\mathbb{Z}_p$ for any odd prime p, while Bar Natan's theory and the Turner spectral sequence work over $\mathbb{Z}_2$. Spectral sequences in Khovanov link homology together with the Bockstein spectral sequence and other homological algebra techniques can be used to determine the structure, support, and properties of Khovanov homology.

For example, the Lee spectral sequence [34] has the Khovanov chain complex as its zeroth page, and Khovanov homology as the first. Lee's differential has bidegree $(1, 4n)$; the spectral sequence for a knot converges to a pawn move piece: two copies of the base ring in bidegrees $(0, s - 1)$ and $(0, s + 1)$, where s is the Rasmussen invariant [52].

According to the Knight Move conjecture [9], Khovanov homology of a knot can be decomposed into a single pawn move piece and Knight move pieces. The Lee spectral sequence can be used to prove this conjecture for all *thin* knots – those whose Khovanov homology is supported in bigradings $2i - j = k \pm 1$ for some integer k, – a class which includes all alternating and quasialternating knots. The Knight Move conjecture also holds for all knots with unknotting number at most 2 [3].

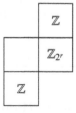

Knight move

Knight Move conjecture was proven to be false by finding a counterexample, a knot whose Lee spectral sequence has nontrivial differential of degree $(1, 8)$ and therefore cannot be decomposed into bidegree $(1, 4)$ Knight moves [37].

Spectral sequences can also be used to analyze the torsion of Khovanov homology not captured by the Jones polynomial. Shumakovitch showed that homologically thin links have only $\mathbb{Z}_{2^k}$ torsion in their Khovanov homology [59], and in [60] he used a relationship between the Turner and Bockstein differentials on $\mathbb{Z}_2$-Khovanov homology to show there is only $\mathbb{Z}_2$ torsion.

Shumakovitch conjectured that all links, besides unknots, Hopf links, and connect sums of these, contain torsion of order two. The proof of this conjecture amounts to alternative proof that Khovanov homology is an unknot detector. There are a number of results concerning the existence, order, distribution, and patterns of torsion in Khovanov homology [44, 47, 56, 60, 54, 1], and constructing infinite families of knots with interesting torsion such as $\mathbb{Z}_{2^k}$ torsion for $k < 24$, [39]. General understanding torsion and its relations with topological properties of knots remains elusive.

70.4 Khovanov Homology and Its Applications

The geometric and topological significance of Khovanov homology is still somewhat of a mystery. Adding to the mystery are the numerous relations of Khovanov homology to other areas of mathematics and the existence of various approaches to constructing it.

Algebraic geometry gives a way to define Khovanov homology using derived categories of coherent sheaves [14] and there is a construction based on symplectic geometry that is conjecturally the same as Khovanov link homology [55].

Classical results from homotopy theory combined with the cube category provide a lifting of Khovanov homology to spectra by canonically associating to each link a space whose cohomology is the Khovanov homology of a link [36, 33].

There are several extensions of Khovanov homology, including odd [43, 58], reduced Khovanov homology [26], colored [27, 12], bordered [25, 53] inspired by the bordered Heegaard Floer homology [35], annular [4], and sutured Khovanov homology which can be used to distinguish braids from other tangles [19].

One of the first remarkable applications of Khovanov link homology comes in the form of a numerical concordance invariant, called the s-invariant. Discovery of the s-invariant and its properties enabled Rasmussen to give a purely combinatorial proof of the Milnor conjecture, also known as the Kronheimer-Mrowka theorem [52].

Theorem 70.4.1 ([52]). *The s-invariant provides a lower bound on the slice genus of a knot.*

Theorem 70.4.2 (Milnor conjecture). *The slice genus of the $T(p, q)$ torus knot is equal to $\frac{1}{2}(p - 1)(q - 1)$.*

In the same spirit as with the s-invariant, Khovanov homology was used to provide bounds on the Thurston–Bennequin number [40] and the self-linking number [45], and to improve the slice-Bennequin inequality for some classes of knots [57]. Khovanov homology can also be used to obtain obstructions to certain Dehn fillings [66].

Around the same time the connections between Khovanov and various Floer homology theories such as instanton and monopole homology were discovered by Ozsvath-Szabo, Bloom, and Kronheimer-Mrowka in the form of spectral sequences. The spectral sequence of Kronheimer-Mrowka abutting the instanton Floer homology of the sutured knot complement yields a proof that Khovanov homology is an unknot-detector.

Theorem 70.4.3 ([32]). *Khovanov homology distinguishes the unknot, that is, any knot K with trivial Khovanov homology is the unknot.*

The corresponding question, if a knot has trivial Jones polynomial does it have to be the unknot, remains open. Another evidence of the advantages of categorification is that although the Jones polynomial does not distinguish the trivial link with two components from other 2-component links, Khovanov homology distinguishes n-component links from the trivial link with the same number of components [21].

Results in contact topology and relations between Khovanov homology, $SU(2)$ representations of the knot group, and instanton Floer homology play a crucial role in proving that Khovanov homology also distinguishes the trefoils [7] and the Hopf links [8].

The rich structure of Bar Natan and Lee Khovanov-type homology theories and their torsion and properties of related spectral sequences provide a new lower bound on the unknotting number, as found by Alishahi and Dowlin [2, 3]. These ideas extend to relating properties of spectral sequences to the Gordian distance, alternation number, and Turaev genus [13].

A recently discovered spectral sequence from Khovanov to knot Floer homology provides further insight into relations between these two theories [16]. It was used to prove a conjecture by Rasmussen stating that for any knot K in S^3, the rank of the reduced Khovanov homology of K is greater than or equal to the rank of the reduced knot Floer homology of K.

Extracting the essential structure and information captured by Khovanov-type homology theories is still an open problem, as well as discovering functorial categorifications of other quantum invariants, and relations between them. The general framework of stratified factorization homology, and in particular the version of the Baez-Dolan Tangle Hypothesis do provide a vector-space-valued invariant of a 3-dimensional manifold with a one-dimension submanifold. Although it is known that factorization homology [6] is not directly related to Khovanov homology, it might be used for construction of related link homology theories. The relation between factorization and Khovanov homology could be understood via the relation between Khovanov homology of $(2, n)$-torus knots [46] and Pirashvili or higher Hochschild homology as a stabilization of the Khovanov-type chromatic graph homology.

70.5   Bibliography

[1] R. Sazdanovic V. Summers A. Chandler, A. M. Lowrance. Torsion in thin regions of Khovanov homology. *arXiv*, 1903.05760, 2019.

[2] Akram Alishahi. Unknotting number and Khovanov homology. *Pacific Journal of Mathematics*, 301(1):15–29, 2019.

[3] Akram Alishahi and Nathan Dowlin. The Lee spectral sequence, unknotting number, and the Knight move conjecture. *Topology and Its Applications*, 254:29–38, 2019.

[4] Marta M Asaeda, Józef H Przytycki, and Adam S Sikora. Categorification of the Kauffman bracket skein module of I–bundles over surfaces. *Algebraic & Geometric Topology*, 4(2):1177–1210, 2004.

[5] Martha M Asaeda and Mikhail Khovanov. Notes on link homology. In *Low-Dimensional Topology*, IAS/Park City Mathematics Series, pages 139–196. AMS, 2009.

[6] David Ayala, John Francis, and Hiro Lee Tanaka. Factorization homology of stratified spaces. *Selecta Mathematica*, 23(1):293–362, 2017.

[7] John A Baldwin and Steven Sivek. Khovanov homology detects the trefoils. *arXiv preprint arXiv:1801.07634*, 2018.

[8] John A Baldwin, Steven Sivek, and Yi Xie. Khovanov homology detects the Hopf links. *arXiv preprint arXiv:1810.05040*, 2018.

[9] Dror Bar-Natan. On Khovanov's categorification of the Jones polynomial. *Algebraic & Geometric Topology*, 2(1):337–370, 2002.

[10] Dror Bar-Natan. Khovanovs homology for tangles and cobordisms. *Geometry & Topology*, 9(3):1443–1499, 2005.

[11] Anna Beliakova, Krzysztof K Putyra, and Stephan M Wehrli. Quantum link homology via trace functor i. *Inventiones mathematicae*, 215(2):383–492, 2018.

[12] Anna Beliakova and Stephan Wehrli. Categorification of the colored Jones polynomial and Rasmussen invariant of links. *Canadian Journal of Mathematics*, 60(6):1240–1266, 2008.

[13] C. Ruey Shan Lee A. Lowrance R. Sazdanovic M. Zhang C. Caprau, N. Gonzalez. On khovanov homology and related invariants. *arXiv*, 2002.05247, 2019.

[14] Sabin Cautis, Joel Kamnitzer, et al. Knot homology via derived categories of coherent sheaves, i: The $\mathfrak{sl}(2)$-case. *Duke Mathematical Journal*, 142(3):511–588, 2008.

[15] David Clark, Scott Morrison, and Kevin Walker. Fixing the functoriality of Khovanov homology. *Geometry & Topology*, 13(3):1499–1582, 2009.

[16] Nathan Dowlin. A spectral sequence from Khovanov homology to knot Floer homology. *arXiv preprint arXiv:1811.07848*, 2018.

[17] Nathan M Dunfield, Sergei Gukov, and Jacob Rasmussen. The superpolynomial for knot homologies. *Experimental Mathematics*, 15(2):129–159, 2006.

[18] Eugene Gorsky, Alexei Oblomkov, and Jacob Rasmussen. On stable Khovanov homology of torus knots. *Experimental Mathematics*, 22(3):265–281, 2013.

[19] J Grigsby and Yi Ni. Sutured Khovanov homology distinguishes braids from other tangles. *Mathematical Research Letters*, 21, 05 2013.

[20] Sergei Gukov, Albert Schwarz, and Cumrun Vafa. Khovanov-Rozansky homology and topological strings. *Letters in Mathematical Physics*, 74(1):53–74, 2005.

[21] M. Hedden and Y. Ni. Khovanov module and the detection of unlinks. *Geometry and Topology*, 17(5):3027–3076, 2013.

[22] Magnus Jacobsson. An invariant of link cobordisms from Khovanov homology. *Algebraic & Geometric Topology*, 4(2):1211–1251, 2004.

[23] Louis H Kauffman. Khovanov homology. *Introductory Lectures in Knot Theory*, 46:248–280, 2011.

[24] Mikhail Khovanov. A categorification of the Jones polynomial. *Duke Math. J.*, 101(3):359–426, 2000.

[25] Mikhail Khovanov. A functor-valued invariant of tangles. *Algebraic & Geometric Topology*, 2(2):665–741, 2002.

[26] Mikhail Khovanov. Patterns in knot cohomology, i. *Experimental Mathematics*, 12(3):365–374, 2003.

[27] Mikhail Khovanov. Categorifications of the colored Jones polynomial. *Journal of Knot Theory and Its Ramifications*, 14(01):111–130, 2005.

[28] Mikhail Khovanov. Link homology and Frobenius extensions. *Fundamenta Mathematicae*, 190:179–190, 2006.

[29] Mikhail Khovanov and Lev Rozansky. Matrix factorizations and link homology. *Fund. Math.*, 199(1):1–91, 2008.

[30] Mikhail Khovanov and Lev Rozansky. Matrix factorizations and link homology ii. *Geometry & Topology*, 12(3):1387–1425, 2008.

[31] Joachim Kock. *Frobenius Algebras and 2D Topological Quantum Field Theories*, volume 59. Cambridge University Press, 2004.

[32] Peter B Kronheimer and Tomasz S Mrowka. Khovanov homology is an unknot-detector. *Publications mathématiques de l'IHÉS*, 113(1):97–208, 2011.

[33] Tyler Lawson, Robert Lipshitz, and Sucharit Sarkar. The cube and the Burnside category. In *Categorification in Geometry, Topology, and Physics*, volume 684 of *Contemp. Math.*, pages 63–85. Amer. Math. Soc., Providence, RI, 2017.

[34] Eun Soo Lee. An endomorphism of the Khovanov invariant. *Advances in Mathematics*, 197(2):554–586, 2005.

[35] Robert Lipshitz, Peter Ozsváth, and Dylan Thurston. *Bordered Heegaard Floer homology*, volume 254. American Mathematical Society, 2018.

[36] Robert Lipshitz and Sucharit Sarkar. A Khovanov stable homotopy type. *J. Amer. Math. Soc.*, 27(4):983–1042, 2014.

[37] Cipiran Manolescu and Marco Marengon. The Knight Move conjecture is false. *Proc. Amer. Math. Soc.*, 148(1):435–439, 2020.

[38] Ciprian Manolescu, Peter Ozsváth, and Sucharit Sarkar. A combinatorial description of knot Floer homology. *Annals of Mathematics*, pages 633–660, 2009.

[39] Sujoy Mukherjee, Józef H Przytycki, Marithania Silvero, Xiao Wang, and Seung Yeop Yang. Search for torsion in Khovanov homology. *Experimental Mathematics*, pages 1–10, 2017.

[40] Lenhard Ng. A Legendrian Thurston–Bennequin bound from Khovanov homology. *Algebraic & Geometric Topology*, 5(4):1637–1653, 2005.

[41] Peter Ozsváth and Zoltán Szabó. Holomorphic disks and knot invariants. *Advances in Mathematics*, 186(1):58–116, 2004.

[42] Peter Ozsváth and Zoltán Szabó. On the Heegaard Floer homology of branched double-covers. *Advances in Mathematics*, 194(1):1–33, 2005.

[43] Peter S Ozsváth, Jacob Rasmussen, and Zoltán Szabó. Odd Khovanov homology. *Algebraic & Geometric Topology*, 13(3):1465–1488, 2013.

[44] Milena D Pabiniak, Józef H Przytycki, and Radmila Sazdanović. On the first group of the chromatic cohomology of graphs. *Geometriae Dedicata*, 140(1):19, 2009.

[45] Olga Plamenevskaya. Transverse knots and Khovanov homology. *Math. Res. Lett.*, 13(4):571–586, 2006.

[46] Jozef H Przytycki. When the theories meet: Khovanov homology as Hochschild homology of links. *Quantum Topology*, 1(2):93–109, 2010.

[47] Józef H. Przytycki and Radmila Sazdanović. Torsion in Khovanov homology of semi-adequate links. *Fund. Math.*, 225(1):277–304, 2014.

[48] Hoel Queffelec and Paul Wedrich. Khovanov homology and categorification of skein modules. *arXiv preprint arXiv:1806.03416*, 2018.

[49] Jacob Rasmussen. Floer homology and knot complements. Ph.D. thesis, Harvard University, arXiv preprint math/0306378, 2003.

[50] Jacob Rasmussen et al. Knot polynomials and knot homologies. *Geometry and Topology of Manifolds*, 47:261–280, 2005.

[51] Jacob Rasmussen et al. Khovanov-Rozansky homology of two-bridge knots and links. *Duke Mathematical Journal*, 136(3):551–583, 2007.

[52] Jacob A Rasmussen. Khovanov homology and the slice genus. *Inventiones Mathematicae*, 182:419–447, 2010.

[53] Lawrence Roberts. A type a structure in Khovanov homology. *Algebraic & Geometric Topology*, 16(6):3653–3719, 2016.

[54] Radmila Sazdanovic and Daniel Scofield. Patterns in Khovanov link and chromatic graph homology. *J. Knot Theory Ramifications*, 27(3):1840007, 38, 2018.

[55] Paul Seidel, Ivan Smith, et al. A link invariant from the symplectic geometry of nilpotent slices. *Duke Mathematical Journal*, 134(3):453–514, 2006.

[56] Alexander Shumakovitch. Torsion of Khovanov homology. *Fundamenta Mathematicae*, 225:343–364, 2014.

[57] Alexander N Shumakovitch. Rasmussen invariant, Slice-Bennequin inequality, and sliceness of knots. *J. Knot Theory and Its Ramifications*, 16(10):1403–1412, 2007.

[58] Alexander N Shumakovitch. Patterns in odd Khovanov homology. *Journal of Knot Theory and Its Ramifications*, 20(01):203–222, 2011.

[59] Alexander N. Shumakovitch. Torsion of Khovanov homology. *Fund. Math.*, 225(1):343–364, 2014.

[60] Alexander N Shumakovitch. Torsion in Khovanov homology of homologically thin knots. *arXiv preprint arXiv:1806.05168*, 2018.

[61] Paul Turner. A hitchhiker's guide to Khovanov homology. *arXiv preprint arXiv:1409.6442*, 2014.

[62] Paul Turner. Five lectures on Khovanov homology. *Journal of Knot Theory and Its Ramifications*, 26(03):1741009, 2017.

[63] Paul Turner et al. A spectral sequence for Khovanov homology with an application to $(3, q)$–torus links. *Algebraic & Geometric Topology*, 8(2):869–884, 2008.

[64] Paul R Turner. Calculating Bar-Natan's characteristic two Khovanov homology. *Journal of Knot Theory and Its Ramifications*, 15(10):1335–1356, 2006.

[65] Oleg Viro. Khovanov homology, its definitions and ramifications. *Fund. Math*, 184:317–342, 2004.

[66] Liam Watson. Surgery obstructions from Khovanov homology. *Selecta Mathematica*, 18(2):417–472, 2012.

[67] Edward Witten. Khovanov homology and gauge theory. *Geom. Topol. Monogr*, 18:291–308, 2012.

Chapter 71

A Short Survey on Knot Floer Homology

András I. Stipsicz, Rényi Institute of Mathematics

71.1 Introduction

Knot Floer homology is part of the Heegaard Floer package discovered by Ozsváth and Szabó around 2000; this package associates algebraic invariants to low dimensional objects (such as three- and four-manifolds with or without boundary, knots in three-manifolds, contact structures on three-manifolds, Legendrian and transverse knots in contact three-manifolds, etc.). In particular, *knot Floer homology* associates invariants to knots and links in three-manifolds; in this chapter we concentrate mostly on knots in the standard three-sphere S^3. There are a number of surveys accessible about this theory in the literature: in Hom's survey [9] the emphasis is on the numerical invariants one can derive from knot Floer homology, in the paper of Juhász [12] the focus is on three-manifold invariants, and knots enter the picture through Sutured Heegaard Floer homology. Manolescu's overview [18] provides a slightly different perspective of the same construction, while the most recent survey of Ozsváth and Szabó [31] stresses their novel construction of computable invariants: bordered knot Floer homology. In this chapter we will sketch the definition of certain concordance invariants, and show some (rather straightforward) consequences and applications.

We start our discussion by recalling the definition of the Alexander polynomial of a smooth knot $K \subset S^3$ using Kauffman states. To this end, consider a diagram D of K, that is, a generic projection of K, together with the extra information at crossings, indicating the lower strand with an interrupted segment. By distinguishing an arc in D with a marking p (distinct from all the crossings) we get a *marked diagram* (D, p). The set of crossings in the diagram will be denoted by C, while the set of domains (i.e., the set of components of the complement of the projection in the plane or in S^2) is $\mathcal{D}$. The domains not containing p on their boundary comprise the set $\mathcal{D}_0 \subset \mathcal{D}$, i.e. we get $\mathcal{D}_0$ from $\mathcal{D}$ by excluding the two domains next to p. It is easy to see that the cardinality of C and of $\mathcal{D}_0$ are equal. A *Kauffman state* κ of the marked diagram (D, p) is a bijection $\kappa \colon C \to \mathcal{D}_0$ such that the domain $\kappa(c)$ for a crossing $c \in C$ contains c in its closure, i.e., $\kappa(c)$ is one of the four quadrants meeting at c. Let $\mathcal{K}(D, p)$ denote the set of all Kauffman states of the marked diagram (D, p).

We associate two values $A(\kappa)$ and $M(\kappa)$ to a Kauffman state κ as follows. Orient the diagram, and at each crossing (depending on its sign) copy the values dictated by the local pictures of Figure 71.1 to the quadrants meeting at the crossing. Then sum the values falling in the quadrants specified by the given Kauffman state κ. It is obvious that $M(\kappa)$ is an integer; it can be shown that so is $A(\kappa)$. (Here M stands for *Maslov*, while A for *Alexander grading*.)

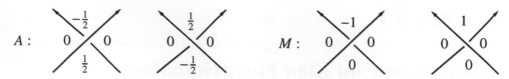

FIGURE 71.1: Local coefficients A and M at a crossing c.

The *(symmetrized) Alexander polynomial* $\Delta_K(t)$ of the knot K is given by

$$\Delta_K(t) = \sum_{\kappa \in \mathcal{K}(D,p)} (-1)^{M(\kappa)} t^{A(\kappa)} \in \mathbb{Z}[t, t^{-1}].$$

As the notation suggests, the above sum depends only on the knot, and is independent of the chosen diagram D (or marking p) of the given knot K.

Remark 71.1.1. For example, for the diagram of the unknot U with no crossing we have $C = \mathcal{D}_0 = \emptyset$, hence there is a unique Kauffman state (the bijection between two empty sets), and $\Delta_U(t) = 1$. The definition verbatim extends to links; if the projection D of a link L is not connected (i.e., L is a split link), then the number $|\mathcal{D}_0|$ of allowed domains and the number $|C|$ of crossings differ, hence $\mathcal{K}(D, p) = \emptyset$, resulting in $\Delta_L(t) = 0$.

Some basic (relevant) properties of Δ_K are as follows:

- The Alexander polynomial Δ_K of the above definition is symmetric (i.e., $\Delta_K(t) = \Delta_K(t^{-1})$), hence can be written as $\Delta_K(t) = a_0 + \sum_{i=1}^{d} a_i(t^i + t^{-i})$ with $a_d \neq 0$; the integer d is the *degree* of Δ_K.

- For the Seifert genus $g(K)$ of K we have $d \leq g(K)$.

- Furthermore, if K is fibered, then $a_d = \pm 1$.

If K is an alternating knot, i.e., admits a diagram with crossings alternating as one traverses through the knot (for an intrinsic characterization of alternating knots see [5, 10]), then the above statements admit sharpenings:

Theorem 71.1.2. *Suppose that K is an alternating knot with Alexander polynomial $\Delta_K(t) = a_0 + \sum_{i=1}^{d} a_i(t^i + t^{-i})$ (and $a_d \neq 0$). Then $d = g(K)$ and K is fibered if and only if $a_d = \pm 1$.* □

Knot Floer homology (in its simplest form) provides an invariant which extends Theorem 71.1.2 to all knots, after substituting Δ_K with the generalization provided by this theory. Indeed, the first idea would be to consider the generating function

$$G_{(D,p)}(s, t) = \sum_{\kappa \in \mathcal{K}(D,p)} s^{M(\kappa)} t^{A(\kappa)} \in \mathbb{Z}[s^{\pm 1}, t^{\pm 1}]$$

of all Kauffman states. It is not hard to see, that this expression is not a knot-invariant; for example, by considering $G_{(D,p)}(1, 1)$ (the number of Kauffman states) for two different (appropriately chosen) diagrams of the unknot, we get different values. Nevertheless, by considering the vector space $C = C(D, p)$ formally generated by the Kauffman states (for simplicity over the field $\mathbb{F}$ of two elements), and equipping this finite-dimensional vector space with the two gradings M and A, the function $G_{(D,p)}(s, t)$ above is simply the graded dimension of this bigraded vector space $C = \oplus_{A,M} C_{A,M}$. In the definition of Δ_K, we considered $G_{(D,p)}(-1, t)$, providing some cancellations in $G_{(D,p)}$ so that the result became a knot invariant. Such cancellation on the vector space C can be introduced by a *boundary operator*, turning C into a chain complex.

Theorem 71.1.3 (Ozsváth-Szabó [26], Rasmussen [34]). *There is a linear map $\widehat{\partial}: C(D, p) \to C(D, p)$ which is of bigrading $(-1, 0)$ (that is, drops M by one and keeps the A-grading), satisfies $(\widehat{\partial})^2 = 0$, and the homology $\widehat{HFK}(K) = H_{*,*}(C(D, p), \widehat{\partial}) = \text{Ker } \widehat{\partial}/\text{Im } \widehat{\partial}$ (as a bigraded vector space over $\mathbb{F}$) is a knot invariant.* □

Consider the Poincaré polynomial of $\widehat{HFK}(K) = \oplus_{A,M} \widehat{HFK}_M(K, A)$ defined as

$$P_K(s, t) = \sum_{A, M \in \mathbb{Z}} \dim_{\mathbb{F}} \widehat{HFK}_M(K, A) \cdot s^M t^A.$$

Simple linear algebra tells us that $\Delta_K(t) = P_K(-1, t)$, while for the polynomial $\mathcal{E}_K(t) = P_K(1, t)$ we have:

Theorem 71.1.4 ([4, 23, 26]). • *The polynomial $\mathcal{E}_K(t) \in \mathbb{Z}[t^{\pm 1}]$ defined above is an invariant of the knot K; it is symmetric, and hence can be written as*

$$\mathcal{E}_K(t) = b_0 + \sum_{i=1}^{e} b_i(t^i + t^{-i}),$$

with $b_e \neq 0$.

• *The Seifert genus $g(K)$ of the knot K is equal to e, the degree of $\mathcal{E}$.*

• *The knot K is fibered if and only if $b_e = \pm 1$.* □

The key step in the construction of $\widehat{HFK}(K)$ is the definition of $\widehat{\partial}$, which we will sketch in Section 71.3. As it will be clear from the geometric picture, the definition allows many variations, leading to a plethora of invariants of knots, of their concordance classes, and of links.

The paper is organized as follows: in Section 71.2 we discuss Heegaard diagrams (and doubly pointed Heegaard diagrams) representing knots and three-manifolds. Section 71.3 is devoted to the introduction of knot Floer homologies, listing several versions of the theory. In Section 71.4 numerical invariants derived from knot Floer homology are presented. We will concentrate on the Upsilon-invariant Υ_K of knots. In Section 71.5 we will use the branched cover construction (in particular, double branched covers) and show a method for further defining (module-valued) invariants of knots. Using these invariants, some simple applications of knot Floer homology are given in Section 71.6. We conclude our survey with a brief outlook to further aspects of the theory.

Acknowledgments: The author was partially supported by *Élvonal (Frontier)* grant KKP 126683 of the NKFIH, Hungary.

71.2   Heegaard diagrams

The definition of the map $\widehat{\partial}$ (and its generalizations) will rest on a purely topological and a symplectic geometric construction. We start with topology, and return to symplectic geometry in the next section.

Suppose that Σ is a closed, oriented surface of genus g. Let $\alpha = \{\alpha_1, \ldots, \alpha_g\}$ be a g-tuple of (smoothly embedded) simple closed curves in Σ, which are disjoint from each other, and their complement $\Sigma \setminus \cup_{i=1}^{g} \alpha_i$ is a connected surface (which is equivalent to being planar). Attaching three-dimensional 2-handles to $\Sigma \times [0, 1]$ along the $\alpha_i \subset \Sigma \times \{0\}$ (and then gluing in a 3-disk), we see that (Σ, α) determines a handlebody. If we take $\beta = \{\beta_1, \ldots, \beta_g\}$ with the same properties, then the triple (Σ, α, β) determines a closed, oriented three-manifold by gluing the two handlebodies together along their common boundary Σ (with the convention that the handlebody determined by Σ and β is taken with its reversed orientation). The resulting three-manifold $Y = Y_{(\Sigma, \alpha, \beta)}$ also comes with a *Heegaard decomoposition*; the triple (Σ, α, β) is called a *Heegaard diagram* of Y. In the following we will always assume that the α- and the β-curves intersect each other transversally.

If we fix two points $w, z \in \Sigma \setminus (\cup_{i=1}^{g} \alpha_i \cup \cup_{j=1}^{g} \beta_j)$ in the complement of these curves, then the five-tuple $(\Sigma, \alpha, \beta, w, z)$ defines an oriented knot in the three-manifold $Y = Y_{(\Sigma, \alpha, \beta)}$ as follows: connect w to z in the complement of the α-curves (which is connected by assumption) and push the interior of the resulting arc into the handlebody defined by the α-curves to get a properly embedded arc in this handlebody. Then connect z to w in the complement of the β-curves (and push the interior of this arc to the β-handlebody); by gluing the two arcs together we get a knot in $Y = Y_{(\Sigma, \alpha, \beta)}$. The *doubly pointed Heegaard diagram* $(\Sigma, \alpha, \beta, w, z)$ then determines (Y, K). (Note that the arcs are uniquely determined up to isotopy by their defining properties.)

It is not hard to see that any pair (Y, K) (with Y a closed, oriented three-manifold and K an oriented knot in it) can be given by this procedure, and indeed the presentation by such a five-tuple is unique

- up to isotopies of the curves in the complement of the points w, z,

- up to handleslides (again in the complement of the points w and z)

- and up to stabilization (i.e., connected sum with the diagram (T^2, α, β) where T^2 is the torus and α intersects β in a unique point) and its inverse operation, destabilization.

Below we give two examples of such diagrams.

Example 71.2.1. Suppose that $(D, p) \subset \mathbb{R}^2$ is a marked diagram of the (oriented) knot $K \subset S^3$. Let νD denote the tubular neighbourhood of the projection (so here we omit the under/overcrossing information) in $\mathbb{R}^3$, and let Σ be the boundary of νD, with its orientation reversed. If the projection determined d domains (connected components of the complement of the projection) then Σ is of genus $d - 1$. The intersection of Σ with the plane $\mathbb{R}^2$ containing the projection is therefore a disjoint collection of d circles; choose those $d - 1$ of them which bound compact domains in the plane. Clearly, by attaching three-dimensional 2-handles along these circles (and a finishing three-disk D^3) we get the handlebody which is the complement of νD (viewed in $S^3 = \mathbb{R}^3 \cup \{\infty\}$). The projection has $d - 2$ double points (these are the crossings of the diagram), where we add the β-curves as dictated by Figure 71.2 — notice that these choices encode the under/over nature of the

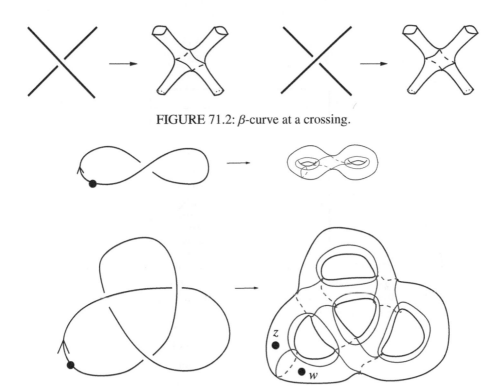

FIGURE 71.2: β-curve at a crossing.

FIGURE 71.3: Doubly pointed Heegaard diagrams from a marked diagram of the unknot and of the trefoil knot. The red curves are the α-curves and the blue curves are the β-curves.

crossing. The final β-curve β_{d-1} is just a meridian of the projection at the marked point $p \in D$. We conclude the construction by putting w and z to the two sides of β_{d-1}. In this last step we decide which side to put w using the orientation of the knot: crossing β_{d-1} from w to z should agree with the orientation of K. It is rather simple to see that by attaching three-dimensional 2-handles along the β_i from the inside (together with a final D^3), we fill νD, and by connecting w to z and z to w in the complements, we recover the original knot K. See Figure 71.3 for the Heegaard diagram given by the above procedure for a diagram of the unknot and for the standard diagram of the trefoil knot.

Example 71.2.2. In certain cases we can find a 5-tuple $(\Sigma, \alpha, \beta, w, z)$ for K where Σ is of genus 1, and hence there is a single α- and a single β-curve. Knots having such doubly pointed Heegaard diagrams are called $(1, 1)$-knots. Not every knot admits such a simple diagram, but surprisingly many do: for instance, each torus knot does. Indeed, the torus knots $T_{2,2n+1}$ and $T_{n,n+1}$ can be given by the diagrams of Figures 71.4 and 71.5. Indeed, a diagram on T^2 with a single α- and a single β-curve represents a knot in S^3 if β can be isotoped (through w and z if needed) to have a single intersection with α, and in this case we have a diagram of a torus knot if we can connect the two points w and z on the torus (from w to z in $T^2 \setminus \alpha$ and from z to w in $T^2 \setminus \beta$) so that the result is embedded in the torus. $(1, 1)$-diagrams can be also used to present knots in lens spaces. For extended discussions of $(1, 1)$-knots see [33, 35].

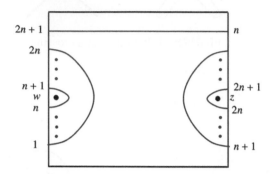

FIGURE 71.4: A $(1, 1)$-diagram of the torus knot $T_{2,2n+1}$. Identifying the right and left, and the top and bottom segments, we get a torus with a single (red) α-curve and a single (blue) β-curve and two points w and z. In the identification of the two sides we apply a shear so that points with equal decorations $\in \{1, \ldots, 2n + 1\}$ match up.

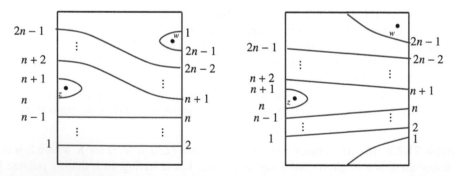

FIGURE 71.5: A $(1, 1)$-diagram for the knot torus knot $T_{n,n+1}$ with $n \geq 3$ in S^3. In general, we apply a shearing in identifying the two sides of the square; on the left diagram this shearing is indicated by numbering the pairs of points which match with each other. The right diagram gives the same picture, presented without the shearing.

71.3 Knot Floer homology

Knot Floer homology is defined as Lagrangian Floer homology of a symplectic manifold with two Lagrangians (and certain further submanifolds) associated to doubly pointed Heegaard diagrams.

Indeed, let $\mathcal{H} = (\Sigma, \alpha, \beta, w, z)$ be a given doubly pointed Heegaard diagram for (Y, K). Fix a complex structure j on Σ and consider the Kähler manifold

$$\mathrm{Sym}^g(\Sigma) = \times_{i=1}^g \Sigma / S_g,$$

the g-fold symmetric power of Σ, where S_g is the symmetric group on g letters, and it acts on the Cartesian product $\times_{i=1}^g \Sigma$ by permuting the coordinates. The fundamental theorem of algebra guarantees that this symmetric power is a smooth manifold (of complex dimension g). Let $\mathbb{T}_\alpha = \times_{i=1}^g \alpha_i$ and $\mathbb{T}_\beta = \times_{j=1}^g \beta_j$ be the two tori specified by the α- and β-curves. Notice that to define the product we need to choose an order on the α-curves (and on the β-curves), but when viewed $\mathbb{T}_\alpha$ (and $\mathbb{T}_\beta$) in $\mathrm{Sym}^g(\Sigma)$, the order will be irrelevant. The transverse intersection among the α- and β-curves implies that $\mathbb{T}_\alpha$ intersects $\mathbb{T}_\beta$ transversally in finitely many points.

The two points w and z define two submanifolds (indeed, divisors) of $\mathrm{Sym}^g(\Sigma)$: V_w consists of those g-tuples in $\mathrm{Sym}^g(\Sigma)$ which contain w (i.e., $V_w = \{w\} \times \mathrm{Sym}^{g-1}(\Sigma)$), and similarly $V_z = \{z\} \times \mathrm{Sym}^{g-1}(\Sigma)$. When defining the chain complex associated to $(\mathcal{H}, j)$, we will apply variations of the *Lagrangian Floer homology construction*. The various versions will depend on our way of using the divisors V_w and V_z.

Let us consider the vector space $\widehat{\mathrm{CFK}}(\mathcal{H})$ (over the field $\mathbb{F}$ of two elements) generated by the intersection points $\mathbb{T}_\alpha \cap \mathbb{T}_\beta$. It is not hard to see that if $\mathcal{H}$ is derived from a marked diagram (D, p) of a knot $K \subset S^3$ as in Example 71.2.1, then the points of $\mathbb{T}_\alpha \cap \mathbb{T}_\beta$ correspond exactly to the Kauffman states of (D, p), hence the above vector space $\widehat{\mathrm{CFK}}(\mathcal{H})$ is a natural generalization of the vector space $C(D, p)$ considered in Section 71.1. The linear map $\widehat{\partial} \colon \widehat{\mathrm{CFK}}(\mathcal{H}) \to \widehat{\mathrm{CFK}}(\mathcal{H})$ is defined by counting certain holomorphic maps.

For two intersection points $\mathbf{x}, \mathbf{y} \in \mathbb{T}_\alpha \cap \mathbb{T}_\beta$ consider first the smooth maps u from the unit disk $\mathbb{D} \subset \mathbb{C}$ to $\mathrm{Sym}^g(\Sigma)$ such that

- the point $i \in \mathbb{D}$ maps to $\mathbf{y}$, and the point $-i \in \mathbb{D}$ maps to $\mathbf{x}$;

- the right semicircle of the boundary of $\mathbb{D}$ maps to $\mathbb{T}_\alpha$ and the left semicircle of the boundary of $\mathbb{D}$ maps to $\mathbb{T}_\beta$.

These smooth maps fall into homotopy classes; the set of these classes is usually denoted by $\pi_2(\mathbf{x}, \mathbf{y})$. The *moduli space* of holomorphic maps $u \colon \mathbb{D} \to \mathrm{Sym}^g(\Sigma)$ with the above properties representing a fixed homotopy class $\varphi \in \pi_2(\mathbf{x}, \mathbf{y})$ will be denoted by $\mathcal{M}(\varphi)$. One such holomorphic map gives rise to an entire $\mathbb{R}$-worth of further such maps by precomposing it with a holomorphic automorphism of $\mathbb{D}$ fixing both i and $-i$. (The precomposition provides different maps once the starting map is non-constant, for example if $\mathbf{x} \neq \mathbf{y}$.)

Now the matrix element $\langle \widehat{\partial}(\mathbf{x}), \mathbf{y} \rangle$ of the linear map $\widehat{\partial}$ will be given by the (mod 2) count of elements in the quotient $\mathcal{M}(\varphi)/\mathbb{R}$ for certain homotopy classes φ (which we will specify later, see

Equation (71.1)). In order to count the points in this space, we need that the space of the above maps is a compact, 0-dimensional manifold. To ensure this we need to have

- a dimension formula for $\mathcal{M}(\varphi)$; this is provided by the Maslov index $\mu(\varphi)$ of φ. We need to take those homotopy classes for which $\mu(\varphi) = 1$, so $\mathcal{M}(\varphi)/\mathbb{R}$ is 0-dimensional.

- a method to show that the moduli space is a smooth manifold. For this step a version of the Sard-Smale theorem is used, together with allowing to perturb the complex structure $\mathrm{Sym}^g(j)$ on $\mathrm{Sym}^g(\Sigma)$ defined by j on Σ to appropriate almost complex structures, and

- the existence of a symplectic form ω on $\mathrm{Sym}^g(\Sigma)$ for which $\mathbb{T}_\alpha$ and $\mathbb{T}_\beta$ are Lagrangian, and $\mathrm{Sym}^g(j)$ is tamed by ω (and so are the almost complex structures in the above perturbation). Then an argument based on a version of Gromov's famous compactness theorem provides the desired compactness of $\mathcal{M}(\varphi)/\mathbb{R}$. (The existence of such ω is provided by a result of Perutz [32].)

Remark 71.3.1. There are further technical difficulties; for example, in the analytic arguments one needs to work with Banach spaces, hence the space of C^∞-maps needs to be completed using some Sobolev norm. Regarding perturbations, we need slightly more than outlined above: we need to allow paths of almost complex structures, and the holomorphic condition should be adapted to this more general setting. Also, in order to have compactness either we need to assume $b_1(Y) = 0$, or we need to restrict our attention to special (***admissible***) Heegaard diagrams. These rather technical fine points of the theory will not be detailed here.

So far the divisors V_w and V_z played no role. For a homotopy class $\varphi \in \pi_2(\mathbf{x}, \mathbf{y})$ let us define $n_w(\varphi) \in \mathbb{Z}$ (and similarly $n_z(\varphi) \in \mathbb{Z}$) as the intersection number of the image of the unit disk under a map u representing φ with V_w (and V_z, respectively). Since holomorphic objects intersect nonnegatively, once $\mathcal{M}(\varphi) \neq \emptyset$, we have $n_w(\varphi) \geq 0$, and the condition $n_w(\varphi) = 0$ means that the holomorphic map u (representing φ) avoids V_w.

Let $n(\varphi)$ denote the mod 2 count of points in the moduli space $\mathcal{M}(\varphi)/\mathbb{R}$ (with $\varphi \in \pi_2(\mathbf{x}, \mathbf{y})$) of holomorphic disks encountered above in the case when the Maslov index of φ is one (so $\dim \mathcal{M}(\varphi)/\mathbb{R} = 0$). Then define

$$\widehat{\partial}(\mathbf{x}) = \sum_{\mathbf{y} \in \mathbb{T}_\alpha \cap \mathbb{T}_\beta} \sum_{\{\varphi \in \pi_2(\mathbf{x}, \mathbf{y}) \mid \mu(\varphi) = 1, n_w(\varphi) = n_z(\varphi) = 0\}} n(\varphi) \cdot \mathbf{y}. \tag{71.1}$$

(If $b_1(Y) = 0$, then there is a unique $\varphi \in \pi_2(\mathbf{x}, \mathbf{y})$ with $\mu(\varphi) = 1$, but when $b_1(Y) > 0$, the set of such homotopy classes is not finite; to have a meaningful sum, we need to appeal to admissible diagrams to ensure that only finitely many homotopy classes have nonempty moduli space.)

The Alexander and Maslov gradings extend naturally to this context provided that $b_1(Y) = 0$. These gradings will be integer valued once Y is an integral homology sphere (so $H_1(Y; \mathbb{Z}) = 0$); in general we will have $\mathbb{Q}$-bigraded vector spaces.

Remark 71.3.2. The definition of the gradings is delicate; there are however simple formulae to compute the ***relative*** gradings: for a homotopy class $\varphi \in \pi_2(\mathbf{x}, \mathbf{y})$

$$M(\mathbf{x}) - M(\mathbf{y}) = \mu(\varphi) - 2n_w(\varphi), \qquad A(\mathbf{x}) - A(\mathbf{y}) = n_z(\varphi) - n_w(\varphi).$$

In the case when $b_1(Y) = 0$, these differences will be independent of the choice of φ in $\pi_2(\mathbf{x}, \mathbf{y})$, and the resulting relative gradings lift to absolute $\mathbb{Q}$-gradings.

Theorem 71.3.3 (Ozsváth-Szabó). *The endomorphism $\widehat{\partial}$ is a boundary operator, that is, $(\widehat{\partial})^2 = 0$. The homology group $\widehat{HFK}(K)$ of the chain complex $(\widehat{CFK}(\mathcal{H}), \widehat{\partial})$ (as a bigraded vector space) is a knot invariant, i.e., independent of the chosen doubly pointed Heegaard diagram, and the almost complex structure.* □

Remark 71.3.4. The proof of $(\widehat{\partial})^2 = 0$ again relies on the Gromov compactness theorem: now we need to consider moduli spaces with $\dim \mathcal{M}(\phi) = 2$, so after taking the quotient we have $\dim(\mathcal{M}(\phi)/\mathbb{R}) = 1$. The ends of these moduli spaces will correspond to contributions of $(\widehat{\partial})^2$; in the verification of this statment we need a further technical tool, called the **gluing** of holomorphic maps. The gluing theorems, together with Gromov's compactness then show $(\widehat{\partial})^2 = 0$. The invariance property mainly follows from general properties of Lagrangian Floer homology, together with invariance under stabilization, which again relies on gluing results.

The theory admits simple variants, depending on the change of the conditions for the holomorphic disks. Obviously, we can only count points in moduli spaces of dimension zero (hence we will keep the assumption $\mu(\varphi) = 1$), but the conditions about n_w and n_z can be relaxed. Indeed, let $\text{CFK}^-(\mathcal{H})$ denote the free $\mathbb{F}[U]$-module generated by the points of $\mathbb{T}_\alpha \cap \mathbb{T}_\beta$ and define

$$\partial^- : \text{CFK}^-(\mathcal{H}) \to \text{CFK}^-(\mathcal{H})$$

by the formula

$$\partial^-(\mathbf{x}) = \sum_{\mathbf{y} \in \mathbb{T}_\alpha \cap \mathbb{T}_\beta} \sum_{\{\varphi \in \pi_2(\mathbf{x}, \mathbf{y}) \mid \mu(\varphi) = 1, n_z(\varphi) = 0\}} n(\varphi) \cdot U^{n_w(\varphi)} \mathbf{y}$$

for $\mathbf{x} \in \mathbb{T}_\alpha \cap \mathbb{T}_\beta$, and extend it to the free module $\text{CFK}^-(\mathcal{H})$ defined above $\mathbb{F}[U]$-linearly. In other words, now we allow our disks to cross the divisor V_w, but introduce a formal parameter U which records the number of these intersections. The homology $\text{HFK}^-(K)$ of the resulting bigraded chain complex $(\text{CFK}^-(\mathcal{H}), \partial^-)$ is again a knot invariant (as a bigraded module over the ring $\mathbb{F}[U]$ of polynomials with coefficients in $\mathbb{F}$).

Remark 71.3.5. In this construction the Alexander grading of U is chosen to be -1, while its Maslov grading is -2, hence the chain complexes and homologies are all modules over this slightly oddly bigraded ring $\mathbb{F}[U]$. In addition, we remark that for $\mathbf{y}$ in $\partial^-(\mathbf{x})$ to have polynomial, rather than power series coefficient, we need a stronger version of admissibility for the Heegaard diagram $\mathcal{H}$. A version of the theory over the ring $\mathbb{F}[[U]]$ of power series can be defined without this further admissibility condition — the admissibility of Remark 71.3.1, however, is still necessary in general.

Finally there is a version which considers all disks. Define $\text{CFK}^\infty(\mathcal{H})$ as the free module over the ring $\mathbb{F}[U, U^{-1}]$ of Laurent polynomails over the field $\mathbb{F}$ of two elements freely generated by the intersection points (the elements of $\mathbb{T}_\alpha \cap \mathbb{T}_\beta$), and define

$$\partial : \text{CFK}^\infty(\mathcal{H}) \to \text{CFK}^\infty(\mathcal{H})$$

by the formula

$$\partial(\mathbf{x}) = \sum_{\mathbf{y} \in \mathbb{T}_\alpha \cap \mathbb{T}_\beta} \sum_{\{\varphi \in \pi_2(\mathbf{x},\mathbf{y}) | \mu(\varphi)=1\}} n(\varphi) \cdot U^{n_w}(\varphi)\mathbf{y}.$$

In this case the point z comes into the picture by taking the filtration (the *Alexander* filtration) defined by the Alexander grading. Furthermore, the module comes with a prescribed basis, which can be coded by another filtration (the *algebraic* filtration) counting (the negative of) the U-power of a monomial.

This construction therefore provides a bifiltered, graded chain complex $\mathrm{CFK}^\infty(\mathcal{H})$ over the ring $\mathbb{F}[U, U^{-1}]$.

Theorem 71.3.6 (Ozsváth-Szabó). *The bifiltered, graded chain homotopy equivalence class of the (bifiltered, graded) chain complex $\mathrm{CFK}^\infty(\mathcal{H})$ is an isotopy invariant of the knot $K \subset Y$.* □

It is easy to see how to derive $(\widehat{\mathrm{CFK}}(\mathcal{H}), \widehat{\partial})$ from $(\mathrm{CFK}^-(\mathcal{H}), \partial^-)$: take $U = 0$. In a similar manner, the bifiltered chain complex $\mathrm{CFK}^\infty(\mathcal{H})$ determines $\mathrm{CFK}^-(\mathcal{H})$ by taking the $\mathbb{F}[U]$-submodule generated by elements of nonpositive algebraic filtration and taking the Alexander grading preserving component of ∂ (i.e. considering the associated graded object).

Indeed, by dropping one of the basepoints, say z, the Heegaard diagram $(\Sigma, \alpha, \beta, w)$ defines only the three-manifold Y, and the constructions above provide the Heegaard Floer invariants $\widehat{\mathrm{HF}}(Y)$, $\mathrm{HF}^-(Y)$ and $\mathrm{HF}^\infty(Y)$ of the closed, oriented three-manifold Y. (By dropping z we lose the Alexander grading of the intersection points, hence $\widehat{\mathrm{HF}}(Y)$ and $\mathrm{HF}^-(Y)$ are graded, while $\mathrm{HF}^\infty(Y)$ is graded and filtered.)

The pictorial presentation of $\mathrm{CFK}^\infty(\mathcal{H})$ goes as follows: Consider the basis of the module given by the intersection points $\mathbb{T}_\alpha \cap \mathbb{T}_\beta$, and represent each generator by a dot in $\mathbb{R}^2$, placed on the y-axis, with y-coordinate being equal to the Alexander grading A of the generator. The set

$$\{U^k\mathbf{x} \mid k \in \mathbb{Z}, \mathbf{x} \in \mathbb{T}_\alpha \cap \mathbb{T}_\beta\}$$

provides a basis of $\mathrm{CFK}^\infty(\mathcal{H})$ over $\mathbb{F}$, and we symbolize them by placing a dot for $U^k\mathbf{x}$ at the point $(-k, A(\mathbf{x}) - k)$. We complete this picture with symbolizing ∂ with arrows: if $U^n\mathbf{y}$ is a nontrivial component of $\partial(U^k\mathbf{x})$, then we draw an arrow pointing from the dot (at $(-k, A(\mathbf{x}) - k)$) representing $U^k\mathbf{x}$ to the dot (at $(-n, A(\mathbf{y}) - n)$) representing $U^n\mathbf{y}$. It is worth pointing out that all the arrows point to the left (including vertical) and down (including horizontal), expressing the facts that the negative of the U-power, as well as the Alexander grading induce filtrations respected by the boundary map ∂. Notice that the resulting picture will be invariant under shifting by vectors of the form (k, k) $(k \in \mathbb{Z})$.

In this picture $\widehat{\mathrm{CFK}}(\mathcal{H})$ can be seen by taking only those dots which are on the y-axis and considering only those arrows which point from a position in $\mathbb{R}^2$ to itself, and $\mathrm{CFK}^-(\mathcal{H})$ is given by taking dots on the left half-plane $\{x \leq 0\}$ and consider only horizontal arrows.

The determination of the chain complexes above are typically quite involved. As it is rather transparent, in general the module can be found easily (for example, for a knot $K \subset S^3$ and the Heegaard diagram derived from a knot diagram by identifying Kauffman states), but the boundary operator

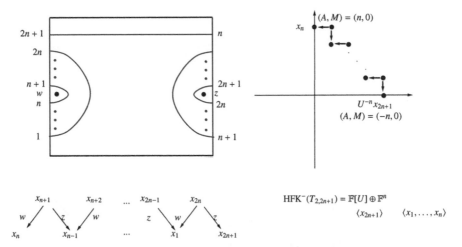

FIGURE 71.6: Computations in $(1, 1)$-diagrams. The doubly pointed Heegaard diagram on the left represents the torus knot $T_{2,2n+1}$; on the right the complex $\mathrm{CFK}^\infty(T_{2,2n+1})$ is shown (to get all points and arrows, one needs to apply translations of this picture by all vectors of the form (k, k) with $k \in \mathbb{Z}$). We also show the maps in the lower portion of the figure; the intersection point i in the diagram gives rise to the generator x_i, and the arrows representing the boundary maps come with the decoration indicating which divisor from $V_w = \{w\}$ and $V_z = \{z\}$ is intersected by the holomorphic disk. A simple computation then determines $\mathrm{HFK}^-(T_{2,2n+1})$ which is also given on the figure (with generators).

involves rather delicate complex analytic considerations. Nevertheless, in some cases the invariants can be explicitly calculated. For example, if $K \subset S^3$ is an alternating knot, then the invariants defined above are determined (through rather simple formulae) by the Alexander polynomial and the signature of K.

As another example, for $(1, 1)$-knots the invariants can be easily determined. Notice that in this case $g = 1$, so $\mathrm{Sym}^g(\Sigma)$ is simply the torus $\mathbb{T}$, and $\mathbb{T}_\alpha \cap \mathbb{T}_\beta$ is simply the intersection $\alpha \cap \beta$. In order to determine the boundary map, we need to examine holomorphic maps $\mathbb{D} \to \mathbb{T}$. These maps lift to $\mathbb{D} \to \mathbb{C}$, and therefore we need to understand bigons in $\mathbb{C}$ (equipped with the lifts of the α- and the β-curve) with boundary partly on a lift of α and partly on a lift of β. The Riemann mapping theorem then determines the relevant moduli spaces (consisting of single points), providing the full invariants. As examples, for the torus knot $T_{2,2n+1}$ see Figure 71.6, and in Figure 71.7 we show how to determine $\mathrm{CFK}^\infty(T_{3,4})$.

71.4 Knot invariants

The homologies (or rather the chain homotopy equivalence classes of the chain complexes) are isotopy invariants of the knots. For concordance invariants, we need to introduce another equivalence

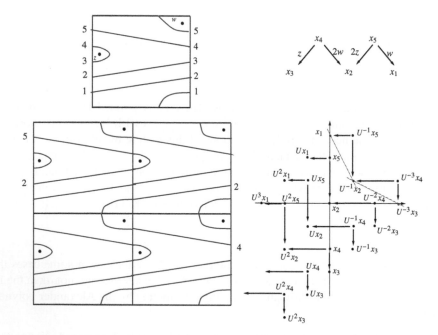

FIGURE 71.7: Determining the chain complex of the torus knot $T_{3,4}$. The genus-1 Heegaard diagram on the top is a special case of the diagrams of Figure 71.5 for $n = 3$. The boundary map (on the top right) can be computed by considering the four fundamental domains in the universal cover depicted on the lower left, and identifying the relevant bigons. (As always, w and z over an arrow indicates whether the corresponding disk intersects V_w or V_z, the multiple in front of the basepoint indicates the intersection multiplicity.) The chain complex $\mathrm{CFK}^{\infty}(T_{3,4})$ is given on the lower right; it also provides the Υ-function of $T_{3,4}$.

relation. Suppose that C_1, C_2 are two graded, bifiltered chain complexes over the ring $\mathbb{F}[U, U^{-1}]$. Suppose furthermore that $H_*(C_i) = \mathbb{F}[U, U^{-1}]$.

Definition 71.4.1. • *The chain complexes C_1, C_2 are **stably equivalent** if there are graded, bifiltered chain complexes A_1, A_2, which are acyclic (that is, $H_*(A_i) = 0$) and $C_1 \oplus A_1$ is (graded, bifiltered) chain homotopy equivalent to $C_2 \oplus A_2$.*

• *C_1 and C_2 are **locally equivalent** if there are (grading and bifiltration respecting) chain maps $f : C_1 \to C_2$ and $g : C_2 \to C_1$ such that the induced maps on the homologies are isomorphisms.*

It can be shown that the two equivalence relations defined above are actually equivalent. In order to see that the above concepts are relevant in our context, we need to show that

Proposition 71.4.2. *Suppose that $K \subset S^3$ is a given knot. Then the chain complex $C = \mathrm{CFK}^\infty(K)$ satisfies $H_*(C) = \mathbb{F}[U, U^{-1}]$.* □

Remark 71.4.3. Since in this case z plays no role in computing $H_*(\mathrm{CFK}^\infty(K))$, the homology is obviously independent of the knot. Computing this homology module for the unknot, the proof of the above proposition follows at once.

Theorem 71.4.4 (Hom [9]). *Suppose that K_1, K_2 are concordant knots. Then the (graded, bifiltered) chain complexes $\mathrm{CFK}^\infty(K_1)$ and $\mathrm{CFK}^\infty(K_2)$ are stably equivalent.* □

The property of CFK^∞ formulated in Proposition 71.4.2 is true for knots in rational homology spheres only after a slight refinement: if Y is a rational homology sphere (that is, $b_1(Y) = 0$) and $K \subset Y$ is a knot, then the chain complex $\mathrm{CFK}^\infty(K)$ splits as a direct sum $\mathrm{CFK}^\infty(K) = \oplus_{\mathfrak{s} \in \mathrm{Spin}^c(Y)} \mathrm{CFK}^\infty(K, \mathfrak{s})$ according to spin^c structures on Y, and each subcomplex $\mathrm{CFK}^\infty(K, \mathfrak{s})$ satisfies the conditions demanded above (i.e., $H_*(\mathrm{CFK}^\infty(K, \mathfrak{s})) = \mathbb{F}[U, U^{-1}]$). We just remark here that the set of spin^c structures on Y can be parametrized by $H_1(Y; \mathbb{Z})$.

The vector space $\widehat{\mathrm{HFK}}(K)$ is easy to handle (and easy to compare for different knots, especially through its Poincaré polynomial), but the concordance invariant (the stable equivalence class of CFK^∞) is hard to work with. There are several methods for getting numerical invariants from this complex; for a beautiful survey see [9]. Here we will recall one invariant, the Upsilon-function Υ_K of a knot $K \subset S^3$. (This invariant has been introduced in [25], the presentation we describe here is based on Livingston's work in [17].) For $t \in [0, 2]$ let us consider the half-plane

$$H_t = \{(x, y) \in \mathbb{R}^2 \mid y \leq \frac{t}{t - 2} x\}$$

through the origin. For $s \in \mathbb{R}$ take its s-shift $H_t(s)$, defined as

$$H_t(s) = \{(x, y) \in \mathbb{R}^2 \mid (x - s, y - s) \in H_t\}.$$

Let $C_t(s) \subset \mathrm{CFK}^\infty(K)$ be the $\mathbb{F}[U]$-subcomplex generated by those $U^k \mathbf{x}$ for which the corresponding dot $(-k, A(\mathbf{x}) - k)$ is in $H_t(s)$. Define $\Upsilon_K(t) \in \mathbb{R}$ as

$$\Upsilon_K(t) = -2 \cdot \inf\{s \mid H_0(C_t(s)) \to H_0(\mathrm{CFK}^\infty(K)) = \mathbb{F} \text{ is onto}\}.$$

It is not hard to check that the above formula is well defined (i.e., the set of s's is non-empty and bounded from below), the inf can be replaced by min, and the resulting function

$$\Upsilon_K \colon [0, 2] \to \mathbb{R}$$

is continuous and piecewise linear. Since $H_0(\mathrm{CFK}^\infty(K))$ does not change within a stable equivalence class, from Theorem 71.4.4 it is clear that Υ_K is a concordance invariant.

From the pictorial presentation of $\mathrm{CFK}^\infty(K)$ (and the identification of elements of Maslov grading 0), the description of Υ_K is not hard. We will present two examples below.

Example 71.4.5. In Figure 71.6 we see all the generators of $\mathrm{CFK}^\infty(T_{2,2n+1})$ with Maslov index 0 and 1, and this information is enough to determine $\Upsilon_{T_{2,2n+1}}$. Indeed, for $t \in [0, 1]$ the first element with $M = 0$ getting into the half-plane $H_t(s)$ as s increases is $U^{-n} x_{2n+1}$; at $t = 1$ we get all $U^k x_i$ with $M(U^k x_i) = 0$ and $1 \leq i \leq n$, $i = 2n + 1$ in the half-plane at the same time; and for $t \in [1, 2]$ we have to identify the value of s when x_n enters $H_t(s)$ (as s increases). Simple calculation then produces $\Upsilon_{T_{2,2n+1}} = -(1 - |t - 1|)n$.

Example 71.4.6. The computation of the chain complex for the torus knot $T_{n,n+1}$ follows a similar, slightly more complicated manner, cf. the explanation in the caption of Figure 71.7 for $T_{3,4}$. In the general case, we need to pass to some covers of the torus to explicitly see the embedded disks in the computation of CFK^∞, and the number of intersections of these disks with the divisors will be typically bigger than 1. We have shown these disks and the chain complex for $T_{3,4}$ in Figure 71.7. Inspecting Figure 71.7, we see that for $t \in [\frac{2}{3}, \frac{4}{3}]$ we have $\Upsilon_{T_{3,4}}(t) = -2$, for $t \in [0, \frac{2}{3}]$ we have $\Upsilon_{T_{3,4}}(t) = -3t$ and for $t \in [\frac{4}{3}, 2]$ we have $\Upsilon_{T_{3,4}}(t) = 6 - 3t$. Similar formulae for $\Upsilon_{T_{n,n+1}}$ can be derived for all $n \in \mathbb{N}$.

It is not hard to see that the unknot has vanishing Υ-invariants, for the mirror $m(K)$ of K we have $\Upsilon_{m(K)} = -\Upsilon_K$, and Υ is additive under connected sum: $\Upsilon_{K_1 \# K_2} = \Upsilon_{K_1} + \Upsilon_{K_2}$. In short, the function

$$K \mapsto \Upsilon_K$$

descends to a homomorphism $\Upsilon \colon C \to C[0, 2]$, where C is the smooth concordance group of knots in S^3 and $C[0, 2]$ is the vector space of continuous real-valued functions on the interval $[0, 2] \subset \mathbb{R}$.

The first breaking point of Υ_K is a knot invariant, which is sometimes easy to determine; for example for the torus knot $T_{n,n+1}$ it is $\frac{2}{n}$. Also, for alternating knots the function Υ_K is determined by the knot signature $\sigma(K)$ as

$$\Upsilon_K(t) = (1 - |t - 1|)\frac{\sigma(K)}{2},$$

hence the first breaking point of the Upsilon-invariant of an alternating knot with nonzero signature is at 1. (An example of such calculation is given in Example 71.4.5 for the alternating torus knot $T_{2,2n+1}$.) As an application we can easily show the following:

Lemma 71.4.7. *The torus knots $\{T_{n,n+1} \mid n \geq 2\}$ are linearly independent in the smooth concordance group C. Indeed, the same independence result holds in the quotient $C/\mathcal{A}$, where $\mathcal{A}$ is the subgroup generated by the equivalence classes of alternating knots, if we also assume $n \geq 3$ for the torus knots $T_{n,n+1}$.*

Proof. In a linear combination of torus knots concordant to the trivial knot (or to an alternating knot in the second case of the lemma) there is a knot $T_{n,n+1}$ with nonzero coefficient and highest n-value. Then at $\frac{2}{n}$ the Υ-function of the connected sum has a breaking point, since the break due to $T_{n,n+1}$ will not be compensated by any of the other components (which provide linear summands at $\frac{2}{n}$). $\qquad\square$

A beautiful argument is given in [3] to derive a very useful, simple inductive formula for $\Upsilon_{T_{p,q}}$ of the (p,q) torus knot $T_{p,q}$ (with the convention that $p < q$):

$$\Upsilon_{T_{p,q}} = \Upsilon_{T_{p,q-p}} + \Upsilon_{T_{p,p+1}}.$$

Using this formula, the argument of Lemma 71.4.7 works for any family $\{T_{p_i,q_i}\}$ of torus knots provided $2 < p_i < q_i$ and all p_i are distinct.

There is a concordance invariant which can be already derived from the $\mathbb{F}[U]$-module $\mathrm{HFK}^-(K)$: it can be shown that this module (for a knot $K \subset S^3$) is of rank one, that is of the form

$$\mathrm{HFK}^-(K) = \mathbb{F}[U] \oplus A,$$

where the $\mathbb{F}[U]$-module A is torsion, i.e. there is $n \in \mathbb{N}$ such that $U^n A = 0$. (This claim indeed follows from Proposition 71.4.2 by algebraic considerations.) Now the Alexander grading of the generator of a free summand $\mathbb{F}[U] \subset \mathrm{HFK}^-(K)$ is a knot invariant. (Note that the generator of a free summand is not well defined, but its grading is. Also, we could consider the Maslov grading of the generator, but for elements in this free summand we have $M = 2A$, hence we would get the same information.) It is not hard to see that this grading, the (negative of the) Ozsváth-Szabó τ-invariant of the knot [27], is a concordance invariant. Indeed, it provides a concordance homomorphism $\tau \colon C \to \mathbb{Z}$, and since it gives a lower bound on the slice genus, and for $T_{p,q}$ it can be computed as

$$\tau(T_{p,q}) = \frac{1}{2}(p-1)(q-1),$$

we can use this invariant to prove Milnor's conjecture on the unknotting number (and slice genus) of torus knots. Incidentally, τ is already encoded in Υ, as the slope of the piecewise linear graph of Υ_K at zero is $-\tau(K)$.

There are further numerical invariants one can derive from knot Floer homology (such as the ν or ν^+ invariants) — for details about them see [9]. In a similar manner, many of the above discussions extend for knots in rational homology three-spheres (i.e., manifolds with vanishing first Betti number).

71.5 Branched Covers

The methods and invariants introduced above admit natural enhanced versions when combined with additional topological methods. One of the most prominent topological constructions using knots (besides surgeries) is to take branched covers. Indeed, since the complement of a knot in S^3

has $\mathbb{Z}$ as its first homology, cyclic branched covers of any degree can be considered, and those manifolds are uniquely determined by the knot and the branch index. Below we restrict our attention to the double branched cover $\varphi\colon \Sigma(K) \to S^3$ of $K \subset S^3$, i.e., φ is 2:1 away from K and 1:1 along K. It has been known for a long time that this three-manifold can be used to understand topological questions about the knot. For example, the torus knot $T_{3,5}$ is not smoothly slice, since the double branched cover $\Sigma(T_{3,5})$, which is the Poincaré homology sphere $\Sigma(2,3,5)$, would be the boundary of the double branched cover of D^4 along the slice disk, giving a four-manifold with rational homology of the four-disk and with boundary the Poincaré homology sphere. Such a four-manifold, however could be glued to the E_8-plumbing, producing a four-manifold violating Donaldson's celebrated diagonalizability theorem. (Of course there are other, less technical ways to show that $T_{3,5}$ is not slice.) Essentially the same argument shows that none of the knots $T_{3,5}\#K_1\#\ldots\#K_n$ are slice for K_i two-bridge knots. This approach was developed to perfection by Lisca in his study of sliceness properties of two-bridge knots [15].

In these arguments only the three-manifold $\Sigma(K)$ has been used, although it carries further information about K: the branch set $\widetilde{K} \subset \Sigma(K)$ for example, or the covering involution $\phi\colon \Sigma(K) \to \Sigma(K)$ where points of the same φ-image downstairs in S^3 are interchanged.

In [1] the Upsilon-invariants of $(\Sigma(K), \widetilde{K})$ have been studied in various spin^c structures of $\Sigma(K)$, and an obstruction of sliceness for K has been identified in terms of these functions. This approach allows the use of Heegaard Floer theoretic methods in studying concordance questions for alternating knots.

In another approach, we investigated the properties of the pair $(\Sigma(K), \phi)$, i.e., the three-manifold $\Sigma(K)$ equipped with the covering involution. In this work [2] we utilized ideas of Hendricks-Manolescu [8] and Hendricks-Hom-Lidman [7] as follows.

The involution $\phi\colon \Sigma(K) \to \Sigma(K)$ induces a homotopy involution $\phi_{\#}\colon \mathrm{CF}^-(\Sigma(K)) \to \mathrm{CF}^-(\Sigma(K))$ on the Heegaard Floer chain complex of $\Sigma(K)$, which maps the summand corresponding to the unique spin structure $\mathfrak{s}_0$ to itself. Consider now the local self-equivalences $f\colon \mathrm{CF}^-(\Sigma(K), \mathfrak{s}_0) \to \mathrm{CF}^-(\Sigma(K), \mathfrak{s}_0)$ which homotopy commutes with $\phi_{\#}$. Through the double branched cover construction, a concordance between K_0 and K_1 provides a homology cobordism W between $\Sigma(K_0)$ and $\Sigma(K_1)$ with a distinguished spin structure and a covering involution Φ. Using this cobordism it is easy to show that the following construction provides a concordance invariant: take $f_{\max}$ among the local self-equivalences of $\mathrm{CF}^-(\Sigma(K), \mathfrak{s}_0)$ homotopy commuting with $\phi_{\#}$ with maximal kernel, and define

$$\mathrm{HFB}^-_{\mathrm{conn}}(K) = H_*(\mathrm{Im}f_{\max}) \subset \mathrm{HF}^-(\Sigma(K), \mathfrak{s}_0).$$

It is not hard to see that this $\mathbb{F}[U]$-module is well defined up to (graded) module isomorphism, and is of rank one, decomposing as

$$\mathrm{HFB}^-_{\mathrm{conn}}(K) = \mathbb{F}[U] \oplus \mathrm{HFB}^-_{\mathrm{red-conn}}(K).$$

The torsion submodule $\mathrm{HFB}^-_{\mathrm{red-conn}}(K)$ is then a concordance invariant. Although it is rather challenging to determine the chain complex together with the action of $\phi_{\#}$, in certain cases it is possible, providing the following results.

Theorem 71.5.1.   • *If $K \subset S^3$ is an alternating knot, then* $\mathrm{HFB}^-_{\mathrm{red-conn}}(K) = 0$.

• *If $K \subset S^3$ is a torus knot, then* $\mathrm{HFB}^-_{\mathrm{red-conn}}(K) = 0$.

- *For the pretzel knot $P(-2, 3, q)$ with $q \geq 7$ and odd, we have* $\mathrm{HFB}^{-}_{\text{red-conn}}(P(-2, 3, q)) = \mathbb{F}$.

- *For the pretzel knot $P(4q + 3, -2q - 1, 4q + 1)$ we have*

$$\mathrm{HFB}^{-}_{\text{red-conn}}(P(4q + 3, -2q - 1, 4q + 1)) = \mathbb{F}[U]/(U^q).$$

The first statement easily follows from the fact that $\mathrm{HFB}^{-}_{\text{conn}}(K)$ is a submodule of $\mathrm{HF}^{-}(\Sigma(K), \mathfrak{s}_0)$, and for an alternating (or even for a quasi-alternating) knot we have $\mathrm{HF}^{-}(\Sigma(K), \mathfrak{s}_0) = \mathbb{F}[U]$. The second statement is particularly simple for a torus knot $T_{p,q}$ with pq odd, since in this case on the double branched cover

$$\Sigma(T_{p,q}) = \{(x, y, z) \in \mathbb{C}^3 \mid x^2 + y^p + z^q, \|(x, y, z)\| = 1, \}$$

the involution ϕ is simply $(x, y, z) \mapsto (-x, y, z)$, and (through the one-parameter family $\psi_t(x, y, z) = (t^{pq}x, t^{2q}y, t^{2p}z)$ for $t \in S^1$) this map is homotopic to the identity.

The third and fourth statements are proved in [2] using lattice homology arguments, see [22]. The third statement implies that the pretzel knots $P(-2, 3, q)$ are nontrivial even in the quotient group $C/(Q + \mathcal{T})$, when we factor out by the subgroup of the smooth concordance group C generated by quasi-alternating (Q) and torus ($\mathcal{T}$) knots. Finally, the fourth statement implies that the pretzel knots of the statement are linearly independent in the quotient group $C/(Q + \mathcal{T})$, providing a subgroup (indeed a direct summand) in this quotient group isomorphic to $\mathbb{Z}^\infty$.

71.6   Some Applications and Further Remarks

Knot Floer homology is rather nicely adaptable to surgery problems. Indeed, suppose that $K \subset S^3$ is a given knot, and let $S^3_r(K)$ denote the three-manifold we get by applying Dehn surgery with coefficient r along K. Then the invariant $\mathrm{HF}^{-}(S^3_r(K))$ of this new three-manifold can be algebraically determined from the chain complex $\mathrm{CFK}^\infty(K)$ through a straightforward algorithm [28, 29]. This feature and further results for knot Floer homology provide proofs of the two statements given below. (These two theorems just highlight some aspects of low-dimensional topology where knot Floer homology can be successfully applied.)

Theorem 71.6.1 ([4, 13]). *Suppose that surgery along the knot K provides the Poincaré homology sphere or the three-dimensional projective space. Then the knot is the left-handed trefoil knot (with surgery coefficient -1) in the first case, and the unknot in the second.*   □

The statement admits various generalizations regarding lens space surgeries, cf. [13].

Another (similar) circle of ideas is to determine which non-trivial knot K admits (truly) cosmetic surgeries, i.e., two distinct surgery coefficients $r \neq s$ with the property that $S^3_r(K)$ and $S^3_s(K)$ are orientation-preserving diffeomorphic. A recent result puts rather strong constraints on knots with such cosmetic surgeries (and the conjecture is that, indeed, there are no such non-trivial knots). A sample of these recent results is the following simple statement:

Theorem 71.6.2 (Hanselman [6]). *Suppose that K is an alternating knot which admits truly cosmetic surgeries. Then K is of genus 2, has a vanishing signature, and its Alexander polynomial is of the form $kt^2 - 4kt + 6k + 1 - 4kt^{-1} + kt^{-2}$ for some integer $k \in \mathbb{Z}$.*   □

Further Remarks

The theories outlined in this paper admit various extensions and variants.

- The definition of knot Floer homology naturally extends to links. In this case (in the most general setting) the link Floer homology module of the link $L \subset S^3$ of n components is a finitely generated $\mathbb{F}[U_1^{\pm 1}, \ldots, U_n^{\pm 1}]$-module, equipped with a grading and $2n$ filtrations. This module can be used to determine the Thurston norm of the complement and whether L is fibered. There is also a (rather complicated) algorithm [19] which determines the invariants of the surgeries along L from its link Floer complex. Notice that since any (closed, oriented) three-manifold can be given by integer surgery along a link in S^3, this method provides an algorithmic way of computing Heegaard Floer homology of three-manifolds from link Floer homology (which, in turn, admits a combinatorial description through grid diagrams, see below).

- Computation and definition of knot Floer homology admits several combinatorial/algebraic approaches. Grid homology [20, 21] provides a conceptually rather simple, algorithmic approach to determine all versions of knot Floer homology, cf. [24]. The problem is that the chain complexes encountered in this approach are typically large, and their computational complexity (even for relatively small knots) is beyond practical.

- Ozsváth and Szabó adapted the structures and ideas of Bordered Floer homology to the knot context, and defined algebraically computable invariants of knots, which are isomorphic to a certain variant of $\mathrm{CFK}^\infty(K)$, see [30]. There is a rather effective computer program which determines these invariants.

- Heegaard Floer homology extends to manifolds with boundary; either we assume that the boundary admits sutures, and get Sutured Floer Homology (defined and applied by Juhász [11]), or we assume that the boundary admits a fixed parameterization and get Bordered Floer homology of Lipshitz-Ozsváth-Thurston [14].

- Although we chose our coefficients always from the field $\mathbb{F}$ of two elements, many aspects of the theory have been developed with coefficients in $\mathbb{Z}$.

- Knot Floer theoretic methods can be used in the context of Legendrian and transverse knots, see [16, 21].

71.7   Bibliography

[1] A. Alfieri, D. Celoria, and A. Stipsicz. Upsilon invariants from cyclic branched covers. arXiv:1809.08269, 2018.

[2] A. Alfieri, S. Kang, and A. Stipsicz. Connected Floer homology of covering involutions. arXiv:1905.11613, 2019.

[3] P. Feller and D. Krcatovich. On cobordisms between knots, braid index, and the upsilon-invariant. *Math. Ann.*, 369(1-2):301–329, 2017.

[4] P. Ghiggini. Knot Floer homology detects genus-one fibred knots. *Amer. J. Math.*, 130(5):1151–1169, 2008.

[5] J. Greene. Alternating links and definite surfaces. *Duke Math. J.*, 166(11):2133–2151, 2017. With an appendix by András Juhász and Marc Lackenby.

[6] J. Hanselman. Heegaard Floer homology and cosmetic surgeries in S^3. arXiv:1906.06773, 2019.

[7] K. Hendricks, J. Hom, and T. Lidman. Applications of involutive Heegaard Floer homology. arXiv:1708.06389, 2018.

[8] K. Hendricks and C. Manolescu. Involutive Heegaard Floer homology. *Duke Math. J.*, 166(7):1211–1299, 2017.

[9] J. Hom. A survey on Heegaard Floer homology and concordance. *J. Knot Theory Ramifications*, 26(2):1740015, 24, 2017.

[10] J. Howie. A characterisation of alternating knot exteriors. *Geom. Topol.*, 21(4):2353–2371, 2017.

[11] A. Juhász. Holomorphic discs and sutured manifolds. *Algebr. Geom. Topol.*, 6:1429–1457, 2006.

[12] A. Juhász. A survey of Heegaard Floer homology. In *New Ideas in Low Dimensional Topology*, volume 56 of *Ser. Knots Everything*, pages 237–296. World Sci. Publ., Hackensack, NJ, 2015.

[13] P. Kronheimer, T. Mrowka, P. Ozsváth, and Z. Szabó. Monopoles and lens space surgeries. *Ann. of Math. (2)*, 165(2):457–546, 2007.

[14] R. Lipshitz, P. Ozsváth, and D. Thurston. Bordered Heegaard Floer homology. *Mem. Amer. Math. Soc.*, 254(1216):viii+279, 2018.

[15] P. Lisca. Sums of lens spaces bounding rational balls. *Algebr. Geom. Topol.*, 7:2141–2164, 2007.

[16] P. Lisca, P. Ozsváth, A. Stipsicz, and Z. Szabó. Heegaard Floer invariants of Legendrian knots in contact three-manifolds. *J. Eur. Math. Soc. (JEMS)*, 11(6):1307–1363, 2009.

[17] C. Livingston. Notes on the knot concordance invariant upsilon. *Algebr. Geom. Topol.*, 17(1):111–130, 2017.

[18] C. Manolescu. An introduction to knot Floer homology. In *Physics and Mathematics of Link Homology*, volume 680 of *Contemp. Math.*, pages 99–135. Amer. Math. Soc., Providence, RI, 2016.

[19] C. Manolescu and P. Ozsváth. Heegaard Floer homology and integer surgeries on links. arXiv:1011.1317, 2015.

[20] C. Manolescu, P. Ozsváth, and S. Sarkar. A combinatorial description of knot Floer homology. *Ann. of Math. (2)*, 169(2):633–660, 2009.

[21] C. Manolescu, P. Ozsváth, Z. Szabó, and D. Thurston. On combinatorial link Floer homology. *Geom. Topol.*, 11:2339–2412, 2007.

[22] A. Némethi. Lattice cohomology of normal surface singularities. *Publ. Res. Inst. Math. Sci.*, 44(2):507–543, 2008.

[23] Y. Ni. Knot Floer homology detects fibred knots. *Invent. Math.*, 170(3):577–608, 2007.

[24] P. Ozsváth, A. Stipsicz, and Z. Szabó. *Grid Homology for Knots and Links*, volume 208 of *Mathematical Surveys and Monographs*. American Mathematical Society, Providence, RI, 2015.

[25] P. Ozsváth, A. Stipsicz, and Z. Szabó. Concordance homomorphisms from knot Floer homology. *Adv. Math.*, 315:366–426, 2017.

[26] P. Ozsváth and Z. Szabó. Holomorphic disks and knot invariants. *Adv. Math.*, 186(1):58–116, 2004.

[27] P. Ozsváth and Z. Szabó. Knot Floer homology, genus bounds, and mutation. *Topology Appl.*, 141(1-3):59–85, 2004.

[28] P. Ozsváth and Z. Szabó. Knot Floer homology and integer surgeries. *Algebr. Geom. Topol.*, 8(1):101–153, 2008.

[29] P. Ozsváth and Z. Szabó. Knot Floer homology and rational surgeries. *Algebr. Geom. Topol.*, 11(1):1–68, 2011.

[30] P. Ozsváth and Z. Szabó. Kauffman states, bordered algebras, and a bigraded knot invariant. *Adv. Math.*, 328:1088–1198, 2018.

[31] P. Ozsváth and Z. Szabó. An overview of knot Floer homology. In *Modern Geometry: A Celebration of the Work of Simon Donaldson*, volume 99 of *Proc. Sympos. Pure Math.*, pages 213–249. Amer. Math. Soc., Providence, RI, 2018.

[32] T. Perutz. Hamiltonian handleslides for Heegaard Floer homology. In *Proceedings of Gökova Geometry-Topology Conference 2007*, pages 15–35. Gökova Geometry/Topology Conference (GGT), Gökova, 2008.

[33] B. Rácz. *Geometry of (1,1)-Knots and Knot Floer Homology*. ProQuest LLC, Ann Arbor, MI, 2015. Thesis (Ph.D.)–Princeton University.

[34] J. Rasmussen. *Floer Homology and Knot Complements*. ProQuest LLC, Ann Arbor, MI, 2003. Thesis (Ph.D.)–Harvard University.

[35] J. Rasmussen. Knot polynomials and knot homologies. In *Geometry and Topology of Manifolds*, volume 47 of *Fields Inst. Commun.*, pages 261–280. Amer. Math. Soc., Providence, RI, 2005.

Chapter 72

An Introduction to Grid Homology

András I. Stipsicz, Rényi Institute of Mathematics, Hungary

72.1   Introduction

Knot Floer homology provides a powerful tool for studying three- and four-dimensional properties of knots in three-manifolds, in particular, in the standard three-sphere S^3. The definition of this homology theory is, however, quite involved, and rests on delicate ideas and methods of infinite-dimensional analysis. The discovery of Sarkar and Wang [21], finding that under certain specific circumstances these analytic difficulties can be bypassed by rather simple combinatorics and planar geometry, opened up the possibility to have an algorithmically computable theory. Indeed, in [12] a combinatorial version of knot Floer homology for knots and links in S^3 has been described, and in [13] the entire theory (i.e. the proof of basic properties) has been put into a combinatorial context. This combinatorial version of the theory is now called *grid homology*. A detailed introduction to this theory has been given in [14].

The main advantage of grid homology is that its presentation requires very little background, yet produces remarkable results and invariants. As an example, it can be used to show how Milnor's famous conjecture about the unknotting number of torus knots can be verified. This conceptual simplicity, however, comes with a price: this theory is defined only for knots and links in S^3 (and admits an extension to knots and links in lens spaces [1]), and hence cannot provide the TQFT features knot Floer homology possesses. This property limits the applicability of the theory. In addition, the computational complexity of the chain complexes involved in this combinatorial definition is enormous and way beyond practical. In that respect, Ozsváth and Szabó's bordered knot Floer homology (developed in [18] for knots in S^3) is the most useful approach at the moment. With that said, grid homology is a beautiful piece of mathematics, providing a glimpse at a very powerful and technical tool (knot Floer homology) through a rather simple and tame lens. Indeed, familiarity with the concepts and constructions of grid homology provides an excellent start in the study of knot Floer homology.

In a nutshell, grid homology is defined as follows. Consider a table $\mathbb{G}$ of n^2 small squares in a bigger square, equipped with n X-markings and n O-markings in a manner that

- each row and each column of $\mathbb{G}$ contains exactly one X and exactly one O, and

- no small square contains both an X and an O.

Such a table $\mathbb{G}$ is called a *grid*. By connecting each X to the O-marking in its column vertically (directed from X to O) and then from O to X horizontally, in such a way that the horizontal segment passes under the vertical at a crossing, we get an oriented knot/link diagram, defining a knot or link $K_{\mathbb{G}} \subset S^3$. It is not hard to see that any knot (and link) occurs in this way: for a diagram of a knot

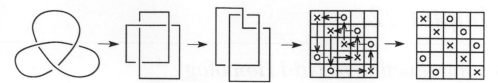

FIGURE 72.1: Turning the usual diagram of the trefoil knot into a grid diagram.

$K \subset S^3$, isotope it until all segments are either horizontal or vertical (as in the second diagram of Figure 72.1), and at the crossings verticals are over horizontal, as shown by the third diagram in Figure 72.1. From this picture the grid defining the given diagram is easy to find (the markings will be located at the turning points), as shown by Figure 72.1.

The chain complex associated to a grid diagram $\mathbb{G}$ providing grid homology is generated by the bijections between the horizontal and the vertical intervals in the large square (over some appropriately chosen polynomial ring) and the boundary map counts rectangles with certain properties. The various versions of the homologies stem from the fact that the markings can be used in many ways to block some rectangles to appear in the count defining the boundary map. We will mainly concentrate on the version $GH^-(K)$ (called the *simply blocked theory* in [14]), corresponding to the minus version $HFK^-(K)$ of knot Floer homology.

Indeed, grid homology can be interpreted as categorification of the Alexander polynomial of knots. To this end, let us describe a definition of the Alexander polynomial relying on grids. Consider the grid $\mathbb{G}$ representing $K \subset S^3$, together with the knot projection (an immersed curve) determined by the markings as above. To each crossing of the horizontal and vertical intervals defining the grid, write the monomial we get by raising t to the (negative of the) winding number of the curve around that point. For an example reflecting the exact conditions, see Figure 72.2. The matrix of the grid of Figure 72.2 is

$$\begin{pmatrix} 1 & 1 & t & t & 1 & 1 \\ 1 & t^{-1} & 1 & t & 1 & 1 \\ 1 & t^{-1} & t^{-1} & 1 & t^{-1} & 1 \\ 1 & t^{-1} & t^{-1} & 1 & 1 & t \\ 1 & 1 & 1 & t & t & t \\ 1 & 1 & 1 & 1 & 1 & 1 \end{pmatrix}.$$

The determinant of the resulting matrix is a function of t, which can be normalized by some expressions of t to get the symmetrized Alexander polynomial of K. Grid homology is the 'categorification' of this construction: the generators of the theory are the summands of the determinant (i.e. bijections between the horizontal and vertical intervals in $\mathbb{G}$), one of the gradings (called the *Alexander grading*) comes from a suitable normalization of the winding numbers, while the other grading (the *Maslov grading*) is a specific integral lift of the sign the summand admits in the determinant.

Grid homology provides a self-contained theory for knots and links in S^3. The definition of the theory, however, comes from a more general theory: for $K \subset S^3$ the invariant $GH^-(K)$ is isomorphic to the knot Floer homology group $HFK^-(K)$ of K (discovered by Ozsváth-Szabó [17] and independently by Rasmussen [19]). Indeed, a grid diagram can be viewed as a (multipointed) Heegaard diagram, and then the usual definition of knot Floer homology immediately provides the chain

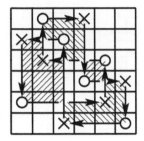

FIGURE 72.2: Illustration of winding numbers of regions; hatchings from lower left to upper right indicate regions with winding number 1, while the other hatchings indicate winding number −1. (No hatching is applied in regions with winding number 0.)

complex $GC^-(\mathbb{G})$: The generators of the chain complexes GC^- and CFK^- defining grid and knot Floer homologies are naturally the same, while for the identification of the boundary maps one needs a relatively simple complex analytic argument. Grid homology also extends further variants of knot Floer homology, such as the graded, bifiltered chain complex $CFK^\infty(K)$, or the Legendrian invariants from [10] (cf. also [14]). Indeed, when combined with the result of Manolescu and Ozsváth [11], describing how Heegaard Floer homologies of the result of a surgery along a link can be determined from the link Floer homology of the link, we get a combinatorial description of Heegaard Floer homology groups of closed, oriented three-manifolds.

The basics of grid homology, and all the results and proofs discussed in this chapter have been explained in detail in [14]. This work is intended to provide a gentle introduction to the topic by sketching the main ideas and steps of the most important results given in the first eight chapters of [14].

The chapter is organized as follows. In Section 72.2 we discuss grids, and explain grid moves which transform grids defining isotopic knots/links into each other. Section 72.3 gives the definition of the grid chain complex and its grading, leading to the definition of grid homology. In this section we also sketch the main ideas of the proof of independence from the chosen grid presenting a given knot. Some illustrative examples are given in Section 72.4, and in Section 72.5 we examine the effect of crossing changes. The resulting construction also provides a structure theorem for grid homology, and an invariant which then gives a proof of Milnor's conjecture on the unknotting number of torus knots. In Section 72.6 we sketch the extension of the theory to links and provide a slice genus bound. This reasoning (together with groundbreaking results of Freedman and Quinn) provides a construction for exotic $\mathbb{R}^4$'s. In Section 72.7 bounds on the unoriented (four-dimensional) genus are discussed.

Acknowledgements: The author was partially supported by *Élvonal (Frontier)* grant KKP 126683 of the NKFIH, Hungary.

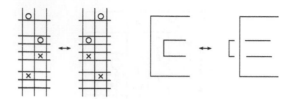

FIGURE 72.3: Commutation of two columns. The right portion of the diagram shows the effect of this particular commutation move on the knot diagram.

72.2 Grid Diagrams

As it was described in the previous section, any knot $K \subset \mathbb{R}^3$ (and in fact any link $L \subset \mathbb{R}^3$) can be described by a grid diagram. Such a diagram $\mathbb{G}$ is usually viewed as part of the plane, but by identifying the top and bottom edges as well as the right and left edges, we can view it on the standard torus $\mathbb{T}^2 \subset \mathbb{R}^3$. In this way the vertical and horizontal segments glue to vertical and horizontal circles. We will mostly view our diagrams in the torus, but it will be convenient sometimes to consider the fundamental domain in $\mathbb{R}^2$, in which case we talk about a *planar grid*.

The grid diagram of a knot is obviously not unique; as knot diagrams are unique only up to Reidemeister moves, grid diagrams are unique up to grid moves: *commutation* and *stabilization/destabilization*, cf. [3]. There are two types of commutations: we can commute columns and rows. (We will discuss only column commutations; the definition of row commutation needs only slight and obvious changes.) Consider the two columns C_1, C_2 next to each other, and the circles in the middle of both passing through the markings. The markings cut these circles into two intervals. (We imagine that the markings are exactly in the middle of their squares.) We can commute the two columns if the resulting intervals (when projecting one to the other) contain each other (i.e., one of the intervals in C_1 contains an interval from C_2 after projection, while the other interval in C_1 is contained in the other interval of C_2, again after projection). See Figure 72.3 for an example (and note that the configuration of Figure 72.14, for example, cannot be commuted). It is easy to see that the commutation of two such columns turning a grid diagram $\mathbb{G}$ to $\mathbb{G}'$ changes only the diagram of the knot associated to these grids, but not its isotopy type, cf. the right-hand diagram of Figure 72.3 for an example.

Stabilization can be performed along a square containing a marking, hence there are two types of stabilizations: X- and O-stabilizations. (Once again, we will discuss only one of those.) Consider a small square which contains an X-marking. Split the column and the row passing through this square of our chosen marking into two. The square containing our X then splits into four small squares. We put two X-markings and one O-marking into three of these four squares, and this choice then determines the position of the further two O's in the column and row of the original X-marking at which we perform the stabilization; see Figure 72.4. It is easy to see that there are four different X-stabilizations (distinguished by the small square left empty in cutting the marked small square into four), as shown by Figure 72.4. The resulting four diagrams differ by commutation moves, and furthermore, for any O-stabilization, there is an X-stabilization differing from it by

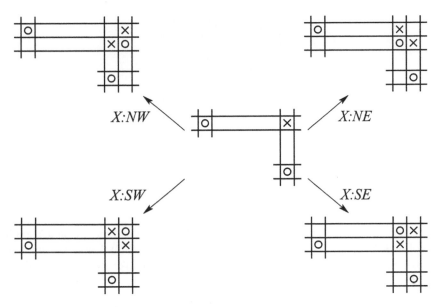

FIGURE 72.4: The four types of stabilizations at an X-marking. The different stabilizations are named after the geographic position of the empty square after splitting the original X-marked square into four.

commutation moves. The appropriate adaptation of Reidemeister's Theorem on knot diagrams for the grid context provides:

Theorem 72.2.1 (Cromwell [3]). *Two grid diagrams $\mathbb{G}_0$ and $\mathbb{G}_1$ define isotopic links if and only if $\mathbb{G}_0$ can be transformed to $\mathbb{G}_1$ by a finite sequence of row and column commutations and X-stabilizations/destabilizations.* □

As in the case of the Reidemeister Theorem, one implication of the above theorem is rather trivial (since the moves obviously do not change the knot or link type); the converse, however, is somewhat tedious, see [14, Appendix B.4].

72.3 Grid Homology

Suppose now that $\mathbb{G}$ is a given $n \times n$ grid. A *grid state* $\mathbf{x}$ is a bijection between the vertical and horizontal lines of the toroidal grid. When viewing the grid on the plane (through one of its fundamental domains), $\mathbf{x}$ can be depicted by n dots, each vertical and horizontal line containing exactly one of them. (Recall that the top and bottom, and right and left intervals are identified.) In particular, there are $n!$ grid states associated to $\mathbb{G}$. (Notice that here only the size of the grid matters, the positions of the markings are irrelevant.) The set of grid states of $\mathbb{G}$ will be denoted by $S(\mathbb{G})$; Figure 72.5 shows an example of a grid state.

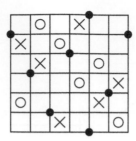

FIGURE 72.5: An example of a grid state. Recall that the top and bottom, as well as the left and right segments are identified.

Each grid state $\mathbf{x}$ admits two gradings, its Maslov grading $M(\mathbf{x})$ and its Alexander grading $A(\mathbf{x})$, defined as follows. Suppose that $P, Q \subset \mathbb{R}^2$ are finite sets. Let $\mathcal{I}(P, Q)$ denote the number of those pairs $((p_1, p_2), (q_1, q_2))$ in $P \times Q$ where $p_1 < q_1$ and $p_2 < q_2$, and take its symmetrized version

$$\mathcal{J}(P, Q) = \frac{1}{2}(\mathcal{I}(P, Q) + \mathcal{I}(Q, P)).$$

Let $\mathbb{O}$ and $\mathbb{X}$ denote the set of O- and X-markings. Then

$$M(\mathbf{x}) = \mathcal{J}(\mathbf{x}, \mathbf{x}) - 2\mathcal{J}(\mathbf{x}, \mathbb{O}) + \mathcal{J}(\mathbb{O}, \mathbb{O}) + 1,$$

and

$$A(\mathbf{x}) = \mathcal{J}(\mathbf{x}, \mathbb{X}) - \mathcal{J}(\mathbf{x}, \mathbb{O}) + \frac{1}{2}(\mathcal{J}(\mathbb{O}, \mathbb{O}) - \mathcal{J}(\mathbb{X}, \mathbb{X}) - n + 1).$$

Although in the definition of these quantities we use a planar representation of the toroidal grid, it can be shown that $M(\mathbf{x})$ and $A(\mathbf{x})$ are associated to the grid state $\mathbf{x}$ in the toroidal grid (and independent of the choice of the fundamental domain).

Number the O-markings from 1 to n, consider the polynomial ring $R = \mathbb{F}[V_1, \ldots, V_n]$ of n variables over the field $\mathbb{F}$, and define the module $GC^-(\mathbb{G})$ as the free R-module generated by the grid states. Define $M(V_i) = -2$ and $A(V_i) = -1$ for all $i \in \{1, \ldots, n\}$.

The bigraded module $GC^-(\mathbb{G})$ will now be equipped with a boundary map as follows. Consider two grid states $\mathbf{x}, \mathbf{y}$ in $\mathbb{G}$. We define the set of rectangles $\mathrm{Rect}(\mathbf{x}, \mathbf{y})$ connecting $\mathbf{x}$ and $\mathbf{y}$ as follows. If $\mathbf{x}$ and $\mathbf{y}$ differ on more than two horizontal (or equivalently, vertical) segments, then we define $\mathrm{Rect}(\mathbf{x}, \mathbf{y}) = \emptyset$. If $\mathbf{x}$ and $\mathbf{y}$ differ on exactly two horizontal segments, then the vertical and horizontal circles containing the differences of $\mathbf{x}$ and $\mathbf{y}$ cut the torus into four rectangles, two of which will contribute to $\mathrm{Rect}(\mathbf{x}, \mathbf{y})$ and the other two to $\mathrm{Rect}(\mathbf{y}, \mathbf{x})$. A rectangle from these four elements is in $\mathrm{Rect}(\mathbf{x}, \mathbf{y})$ if the orientation of the torus induces the orientation on the rectangle so that its boundary orientation on the horizontal circle points from the $\mathbf{x}$-component to the $\mathbf{y}$-component. Finally, $\mathrm{Rect}^\circ(\mathbf{x}, \mathbf{y})$ is the set of those rectangles from $\mathrm{Rect}(\mathbf{x}, \mathbf{y})$ which do not contain any further component of $\mathbf{x}$ (and so of $\mathbf{y}$) in their interior. (In particular, $\mathrm{Rect}^\circ(\mathbf{x}, \mathbf{y})$ has zero, one or two elements, depending on $\mathbf{x}$ and $\mathbf{y}$. For examples see Figure 72.6.) Suppose now that $r \in \mathrm{Rect}(\mathbf{x}, \mathbf{y})$ and O_i is an O-marking. Then $O_i(r)$ is one or zero depending on whether r contains O_i or it does not. With these definitions in place, we can define the boundary operator $\partial^- : GC^-(\mathbb{G}) \to GC^-(\mathbb{G})$ as follows: for

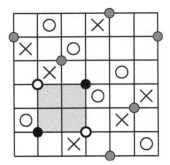 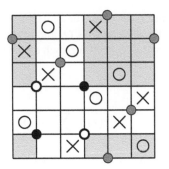

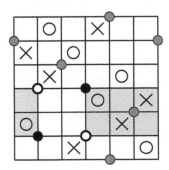 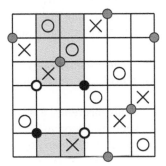

FIGURE 72.6: The diagrams show the four examples of rectangles between the grid states **x** (represented by full circles) and **y** (for which the two differing coordinates are shown by hollow circles, the common coordinates of **x** and **y** are denoted by gray full circles). The upper two rectangles are in Rect(**x**, **y**), while the lower two are in Rect(**y**, **x**). Only the upper left rectangle is empty, i.e. contains no gray full circles.

a grid state $\mathbf{x} \in \mathcal{S}(\mathbb{G})$ take

$$\partial^{-}(\mathbf{x}) = \sum_{\mathbf{y} \in \mathcal{S}(\mathbb{G})} \sum_{\{r \in \text{Rect}^{\circ}(\mathbf{x},\mathbf{y}) | r \cap \mathbb{X} = \emptyset\}} V_1^{O_1(r)} \cdots V_n^{O_n(r)} \mathbf{y},$$

and extend this map R-linearly to an endomorphism of $\text{GC}^{-}(\mathbb{G})$.

Theorem 72.3.1. *The map ∂^{-} is homogeneous of bidegree $(-1, 0)$ with respect to the Maslov-Alexander bigrading of $\text{GC}^{-}(\mathbb{G})$, and satisfies $(\partial^{-})^2 = 0$.* □

The claim on bidegree is a simple calculation, while the fact that ∂^{-} is a boundary map follows from some elementary planar geometry: the composition of two rectangles (contributing to $(\partial^{-})^2(\mathbf{x})$) admits two such decompositions, hence over the field $\mathbb{F}$ of two elements they cancel.

A similar argument shows that if $\mathbb{G}$ gives rise to a knot (and not a link), then multiplications by V_i and by V_j for any pair i, j are chain homotopic. Indeed, it is enough to prove this claim for 'neighbouring pairs', i.e. for pairs of indices with the property that there is an X-marking X_{ij} which is in the same row as O_i and in the same column as O_j. If $V_i \colon \mathrm{GC}^-(\mathbb{G}) \to \mathrm{GC}^-(\mathbb{G})$ denotes multiplication by V_i, then for such a pair i, j we have

$$V_i - V_j = \partial^- \circ h + h \circ \partial^-,$$

where

$$h(\mathbf{x}) = \sum_{} = \sum_{\mathbf{y} \in S(\mathbb{G})} \sum_{\{r \in \mathrm{Rect}^\circ(\mathbf{x}, \mathbf{y}) | r \cap \mathbb{X} = \{X_{ij}\}\}} V_1^{O_1(r)} \cdots V_n^{O_n(r)} \mathbf{y}.$$

This shows that by defining the U-action as multiplication by any of the V_i, the homology $H(\mathrm{GC}^-(\mathbb{G}), \partial^-)$ is naturally an $\mathbb{F}[U]$-module, and this structure is independent of the choice of the above index i.

In conclusion, the homology $\mathrm{GH}^-(\mathbb{G})$ of the chain complex $(\mathrm{GC}^-(\mathbb{G}), \partial^-)$ associated to $\mathbb{G}$ is a bigraded module over the ring $\mathbb{F}[U]$ of polynomials in one variable (and coefficients from the field $\mathbb{F}$ of two elements). The first main result of the subject is

Theorem 72.3.2. *The grid homology* $\mathrm{GH}^-(\mathbb{G})$, *as a bigraded* $\mathbb{F}[U]$-*module, is a knot invariant, i.e. for two toroidal grids* $\mathbb{G}_0$, $\mathbb{G}_1$ *both representing the knot* K *we have* $\mathrm{GH}^-(\mathbb{G}_0) \cong \mathrm{GH}^-(\mathbb{G}_1)$. $\qquad\square$

In order to prove this theorem, i.e. to show that the homology does not depend on the grid presentation of the knot, by Theorem 72.2.1 we need to verify that a commutation and a stabilization move on $\mathbb{G}$ does not alter the homology.

72.3.1 Commutation Invariance

Suppose that $\mathbb{G}$ and $\mathbb{G}'$ differ by a commutation, which we assume to be a commutation of columns. The two diagrams can be drawn on the same grid, by allowing one more vertical line, and two distinguished verticals intersecting each other in two points, see Figure 72.3. In this diagram there are pentagons, where one vertex of the pentagon is one of the two intersection points of the two distinguished vertical curves, two further vertices are parts of a grid state in $\mathbb{G}$, and the final two vertices are part of a grid state in $\mathbb{G}'$. By counting empty pentagons (as we have counted rectangles in the definition of ∂^-) as shown by Figure 72.8, disjoint from the X-markings, and by counting the O-markings contained by them in the exponents of the V_i variables, we get two maps

$$P \colon \mathrm{GC}^-(\mathbb{G}) \to \mathrm{GC}^-(\mathbb{G}'), \qquad P' \colon \mathrm{GC}^-(\mathbb{G}') \to \mathrm{GC}^-(\mathbb{G}).$$

It can be shown that these two maps are chain maps, i.e. $P \circ \partial^-_{\mathbb{G}} = \partial^-_{\mathbb{G}'} \circ P$ (and similarly for P') and preserve the bigradings. (The proof of the property that these maps are chain maps follows the same rough idea we used in proving $(\partial^-)^2 = 0$: apply combinations of pentagons and rectangles and verify that they can be always decomposed in two different ways.) Furthermore, these maps are *homotopy inverses* of each other, i.e. the compositions $P \circ P'$ and $P' \circ P$ are homotopic to the respective identities. This amounts to showing that there is a map $H \colon \mathrm{GC}^-(\mathbb{G}) \to \mathrm{GC}^-(\mathbb{G})$ which satisfies

$$P \circ P' + \mathrm{id}_{\mathrm{GC}^-(\mathbb{G})} = \partial^-_{\mathbb{G}} \circ H + H \circ \partial^-_{\mathbb{G}}$$

(and a similar map H' for $P' \circ P$). This map H counts hexagons with two vertices being the two intersection points of the two distinguished vertical curves, and the other four vertices are parts of

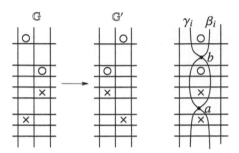

FIGURE 72.7: The commutation move from $\mathbb{G}$ to $\mathbb{G}'$ can be depicted in one diagram (shown on the right) by allowing two verticals to be curved; choosing β_i we get $\mathbb{G}$, while choosing γ_i we get $\mathbb{G}'$.

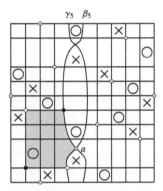

FIGURE 72.8: An empty pentagon from $\mathbb{G}$ to $\mathbb{G}'$, containing a unique O-marking. The grid state $\mathbf{x}$ is denoted by full circles, while $\mathbf{y}$ by hollow circles, and the common coordinates are grey circles.

two grid states in $\mathbb{G}$, see Figure 72.9. The combination of two empty pentagons can be decomposed as a hexagon and rectangle (in some order) with one exception, when the two pentagons provide a hexagon of width one running around the entire toroidal grid — this case being responsible for $\mathrm{id}_{\mathrm{GC}^-(\mathbb{G})}$ in the formula. Since homotopy inverses on the chain level induce actual inverses in homology, the commutation invariance of GH$^-$ follows.

72.3.2 Stabilization Invariance

The independence from stabilization is slightly more involved, since in this case the ring $\mathbb{F}[V_1, \ldots, V_n]$ over which the chain complex is defined also changes. The rough idea of the proof in this case goes as follows.

Suppose that $\mathbb{G}'$ is given by an X-stabilization of the grid $\mathbb{G}$ (for simplicity we assume that it is an $X : SW$ stabilization), and let $O_2, \ldots, O_n$ be the O-markings in $\mathbb{G}$, hence $\mathrm{GC}^-(\mathbb{G})$ is defined over $\mathbb{F}[V_2, \ldots, V_n]$. We can consider $\mathrm{GC}^-(\mathbb{G})[V_1]$, which is now a chain complex over the ring $\mathbb{F}[V_1, \ldots, V_n]$. Obviously it is a much larger chain complex, having larger homology as well. On

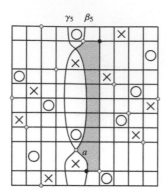
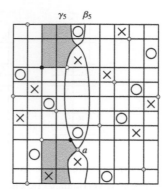

FIGURE 72.9: Two examples of hexagons.

the other hand, some standard algebraic arguments show that the multiplication map by $V_1 - V_2$,

$$V_1 - V_2 \colon \mathrm{GC}^-(\mathbb{G})[V_1] \to \mathrm{GC}^-(\mathbb{G})[V_1]$$

has the property that the homology of its mapping cone $\mathrm{Cone}(V_1 - V_2)$ is isomorphic to the homology of $\mathrm{GC}^-(\mathbb{G})$. (Recall that the mapping cone of a chain map $f \colon C_1 \to C_2$ is the chain complex $(C_1 \oplus C_2, \partial)$, where $\partial(x_1, x_2) = (\partial_{C_1}(x_1), \partial_{C_2}(x_2) + f(x_1))$.) Intuitively, the mapping cone of $V_1 - V_2$ provides that V_1 becomes chain homotopic to V_2.

Therefore we will compare $\mathrm{GC}^-(\mathbb{G}')$ (a chain complex over $\mathbb{F}[V_1, \ldots, V_n]$) with the mapping cone $\mathrm{Cone}(V_1 - V_2)$. To this end, we write $\mathrm{GC}^-(\mathbb{G}')$ as a mapping cone itself. Indeed, in the stabilization, when cutting the row and the column through the distinguished X-marking into two, the horizontal and vertical bisecting circles intersect each other in a point c; the new markings are denoted as shown by the diagram $\dfrac{X_1 \mid O_1}{\mid X_2}$.

We can use this point c to distinguish two types of grid states in the stabilized grid diagram $\mathbb{G}'$: $\mathbf{x}$ belongs to class $\mathbf{I}$ if it contains c and to $\mathbf{N}$ if it does not. Let the submodules generated by these grid states also be denoted by $\mathbf{I}$ and $\mathbf{N}$; obviously $\mathrm{GC}^-(\mathbb{G}') = \mathbf{I} \oplus \mathbf{N}$ as a module, and since we are performing an $X : SW$-stabilization, $\mathbf{N}$ is indeed a subcomplex; the boundary map on $\mathbf{N}$ will be denoted by $\partial_{\mathbf{N}}^{\mathbf{N}}$. Although $\mathbf{I}$ is not a subcomplex, we can consider the component of the restriction of ∂ to $\mathbf{I}$ which points to $\mathbf{I}$ (and denote it by $\partial_{\mathbf{I}}^{\mathbf{I}}$), and similarly consider the component of the restriction of ∂ to $\mathbf{I}$ pointing to $\mathbf{N}$ (and denote it by $\partial_{\mathbf{N}}^{\mathbf{I}}$). With these notations in place, $\mathrm{GC}^-(\mathbb{G}')$ can be written as the mapping cone of the restriction map $\partial_{\mathbf{N}}^{\mathbf{I}} \colon (\mathbf{I}, \partial_{\mathbf{I}}^{\mathbf{I}}) \to (\mathbf{N}, \partial_{\mathbf{N}}^{\mathbf{N}})$.

The natural identification of grid states in $\mathbf{I}$ with the grid states of $\mathbb{G}$ provides an isomorphism $(\mathbf{I}, \partial_{\mathbf{I}}^{\mathbf{I}}) \to \mathrm{GC}^-(\mathbb{G})[V_1]$. The map $\mathcal{H} \colon (\mathbf{N}, \partial_{\mathbf{N}}^{\mathbf{N}}) \to (\mathbf{I}, \partial_{\mathbf{I}}^{\mathbf{I}})$ defined by counting rectangles containing the X_2-marking for $\mathbf{x} \in \mathbf{N}$ as

$$\mathcal{H}(\mathbf{x}) = \sum_{y \in \mathbf{I}} \ \sum_{\{r \in \mathrm{Rect}^\circ(\mathbf{x}, \mathbf{y}) \mid \mathrm{Int}(r) \cap \mathbb{X} = X_2\}} V_1^{O_1(r)} \cdots V_n^{O_n(r)} \mathbf{y}$$

shows that $(\mathbf{N}, \partial_{\mathbf{N}}^{\mathbf{N}})$ and $(\mathbf{I}, \partial_{\mathbf{I}}^{\mathbf{I}})$ are homotopy equivalent chain complexes. From this observation standard algebra provides that $H_*(\mathrm{Cone}(V_1 - V_2))$ and $H_*(\mathrm{GC}^-(\mathbb{G}'))$ are isomorphic, concluding the proof of stabilization invariance.

FIGURE 72.10: Grid diagram for the unknot. The two grid states of this diagram are indicated by full ($\mathbf{x}$) and hollow ($\mathbf{y}$) circles. Recall that top and bottom, as well as left and right sides are identified, hence there is a full circle at each corner, and a hollow at each bisector, which we did not draw.

Combining the results of the last two subsections, the outline of the proof of Theorem 72.3.2 (the invariance of grid homology) is complete.

72.4 Examples

Before proceeding further, we work out some simple examples, which are instructive in understanding the above arguments, and also will be used in our later discussions.

72.4.1 The Unknot

The unknot K_U can be represented by the 2×2 grid diagram of Figure 72.10, hence we work over the ring $\mathbb{F}[V_1, V_2]$ and the module $GC^-(\mathbb{G})$ is generated by the two grid states $\mathbf{x}$ and $\mathbf{y}$ shown by Figure 72.10. It is not hard to see that $\partial^- \mathbf{x} = 0$ and $\partial^- \mathbf{y} = (V_1 + V_2)\mathbf{x}$, hence the class of $\mathbf{x}$ generates $GH^-(\mathbb{G})$, and since $V_1 \mathbf{x}$ is homologous to $V_2 \mathbf{x}$, we have $GH^-(K_U) \cong \mathbb{F}[U]$ (where U could be any of V_1 or V_2). The gradings are also easy to determine: $M(\mathbf{x}) = A(\mathbf{x}) = 0$ while $M(\mathbf{y}) = -1$, $A(\mathbf{y}) = 0$.

72.4.2 Torus Knots

The (p, q) torus knot $T_{p,q}$, or rather, its mirror image $T_{-p,q}$ can be described by the following $(p + q) \times (p + q)$ grid diagram. Put O's in the skew-diagonal, and X's on the off-skew-diagonal p steps away from the skew diagonal (and view the diagram on the torus), resulting, for example, in the diagram of Figure 72.11 for $T_{-3,7}$. In the diagram we also indicate a grid state $\mathbf{x}$, which is given by the lower left corners of the small squares occupied by X-markings. Obviously every rectangle starting at $\mathbf{x}$ contains an X-marking, hence $\partial^-(\mathbf{x}) = 0$.

A somewhat tedious computation shows that $\mathbf{x}$ has Alexander grading

$$A(\mathbf{x}) = \frac{1}{2}(p - 1)(q - 1),$$

$\mathbf{x}$ has the highest Alexander grading among all the grid states of the given grid, and indeed in that Alexander grading $\mathbf{x}$ is the only grid state. These findings immediately imply that the homology class $[\mathbf{x}]$ is nonzero in $GH^-(\mathbb{G})$. Indeed, more is true: $[\mathbf{x}]$ is nontorsion, that is, $U^n[\mathbf{x}] \neq 0$ in $GH^-(\mathbb{G})$

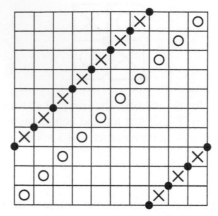

FIGURE 72.11: This grid diagram represents the negative torus knot $T_{-3,7}$, together with a distinguished grid state **x**, occupying the lower left corners of the small squares containing X-markings.

for any $n \in \mathbb{N}$. To prove this claim, we need to introduce another version of the homology theory, where we consider the $\mathbb{F}$-vector space $C'(\mathbb{G})$ generated by the grid states of $\mathbb{G}$ and

$$\partial' \mathbf{x} = \sum_{\mathbf{y} \in S(\mathbb{G})} \sum_{\{r \in \mathrm{Rect}^{\circ}(\mathbf{x},\mathbf{y}) \mid r \cap \mathbb{X} = \emptyset\}} \mathbf{y}.$$

Note that this chain complex does not depend on the O-markings, but does depend on the size of the grid. There is a natural chain map $\mathrm{GC}^{-}(\mathbb{G}) \to C'(\mathbb{G})$ (by setting $V_i = 1$ for all i) inducing a homomorphism $\phi \colon \mathrm{GH}^{-}(\mathbb{G}) \to H_*(C'(\mathbb{G}))$. It is easy to see that $\phi(U^n \mathbf{x}) = \phi(\mathbf{x})$ for all $n \in \mathbb{N}$, hence to prove that a certain class $[\mathbf{x}]$ is nontorsion in $\mathrm{GH}^{-}(\mathbb{G})$, it suffices to show that the class $[\mathbf{x}]$ generated by the same grid state is nonzero in $H_*(C'(\mathbb{G}))$.

For the grid state given by Figure 72.11 this is quite straightforward: if there is an empty rectangle $r \in \mathrm{Rect}^{\circ}(\mathbf{y}, \mathbf{x})$ for some grid state $\mathbf{y}$, then the symmetry of the position of the X-markings show that there is another such rectangle from $\mathbf{y}$ to $\mathbf{x}$ as well. The two rectangles cancel in the boundary map, and so $\mathbf{x}$ is not a boundary in $C'(\mathbb{G})$, hence $[\mathbf{x}] \in \mathrm{GH}^{-}(\mathbb{G})$ is nontorsion.

72.4.3   Negative Whitehead Double of the Left-Handed Trefoil

Consider $W_0^-(T_{-2,3})$, the 0-framed negative clasped Whitehead double of the left-handed trefoil knot $T_{-2,3}$, as shown by Figure 72.12. A grid diagram of that knot is given by Figure 72.13, together with a distinguished grid state **x**. It is not hard to see that $A(\mathbf{x}) = 1$ and (as in our previous example for torus knots) **x** is a cycle. It requires a slightly longer argument to show that **x** is nonzero in homology; indeed, in $\mathrm{GH}^{-}(\mathbb{G})$ we have $U^n \cdot [\mathbf{x}] \neq 0$ for any $n \in \mathbb{N}$.

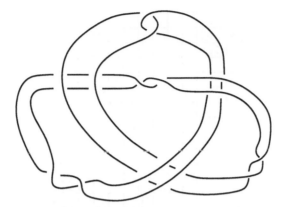

FIGURE 72.12: A knot diagram of the 0-framed, negatively clasped Whitehead double of the left-handed trefoil knot $T_{-2,3}$.

72.5 The Milnor Conjecture

Recall that the Milnor conjecture asserts that the unknotting number of the torus knot $T_{p,q}$ (with $p > q > 0$ relatively prime) is $\frac{1}{2}(p-1)(q-1)$. The first proof of this result (due to Kronheimer and Mrowka [9]) uses gauge theory and delicate arguments from infinite-dimensional analysis, while Rasmussen's proof [20] (based on Khovanov homology and Rasmussen's s-invariant) is combinatorial in nature.

Below we outline the argument using grid homology. At the same time, the proof provides a structure theorem for $GH^-(K)$ of a knot K leading to a numerical invariant. Suppose now that two neighbouring columns (or rows) have marking such that the intervals defined by these markings (as they were defined in the commutation move) now intertwine, see Figure 72.14 for an example. If we commute two such columns, the resulting diagram $\mathbb{G}'$ will define a knot K' which is not necessarily isotopic to the original K given by $\mathbb{G}$, rather K' differs from K by a crossing change. Although this move is not a commutation, we can still draw the two diagrams in one grid, with the difference that the two curves (one for $\mathbb{G}$ and one for $\mathbb{G}'$) now intersect in four points, as shown by the right part of Figure 72.15.

By carefully choosing one of the four crossings (in Figure 72.15 denoted by s) and considering pentagons having this chosen point as one of the vertices, the count of empty pentagons (similar to the commutation move) provides a chain map $c_-: GC^-(\mathbb{G}_+) \to GC^-(\mathbb{G}_-)$, inducing a bigraded map

$$C_-: GH^-(\mathbb{G}_+) \to GH^-(\mathbb{G}_-).$$

Another intersection point t of Figure 72.15 and a similar pentagon counting chain map $c_+: GC^-(\mathbb{G}_-) \to GC^-(\mathbb{G}_+)$, now in the opposite direction, gives rise to a module homomorphism

$$C_+: GH^-(\mathbb{G}_-) \to GH^-(\mathbb{G}_+),$$

which is of (Maslov-Alexander) bidegree $(-2, -1)$, and more importantly the compositions of the

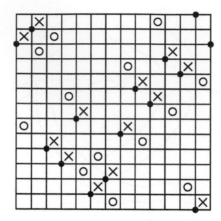

FIGURE 72.13: A grid diagram for the 0-framed, negatively clasped Whitehead double of the left-handed trefoil knot $T_{-2,3}$ of Figure 72.12. In the diagram we also show the distinguished grid state **x**, for which the coordinates are the lower left corners of the small squares occupied by the X-markings.

two maps are simply multiplications by U:

$$C_- \circ C_+ = U \qquad C_+ \circ C_- = U.$$

The existence of such maps has two important consequences. First, notice that by the above composition property, we get that the rank of $\mathrm{GH}^-(K_+)$ (as an $\mathbb{F}[U]$-module) is equal to the rank of $\mathrm{GH}^-(K_-)$. Since any knot can be unknotted (and as we saw $\mathrm{GH}^-(K_U) = \mathbb{F}[U]$ for the unknot K_U), this implies the following:

Theorem 72.5.1. *The grid homology* $\mathrm{GH}^-(K)$ *of a knot K is of rank one, that is,*

$$\mathrm{GH}^-(K) = \mathbb{F}[U] \oplus A,$$

where A is a finitely generated U-torsion module: there is $n \in \mathbb{N}$ with $U^n A = 0$. □

The Alexander grading of the generator of the free part in the decomposition is a knot invariant, which is usually denoted by $-\tau(K)$. (The negative sign is due to another definition of the same quantity.)

Example 72.5.2. Based on our discussion in Subsection 72.4.2, the tau-invariant $\tau(T_{-p,q})$ of the torus knot $T_{-p,q}$ is equal to $-\frac{1}{2}(p-1)(q-1)$, while Subsection 72.4.3 shows that the tau-invariant $\tau(Wh^+(T_{-2,3}))$ is at most -1.

Since both C_+ and C_- shift degree by at most 1, the proof of the Milnor conjecture reduces to the observation that a crossing change can alter the τ-invariant of the underlying knot by at most one, hence

$$|\tau(K)| \le u(K), \tag{72.1}$$

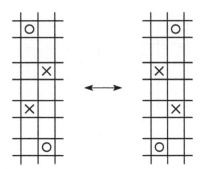

FIGURE 72.14: Exchanging two neighbouring columns, resulting in a crossing change of the underlying knot diagram.

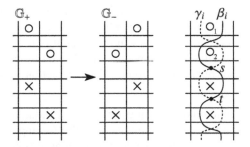

FIGURE 72.15: Crossing change depicted in one diagram. Notice that now the two curved segments intersect in four points rather than two (as was the case for commutation). We distinguished two intersection points s and t on the right-hand diagram.

where $u(K)$ is the unknotting number of K. The calculation of Example 72.5.2 then immediately implies that the unknotting number of the torus knot $T_{-p,q}$ (and therefore of $T_{p,q}$) is at least $\frac{1}{2}(p-1)(q-1)$. It is not hard to see that $T_{p,q}$ can be unknotted using $\frac{1}{2}(p-1)(q-1)$ steps, hence we conclude that

$$u(T_{p,q}) = \frac{1}{2}(p-1)(q-1),$$

verifying Milnor's famous conjecture.

72.6 Knot and Link Invariants

Inequality (72.1) can be strengthened to a bound on the slice genus $g_s(K)$ of a knot K by its τ-invariant. This improvement, however, requires the extension of grid homology from knots to links. The Alexander polynomial of a link has several versions (one variable, or multivariable versions), and the situation is similar for grid homologies. Indeed, the arguments for gradings and invariance of the previous sections work well in case the grid diagram represents a link rather than a knot,

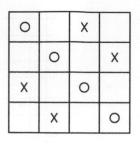

FIGURE 72.16: The (positive) Hopf link, a grid diagram for it and the grid matrix with two variables. The variable t_1 corresponds to the component L_1, the variable t_2 to L_2.

with one notable exception. Recall that in the proof of the fact that on homologies the various variables V_i act identically, we found a chain homotopy between V_i and V_j provided there was an X-marking in the intersection of the row of O_i and the column of O_j. Induction then extended the identification of the actions of all V_i's corresponding to O_i's on the knot. If $\mathbb{G}$ presents a link L, however, this method does not relate the action of V_i to the action of V_j once O_i and O_j are on different components of L.

There are two versions of link homology coming out of this discussion: if we choose a variable V_{i_j} corresponding to an O-marking for the ith component of the ℓ-component link L (and denote the chosen V_{i_j} by U_j with $j \in \{1, \ldots, \ell\}$), then we get a homology theory over $\mathbb{F}[U_1, \ldots, U_\ell]$, which admits ℓ different Alexander gradings, categorifying the multivariable Alexander polynomial of L. (This last statement is also visible through the definition of $\Delta_L(t_1, \ldots, t_\ell)$ as the determinant of the grid matrix introduced in Section 72.1, after the refinement that the winding numbers of the various components are exponents of different variables, see a simple example in Figure 72.16.)

The other method is to take the representative V_{i_j}'s for all the components, and then consider the quotient of the chain complex with the relation $V_{i_1} = \ldots = V_{i_\ell}$ (and denote this variable by U); call the resulting chain complex the *collapsed* complex and denote it by $\mathrm{cGC}^-(\mathbb{G})$. In this way we get that the actions of all the V_i are homotopic, and the resulting homology group $\mathrm{cGH}^-(\mathbb{G})$ will be a link invariant, which is again a (finitely generated) module over the polynomial ring $\mathbb{F}[U]$. Notice that in this case the collapse of the variables forces the Alexander gradings to collapse as well (by adding the individual Alexander gradings), hence $\mathrm{cGH}^-(\mathbb{G})$ is a (Maslov and Alexander) bigraded $\mathbb{F}[U]$-module, which is a link invariant. Most of the basic properties of knot grid homology extend to the collapsed version for links, with the exception that the collapse of the variables implies that the rank of $\mathrm{cGH}^-(L)$ (as an $\mathbb{F}[U]$-module) for an ℓ-component link is $2^{\ell-1}$. In particular, we can imitate the definition of the τ-invariant, but now for a link this invariant will be a set $\{\tau_1(L), \ldots, \tau_{2^{\ell-1}}(L)\}$ of integers, and we will be mainly interested in the smallest and largest of those, $\tau_{\min}(L)$ and $\tau_{\max}(L)$.

Using this extension, we can strengthen Inequality (72.1) to a bound on the slice genus $g_s(K)$ of K:

$$|\tau(K)| \le g_s(K) \tag{72.2}$$

as follows. Suppose that $\Sigma \subset S^3 \times [0, 1]$ is a smoothly and properly embedded genus-g surface providing a cobordism between the knots $K_i \subset S^3 \times \{i\}$ ($i = 0, 1$). Then after a small isotopic

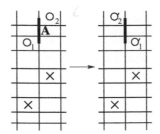

FIGURE 72.17: Grid presentation of an oriented band attachment. The circle between the two markings is divided into two intervals by the O_i markings, one of them is called A (shown in the figure, and not passing by the X-markings), the complement is called B.

perturbation we can assume that projection to the second factor restricts to a Morse function on Σ, and by further isotopy we can also assume that Σ is in normal position: as t grows from 0 to 1, on the critical level sets $S^3 \times \{t\}$ we encounter first 0-handles, then those 1-handles which connect different components (hence after adding those, we have a knot K_0' as boundary), then the $2g$ 1-handles making up the genus of Σ, having the knot K_1' as boundary, and finally the remaining 1-handles and 2-handles. (In many of the intermediate levels the slices are links rather than knots; for this reason we need the extension of the theory to links.) A surface in such a position is called a *surface in normal form*.

Ideas similar to the ones used in the proof of stabilization invariance imply that for a link L' which is the union of another link L and an unlinked unknot, we have

$$\mathrm{cGC}^-(L') \cong \mathrm{cGC}^-(L) \oplus \mathrm{cGC}^-(L)$$

(with some degree shift on one of the summands). This proof allows us to determine the collapsed link homology once we pass all the 0-handles of a cobordism Σ connecting K_0 and K_1.

Then we would like to have a formula determining the changes when passing through a 1-handle. The geometric picture in this case is that we attach an oriented band to the link L; notice that such a band can be attached to two distinct components of L, in which case it merges them, or to one component, in which case (since it is an oriented band) it splits that component into two. Diagrammatically such a band attachment can be depicted in the grid context by finding an appropriate grid diagram of L where the move is simply exchanging the two O-markings of two neighbouring columns, see Figure 72.17.

This picture can be used to find simple maps corresponding to the band attachment: suppose that L' is given by attaching a band to L and it splits a component of L into two, and let W denote the vector space $W = \mathbb{F}^2$, with one generator in bidegree $(0,0)$ and another in $(-1,-1)$. Then there are maps

$$\sigma : \mathrm{cGC}^-(L) \otimes W \to \mathrm{cGC}^-(L') \qquad \mu : \mathrm{cGC}^-(L') \to \mathrm{cGC}^-(L) \otimes W$$

such that σ is of bidegree $(-1,0)$, μ is of bidegree $(-1,-1)$, and both compositions $\mu \circ \sigma$ and $\sigma \circ \mu$ are multiplications by U. Indeed, these maps are rather easy to define: suppose that $\mathbf{x}$ is a grid state with coordinate c on the circle separating the two O-markings to be switched. Depending on the position of c with respect the two O-markings O_1 and O_2, $\mathbf{x}$ is of type A or B (see Figure 72.17).

Then define $\sigma(\mathbf{x}) = U \cdot \mathbf{x}$ if $\mathbf{x}$ is of type A and $\sigma(\mathbf{x}) = \mathbf{x}$ otherwise, while $\mu(\mathbf{x}) = \mathbf{x}$ if $\mathbf{x}$ is of type A and $\mu(\mathbf{x}) = U \cdot \mathbf{x}$ if $\mathbf{x}$ is of type B. Obviously both compositions are equal to multiplication by U.

Using these maps, now it follows at once that if L' is given as a split band attachment to L, then

$$\tau_{\min}(L) - 1 \leq \tau_{\min}(L') \leq \tau_{\min}(L)$$

and

$$\tau_{\max}(L) \leq \tau_{\max}(L') \leq \tau_{\max}(L) + 1.$$

From these inequalities it follows that for a surface $\Sigma \subset S^3 \times [0,1]$ in normal form between the knots K_0, K_1 (and K_0', K_1' as given by the definition of normal form) we have $\tau(K_0) = \tau(K_0')$, $\tau(K_1) = \tau(K_1')$, and $|\tau(K_0') - \tau(K_1')| \leq g(\Sigma)$, ultimately implying

$$|\tau(K_0) - \tau(K_1)| \leq g(\Sigma).$$

Applying this principle to a punctured slice surface (i.e. a cobordism between a knot K and the unknot K_U), and using the fact that $\tau(K_U) = 0$ we get the desired genus bound

$$|\tau(K)| \leq g_s(K).$$

72.6.1   Application

As an application, we see that the slice genus $g_s(T_{p,q})$ of the torus knot $T_{p,q}$ is at least $\frac{1}{2}(p-1)(q-1)$, and since $g_s(K) \leq u(K) = \frac{1}{2}(p-1)(q-1)$, it follows that

$$g_s(T_{p,q}) = \frac{1}{2}(p-1)(q-1).$$

In a similar manner, we get that $\tau(W_0^-(T_{-2,3})) \leq -1$, hence $g_s(W_0^-(T_{-2,3})) \geq 1$. (Indeed, since the unknotting number of a Whitehead double is always at most 1, we also get that $g_s(W_0^-(T_{-2,3})) = 1$.) This computation then shows that the knot $W_0^-(T_{-2,3})$ is not smoothly slice. On the other hand, by a simple calculation the Alexander polynomial $\Delta_{W_0^-(T_{-2,3})}(t)$ of this knot is (as a 0-framed Whitehead double) equal to 1, hence by Freedman's groundbreaking result, this knot is topologically slice [4, 6] conclusion, we get an example of a topologically slice, smoothly nonslice knot.

Topologically slice but smoothly nonslice knots can be used to show the existence of exotic $\mathbb{R}^4$'s, i.e. the existence of smooth four-manifolds which are homeomorphic but not diffeomorphic to $\mathbb{R}^4$. Indeed, suppose that K is a topologically slice knot. View S^4 as the union of two disks: $S^4 = D^4 \cup D^4$ and take the complement of the union of (the interior of) one of them, together with the (locally flat) open neighbourhood of the topological slice disk of K in the other D^4. The result is a compact topological four-manifold with boundary. Delete a point from this compact four-manifold and put a smooth structure on the resulting noncompact manifold Z (which is possible by a deep result of Quinn [5]). Gluing back the smooth four-manifold $X_0(K)$ we get by attaching a 0-framed 2-handle along K to the four-dimensional 0-handle, we get a smooth manifold $R_K = Z \cup X_0(K)$, which is obviously homeomorphic to $\mathbb{R}^4$. This manifold R_K is not diffeomorphic to $\mathbb{R}^4$, since if it were, then R_K smoothly embeds into S^4, hence $X_0(K) \subset R_K$ smoothly embeds into S^4, which then easily shows that K is smoothly slice, which contradicts our choice of K.

72.7 Unoriented Genus

In the previous sections we examined the effect of changing markings in neighbouring columns in various circumstances.

- If we switch the two columns and the markings are not intertwining (as in Figure 72.3), then the geometric effect is a Reidemeister move on the knot diagram defined by the grid, and grid homology remains unchanged.

- If the markings are intertwined (as in Figure 72.14), then the change of the columns creates a crossing change, inducing maps $C_\pm$ on grid homology (cf. Section 72.5). This idea provided the proof of Milnor's conjecture.

- Only switching the O-markings (as in Figure 72.17), we get a grid representation of oriented band attachment, and the corresponding maps σ, μ have been discussed in Section 72.6. (The exchange of two X-markings can be decomposed into one of the column changes and the switch of the O-markings.) This approach gave the key step to the proof of the lower bound on g_s by τ.

There is one more possibility: exchange an O- and an X-marking in the two columns. Notice that the result will not be a grid diagram anymore, since in one column we will have two O's, while in a neighbouring column two X's. This information still provides a knot (or link), but the orientation information coded in the types of the markings has been lost, see Figure 72.18. Such a change corresponds to an *unoriented* band attachment. This leads us to a variant of grid homology (and indeed of knot Floer homology): unoriented grid homology.

The concept of *unoriented knot Floer homology* has been introduced in [16] (and the idea has been extended to further deformations in [15]). The construction goes as follows. Represent the knot $K \subset S^3$ again with a grid diagram $\mathbb{G}$, and define the δ-grading of a grid state $\mathbf{x}$ as

$$\delta(\mathbf{x}) = M(\mathbf{x}) - A(\mathbf{x}).$$

Consider the $\mathbb{F}[U]$-module generated by the grid states, resulting in a graded (and not bigraded!) module $GC'(\mathbb{G})$; the grading of U in this theory is -1. Now in defining the (unoriented) boundary map ∂_u, we will not make any distinction between the X- and O-markings:

$$\partial_u(\mathbf{x}) = \sum_{\mathbf{y} \in S(\mathbb{G})} \sum_{r \in \mathrm{Rect}^\circ(\mathbf{x}, \mathbf{y})} U^{O(r) + X(r)} \mathbf{y},$$

where $O(r) + X(r)$ is the total number of markings (O or X) contained by the rectangle. Notice that now we do not distinguish the O-markings with separate variables.

The proof of $(\partial_u)^2 = 0$ and commutation invariance proceeds as before, in the proof of stabilization one can notice that (since all variables are already collapsed to one single U) we have that

$$GH'(\mathbb{G}') \cong GH'(\mathbb{G}) \otimes V,$$

where V is the 2-dimensional vector space $\mathbb{F}^2$ in δ-grading 0 and $\mathbb{G}'$ is the stabilization of $\mathbb{G}$. It can be shown that for an $n \times n$ grid representing the knot K we have

$$GH'(\mathbb{G}) \cong HFK'(K) \otimes V^{n-1}.$$

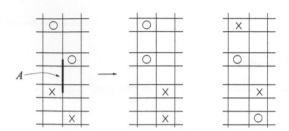

FIGURE 72.18: By switching an X- and an O-marking, an unoriented band attachment can be depicted through grid diagrams. Notice that the result is, strictly speaking, not a grid diagram, and one needs to interchange a number of markings to recover a grid diagram from the picture, shown on the right.

for a knot invariant (the unoriented knot Floer homology) $\mathrm{HFK}'(K)$. Once again, a simple modification of our earlier arguments show that $\mathrm{HFK}'(K)$ (as an $\mathbb{F}[U]$-module) is of rank one, hence the δ-grading of the generator of the free part is a knot invariant $\upsilon(K)$ (upsilon of K). As before, the definition of υ extends to links (and again we get an entire set of integers as invariant); furthermore, $|\upsilon(K)|$ provides a lower bound on the unknotting number and the slice genus — just like $\tau(K)$ does.

The new phenomenon (related to the fact that in this version the orientation of the knot is lost) appeard when we try to bound the *unoriented genus* of K. Let us define

$$\gamma_4(K) = \min\{b_1(F^2) \mid F^2 \subset D^4 \text{ smoothly and properly embedded, compact, } \partial F^2 = K\}.$$

By assuming F^2 orientable, we get the slice genus $g_s(K)$ of K, while above we also allow F^2 to be a nonorientable surface. Note that for the alternating torus knots $T_{2,2n+1}$ we have $g_s(K) = n$, while all these knots are boundaries of Möbius bands, hence $\gamma_4(T_{2,2n+1}) = 1$ for all n. Therefore the two numbers $g_s(K)$ and $\gamma_4(K)$ can drastically differ.

When imitating the proof of slice bound now for nonorientable surfaces, in the case of 1-handles we switch X- and O-markings, and the direct adaptations of the maps σ and μ provide similar bounds. The only difference is that the grading shift is harder to determine, since after the switch the result is not a real grid diagram, while in the computation of δ the classification of markings into X- and O-markings plays an important role. Therefore to compute δ, we need to flip some of the markings, bringing us back to the world of grid diagrams, see the right part of Figure 72.18.

This change then brings in a correction term, leading to the formula

$$|\upsilon(K) - \frac{e(F)}{4}| \le \frac{b_1(F)}{2},$$

where $e(F)$ is the Euler number of the embedded, unoriented surface in D^4 (relative to the Seifert framing on the boundary). Combining this result with the Gordon-Litherland formula [7]

$$|\sigma(K) - \frac{e(F)}{2}| \le b_1(F),$$

(where $\sigma(K)$ is the signature of the knot K) we get

$$|\upsilon(K) - \frac{\sigma(K)}{2}| \le \gamma_4(K),$$

a lower bound on the unoriented genus. (A similar bound has been found earlier by Batson [2], using Heegaard Floer homology and correction terms.)

The map $K \mapsto \upsilon(K)$ descends to a concordance homomorphism, and simple computation shows that for the n-fold connected sum $\#_n T_{3,4}$ of the $(3, 4)$-torus knot with itself we have $\upsilon(\#_n T_{3,4}) = -2n$ and $\sigma(\#_n T_{3,4}) = -6n$, hence $\gamma_4(\#_n T_{3,4}) \geq n$. Indeed, it is not hard to see that

$$\gamma_4(\#_n T_{3,4}) = n,$$

implying that the unoriented smooth four-dimensional genus is not bounded. (Further applications of the υ-invariant in studying the unoriented genus has been given, for example in [8].) We just note here that $\upsilon(K)$ is a special case of a wider construction of the Upsilon-function $\Upsilon_K : [0, 2] \to \mathbb{R}$ of a knot; in fact, $\upsilon(K) = \Upsilon_K(1)$.

72.8 Bibliography

[1] K. Baker, E. Grigsby, and M. Hedden. Grid diagrams for lens spaces and combinatorial knot Floer homology. *Int. Math. Res. Not. IMRN*, (10):Art. ID rnm024, 39, 2008.

[2] J. Batson. Nonorientable slice genus can be arbitrarily large. *Math. Res. Lett.*, 21(3):423–436, 2014.

[3] P. Cromwell. Embedding knots and links in an open book. I. Basic properties. *Topology Appl.*, 64(1):37–58, 1995.

[4] M. Freedman. A surgery sequence in dimension four; the relations with knot concordance. *Invent. Math.*, 68(2):195–226, 1982.

[5] M. Freedman and F. Quinn. *Topology of 4-Manifolds*, volume 39 of *Princeton Mathematical Series*. Princeton University Press, Princeton, NJ, 1990.

[6] S. Garoufalidis and P. Teichner. On knots with trivial Alexander polynomial. *J. Differential Geom.*, 67(1):167–193, 2004.

[7] C. Gordon and R. Litherland. On the signature of a link. *Invent. Math.*, 47(1):53–69, 1978.

[8] S. Jabuka and T. Kelly. The nonorientable 4-genus for knots with 8 or 9 crossings. *Algebr. Geom. Topol.*, 18(3):1823–1856, 2018.

[9] P. Kronheimer and T. Mrowka. Gauge theory for embedded surfaces. I. *Topology*, 32(4):773–826, 1993.

[10] P. Lisca, P. Ozsváth, A. Stipsicz, and Z. Szabó. Heegaard Floer invariants of Legendrian knots in contact three-manifolds. *J. Eur. Math. Soc. (JEMS)*, 11(6):1307–1363, 2009.

[11] C. Manolescu and P. Ozsváth. Heegaard Floer homology and integer surgeries on links. arXiv:1011.1317, 2015.

[12] C. Manolescu, P. Ozsváth, and S. Sarkar. A combinatorial description of knot Floer homology. *Ann. of Math. (2)*, 169(2):633–660, 2009.

[13] C. Manolescu, P. Ozsváth, Z. Szabó, and D. Thurston. On combinatorial link Floer homology. *Geom. Topol.*, 11:2339–2412, 2007.

[14] P. Ozsváth, A. Stipsicz, and Z. Szabó. *Grid Homology for Knots and Links*, volume 208 of *Mathematical Surveys and Monographs*. American Mathematical Society, Providence, RI, 2015.

[15] P. Ozsváth, A. Stipsicz, and Z. Szabó. Concordance homomorphisms from knot Floer homology. *Adv. Math.*, 315:366–426, 2017.

[16] P. Ozsváth, A. Stipsicz, and Z. Szabó. Unoriented knot Floer homology and the unoriented four-ball genus. *Int. Math. Res. Not. IMRN*, (17):5137–5181, 2017.

[17] P. Ozsváth and Z. Szabó. Holomorphic disks and knot invariants. *Adv. Math.*, 186(1):58–116, 2004.

[18] P. Ozsváth and Z. Szabó. Kauffman states, bordered algebras, and a bigraded knot invariant. *Adv. Math.*, 328:1088–1198, 2018.

[19] J. Rasmussen. *Floer Homology and Knot Complements*. ProQuest LLC, Ann Arbor, MI, 2003. Thesis (Ph.D.)–Harvard University.

[20] J. Rasmussen. Khovanov homology and the slice genus. *Invent. Math.*, 182(2):419–447, 2010.

[21] S. Sarkar and J. Wang. An algorithm for computing some Heegaard Floer homologies. *Ann. of Math. (2)*, 171(2):1213–1236, 2010.

Chapter 73

Categorification

Volodymyr Mazorchuk, Uppsala University

In 1995, the *Journal of Mathematical Physics* published the paper "Clock and Category: Is Quantum Gravity Algebraic?" by Louis Crane at Kansas State University, cf. [9]. The second subsection in Section 4 of that paper starts with the following passage:

> *"As we stated above, factorizable TQFTs have tiers of analogous structures at higher categorical levels. The structure on surfaces in a 3-D TQFT, for example, is a categorical analog of the structure of a 2-D TQFT. To make this notion precise, we have defined the idea of a categorification, cf. [10].*
>
> *The categorification of any type of algebraic structure is a type of structure with analogous operations one categorical level higher. A categorification of a particular algebraic structure is an example of the categorified type of the structure which when "traced down" gives us back the original structure."*

This seems to be the earliest mentioning of the term **categorification**, despite the reference to [10], a paper by Louis Crane and Igor Frenkel (at Yale) from 1994, where the main ideas behind this term were described.

The above definition from [10] is a very concise, but, at the same time, fairly precise and to-the-point description of the idea of categorification. Below, we will try to explain what categorification is in more detail using some examples. Examples of categorification are many, and a number of them were well known and used long before 1994.

Let us start with a step back to combinatorics, where one can find the idea of a so-called "combinatorial proof," as described by Richard Stanley in [21]. This is a natural ancestor of the idea of categorification. A combinatorial proof is a term used to describe, among other things, a proof of a numerical identity $A = B$ by interpreting both A and B as the cardinalities of certain finite sets X and Y, respectively, and then using either a double counting (if $X = Y$) or construction of a bijection between X and Y to obtain the result.

As an example, consider the following identity:

$$n! = \sum_{\substack{\lambda_1 \geqslant \cdots \geqslant \lambda_k \geqslant 1 \\ \lambda_1 + \cdots + \lambda_k = n}} \left(n! \cdot \det\left[\frac{1}{(\lambda_i - i + j)!} \right]_{i,j=1}^k \right)^2, \tag{73.1}$$

where the bracket on the right-hand side contains the determinant of a $k \times k$ matrix with the entry

$\frac{1}{(\lambda_i-i+j)!}$ in the intersection of row i and column j. A combinatorial proof of (73.1) would be to observe that the left-hand side enumerates the set S_n of all permutations of $\{1, 2, \ldots, n\}$ while the expression

$$n! \cdot \det\left[\frac{1}{(\lambda_i - i + j)!}\right]_{i,j=1}^k$$

on the right-hand side enumerates the set SYT_λ of the so-called *standard Young tableaux* of shape $\lambda = (\lambda_1, \lambda_2, \ldots, \lambda_k)$, and then use the classical *Schensted algorithm*, see [20], which provides a bijection

$$S_n \overset{1:1}{\longleftrightarrow} \coprod_\lambda \mathrm{SYT}_\lambda \times \mathrm{SYT}_\lambda. \tag{73.2}$$

In this way, we interpret (73.1) as a numerical shadow of a certain combinatorial phenomenon in the theory of finite sets.

In fact, we can do even better. In its turn, our set-theoretic interpretation (73.2) of (73.1) is a shadow of a fancier phenomenon in representation theory (which, in reality, is the origin of (73.1)). The set S_n has the natural structure of a group, given by composition of maps. The left regular representation of S_n (say, over the complex numbers), has a basis naturally indexed by the elements of S_n. This representation decomposes into a direct sum of simple representations, each appearing with multiplicity equal to its dimension. Further, simple representations of S_n are parameterized by λ as above, and the representation S_λ corresponding to λ has a natural basis indexed by the elements in SYT_λ (see [19]). In this way, we obtain an isomorphism

$$\mathbb{C}[S_n] \cong \bigoplus_\lambda S_\lambda^{\oplus \dim(S_\lambda)} \tag{73.3}$$

from which the bijection (73.2) follows using basic linear algebra. Thus, (73.2) becomes a set-theoretic shadow of a certain phenomenon in the category of complex representations of S_n. Put briefly, one simply says that (73.3) categorifies (73.2).

Historically, the above example is misleading: the numerical equality (73.1) was originally motivated by representation theory. The above example is just an attempt to explain the concept. However, here is a more accurate example. In representation theory of finite groups, there is a famous conjecture, called *Alperin's weight conjecture*, formulated by Alperin in [1] in 1986. Let G be a finite group and $\Bbbk$ an algebraically closed field of positive characteristic p. The finite-dimensional algebra $\Bbbk G$ decomposes into a direct sum of indecomposable subalgebras, called *blocks*. These subalgebras are not semi-simple, in general. Given such a block A, the defect group of this block is a minimal subgroup D of G such that every simple A-module is a direct summand of a module induced from $\Bbbk D$. The defect group D is a p-subgroup of G and is determined uniquely up to conjugacy. Brauer's first main theorem asserts that there is a natural bijection between blocks of $\Bbbk G$ with defect group D and blocks of $\Bbbk N_G(D)$ with defect group D, where $N_G(D)$ is the normalizer of D in G. The block B of $\Bbbk N_G(D)$ corresponding to A under this bijection is called the *Brauer correspondent* of A. Alperin's conjecture is based on a large number of empirical observations and can be formulated, for example, as follows:

Let G be a finite group and $\Bbbk$ an algebraically closed field of positive characteristic. Let, further, A be a block of $\Bbbk G$ with abelian defect group and B its Brauer correspondent. Then A and B have the same number of isomorphism classes of simple modules.

From the categorical point of view, the easiest way to prove Alperin's conjecture would be to prove that A and B are Morita equivalent, however, there are examples for which this is not the case. In 1990, in [6], Broué proposed the following upgrade of this conjecture, now known as *Broué's conjecture*:

Let G be a finite group and $\Bbbk$ an algebraically closed field of positive characteristic. Let, further, A be a block of $\Bbbk G$ with abelian defect group and B its Brauer correspondent. Then the bounded derived categories $\mathcal{D}^b(A)$ and $\mathcal{D}^b(B)$ are equivalent.

Broué's conjecture implies Alperin's conjecture by noting that equivalence of $\mathcal{D}^b(A)$ and $\mathcal{D}^b(B)$ implies isomorphism of the Grothendieck groups of these two categories and that the number of isomorphism classes of simple modules for A or B is exactly the rank of the Grothendieck group of $\mathcal{D}^b(A)$ or $\mathcal{D}^b(B)$, respectively. In this way, the numerical content of Alperin's conjecture becomes a shadow of a much deeper categorical fact predicted in Broué's conjecture and hence, in this way, Broué's conjecture is a categorification of Alperin's conjecture. Of course, this story predates establishment of categorification terminology.

The most famous example of categorification is Khovanov homology for oriented knots and links, introduced in [16]. It categorifies the Jones polynomial from [14, 15]. In low-dimensional topology, a basic problem is to tell apart two knots (or links), that is two "sufficiently nice" embeddings of the circle, that is a 1-dimensional sphere (resp. a finite number of such spheres) into $\mathbb{R}^3$. The usual mathematical approach to such problems is invariant theory, which is the construction of invariants for knots and links. The Jones polynomial is one of the most famous such invariants.

One of the easiest ways to represent a link is via its "sufficiently nice" projection L onto a plane. In such a projection, the essential information about the link is contained in so-called *crossings* of L which can be of the following two kinds (for oriented knots and links):

right crossing   left crossing

The *Kauffman bracket polynomial* $\langle L \rangle \in \mathbb{Z}[v, v^{-1}]$ of L is defined via the following elementary combinatorial manipulation of crossings:

$$\left\langle \times \right\rangle = \left\langle \smile\hspace{-0.3em}\frown \right\rangle - v \left\langle)(\right\rangle \qquad (73.4)$$

together with the convention $\langle \bigcirc L \rangle = (v + v^{-1})\langle L \rangle$ and normalization $\langle \emptyset \rangle = 1$.

The (unnormalized) *Jones polynomial* $\hat{\mathrm{J}}(L)$ of L is given by

$$\hat{\mathrm{J}}(L) := (-1)^{n_-} v^{n_+ - 2n_-} \langle L \rangle \in \mathbb{Z}[v, v^{-1}],$$

where n_+ and n_- denote the numbers of right and left crossings in L, respectively. The usual Jones polynomial $\mathrm{J}(L)$ is defined via $(v + v^{-1})\mathrm{J}(L) = \hat{\mathrm{J}}(L)$. The celebrated theorem of Jones asserts that $\mathrm{J}(L)$ is an invariant of an oriented link.

In [16], Khovanov proposes to "upgrade" the Kauffman bracket $\langle \cdot \rangle$ to a new, much fancier, bracket $[\![\cdot]\!]$, which takes values in complexes of graded complex vector spaces. The definitions also work over $\mathbb{Z}$, Khovanov originally worked over $\mathbb{Z}[c]$ with $\deg c = 2$. Let V be the two-dimensional graded vector space such that the nonzero homogeneous parts of V are of degrees 1 and -1. The normalization conditions for $\langle \cdot \rangle$ are easy to upgrade:

$$[\![\varnothing]\!] = 0 \to \mathbb{C} \to 0, \qquad [\![\bigcirc L]\!] = V \otimes [\![L]\!].$$

As the main technical point of [16], Khovanov proposes the following "upgrade" of the crossings manipulation rule (73.4):

$$\left[\!\!\left[\times \right]\!\!\right] = \mathrm{Total}\left(0 \to \left[\!\!\left[\asymp \right]\!\!\right] \xrightarrow{d} \left[\!\!\left[ )(\right]\!\!\right] \langle -1 \rangle \to 0 \right),$$

for a certain highly non-trivial differential d. We refer the reader to [16, 2] for details on how this works.

The above results in a chain complex that heavily depends on the diagram L that one starts with. However, the main theorem of [16] asserts that the *homology* of this complex is, in fact, an invariant of an oriented link. This is what is now known as *Khovanov homology*. The connection to the Jones polynomial is given by the observation that the Kauffman bracket polynomial $\langle L \rangle$ is just the graded Euler characteristic of $[\![L]\!]$. In this way, $\hat{J}(L)$ becomes the graded Euler characteristic of the Khovanov homology of L and therefore Khovanov homology can be viewed as a *categorification* of the Jones polynomial.

This example also clearly shows the strength of the categorification approach. It is easy to guess and then verify by computations that the homology of $[\![L]\!]$ contains significantly more information than its graded Euler characteristic and hence Khovanov homology provides a significantly stronger invariant than the Jones polynomial. A slight disclaimer is that this does come with a very heavy computational cost (and a way more sophisticated theoretical machinery required to establish the theory).

After the examples, let us discuss Crane's definition of *categorification* in more detail. To start with, it is convenient to compare various notions in set theory and category theory. This can be summarized in the following table:

Set Theory	Category Theory
set	category
element	object
function	functor
relation between elements	morphism between objects
relation between functions	natural transformation of functors

So, a *categorification* of some algebraic structure expressed in set-theoretic terms should be a categorical "upgrade" of this structure in which the set-theoretic ingredients of the original structure are replaced by their category-theoretic counterparts as indicated by the above table.

A slightly different way to look at this is, rather than defining categorification, to define the inverse procedure, usually referred to as *decategorification*. Definition of decategorification is easier due to the fact that it is always easier to forget information than to invent some. In algebra and topology, there are two slightly different classical ways to define decategorification.

The first, most common, one is via the so-called *Grothendieck group*. Given an essentially small abelian category $\mathcal{A}$, the Grothendieck group $\mathrm{Gr}(\mathcal{A})$ is the quotient of the free abelian group generated by the isomorphism classes of objects in $\mathcal{A}$ modulo the subgroup given by all possible relations of the form $[Y] - [X] - [Z]$, where $0 \to X \to Y \to Z \to 0$ is a short exact sequence in $\mathcal{A}$. For example, if we consider the category $\mathcal{A} = A$-mod of finite-dimensional modules over a finite-dimensional algebra A over some field, then $\mathrm{Gr}(\mathcal{A})$ is a free abelian group whose rank equals the number of isomorphism classes of simple A-modules.

In case $\mathcal{A}$ is an essentially small, additive, idempotent split and Krull-Schmidt category, one considers instead the *split Grothendieck group* $\mathrm{Gr}_\oplus(\mathcal{A})$, which is the quotient of the free abelian group generated by the isomorphism classes of objects in $\mathcal{A}$ modulo the subgroup given by all possible relations of the form $[Y] - [X] - [Z]$, whenever $Y \cong X \oplus Z$ in $\mathcal{A}$. For example, if $\mathcal{A} = A$-proj, the category of finitely generated projective modules over a finite-dimensional algebra A over some field, then $\mathrm{Gr}_\oplus(\mathcal{A})$ is a free abelian group whose rank equals the number of isomorphism classes of indecomposable projective A-modules. In particular, $\mathrm{Gr}(A\text{-mod}) \cong \mathrm{Gr}_\oplus(A\text{-proj})$.

In each of the above cases, if, additionally, $\mathcal{A}$ has a tensor product (which is assumed to be biexact in case $\mathcal{A}$ is abelian or biadditive in case $\mathcal{A}$ is additive), this tensor product induces on $\mathrm{Gr}(\mathcal{A})$ and $\mathrm{Gr}_\oplus(\mathcal{A})$, respectively, the structure of a ring, called the *Grothendieck ring* of $\mathcal{A}$. One usually refers to $\mathrm{Gr}(\mathcal{A})$ or $\mathrm{Gr}_\oplus(\mathcal{A})$ as *Grothendieck decategorification* of $\mathcal{A}$. Turning the terminology around, one says that $\mathcal{A}$ is a *Grothendieck categorification* of $\mathrm{Gr}(\mathcal{A})$ (resp. $\mathrm{Gr}_\oplus(\mathcal{A})$).

For example, let $\mathcal{A}$ be the category vect($\Bbbk$) of finite-dimensional vector spaces over a field $\Bbbk$. We have $\mathcal{A} \cong \Bbbk$-mod $\cong \Bbbk$-proj. The category $\mathcal{A}$ contains a unique, up to isomorphism, simple object, the one-dimensional $\Bbbk$-vector space $_\Bbbk\Bbbk$. Therefore we have $\mathrm{Gr}(\mathcal{A}) \cong \mathrm{Gr}_\oplus(\mathcal{A}) \cong \mathbb{Z}$, where the class of the vector space V is represented by $\dim(V)$. Here we see how a number (i.e. an element of the set $\mathbb{Z}$) is replaced by an object V in $\mathcal{A}$ (in fact, by many isomorphic objects). As $\dim(V \otimes W) = \dim(V) \cdot \dim(W)$, the obvious tensor structure on $\mathcal{A}$ given by the usual tensor product of vector spaces defines on $\mathbb{Z}$ the structure of a ring which coincides with the usual ring structure on integers.

An exact (resp. additive) functor F between abelian (resp. additive) categories $\mathcal{A}$ and $\mathcal{B}$ induces a homomorphism $\overline{\mathrm{F}}$ between the abelian groups $\mathrm{Gr}(\mathcal{A})$ and $\mathrm{Gr}(\mathcal{B})$ (resp. $\mathrm{Gr}_\oplus(\mathcal{A})$ and $\mathrm{Gr}_\oplus(\mathcal{B})$). This shows that *Grothendieck decategorification* is even functorial.

The second, much younger, classical approach to categorification uses *trace*, see [3]. Let $\Bbbk$ be a commutative ring and C a small $\Bbbk$-linear category. Then the *trace* of C is defined as the $\Bbbk$-module

$$\mathrm{Tr}(C) := \left(\bigoplus_{X \in C} C(X, X) \right) / \mathrm{Span}_\Bbbk(fg - gf),$$

where f and g run through all possible pairs (f, g) of morphisms $f : X \to Y$ and $g : Y \to X$, where $X, Y \in C$. Note that, in case $f = g^{-1}$, we obtain the relation $\mathrm{id}_X = \mathrm{id}_Y$ which allows one to extend the above definition to the case of a skeletally small $\Bbbk$-linear category.

For example, if $C = $ vect($\Bbbk$), where $\Bbbk$ is a field, then the class in $\mathrm{Tr}(C)$ of an endomorphism f of a vector space V equals the class of $\mathrm{tr}(f)\mathrm{id}_V$, where tr is the usual trace of a linear endomorphism. In fact, $\mathrm{Tr}(C)$ is isomorphic to $\Bbbk$, see [3, Section 3].

The relation between the Grothendieck and the trace decategorifications is studied in [3, Section 3.5]. As explained there, for any small, additive category C, there is a natural homomorphism of abelian groups $\mathbf{h} : \mathrm{Gr}_\oplus(C) \to \mathrm{Tr}(C)$, however, this homomorphism is neither injective nor surjective in general.

We refer the reader to [3, 4] for further details and examples of trace decategorification and also to surveys [17, 18] for further details on the rest of the material.

Given all the definitions, a natural question is: Why bother? Why is categorification interesting or useful? There are many different aspects which one could discuss in a potential answer. The most obvious one: categorification is, clearly, useful, as it allows one to prove stronger results. For example, Khovanov homology is a strictly stronger invariant than the Jones polynomial, which Khovanov homology categorifies. The reason why this works is the fact that any kind of categorification introduces more structure and hence creates more opportunities to analyze, preserve and compare information.

Another important aspect of categorification is that it provides external tools to connect different objects and also different areas. For instance, the same categorical structure can be decategorified in different ways thus establishing a non-trivial connection between the outcomes of these different categorifications. Also, categorification techniques developed in one area might turn out to be very useful for applications in other areas. For examples, categorification ideas and techniques inspired by topology turned out to be very useful in representation theory, see, for instance, [5, 22] for a reformulation of Khovanov homology in terms of functorial actions on BGG category O for the Lie algebra $\mathfrak{sl}_n$, which led to very significant progress in understanding the structure of category O.

Finally, yet another important aspect of the general appeal of categorification is its intrinsic beauty. A discovery of a higher categorical symmetry which underlines a classical algebraic or set-theoretic structure or statement naturally comes together with a significant feeling of a profound aesthetic admiration.

Let us finish with mentioning yet another interesting and influential example of categorification. In 2002, Fomin and Zelevinsky published their first foundational paper [11] on what is now known as the theory of cluster algebras. The principal novel invention of [11] is the combinatorial notion of a *mutation* of a *seed*. Fix a positive integer n and let R denote the field $\mathbb{Q}(x_1, x_2, \ldots, x_n)$ of rational functions in n variables. A *seed* is a tuple $(\mathbf{v}, \Gamma)$ consisting of a vector $\mathbf{v} = (v_1, v_2, \ldots, v_n)$, called a *cluster*, where $\{v_1, \ldots, v_n\}$ is an algebraically independent set of generators in R, and a finite oriented graph Γ with vertex set $\{1, 2, \ldots, n\}$ which has no loops and no 2-cycles. Each seed can be mutated at any coordinate. The mutation μ_i of a seed $(\mathbf{v}, \Gamma)$ at the coordinate i produces a new seed $(\mathbf{v}', \Gamma') = \mu_i((\mathbf{v}, \Gamma))$. The only difference between the clusters $\mathbf{v}'$ and $\mathbf{v}$ is the i-th coordinate which is computed by the rule

$$v_i' v_i = \prod_{\alpha : j \to i} v_j + \prod_{\beta : i \to j} v_j,$$

where the first product is taken over all arrows in Γ which terminate at i and the second product is taken over all arrows in Γ which start at i. The graph Γ' is obtained from Γ in three steps: first, for each $\alpha : j \to i$ and $\beta : i \to k$ in Γ, one adds a new arrow from j to k in Γ', then one reverses the orientation of all arrows adjoint to i, and then one deletes a maximal set of 2-cycles (which might be potentially created at the first step). One can check that such a mutation is well defined in the sense that it produces a seed and that every mutation is involutive, that is $\mu_i^2((\mathbf{v}, \Gamma)) = (\mathbf{v}, \Gamma)$.

The above defines an action of the free product of n copies of $\mathbb{Z}/2\mathbb{Z}$ on the set of all seeds. A natural question to ask is: Are there any finite orbits of this action, and, if there are, can one classify them? A very surprising result in [12] asserts that such finite orbits can be classified using orientations of simply laced Dynkin diagrams.

Orientations of simply laced Dynkin diagrams are classical objects in representation theory. The famous theorem of Gabriel, see [13], asserts that path algebras of oriented, simply laced Dynkin diagrams are exactly those hereditary finite-dimensional algebras (over an algebraically closed field) which have *finite representation type*, that is, only finitely many indecomposable modules, up to isomorphism. This observation suggests the possibility of interpretation of cluster combinatorics in terms of representations of finite-dimensional algebras. Such a connection was established in [7]. Given the path algebra A of an oriented, simply laced Dynkin diagram Γ, we can consider the category A-mod of all finite-dimensional A-modules and the corresponding bounded derived category $\mathcal{D}^b(A)$. The paper [7] introduces a certain quotient C of $\mathcal{D}^b(A)$ called the *cluster category* and shows that C is a natural model for categorification of the mutation combinatorics of the seed associated to Γ. In this categorification, clusters of Fomin and Zelevinsky correspond to tilting objects (i.e. maximal, self-orthogonal object) and mutations correspond to replacing an indecomposable direct summand of a tilting object by another indecomposable self-orthogonal module. The striking fact of the theory is that the replacement possibility is unique, up to isomorphism, if we want to obtain a tilting object. So, in this way, the analogy between the set-theoretic mutation and its categorification becomes very clean. In [8] it was further shown that, in the setup of [7], the oriented graph of a seed has an interpretation as the Gabriel quiver of the endomorphism algebra of the tilting object associated with the cluster of that seed.

Connection to cluster combinatorics provides a very strong motivation to study the category C which gives an obvious example of categorification. The knowledge and insight obtained by understanding C led to development of and significant progress in several new directions in representation theory of finite-dimensional algebras, including, in particular, higher Auslander-Reiten theory. So, again, one can clearly see the impact of the idea of categorification on the development of modern mathematics.

73.1 Bibliography

[1] J. Alperin. Weights for finite groups. The Arcata Conference on Representations of Finite Groups (Arcata, Calif., 1986), 369–379, *Proc. Sympos. Pure Math.*, 47, Part 1, Amer. Math. Soc., Providence, RI, 1987.

[2] D. Bar-Natan. On Khovanov's categorification of the Jones polynomial. *Algebr. Geom. Topol.* **2** (2002), 337–370.

[3] A. Beliakova, Z. Guliyev, K. Habiro, A. Lauda. Trace as an alternative decategorification functor. *Acta Math. Vietnam.* **39** (2014), no. 4, 425–480.

[4] A. Beliakova; K. Habiro, A. Lauda, M. Živković. Trace decategorification of categorified quantum $\mathfrak{sl}_2$. *Math. Ann.* **367** (2017), no. 1-2, 397–440.

[5] J. Bernstein, I. Frenkel, M. Khovanov. A categorification of the Temperley-Lieb algebra and Schur quotients of $U(\mathfrak{sl}_2)$ via projective and Zuckerman functors. *Selecta Math.* (N.S.) **5** (1999), no. 2, 199–241.

[6] M. Broué. Isométries parfaites, types de blocs, catégories dérivées. Astérisque No. **181-182** (1990), 61–92.

[7] A. Buan. R. Marsh, M. Reineke, I. Reiten, G. Todorov. Tilting theory and cluster combinatorics. *Adv. Math.* **204** (2006), no. 2, 572–618.

[8] A. Buan, R. Marsh, I. Reiten. Cluster mutation via quiver representations. Comment. *Math. Helv.* **83** (2008), no. 1, 143–177.

[9] L. Crane. Clock and category: is quantum gravity algebraic? *J. Math. Phys.* **36** (1995), no. 11, 6180–6193.

[10] L. Crane; I. Frenkel. Four-dimensional topological quantum field theory, Hopf categories, and the canonical bases. Topology and physics. *J. Math. Phys.* **35** (1994), no. 10, 5136–5154.

[11] S. Fomin, A. Zelevinsky. Cluster algebras. I. Foundations. *J. Amer. Math. Soc.* **15** (2002), no. 2, 497–529.

[12] S. Fomin, A. Zelevinsky. Cluster algebras. II. Finite type classification. *Invent. Math.* **154** (2003), no. 1, 63–121.

[13] P. Gabriel. Unzerlegbare Darstellungen. I. *Manuscripta Math.* **6** (1972), 71–103; correction, ibid. **6** (1972), 309.

[14] V. Jones, A polynomial invariant for knots via von Neumann algebras. *Bull. Amer. Math. Soc.* (N.S.) **12** (1985), 103–111.

[15] V. Jones. Hecke algebra representations of braid groups and link polynomials. Ann. Math. **126** (1987), no. 2, 335–388.

[16] M. Khovanov. A categorification of the Jones polynomial. *Duke Math. J.* **101** (2000), no. 3, 359–426.

[17] M. Khovanov, V. Mazorchuk, C. Stroppel. A brief review of abelian categorifications. *Theory Appl. Categ.* **22** (2009), No. 19, 479–508.

[18] V. Mazorchuk. *Lectures on algebraic categorification.* QGM Master Class Series. European Mathematical Society (EMS), Zürich, 2012. x+119 pp.

[19] B. Sagan. *The symmetric group. Representations, combinatorial algorithms, and symmetric functions.* Second edition. Graduate Texts in Mathematics, **203**. Springer-Verlag, New York, 2001. xvi+238 pp.

[20] C. Schensted. Longest increasing and decreasing subsequences. *Canadian J. Math.* **13** (1961), 179–191.

[21] R. Stanley. *Enumerative combinatorics.* Vol. I. Wadsworth & Brooks/Cole Advanced Books & Software, Monterey, CA, 1986. xiv+306 pp.

[22] C. Stroppel. Categorification of the Temperley-Lieb category, tangles, and cobordisms via projective functors. *Duke Math. J.* **126** (2005), no. 3, 547–596.

Chapter 74

Khovanov Homology and the Jones Polynomial

Alexander N. Shumakovitch, The George Washington University

74.1 Introduction

Khovanov homology is a special case of *categorification*, a novel approach to construction of knot (or link) invariants that was introduced by Mikhail Khovanov at the turn of the millennium in his seminal paper [15] and is being actively developed ever since. The idea of categorification is to replace a known polynomial knot (or link) invariant with a family of chain complexes, such that the coefficients of the original polynomial are the Euler characteristics of these complexes. Although the chain complexes themselves depend heavily on a diagram that represents the link, their homology depends on the isotopy class of the link only. Khovanov homology categorifies the Jones polynomial [12], see also Chapter 61.

More specifically, let L be an oriented link in $\mathbb{R}^3$ represented by a planar diagram D and let $J_L(q)$ be a version of the Jones polynomial of L that is defined by the following skein relation and normalization:

$$-q^{-2}J_{\diagup\!\!\!\!\diagdown_+}(q) + q^2 J_{\diagdown\!\!\!\!\diagup_-}(q) = (q - 1/q)J_{\smile^0\frown}(q); \qquad J_{\bigcirc}(q) = q + 1/q. \qquad (74.1)$$

$J_L(q)$ is a Laurent polynomial in q for every link L.

In [15] Khovanov assigned to D a family of Abelian groups $\mathcal{H}^{i,j}(L)$ whose isomorphism classes depend on the isotopy class of L only. These groups are defined as homology groups of an appropriate (graded) chain complex $C^{i,j}(D)$ with integer coefficients. Groups $\mathcal{H}^{i,j}(L)$ are nontrivial for finitely many values of the pair (i, j) only. The gist of the categorification is that the graded Euler characteristic of the Khovanov chain complex equals $J_L(q)$:

$$J_L(q) = \sum_{i,j} (-1)^i q^j h^{i,j}(L), \qquad (74.2)$$

where $h^{i,j}(L) = \mathrm{rk}(\mathcal{H}^{i,j}(L))$, the Betti numbers of $\mathcal{H}$. The reader is referred to Section 74.2 for detailed treatment, see also [3, 15].

In this chapter, we make use of another version of the Jones polynomial, denoted $\widetilde{J}_L(q)$, that satisfies the same skein relation (74.1) but is normalized to equal 1 on the unknot. For the sake of completeness, we also list the skein relation for the original Jones polynomial, $V_L(t)$, from [12]:

$$t^{-1}V_{\diagup\!\!\!\!\diagdown}(t) - tV_{\diagdown\!\!\!\!\diagup}(t) = (t^{1/2} - t^{-1/2})V_{\smile\frown}(t); \qquad V_{\bigcirc}(t) = 1. \qquad (74.3)$$

We note that $J_L(q) \in \mathbb{Z}[q, q^{-1}]$ while $V_L(t) \in \mathbb{Z}[t^{1/2}, t^{-1/2}]$. In fact, the terms of $V_L(t)$ have half-integer (resp. integer) exponents if L has an even (resp. odd) number of components. This is one of

	0	1	2	3
9				1
7				1_2
5			1	
3	1			
1	1			

FIGURE 74.1: Right trefoil and its Khovanov homology.

the main motivations for our convention (74.1) to be different from (74.3). The different versions of the Jones polynomial are related as follows:

$$J_L(q) = (q + 1/q)\widetilde{J}_L(q), \qquad \widetilde{J}_L(-t^{1/2}) = V_L(t), \qquad V_L(q^2) = \widetilde{J}_L(q) \tag{74.4}$$

Another way to look at Khovanov's identity (74.2) is via the *Poincaré polynomial* of the Khovanov homology:

$$Kh_L(t, q) = \sum_{i,j} t^i q^j h^{i,j}(L). \tag{74.5}$$

With this notation, we get

$$J_L(q) = Kh_L(-1, q). \tag{74.6}$$

Example 74.1.1. Consider the right trefoil T. Its non-zero homology groups are tabulated in Figure 74.1, where the i-grading is represented horizontally and the j-grading vertically. The homology is non-trivial for odd j-grading only, and hence, even rows are not shown in the table. A table entry of **1** or **1_2** means that the corresponding group is $\mathbb{Z}$ or $\mathbb{Z}_2$, respectively (one can find a more interesting example in Figure 74.7). In general, an entry of the form **a**, **b_2** corresponds to the group $\mathbb{Z}^a \oplus \mathbb{Z}_2^b$. For the trefoil T, we have that $\mathcal{H}^{0,1}(T) \simeq \mathcal{H}^{0,3}(T) \simeq \mathcal{H}^{2,5}(T) \simeq \mathcal{H}^{3,9}(T) \simeq \mathbb{Z}$ and $\mathcal{H}^{3,7}(T) \simeq \mathbb{Z}_2$. Therefore, $Kh_T(t, q) = q + q^3 + t^2 q^5 + t^3 q^9$. On the other hand, the Jones polynomial of T equals $V_T(t) = t + t^3 - t^4$. Relation (74.4) implies that $J_T(q) = (q + 1/q)(q^2 + q^6 - q^8) = q + q^3 + q^5 - q^9 = Kh_T(-1, q)$.

The importance of the Khovanov homology became apparent after a seminal result by Jacob Rasmussen [35], who used the Khovanov chain complex to give the first purely combinatorial proof of the Milnor conjecture. This conjecture states that the 4-dimensional (slice) genus (and, hence, the genus) of a (p, q)-torus knot equals $\frac{(p-1)(q-1)}{2}$. It was originally proved by Kronheimer and Mrowka [25] using the gauge theory in 1993.

Without going into details, we note that the initial categorification of the Jones polynomial by Khovanov was followed with a flurry of activity. Categorifications of the colored Jones polynomial [6, 18] and skein $\mathfrak{sl}(3)$ polynomial [19] were based on the original Khovanov construction. The matrix factorization technique was used to categorify the $\mathfrak{sl}(n)$ skein polynomials [22], HOMFLY-PT polynomial [23], Kauffman 2-variable polynomial [24], and colored $\mathfrak{sl}(n)$ polynomials [46, 47].

Ozsváth, Szabó and, independently, Rasmussen used a completely different method of Floer homology to categorify the Alexander polynomial [31, 34], see also Chapters 71 and 73. Ideas of categorification were successfully applied to tangles [4, 17], virtual links [8, 28, 29], skein modules [2, 33], and polynomial invariants of graphs [9, 10, 32].

The scope of this article is limited to the categorifications of the Jones polynomial only. The reader is referred to expository papers on the subject [1, 14, 21, 36, 41, 42] to learn more about the interrelations between different types of categorifications.

74.2 Definition of the Khovanov homology

In this section we give a brief outline of the original Khovanov construction. Our setting is slightly more general than the one in the Introduction as we allow different coefficient rings, not only $\mathbb{Z}$.

74.2.1 Algebraic preliminaries

Let R be a commutative ring with unity. In this chapter, we are mainly interested in the cases of $R = \mathbb{Z}$, $\mathbb{Q}$, or $\mathbb{Z}_2$.

Definition 74.2.1. A $\mathbb{Z}$-graded (or simply *graded*) R-module M is an R-module decomposed into a direct sum $M = \bigoplus_{j \in \mathbb{Z}} M_j$, where each M_j is an R-module itself. The summands M_j are called *homogeneous components* of M and elements of M_j are called the *homogeneous elements of degree j*.

Definition 74.2.2. Let $M = \bigoplus_{j \in \mathbb{Z}} M_j$ be a graded free R-module. The *graded dimension* of M is the power series $\dim_q(M) = \sum_{j \in \mathbb{Z}} q^j \dim(M_j)$ in variable q. If $k \in \mathbb{Z}$, the *shifted module* $M\{k\}$ is defined as having homogeneous components $M\{k\}_j = M_{j-k}$.

Definition 74.2.3. Let M and N be two graded R-modules. A map $\varphi : M \to N$ is said to be *graded of degree k* if $\varphi(M_j) \subset N_{j+k}$ for each $j \in \mathbb{Z}$.

Note 74.2.4. It is an easy exercise to check that $\dim_q(M\{k\}) = q^k \dim_q(M)$, $\dim_q(M \oplus N) = \dim_q(M) + \dim_q(N)$, and $\dim_q(M \otimes_R N) = \dim_q(M) \dim_q(N)$, where M and N are graded R-modules. Moreover, if $\varphi : M \to N$ is a graded map of degree k', then the *shifted map* $\varphi : M \to N\{k\}$ is graded of degree $k' + k$. We slightly abuse the notation here by denoting the shifted map in the same way as the map itself.

Definition 74.2.5. Let $(C, d) = \cdots \leftrightarrow C^{i-1} \overset{d^{i-1}}{\leftrightarrow} C^i \overset{d^i}{\leftrightarrow} C^{i+1} \leftrightarrow \cdots$ be a (co)chain complex of graded free R-modules with graded differentials d^i of degree 0 for all $i \in \mathbb{Z}$. Then the *graded Euler characteristic* of C is defined as $\chi_q(C) = \sum_{i \in \mathbb{Z}} (-1)^i \dim_q(C^i)$.

Remark 1. One can think of a graded (co)chain complex of R-modules as a *bigraded R-module* where the homogeneous components are indexed by pairs of numbers $(i, j) \in \mathbb{Z}^2$.

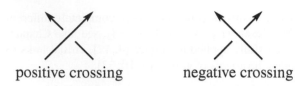

positive crossing   negative crossing

FIGURE 74.2: Positive and negative crossings.

positive marker   negative marker

FIGURE 74.3: Positive and negative markers and the corresponding resolutions of a diagram.

Let $A = R[X]/X^2$ be the algebra of truncated polynomials. As an R-module, A is freely generated by 1 and X. We put grading on A by specifying that $\deg(1) = 1$ and $\deg(X) = -1$[2]. In other words, $A \simeq R\{1\} \oplus R\{-1\}$ and $\dim_q(A) = q + q^{-1}$. At the same time, A is a (graded) commutative algebra with the unit 1 and multiplication $m : A \otimes A \to A$ given by

$$m(1 \otimes 1) = 1, \qquad m(1 \otimes X) = m(X \otimes 1) = X, \qquad m(X \otimes X) = 0. \tag{74.7}$$

The algebra A can also be equipped with a coalgebra structure with comultiplication $\Delta : A \to A \otimes A$ and counit $\varepsilon : A \to R$ defined as

$$\Delta(1) = 1 \otimes X + X \otimes 1, \qquad\qquad \Delta(X) = X \otimes X; \tag{74.8}$$
$$\varepsilon(1) = 0, \qquad\qquad \varepsilon(X) = 1. \tag{74.9}$$

The comultiplication Δ is coassociative and cocommutative and satisfies

$$(m \otimes \mathrm{id}_A) \circ (\mathrm{id}_A \otimes \Delta) = \Delta \circ m \tag{74.10}$$
$$(\varepsilon \otimes \mathrm{id}_A) \circ \Delta = \mathrm{id}_A . \tag{74.11}$$

Together with the unit map $\iota : R \to A$ given by $\iota(1) = 1$, this makes A into a commutative Frobenius algebra over R, see [20].

It follows directly from the definitions that ι, ε, m, and Δ are graded maps with

$$\deg(\iota) = \deg(\varepsilon) = 1 \quad \text{and} \quad \deg(m) = \deg(\Delta) = -1. \tag{74.12}$$

74.2.2 Khovanov chain complex

Let L be an oriented link and D its planar diagram. We assign a number ± 1, called the *sign*, to every crossing of D according to the rule depicted in Figure 74.2. The sum of these signs over all the crossings of D is called the *writhe number* of D and is denoted by $w(D)$.

---

[2]We follow the original grading convention from [15] and [3] here. It is different by a sign from the one in [1].

Every crossing of D can be *resolved* in two different ways according to a choice of a *marker*, which can be either *positive* or *negative*, at this crossing (see Figure 74.3). A collection of markers chosen at every crossing of a diagram D is called a *(Kauffman) state* of D (see also Chapter 61). For a diagram with n crossings, there are, obviously, 2^n different states. Denote by $\sigma(s)$ the difference between the numbers of positive and negative markers in a given state s. Define

$$i(s) = \frac{w(D) - \sigma(s)}{2}, \qquad j(s) = \frac{3w(D) - \sigma(s)}{2}. \qquad (74.13)$$

Since both $w(D)$ and $\sigma(s)$ are congruent to n modulo 2, $i(s)$ and $j(s)$ are always integer. For a given state s, the result of the resolution of D at each crossing according to s is a family D_s of disjointly embedded circles. Denote the number of these circles by $|D_s|$.

For each state s of D, let $\mathcal{A}(s) = A^{\otimes|D_s|}\{j(s)\}$. One should understand this construction as assigning a copy of algebra A to each circle from D_s, taking the tensor product of all of these copies, and shifting the grading of the result by $j(s)$. By construction, $\mathcal{A}(s)$ is a graded free R-module of graded dimension $\dim_q(\mathcal{A}(s)) = q^{j(s)}(q + q^{-1})^{|D_s|}$, see Note 74.2.4. Let $C^i(D) = \bigoplus_{i(s)=i} \mathcal{A}(s)$ for each $i \in \mathbb{Z}$. In order to make $C(D)$ into a graded complex, we need to define a (graded) differential $d^i : C^i(D) \to C^{i+1}(D)$ of degree 0. But even before this differential is defined, the (graded) Euler characteristic of $C(D)$ makes sense.

Lemma 74.2.6. *The graded Euler characteristic of $C(D)$ equals the Jones polynomial of the link L. That is, $\chi_q(C(D)) = J_L(q)$.*

Proof.
$$\chi_q(C(D)) = \sum_{i \in \mathbb{Z}} (-1)^i \dim_q(C^i(D))$$
$$= \sum_{i \in \mathbb{Z}} (-1)^i \sum_{i(s)=i} \dim_q(\mathcal{A}(s))$$
$$= \sum_s (-1)^{i(s)} q^{j(s)} (q + q^{-1})^{|D_s|}$$
$$= \sum_s (-1)^{\frac{w(D) - \sigma(s)}{2}} q^{\frac{3w(D) - \sigma(s)}{2}} (q + q^{-1})^{|D_s|}.$$

Let us forget for a moment that A denotes an algebra and (temporarily) use this letter for a variable. Substituting $(-A^{-2})$ instead of q and noticing that $w(D) \equiv \sigma(s) \pmod 2$, we arrive at

$$\chi_q(C(D)) = (-A)^{-3w(D)} \sum_s A^{\sigma(s)}(-A^2 - A^{-2})^{|D_s|} = (-A^2 - A^{-2})\langle L \rangle_N,$$

where $\langle L \rangle_N$ is the normalized Kauffman bracket polynomial of L (see [13] and Chapter 61 for details). The normalized bracket polynomial of a link is related to the bracket polynomial of its diagram as $\langle L \rangle_N = (-A)^{-3w(D)}\langle D \rangle$. Kauffman proved in [13] that $\langle L \rangle_N$ equals the Jones polynomial $V_L(t)$ of L after substituting $t^{-1/4}$ instead of A. The relation (74.4) between $V_L(t)$ and $J_L(q)$ completes our proof. □

Let s_+ and s_- be two states of D that differ at a single crossing, where s_+ has a positive marker while s_- has a negative one. We call two such states *adjacent*. In this case, $\sigma(s_-) = \sigma(s_+) - 2$ and,

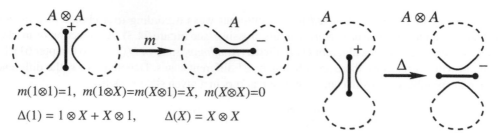

$m(1\otimes1)=1,\ m(1\otimes X)=m(X\otimes1)=X,\ m(X\otimes X)=0$

$\Delta(1) = 1 \otimes X + X \otimes 1, \qquad \Delta(X) = X \otimes X$

FIGURE 74.4: Diagram resolutions corresponding to adjacent states and maps between the algebras assigned to the circles.

consequently, $i(s_-) = i(s_+)+1$ and $j(s_-) = j(s_+)+1$. Consider now the resolutions of D corresponding to s_+ and s_-. One can readily see that D_{s_-} is obtained from D_{s_+} by either merging two circles into one or splitting one circle into two (see Figure 74.4). All the circles that do not pass through the crossing at which s_+ and s_- differ, remain unchanged. We define $d_{s_+:s_-} : \mathcal{A}(s_+) \to \mathcal{A}(s_-)$ as either $m \otimes \mathrm{id}$ or $\Delta \otimes \mathrm{id}$ depending on whether the circles merge or split. Here, the multiplication or comultiplication is performed on the copies of A that are assigned to the affected circles, as in Figure 74.4, while $d_{s_+:s_-}$ acts as an identity on all the A's corresponding to the unaffected ones. The difference in grading shift between $\mathcal{A}(s_+)$ and $\mathcal{A}(s_-)$ and (74.12) ensures that $\deg(d_{s_+:s_-}) = 0$ by Note 74.2.4.

We need one more ingredient in order to finish the definition of the differential on $C(D)$, namely, an ordering of the crossings of D. For an adjacent pair of states (s_+, s_-), define $\xi(s_+, s_-)$ to be the number of the *negative* markers in s_+ (or s_-) that appear in the ordering of the crossings *after* the crossing at which s_+ and s_- differ. Finally, let $d^i = \sum_{(s_+,s_-)}(-1)^{\xi(s_+,s_-)}d_{s_+:s_-}$, where (s_+, s_-) runs over all adjacent pairs of states with $i(s_+) = i$. It is straightforward to verify [15] that $d^{i+1} \circ d^i = 0$ and, hence, $d : C(D) \to C(D)$ is indeed a differential.

Definition 74.2.7 (Khovanov, [15]). The resulting (co)chain complex $C(D) = \cdots \leftrightarrow C^{i-1}(D) \overset{d^{i-1}}{\leftrightarrow} C^i(D) \overset{d^i}{\leftrightarrow} C^{i+1}(D) \leftrightarrow \cdots$ is called the *Khovanov chain complex* of the diagram D. The homology of $C(D)$ with respect to d is called the *Khovanov homology* of L and is denoted by $\mathcal{H}(L)$. We write $C(D; R)$ and $\mathcal{H}(L; R)$ if we want to emphasize the ring of coefficients that we work with. If R is omitted from the notation, integer coefficients are assumed.

Theorem 74.2.8 (Khovanov, [15], see also [3]). *The isomorphism class of $\mathcal{H}(L; R)$ depends on the isotopy class of L only and, hence, is a link invariant. In particular, it does not depend on the ordering chosen for the crossings of D. $\mathcal{H}(L; R)$ categorifies $J_L(q)$, a version of the Jones polynomial defined by (74.1).*

Remark 2. One can think of $C(D; R)$ as a bigraded (co)chain complex $C^{i,j}(D; R)$ with a differential of bidegree $(1, 0)$. In this case, i is the homological grading of this complex, and j is its q-grading, also called the *Jones grading*. Correspondingly, $\mathcal{H}(L; R)$ can be considered to be a bigraded R-module as well.

Note 74.2.9. Let $\#L$ be the number of components of a link L. One can check that $j(s) + |D_s|$ is

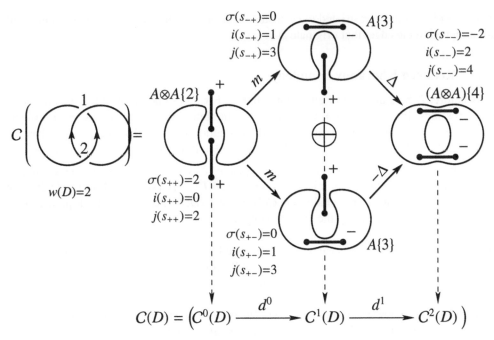

FIGURE 74.5: Khovanov chain complex for the Hopf link.

congruent modulo 2 to $\#L$ for every state s. It follows that $C(D; R)$ has non-trivial homogeneous components only in the degrees that have the same parity as $\#L$. Consequently, $\mathcal{H}(L; R)$ is non-trivial only in the q-gradings with this parity (see Example 74.1.1).

Example 74.2.10. Figure 74.5 shows the Khovanov chain complex for the Hopf link with the indicated orientation. The diagram has two positive crossings, so its writhe number is 2. Let $s_{\pm\pm}$ be the four possible resolutions of this diagram, where each "+" or "−" describes the sign of the marker at the corresponding crossings. The chosen ordering of crossings is depicted by numbers placed next to them. By looking at Figure 74.5, one easily computes that $\mathcal{A}(s_{++}) = A^{\otimes 2}\{2\}$, $\mathcal{A}(s_{+-}) = \mathcal{A}(s_{-+}) = A\{3\}$, and $\mathcal{A}(s_{--}) = A^{\otimes 2}\{4\}$. Correspondingly, $C^0(D) = \mathcal{A}(s_{++}) = A^{\otimes 2}\{2\}$, $C^1(D) = \mathcal{A}(s_{+-}) \oplus \mathcal{A}(s_{-+}) = (A \oplus A)\{3\}$, and $C^2(D) = \mathcal{A}(s_{--}) = A^{\otimes 2}\{4\}$. It is convenient to arrange the four resolutions in the corners of a square placed in the plane in such a way that its diagonal from s_{++} to s_{--} is horizontal. Then the edges of this square correspond to the maps between the adjacent states (see Figure 74.5). We notice that only one of these maps, namely the one corresponding to the edge from s_{+-} to s_{--}, comes with the negative sign.

In general, 2^n resolutions of a diagram D with n crossings can be arranged into an n-dimensional *cube of resolutions*, where vertices correspond to the 2^n states of D. The edges of this cube connect adjacent pairs of states and can be oriented from s_+ to s_-. Every edge is assigned either m or Δ with the sign $(-1)^{\xi(s_+, s_-)}$, as described above. It is easy to check that this makes each square (that is, a 2-dimensional face) of the cube anti-commutative (all squares are commutative without the signs).

Finally, the differential d^i restricted to each summand $\mathcal{A}(s)$ with $i(s) = i$ equals the sum of all the maps assigned to the edges that originate at s.

74.2.3   Reduced Khovanov homology

Let, as before, D be a diagram of an oriented link L. Fix a base point on D that is different from all the crossings. For each state s, we define $\widetilde{\mathcal{A}}(s)$ in almost the same way as $\mathcal{A}(s)$, except that we assign XA instead of A to the circle from the resolution D_s of D that contains that base point. That is, $\widetilde{\mathcal{A}}(s) = \left((XA) \otimes A^{\otimes(|D_s|-1)}\right)\{j(s)\}$. We can now build the *reduced Khovanov chain complex* $\widetilde{\mathcal{A}}(D; R)$ in exactly the same way as $C(D; R)$ by replacing $\mathcal{A}$ with $\widetilde{\mathcal{A}}$ everywhere. The grading shifts and differentials remain the same. It is easy to see that $\widetilde{\mathcal{A}}(D; R)$ is a subcomplex of $C(D; R)$ of index 2. In fact, it is the image of the chain map $C(D; R) \to C(D; R)$ that acts by multiplying elements assigned to the circle containing the base point by X.

Definition 74.2.11 (Khovanov [16], cf. 74.2.7). The homology of $\widetilde{\mathcal{A}}(D; R)$ is called the *reduced Khovanov Homology of L* and is denoted by $\widetilde{\mathcal{A}}(L; R)$. It is clear from the construction of $\widetilde{\mathcal{A}}(D; R)$ that its graded Euler characteristic equals $\overline{J}_L(q)$.

Theorem 74.2.12 (Khovanov [16], cf. 74.2.8). *The isomorphism class of $\widetilde{\mathcal{A}}(L; R)$ is a link invariant that categorifies $\overline{J}_L(q)$, a version of the Jones polynomial defined by (74.1) and (74.4). Moreover, if two base points are chosen on the same component of L, then the corresponding reduced Khovanov homologies are isomorphic. On the other hand, $\widetilde{\mathcal{A}}(L; R)$ might depend on the component of L that the base point is chosen on.*

Although $\widetilde{\mathcal{A}}(D; R)$ can be determined from $C(D; R)$, it is in general not clear how $\mathcal{H}(L; R)$ and $\widetilde{\mathcal{A}}(L; R)$ are related. There are several examples of pairs of knots (the first one being 14^n_{9933} and $\overline{15}^n_{129763}$[3]) that have the same rational Khovanov homology, but different rational reduced Khovanov homology. No such examples are known for homologies over $\mathbb{Z}$ among all prime knots with at most 15 crossings. On the other hand, it is proved that $\mathcal{H}(L; \mathbb{Z}_2)$ and $\widetilde{\mathcal{A}}(L; \mathbb{Z}_2)$ determine each other completely.

Theorem 74.2.13 ([38]). $\mathcal{H}(L; \mathbb{Z}_2) \simeq \widetilde{\mathcal{A}}(L; \mathbb{Z}_2) \otimes_{\mathbb{Z}_2} A_{\mathbb{Z}_2}$. *In particular, $\widetilde{\mathcal{A}}(L; \mathbb{Z}_2)$ does not depend on the component that the base point is chosen on.*

Remark 3. $XA \simeq R\{0\}$ as a graded R-module. It follows that $\widetilde{\mathcal{A}}$ and $\widetilde{\mathcal{A}}$ are non-trivial only in the q-gradings with parity different from that of #L, the number of components of L (cf. 74.2.9).

---

[3]Here, 14^n_{9933} denotes the non-alternating knot number 9933 with 14 crossings from the Knotscape knot table [11] and $\overline{15}^n_{129763}$ is the mirror image of the knot 15^n_{129763}. See also remark on page 742.

74.3 Properties of the Khovanov homology

In this section we summarize the main properties of the Khovanov homology and list related constructions. We emphasize similarities and differences in properties exhibited by different versions of the Khovanov homology. Some of them were already mentioned in the previous sections.

Note 74.3.1. Let L be an oriented link and D its planar diagram. Then the following hold:

- $\widetilde{\mathcal{A}}(D; R)$ is a subcomplex of $C(D; R)$ of index 2.

- $\chi_q(\mathcal{H}(L; R)) = J_L(q)$ and $\chi_q(\widetilde{\mathcal{A}}(L; R)) = \widetilde{J}_L(q)$.

- $\mathcal{H}(L; \mathbb{Z}_2) \simeq \widetilde{\mathcal{A}}(L; \mathbb{Z}_2) \otimes_{\mathbb{Z}_2} A_{\mathbb{Z}_2}$ (see [38]). On the other hand, $\mathcal{H}(L; \mathbb{Z})$ and $\mathcal{H}(L; \mathbb{Q})$ do not split in general.

- For links, $\widetilde{\mathcal{A}}(L; \mathbb{Z}_2)$ does not depend on the choice of a component with the base point. This is, in general, not the case for $\widetilde{\mathcal{A}}(L; \mathbb{Z})$ and $\widetilde{\mathcal{A}}(L; \mathbb{Q})$.

- If L is a non-split alternating link, then $\mathcal{H}(L; R)$ and $\widetilde{\mathcal{A}}(L; R)$ are completely determined by the Jones polynomial and signature of L, see [16, 26, 38, 39].

- $\mathcal{H}(L; \mathbb{Z}_2)$ is invariant under the component-preserving link mutations [45]. It is unclear whether the same holds true for $\mathcal{H}(L; \mathbb{Z})$, though. On the other hand, $\mathcal{H}(L; \mathbb{Z})$ can change under a mutation that exchanges components of a link [44] and under a cabled mutation [7].

- $\mathcal{H}(L; \mathbb{Z})$ almost always has torsion (except for several special cases), but mostly of order 2. The first prime knot with 4-torsion is the $(4, 5)$-torus knot that has 15 crossings. The first known knot with 3-torsion has 22 crossings [30]. The $(5, 6)$-torus knot with 24 crossings has 3-torsion as well. On the other hand, $\widetilde{\mathcal{A}}(L; \mathbb{Z})$ was observed to have very little torsion. The first knot with torsion in the reduced Khovanov homology has 13 crossings.

74.3.1 Homological thickness

Definition 74.3.2. Let L be a link. The *homological width* of L over a ring R is the minimal number of adjacent diagonals $j - 2i = const$ such that $\mathcal{H}(L; R)$ is zero outside of these diagonals. It is denoted by $hw_R(L)$. The *reduced homological width*, $\widetilde{hw}_R(L)$ of L is defined similarly.

Note 74.3.3. It follows from 74.2.13 that $\widetilde{hw}_{\mathbb{Z}_2}(L) = hw_{\mathbb{Z}_2}(L) - 1$. The same holds true in the case of rational Khovanov homology: $\widetilde{hw}_{\mathbb{Q}}(L) = hw_{\mathbb{Q}}(L) - 1$ (see [16]). It is currently unknown whether $\widetilde{hw}_{\mathbb{Z}}(L) = hw_{\mathbb{Z}}(L) - 1$ for every link L.

Definition 74.3.4. A link L is said to be *homologically thin* over a ring R, or simply RH-thin, if $hw_R(L) = 2$. L is *homologically thick*, or RH-thick, otherwise.

Theorem 74.3.5 (Lee). *Non-split alternating links are RH-thin for every R.*

Theorem 74.3.6 (Khovanov [16]). *Adequate links are RH-thick for every R.*

Note 74.3.7. Homological thickness of a link L often does not depend on the base ring. The first prime knot with $hw_{\mathbb{Q}}(L) < hw_{\mathbb{Z}_2}(L)$ and $hw_{\mathbb{Q}}(L) < hw_{\mathbb{Z}}(L)$ is 15^n_{41127} with 15 crossings (see Figure 74.6). The first prime knot that is $\mathbb{Q}H$-thin but $\mathbb{Z}H$-thick, 16^n_{197566}, has 16 crossings (see Figure 74.7). Its mirror image, $\overline{16}^n_{197566}$ is both $\mathbb{Q}H$- and $\mathbb{Z}H$-thin but is $\mathbb{Z}_2H$-thick with $\mathcal{H}^{-8,-21}(\overline{16}^n_{197566}; \mathbb{Z}_2) \simeq \mathbb{Z}_2$, for example, because of the Universal Coefficient Theorem. Also observe that $\mathcal{H}^{9,25}(16^n_{197566}; \mathbb{Z})$ and $\mathcal{H}^{-8,-25}(\overline{16}^n_{197566}; \mathbb{Z})$ have 4-torsion, shown in a small box in the tables.

Remark 4. Throughout this chapter we use the following notation for knots: knots with 10 crossings or less are numbered according to Rolfsen's knot table [37] and knots with 11 crossings or more are numbered according to the knot table from Knotscape [11]. Mirror images of knots from either table are denoted with a bar on top. For example, $\overline{9}_{46}$ is the mirror image of the knot number 46 with 9 crossings from Rolfsen's table and 16^n_{197566} is the non-alternating knot number 197566 with 16 crossings from the Knotscape one.

74.3.2   Lee spectral sequence and the Knight-Move Conjecture

In [26], Eun Soo Lee introduced a structure of a spectral sequence on the rational Khovanov chain complex $C(D; \mathbb{Q})$ of a link diagram D. Namely, Lee defined a differential $d' : C(D; \mathbb{Q}) \to C(D; \mathbb{Q})$ of bidegree $(1, 4)$ by setting

$$
\begin{aligned}
m' : A \otimes A \to A : \quad & m'(1 \otimes 1) = m'(1 \otimes X) = m'(X \otimes 1) = 0, \quad m'(X \otimes X) = 1 \\
\Delta' : A \to A \otimes A : \quad & \Delta'(1) = 0, \qquad \Delta'(X) = 1 \otimes 1.
\end{aligned}
\tag{74.14}
$$

It is straightforward to verify that d' is indeed a differential and that it anti-commutes with d, that is, $d \circ d' + d' \circ d = 0$. This makes $(C(D; \mathbb{Q}), d, d')$ into a double complex. Let d'_* be the differential induced by d' on $\mathcal{H}(L; \mathbb{Q})$. Lee proved that d'_* is functorial, that is, it commutes with isomorphisms induced on $\mathcal{H}(L; \mathbb{Q})$ by isotopies of L. It follows that there exists a spectral sequence with $(E_1, d_1) = (C(D; \mathbb{Q}), d)$ and $(E_2, d_2) = (\mathcal{H}(L; \mathbb{Q}), d'_*)$ that converges to the homology of the total (filtered) complex of $C(D; \mathbb{Q})$ with respect to the differential $d + d'$. It is called the *Lee spectral sequence*. The differentials d_n in this spectral sequence have bidegree $(1, 4(n - 1))$.

Theorem 74.3.8 (Lee [26]). *If L is an oriented link with $\#L$ components, then $H(\text{Total}(C(D; \mathbb{Q})), d + d')$, the limit of the Lee spectral sequence, consists of $2^{\#L-1}$ copies of $\mathbb{Q} \oplus \mathbb{Q}$, each located in a specific homological grading that is explicitly defined by linking numbers of the components of L. In particular, if L is a knot, then the Lee spectral sequence converges to $\mathbb{Q} \oplus \mathbb{Q}$ located in homological grading 0.*

Theorem 74.3.9 (Rasmussen [35]). *If K is a knot, then the two copies of $\mathbb{Q}$ in the limiting term of the Lee spectral sequence for K are "neighbors," that is, their q-gradings are different by 2.*

Definition 74.3.10 (Rasmussen [35]). Let K be a knot. Then its *Rasmussen invariant* $s(K)$ is defined as the average of the two q-gradings from Theorem 74.3.9. Since these q-gradings are odd and are different by 2, the Rasmussen invariant is always an even integer.

	-7	-6	-5	-4	-3	-2	-1	0	1	2	3
9											1
7											1_2
5									1	1	
3							1	1	1_2		
1							1_2	$2, 1_2$			
-1					1	2		1_2			
-3				1	1_2	1_2					
-5			$1, 1_2$	1							
-7		1	1								
-9		1_2									
-11	1										

$$\mathrm{hw}_{\mathbb{Q}} = 3, \qquad \mathrm{hw}_{\mathbb{Z}} = 4, \qquad \mathrm{hw}_{\mathbb{Z}_2} = 4$$

	-7	-6	-5	-4	-3	-2	-1	0	1	2	3
8											1
6										1	
4									1		
2							1	2			
0						1		1_2			
-2					1	1					
-4				2	1						
-6			1								
-8		1									
-10	1										

$$\widetilde{\mathrm{hw}}_{\mathbb{Q}} = 2, \qquad \widetilde{\mathrm{hw}}_{\mathbb{Z}} = 3, \qquad \widetilde{\mathrm{hw}}_{\mathbb{Z}_2} = 3$$

FIGURE 74.6: Integral Khovanov homology (above) and integral reduced Khovanov Homology (below) of the knot 15^n_{41127}.

	-2	-1	0	1	2	3	4	5	6	7	8	9	10
29													1
27												4	1_2
25											7	$1, 3_2\boxed{1}_4$	
23										12	$4, 8_2$		
21									15	$7, 12_2$	1_2		
19								17	$12, 16_2$				
17							16	$15, 18_2$	1_2				
15						15	$17, 16_2$	1_2					
13					10	$16, 15_2$							
11				6	$15, 10_2$								
9			3	$10, 6_2$									
7		1	$7, 2_2$										
5		$2, 1_2$											
3	1												

The free part of $\mathcal{H}(16^n_{197566}; \mathbb{Z})$ is supported on diagonals $j - 2i = 7$ and $j - 2i = 9$. On the other hand, there is 2-torsion on the diagonal $j - 2i = 5$. Therefore, 16^n_{197566} is $\mathbb{Q}$H-thin, but $\mathbb{Z}$H-thick and $\mathbb{Z}_2$H-thick.

	-10	-9	-8	-7	-6	-5	-4	-3	-2	-1	0	1	2
-3													1
-5												2	1_2
-7											7	$1, 2_2$	
-9										10	$3, 6_2$		
-11									15	$6, 10_2$			
-13								16	$10, 15_2$				
-15							$17, 1_2$	$15, 16_2$					
-17						$15, 1_2$	$16, 18_2$						
-19					12	$17, 16_2$							
-21				$7, 1_2$	$15, 12_2$								
-23			4	$12, 8_2$									
-25		1	$7, 3_2\boxed{1}_4$										
-27		$4, 1_2$											
-29	1												

$\mathcal{H}(\overline{16}^n_{197566}; \mathbb{Z})$ is supported on diagonals $j - 2i = -7$ and $j - 2i = -9$.
But there is 2-torsion on the diagonal $j - 2i = -7$.
Therefore, $\overline{16}^n_{197566}$ is $\mathbb{Q}$H-thin and $\mathbb{Z}$H-thin, but $\mathbb{Z}_2$H-thick.

FIGURE 74.7: Integral Khovanov homology of the knots 16^n_{197566} and $\overline{16}^n_{197566}$.

Remark 5. The Rasmussen invariant gives rise to some of the main applications of the Khovanov homology.

Corollary 74.3.11. *If the Lee spectral sequence for a knot K collapses after the second page (that is, $d_n = 0$ for $n \geq 3$), then $\mathcal{H}(K; \mathbb{Q})$ consists of one "pawn-move" pair in homological grading 0 and multiple "knight-move" pairs, shown below, with appropriate grading shifts.*

Pawn-move pair:
| $\mathbb{Q}$ |
| $\mathbb{Q}$ |

Knight-move pair:
| | $\mathbb{Q}$ |
| $\mathbb{Q}$ | |

Corollary 74.3.12. *Since d_3 has bidegree $(1, 8)$, the Lee spectral sequence collapses after the second page for all knots with homological width 2 or 3, in particular, for all alternating knots. Hence, Corollary 74.3.11 can be applied to such knots.*

Knight-Move Conjecture 74.3.13. *[Garoufalidis–Khovanov–Bar-Natan [3, 15]] The conclusion of Corollary 74.3.11 holds true for every knot.*

Note 74.3.14. The Knight-Move Conjecture was recently proved to be false by Ciprian Manolescu and Marco Marengon, see [27]. Their counter-example has 38 crossings and homological width 8.

Remark 6. While the Lee spectral sequence exists over any ring R, the statement of Theorem 74.3.8 does not hold true for all of them. In particular, it is wrong over $\mathbb{Z}_2$. In this case, though, a similar theory was constructed by Paul Turner [40]. His construction works for reduced Khovanov homology over $\mathbb{Z}_2$ as well because of 74.2.13. While Theorems 74.3.8 and 74.3.9 are true over $\mathbb{Z}_p$ with odd prime p, Knight-Move Conjecture is known to be false over such rings [5].

Remark 7. The Lee spectral sequence has no analog in the reduced Khovanov homology theory, except over $\mathbb{Z}_2$, as noted above. The Knight-Move Conjecture has no analog in these theories either.

Corollary 74.3.11 can be extended to integer Khovanov homology for non-split alternating links.

Theorem 74.3.15 ([26, 38, 39]). *Let L be a $\mathbb{Z}_2H$-thin link with #L components. Then $\mathcal{H}(K; \mathbb{Z})$ consists of $2^{\#L-1}$ "integer pawn-move" pairs and multiple "integer knight-move" pairs, shown below, with appropriate grading shifts. For example, the trefoil has a single (integer) knight-move pair, see Figure 74.1.*

Integer pawn-move pair:
| $\mathbb{Z}$ |
| $\mathbb{Z}$ |

Integer knight-move pair:
	$\mathbb{Z}$
	$\mathbb{Z}_2$
$\mathbb{Z}$	

74.3.3 Long exact sequence of the Khovanov homology

One of the most useful tools in studying Khovanov homology is the long exact sequence that categorifies the Kauffman's *unoriented* skein relation for the Jones polynomial, see Chapters 61 and 67. If we forget about the grading, then it is clear from the construction from Section 74.2.2 that $C(\asymp)$ is a subcomplex of $C(\times)$ and $C()() \simeq C(\ )/C(\ )$ (see also Figure 74.5). Here, and $\rangle\langle$ depict link diagrams where a single crossing $\times$ is resolved in a negative or, respectively, positive direction. This results in a short exact sequence of non-graded chain complexes:

$$0 \leftrightarrow C(\times) \overset{in}{\leftrightarrow} C(\ ) \overset{p}{\leftrightarrow} C(\ ) \leftrightarrow 0, \qquad (74.15)$$

where *in* is the inclusion and *p* is the projection.

In order to introduce grading into (74.15), we need to consider the cases when the crossing to be resolved is either positive or negative. We get (see [36]):

$$0 \leftrightarrow C(\ )\{2+3\omega\}[1+\omega] \overset{in}{\leftrightarrow} C(\underset{+}{\times}) \overset{p}{\leftrightarrow} C(\rangle\langle)\{1\} \leftrightarrow 0,$$

$$0 \leftrightarrow C(\ )\{-1\} \overset{in}{\leftrightarrow} C(\underset{-}{\times}) \overset{p}{\leftrightarrow} C(\ )\{1+3\omega\}[\omega] \leftrightarrow 0, \qquad (74.16)$$

where ω is the difference between the numbers of negative crossings in the unoriented resolution (it has to be oriented somehow in order to define its Khovanov chain complex) and in the original diagram. The notation $C[k]$ is used to represent a shift in the homological grading of a complex C by k. The graded versions of *in* and *p* are both homogeneous, that is, have bidegree $(0, 0)$.

By passing to homology in (74.16), we get the following result.

Theorem 74.3.16 (Khovanov, Viro, Rasmussen [15, 43, 36]). *The Khovanov homology is subject to the following long exact sequences:*

$$\cdots \leftrightarrow \mathcal{H}(\ )\{1\} \overset{\partial}{\leftrightarrow} \mathcal{H}(\ )\{2+3\omega\}[1+\omega] \overset{in_*}{\leftrightarrow} \mathcal{H}(\ ) \overset{p_*}{\leftrightarrow} \mathcal{H}(\ )\{1\} \leftrightarrow \cdots$$

$$\cdots \leftrightarrow \mathcal{H}(\ )\{-1\} \overset{in_*}{\leftrightarrow} \mathcal{H}(\ ) \overset{p_*}{\leftrightarrow} \mathcal{H}(\ )\{1+3\omega\}[\omega] \overset{\partial}{\leftrightarrow} \mathcal{H}(\ )\{-1\} \leftrightarrow \cdots \qquad (74.17)$$

where in_* and p_* are homogeneous and ∂ is the connecting differential and has bidegree $(1, 0)$.

74.4   Bibliography

[1] M. Asaeda and M. Khovanov, *Notes on link homology*, in *Low dimensional topology*, 139–195, IAS/Park City Math. Ser., 15, Amer. Math. Soc., Providence, RI, 2009; arXiv:0804.1279.

[2] M. Asaeda, J. Przytycki, and A. Sikora. *Categorification of the Kauffman bracket skein module of I-bundles over surfaces*, Alg. Geom. Topol., **4** (2004), no. 2, 1177–1210; arXiv:math/0409414.

[3] D. Bar-Natan, *On Khovanov's categorification of the Jones polynomial*, Alg. Geom. Topol., **2** (2002) 337–370; arXiv:math.QA/0201043.

[4] D. Bar-Natan, *Khovanov's homology for tangles and cobordisms*, Geom. Topol., **9** (2005), 1465–1499; arXiv:math.GT/0410495.

[5] D. Bar-Natan, *Fast Khovanov homology computations*, J. Knot Th. and Ramif. **16** (2007), no. 3, 243–255; arXiv:math.GT/0606318.

[6] A. Beliakova and S. Wehrli, *Categorification of the colored Jones polynomial and Rasmussen invariant of links*, Canad. J. Math. **60** (2008), no. 6, 1240–1266; arXiv:math.GT/0510382.

[7] N. Dunfield, S. Garoufalidis, A. Shumakovitch, and M. Thistlethwaite, *Behavior of knot invariants under genus 2 mutation*, New York J. Math. **16** (2010), 99–123; arXiv:math.GT/0607258.

[8] H. Dye, L. Kauffman, and V. Manturov, *On two categorifications of the arrow polynomial for virtual knots*, in The Mathematics of Knots, Contributions in Mathematical and Computational Sciences **1**, 95–124, Springer, 2011; arXiv:0906.3408.

[9] L. Helme-Guizon and Y. Rong, *A categorification for the chromatic polynomial*, Algebr. Geom. Topol. **5** (2005), 1365–1388; arXiv:math.CO/0412264.

[10] L. Helme-Guizon and Y. Rong, *Khovanov type homologies for graphs*, Kobe J. Math. **29** (2012), no. 1–2, 25–43; arXiv:math.QA/0506023.

[11] J. Hoste and M. Thistlethwaite, Knotscape — a program for studying knot theory and providing convenient access to tables of knots,
http://www.math.utk.edu/~morwen/knotscape.html

[12] V. Jones, *A polynomial invariant for knots via von Neumann algebras*, Bull. Amer. Math. Soc. **12** (1985), 103–111.

[13] L. Kauffman, *State models and the Jones polynomial*, Topology **26** (1987), no. 3, 395–407.

[14] L. Kauffman, *Khovanov homology* in Introductory Lectures in Knot Theory, Series on Knots and Everything **46**, 248–280, World Scientific, 2012; arXiv:1107.1524.

[15] M. Khovanov, *A categorification of the Jones polynomial*, Duke Math. J. **101** (2000), no. 3, 359–426; arXiv:math.QA/9908171.

[16] M. Khovanov, *Patterns in knot cohomology I*, Experiment. Math. **12** (2003), no. 3, 365–374; arXiv:math.QA/0201306.

[17] M. Khovanov, *An invariant of tangle cobordisms*, Trans. Amer. Math. Soc. **358** (2006), 315–327; arXiv:math.QA/0207264.

[18] M. Khovanov, *Categorifications of the colored Jones polynomial*, J. Knot Th. Ramif. **14** (2005), no. 1, 111–130; arXiv:math.QA/0302060.

[19] M. Khovanov, $\mathfrak{sl}(3)$ *link homology I*, Algebr. Geom. Topol. **4** (2004), 1045–1081; arXiv:math.QA/0304375.

[20] M. Khovanov, *Link homology and Frobenius extensions*, Fundamenta Mathematicae, **190** (2006), 179–190; arXiv:math.QA/0411447.

[21] M. Khovanov, *Link homology and categorification*, ICM–2006, Madrid, Vol. II, 989–999, Eur. Math. Soc., Zürich, 2006; arXiv:math/0605339.

[22] M. Khovanov and L. Rozansky, *Matrix factorizations and link homology*, Fund. Math. **199** (2008), no. 1, 1–91; arXiv:math.QA/0401268.

[23] M. Khovanov and L. Rozansky, *Matrix factorizations and link homology II*, Geom. Topol. **12** (2008), no. 3, 1387–1425; arXiv:math.QA/0505056.

[24] M. Khovanov and L. Rozansky, *Virtual crossings, convolutions and a categorification of the SO(2N) Kauffman polynomial*, J. Gökova Geom. Topol. GGT **1** (2007), 116–214; arXiv:math/0701333.

[25] P. Kronheimer and T. Mrowka, *Gauge theory for embedded surfaces I*, Topology **32** (1993), 773–826.

[26] E. S. Lee, *An endomorphism of the Khovanov invariant*, Adv. Math. **197** (2005), no. 2, 554–586; arXiv:math.GT/0210213.

[27] C. Manolescu and M. Marengon, *The Knight Move Conjecture is false*, arXiv:1809.09769.

[28] V. Manturov, *The Khovanov complex for virtual links*, Fund. and Appl. Math., **11** (2005), no. 4, 127–152; arXiv:math/0501317.

[29] V. Manturov, *Khovanov homology for virtual knots with arbitrary coefficients*, J. Knot Th. Ramif. **16** (2007), no. 03, 345–377; arXiv:math/0601152.

[30] S. Mukherjee, J. H. Przytycki, M. Silvero, X. Wang, and S. Y. Yang, *Search for Torsion in Khovanov Homology*, Exp. Math. **27** (2018), no. 4, 488–497; arXiv:1701.04924.

[31] P. Ozsváth and Z. Szabó, *Holomorphic disks and knot invariants*, Adv. Math. **186** (2004), no. 1, 58–116; arXiv:math.GT/0209056.

[32] M. Pabiniak, J. Przytycki, and R. Sazdanović, *On the first group of the chromatic cohomology of graphs*, Geom. Dedicata, **140** (2009), no. 1, 19–48; arXiv:math.GT/0607326.

[33] H. Quelec and P. Wedrich *Khovanov homology and categorification of skein modules*; arXiv:1806.03416.

[34] J. Rasmussen, *Floer homology and knot complements*, Ph.D. Thesis, Harvard U.; arXiv:math.GT/0306378.

[35] J. Rasmussen, *Khovanov homology and the slice genus*, Invent. Math. **182** (2010), no. 2, 419–447; arXiv:math.GT/0402131.

[36] J. Rasmussen, *Knot polynomials and knot homologies*, Geometry and Topology of Manifolds (Boden et al., eds.), Fields Institute Communications **47** (2005) 261–280, AMS; arXiv:math.GT/0504045.

[37] D. Rolfsen, *Knots and Links*, Publish or Perish, Mathematics Lecture Series 7, Wilmington 1976.

[38] A. Shumakovitch, *Torsion of the Khovanov homology*, Fund. Math. **225** (2014), 343–364; arXiv:math.GT/0405474.

[39] A. Shumakovitch, *Torsion in Khovanov homology of homologically thin knots*, to appear J. Knot Th. Ramif.; arXiv:1806.05168.

[40] P. Turner, *Calculating Bar-Natan's characteristic two Khovanov homology*, J. Knot Th. Ramif. **15** (2006), no. 10, 1335–1356; arXiv:math.GT/0411225.

[41] P. Turner, *Five lectures on Khovanov homology*, J. Knot Th. Ramif., **26** (2017), no. 03, 1741009; arXiv:math/0606464.

[42] P. Turner, *A hitchhiker's guide to Khovanov homology*, arXiv:1409.6442.

[43] O. Viro, *Khovanov homology, its definitions and ramifications*, Fund. Math. **184** (2004), 317–342; arXiv:math.GT/0202199.

[44] S. Wehrli, *Khovanov homology and Conway mutation*, arXiv:math/0301312.

[45] S. Wehrli, *Mutation invariance of Khovanov homology over* $\mathbb{F}_2$, Quantum Topol. **1** (2010), no. 2, 111–128; arXiv:0904.3401.

[46] H. Wu, *A colored* $\mathfrak{sl}(N)$*-homology for links in* S^3, Diss. Math. **499** (2014), 217 pp.; arXiv:0907.0695.

[47] Y. Yonezawa, *Quantum* $(\mathfrak{sl}_n, \wedge V_n)$ *link invariant and matrix factorizations*, Nagoya Math. J. **204** (2011), 69–123; arXiv:0906.0220.

Chapter 75

Virtual Khovanov Homology

William Rushworth, McMaster University

75.1 Introduction

Since its introduction at the turn of the century, *Khovanov homology* [11] has assumed a prominent position in the modern knot theorist's arsenal of invariants. Due to its practical power and mysterious origins, it is a natural task to define Khovanov homology for classes of links other than those in S^3. This expository chapter concerns the problem of extending Khovanov homology to one such class, known as *virtual links*. We refer to links in S^3 as *classical links*.

Virtual links form a proper superset of the classical links, and may be defined via classical diagrams with an additional type of crossing. While we work with this combinatorial definition here, there is an equivalent realization of virtual links as links in thickened orientable surfaces, up to diffeomorphism and certain handle (de)stabilizations of the surface. In this light, extending Khovanov homology to virtual links is seen as an instance of the problem of extending Khovanov homology to 3-manifolds other than S^3.

We begin with a brief overview of virtual knot theory, before describing three extensions of Khovanov homology. We assume familiarity with the construction of Khovanov homology for links in S^3.

75.2 Virtual knot theory

The study of virtual knot theory was initiated by Kauffman in the late 1990s [9]. In this chapter the diagrammatic definition of virtual links is sufficient; however, there is an equivalent topological definition. We direct the interested reader to [9, 4, 15] for details of both.

Definition 75.2.1. A *virtual link diagram* is a 4-valent planar graph, the vertices of which are decorated with either the classical overcrossing and undercrossing decorations, or a new decoration, ⧖, known as a *virtual crossing*.

An example of a virtual link diagram is given in Figure 75.4a.

Definition 75.2.2. A *virtual link* is an equivalence class of virtual link diagrams, up to the *virtual*

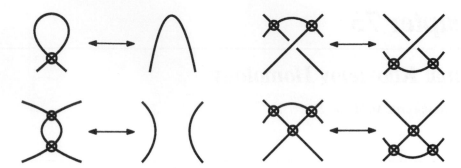

FIGURE 75.1: The non-classical moves making up the virtual Reidemeister moves.

Reidemeister moves. These moves consists of those of classical knot theory, together with four moves involving virtual crossings, depicted in Figure 75.1.

Goussarov, Polyak, and Viro proved that classical links are a proper subset of virtual links [7]. However, one encounters a number of counter-intuitive phenomena when passing from classical to virtual links. For example, unknotting number is no longer necessarily finite, and connected sum is not well-defined. The Jones polynomial extends to virtual links automatically, and there exist infinitely many virtual knots with trivial Jones polynomial. Every such knot is also trivial with respect to the three extensions of Khovanov homology we describe in Section 75.3.

75.3 Extending Khovanov homology

In this section we discuss the difficulties encountered when extending Khovanov homology to virtual links, before describing three such extensions. We aim to illustrate, in broad brush strokes, the problems that arise and the different approaches to solving them. For a detailed account of the source of these problems see [22] (and references therein). As our focus is on the construction of Khovanov homology theories for virtual links, we are not able to discuss applications of such theories in detail.

Khovanov homology associates to an oriented classical link a finitely generated bigraded Abelian group. Likewise, the extensions we describe associate such a group to an oriented virtual link. Let L be an oriented virtual link; we shall discuss

Section 75.3.2 Manturov's extension, denoted $VKh(L)$ [14]

Section 75.3.3 Doubled Khovanov homology, denoted $DKh(L)$ [16]

Section 75.3.2 Tubbenhauer's extension, denoted $UKh(L)$ [19]

Although there are families of virtual links on which all three extensions contain equivalent information, they differ in essential ways. We shall see an example confirming that DKh is distinct from both VKh and UKh, in general. It is currently unknown if VKh and UKh are distinct.

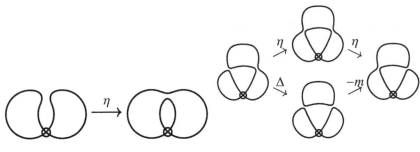

(a) The single-cycle smoothing.   (b) The problem face.

FIGURE 75.2

The extensions we discuss do not provide new information for classical links, and their graded Euler characteristics yield (a multiple of) the Jones polynomial.

Classical Khovanov homology is an unknot detector: a classical knot has trivial Khovanov homology if and only if it is the unknot [12]. This is not replicated by any of the three extensions we discuss.

A feature of classical Khovanov homology that is replicated by these extensions is the presence of concordance information. Dye, Kaestner, and Kauffman reformulated Manturov's extension in order to define the *virtual Rasmussen invariant* [5]. A distinct extension of the Rasmussen invariant, known as the *doubled Rasmussen invariant*, may be defined using doubled Khovanov homology [16, Section 4]. Recall that there is a realization of virtual links as equivalence classes of links in thickened surfaces. These invariants contain information regarding surfaces bounding such links, embedded in certain 4-manifolds. For details of this and further work on the concordance of virtual links see [10, 3, 17, 20].

While not directly applicable to virtual links, there are extensions of Khovanov homology to the closely related theories of links in I-bundles over surfaces [1, 6].

75.3.1   Obstacles to extension

We denote the module assigned to one circle in the construction of classical Khovanov homology by $\mathcal{A}$, where

$$\mathcal{A} = \langle v_+, v_- \rangle_{\mathbb{Z}}.^1$$

A *smoothing* of a (classical or virtual) link diagram is obtained by arbitrarily resolving all of its classical crossings; in the case of a virtual diagram a smoothing is a collection of circles immersed in the plane. Given an oriented virtual link diagram, the cube of smoothings can be formed exactly as in the classical case, simply leaving the virtual crossings untouched. An example is given in Figure 75.2b. However, problems are encountered when converting the diagrammatic cube into a chain complex.

Any extension of Khovanov homology to virtual links must deal with the fundamental problem

---

[1]In general $\mathbb{Z}$ may be replaced by any commutative unital ring.

presented by the *single-cycle smoothing* (also known as the *one-to-one bifurcation*). This is depicted in Figure 75.2a: altering the resolution of a classical crossing no longer necessarily splits one circle or merges two circles, but can take one circle to one circle. The realization of this as a cobordism is a once-punctured Möbius band. We must associate an algebraic map, η, to such edges. Looking at the quantum grading we notice that

$$
\begin{array}{ccc}
0 & & v_+ \\
v_+ & \xrightarrow{\ \eta\ } & 0 \\
0 & & v_- \\
v_- & & 0
\end{array}
$$

from which it follows that the map $\eta : \mathcal{A} \to \mathcal{A}$ must be the zero map if it is to be quantum grading-preserving (we have arranged the generators vertically by quantum grading).[2]

However, naïvely setting η to be the zero map causes collateral damage. Consider the cube of smoothings known as the *problem face*, depicted in Figure 75.2b (it is the cube of the diagram given in Figure 75.4a). Along the upper two edges we have $\eta \circ \eta = 0$, but along the lower two edges we have $-m \circ \Delta$. The latter is non-zero (as $m \circ \Delta(v_+) = 2v_-$), so that this face fails to anticommute. As a consequence, repeating the classical definition of Khovanov homology fails to yield a chain complex (over any ground ring other than $\mathbb{Z}_2$).

There are at least two ways to overcome these obstacles:

> (1) Declare η to be the zero map, and develop a method to recover a chain complex.

or

> (2) Alter the construction in such a way that η may be both quantum grading-preserving and non-zero.

Both Manturov and Tubbenhauer take the first route, as described in Sections 75.3.2 and 75.3.4, respectively. Recall that when traversing the lower two edges of the problem face we apply the non-zero map $-m \circ \Delta$. This composition must be set to zero if the problem face is to anticommute and a well-defined chain complex obtained. However, this must be done locally, as doing so globally would destroy the link invariance of the theory. Manturov and Tubbenhauer add new diagrammatic technology in order to detect the problem face within the cube of smoothings, allowing the necessary local alteration. While the diagrammatic methods used by Manturov and Tubbenhauer are similar in spirit, it is not clear if they are equivalent.

As described in Section 75.3.3, the construction of doubled Khovanov homology takes the second route. The way in which diagrams are converted into algebra is altered, so that η may be both quantum grading-preserving and non-zero. Unlike Manturov's and Tubbenhauer's constructions, no additional diagrammatic technology is required to recover a well-defined chain complex. The result is a theory which exhibits a number of structural differences to classical Khovanov homology.

---

[2]There are classes of virtual links for which no single-cycle smoothings appear, including alternating virtual links and chequerboard colourable virtual links.

75.3.2 Manturov's extension

In this section we describe the extension of Khovanov homology to virtual links due to Manturov [14]. His theory was reformulated by Dye, Kaestner, and Kauffman in order to define a virtual Rasmussen invariant [5].

As outlined in Section 75.3.1, we proceed by taking $\eta = 0$, and add new diagrammatic technology to recover a chain complex. In fact, there are two pieces of new technology. The first piece allows for the exploitation of a symmetry present in $\mathcal{A}$ (that corresponds to the two possible orientations of S^1), using the following involution.

Definition 75.3.1. Let $\mathcal{A} = \langle v_+, v_- \rangle$. The *barring operator* is the map

$$\overline{} : \mathcal{A} \to \mathcal{A}, \ v_- \mapsto -v_-, \ v_+ \mapsto v_+. \qquad \square$$

How the barring operator is applied within the Khovanov complex is determined using an extra decoration on link diagrams, the *source-sink decoration*.

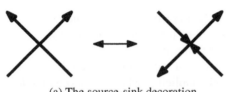

(a) The source-sink decoration.   (b) The source-sink orientation.

FIGURE 75.3

Definition 75.3.2. Let D be an oriented virtual link diagram. Denote by $S(D)$ the diagram formed by replacing the classical crossings of D with the source-sink decoration, depicted in Figure 75.3a. The source-sink decoration at each classical crossing induces an orientation of the arcs of $S(D)$ as in Figure 75.3b. An arc on which the orientations induced by distinct crossings disagree is marked with a *cut locus*. We refer to $S(D)$ as the *source-sink diagram of D*.

An example of a source-sink diagram is given in Figure 75.4b. All smoothings of D are also marked with cut loci, in positions as dictated by those of $S(D)$, as in Figure 75.4c.

Given a smoothing of a virtual link diagram the source-sink decoration at each classical crossing induces an orientation on the (at most two) circles of the smoothing that are incident to it, as in Figure 75.3b. This orientation is known as the *source-sink orientation*, and an example of a smoothing oriented in this way is given in Figure 75.4c. Note that the orientations induced by distinct classical crossings may disagree. This is captured by the cut loci, and it is this disagreement that allows for the detection of the problem face.

The source-sink decoration and the barring operator are used to detect the problem face as follows. The circles within a smoothing have no canonical orientation i.e. the construction of classical Khovanov homology does not depend on their orientation. The source-sink orientation is a way of picking an orientation: between each cut locus there is a well-defind orientation, and passing a cut

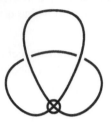

(a) A virtual knot diagram.

(b) The source-sink diagram of the virtual knot diagram on the left, marked with cut loci.

(c) A smoothing of the virtual knot diagram on the far left, marked with cut loci and source-sink orientations.

FIGURE 75.4

locus one moves onto a region of reverse orientation. This is replicated in algebra by applying the barring operator. Specifically, one applies the barring operator to move between two distinct bases of $\mathcal{A}$, $\langle v_+, v_- \rangle_{\mathbb{Z}}$ and $\langle v_+, -v_- \rangle_{\mathbb{Z}}$.

Due to the form of the source-sink decoration, the only non-trivial way in which the source-sink orientations on a circle may disagree is in the presence of the η map. Therefore the barring operator is applied only in these situations. This fact, combined with a careful construction of the differential, can be used to recover the anticommutativity of the problem face without destroying link invariance. This source-sink orientation method is implicit in Manturov's original definition, and explicit in Dye-Kaestner-Kauffman.

The second piece of new technology is used to fix a sign convention. Recall that there is no preferred sign convention in classical Khovanov homology. However, as the barring operator involves multiplication by -1, it is important to fix a sign convention that does not disrupt the detection of the problem face. Manturov replaced the standard tensor product with the exterior product to do so. Dye-Kaestner-Kauffman also use this technique, presented in a different manner.

The result of adding both pieces of technology is a well-defined chain complex, the homology of which is an invariant of virtual links. While the task of ensuring that the complex is well-defined is intricate, the link invariance of its homology may be verified using arguments very similar to those used in the classical case. The result is a well-defined invariant of oriented virtual links.

Definition 75.3.3 ([14, Section 4], [5, Section 3]). Let D be a diagram of an oriented virtual link L. Associated to D is the chain complex $CVKh(D)$, the homology of which is an invariant of L, and is denoted $VKh(L)$.

An example of $VKh(L)$ is given in Figure 75.5. If L is a classical link, then $VKh(L) = Kh(L)$, for $Kh(L)$ the classical Khovanov homology. Given a virtual link L', it is in general not possible to use the homology $VKh(L')$ exclusively to prove that a L' is not classical, but when combined with other invariants it may be used for this purpose (see [5, Theorem 3.4]).

There is a particular case in which VKh may be used exclusively to show that a virtual knot is not classical. A virtual knot diagram is *alternating* if one alternates between overcrossings and undercrossings when traversing the diagram, ignoring virtual crossings. A virtual knot is alternating

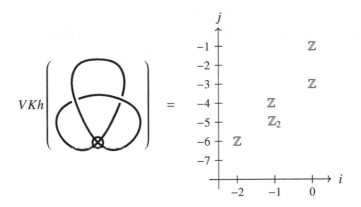

FIGURE 75.5: The bigraded group $VKh(K)$ associated to the virtual knot depicted.

if it possesses an alternating diagram. Recall that the Khovanov homology of a classical alternating knot is supported on 2 diagonals [13]. Karimi proved that given an alternating virtual knot K, the homology $VKh(K)$ is supported on $g + 2$ diagonals, where g depends only on a topological property of K [8]. Therefore, an alternating virtual knot K' for which $VKh(K')$ is not supported on 2 diagonals is not classical.

75.3.3 Doubled Khovanov homology

In this section we describe the extension of Khovanov homology to virtual links due to the author, known as *doubled Khovanov homology* [16]. This extension is distinct from those described in Sections 75.3.2 and 75.3.4. While those theories retain some of the structural features of classical Khovanov homology, doubled Khovanov homology exhibits a number of new phenomena.

Recall from Section 75.3 that if one associates the module $\mathcal{A}$ to a circle, the η map must be the zero map if it is to be quantum grading-preserving. Doubled Khovanov homology is constructed by altering the module assigned to a circle, so that the η map may be non-zero and quantum grading-preserving.

Specifically, we proceed by "doubling up" the complex associated to a link diagram in order to fill the gaps in the quantum grading. Taking the direct sum of the standard Khovanov chain space with itself, with one copy shifted in quantum grading by -1, we obtain $\eta : \mathcal{A} \oplus \mathcal{A}\{-1\} \to \mathcal{A} \oplus \mathcal{A}\{-1\}$, that is

$$
\begin{array}{ccc}
0 & & v_+^u \\
v_+^u & & v_+^l \\
v_+^l & \xrightarrow{\ \eta\ } & v_-^u \\
v_-^u & & v_-^l \\
v_-^l & & 0
\end{array}
$$

where $\mathcal{A} = \langle v_+^u, v_-^u \rangle$ and $\mathcal{A}\{-1\} = \langle v_+^l, v_-^l \rangle$ (u for "upper" and l for "lower") are graded modules and for W a graded module $W_{l-k} = W\{k\}_l$ (we have arranged the generators vertically by quantum grading as before). Thus η may now be non-zero while still quantum grading-preserving.

It remains to determine a form of the map η that yields a well-defined chain complex, whose homology is a virtual link invariant. To do so, we examine the following possible faces of the cube:

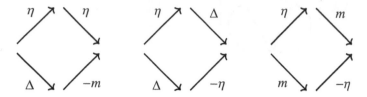

These faces suffice, as all other possible faces anticommute trivially. It can be quickly determined that the unique form of η that causes these faces to anticommute is given by

$$\eta(v_+^u) = v_+^l \qquad \eta(v_+^l) = 2v_-^u$$
$$\eta(v_-^u) = v_-^l \qquad \eta(v_-^l) = 0.$$

The m and Δ maps are left unchanged and do not interact with the upper and lower superscripts (so that $m(v_+^u \otimes v_+^u) = v_+^u$, for example).

It follows that the construction outlined above associates a well-defined chain complex to an oriented virtual link diagram. As in the case of Manturov's extension, the proof that the homology of this chain complex is an invariant of the virtual link represented by the diagram follows the classical proof. Surprisingly, the moves depicted in Figure 75.1 induce the identity on the chain complex level. It remains to verify invariance under the classical Reidemeister moves; here one may repeat *mutatis mutandis* the arguments applied to classical Khovanov homology.

Definition 75.3.4 ([16, Section 2]). Let D be a diagram of an oriented virtual link L. Associated to D is the chain complex $CDKh(D)$, the homology of which is an invariant of L, and is denoted $DKh(L)$. We refer to $DKh(L)$ as the *doubled Khovanov homology of L*.

An example of the doubled Khovanov homology of a virtual knot is given in Figure 75.6. Comparing Figures 75.6 and 75.5 we see that doubled Khovanov homology is distinct to Manturov's extension.

For a classical link L, the quantum degrees in which $Kh(L)$ is non-trivial are either all even, or all odd (which case holds depends on the number of components of L). Figure 75.6 demonstrates that this is not the case for doubled Khovanov homology. This is an example of the structural differences between the classical Khovanov and doubled Khovanov homologies. Despite these differences, doubled Khovanov homology does not yield any new information when restricted to classical links (see [16, Proposition 2.5]).

Unlike Manturov's extension, we can sometimes use doubled Khovanov homology exclusively to detect virtual links that are not classical.

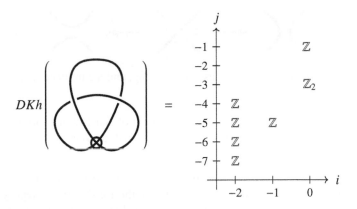

FIGURE 75.6: The doubled Khovanov homology of the virtual knot depicted.

Theorem 75.3.1 ([16, Corollary 2.6]). *Let L be a virtual link. If*

$$DKh(L) \neq G \oplus G\{-1\}$$

for G a non-trivial bigraded Abelian group, then L is not classical.

The connected sum operation is not well-defined on virtual knots. That is, one may connect sum two virtual knot diagrams at distinct sites to obtain distinct virtual knots (see [15, Section 2.3]). In particular, there exist non-trivial virtual knots that can be obtained as a connected sum of two unknots. Doubled Khovanov homology yields a condition met by such knots.

Theorem 75.3.2 ([16, Theorem 5.12]). *Let K be a virtual knot that is a connected sum of two unknots. Then $DKh(K) = DKh(\bigcirc)$.*

For example, it can be concluded from Figure 75.6 that the virtual knot depicted is not classical, nor the connected sum of two unknots.

75.3.4 Tubbenhauer's extension

We conclude this chapter by describing the extension of Khovanov homology due to Tubbenhauer [19]. His construction is the most technical, and as such the description here is more limited than those given in Sections 75.3.2 and 75.3.3.

The construction of classical Khovanov homology utilises a topological quantum field theory (TQFT): a functor from a topological category to an algebraic one. Khovanov originally worked almost entirely post-functor, in the algebraic category. Bar-Natan re-expressed this construction by working pre-functor, in the topological category [2]. Specifically, he constructed a category, *Cob*, whose objects are simple closed curves in the plane and whose morphisms are oriented cobordisms. Working in this category one can, in a very precise sense, define chain complexes of smoothings of classical link diagrams. This allowed Bar-Natan to reprove that Khovanov's construction yields

FIGURE 75.7: A choice of x-marker at a classical crossing, and the induced marker on smoothings.

a link invariant, without applying a TQFT. Another benefit of Bar-Natan's formulation is that it extends to a theory of *tangles* (properly embedded arcs in a 3-ball up to boundary-preserving isotopy). This allows for powerful 'cut-and-paste' arguments.

As noted in Section 75.3.1, the realization of the η map as a cobordism is a once-punctured Möbius band. Tubbenhauer's extension proceeds by generalizing the Bar-Natan's category *Cob* to include such non-orientable cobordisms. We denote this new topological category *uCob*: the objects are immersed curves in the plane (double points are virtual crossings) and the morphisms are possibly non-orientable cobordisms. The boundaries of these cobordisms are also decorated using the formal signs $+$, $-$, known as the *gluing numbers*.

Using this category Tubbenhauer is able to follow Bar-Natan, constructing a purely diagrammatic chain complex. He then verifies that this construction yields an invariant of virtual links, without passing to an algebraic category. The way in which the single-cycle smoothing and problem face are dealt with is similar in spirit to Manturov's extension: one can orient the circles within a smoothing, and use disagreements between these orientations to detect the problem face. However, there is no canonical way to assign such orientations. In Manturov's extension the source-sink orientation is used. Tubbenhauer takes a different approach, and arbitrarily assigns orientations to the circles within smoothings. The construction also requires the 180°-symmetry of classical crossings to be broken. This is done using an *x-marker*, as depicted in Figure 75.7.

With an arbitrary choice of orientation of the circles within smoothings and an x-marker at each classical crossing, the problem face may be detected as follows. Given an edge of the cube of smoothings, the choices of orientation and x-marker are used to decorate the initial and final smoothings with gluing numbers. When composing two edges together, one can carefully use the configuration of these gluing numbers to recover anticommutativity of the problem face, employing an automorphism similar to the barring operator of 75.3.1. That the resulting chain complex yields a virtual link invariant is verified in a manner very similar to Bar-Natan's original proof. Tubbenhauer verifies that the construction is independent of the choice of orientation of the circles within smoothings and the x-markers, so that the invariant is well-defined.

As in the classical case, one must now apply a functor to pass from the topological category to an algebraic one. The relevant functor is now an *unoriented TQFT* [21]. Tubbenhauer classifies all such functors whose application yields an invariant of virtual links (these functors are intimately related to skew-extended Frobenius algebras). He demonstrates that exactly one unoriented TQFT yields a bigraded Abelian group (the others are singly-graded). We denote the resulting homology theory as follows.

Definition 75.3.5 ([19, Section 5]). Let D be a diagram of an oriented virtual link L. Associated to D is the chain complex $CUKh(D)$, the homology of which is an invariant of L, and is denoted $UKh(L)$.

In all known examples the homology groups $VKh(L)$ and $UKh(L)$ are isomorphic. While the constructions of these homologies are similar in spirit – both orient the circles of smoothings and exploit any disagreements – it is not clear if they are compatible. In addition, both theories require a particular sign convention, and it is not clear if the two distinct conventions agree. Both theories recover anticommutativity of the problem face using an automorphism that involves multiplication by -1. As such, it is possible that sign disagreements could cause the theories to be inequivalent.

Tubbenhauer also investigates analogues of Lee homology and Bar-Natan homology for virtual links, using the strength of the categorical framework provided by *uCob*. Further, he constructs an extension of Khovanov homology to *virtual tangles*. This is intricate, and requires the generalization of *uCob* to include cobordisms with more complicated boundaries. The extension to tangles gives access to some of the 'cut-and-paste' arguments prevalent in classical Khovanov homology. Tubbenhauer has written a Mathematica program to compute $UKh(L)$ [18]; in contrast, both Manturov's extension and doubled Khovanov homology must be computed by hand at present.

75.4 Bibliography

[1] M. M. Asaeda, J. H. Przytycki, and A. S. Sikora. Categorification of the Kauffman bracket skein module of *I*-bundles over surfaces. *Algebr. Geom. Topol.*, 4:1177–1210, 2004.

[2] D. Bar-Natan. Khovanov's homology for tangles and cobordisms. *Geom. Topol.*, 9:1443–1499, 2005.

[3] H. U. Boden, M. Chrisman, and R. Gaudreau. Virtual knot cobordism and bounding the slice genus. *Experiment. Math.*, pages 1–17, 2018.

[4] J. S. Carter, S. Kamada, and M. Saito. Stable equivalence of knots on surfaces and virtual knot cobordisms. *J. Knot Theory Ramifications*, 11(3):311–322, 2002.

[5] H. A. Dye, A. Kaestner, and L. H. Kauffman. Khovanov homology, Lee homology and a Rasmussen invariant for virtual knots. *J. Knot Theory Ramifications*, 26(03):1741001, 2017.

[6] Boštjan Gabrovšek. The categorification of the Kauffman bracket Skein module of $\mathbb{RP}^3$. *Bull. Aust. Math. Soc.*, 88(3):407–422, 2013.

[7] M. Goussarov, M. Polyak, and O. Viro. Finite-type invariants of classical and virtual knots. *Topology*, 39(5):1045–1068, 2000.

[8] H. Karimi. The Khovanov homology of alternating virtual links, 2019. `arXiv:1904.12235`.

[9] L. H. Kauffman. Virtual Knot Theory. *European J. Combin.*, 20(7):663–691, 1999.

[10] L. H. Kauffman. Virtual knot cobordism. In *New ideas in low dimensional topology*, pages 335–377. World Scientific, 2015.

[11] M. Khovanov. A categorification of the Jones polynomial. *Duke Mathematical Journal*, 101(03):359–426, 1999.

[12] P. B. Kronheimer and T. S. Mrowka. Khovanov homology is an unknot-detector. *Publ. Math. Inst. Hautes Études Sci.*, (113):97–208, 2011.

[13] E. S. Lee. An endomorphism of the Khovanov invariant. *Adv. Math.*, 197(2):554–586, 2005.

[14] V. O. Manturov. Khovanov homology for virtual links with arbitrary coefficients. *J. Knot Theory Ramifications*, 16(03):343–377, 2007.

[15] V. O. Manturov and D. P. Ilyutko. *Virtual Knots: The State of the Art.* World Scientific, 2013.

[16] W. Rushworth. Doubled Khovanov Homology. *Canad. J. Math.*, 70:1130–1172, 2018.

[17] W. Rushworth. Computations of the slice genus of virtual knots. *Topology Appl.*, 253:57–84, 2019.

[18] D. Tubbenhauer. http://www.dtubbenhauer.com/vKh.html.

[19] D. Tubbenhauer. Virtual Khovanov homology using cobordisms. *J. Knot Theory Ramifications*, 23(09):1450046, 2014.

[20] V. Turaev. Cobordism of knots on surfaces. *J. Topol.*, 1(2):285–305, 2008.

[21] V. Turaev and P. Turner. Unoriented topological quantum field theory and link homology. *Algebr. Geom. Topol.*, 6(3):1069–1093, 2006.

[22] O. Viro. Virtual links, orientations of chord diagrams and Khovanov homology. In *Proceedings of Gökova Geometry-Topology Conference 2005*, pages 187–212. Gökova Geometry/Topology Conference (GGT), Gökova, 2006.

Part XIII

Algebraic and Combinatorial Invariants

Chapter 76

Knot Colorings

Pedro Lopes, Universidade de Lisboa

76.1   Introduction

Knot colorings have to do with associating mathematical objects with arcs of knot diagrams in a way that is insensitive to the Reidemeister moves. Thus, knot colorings are knot invariants. The most popular of these must be Fox colorings. A Fox coloring of a knot diagram is the ability (or otherwise) of associating one of three colors to each arc of the diagram in hand, in such a way that (*i*) at least two colors are used; (*ii*) at each crossing of the diagram, either the three colors meet or there is only one color. What remains invariant here? If there exists an assignment satisfying (*i*) and (*ii*) for the original diagram, then there will also be such an assignment for any diagram obtained from the original one by a Reidemeister move. We leave the proof for the reader noting Reidemeister moves are local transformations. Therefore, outside a certain neighborhood the diagram (and the coloring) remains the same after the Reidemeister move is performed. Conjugating this with condition (*ii*) (which states how colors meet at crossings), an argument is made for each of the Reidemeister moves. We consider the first Reidemeister move. On one side of it there is only one arc and no crossings. Therefore there is only one color involved. On the other side of this move there are two arcs and a crossing. The crossing allows for the existence of three colors, but since one of the arcs provides for both an under-arc and the over-arc at the crossing, there can be at most two colors. But two colors is not an option. Therefore, on this side of the move there is also only one color. On both sides of the diagram, the arc(s) extend to the rest of the diagram outside the neighborhood where the transformation takes place (which remains the same). Therefore, the colors we referred to have to be equal on both sides of the transformation. Therefore, the first Reidemeister move does not affect the colorability i.e., the existence (or not) of a Fox coloring. The other Reidemeister moves are dealt with analogously. We, then, state the following result.

Theorem 76.1.1. *Fox colorings constitute a knot invariant.*

Look at Figures 76.1 for first applications of Fox colorings.

To the best of our knowledge, the first reference in print to Fox colorings is in [3] (Exercises 6 and 7 in Chapter VI, see also Exercises 8 and 10 in Chapter VIII). Also, note that we can use *p* colors instead of three, for any given odd prime *p*, and keep the conditions (i) and (ii); in this way we will have Fox *p*-colorings. The same sort of argument as above will prove Fox *p*-colorings are knot invariants. From now on, Fox 3-colorings are what we called before Fox colorings.

Before we move on to the algebraization of these ideas we state an interesting concept, the minimum number of colors.

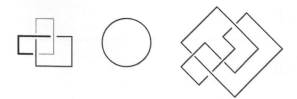

FIGURE 76.1: First Fox colorings. Leftmost: a diagram of the trefoil knot with a Fox coloring (different thickness meaning different color). Center: a diagram of the unknot; only one arc, only one color, i.e., the unknot does not admit Fox colorings. Therefore, the trefoil knot and the unknot are different knots. Rightmost: a diagram of the figure-eight knot. Exercise: the figure-eight knot does not admit Fox colorings but admits Fox 5-colorings, whereas the trefoil knot does not admit 5-Fox colorings. (Thus, the trefoil knot and the figure-eight knot are different knots.)

Definition 76.1.1. *Suppose p is an odd prime and K is a knot admitting Fox p-colorings i.e., for any diagram $\mathcal{D}$ of K there is a Fox p-coloring of it i.e., an assignment of colors from a set of p different colors, to the arcs of $\mathcal{D}$ such that conditions (i) and (ii) above are satisfied, for p = 3 (in the sequel, we will clarify what these conditions are for $p \neq 3$). For each such $\mathcal{D}$, consider the number $m_{\mathcal{D},p}$ which stands for the minimum number of colors it takes to assemble a Fox p-coloring on diagram $\mathcal{D}$. Then, the minimum number of colors of K mod p, notation $mincol_p(K)$, is by definition,*

$$mincol_p(K) := \min\{m_{\mathcal{D},p} \mid \mathcal{D} \text{ is a diagram of } K\}.$$

Look at Table 76.1 to find out where we stand with respect to these minima.

p (odd prime)	$mincol_p(K)$	dependence on K admitting Fox p-colorings
3	3	none
5	4	none
7	4	none
11	5	none
13	5	none
$p > 13$	$mincol_p(K) \geq 2 + \lfloor \log_2 p \rfloor$	not known

TABLE 76.1: Where we stand regarding $mincol_p$.

We emphasize that these minima are known for p in the range $\{3, 5, 7, 11, 13\}$ (see [7], [15], [13], [2], [12], [1]), that for p in this range each of these minima is the same no matter which knot we choose as long as it admits Fox p-colorings, that there is an upper bound which is $2 + \lfloor \log_2 p \rfloor$ (see [11]) which happens to coincide with the minima in the indicated range, and that for prime $p > 13$ we do not know the exact minima nor if it varies over the set of knots admitting Fox p-colorings.

76.2 Algebraization of Fox Colorings

A Fox coloring is an assignment of objects to arcs of a diagram and satisfying conditions at each crossing of the diagram. This is reminiscent of presentations of algebraic structures, generators and relations. What if we could find an algebraic structure which would fit in the current picture? That is, an algebraic structure that, given a knot diagram, would envisage arcs of the diagram as generators with relations read off the crossings of the diagram, with the presentation yielding a knot invariant? For that to happen we need the axioms of said algebraic structure to be insensitive to the Reidemeister moves. Such an algebraic structure was found independently by Joyce ([6]) and Matveev ([10]), the first author calls it a *quandle* while the second author calls it a *distributive groupoid*. We will here use quandle for it is shorter. Here are the quandle axioms.

Definition 76.2.1. *Given a set X equipped with a binary operation denoted $*$, $(X, *)$ is called a quandle, by definition, if*

I. for any $a \in X$, $a * a = a$;

II. for any $a, b \in X$, there exists a unique $x \in X$, such that $x * b = a$;

III. for any $a, b, c \in X$, $(a * b) * c = (a * c) * (b * c)$.

These three axioms are the algebraic consequences of the Reidemeister moves so that given a knot diagram and following the instructions "arcs are generators and crossings give rise to relations" we end up with a presentation which is a knot invariant, up to isomorphism. This is a presentation of a quandle, known as the Fundamental Quandle of the Knot (at hand). Unfortunately, there is no algorithm that in a finite number of steps would tell us if two such presentations are isomorphic or not so, in order to tell knots apart we resort to counting homomorphisms, for instance. For example, consider the so-called dihedral quandle of order 3, notation D_3, whose underlying set is $\mathbf{Z}/3\mathbf{Z} = \{\bar{0}, \bar{1}, \bar{2}\}$, and with multiplication table displayed in Table 76.2.

$*$	0	1	2
0	0	2	1
1	2	1	0
2	1	0	2

TABLE 76.2: Multiplication table of D_3. Bars over representatives are omitted.

The diagonal is $\{0, 1, 2\}$ to comply with the first axiom and when two distinct elements are "'multiplied'" the result is the third element. Let us now calculate the fundamental quandle of the trefoil knot, the unknot, and the figure-eight knot. We will then be able to calculate numbers of homomorphisms, from these fundamental quandles to D_3, and then assess whether these knots are distinct or not. Consider now Figure 76.2.

Now we can calculate how many homomorphisms there are from each of these fundamental quandles to D_3; it is a matter of verifying when the relations of the source quandle are satisfied in D_3. We leave the details to the reader the outcome being the following. For the trefoil knot, any assignment

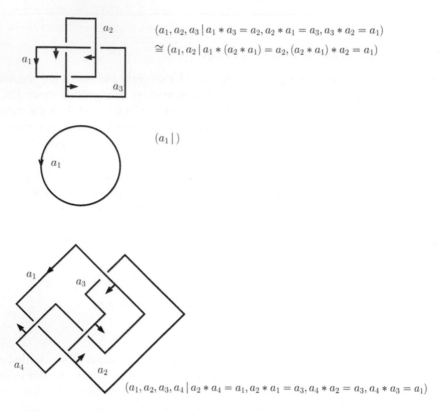

$$(a_1, a_2, a_3 \mid a_1 * a_3 = a_2, a_2 * a_1 = a_3, a_3 * a_2 = a_1)$$

$$\cong (a_1, a_2 \mid a_1 * (a_2 * a_1) = a_2, (a_2 * a_1) * a_2 = a_1)$$

$$(a_1 \mid)$$

$$(a_1, a_2, a_3, a_4 \mid a_2 * a_4 = a_1, a_2 * a_1 = a_3, a_4 * a_2 = a_3, a_4 * a_3 = a_1)$$

FIGURE 76.2: First row: trefoil knot and its fundamental quandle. Second row: unknot and its fundamental quandle. Third row: figure-eight knot and its fundamental quandle.

of elements of D_3 to a_1 and a_2 works. Therefore, there are 9 homomorphisms from the fundamental quandle of the trefoil to D_3. From the fundamental quandle of the unknot to D_3 there are only three homomorphisms, one per element of D_3. Finally, from the fundamental quandle of the figure-eight knot to D_3 there are only three, one per element of D_3. If the fundamental quandles of these knots were isomorphic then the numbers of homomorphisms would be equal. Since they are not we may conclude that the trefoil is distinct from the unknot and from the figure-eight knot. But wait! We already knew this! In fact, what we are doing is the algebraization of the Fox colorings. We now regard the colorings as homomorphisms from the fundamental quandle to a target quandle. If we pick the target quandle to be D_3 then we are translating the Fox 3-colorings to this setting, modulo the fact that now we also consider the trivial colorings i.e., the ones that only involve one color. These are as many as the elements of the target quandle.

Also, D_3 is but one of a family of quandles known as the dihedral quandles. There is one for each integer greater than 2; the underlying set is $\mathbf{Z}/n\mathbf{Z}$ and the quandle operation is $a * b = 2b - a$, mod n. For each odd prime p, the Fox p-colorings translate to the homomorphisms from the fundamental quandle to the dihedral quandle of order p, modulo the monochromatic homomorphisms as mentioned above.

Here is another way of looking at these homomorphisms to the dihedral quandles and how to calculate them. Each crossing in the diagram gives rise to an equation of the sort $2b - a - c = 0$ and ignoring for the moment which modulo to choose. Collecting these equations over the crossings, we obtain a system of linear homogeneous equations over $\mathbf{Z}$. The determinant of the matrix of the coefficients is zero because each row of this matrix has exactly one 2 and two -1's — and this corresponds to the existence of the trivial coloring. If we want more solutions, the polychromatic solutions, we have to require that the first minor determinants of the matrix of the coefficients also be 0. But these minor determinants are always some non-zero integer (for knots). So we work modulo this integer (or any of its prime divisors) in order to also obtain the polychromatic solutions. This explains the choice of the prime modulus. Also note that this minor determinant is known as the *determinant* of the knot under study.

We can now state the following result.

Theorem 76.2.1 (Kauffman-Harari). *Let p be an odd integer and K an alternating knot admitting non-trivial p-colorings. Let $\mathcal{D}$ be an alternating minimal diagram of K. Then each non-trivial p coloring of $\mathcal{D}$ assigns different colors to different arcs.*

This result began as a conjecture stated in [5] and proved in [9]. Does this result generalize? That is, now that we know that colorings correspond to homomorphisms to dihedral quandles, we can also use other target quandles to try to tell knots apart. If we use other target quandles will we also have a corresponding Kauffman-Harari conjecture? For instance, consider the class of linear Alexander quandles. There is one such quandle for each pair of integers, n, m greater than 2 which are relatively prime (in order to satisfy the second quandle axiom). The underlying set of such a quandle is $\mathbf{Z}/n\mathbf{Z}$ and the operation is $a * b := ma + (1 - m)b$, mod n. In order to obtain non-trivial colorings n should factor a certain minor determinant (which depends on m). If we make $m = -1$ we obtain the dihedral quandles. Again, is there a corresponding Kauffman-Harari conjecture for the quandles of this family when $m \neq -1$? The answer appears to be no, although it is not clear why; check [8] for examples and counter-examples. In passing, if we take m to be an invertible algebraic variable, then we would be looking at "Alexander colorings"; for non-trivial colorings we would have to require that a certain minor determinant be null i.e., we would have to work modulo - the Alexander polynomial, see [3]. Reciprocally, evaluating the Alexander polynomial at a given m tells us which modulus we should use when working with linear Alexander quandles for target quandles, see [8]. This will be used below.

Finally, we mention the existence of equivalence classes of non-trivial colorings in some instances when a power of the modulus factors the determinant of the knot, see [4] and [8]. The target quandles we are thinking of stem from the linear Alexander quandles. These equivalence classes are the orbits of the colorings acted upon the permutations of the modulus at hand (i.e., the total number of colors) which are compatible with the quandle operation. The number of these orbits is a knot invariant which depends on the nullity of the coefficient matrix, which is the number of zeros in the diagonal of the Smith Normal Form for the matrix of the coefficients. The formula for the number of these classes, working modulo odd prime p, with p-nullity n_p is

$$\frac{p^{n_p - 1} - 1}{p - 1}.$$

In the tables below we study the knots from Rolfsen's table, [14], whose Fox determinants are prime powers for primes $3, 5, 7, 11, 13$.

Encyclopedia of Knot Theory

K	$S(K)$	Equiv. Classes	1st prime value	Oddities
3_1	diag$\{1, 3, 0\}$	1	Alex(2) = 3	
6_1	diag$\{1, 1, 1, 1, 3^2, 0\}$	1	Alex(3) = 5	
8_{10}	diag$\{1, 1, 1, 1, 1, 1, 3^3, 0\}$	1	Alex(63) = 490558501	
8_{11}	diag$\{1, 1, 1, 1, 1, 1, 3^3, 0\}$	1	Alex(199) = 1213298069	Alex(2) = 0
8_{19}	diag$\{1, 1, 1, 1, 1, 1, 3, 0\}$	1	Alex(44) = 1498451819	
8_{20}	diag$\{1, 1, 1, 1, 1, 1, 3^2, 0\}$	1	Alex(289) = 1662202303	
9_1	diag$\{1, 1, 1, 1, 1, 1, 1, 3^2, 0\}$	1	Alex(26) = 770734061	
9_6	diag$\{1, 1, 1, 1, 1, 1, 1, 3^3, 0\}$	1	Alex(75) = 1271385649	
9_{35}	diag$\{1, 1, 1, 1, 1, 1, 3, 3^2, 0\}$	4	Alex(3) = 31	
9_{46}	diag$\{1, 1, 1, 1, 1, 1, 3, 3, 0\}$	4	Alex(3) = 5	Alex(2) = 0
9_{47}	diag$\{1, 1, 1, 1, 1, 1, 3, 3^2, 0\}$	4	Alex(3) = 151	
9_{48}	diag$\{1, 1, 1, 1, 1, 1, 3, 3^2, 0\}$	4	Alex(3) = 29	
10_4	diag$\{1, 1, 1, 1, 1, 1, 1, 1, 3^3, 0\}$	1	Alex(7) = 5099	
10_{42}	diag$\{1, 1, 1, 1, 1, 1, 1, 1, 3^4, 0\}$	1	Alex(3) = 11	
10_{75}	diag$\{1, 1, 1, 1, 1, 1, 1, 1, 3^4, 0\}$	1	Alex(3) = 11	
10_{87}	diag$\{1, 1, 1, 1, 1, 1, 1, 1, 3^4, 0\}$	1	Alex(95) = 1830711601	Alex(2) = 0
10_{98}	diag$\{1, 1, 1, 1, 1, 1, 1, 1, 3^4, 0\}$	1	Alex(95) = 1830711601	
10_{99}	diag$\{1, 1, 1, 1, 1, 1, 1, 3^2, 3^2, 0\}$	4	Alex(51) = 493236641	Alex(2) = 0
10_{124}	diag$\{1, 1, 1, 1, 1, 1, 1, 1, 1, 0\}$			Alex(−1) = −1
10_{139}	diag$\{1, 1, 1, 1, 1, 1, 1, 1, 3, 0\}$	1	Alex(18) = 1821281807	
10_{140}	diag$\{1, 1, 1, 1, 1, 1, 1, 1, 3^2, 0\}$	1	Alex(289) = 1662202303	
10_{143}	diag$\{1, 1, 1, 1, 1, 1, 1, 1, 3^3, 0\}$	1	Alex(63) = 490558501	
10_{145}	diag$\{1, 1, 1, 1, 1, 1, 1, 1, 3, 0\}$	1	Alex(4) = 277	
10_{147}	diag$\{1, 1, 1, 1, 1, 1, 1, 1, 3^3, 0\}$	1	Alex(199) = 1213298069	Alex(2) = 0
10_{153}	diag$\{1, 1, 1, 1, 1, 1, 1, 1, 1, 0\}$		Alex(45) = 474526411	Alex(−1) = 1

TABLE 76.3: List of knots up to 10 crossings whose (Fox) determinant is a power of 3 (including 3^0). $S(K)$ denotes the Smith Normal Form for the matrix of the coefficients associated with the Fox colorings. Alex(...) denotes the Alexander polynomial of the knot under study.

76.3 Bibliography

[1] F. Bento, P. Lopes, The minimum number of Fox colors modulo 13 is 5, *Topology Appl.* **216** (2017), 85–115.

[2] W. Cheng, X. Jin, and N. Zhao, Any 11-colorable knot can be colored with at most six colors, *J. Knot Theory Ramifications* **23** (2014), no. 11, 1450062, 25pp.

[3] R. H. Crowell, R. H. Fox, *Introduction to Knot Theory*, Dover Publications, 2008 (originally published by Ginn and Co. in 1963)

[4] J. Ge, S. Jablan, L. H. Kauffman, P. Lopes, Equivalence classes of colorings, *Banach Center Publ.*, **103** (2014), 63-76.

[5] F. Harary, L. Kauffman, Knots and graphs. I. Arc graphs and colorings, *Adv. in Appl. Math.* **22** (1999), no. 3, 312–337.

K	$S(K)$	Equiv. Classes	1st prime value	Oddities
4_1	diag$\{1, 1, 5, 0\}$	1	Alex(5) = 11	
5_1	diag$\{1, 1, 5, 0\}$	1	Alex(2) = 11	
8_9	diag$\{1, 1, 1, 1, 1, 1, 5^2, 0\}$	1	Alex(2) = 7	
9_{49}	diag$\{1, 1, 1, 1, 1, 1, 5, 5, 0\}$	6	Alex(2) = 19	
10_3	diag$\{1, 1, 1, 1, 1, 1, 1, 1, 5^2, 0\}$	1	Alex(18921) = 2147187817	
10_{129}	diag$\{1, 1, 1, 1, 1, 1, 1, 1, 5^2, 0\}$	1	Alex(183) = 2088411671	
10_{132}	diag$\{1, 1, 1, 1, 1, 1, 1, 1, 5, 0\}$	1	Alex(2) = 11	
10_{137}	diag$\{1, 1, 1, 1, 1, 1, 1, 1, 5^2, 0\}$	1	Alex(285) = 2130436951	Alex(2) = 1 = Alex(3)
10_{155}	diag$\{1, 1, 1, 1, 1, 1, 1, 5, 5, 0\}$	6	Alex(2) = 7	
10_{161}	diag$\{1, 1, 1, 1, 1, 1, 1, 1, 5, 0\}$	1	Alex(3) = 829	

TABLE 76.4: List of knots up to 10 crossings whose (Fox) determinant is a power of 5. $S(K)$ denotes the Smith Normal Form for the matrix of the coefficients associated with the Fox colorings. Alex(...) denotes the Alexander polynomial of the knot under study.

K	$S(K)$	Equiv. Classes	1st prime value
5_2	diag$\{1, 1, 1, 7, 0\}$	1	Alex(3) = 11
7_1	diag$\{1, 1, 1, 1, 1, 7, 0\}$	1	Alex(2) = 43
9_{27}	diag$\{1, 1, 1, 1, 1, 1, 7^2, 0\}$	1	Alex(2) = 5
9_{41}	diag$\{1, 1, 1, 1, 1, 1, 7, 7, 0\}$	8	Alex(2) = 7
9_{42}	diag$\{1, 1, 1, 1, 1, 1, 7, 0\}$	1	Alex(3) = 31
10_{22}	diag$\{1, 1, 1, 1, 1, 1, 1, 1, 7^2, 0\}$	1	Alex(37) = 438518371
10_{35}	diag$\{1, 1, 1, 1, 1, 1, 1, 1, 7^2, 0\}$	1	Alex(3) = 7
10_{48}	diag$\{1, 1, 1, 1, 1, 1, 1, 1, 7^2, 0\}$	1	Alex(18) = 781537259
10_{157}	diag$\{1, 1, 1, 1, 1, 1, 7, 7, 0\}$	8	Alex(2) = 23

TABLE 76.5: List of knots up to 10 crossings whose (Fox) determinant is a power of 7. $S(K)$ denotes the Smith Normal Form for the matrix of the coefficients associated with the Fox colorings. Alex(...) denotes the Alexander polynomial of the knot under study.

[6] D. Joyce, A classifying invariant of knots, the knot quandle, *J. Pure Appl. Alg.*, **23** (1982), 37–65

[7] L. H. Kauffman, P. Lopes, On the minimum number of colors for knots, *Adv. in Appl. Math.*, **40** (2008), no. 1, 36–53

[8] L. H. Kauffman, P. Lopes, Colorings beyond Fox: The other linear Alexander quandles, *Linear Algebra Appl.*, **548** (2018), no. 1, 221 - 258

[9] T. Mattman, P. Solis, A proof of the Kauffman-Harary conjecture, *Algebr. Geom. Topol.* **9** (2009), 2027–2039.

[10] S. Matveev, Distributive groupoids in knot theory, *Math. USSR Sbornik* **47** (1984), no. 1, 73–83.

[11] T. Nakamura, Y. Nakanishi, S. Satoh, The palette graph of a Fox coloring, *Yokohama Math. J.* **59** (2013), 91–97.

K	$S(K)$	Equiv. Classes	1st prime value
6_2	diag$\{1, 1, 1, 1, 11, 0\}$	1	Alex$(3) = 11$
7_2	diag$\{1, 1, 1, 1, 1, 11, 0\}$	1	Alex$(2) = 43$
10_{123}	diag$\{1, 1, 1, 1, 1, 1, 1, 11, 11, 0\}$	12	Alex$(33) = 933171329$
10_{125}	diag$\{1, 1, 1, 1, 1, 1, 1, 1, 11, 0\}$	1	Alex$(2) = 29$
10_{128}	diag$\{1, 1, 1, 1, 1, 1, 1, 1, 11, 0\}$	1	Alex$(3) = 839$
10_{130}	diag$\{1, 1, 1, 1, 1, 1, 1, 1, 11, 0\}$	1	Alex$(3) = 89$
10_{152}	diag$\{1, 1, 1, 1, 1, 1, 1, 1, 11, 0\}$	1	Alex$(2) = 139$

TABLE 76.6: List of knots up to 10 crossings whose (Fox) determinant is a power of 11. $S(K)$ denotes the Smith Normal Form for the matrix of the coefficients associated with the Fox colorings. Alex(...) denotes the Alexander polynomial of the knot under study.

K	$S(K)$	Equiv. Classes	1st prime value
6_3	diag$\{1, 1, 1, 1, 13, 0\}$	1	Alex$(2) = 7$
7_3	diag$\{1, 1, 1, 1, 1, 13, 0\}$	1	Alex$(3) = 101$
8_1	diag$\{1, 1, 1, 1, 1, 1, 13, 0\}$	1	Alex$(4) = 23$
9_{43}	diag$\{1, 1, 1, 1, 1, 1, 1, 13, 0\}$	1	Alex$(2) = 7$
10_{154}	diag$\{1, 1, 1, 1, 1, 1, 1, 13, 0\}$	1	Alex$(2) = 41$

TABLE 76.7: List of knots up to 10 crossings whose (Fox) determinant is a power of 13. $S(K)$ denotes the Smith Normal Form for the matrix of the coefficients associated with the Fox colorings. Alex(...) denotes the Alexander polynomial of the knot under study.

[12] T. Nakamura, Y. Nakanishi, S. Satoh, 11-colored knot diagram with 5 colors, *J. Knot Theory Ramifications* **25** (2016), no. 4, 1650017

[13] K. Oshiro, Any 7-colorable knot can be colored by four colors, *J. Math. Soc. Japan* **62**, no. 3 (2010), 963–973

[14] D. Rolfsen, *Knots and Links*, AMS Chelsea Publishing, 2003

[15] S. Satoh, 5-colored knot diagram with four colors, *Osaka J. Math.*, **46**, no. 4 (2009), 939–948

Chapter 77

Quandle Cocycle Invariants

J. Scott Carter, University of South Alabama

77.1 Introduction

Matveev [19] and, independently, Joyce [13] introduced an algebraic structure that mimics the Reidemeister moves for classical knots. The structure consists of a set Q with a binary operation $\triangleleft$ that satisfies the following three axioms.

I. $\forall a \in Q, a \triangleleft a = a$;

II. $\forall a, b \in Q, \exists! c \in Q$ such that $c \triangleleft a = b$;

III. $\forall a, b, c \in Q, (a \triangleleft b) \triangleleft c = (a \triangleleft c) \triangleleft (b \triangleleft c)$.

Following Joyce, call $(Q, \triangleleft)$ a *quandle*. Several aspects of this structure were rediscovered during the twentieth century. Takasaki [21] introduced the notion of *Kei* in which $(x \triangleleft y) \triangleleft y = x$. As reported in [6], the notion of a (w)rack was discussed in correspondences between Conway and Wraith to describe a group with conjugation as the operation. In modern usage, a *rack* satisfies axioms II and III. In [1], the concept of a shelf was introduced as a set for which axiom III holds. This is also called a right distributive system (RDS). A number of other generalizations exist including that of a biquandle [7], a qualgebra [17], and multiple conjugation quandles [11].

Here the focus is upon quandles and racks, homology theories thereof, and invariants of codimension 2 embeddings that can be formed via these. However, generalizations of these homology theories exist for many of the contexts defined above which also have applications to knot theory in a variety of contexts.

77.2 Quandle Homology

Let a quandle Q be given. For $n = 1, 2, \ldots$, consider the chain group $C_{n+1}(Q)$ to be the free abelian group generated by $(n + 1)$-tuples of quandle elements $(a_1, a_2, \ldots, a_{n+1}) \in Q^{n+1}$. The chain group $C_0(Q)$ is trivial. Define differentials $C_n(Q) \xleftarrow{\partial_{n+1}} C_{n+1}(Q)$ by

$$
\begin{aligned}
\partial_{n+1}(a_1, a_2, \ldots, a_{n+1}) &= \sum_{j=1}^{n+1} (-1)^j \Big[(a_1, \ldots, \widehat{a_j}, \ldots, a_{n+1}) \\
&\quad - (a_1 \triangleleft a_j, \ldots, a_{j-1} \triangleleft a_j, a_{j+1} \ldots, a_{n+1}) \Big].
\end{aligned}
$$

That the composition $\partial_n \circ \partial_{n+1} = 0$ follows from the self-distributivity axiom III: $(a \triangleleft b) \triangleleft c = (a \triangleleft c) \triangleleft (b \triangleleft c)$. Let $Z_n(Q)$ denote the kernel of ∂_n, and let $B_n(Q)$ denote the image of ∂_{n+1}. The (rack) homology of Q is defined to be $H_n^R(Q) = Z_n(Q)/B_n(Q)$. There is a subcomplex of degeneracies: $D_n(Q)$ is the submodule that is generated by chains of the form $(x_1, \ldots, x_i, x_i, x_{i+2}, \ldots, x_{n-1})$ for some $i = 1, \ldots, n-1$. The idempotence axiom I: $a \triangleleft a = a$ is used to show that the boundary of a degenerate chain is degenerate. Define the chain complex $C^Q(Q) = (\{C_n(Q)/D_n(Q) : n = 0, 1, \ldots\}, \partial_n)$, and let $H_n^Q(Q)$ denote its homology. This is the *quandle homology of a quandle Q*.

Litherland and Nelson [18] showed that the rack homology splits as a direct sum of the quandle homology and the degenerate homology.

Cohomology is defined dually, so that a quandle 2-cocycle is a function ϕ that takes values in an abelian group, A, and that satisfies

$$\phi(a, b) + \phi(a \triangleleft b, c) = \phi(a, c) + \phi(a \triangleleft c, b \triangleleft c), \quad \phi(a, a) = 0,$$

for all $a, b, c \in Q$. A 3-cocycle also takes values in A and satisfies

$$\theta(a, b, c) + \theta(a \triangleleft c, b \triangleleft c, d) + \theta(a, c, d) = \theta(a, b, d) + \theta(a \triangleleft b, c, d) + \theta(a \triangleleft d, b \triangleleft d, c \triangleleft d),$$

$$\theta(a, a, c) = \theta(a, b, b) = 0,$$

for all $a, b, c, d \in Q$. Quandle extensions are controlled by 2-cocycles. The product set $Q \times A$ has the structure of a quandle with $(a, s) \triangleleft (b, t) = (a \triangleleft b, s + \phi(a, b))$.

77.3 Codimension 2 Embeddings

When a smooth closed oriented n-dimensional manifold M is embedded into the $(n + 2)$-dimensional sphere S^{n+2}, it can be knotted. On first reading, the reader may assume $n = 1, 2$ and subsequently generalize to higher dimension as needed. We let K denote the image of M in S^{n+2}. The target space is identified with the one point compactification of $\mathbb{R}^{n+2}$; the point at infinity is chosen away from the image K. The *knot exterior* is defined to be the closure of the complement of a tubular neighborhood: $E(K) = \mathrm{Cl}(S^{n+2} \setminus N(K))$. While it is true that the exterior is not necessarily a complete invariant unless $M = S^1$ is a circle embedded in 3-dimensional space [10], the exterior and secondary invariants are often powerful. In particular, the fundamental group and the fundamental quandle are defined via homotopy classes of paths in the exterior. Both of these can be described diagrammatically using a generic projection of K into an $(n + 1)$-dimensional hyperplane $\mathbb{R}^{n+1}$.

For most points in K, the rank of the derivative of the restriction of the projection is n. Thus in a neighborhood of the point, the projection from K onto $\mathbb{R}^{n+1}$ appears as a coordinate n-plane. At other points in K the derivative drops in rank; the *branch point set* is an $(n - 2)$-dimensional subset upon which derivative of the projection is singular. At non-singular points, the image in $\mathbb{R}^{n+1}$ resembles $\{(x_1, x_2, \ldots, x_{n+1}) \in [-1, 1]^{n+1} : x_1 \cdot x_2 \cdot \cdots \cdot x_k = 0\}$. Such points will be called *k-tuple points* (non-singular, double, triple, quadruple, etc.) of the projection. We will use this local model to describe the fundamental quandle. Define

$$\pi_j = \{(x_1, \ldots, x_{j-1}, 0, x_{j+1}, \ldots, x_{n+1})\},$$

and let the standard basis vector e_j denote its oriented normal direction in $\mathbb{R}^{n+1}$. The normal vector e_j is a stand-in for a meridional loop in R^{n+2} that lies in the boundary of a small tubular neighborhood of the knotting K (See Fig. 1 of Chapter 43).

A local picture of the image of K in $\mathbb{R}^{n+2}$ can be reconstructed by lifting the hypoplane in which the jth coordinate is 0 to the location

$$(x_1, \ldots, x_{j-1}, 0, x_{j+1}, \ldots, x_{n+1}, j) \subset [-1,1]^{n+1} \times [0, n+2] \subset \mathbb{R}^{n+2}$$

for $j = 1, \ldots, n + 1$. Within $\mathbb{R}^{n+1}$ a broken sheeted diagram can be constructed that imitates the crossing data that are contained by considering the projection of a local picture $\cup_{j=1}^{k}\{(x_1, \ldots, x_{j-1}, 0, x_{j+1}, \ldots, x_{n+1}, j) \subset [-1,1]^{n+1} \times [0, n+2]\}$ of a k-tuple point. It is important to note that both the broken sheeted diagram and the configuration that is lifted to $\mathbb{R}^{n+2}$ are local depictions, and they do not inform how to glue the pictures together across patches of the manifold M that is being knotted.

Let $k \in \{1, 2, \ldots, n + 1\}$ be fixed, and consider a k-tuple point. Associate a label a_k to the vector e_k. The label is meant to indicate a path from the boundary of a tubular neighborhood of the kth sheet, π_k, to a base point in the exterior $E(K)$. In the penultimate sheet, π_{k-1}, introduce a pair of labels: when $x_k < 0$, the label associated to e_{k-1} is a_{k-1}, but when $0 < x_k$, the label associated to e_{k-1} is $a_{k-1} \triangleleft a_k$. Continue inductively. If $\ell < k$, label the region of π_ℓ for which $x_{\ell+1}, \ldots, x_k < 0$ with a_ℓ. By crossing a sheet (from $x_j < 0$ to $0 < x_j$) that is labeled a_j, the label associated to e_ℓ on the positive side of π_j is $a_\ell \triangleleft a_j$. If two coordinates change from negative to positive, then one can move via first crossing π_j and subsequently crossing π_m with $j < m$, or by first crossing π_m and then π_j. In the first case the label where $x_j, x_m > 0$ is $(a_\ell \triangleleft a_j) \triangleleft a_m$. In the other case, the label is $(a_\ell \triangleleft a_m) \triangleleft (a_j \triangleleft a_m)$. By quandle axiom III these labels coincide.

Proceed with caution. In this local picture, the chosen normal to π_ℓ is e_ℓ. The oriented normal to π_ℓ at an arbitrary k-tuple point that is induced from the orientation of the knotting K may, in fact, be $-e_\ell$. A change in source and target regions in all sheets π_j when $j < \ell$ occurs under this orientation change.

In general and with respect to any induced set of orientation normals, at a k-tuple point, there are $2^k - 1$ labels on the broken sheets. The labels coincide with the non-empty subsets of $\{a_1, \ldots, a_k\}$. Specifically, given $j_1 < j_2 < \cdots < j_\ell$, a label $(\cdots(a_{j_1} \triangleleft a_{j_2}) \triangleleft \cdots) \triangleleft a_{j_\ell}$ appears on the region of the hypoplane π_{j_1} towards which the normals to $\pi_{j_2}, \ldots, \pi_{j_\ell}$ point. In each hypoplane π_j, the region from which subsequent normals point are called the *source regions*.

A *broken sheeted* diagram of the knotting K consists of the projection into $\mathbb{R}^{n+1}$ together with breaks along the k-tuple set that indicate the relative height in $\mathbb{R}^{n+2}$. The labels on the local pictures of crossings are chosen to be consistent along the unbroken sheets of the diagram. The *fundamental quandle*, then, is presented via the labels on the unbroken sheets with relations of the form $c = a \triangleleft b$ that occur by moving across a broken sheet in the direction of the normal arrow to the sheet that is above and that is labeled b. Here the normal arrow is chosen so that the oriented tangent to M followed by the normal coincides with the orientation of $\mathbb{R}^{n+1}$. We remark that along branch points two sheets merge; the labels on the sheets coincide since there is no breaking along the branch set.

One of the principal theorems that grew out of the work by Joyce and Matveev is that the fundamental quandle of a knotted circle in 3-space classifies the knot up to orientation-reversing orientation. In particular, the fundamental quandle contains information about the fundamental group and its peripheral subgroup. It is easy to pass from the presentation of the fundamental quandle to the fundamental group by using the same set of generators — the labels on the connected sheets of the diagram — and declaring relations such as $a \triangleleft b = c = b^{-1}ab$ along the double point set.

77.4   Counting Invariants

The *three-element dihedral quandle* R_3 is the set $\{1, 2, 3\}$ with the operation $1 \triangleleft 2 = 3 = 2 \triangleleft 1$, $2 \triangleleft 3 = 1 = 3 \triangleleft 2$, $3 \triangleleft 1 = 2 = 3 \triangleleft 1$. It is easy to assign the labels $1, 2, 3$ to the arcs of a trefoil diagram so that the crossing conditions that were described above hold. Each arc receives a color and all colors are assigned. There are six ways to do so. There are also three ways to assign the same element to all three arcs. This means that the number of 3-colorings of the trefoil knot is nine.

In general given a finite quandle Q, a counting invariant for a knot K (whether it be classical or higher dimensional) is the number of ways of assigning labels in Q to the sheets of K such that the quandle condition holds along the double point set. That this is an invariant of the knot K (rather than the chosen diagram) depends upon the fact that the fundamental quandle is an invariant of the exterior $E(K)$. When a quandle is constructed from a group, the number of colorings is related to the cohomology of the infinite cyclic cover of the exterior with coefficients in the abelianization of the group.

The quandle cocycle invariant is a refinement of the counting invariant in that it associates to each quandle coloring an element in an abelian group, and thereby filters the colorings according to the group elements that are assigned.

77.5   Colored Knots as Chains and Cocycle Invariants

For any integer n, quandle n-cocycles can be used to define invariants of knottings $K : M^n \subset S^{n+2}$. First, consider the local model of an $(n + 1)$-tuple point in $\mathbb{R}^{n+1}$. Associate to this model the $(n + 1)$-tuple $(a_1, a_2, \ldots, a_{n+1})$ of quandle elements where a_{n+1} is the quandle element on the top sheet and a_n is the quandle element on the source sheet (of the normal to x_{n+1}) at which $x_n = 0$. In general a_k is the quandle label on the source sheet ($x_k = 0$ and the normals to $\pi_{k+1}, \ldots, \pi_{n+1}$ point away). When a given oriented knot $K : M^n \subset S^{n+2}$ is projected into $\mathbb{R}^{n+1}$, each $(n + 1)$-tuple point inherits an ordering of sheets from K, and orientation from M. Thus associated to the $(n + 1)$-tuple point is a signed quandle chain $\pm(a_1, a_2, \ldots, a_{n+1})$ where the sign is positive if the choice of the sequence of normals agrees with the orientation of $\mathbb{R}^{n+1}$.

Since M is closed, the sum of all the signed chains associated to the $(n + 1)$-tuple points is a cycle. The cycle itself represents a homology class in the homology of the fundamental quandle of the knot. In the classical case, Eisermann [4] showed that an embedding $S^1 \subset S^3$ is unknotted if and only if this cycle is a boundary.

Let a finite quandle Q be given. A *(quandle) coloring* of K is an assignment of elements of Q to the unbroken sheets of the diagram in such a way that along the double point set, the target sheet receives the quandle product $a \triangleleft b$ where a is the color on the source region of the under crossing and b is the color on the over sheet. Let $\omega \in Z_Q^{n+1}(Q)$ denote a quandle cocycle (a function that takes values in an abelian group A and that vanishes on the boundary of every $(n + 2)$-chain). Consider an $(n + 1)$-tuple point of the projection of K into $\mathbb{R}^{n+1}$, and evaluate ω on the $(n + 1)$-tuple of source colors $(q_1, q_2, \ldots, q_{n+1})$ on the multiple point. Here q_1 is the quandle color in Q that labels the source region in the bottom-most sheet, q_2 lies on the next highest sheet, and so forth, with q_{n+1} indicating the color on the unbroken sheet at the multiple point. Assign a sign $\sigma(t) = (\pm)$ to the multiple point t according to whether the oriented frame of normals induced from the orientation of M agrees with the orientation of $\mathbb{R}^{n+1}$ or not. In this way, an element in the abelian group A is obtained as

$$V(C) = \sum_{t \in (n+1)\text{-tuple points}} \sigma(t)\omega(q_1, \ldots, q_{n+1}).$$

The value depends upon the coloring C that is chosen. The multiset

$$\mathrm{QCI}(K) = \{\{V(C) : C \text{ is a coloring of } K\}\}$$

is the quandle cocycle invariant [2] of the knotting K.

Theorem 77.5.1 ([2]). *Let $n = 1, 2$. The quandle cocycle invariant $\mathrm{QCI}(K)$ is invariant under the Reidemeister or Roseman moves, and therefore defines an invariant of the isotopy class of the embedding $K : M \subset \mathbb{R}^{n+2}$.*

Results of [8] may be used to extend this result to oriented dimensions n. The proof for dimension $n = 1, 2$ involves showing that the multiset $\mathrm{QCI}(K)$ remains unchanged under the Reidemeister or Roseman moves [3]. This invariance is easy to show since the quandle cocycle conditions mirror the type III or the tetrahedral move.

The quandle cocycle invariants have found applications in a wide variety of contexts. In addition to Eisermann's result [5] which uses quandle homology to characterize the unknot, we mention a few other results. In [20], the 2-twist spun trefoil is shown to have at least four triple points in any projection to $\mathbb{R}^3$. In [2], we showed that the 2-twist spun trefoil is non-invertable. Generalization of the cocycle invariants have been used to distinguish handlebody knots [12]. The marked vertex diagrams of Yoshikawa [22] that encode ideas from [9] and later [15],[16] can be used to compute the invariants [14].

77.6 Bibliography

[1] John C. Baez and Alissa S. Crans. Higher-dimensional algebra. VI. Lie 2-algebras. *Theory Appl. Categ.*, 12:492–538, 2004.

[2] J. Scott Carter, Daniel Jelsovsky, Seiichi Kamada, Laurel Langford, and Masahico Saito. Quandle cohomology and state-sum invariants of knotted curves and surfaces. *Trans. Amer. Math. Soc.*, 355(10):3947–3989, 2003.

[3] J. Scott Carter and Masahico Saito. Broken surface diagrams and Roseman moves. This volume, Chapter 43.

[4] Michael Eisermann. Homological characterization of the unknot. *J. Pure Appl. Algebra*, 177(2):131–157, 2003.

[5] Michael Eisermann. Yang–Baxter deformations of quandles and racks. *Algebr. Geom. Topol.*, 5:537–562 (electronic), 2005.

[6] Roger Fenn and Colin Rourke. Racks and links in codimension two. *J. Knot Theory Ramifications*, 1(4):343–406, 1992.

[7] Roger Fenn, Colin Rourke, and Brian Sanderson. An introduction to species and the rack space. In *Topics in Knot Theory (Erzurum, 1992)*, volume 399 of *NATO Adv. Sci. Inst. Ser. C Math. Phys. Sci.*, pages 33–55. Kluwer Acad. Publ., Dordrecht, 1993.

[8] Roger Fenn, Colin Rourke, and Brian Sanderson. The rack space. *Trans. Amer. Math. Soc.*, 359(2):701–740, 2007.

[9] R. H. Fox. A quick trip through knot theory. In *Topology of 3-Manifolds and Related Topics (Proc. The Univ. of Georgia Institute, 1961)*, pages 120–167. Prentice-Hall, Englewood Cliffs, N.J., 1962.

[10] C. McA. Gordon and J. Luecke. Knots are determined by their complements. *J. Amer. Math. Soc.*, 2(2):371–415, 1989.

[11] Atsushi Ishii. A multiple conjugation quandle and handlebody-knots. *Topology Appl.*, 196(part B):492–500, 2015.

[12] Atsushi Ishii, Masahide Iwakiri, Yeonhee Jang, and Kanako Oshiro. A G-family of quandles and handlebody-knots. *Illinois J. Math.*, 57(3):817–838, 2013.

[13] David Joyce. A classifying invariant of knots, the knot quandle. *J. Pure Appl. Algebra*, 23(1):37–65, 1982.

[14] Seiichi Kamada, Jieon Kim, and Sang Youl Lee. Computations of quandle cocycle invariants of surface-links using marked graph diagrams. *J. Knot Theory Ramifications*, 24(10):1540010, 35, 2015.

[15] Akio Kawauchi, Tetsuo Shibuya, and Shin'ichi Suzuki. Descriptions on surfaces in four-space. I. Normal forms. *Math. Sem. Notes Kobe Univ.*, 10(1):75–125, 1982.

[16] Akio Kawauchi, Tetsuo Shibuya, and Shin'ichi Suzuki. Descriptions on surfaces in four-space. II. Singularities and cross-sectional links. *Math. Sem. Notes Kobe Univ.*, 11(1):31–69, 1983.

[17] Victoria Lebed. Qualgebras and knotted 3-valent graphs. *Fund. Math.*, 230(2):167–204, 2015.

[18] R. A. Litherland and Sam Nelson. The Betti numbers of some finite racks. *J. Pure Appl. Algebra*, 178(2):187–202, 2003.

[19] Sergei Matveev. Distributive groupoids in knot theory. *Mat. Sb. (N.S.)*, 119(161)(1):78–88, 160, 1982.

[20] Shin Satoh and Akiko Shima. The 2-twist-spun trefoil has the triple point number four. *Trans. Amer. Math. Soc.*, 356(3):1007–1024, 2004.

[21] Mituhisa Takasaki. Abstraction of symmetric transformations. *Tôhoku Math. J.*, 49:145–207, 1943.

[22] Katsuyuki Yoshikawa. An enumeration of surfaces in four-space. *Osaka J. Math.*, 31(3):497–522, 1994.

Chapter 78

Kei and Symmetric Quandles

Kanako Oshiro, Sophia University

In 1942, M. Takasaki [12] introduced the notion of a *kei* (圭), that is a quandle satisfying a condition. In [6], S. Kamada introduced the notion of a symmetric quandle, which is a quandle equipped with an involution satisfying certain identities motivated by knot diagrams with local orientations. By using a kei or a symmetric quandle, we can give a coloring invariant for links and surface-links which are not necessarily orientated or orientable. By using a symmetric quandle, we can also give a cocycle invariant for (unoriented or nonorientable) links and surface-links.

78.1   Definitions of a Kei and a Symmetric Quandle

Definition 1. ([5, 9]) A *quandle* is a set Q equipped with a binary operation $* : Q \times Q \to Q$ satisfying the following conditions:

(Q1) For any $a \in Q$, $a * a = a$.

(Q2) For any $a, b \in Q$, there exists a unique $c \in Q$ such that $c * b = a$.

(Q3) For any $a, b, c \in Q$, $(a * b) * c = (a * c) * (b * c)$.

We denote by $a * b^{-1}$ the unique $c \in Q$ for the second axiom (Q2). We write a quandle $(Q, *)$ simply as Q when no confusion can arise.

A quandle $(Q, *)$ is called a *kei* if it satisfies the condition:

(Q2′) For any $a, b \in Q$, $(a * b) * b = a$.

We can easily see that "(Q2′) $\Rightarrow$ (Q2)" holds. A kei is also known as an *involutory quandle* since its right translations $*a : Q \to Q; x \mapsto x * a$ are involutory.

Example 1. For an integer $n \geq 3$, we set a map $* : \mathbb{Z}_n \times \mathbb{Z}_n \to \mathbb{Z}_n$ by $a * b = 2b - a$. Then $(\mathbb{Z}_n, *)$ is a quandle which is a kei. We call it a *dihedral quandle* (or a *dihedral kei*) of order n and we denote it by R_n.

Definition 2. ([6, 8]) A *symmetric quandle* is a quandle $(Q, *)$ equipped with an involution $\rho : Q \to Q$ satisfying the following conditions:

(SQ1) For any $a, b \in Q$, $\rho(a * b) = \rho(a) * b$.

(SQ2) For any $a, b \in Q$, $a * \rho(b) = a * b^{-1}$.

We call the involution ρ a *good involution* of $(Q, *)$.

Example 2. Let Q be a kei. Then the identity map id_Q is a good involution of Q. In particular, the identity map of the dihedral quandle R_n is a good involution.

78.2   Kei Colorings

A kei is a quandle, and thus, a kei coloring for a link (or surface-link) diagram is a kind of quandle coloring. Here we focus on the property that kei colorings can be defined without considering orientations of links (or surface-links).

Let $(Q, *)$ be a kei. Let D be a diagram of a (unoriented) link.

Definition 3. (cf. [1, 3]) A Q-*coloring* of D is an assignment of an element of Q to each arc of D satisfying the following crossing condition:

- For a crossing v of D, let a_1, a_2 and a_3 denote the assigned elements to the under-arcs and the over-arc of v as depicted in the left of Figure 78.1. Then $a_1 * a_3 = a_2$ holds.

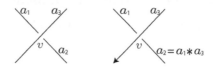

FIGURE 78.1: The crossing conditions.

We denote by $\mathrm{Col}_Q(D)$ the set of Q-colorings of D.

We remark that in the case of quandle colorings, the orientation of the over-arc and the location of the under-arcs are important for the crossing condition, see the right of Figure 78.1. However in the case of kei colorings, there is no need to consider such information since "$a_1 * a_3 = a_2 \Leftrightarrow a_1 = a_2 * a_3$" holds.

By considering the cardinality of the set of kei colorings, we have an invariant for (unoriented) links:

Proposition 1. *Let D and D' be diagrams of (unoriented) links. If D and D' represent the same link, then* $\#\mathrm{Col}_Q(D) = \#\mathrm{Col}_Q(D')$.

For the proof of this proposition, refer to the proof of Proposition 2 and Remark 1.

78.3   Symmetric Quandle Colorings

A symmetric quandle coloring is defined similar to the case of quandle colorings by considering "local orientations" instead of the orientations of links. That is, the labels of the arcs now come with a choice of normal direction indicated by an arrow.

Let $(Q, *)$ be a quandle and ρ a good involution of Q. Let D be a diagram of a (unoriented) link.

Definition 4. ([6, 8]) A (Q, ρ)-*coloring* of D is the Reidemeister equivalence class of an assignment of an element of Q and a normal orientation to each semi-arc of D satisfying the following crossing condition:

- Let a_1 and a_2 denote the assigned elements to two diagonal semi-arcs coming from an over-arc of D at a crossing v. If the normal orientations are coherent, then $a_1 = a_2$, otherwise $a_1 = \rho(a_2)$, see the first and second parts of Figure 78.2.

- Let a_1 and a_2 denote the assigned elements to two diagonal semi-arcs, say e_1 and e_2, coming from an under-arc of D at a crossing v. Suppose that one of the semi-arcs coming from an over-arc of D at v, say e_3, is labeled as a_3. We assume that the normal orientation of the over semi-arc e_3 points from e_1 to e_2. If the normal orientations of e_1 and e_2 are coherent, then $a_1 * a_3 = a_2$, otherwise $a_1 * a_3 = \rho(a_2)$, see the third and fourth parts of Figure 78.2.

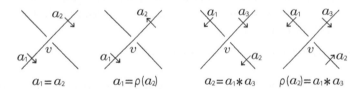

FIGURE 78.2: The crossing conditions.

$$\frac{a}{}\downarrow \quad \sim \quad \frac{\rho(a)}{}\uparrow$$

FIGURE 78.3: A basic inversion.

A *basic inversion* is an operation which reverses the normal orientation of a semi-arc and replaces the assigned element a of the semi-arc with $\rho(a)$, see Figure 78.3. The equivalence relation is generated by basic inversions.

We denote by $\mathrm{Col}_{(Q,\rho)}(D)$ the set of (Q, ρ)-colorings of D.

By considering the cardinality of the set of symmetric quandle colorings, we have an invariant for (unoriented) links:

Proposition 2. Let D and D' be diagrams of (unoriented) links. If D and D' represent the same link, then $\#\mathrm{Col}_{(Q,\rho)}(D) = \#\mathrm{Col}_{(Q,\rho)}(D')$.

Proof. Let D and D' be diagrams of a link L which differ by a Reidemeister move I, II or III. Let C be a (Q, ρ)-coloring of D. Then there exists a unique (Q, ρ)-coloring C' of D' such that $C = C'$ holds outside the disk where the Reidemeister move is applied. Thus we have a bijection $\mathrm{Col}_{(Q,\rho)}(D) \to \mathrm{Col}_{(Q,\rho)}(D'); C \mapsto C'$. See [8] for more details. $\qquad\square$

Remark 1. Any kei coloring can be regarded as a symmetric quandle coloring. More precisely, for a kei Q, a Q-coloring C of a diagram D can be regarded as a (Q, id)-coloring C of D by considering arbitrary assignments of normal orientations to semi-arcs of D.

For a symmetric quandle (Q, ρ), the *associated group* $G_{(Q,\rho)}$ is defined by

$$G_{(Q,\rho)} = \langle a \in Q \mid a * b = b^{-1}ab \ (a, b \in Q), \ \rho(a) = a^{-1} \ (a \in Q) \rangle.$$

A (Q, ρ)-*set* is a set Y equipped with a right action by the associated group $G_{(Q,\rho)}$. We denote by $y * g$ the image of an element $y \in Y$ by the action of $g \in G_{(Q,\rho)}$. We note that Q can be regarded as a (Q, ρ)-set with the right action which coincides with the quandle operation $*$.

Let (Q, ρ) be a symmetric quandle and Y a (Q, ρ)-set. Let D be a diagram of a (unoriented) link.

Definition 5. ([6, 8]) A $(Q, \rho)_Y$-*coloring* of D is a (Q, ρ)-coloring of D equipped with an assignment of an element of Y to each complementary region of D satisfying the following semi-arc condition:

- Let y_1 and y_2 denote the assigned elements to two adjacent regions f_1 and f_2 which are separated by a semi-arc e. Suppose that the semi-arc e is labeled as a. If the normal orientation of e points from f_1 to f_2, then $y_1 * a = y_2$.

We denote by $\mathrm{Col}_{(Q,\rho)_Y}(D)$ the set of $(Q, \rho)_Y$-colorings of D.

Proposition 3. Let D and D' be diagrams of (unoriented) links. If D and D' represent the same link, then $\#\mathrm{Col}_{(Q,\rho)_Y}(D) = \#\mathrm{Col}_{(Q,\rho)_Y}(D')$.

Proof. As in the proof of Proposition 2, we have a bijection $\mathrm{Col}_{(Q,\rho)_Y}(D) \to \mathrm{Col}_{(Q,\rho)_Y}(D'); C \mapsto C'$ such that $C = C'$ outside the disk where the Reidemeister move is applied. $\qquad\square$

78.4 Symmetric Quandle Cocycle Invariants

Let D be a diagram of a (unoriented) link. Let (Q, ρ) be a symmetric quandle and fix a $(Q, \rho)_Y$-coloring C of D. For an abelian group A, we set a map $\theta : Y \times Q^2 \to A$ satisfying

- $\theta(y, a, a) = 0$ for $y \in Y$ and $a \in Q$,

- $-\theta(y, b, c) + \theta(y * a, b, c) + \theta(y, a, c) - \theta(y * b, a * b, c) - \theta(y, a, b) + \theta(y * c, a * c, b * c) = 0$ for $y \in Y$ and $a, b, c \in Q$, and

- $\theta(y, a, b) + \theta(y * a, \rho(a), b) = 0$ and $\theta(y, a, b) + \theta(y * b, a * b, \rho(b)) = 0$ for $y \in Y$ and $a, b \in Q$.

We call such a map θ a 2-*cocycle* of (Q, ρ) and Y. We note that we have a symmetric quandle cohomology theory, see [6, 8]. The above conditions for θ coincide with the 2-cocycle conditions of the cohomology theory.

For a crossing v of D, there are locally four complementary regions of D around v. Choose one of them, say f, which we call a *specified region* for v. Let $y \in Y$ be the element assigned to f. Let e_1 and e_2 be the under semi-arc and the over semi-arc at v, respectively, which face the region f. By basic inversions, we may assume that the normal orientations n_1 and n_2 of e_1 and e_2 point from f. Let $a_1, a_2 \in Q$ be the assigned elements of e_1 and e_2, respectively. The weight of v is defined to be $\varepsilon\theta(y, a_1, a_2)$, where ε is the sign of v and the sign of v with respect to the region f is $+1$ (or -1) if the pair of normal orientations (n_2, n_1) does (or does not) match the orientation of $\mathbb{R}^2$. See Figure 78.4.

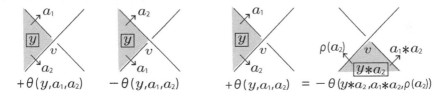

FIGURE 78.4: Weights.   FIGURE 78.5: Weights.

Lemma 1. The weight $\varepsilon\theta(y, a_1, a_2)$ does not depend on the choice of the specified region.

Proof. When we change the specified region as in Figure 78.5, the weight $\theta(y, a_1, a_2)$ is changed to $-\theta(y * a_2, a_1 * a_2, \rho(a_2))$. By the 2-cocycle conditions for θ, $\theta(y, a_1, a_2) = -\theta(y * a_2, a_1 * a_2, \rho(a_2))$ holds. See [8] for more details. $\qquad\square$

Define $W_\theta(D, C)$ by $W_\theta(D, C) = \sum_v$ (the weight of v), where v runs over all crossings of D. Define $\Phi_\theta(D)$ by the multiset

$$\Phi_\theta(D) = \{W_\theta(D, C) \mid C \in \mathrm{Col}_{(Q,\rho)_Y}(D)\}.$$

Theorem 1. The multiset $\Phi_\theta(D)$ is an invariant for (unoriented) links.

By Theorem 1, we may denote the invariant by $\Phi_\theta(L)$, where L is the unoriented link which D represents.

Proof. Let (D, C) and (D', C') be $(Q, \rho)_Y$-colored diagrams of a link L which differ by a Reidemeister move I, II or III, where C and C' are related by the bijection shown in the proof of Proposition 3. Then $W_\theta(D, C) = W_\theta(D', C')$, and thus, we have $\Phi_\theta(D) = \Phi_\theta(D')$ by Proposition 3. See [8] for more details. $\qquad\square$

Example 3. Let Q be the dihedral quandle R_3 and ρ the identity map of Q. As shown in Example 2, (Q, ρ) is a symmetric quandle. Set $\theta : Q \times Q^2 \to \mathbb{Z}_3$ by

$$\theta(y, a_1, a_2) = (y - a_1)(a_1 - a_2)^2 a_2.$$

The map θ is a 2-cocycle of (Q, ρ) and Q.

Let K and K^* be the left- and right-handed trefoil knots, respectively, see Figure 78.6 for a diagram D of K. Then we have

$$\Phi_\theta(K) = \{\underbrace{0, \ldots, 0}_{9}, \underbrace{1, \ldots, 1}_{18}\} \text{ and } \Phi_\theta(K^*) = \{\underbrace{0, \ldots, 0}_{9}, \underbrace{-1, \ldots, -1}_{18}\}.$$

Thus we can distinguish the left- and right-handed trefoil knots by using the symmetric quandle cocycle invariants.

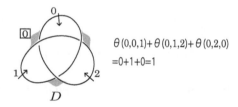

FIGURE 78.6: A diagram of the left-handed trefoil knot.

78.5 Studies for Surface-Links

Symmetric quandle cocycle invariants can be defined for surface-links. In particular, the invariants are advanced in that they give quandle cocycle invariants for "nonorientable" surface-links. Here we introduce some applications of the invariants for surface-links.

In [11], a relationship between symmetric quandle cocycle invariants and triple linking numbers was studied. More precisely, the symmetric quandle cocycle invariants obtained from trivial quandles with good involutions are determined by the triple linking numbers of surface-links. Conversely, each of the triple linking numbers is determined by the symmetric quandle cocycle invariant obtained from a trivial quandle with a good involution when we choose a particular coloring for a given surface-link diagram.

In [10], an evaluation formula with respect to triple point numbers for surface-links was introduced. In particular, the author constructed some examples of 2-component nonorientable surface-links whose triple point numbers are exactly $2n$ for a given positive integer n. See also [2, 8].

For other studies using symmetric quandles, see [4, 7] for example.

78.6 Bibliography

[1] J. S. Carter, D. Jelsovsky, S. Kamada, L. Langford and M. Saito, Quandle cohomology and state-sum invariants of knotted curves and surfaces, *Trans. Amer. Math. Soc.* **355** (2003) 3947–3989.

[2] J. S. Carter, Kanako Oshiro and Masahico Saito, Symmetric extensions of dihedral quandles and triple points of non-orientable surfaces, *Topology Appl.* **157** (2010), 857–869.

[3] R. Fenn, C. Rourke, Racks and links in codimension two, *J. Knot Theory Ramifications* **1** (1992), no. 4, 343–406.

[4] Y. Jang and K. Oshiro, Symmetric quandle colorings for spatial graphs and handlebody-links, *J. Knot Theory Ramifications* **21** (2012), no. 4, 1250050, 16 pp.

[5] D. Joyce, A classifying invariant of knots, the knot quandle, *J. Pure Appl. Alg.* **23** (1982) 37–65.

[6] S. Kamada, Quandles with good involutions, their homologies and knot invariants, Intelligence of low dimensional topology 2006, 101–108, *Ser. Knots Everything* **40** (2007), World Sci. Publ., Hackensack, NJ.

[7] S. Kamada, Quandles and symmetric quandles for higher dimensional knots, Knots in Poland III. Part III, 145–158, *Banach Center Publ.* **103** (2014), Polish Acad. Sci. Inst. Math., Warsaw.

[8] S. Kamada and K. Oshiro, Homology groups of symmetric quandles and cocycle invariants of links and surface-links, *Trans. Amer. Math. Soc.* **362** (2010), no. 10, 5501–5527.

[9] S. V. Matveev, Distributive groupoids in knot theory, *Mat. Sb. (N.S.)* **119(161)** (1982) 78–88.

[10] K. Oshiro, Triple point numbers of surface-links and symmetric quandle cocycle invariants, *Algebr. Geom. Topol.* **10** (2010), 853–865.

[11] K. Oshiro, Homology groups of trivial quandles with good involutions and triple linking numbers of surface-links, *J. Knot Theory Ramifications* **20** (2011), no.4, 595–608.

[12] M. Takasaki, Abstractions of symmetric functions, *Tohoku Math. J.* **49**(1942), 143–207.

Chapter 79

Racks, Biquandles and Biracks

Sam Nelson, Claremont McKenna College

The idea of combinatorially generalizing the knot group to encode knot colorings preserved under Reidemeister moves in the form of an algebraic structure, e.g., like Joyce did to define quandles in [7], has been independently rediscovered many times and applied to many different types of knotted objects. In this entry we will consider three generalizations of the quandle concept: racks, biquandles and biracks.

79.1 Racks

A *rack* is a set X with a binary operation $\triangleright : X \times X \to X$ satisfying the axioms:

(i) for all $y \in X$, the map $f_y : X \to X$ defined by $f_y(x) = x \triangleright y$ is bijective, and

(ii) for all $x, y, z \in X$, we have $(x \triangleright y) \triangleright z = (x \triangleright z) \triangleright (y \triangleright z)$.

Racks are the analogue of quandles for oriented framed knots, with the same coloring convention from quandles, i.e.,

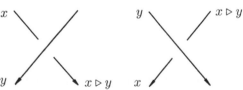

but with the usual Reidemeister I move replaced with the *framed Reidemeister I* move:

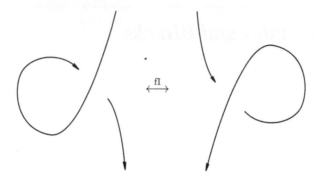

We can think of framed isotopy as describing knotted solid tori, i.e., physical closed knots, since doing a standard Reidemeister I move on a knotted solid torus requires a meridional Dehn twist. The name "rack" was introduced in [5], a modification of earlier work by Conway and Wraith in which the term "wrack" from the phrase "wrack and ruin" was used to describe what remains of the knot group after keeping only conjugation. Dropping the "w" makes sense since the new structure drops the writhe invariance resulting from the unframed Reidemeister I move.

Racks have a bijective *kink map* $\pi : X \to X$ which we can think of as the result of going through a positive kink:

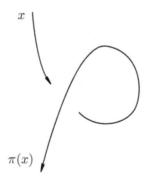

For finite racks, the exponent of the kink map (i.e., the smallest integer N such that $\pi^N(x) = x$ for all $x \in X$) is called the *rack rank* or *rack characteristic*; colorings of oriented knots and links by a rack of characteristic N are in one-to-one correspondence before and after the *N-phone cord move*:

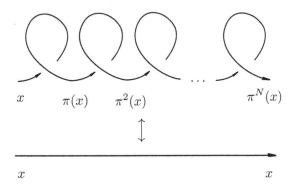

In particular, a quandle is a rack of characteristic 1.

Many examples of non-quandle racks are known; one class of examples is $\mathbb{Z}_n$ with

$$x \triangleright y = tx + sy$$

for any choice of t coprime to n and $s \in \mathbb{Z}_n$ such that $s^2 = (1 - t)s$, e.g., $\mathbb{Z}_8$ with $t = 7$ and $s = 2$.

Racks appear in the literature under a few alternative names, such as *automorphic sets* [1], with some alternative notations such as $x^y = x \triangleright y$ [5]. These structures have been studied by algebraists independently of knot theory (e.g., [2]), with motivations such as the study of self-distributivity and the observation that a rack acts on itself by automorphisms, since the maps $f_y : X \to X$ are bijections satisfying

$$f_z(x \triangleright y) = (x \triangleright y) \triangleright z = (x \triangleright z) \triangleright (y \triangleright z) = f_z(x) \triangleright f_z(y).$$

For a given framed knot, the number of colorings by a finite rack X is an invariant of framed isotopy, and the multiset of numbers of colorings over a complete set of framings modulo the rack rank is an invariant of the unframed knot. See [3, 9] for more.

79.2 Biquandles

To generalize quandles to *biquandles*, instead of looking at a different Reidemeister equivalence relation like we did with racks, we look at a different labeling rule for diagrams. For quandle colorings of knots, quandle elements are attached to *arcs* in a knot diagram, portions of the diagram running from one undercrossing to another. For biquandles we'll attach elements to *semiarcs*, portions of the diagram going from one crossing point (over or under) to the next crossing point (over or under) and let the two labels x, y on the left side interact to form the labels on the right side

$x \underline{\triangleright} y, y \overline{\triangleright} x.$

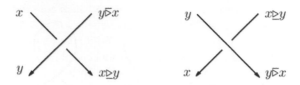

The biquandle axioms are then the conditions resulting from the requirement that for every coloring of an oriented knot or link diagram before a Reidemeister move, there is a *unique* coloring of the diagram after the move which agrees with the original coloring outside the neighborhood of the move. We note that contrary to possible expectation, a biquandle is not simply a set with two compatible quandle structures.

A *biquandle* is a set X with two binary operations $\underline{\triangleright}, \overline{\triangleright} : X \times X \to X$ satisfying:

(i) for all $x \in X$, $x \overline{\triangleright} x = x \underline{\triangleright} x$,

(ii) for all $y \in X$, the maps $\alpha_y, \beta_y : X \to X$ and the map $S : X \times X \to X \times X$ defined by

$$\alpha_y(x) = x \overline{\triangleright} y, \ \beta_y(x) = x \underline{\triangleright} y \text{ and } S(x,y) = (y \overline{\triangleright} x, x \underline{\triangleright} y)$$

are invertible, and

(iii) for all $x, y, z \in X$, the *exchange laws* are satisfied:

$$\begin{aligned}
(x \overline{\triangleright} y) \overline{\triangleright} (z \overline{\triangleright} y) &= (x \overline{\triangleright} z) \overline{\triangleright} (y \underline{\triangleright} z) \\
(x \underline{\triangleright} y) \overline{\triangleright} (z \underline{\triangleright} y) &= (x \overline{\triangleright} z) \underline{\triangleright} (y \overline{\triangleright} z) \\
(x \underline{\triangleright} y) \underline{\triangleright} (z \underline{\triangleright} y) &= (x \underline{\triangleright} z) \underline{\triangleright} (y \overline{\triangleright} z).
\end{aligned}$$

Thus, we can think of quandles as biquandles in which the over-crossing operation $\overline{\triangleright}$ is trivial, i.e., $x \overline{\triangleright} y = x$ for all x, y.

Many examples of non-quandle biquandles are known; a simple type of example is $\mathbb{Z}_n$ with

$$x \underline{\triangleright} y = tx + (s - t)y \quad \text{and} \quad x \overline{\triangleright} y = sy$$

where t, s are any elements coprime to n, e.g., $\mathbb{Z}_5$ with $t = 3$ and $s = 2$.

As with racks, the literature includes a few alternative notations for the biquandle operations, such as $x \underline{\triangleright} y = x^y$ and $x \overline{\triangleright} y = x_y$. Perhaps more importantly, the literature includes different choices for biquandle operations, with some papers such as [4, 8] using what we might call the "downward operations"

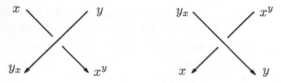

and others such as [6, 10] using the "sideways" operations as defined above. The strong invertibility conditions required by the Reidemeister II move in axiom (ii) give us the freedom to switch between these and other choices of operations; see [3] for more details.

As with quandles, we can count colorings of oriented knots and links by a finite biquandle and enhance this invariant with cocycles in a cohomology theory. Such colorings can be understood as homomorphisms to the coloring biquandle from the *fundamental biquandle* of the knot, an object defined combinatorially by generators and relations [3]. While the fundamental quandle is known to be complete invariant up to reflection for classical knots and links, the fundamental biquandle is at the time of this writing, conjectured to be a complete invariant up to a type of orientation-reversed mirror for virtual knots and links.

79.3 Biracks

We have seen two ways of generalizing quandles: using framings gives us racks, while using semi-arcs instead of arcs yields biquandles. If we do both, we obtain *biracks*.

More precisely, a *birack* is a set X with two binary operations satisfying the second two biquandle axioms with a modified version of the first:

(i) there is a bijection $\pi : X \to X$ such that for all $x \in X$, $\pi(x) \,\overline{\triangleright}\, x = x \,\underline{\triangleright}\, \pi(x)$,

(ii) for all $y \in X$, the maps $\alpha_y, \beta_y : X \to X$ and the map $S : X \times X \to X \times X$ defined by

$$\alpha_y(x) = x \,\overline{\triangleright}\, y, \quad \beta_y(x) = x \,\underline{\triangleright}\, y \quad \text{and} \quad S(x, y) = (y \,\overline{\triangleright}\, x, x \,\underline{\triangleright}\, y)$$

are invertible, and

(iii) for all $x, y, z \in X$, the *exchange laws* are satisfied:

$$
\begin{aligned}
(x \,\overline{\triangleright}\, y) \,\overline{\triangleright}\, (z \,\overline{\triangleright}\, y) &= (x \,\overline{\triangleright}\, z) \,\overline{\triangleright}\, (y \,\underline{\triangleright}\, z) \\
(x \,\underline{\triangleright}\, y) \,\overline{\triangleright}\, (z \,\underline{\triangleright}\, y) &= (x \,\overline{\triangleright}\, z) \,\underline{\triangleright}\, (y \,\overline{\triangleright}\, z) \\
(x \,\underline{\triangleright}\, y) \,\underline{\triangleright}\, (z \,\underline{\triangleright}\, y) &= (x \,\underline{\triangleright}\, z) \,\underline{\triangleright}\, (y \,\overline{\triangleright}\, z).
\end{aligned}
$$

As with racks and biquandles, many examples of biracks which are neither racks nor biquandles are known. One simple family of examples is $X = \mathbb{Z}_n$ with operations

$$x \,\underline{\triangleright}\, y = tx + sy \quad \text{and} \quad x \,\overline{\triangleright}\, y = ry$$

where $t, r \in X$ are coprime to n and s satisfies $s^2 = s(r - t)$, e.g., $\mathbb{Z}_6$ with $t = 1$, $r = 5$ and $s = 4$.

As with racks, the exponent of the kink map π is the *birack characteristic*, and the multiset of birack colorings over a period of framings modulo this characteristic is an invariant of unframed knots. As with biquandles, a birack with trivial over-operation so that $x \,\overline{\triangleright}\, y = x$ for all $x, y \in X$ is a rack, and a birack of characteristic 1 is a biquandle. We can summarize the relationship between these algebraic structures with a Venn diagram:

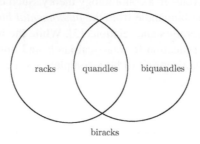

79.4   Bibliography

[1] E. Brieskorn. Automorphic sets and braids and singularities. In *Braids (Santa Cruz, CA, 1986)*, volume 78 of *Contemp. Math.*, pages 45–115. Amer. Math. Soc., Providence, RI, 1988.

[2] Patrick Dehornoy. A canonical ordering for free self-distributive systems. *Proc. Amer. Math. Soc.*, 122(1):31–36, 1994.

[3] Mohamed Elhamdadi and Sam Nelson. *Quandles—an Introduction to the Algebra of Knots*, volume 74 of *Student Mathematical Library*. American Mathematical Society, Providence, RI, 2015.

[4] Roger Fenn, Mercedes Jordan-Santana, and Louis Kauffman. Biquandles and virtual links. *Topology Appl.*, 145(1-3):157–175, 2004.

[5] Roger Fenn and Colin Rourke. Racks and links in codimension two. *J. Knot Theory Ramifications*, 1(4):343–406, 1992.

[6] Roger Fenn, Colin Rourke, and Brian Sanderson. Trunks and classifying spaces. *Appl. Categ. Structures*, 3(4):321–356, 1995.

[7] David Joyce. A classifying invariant of knots, the knot quandle. *J. Pure Appl. Algebra*, 23(1):37–65, 1982.

[8] Louis H. Kauffman and David Radford. Bi-oriented quantum algebras, and a generalized Alexander polynomial for virtual links. In *Diagrammatic morphisms and applications (San Francisco, CA, 2000)*, volume 318 of *Contemp. Math.*, pages 113–140. Amer. Math. Soc., Providence, RI, 2003.

[9] Sam Nelson. Link invariants from finite racks. *Fund. Math.*, 225:243–258, 2014.

[10] Sam Nelson, Michael E. Orrison, and Veronica Rivera. Quantum enhancements and biquandle brackets. *J. Knot Theory Ramifications*, 26(5):1750034, 24, 2017.

Chapter 80

Quantum Invariants via Hopf Algebras and Solutions to the Yang-Baxter Equation

Leandro Vendramin, Universidad de Buenos Aires

A Carloncha

The fundamental problem of knot theory is to know whether two knots are equivalent or not. As a tool to prove that two knots are different, mathematicians have developed various invariants. Knots invariants are just functions that can be computed from the knot and depend only on the topology of the knot. Here we describe quantum invariants, a powerful family of invariants related to the celebrated Yang–Baxter equation.

The basic idea of quantum invariants belongs to Witten [8]. It was then developed by Reshetikhin and Turaev in [6] and [7]. In modern language, Reshetikhin and Turaev's construction is essentially the observation that finite-dimensional representations of a quantum group $\mathcal{U}_q(\mathfrak{g})$ form a ribbon category.

80.1 Braidings and Yetter-Drinfeld Modules

Recall that a *braided vector space* is a pair (V, c), where V is a vector space and $c \colon V \otimes V \to V \otimes V$ is a solution of the *braid equation*, that is a linear isomorphism such that

$$(c \otimes \mathrm{id})(\mathrm{id} \otimes c)(c \otimes \mathrm{id}) = (\mathrm{id} \otimes c)(c \otimes \mathrm{id})(\mathrm{id} \otimes c).$$

There is a procedure that produces braided vector spaces from Hopf algebras. We will present this method in the language of Yetter-Drinfeld modules. Let H be a Hopf algebra with invertible antipode S. A *Yetter-Drinfeld module* over H is triple $(V, \cdot, \delta)$, where $(V, \cdot)$ is a left H-module, (V, δ) is a left H-comodule and such that the compatibility

$$\delta(h \cdot v) = h_1 v_{-1} S(h_3) \otimes h_2 v_0$$

holds for all $h \in H$ and $v \in H$.

Yetter-Drinfeld modules over H form a category (the objects are the Yetter-Drinfeld modules over

H and the morphisms are linear maps that are module and comodule homomorphisms). If V and W are Yetter-Drinfeld modules over H, then $V \otimes W$ is a Yetter-Drinfeld module over H with

$$h \cdot v \otimes w = h_1 \cdot v \otimes h_2 \cdot w, \quad \delta(v \otimes w) = v_{-1}w_{-1} \otimes (v_0 \otimes w_0).$$

Yetter-Drinfeld modules produce braided vector spaces. More generally, if U, V and W are Yetter-Drinfeld modules over H, the map

$$c_{V,W} \colon V \otimes W \to W \otimes V, \quad c_{V,W}(v \otimes w) = v_{-1} \cdot w \otimes v_0,$$

is an invertible morphism of Yetter-Drinfeld modules over H such that

$$(c_{W,X} \otimes \mathrm{id}_V)(\mathrm{id}_W \otimes c_{V,X})(c_{V,W} \otimes \mathrm{id}_X) = (\mathrm{id}_X \otimes c_{V,W})(c_{V,X} \otimes \mathrm{id}_W)(\mathrm{id}_V \otimes c_{W,X})$$

In particular, when $X = V = W$ one obtains braided vector spaces.

Remark 80.1.1. *Yetter-Drinfeld modules and their braided vector spaces appear in the classification of Hopf algebras, see for example [1].*

Finite-dimensional Yetter-Drinfeld modules have duals. Let V be a finite-dimensional Yetter-Drinfeld module over H with basis $\{v_1, \ldots, v_n\}$. Let V^* be the vector space with basis $\{f_1, \ldots, f_n\}$, where $f_i(v_j) = \delta_{ij}$. Then V^* is a Yetter-Drinfeld module over H with

$$(h \cdot f)(v) = f(S(h) \cdot v), \quad \delta(f) = \sum_{k=1}^{n} S^{-1}((f_i)_{-1}) \otimes f((v_i)_0) f_i$$

for all $h \in H$, $f \in V^*$ and $v \in V$. Recall that there are two canonical maps $b_V \colon \mathbb{C} \to V \otimes V^*$, $e_V \colon V^* \otimes V \to \mathbb{C}$, such that

$$(\mathrm{id}_V \otimes e_V)(b_V \otimes \mathrm{id}_V) = \mathrm{id}_V, \quad (e_V \otimes \mathrm{id}_{V^*})(\mathrm{id}_{V^*} \otimes b_V) = \mathrm{id}_{V^*}.$$

The map $b \colon \mathbb{C} \to V \otimes V^*$ is given by

$$b(1) = \sum_{j=1}^{n} v_i \otimes f_i$$

and $e \colon V^* \otimes V \to \mathbb{C}$, $f \otimes v \mapsto f(v)$, is the usual evaluation map. We leave to the reader the exercise of proving that

$$c_{V^*,V} = (e \otimes \mathrm{id}_V \otimes \mathrm{id}_{V^*})(\mathrm{id}_{V^*} \otimes c_{V,V} \otimes \mathrm{id}_{V^*})(\mathrm{id}_{V^*} \otimes \mathrm{id}_V \otimes b).$$

Similarly one computes c_{V,V^*} and c_{V^*,V^*}.

The category of finite-dimensional Yetter-Drinfeld modules over H form a *ribbon* category, this means that there is a family of natural isomorphisms $\theta_V \colon V \to V$ such that

$$\theta_{\mathbb{C}} = \mathrm{id}_{\mathbb{C}}, \quad \theta_{V^*} = (\theta_V)^*, \quad \theta_{V \otimes W} = c_{W,V} c_{V,W}(\theta_V \otimes \theta_W).$$

See [5] for a classification of all such θ.

80.2 Braids and Knots

The *braid group* $\mathbb{B}_n$ is the group with generators $\sigma_1, \ldots, \sigma_{n-1}$ and relations

$$\sigma_i \sigma_j = \sigma_j \sigma_i \qquad\qquad |i - j| \geq 2,$$
$$\sigma_i \sigma_{i+1} \sigma_i = \sigma_{i+1} \sigma_i \sigma_{i+1} \qquad 1 \leq i \leq n - 2.$$

One can draw pictures to represent elements of the braid group, see for example Figure 80.1. The group operation is just composition of braids.

FIGURE 80.1: The braid $\sigma_1^{-1} \sigma_2 \sigma_1^{-1} \in \mathbb{B}_3$.

Braided vector spaces produce representations of the braid group: if (V, c) is a braided vector space, the map $\rho_n \colon \mathbb{B}_n \to \mathbf{GL}(V^{\otimes})$, $\sigma_j \mapsto c_j$, where

$$c_j(v_1 \otimes \cdots \otimes v_n) = v_1 \otimes \cdots \otimes v_{j-1} \otimes c(v_j \otimes v_{j+1}) \otimes v_{j+2} \otimes \cdots \otimes v_n,$$

extends to a group homomorphism. This ρ_n yields a graphical calculus where certain morphisms are represented by pictures.

One can use braids to represent knots. Alexander's theorem states that each knot is the closure of a braid, see Figure 80.2. This gives a way to move between braids and knots. Two equivalent braids will become equivalent knots. However, the closure of two non-equivalent braids could produce equivalent knots. This means that one needs a method to determine if two closed braids are equivalent. The answer to this fundamental method is now known as Markov's theorem.

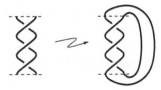

FIGURE 80.2: The closure of a braid.

To obtain invariants of knots from quantum groups, one needs to accept that the invariants on the diagrams of Figure 80.3 could receive different values. Without this it would be very hard to construct good invariants. A *framed knot* is a knot equipped with a smooth family of non-zero vectors orthogonal to the knot. Once a plane projection is given, one way to choose a framing is to use a vector field everywhere parallel to the projection plane. This is the *blackboard framing*. Therefore a framed knot can be viewed as a *ribbon knot*. Framed knots with blackboard framing are not invariant under the first Reidemeister move.

FIGURE 80.3: Trivial knots.

We can use morphisms of the category of finite-dimensional Yetter-Drinfeld modules over H to color the arcs of our knot. We use c, c^{-1}, e, b,

$$b^- : \mathbb{C} \to V^* \otimes V, \qquad\qquad b^- = (\mathrm{id}_V \otimes \theta_V^{-1}) c_{V,V^*}^{-1} . b,$$

and

$$e^- : V \otimes V^* \to \mathbb{C}, \qquad\qquad e^- = e c_{V,V^*} (\theta_V \otimes \mathrm{id}_{V^*})$$

as follows:

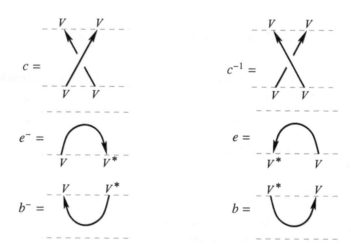

It is an exercise to check that this is invariant under the second and third Reidemeister moves. In some cases this construction yields a knot invariant. Let us assume for example that the twist θ acts on V and V^* by scalar multiplication. To normalize the diagram one multiplies by $\theta^{-\omega(K)}$, where $\omega(K)$ is the writhe of the knot K (that is the number of positive crossings of K minus the number of negative crossings of K).

80.3   An Example: The Taft Algebra

Let m be an odd positive integer and let $q \in \mathbb{C}^\times$ be a primitive m-root of 1. Let H be the algebra generated by x and g with relations

$$gx = qxg, \quad x^m = 0, \quad g^m = 1.$$

Then $\dim H = m^2$ and $\{x^i g^j : 1 \le i, j < m\}$ is a basis of H. Furthermore H is a Hopf algebra with

$$\Delta(g) = g \otimes g, \qquad \Delta(x) = g \otimes x + x \otimes 1, \qquad \epsilon(g) = 1,$$
$$\epsilon(x) = 0, \qquad S(g) = g^{-1}, \qquad S(x) = -g^{-1}x.$$

Let $n \ge 0$ and V_n be a vector space with basis $\{v_{-n}, v_{-n+2}, \ldots, v_{n-2}, v_n\}$. Then V_n is a simple left H-module with

$$g \cdot v_k = q^{-k/2}, \qquad x \cdot v_k = v_{k-2},$$

and a left H-comodule with

$$\delta(v_k) = \sum_{i \ge 0} \frac{1}{(i)!_q} \prod_{j=0}^{i-1} \alpha_{k+2j} x^i g^{-\frac{k+2i}{2}} \otimes v_{k+2i},$$

where

$$\alpha_k = q^{-\frac{k+n+2}{2}} (q-1) \left(\frac{k+n+2}{2} \right)_q \left(\frac{n-k}{2} \right)_q$$

and $(k)_q = 1 + q + \cdots + q^{k-1}$.

One proves that the V_n are simple Yetter-Drinfeld modules over H. The simplicity of V_n implies that θ acts on V_n by scalar multiplication by some number θ_n. Since

$$c_{V_n, V_m}(v_n \otimes v_m) = q^{nm/4} v_m \otimes v_n,$$

one needs $\theta_{n+m} = q^{nm/2} \theta_n \theta_m$ and $\theta_0 = 1$. By using that $V_1 \otimes V_1 \simeq V_2 \oplus V_0$, one proves that $\theta_n = q^{\frac{n^2+2n}{4}}$.

Let us now study the case $n = 1$. The matrix of the braiding is

$$\begin{pmatrix} q^{1/4} & 0 & 0 & 0 \\ 0 & q^{1/4} - q^{-3/4} & q^{-1/4} & 0 \\ 0 & q^{-1/4} & 0 & 0 \\ 0 & 0 & 0 & q^{1/4} \end{pmatrix}.$$

We leave to the reader the exercise to compute the matrices of c_{V^*, V^*}, $c_{V^*, V}$ and $c_{V^*, V}$. Then

$$b^-(1) = q^{1/2} v_{-1} \otimes v_{-1} + q^{-1/2} v_1 \otimes v_1,$$
$$e^-(v_{-1} \otimes v_{-1}, v_{-1} \otimes v_1, v_1 \otimes v_{-1}, v_1 \otimes v_1) = (q^{1/2}, 0, 0, q^{1/2}).$$

Let us apply this to prove that the left and right trefoil knots are different. These knots are the closure of $\sigma_1^{-3} \in \mathbb{B}_2$ and $\sigma_1^3 \in \mathbb{B}_2$. The invariant for the left trefoil knot of Figure 80.4 is

$$\theta_2^{-3}(e \otimes e^-)(\mathrm{id}_{V^*} \otimes c_{V,V}^{-3} \otimes \mathrm{id}_{V^*})(b^- \otimes b) = -q^{9/2} + q^{5/2} + q^{3/2} + q^{1/2}.$$

The invariant for the right trefoil knot is

$$\theta_2^3(e \otimes e^-)(\mathrm{id}_{V^*} \otimes c_{V,V}^3 \otimes \mathrm{id}_{V^*})(b^- \otimes b) = q^{-1/2} + q^{-3/2} + q^{-5/2} + q^{1/2}.$$

Therefore that these knots are different.

Remark 80.3.1. *The modules V_n can be used to recover the Jones polynomial.*

Remark 80.3.2. *In [4], Graña showed that the quandle cocycle invariants in the sense of [2] are quantum invariants related to Yetter-Drinfeld modules over group algebras.*

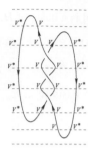

FIGURE 80.4: The left trefoil knot.

Acknowledgments

This chapter is based on a minicourse given by Matías Graña in Córboba in 2003. Thanks to Sergei Chmutov, the figures were taken from [3]. The author is partially supported by PICT-201-0147 and MATH-AmSud 17MATH-01.

80.4 Bibliography

[1] Nicolás Andruskiewitsch and Matías Graña. From racks to pointed Hopf algebras. *Adv. Math.*, 178(2):177–243, 2003.

[2] J. Scott Carter, Daniel Jelsovsky, Seiichi Kamada, Laurel Langford, and Masahico Saito. Quandle cohomology and state-sum invariants of knotted curves and surfaces. *Trans. Amer. Math. Soc.*, 355(10):3947–3989, 2003.

[3] S. Chmutov, S. Duzhin, and J. Mostovoy. *Introduction to Vassiliev knot invariants*. Cambridge University Press, Cambridge, 2012.

[4] Matías Graña. Quandle knot invariants are quantum knot invariants. *J. Knot Theory Ramifications*, 11(5):673–681, 2002.

[5] Louis H. Kauffman and David E. Radford. A necessary and sufficient condition for a finite-dimensional Drinfeld double to be a ribbon Hopf algebra. *J. Algebra*, 159(1):98–114, 1993.

[6] N. Yu. Reshetikhin and V. G. Turaev. Ribbon graphs and their invariants derived from quantum groups. *Comm. Math. Phys.*, 127(1):1–26, 1990.

[7] V. G. Turaev. The Yang-Baxter equation and invariants of links. *Invent. Math.*, 92(3):527–553, 1988.

[8] Edward Witten. Quantum field theory and the Jones polynomial. *Comm. Math. Phys.*, 121(3):351–399, 1989.

Chapter 81

The Temperley-Lieb Algebra and Planar Algebras

Stephen Bigelow, University of California, Santa Barbara

81.1 Two Definitions of the Temperley-Lieb Algebra

The Temperley-Lieb algebra was introduced by Temperley and Lieb in 1971 [7] in connection to a problem in statistical mechanics. It has since found applications in diverse fields such as knot theory, representation theory, and von Neumann algebras. It is arguably the first and simplest example of a *diagrammatic algebra*.

When we speak of "the" Temperley-Lieb algebra, we really refer to a family of algebras $TL_n(\delta)$, indexed by a non-negative integer n and a complex number δ. The easiest way to define $TL_n(\delta)$ is by generators and relations.

Definition 1. $TL_n(\delta)$ *is the unital associative $\mathbb{C}$-algebra generated by*

$$e_1, \ldots, e_{n-1}$$

subject to the relations:

- $e_i^2 = \delta e_i$.
- $e_i e_j e_i = e_i$ *if* $|i - j| = 1$.
- $e_i e_j = e_j e_i$ *if* $|i - j| > 1$.

(We could use an arbitrary ring of scalars, but we will stick with $\mathbb{C}$.)

What makes $TL_n(\delta)$ a diagrammatic algebra is an alternative definition in terms of *Temperley-Lieb diagrams*. Fix a rectangle with n marked points on the bottom edge and n marked points on the top edge. An n-strand Temperley-Lieb diagram is a collection of n disjoint arcs in the rectangle, each connecting a pair of marked points. The arcs are called *strands*.

Two n-strand Temperley-Lieb diagrams are considered to be the same if there is an isotopy that goes from one diagram to the other, while remaining a Temperley-Lieb diagram at all times. Thus, all that matters is which pairs of marked points are connected.

The set of n-strand Temperley-Lieb diagrams is a basis for $TL_n(\delta)$, as a vector space. There is no diagrammatic interpretation of addition or scalar multiplication: a typical element of $TL_n(\delta)$ is a purely formal linear combination of diagrams.

The product of Temperley-Lieb diagrams is defined by the following procedure of "stacking and bursting bubbles." Given two n-strand Temperley-Lieb diagrams a and b, place a on top of b, and connect the marked points on the top of a to the corresponding marked points on the bottom of b. The resulting diagram might contain closed loops made up of arcs from a and b. Let ℓ be the number of closed loops, and let c be the Temperley-Lieb diagram obtained by removing these closed loops. The product of a and b is then defined to be $ab = \delta^\ell c$.

Extend the above multiplication of diagrams to a bilinear operation on $TL_n(\delta)$. Thus, to multiply two linear combinations of diagrams, use the distributive law, and then the above rule for multiplication of diagrams. In summary, the diagrammatic definition of $TL_n(\delta)$ is as follows.

Definition 2. $TL_n(\delta)$ *is the algebra of formal linear combinations of n-strand Temperley-Lieb diagrams. Multiplication is the unique bilinear operation that extends the above "stacking and bursting bubbles" operation on diagrams.*

It is not immediately obvious, but Definitions 1 and 2 do in fact give the same algebra. This was proved by Kauffman [5].

The identity element **1** of $TL_n(\delta)$ is the diagram in which every strand is vertical. The generator e_i is the diagram in which every strand is vertical except for a strand connecting marked points number i and $i + 1$ on the bottom, and a strand connecting marked points number i and $i + 1$ on the top. See Figure 81.1.

See Figure 81.2 for an example of a Temperley-Lieb diagram rewritten as a product of diagrams e_i. Kauffman proved that this is always possible [5]. See also his survey article [6], which fills in some of the details of the proof that were previously left to the reader. Ernst, Hastings and Salmon [1] used a similar procedure to go from a diagram to a word of shortest possible length.

The diagrams e_i satisfy the relations in Definition 1, as can be verified by simply drawing the diagrams representing each side of each relation. The fact that no additional relations are needed

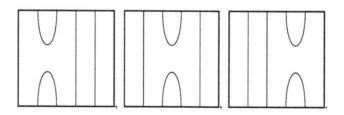

FIGURE 81.1: Generators e_1, e_2, e_3 of $TL_4(\delta)$.

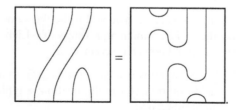

FIGURE 81.2: The element $e_1 e_2 e_3$ of $TL_4(\delta)$.

can be proved by a dimension count. In the diagrammatic definition, the dimension of $TL_n(\delta)$ is the number of Temperley-Lieb diagrams, which is known to equal the *Catalan number*

$$\dim TL_n(\delta) = C_n = \frac{1}{n+1}\binom{2n}{n}.$$

Jones [2] showed that the relations given in the algebraic definition of $TL_n(\delta)$ can be used to put any word in the generators into a certain normal form, and the number of words in normal form is also C_n.

81.2 The Temperley-Lieb Planar Algebra

The Temperley-Lieb planar algebra takes the idea of multiplying by stacking diagrams, and extends it to allow for infinitely many other ways of connecting diagrams. This uses the two-dimensional nature of the diagrams to give extra structure, whereas a traditional algebra only uses the one-dimensional operation of concatenating strings of letters. We use *planar tangles* to specify operations; see, for example, Figure 81.3.

A planar tangle consists of the following data:

- an *output disk D*,
- *input disks* $D_1, \ldots, D_k$ in the interior of D,
- an even number of *marked points* on the boundary of each input and output disk,
- finitely many *strands*, and
- a *marked interval* touching the boundary of each input or output disk.

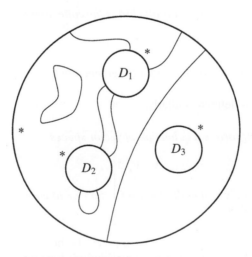

FIGURE 81.3: A planar tangle with three input disks.

The marked points are connected in pairs by strands that are disjoint from each other and from the interior of the input disks. We also allow strands that are closed loops. Each input and output disk has one marked interval between adjacent marked points, usually indicated by a star near the boundary. If a disk has no marked points on the boundary then the marked interval is the entire boundary. Two planar tangles are considered to be the same if there is an isotopy that goes from one to the other, while remaining a planar tangle at all times.

For each planar tangle T, there is a corresponding *partition function* Z_T. Suppose T has k input disks with $2n_1, \ldots, 2n_k$ marked points, and $2n$ marked points on the boundary of the output disk. Then Z_T will be a linear map

$$Z_T : \mathrm{TL}_{n_1} \otimes \cdots \otimes \mathrm{TL}_{n_k} \to \mathrm{TL}_n.$$

To define Z_T, it suffices to define its action on

$$v_1 \otimes \cdots \otimes v_k,$$

where v_i is an n_i-strand Temperley-Lieb diagram.

The idea is to insert the diagrams v_i into the input disks of T. It helps to think of them as diagrams in a disk, by "rounding off the corners" of the rectangles, but keeping a marked interval at what was the left edge of the rectangle. Insert each diagram v_i into the input disk D_i, rotated so that their marked intervals align, and then join each marked point on the boundary of v_i to the corresponding marked point on the boundary of D_i.

After inserting a diagram into each input disk, we obtain a diagram in D. Let ℓ be the number of closed loops in this diagram, and let v be the Temperley-Lieb diagram obtained by deleting these closed loops. Then

$$Z_T(v_1 \otimes \cdots \otimes v_k) = \delta^\ell v.$$

Extend this to a linear function on all of $\mathrm{TL}_{n_1} \otimes \cdots \otimes \mathrm{TL}_{n_k}$.

Definition 3. *The Temperley-Lieb planar algebra is the collection of vector spaces*

$$\mathrm{TL}_0, \mathrm{TL}_1, \mathrm{TL}_2, \ldots$$

together with the above rule that assigns a partition function to each planar tangle.

A general planar algebra is defined similarly.

Definition 4. *A planar algebra is a collection of vector spaces*

$$V_0, V_1, V_2, \ldots$$

together with a rule that assigns a multilinear operation Z_T to each planar tangle T, and satisfies a certain naturality condition.

The naturality condition is cumbersome to state precisely, but comes for free in most examples where the vector spaces V_n are spanned by some kind of diagrams. Here, the diagrams include Temperley-Lieb diagrams, but may also allow extra features like crossings, orientations, or vertices.

The function Z_T is defined by inserting diagrams into input disks of T, and perhaps performing local operations that simplify the resulting diagram. The naturality condition then says that if you insert diagrams into planar tangles, and insert those planar tangles into other planar tangles, then the order of those two operations should not matter.

Planar algebras were originally defined by Jones [3] as a way to axiomatize the standard invariant of a subfactor. This original definition, now called a *subfactor planar algebra*, has additional features and axioms that we have left out. (The relevant adjectives are: shaded, evaluable, unital, involutive, spherical, and positive.) Conversely, a less restrictive definition than ours might allow odd number of marked points on boundaries of a disk, or might not require Z_T to be invariant under isotopy of T. A survey of variations on the notion of planar algebra can be found in Jones's lecture notes from 2011 [4].

81.3 Bibliography

[1] Dana C. Ernst, Michael G. Hastings, and Sarah K. Salmon. Factorization of Temperley-Lieb diagrams. *Involve*, 10(1):89–108, 2017.

[2] V. F. R. Jones. Index for subfactors. *Invent. Math.*, 72(1):1–25, 1983.

[3] V. F. R. Jones. Planar algebras, I. ArXiv Mathematics e-prints, September 1999.

[4] Vaughan Jones. Lecture notes. `http://math.berkeley.edu/~vfr/VANDERBILT/pl21.pdf`, 2011. [Online; accessed 2018-03-09].

[5] Louis H. Kauffman. An invariant of regular isotopy. *Trans. Amer. Math. Soc.*, 318(2):417–471, 1990.

[6] Louis H. Kauffman. Knot diagrammatics. In *Handbook of Knot Theory*, pages 233–318. Elsevier B. V., Amsterdam, 2005.

[7] H. N. V. Temperley and E. H. Lieb. Relations between the "percolation" and "colouring" problem and other graph-theoretical problems associated with regular planar lattices: Some exact results for the "percolation" problem. *Proc. Roy. Soc. London Ser. A*, 322(1549):251–280, 1971.

Chapter 82

Vassiliev/Finite-Type Invariants

Sergei Chmutov, Ohio State University at Mansfield

Alexander Stoimenow, Daedook Innopolis

Vassiliev knot invariants, also known as *finite-type invariants*, form a class of knot invariants introduced independently by Victor Vassiliev (Moscow) [41] and by Mikhail Goussarov (St. Petersburg) [17] at the end of the 1980s. This class includes (the Taylor coefficients of) all classical and quantum polynomial invariants [7]. The greatest still open problem of this theory is whether Vassiliev invariants distinguish knots. If this were true, then it would provide another proof of the Poincaré conjecture [15].

82.1 The Filtration of the Knot Space

Consider the linear (infinite dimensional) space $\mathcal{V}$ over a field, for example $\mathbb{C}$ or $\mathbb{Q}$, (freely) generated by all the (isotopy classes of) knot embeddings. Let $\mathcal{V}^p$ be the space of singular knots with exactly p double points (up to isotopy).

$\mathcal{V}^p$ can be identified with a linear subspace of $\mathcal{V}$ by resolving the singularities into the difference of an over- and an undercrossing via the rule

$$\left(\!\!\!\includegraphics{}\!\!\!\right) = \left(\!\!\!\includegraphics{}\!\!\!\right) - \left(\!\!\!\includegraphics{}\!\!\!\right), \tag{82.1}$$

where all the rest of the knot projections are assumed to be equal. (One can show that the result does not depend on the order in which all the double points are resolved.) This yields a filtration of $\mathcal{V}$

$$\mathcal{V} = \mathcal{V}^0 \supseteq \mathcal{V}^1 \supseteq \mathcal{V}^2 \supseteq \mathcal{V}^3 \supseteq \cdots \tag{82.2}$$

There exists a combinatorial description of the graded vector space

$$\bigoplus_{i=0}^{\infty} \left(\mathcal{V}^i / \mathcal{V}^{i+1} \right), \tag{82.3}$$

associated to this filtration, namely

$$\mathcal{V}^i / \mathcal{V}^{i+1} \; \simeq \; Lin\{\text{chord diagrams of degree } i\} / \left\{ \begin{matrix} 4T \text{ relation} \\ FI \text{ relation} \end{matrix} \right\}. \tag{82.4}$$

Here the chord diagrams (CDs) are objects like this (an oriented counterclockwise circle with finitely many chords in it)

(up to isotopy) and are graded by the number of chords. The $4T$ (4-term) relations have the form

and the FI (framing independence) relation requires that each CD with an *isolated* chord (i.e., a chord whose endpoints are consecutive on the circle as on the figure below) is zero:

The map which yields the isomorphism (82.4) is a simple way to assign a CD D_K to a singular knot K. Connect in the parameter space of K (which is an oriented S^1) pairs of points with the same image by a chord. Actually, the idea to describe singular knots in this way led to the representation (82.4), and, more generally, many of the further algebraic statements are based on this idea. With the CDs once introduced, the $4T$ and FI relations become self-suggesting. Look, e.g., at the FI relation. If one takes a singular knot corresponding to a diagram with an isolated chord and resolves the singularity corresponding to this chord, one gets exactly an ambient isotopy relation of (singular) knots.

Definition 82.1.1. *A* Vassiliev *invariant of degree* m *is an element* $v \in \mathcal{V}^*$*, s.t.*

$$V|_{\mathcal{V}^{m+1}} \equiv 0 \quad \text{and} \quad V|_{\mathcal{V}^m} \not\equiv 0.$$

It is possible, in some sense, to consider (82.1) as a way to "differentiate" a knot, and in this language a Vassiliev invariant corresponds to a polynomial invariant, i.e., a function with a vanishing derivative.

This notion is interesting in connection with many other known knot invariants, such as the Conway and the HOMFLY polynomials, which have originally come satisfying certain (*skein*) relations closely connected to (82.1). E.g. the Conway polynomial $\nabla_K(z)$ (see Chapter 64) of a knot K satisfies the (skein) relation

$$\nabla(\text{⤬})(z) := \nabla(\text{⤬})(z) - \nabla(\text{⤬})(z) = z\nabla(\text{)(})), \tag{82.5}$$

from which one can obtain the following standard fact.

Lemma 82.1.2. *The coefficient* $coeff_{z,k} \nabla_\star(z)$ *is a Vassiliev invariant of degree at most k.*

By similar arguments, the same is true for the HOMFLY polynomial in his reparameterization introduced by Jones [22] $F_K(N, e^z)$ (and therefore, setting $N = 2$, for the Jones polynomial [21] as well) and the Kauffman polynomial [24] in its version called by Kauffman the Dubrovnik polynomial [25], and more generally for any quantum group invariant [7] (see §82.3).

Analogous theories of (finite type) Vassiliev invariants can be developed for braids, tangles [4, 3], links, spatial graphs [34], virtual knots [19], knots in 3-manifolds [23], 3-manifolds themselves [29, 13], and immersed curves [1].

By definition, all Vassiliev invariants of degree $\leqslant m$ are sensitive with respect to knot classification only *maximally* up to $\mathcal{V}^{m+1}$. However, one does not know yet whether *all* Vassiliev invariants are capable of a *complete* topological classification of knots.

Conjecture 82.1.3. *Vassiliev invariants separate knots.*

By now conjecture 82.1.3 has been proved at least for pure braids [4, 3], but for knots we do not know much about it. An affirmative answer to this conjecture for all homotopy spheres would imply the Poincaré conjecture [15].

Question 82.1.1. *Do Vassiliev invariants distinguish knot orientation?*

This is one of the hardest problems in knot theory. Yet, there are no easily definable invariants, as quantum and skein invariants, which distinguish knot orientation. As pointed out by J. Birman [6], the fact that non-invertible knots exist was proved only in the 60s by Trotter [40]. G. Kuperberg [28] showed that a negative answer to this question would invalidate Conjecture 82.1.3 even for unoriented knots.

82.2   Weight Systems

Let us for a moment forget about the *FI* relation. The grading of the CDs is preserved by $4T$, so we can define the linear space $\mathcal{A}^r$ spanned by chord diagrams with r chords modulo $4T$. The graded vector space $\mathcal{A} := \bigoplus_{r=0}^{\infty} \mathcal{A}^r$ has a structure of a graded commutative and cocommutative HOPF algebra with multiplication given by the connected sum of CDs, which (modulo $4T$) does not depend on the specific arcs where the connected sum is performed. The classical Milnor-Moore [30] theorem states that $\mathcal{A}$ is isomorphic to the algebra of polynomials on the primitive space.

Definition 82.2.1. *A weight system of degree m is an element* $W \in (\mathcal{A}^m)^*$. *Let* $\mathcal{W}_m$ *be the linear space of all weight systems of degree m.*

It is easy to assign to a Vassiliev invariant v of deg $\leqslant m$ a weight system w of degree m via the (graded) map W_* given by

$$\mathcal{V}_m := (\mathcal{V}^m)^* \xrightarrow{\ W_m\ } (\mathcal{A}^m)^* =: \mathcal{W}_m \qquad (82.6)$$

$$W_m(v)\,(\text{CD of deg } m) \quad := \quad v\!\left(\begin{array}{l}\text{a singular knot with } m \text{ double points}\\ \text{whose chord diagram is equal to the CD}\end{array}\right). \qquad (82.7)$$

The kernel of this map is by definition $\mathcal{V}_{m-1} = (\mathcal{V}^{m-1})^*$, i.e., the space of Vassiliev invariants of deg $\leqslant m - 1$, so we have the short exact sequence

$$0 \longrightarrow \mathcal{V}_{m-1} \lhook\joinrel\longrightarrow \mathcal{V}_m \xrightarrow{\ W_m\ } \mathcal{W}_m \longrightarrow 0 .$$

Theorem 82.2.2 (Fundamental Theorem of Vassiliev invariants. [27])**.** *This exact sequence splits, i.e., there is a (graded) map V*

$$V_m : \mathcal{W}_m \longrightarrow \mathcal{V}_m ,$$

such that

$$Im(\,Id_{\mathcal{V}_m} - V_m \circ W_m\,) \subset \mathcal{V}_{m-1} \quad and \quad W_m \circ V_m = Id_{\mathcal{W}_m} .$$

In some sense the map V can be considered as "integrating" a weight system. In [27] it is given by the famous KONTSEVICH INTEGRAL.

82.3   Vassiliev Invariants and Lie Algebras

RESHETIKHIN and TURAEV [33] introduced a way to construct a framing-dependent weight system $W_{\mathfrak{g},R}$ (an element in $\mathcal{A}^*$ instead of $(\mathcal{A}^r)^*$) given an irreducible representation R of a semi-simple Lie algebra $\mathfrak{g}$. It is easy to "untwist" such a $W_{\mathfrak{g},R}$, i.e., to make it into a (framing-independent) weight system $\hat{W}_{\mathfrak{g},R}$. Integrating $\hat{W}_{\mathfrak{g},R}$ one obtains special Vassiliev invariants, called *Reshetikhin-Turaev* or quantum invariants. It seemed for a while, that these invariants, constructed to the most common Lie algebras, already exhaust all Vassiliev invariants. Some explicit computations in lower degrees led Bar-Natan [2] to the conjecture that all Vassiliev invariants are Reshetikhin-Turaev invariants.

However, it was already known that this conjecture contradicts the no less plausible looking conjecture 82.1.3, because quantum invariants are known not to detect orientation (see Question 82.1.1). Later, Vogel [42] indeed proved that there are Vassiliev invariants which do not come out of the Reshetikhin-Turaev construction for semi-simple Lie algebras.

82.4   Dimension Bounds, Modular Forms and Hyperbolic Volume

Though we have a nice description of the spaces introduced above, some simple algebraic features seem hopelessly hard to examine.

As a remarkable example in this regard, Bar-Natan observed [5], that a certain degeneracy statement of the sl_n weight system is equivalent to the Four Color Theorem, one of the most complicatedly proven theorems in mathematical history!

Another illustration: one does not know any way to compute the graded dimension of $\mathcal{A}$. (This has only been done for degrees up to 12 at the cost of a huge computer effort [26].)

There only exist some crude bounds about the asymptotical behaviour of these dimensions as degree $\to \infty$, mainly due to O. Dasbach [14] and the first author with Duzhin [8, 9] and the second author [35]. This shows the fascinating complexity of these simple-looking objects.

Nevertheless, the enumeration problems of [35] turned out to be interesting in a different context. D. Zagier [43] discovered that they give the coefficients of an asymptotic expansion near 1 of a "derivative" of the Dedekind eta-function, a classical modular form of weight 1/2. Using the theory of Dirichlet series, Zagier found the exact asymptotical behaviour of the numbers of [35], thus establishing the currently best upper bound for the graded dimension of $\mathcal{A}$.

A similar phenomenon occurs in the asymptotic expansion of Witten-Reshetikhin-Turaev-invariants of certain 3-manifolds, and is therefore of great number theoretic interest. The existence of such expansions was established by Ohtsuki, and they form the 3-manifold counterpart of Vassiliev theory for knots [29].

The same type of power series appeared in yet another context related to Vassiliev invariants. They give limits of evaluations at roots of unity of (some projective version of) the colored Jones polynomials. These cablings of the Jones polynomial can be transformed by the method of [7] into power series whose coefficients are Vassiliev invariants with weight systems coming via the construction in §82.3 from the irreducible representations of sl_2.

These evaluations have caused much excitement, because Murakami and Murakami [31] showed that they coincide with Kashaev's quantum dilogarithm invariants, for which he conjectured that the asymptotical behaviour of their evaluations at $e^{2\pi i/N}$ as $N \to \infty$ can be used to obtain the hyperbolic volume (or, if the knot is non-hyperbolic, the Gromov norm) of the complement of the knot. The KASHAEV-MURAKAMI-MURAKAMI conjecture has been established for some narrow classes of knots, but seems hard to approach in general. It has been noted by Murakami and Murakami [31], that the conjecture implies that a knot is trivial if and only if all of its Vassiliev invariants are trivial, that is, a partial case of conjecture 82.1.3 for the unknot.

82.5   The Approach of Small State Sums

Before Vassiliev, the common approach for finding knot invariants, as applied for the polynomials, is to fix certain initial values and a relation of the values of the invariant on links that are equal except near one crossing. Except for the Kauffman polynomial this relation involves triples as depicted in (82.5) and such a relation is called a *skein relation*.

An alternative approach was initiated by Fiedler [16] and Polyak-Viro [32]. They look at an n-tuple of crossings in a fixed knot projection satisfying certain conditions. Then they sum over all such n-tuples certain weights that can be calculated from the crossings in the tuple. If the conditions are well chosen, then the sum is invariant under the Reidemeister moves, and thus gives a knot invariant. It was proved [32], see also [36], that the Polyak-Viro invariants are Vassiliev invariants. In [19] it was proved that an arbitrary Vassiliev invariant can be represented by a Polyak-Viro formula. It is interesting that this proof requires an extension of the invariants to *virtual* knots. Concrete such formulas for the coefficients of the CONWAY and the HOMFLY polynomials can be found in [11] and [12], respectively.

The Polyak-Viro invariants, also called *Gauß sum*, *Gauss diagram formulas*, or *state sum* invariants, lead to a series of new results on the values of the Jones polynomial [39] and on positive [37] and almost positive knots [38].

82.6   The Goussarov–Habiro Theory

This theory describes certain moves on knot diagrams which preserve all Vassiliev invariants up to order m (and can be generalized to any class of knotted objects). It was announced by M. Goussarov in 1995 at a conference in Oberwolfach and published several years later in [18]. About the same time, K. Habiro independently found the same theorem [20], stating that the converse holds as well for knots.

Theorem 82.6.1 (Goussarov–Habiro). *Suppose that $v(K_1) = v(K_2)$ for all $\mathbb{Z}$-valued Vassiliev invariants v of order $\leq m$. Then arbitrary diagrams of K_1 and K_2 are related by a finite sequence of the Reidemeister moves and of moves $\mathcal{M}_m$:*

It is not difficult to prove that the move $\mathcal{M}_m$ preserves all Vassiliev invariants up to degree m. The tangle on the left side is an example of a *Brunnian tangle*: removing any of its strands makes it the trivial tangle with $m + 1$ components.

The book [10] is entirely devoted to the detailed study of Vassiliev knot invariants.

82.7   Bibliography

[1] V.I. Arnold. *Topological Invariants of Plane Curves and Caustics*. AMS, 1994.

[2] D. Bar-Natan. On the Vassiliev knot invariants. *Topology*, 34:423–472, 1995.

[3] D. Bar-Natan. Vassiliev homotopy string link invariants. *Jour. of Knot Theory and Its Ramifications*, 4(1):13–32, 1995.

[4] D. Bar-Natan. Vassiliev and quantum invariants for braids. In *Proceedings of Symposia in Applied Mathematics*, pages 129–144, 1996.

[5] D. Bar-Natan. Lie algebras and the four color theorem. *Combinatorica*, 17(1):43–52, 1997.

[6] J.S. Birman. New points of view in knot theory. *Bull. Amer. Math. Soc.*, 28:253–287, 1993.

[7] J.S. Birman and X-S. Lin. Knot polynomials and Vassiliev's invariants. *Invent. Math.*, 111:225–270, 1993.

[8] S. Chmutov and S. Duzhin. An upper bound for the number of Vassiliev knot invariants. *Jour. of Knot Theory and Its Ramifications*, 3(2):141–151, 1994.

[9] S. Chmutov and S. Duzhin. A lower bound for the number of Vassiliev knot invariants. *Topol. Appl.*, 92(3):201–223, 1999.

[10] S. Chmutov, S. Duzhin, and Y. Mostovoy. *Introduction to Vassiliev Knot Invariants*. Cambridge University Press, 2012.

[11] S. Chmutov, M. Khoury, and A. Rossi. Polyak-Viro formulas for coefficients of the Conway polynomial. *Jour. of Knot Theory and Its Ramifications*, 18(6):773–783, 2009.

[12] S. Chmutov and M. Polyak. Elementary combinatorics of the HOMFLYPT polynomial. *Int. Math. Res. Notes*, 2010(3):480–495, 2010.

[13] T. Cochran and P. Melvin. Finite type invariants of 3-manifolds. *Invent. Math.*, 140(1):45–100, 2000.

[14] O. Dasbach. On the combinatorial structure of primitive Vassiliev invariants III — a lower bound. *Communications in Contemporary Mathematics*, 2(4):579–590, 2000.

[15] M. Eisermann. Vassiliev invariants and the Poincaré conjecture. *Topology*, 43:1211–1229, 2004.

[16] Th. Fiedler. A small state sum for knots. *Topology*, 32(2):281–294, 1993.

[17] M. Goussarov. On *n*-equivalence of knots and invariants of finite degree. In *Topology of Manifolds and Varieties (O. Viro, editor), Adv. in Soviet Math.*, volume 18, pages 173–192. Amer. Math. Soc., Providence, RI, 1994.

[18] M. Goussarov. Variations of knotted graphs, geometric technique of *n*-equivalence. *St. Petersburg Math. J.*, 12(4):569–604, 2001.

[19] M. Goussarov, M. Polyak, and O. Viro. Finite type invariants of classical and virtual knots. *Topology*, 39:1045–1068, 2000.

[20] K. Habiro. Claspers and finite type invariants of links. *Geom. Topol.*, 4:1–83, 2000.

[21] V.F.R. Jones. A polynomial invariant of knots and links via von Neumann algebras. *Bull. Amer. Math. Soc.*, 12:103–111, 1985.

[22] V.F.R. Jones. On knot invariants related to some statistical mechanical models. *Pacific. J. Math.*, 137(2):311–334, 1989.

[23] E. Kalfagianni. Finite type invariants for knots in 3-manifolds. *Topology*, 37(3):673–707, 1998.

[24] L.H. Kauffman. An invariant of regular isotopy. *Trans. Amer. Math. Soc.*, 318:417–471, 1990.

[25] L.H. Kauffman. *Knots and Physics (Second Edition)*. World Scientific, Singapore, 1993.

[26] J. Kneissler. The number of primitive Vassiliev invariants up to degree twelve. arXiv:math.QA/9706022, June 1997.

[27] M. Kontsevich. Vassiliev's knot invariants. *Adv. in Soviet Math.*, 16:137–150, 1993.

[28] G. Kuperberg. Detecting knot invertibility. *Jour. of Knot Theory and Its Ramifications*, 5(2):173–181, 1996.

[29] J. Le, T. Q. T. Murakami and T. Ohtsuki. On a universal perturbative invariant of 3-manifolds. *Topology*, 37(3):271–291, 1998.

[30] J. Milnor and J. Moore. On the structure of Hopf algebras. *Annals of Math.*, 81:211–264, 1965.

[31] H. Murakami and J. Murakami. The colored Jones polynomials and the simplicial volume of a knot. *Acta Math.*, 186(1):85–104, 2001.

[32] M. Polyak and O. Viro. Gauss diagram formulas for Vassiliev invariants. *Int. Math. Res. Notes*, 11:445–454, 1994.

[33] N. Yu. Reshetikhin and V. G. Turaev. Ribbon graphs and their invariants derived from quantum groups. *Commun. Math. Phys.*, 127:1–26, 1990.

[34] T. Stanford. Finite-type invariants of knots, links, and graphs. *Topology*, 35:1027–1050, 10 1996.

[35] A. Stoimenow. Enumeration of chord diagrams and an upper bound for Vassiliev invariants. *J. of Knot Theory and Its Ram.*, 7(1):93–114, 1998.

[36] A. Stoimenow. Gauß sum invariants, Vassiliev invariants and braiding sequences. *Jour. of Knot Theory and Its Ramifications*, 9(2):221–269, 2000.

[37] A. Stoimenow. Positive knots, closed braids and the Jones polynomial. *Ann. Scuola Norm. Sup. Pisa Cl. Sci.*, 2(2):237–285, 2003.

[38] A. Stoimenow. Gauss sums on almost positive knots. *Compositio Mathematica*, 140(1):228–254, 20040.

[39] A. Stoimenow. On some restrictions to the values of the Jones polynomial. *Indiana Univ. Math. J.*, 54(2):557–574, 2005.

[40] H.F. Trotter. Non-invertible knots exist. *Topology*, 2:341–358, 1964.

[41] V.A. Vassiliev. Cohomology of knot spaces. In *Theory of Singularities and Its Applications (ed. V. I. Arnold), Adv. in Soviet Math.*, volume 1, pages 23–69. Amer. Math. Soc., Providence, RI, 1990.

[42] P. Vogel. Algebraic structures on modules of diagrams. *J. Pure Appl. Algebra*, 215(6):1292–1339, 2011. Université Paris VII preprint, July 1995.

[43] D. Zagier. Vassiliev invariants and a strange identity related to the Dedekind eta-function. *Topology*, 40(5):945–960, 2001.

Chapter 83

Linking Number and Milnor Invariants

Jean-Baptiste Meilhan, Université Grenoble Alpes

83.1 Introduction

The linking number is probably the oldest invariant of knot theory. In 1833, several decades before the seminal works of physicists Tait and Thompson on knot tabulation, the work of Gauss on electrodynamics led him to formulate the number of "intertwinings of two closed or endless curves" as

$$\frac{1}{4\pi} \int \int \frac{(x'-x)(dydz' - dzdy') + (y'-y)(dzdx' - dxdz') + (z'-z)(dxdy' - dydx')}{[(x'-x)^2 + (y'-y)^2 + (z'-z)^2]^{3/2}}, \quad (83.1)$$

where x, y, z (resp. x', y', z') are coordinates on the first (resp. second) curve. This number, now called the *linking number*, is an invariant of 2–component links,[1] which typically distinguishes the trivial link and the Hopf link, shown below.

Alternatively, the linking number can be simply defined in terms of the first homology group of the link complement (see Thm. 83.2.1). But, rather than the abelianization of the fundamental group, one can consider finer quotients, for instance, by the successive terms of its lower central series, to construct more subtle link invariants. This is the basic idea upon which Milnor based his work on higher-order linking numbers, now called *Milnor $\overline{\mu}$-invariants*.

We review in Section 83.2 several definitions and key properties of the linking number. In Section 83.3, we shall give a precise definition of Milnor invariants and explore some of their properties, generalizing those of the linking number. In the final Section 83.4, we briefly review further known results on Milnor invariants.

83.2 The Linking Number

The linking number has been studied from multiple angles and has thus been given many equivalent definitions of various natures. We already saw in (83.1) the original definition due to Gauss; let us review below a few others.

---

[1] In this chapter, all links will be ordered and oriented tame links in the 3-sphere.

83.2.1 Definitions

For the rest of this section, let $L = L_1 \cup L_2$ be a 2-component link, and let $\mathrm{lk}(L)$ denote the linking number of L.

Recall that the first homology group of the complement $S^3 \setminus L_1$ of L_1 is the infinite cyclic group generated by the class $[m_1]$ of a meridian m_1, as shown on the right.

Theorem 83.2.1. *The linking number of L is the integer k such that $[L_2] = k[m_1]$ in $H_1(S^3 \setminus L_1; \mathbb{Z})$.*

Recall now that any knot bounds orientable surfaces in S^3, called Seifert surfaces. Now, given a Seifert surface S for L_2, one may assume up to isotopy that L_1 intersects S at finitely many transverse points. At each intersection point, the orientation of L_1 and that of S form an oriented 3-frame, hence a sign using the right-hand rule; the *algebraic intersection* of L_1 and S is the sum of these signs over all intersection points.

Theorem 83.2.2. *The linking number of L is the algebraic intersection of L_1 with a Seifert surface for L_2.*

Given a regular diagram of L, the linking number is simply given by the number of crossings where L_1 passes over L_2, counted with signs as follows.

Theorem 83.2.3.

$$\mathrm{lk}(L) = \left(number\ of\ {}_1\!\!\diagdown\!\!{}_2 \right) - \left(number\ of\ {}_2\!\!\diagup\!\!{}_1 \right).$$

The proof of the above results, *i.e.* the fact that these definitions are all equivalent and equivalent to (83.1), can be found in [41, §. 5.D], along with several further definitions. See also [40] for details.

Remark 83.2.4. As part of Theorems 83.2.2 and 83.2.3, the linking number does not depend on the choices involved in these definitions, namely the choice of a Seifert surface for L_2 and of a diagram of L, respectively.

83.2.2 Basic Properties

It is rather clear, using any of the above definitions, that reversing the orientation of either component of L changes the sign of $\mathrm{lk}(L)$.

Also clear from the Gauss formula (83.1) is the fact that the linking number is symmetric:

$$\mathrm{lk}(L_1, L_2) = \mathrm{lk}(L_2, L_1). \tag{83.2}$$

This can also be seen, for example, from the diagrammatic definition (Thm. 83.2.3), by considering two projections, on two parallel planes which are 'on either side' of L, so that crossings where 1 overpasses 2 in one diagram are in one-to-one correspondence with crossings where 2 overpasses 1 in the other, with the same sign.

The symmetry property (83.2) allows for a symmetrized version of Theorem 83.2.3:

$$\mathrm{lk}(L) = \frac{1}{2}\left(\text{(number of } \diagup\kern-0.6em\nwarrow\kern-0.4em\nearrow\text{)} - (\text{ number of } \diagup\kern-0.6em\nearrow\kern-0.4em\nwarrow\text{)}\right),$$

where the sum runs over *all* crossings involving a strand of component 1 and a strand of component 2.

83.2.3 Two Classification Results

The following notion was introduced by Milnor in the fifties [29].

Definition 83.2.5. Two links are *link-homotopic* if they are related by a sequence of ambient isotopies and self-crossing changes, *i.e.* crossing changes involving two strands of the same component (see Figure 83.1).

The idea behind this notion is that, as a first approximation to the general study of links, working up to link-homotopy allows us to unknot each individual component of a link, and only records their 'mutual interactions'—this is, in a sense, studying 'linking modulo knotting'.

Observe that the linking number is invariant under link-homotopy. Furthermore, it is not too difficult to check that any 2–component link is link-homotopic to an iterated Hopf link, *i.e.* the closure of the pure braid σ_1^{2n} for some $n \in \mathbb{Z}$. This number n is precisely the linking number of the original link, thus showing:

Theorem 83.2.6. *The linking number classifies 2-component links up to link-homotopy.*

Generalizing this result to a higher number of components, however, requires additional invariants, and this was one of the driving motivations that led Milnor to develop his $\bar{\mu}$-invariants, see Section 83.3.3.2.

FIGURE 83.1: A self-crossing change (left) and a Δ-move (right).

The equivalence relation on links with an arbitrary number of components, which is classified by the linking number, is the Δ-*equivalence*, which is generated by the Δ-move, shown on the right-hand side of Figure 83.1. Indeed, the following is due H. Murakami and Y. Nakanishi [32], see also [26].

Theorem 83.2.7. *Two links $\cup_{i=1}^n L_i$ and $\cup_{i=1}^n L_i'$ ($n \geq 2$) are Δ-equivalent if, and only if $\mathrm{lk}(L_i, L_j) = \mathrm{lk}(L_i', L_j')$ for all i, j such that $1 \leq i < j \leq n$.*

Notice in particular that the Δ-move is an unknotting operation, meaning that any knot can be made trivial by a finite sequence of isotopies and Δ-moves.

83.3   Milnor Invariants

In this whole section, let $L = L_1 \cup \cdots \cup L_n$ be an n-component link, for some fixed $n \geq 2$.

83.3.1   Definition of Milnor $\overline{\mu}$-Invariants

In his master's and doctoral theses, supervised by R. Fox, Milnor defined numerical invariants extracted from the peripheral system of a link, which widely generalize the linking number.

Denote by X the complement of an open tubular neighborhood of L. Pick a point in the interior of X, and denote by π the fundamental group of X based at this point. It is well known that, given a diagram of L, we can write an explicit presentation of π, called the Wirtinger presentation, where each arc in the diagram provides a generator, and each crossing yields a relation. Such presentations, however, are in general difficult to work with, owing to their large number of generators. We introduce below a family of quotients of π for which a much simpler presentation can be given, and which still retain rich topological information on the link L.

The *lower central series* $(\Gamma_k G)_{k \geq 1}$ of a group G is the nested family of subgroups defined inductively by

$$\Gamma_1 G = G \quad \text{and} \quad \Gamma_{k+1} G = [G, \Gamma_k G].$$

The *kth nilpotent quotient* of the group G is the quotient $G/\Gamma_k G$.

Let us fix, from now on, a value of $k \geq 2$. For each $i \in \{1, \cdots, n\}$, consider two elements $m_i, \lambda_i \in \pi/\Gamma_k \pi$, where m_i represents a *choice* of a meridian for L_i in X, and where λ_i is a word representing the ith *preferred longitude* of L_i, *i.e.* a parallel copy of L_i in X having linking number zero with L_i. Denote also by F the free group $F(m_1, \cdots, m_n)$.

Theorem 83.3.1 (Chen-Milnor Theorem). *The kth nilpotent quotient of π has a presentation given by*

$$\pi/\Gamma_k \pi = \left\langle m_1, \ldots, m_n \mid m_i \lambda_i m_i^{-1} \lambda_i^{-1} \ (i = 1, \cdots, n) ; \Gamma_k F \right\rangle.$$

In order to extract numerical invariants from the nilpotent quotients of π, we consider the *Magnus expansion* [25] of each λ_j, which is an element $E(\lambda_j)$ of $\mathbb{Z}\langle\langle X_1, ..., X_n \rangle\rangle$, the ring of formal power series in non-commuting variables $X_1, \ldots, X_n$, obtained by the substitution

$$m_i \mapsto 1 + X_i \quad \text{and} \quad m_i^{-1} \mapsto 1 - X_i + X_i^2 - X_i^3 + \ldots$$

We denote by $\mu_L(i_1 i_2 \ldots i_p j)$ the coefficient of $X_{i_1} X_{i_2} \ldots X_{i_p}$ in the Magnus expansion of λ_j:

$$E(\lambda_j) = 1 + \sum_{i_1 i_2 \ldots i_p} \mu_L(i_1 i_2 \ldots i_p j) X_{i_1} X_{i_2} \ldots X_{i_p}. \tag{83.3}$$

This coefficient is *not*, in general, an invariant of the link, as it depends upon the choices made in this construction (essentially, picking a system of based meridians for L). We can, however, promote these numbers to genuine invariants by regarding them modulo the following indeterminacy.

Definition 83.3.2. Given a sequence I of indices in $\{1, ..., n\}$, let $\Delta_L(I)$ be the greatest common divisor of the coefficients $\mu_L(I')$, for all sequences I' obtained from I, by deleting at least one index and permuting cyclically.

For example, $\Delta_L(123) = \gcd\{\mu_L(12), \mu_L(21), \mu_L(13), \mu_L(31), \mu_L(23), \mu_L(32)\}$.

Theorem–Definition 1 (Milnor [30]). *The residue class*

$$\overline{\mu}_L(I) \equiv \mu_L(I) \bmod \Delta_L(I)$$

is an invariant of ambient isotopy of L, for any sequence I of $\leq k$ integers, called a Milnor *invariant of L. The number of indices in I is called the* length.

Before giving some examples in Section 83.3.2, a few remarks are in order.

Remark 83.3.3. Observe from (83.3) that $\mu_L(I)$ is only defined for a sequence I of *two* or more indices. We set $\mu_L(i) = 0$ for all i as a convention.

Remark 83.3.4. If all Milnor invariants of L of length $< k$ are zero, then those of length k are well-defined integers. These *first non-vanishing Milnor invariants* are thus much easier to study in practice and, as a matter of fact, a large proportion of the literature on Milnor link invariants focusses on them.

Remark 83.3.5. Working with the kth nilpotent quotient of π only allows us to define Milnor invariants up to length k, but this k can be chosen arbitrarily large, so this is no restriction.

83.3.2 First Examples

By Remark 83.3.3, length 2 Milnor invariants are well defined over $\mathbb{Z}$. In order to define them, it suffices to work in the second nilpotent quotient, *i.e.* in $\pi/\Gamma_2\pi = H_1(X; \mathbb{Z})$. But the jth preferred longitude in $H_1(X; \mathbb{Z})$ is given by $\lambda_j = \sum_{i \neq j} \text{lk}(L_i, L_j)m_i$, and we have

$$E(\lambda_j) = 1 + \sum_{i \neq j} \text{lk}(L_i, L_j)X_i + \text{terms of degree} \geq 2.$$

Hence length 2 Milnor invariants are exactly the pairwise linking numbers of a link. This justifies regarding length ≥ 3 Milnor invariants as 'higher- order linking numbers'. In order to get a grasp on these, let us consider an elementary example.

Example 83.3.6. Consider the Boromean rings B, as illustrated below.[2] The Wirtinger presentation for π is

---

[2]There, we use the usual convention that a small arrow underpassing an arc of the diagram represents a loop, based at the reader's eye and going straight down to the projection plane, positively enlacing the arc and going straight back to the basepoint.

$$\pi = \left\langle \begin{array}{l} m_1, m_2, m_3 \\ n_1, n_2, n_3 \end{array} \right| \left. \begin{array}{lll} m_3 m_1 m_3^{-1} n_1^{-1} ; & m_1 m_2 m_1^{-1} n_2^{-1} ; & m_2 m_3 m_2^{-1} n_3^{-1} \\ n_1^{-1} n_2 n_1 m_2^{-1} ; & n_2^{-1} n_3 n_2 m_3^{-1} ; & n_3^{-1} n_1 n_3 m_1^{-1} \end{array} \right\rangle.$$

Consider the third nilpotent quotient of π, *i.e.* assume $k = 3$. We *choose* the meridians m_i as (representatives of) generators, and express preferred longitudes λ_i: we have $\lambda_3 = m_2^{-1} n_2 = m_2^{-1} m_1 m_2 m_1^{-1}$ (the other two are obtained by cyclic permutation of the indices, owing to the symmetry of the link). Taking the Magnus expansion gives

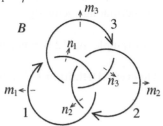

$$E(\lambda_3) = 1 + X_1 X_2 - X_2 X_1 + \text{terms of degree} \geq 3.$$

We thus obtain that $\overline{\mu}_B(I) = 0$ for any sequences I of two indices (*i.e.* all linking numbers are zero), and

$$\overline{\mu}_B(123) = \overline{\mu}_B(231) = \overline{\mu}_B(312) = 1 \quad \text{and} \quad \overline{\mu}_B(132) = \overline{\mu}_B(213) = \overline{\mu}_B(321) = -1.$$

Example 83.3.6 illustrates the fact that, just like the linking number detects (and actually, counts copies of) the Hopf link, the *triple linking number* $\overline{\mu}(123)$ detects the Borromean rings. This generalizes to the following realization result, due to Milnor [29].

Lemma 83.3.7. *Let M_n be the n-component link shown on the right ($n \geq 2$), and let σ be any permutation in S_{n-1}. Then*

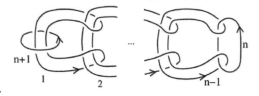

$$\overline{\mu}_{M_n} (\sigma(1) \cdots \sigma(n-1) n(n+1)) = \begin{cases} 1 & \text{if } \sigma = Id, \\ 0 & \text{otherwise.} \end{cases}$$

83.3.3 Some Properties

We gather here several well-known properties of Milnor $\overline{\mu}$-invariants, most being due to Milnor himself [29, 30] (unless otherwise specified).

83.3.3.1 Symmetry and Shuffle

It is rather clear that, if L' is obtained from L by reversing the orientation of the ith component, then $\overline{\mu}_{L'}(I) = (-1)^{i(I)} \overline{\mu}_L$, where $i(I)$ denotes the number of occurrences of the index i in the sequence I. The next two relations were shown by Milnor [30], using properties of the Magnus expansion:

Cyclic symmetry: For any sequence $i_1 \cdots i_m$ of indices in $\{1, \cdots, n\}$, we have

$$\overline{\mu}_L(i_1 \cdots i_m) = \overline{\mu}_L(i_m i_1 \cdots i_{m-1}).$$

Shuffle: For any $k \in \{1, \cdots, n\}$ and any two sequences I, J of indices in $\{1, \cdots, n\}$, we have

$$\sum_{H \in S(I,J)} \overline{\mu}_L(Hk) \equiv 0 \pmod{\gcd\{\Delta(Hk) ; H \in S(I,J)\}},$$

where $S(I, J)$ denotes the set of all sequences obtained by inserting the indices of J into I, preserving order.

It follows in particular that there is essentially only one triple linking number: for any 3-component algebraically split link L, we have $\overline{\mu}_L(123) = \overline{\mu}_L(231) = \overline{\mu}_L(312) = -\overline{\mu}_L(132) = -\overline{\mu}_L(213) = -\overline{\mu}_L(321)$.

In general, the number of independent $\overline{\mu}$-invariants was given by K. Orr, see [36, Thm. 15].

83.3.3.2 Link-Homotopy and Concordance

Milnor invariants are not only invariants of ambient isotopy: they are actually invariants of *isotopy*, *i.e.* homotopy through embeddings. In particular that Milnor invariants do not see 'local knots'.

As mentioned in Section 83.2.3, the notion of link-homotopy was introduced by Milnor himself, who proved the following in [30].

Theorem 83.3.8. *For any sequence I of pairwise distinct indices, $\overline{\mu}(I)$ is a link-homotopy invariant.*

Milnor invariants are sharp enough to detect link-homotopically trivial links, and to classify links up to link-homotopy for ≤ 3 components [30]. The case of 4-component links was only completed thirty years later by J. Levine, using a refinement of Milnor's construction [20]. The general case is discussed in Section 83.4.1. Milnor link-homotopy invariants also form a complete set of link-homotopy invariants for *Brunnian* links, *i.e.* links which become trivial after removal of any component (see Section 83.3.2 for examples) [29].

Recall that two n-component links, L and L', are *concordant* if there is an embedding $f : \sqcup_{i=1}^n (S^1 \times [0,1])_i \longrightarrow S^3 \times [0,1]$ of n disjoint copies of the annulus $S^1 \times [0,1]$, such that $f\left((\sqcup_{i=1}^n S_i^1) \times \{0\}\right) = L \times \{0\}$ and $f\left((\sqcup_{i=1}^n S_i^1) \times \{1\}\right) = L' \times \{1\}$. The following was essentially shown by J. Stallings [43] (and also by A. Casson [6] for the more general relation of cobordism).

Theorem 83.3.9. *Milnor invariants are concordance invariants.*

Let us also mention here that Milnor invariants are all zero for a *boundary link*, that is, for a link whose components bound mutually disjoint Seifert surfaces in S^3 [42]. Characterizing geometrically links with vanishing of Milnor invariants is the subject of the k-slice conjecture, proved by K. Igusa and K. Orr in [14], which roughly states that all invariants of length $\leq 2k$ vanish if and only if the link bounds a surface which "looks like slice disks" modulo k-fold commutators of the fundamental group of the surface complement.

83.3.3.3 Cabling Formula

Milnor invariants indexed by non-repeated sequences are not only topologically relevant, by Theorem 83.3.8; they also 'generate' all Milnor invariants, by the following.

Theorem 83.3.10. *Let I be a sequence of indices in $\{1, \cdots, n\}$, such that the index i appears twice in I. Then $\overline{\mu}_L(I) = \overline{\mu}_{\tilde{L}}(\tilde{I})$, where $\tilde{I}$ is obtained from I by replacing the second occurrence of i by $n + 1$, and where $\tilde{L}$ is obtained from L by adding an $(n + 1)$th component, which is a parallel copy of L_i having linking number zero with it.*

Example 83.3.11. A typical example of a link-homotopically trivial link is the Whitehead link W, shown below on the left-hand side. But Milnor concordance invariants do detect W; specifically, we have $\overline{\mu}_W(1122) = 1$.

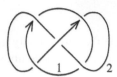 

FIGURE 83.2: The Whitehead link W (left) and the 4-component link $\tilde{W}$ (right).

As a matter of fact, one can check that $\overline{\mu}_W(1122) = \overline{\mu}_{\tilde{W}}(1324) = 1$, where $\tilde{W}$ is the 4-component link shown on the right of figure 83.2.

83.3.3.4 Additivity

Given two n-component links L and L', lying in two disjoint 3-balls of S^3, a *band sum* $L\sharp L'$ of L an L' is any link made of pairwise disjoint connected sums of the ith components of L and L' ($1 \le i \le n$). Although this operation is not well defined, V. Krushkal showed that, for any sequence I,

$$\overline{\mu}_{L\sharp L'}(I) \equiv \overline{\mu}_L(I) + \overline{\mu}_{L'}(I) \pmod{\gcd(\Delta_L(I), \Delta_{L'}(I))},$$

meaning that Milnor invariants are independent of the choice of bands defining $L\sharp L'$, see [19].

83.4 Further Properties

Since their introduction in the fifties, Milnor invariants have been the subject of numerous works. In this section, which claims neither exhaustivity nor precision, we briefly overview some of these results.

83.4.1 Refinements and Generalizations

The main difficulty in understanding Milnor $\overline{\mu}$-invariants lies in these intricate indeterminacies $\Delta(I)$. Multiple attempts have been made to refine this indeterminacy, in order to get more subtle invariants.

T. Cochran defined, by considering recursive intersection curves of Seifert surfaces, link invariants which recover (and shed beautiful geometric lights on) $\overline{\mu}$-invariants, and sometimes refine them [8].

K. Orr defined invariants, as the class of the ambient S^3 in some third homotopy group, which refines significantly the indeterminacy of Milnor invariants [35, 36]. Orr's invariant is also the first attempt towards a transfinite version of Milnor invariants, a problem posed in [30]; see also the work of J. Levine in [22].

A decisive step was taken by Habegger and Lin, who showed that the indeterminacies $\Delta_L(I)$ are equivalent to the indeterminacy in representing L as the closure of a *string link*, *i.e.* a pure tangle without closed component [11] (see also [21]). This led them to a full link-homotopy classification of (string) links [11].

83.4.2 Link Maps and Higher Dimensional Links

There are several higher-dimensional versions of the linking number; see [5, § 3.4] for a good survey. Some are invariants of *link maps*, *i.e.* maps from a union of two spheres (of various dimensions) to a sphere with disjoint images – these are the natural objects to consider when working up to link-homotopy. In particular, the first example of a link map which is not link-homotopically trivial was given in [9] using an appropriate generalization of the linking number.

Higher-dimensional generalizations of Milnor invariants were defined and extensively studied by U. Koschorke for link maps with many components, see [16, 17] and references therein. Note also that Orr's invariants, which generalize Milnor invariants (see Section 83.4.1), are also defined in any dimensions. But there does not seem to be nontrivial analogues of Milnor invariants for n-dimensional links ($n \geq 2$), *i.e.* for embedded n-spheres in codimension 2; in fact, all are link-homotopically trivial [4]. However, Milnor invariants generalize naturally to 2-string links, *i.e.* knotted annuli in the 4-ball bounded by a prescribed unlink in the 3-sphere, and classify them up to link-homotopy [2], thus providing a higher-dimensional version of [11].

83.4.3 Relations with Other Invariants

Milnor invariants are not only natural generalizations of the linking number, but are also directly related to this invariant. Indeed, K. Murasugi expressed Milnor $\bar{\mu}$-invariants of a link as a linking number in certain branched coverings of S^3 along this link [34]. The Alexander polynomial is also rather close in nature to Milnor invariants, being extracted from the fundamental group of the complement. As a matter of fact, there are a number of results relating these two invariants: see for example [7, 23, 33, 37, 44]. On the other hand, the relation to the Konstevich integral [12] hints to a potential connections to quantum invariants. Such relations were given with the HOMFLY-PT polynomial [28] and the (colored) Jones polynomial [27]. Milnor string link invariants also satisfy a skein relation [38], which is a typical feature of polynomial and quantum invariants.

There also are known relations outside knot theory. Milnor $\bar{\mu}$-invariants can be expressed in terms of Massey products of the complement, which are higher-order cohomological invariants generalizing the cup product [39, 45]. Also, Milnor string link invariants are in natural correspondence with Johnson homomorphisms of homology cylinders, which are 3-dimensional extensions of certain abelian quotients of the mapping class group [10]. Finally, as part of the deep analogies he established between knots and primes, M. Morishita defined and studied arithmetic analogues of Milnor invariants for prime numbers [31].

83.4.4   Finite Type Invariants

The linking number is (up to scalar) the unique degree 1 finite-type link invariant. This was generalized independently by D. Bar Natan [3] and X.S. Lin [24], who showed that any Milnor *string link* invariant of length k is a finite-type invariant of degree $k - 1$. As a consequence, Milnor string link invariants can be, at least in principle, extracted from the Kontsevich integral, which is universal among finite-type invariants. This was made completely explicit by G. Masbaum and N. Habegger in [12].

Note that the finite-type property does not make sense for higher-order $\bar{\mu}$-invariants (of *links*), since the indeterminacy $\Delta(I)$ is, in general, not the same for two links which differ by a crossing change. Nonetheless, $\bar{\mu}$-invariants of length k are invariants of C_k-equivalence [13], a property shared by all degree $k - 1$ invariants.

83.4.5   Virtual Theory

There are two distinct extensions of the linking number for virtual 2-component links, namely $\mathrm{lk}_{1,2}$ and $\mathrm{lk}_{2,1}$, where $\mathrm{lk}_{i,j}$ is the sum of signs of crossings where i passes over j. Notice that these virtual linking numbers are actually invariants of *welded links*, *i.e.* are invariant under the forbidden move allowing a strand to pass *over* a virtual crossing. Indeed, the linking number is extracted from (a quotient of) the fundamental group, which is a welded invariant [15]. Likewise, extensions of Milnor invariants shall be welded invariants. A general welded extension of Milnor string link invariants is given in [1], which classifies welded string links up to self-virtualization, generalizing the classification of [11]. This extension recovers and extends that of [18], which gives general Gauss diagram formulas for virtual Milnor invariants.

Acknowledgments

The author would like to thank Benjamin Audoux and Akira Yasuhara, for their comments on a preliminary version of this note.

83.5   Bibliography

[1] B. Audoux, P. Bellingeri, J.-B. Meilhan, and E. Wagner. Homotopy classification of ribbon tubes and welded string links. *Ann. Sc. Norm. Super. Pisa Cl. Sci.*, XVII(2):713–761, 2017.

[2] B. Audoux, J.-B. Meilhan, and E. Wagner. On codimension 2 embeddings up to link-homotopy. *J. Topology*, 10(4):1107–1123, 2017.

[3] D. Bar-Natan. Vassiliev homotopy string link invariants. *J. Knot Theory Ramifications*, 4(1):13–32, 1995.

[4] A. Bartels and P. Teichner. All two dimensional links are null homotopic. *Geom. Topol.*, 3:235–252, 1999.

[5] J. Carter, S. Kamada, and M. Saito. *Surfaces in 4-Space*, volume 142 of *Encyclopaedia of Mathematical Sciences*. Springer-Verlag, Berlin, 2004. Low-Dimensional Topology, III.

[6] A. Casson. Link cobordism and Milnor's invariant. *Bull. Lond. Math. Soc.*, 7:39–40, 1975.

[7] T. D. Cochran. Concordance invariance of coefficients of Conway's link polynomial. *Invent. Math.*, 82(3):527–541, 1985.

[8] T. D. Cochran. Derivatives of links: Milnor's concordance invariants and Massey's products. *Mem. Am. Math. Soc.*, 427:73, 1990.

[9] R. Fenn and D. Rolfsen. Spheres may link homotopically in 4-space. *J. London Math. Soc.*, 34(2):177–184, 1986.

[10] N. Habegger. Milnor, Johnson, and tree level perturbative invariants. Preprint, 2000.

[11] N. Habegger and X.-S. Lin. The classification of links up to link-homotopy. *J. Amer. Math. Soc.*, 3:389–419, 1990.

[12] N. Habegger and G. Masbaum. The Kontsevich integral and Milnor's invariants. *Topology*, 39(6):1253–1289, 2000.

[13] K. Habiro. Claspers and finite type invariants of links. *Geom. Topol.*, 4:1–83 (electronic), 2000.

[14] K. Igusa and K. E. Orr. Links, pictures and the homology of nilpotent groups. *Topology*, 40(6):1125–1166, 2001.

[15] L. H. Kauffman. Virtual knot theory. *European J. Combin.*, 20(7):663–690, 1999.

[16] U. Koschorke. Higher order homotopy invariants for higher dimensional link maps. Algebraic topology, Proc. Conf., Göttingen/Ger. 1984, Lect. Notes Math. 1172, 116–129, 1985.

[17] U. Koschorke. A generalization of Milnor's μ-invariants to higher dimensional link maps. *Topology*, 36(2):301–324, 1997.

[18] O. Kravchenko and M. Polyak. Diassociative algebras and Milnor's invariants for tangles. *Lett. Math. Phys.*, 95(3):297–316, 2011.

[19] V. S. Krushkal. Additivity properties of Milnor's $\bar{\mu}$-invariants. *J. Knot Theory Ramifications*, 7(5):625–637, 1998.

[20] J. Levine. An approach to homotopy classification of links. *Trans. Am. Math. Soc.*, 306(1):361–387, 1988.

[21] J. Levine. The μ-invariants of based links. In *Differential Topology (Siegen, 1987)*, volume 1172 of *Lecture Notes in Mathematics*, pages 87–103. Springer, Berlin, 1988.

[22] J. Levine. Link concordance and algebraic closure of groups. *Comment. Math. Helv.*, 64(2):236–255, 1989.

[23] J. Levine. A factorization of the Conway polynomial. *Comment. Math. Helv.*, 74(1):27–52, 1999.

[24] X.-S. Lin. Power series expansions and invariants of links. In *Geometric topology (Athens, GA, 1993)*, volume 2 of *AMS/IP Stud. Adv. Math.*, pages 184–202. Amer. Math. Soc., Providence, RI, 1997.

[25] W. Magnus, A. Karrass, and D. Solitar. Combinatorial Group Theory. Dover Books on Advanced Mathematics. New York: Dover Publications, Inc. XII, 444p., 1976.

[26] S. V. Matveev. Generalized surgeries of three-dimensional manifolds and representations of homology spheres. *Mat. Zametki*, 42(2):268–278, 345, 1987.

[27] J.-B. Meilhan and S. Suzuki. The universal sl_2 invariant and Milnor invariants. *Int. J. Math.*, 27(11):37, 2016.

[28] J.-B. Meilhan and A. Yasuhara. Milnor invariants and the HOMFLYPT polynomial. *Geom. Topol.*, 16(2):889–917, 2012.

[29] J. Milnor. Link groups. *Ann. of Math. (2)*, 59:177–195, 1954.

[30] J. Milnor. Isotopy of links. Algebraic geometry and topology. In *A Symposium in Honor of S. Lefschetz*, pages 280–306. Princeton University Press, Princeton, N. J., 1957.

[31] M. Morishita. On certain analogies between knots and primes. *J. Reine Angew. Math.*, 550:141–167, 2002.

[32] H. Murakami and Y. Nakanishi. On a certain move generating link-homology. *Math. Ann.*, 284(1):75–89, 1989.

[33] K. Murasugi. On Milnor's invariant for links. *Trans. Am. Math. Soc.*, 124:94–110, 1966.

[34] K. Murasugi. Nilpotent coverings of links and Milnor's invariant. Low dimensional topology, 3rd Topology Semin. Univ. Sussex 1982, *Lond. Math. Soc. Lect. Note Ser.* 95, 106–142, 1985.

[35] K. E. Orr. New link invariants and applications. *Comment. Math. Helv.*, 62:542–560, 1987.

[36] K. E. Orr. Homotopy invariants of links. *Invent. Math.*, 95(2):379–394, 1989.

[37] M. Polyak. On Milnor's triple linking number. *C. R. Acad. Sci., Paris, Sér. I, Math.*, 325(1):77–82, 1997.

[38] M. Polyak. Skein relations for Milnor's μ-invariants. *Algebr. Geom. Topol.*, 5:1471–1479, 2005.

[39] R. Porter. Milnor's $\bar{\mu}$-invariants and Massey products. *Trans. Am. Math. Soc.*, 257:39–71, 1980.

[40] R. L. Ricca and B. Nipoti. Gauss' linking number revisited. *J. Knot Theory Ramifications*, 20(10):1325–1343, 2011.

[41] D. Rolfsen. Knots and Links. Mathematical Lecture Series. 7. Berkeley, Ca.: Publish or Perish, Inc. 439p., 1976.

[42] N. Smythe. Boundary links. *Ann. Math. Stud.*, 60:69–72, 1966.

[43] J. Stallings. Homology and central series of groups. *J. Algebra*, 2:170–181, 1965.

[44] L. Traldi. Milnor's invariants and the completions of link modules. *Trans. Amer. Math. Soc.*, 284(1):401–424, 1984.

[45] V. Turaev. Milnor invariants and Massey products. *J. Sov. Math.*, 12:128–137, 1979.

Part XIV

Physical Knot Theory

Part XIV

Physical Knot Theory

Chapter 84

Stick Number for Knots and Links

Colin Adams, Williams College

84.1 Introduction

Given a knot K, one can ask for the least number of sticks glued end-to-end in order to construct the knot, as in Figure 84.1.

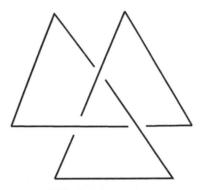

FIGURE 84.1: A trefoil knot realized with six sticks.

This is the *stick number $s(K)$*, also sometimes called the *polygon number $p(K)$* or *edge number $e(K)$*. Given its simple definition, $s(K)$ is a natural invariant to consider. Moreover, it is relevant to questions about knotted molecules. Specifically, if we model a knotted molecule using vertices for the atoms and sticks for the bonds between the atoms, which is not a bad first approximation for a model, then the number of sticks needed to make a given knot may impact the number of atoms required to make a knotted molecule in the shape of that knot. In this initial section, we discuss what is known about the stick number. In the second section, we consider various other stick invariants. And in the final section, we turn to stick realizations of projections.

The obvious first question one might ask is how many sticks it takes to make a nontrivial knot. Three sticks yield a triangle, which clearly represents a trivial knot. With four sticks, a projection yields at most one crossing, which is not enough to generate a nontrivial knot. If we construct a knot from five sticks, we can project down one of the sticks and see only four sticks in the plane, which again yields at most one crossing. So again this cannot be a nontrivial knot. However, in Figure 84.1, we see a trefoil knot constructed with six sticks. So 6 is the least stick number of a nontrivial knot. It is not difficult to prove that the only nontrivial knot with stick number 6 is the trefoil.

Note that even if we have a projection showing some number of sticks, that does not mean the knot can actually be constructed with that number of sticks. It could be that the sticks are bent in 3-space, but their projections are straight. In the case of the six-stick trefoil, we can label the vertices by points in 3-space to show that the stick configuration can be realized in 3-space. In Figure 84.2 we see a projection of the knot 5_2 made with five sticks. However, the sticks cannot be realized as straight in space. We can see this by picking two adjacent sticks and labelling their vertices with P to represent their lying in the plane of the page. Then we can label the remaining vertices of the adjacent sticks with H for high (above the plane) and L for low (below the plane) in order to achieve their crossings with the first two sticks. However, the stick with vertices P and H must then pass beneath the stick labelled P and L, which cannot occur.

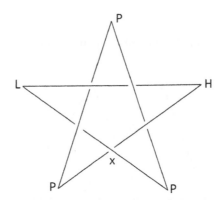

FIGURE 84.2: A five-stick projection that cannot be realized in 3-space with straight sticks.

One of the first papers investigating the stick number was [27], which appeared in 1994. There, Randell determined the stick number for a few small crossing knots and discussed how certain invariants are impacted for low-stick-number knots.

In general, determining stick numbers for knots can be difficult. If we construct a knot K from n sticks, we certainly know $s(K) \leq n$, but obtaining a lower bound is almost always the tricky part.

An obvious fundamental question is how the stick number behaves under composition. We do know the answer in the case of the composition of a collection of n trefoil knots, denoted nT.

Theorem 84.1.1. *([1])* $s(nT) = 2n + 4$.

This is somewhat surprising. Although the first trefoil costs us six sticks, each subsequent trefoil knot only requires an additional two sticks. The general case of how the stick number behaves under composition is still open. We do have the upper bound of $s(K_1 \# K_2) \leq s(K_1) + s(K_2) - 3$.

There are very few infinite families of knots for which we know the stick number. But we do know it for certain torus knots.

Theorem 84.1.2. *If $2 \leq p < q < 2p$, the stick number of the (p, q)-torus knot is $2q$.*

To show the stick number is at most $2q$, (see either [1] or [19]), we construct a model of the corresponding knot out of sticks on a pair of hyperboloids, as in Figure 84.3. A given hyperboloid is a ruled surface, so there exist two families of lines, one the reflection of the other that make up the surface of the hyperboloid. By choosing two hyperboloids that intersect in unit circles in the planes given by $z + 1$ and $z = -1$, but that have different waist sizes, we can construct a torus with crescent-shaped cross section from the two annuli on the hyperboloids between their shared circles. We can then choose line segments on each that form the sticks making up our torus knots. This gives an upper bound on the number of sticks.

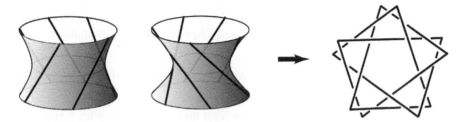

FIGURE 84.3: Constructing a $(5, 3)$-torus knot from sticks on two hyperboloids.

A proof that there is a lower bound on the stick number which matches this upper bound utilizing the total curvature of the knot works in the case of $(p, p + 1)$-torus knots(c.f. [1]). The lower bound in the more general case is achieved using the superbridge number (cf. [19]).

In [2], the stick number of the $(p, 2p+1)$-torus knots was shown to be $4p$. In [20], this was extended to show the stick number of all (p, q)-torus knots with $2p < q < 3p$ is $4p$. In [1], the exact stick number was further determined for the composition of any collection of torus knots of the form $T_{(p,p+1)}$, with p allowed to vary over the collection.

We can obtain bounds on the stick number in terms of the crossing number:

$$\frac{1}{2}(7 + \sqrt{8c(K) + 1}) \le s(K) \le \frac{3}{2}(c(K) + 1).$$

Negami first determined lower and upper bounds in [26]. His lower bound was improved to this one by Calvo ([6]) and his upper bound was improved to this one by Huh and Oh[14]. Note that both bounds are tight for the trefoil knot.

Various authors have found upper bounds on stick numbers for 2-bridge knots and links. In particular, McCabe([25]) showed $s(K) \le c(K)+3$ for all 2-bridge knots and links other than the unlink and Hopf link. Huh, No and Oh ([12]) improved this to $s(K) \le c(K) + 2$ for all 2-bridge knots and links of crossing number at least 6. In [17], the authors further improved this bound for many 2-bridge links, including all that have eight or more crossings per integer tangle. Moreover, they showed that for many 2-bridge knots, including for instance every two-braid with 15 or more crossings and every 2-bridge knot with at least 11 crossings per integer tangle, the minimal stick configuration does not correspond to a minimal crossing projection.

84.2   Other Stick-Type Invariants

Here we list some additional stick-type invariants. In all cases, the sticks are glued end-to-end to create the knot.

1. The equal-length stick number $s_=(K)$ is the least number of equal-length sticks to make the knot K.

2. The lattice stick number $s_L(K)$ is the least number of sticks, each parallel to the x, y and z axes, to construct a knot.

3. The unit-length lattice stick number $e_L(K)$ is the least number of unit-length sticks parallel to the x, y and z axes to make the knot.

4. The unit sphere stick number $s_u(K)$ is the least number of sticks with vertices on the unit sphere to construct the knot.

5. The cube number $C_L(K)$ is the least number of unit cubes glued face to face with centers in the cubic lattice to create a given knot as the core curve of the resultant solid torus.

Question: Is $s_=(K) = s(K)$ for all K?

This is a very natural question. In [28], algorithms from the software KnotPlot were utilized to determine upper bounds for the stick number for prime knots of ten or fewer crossings. The authors were able to show that for all but seven of these knots, $s_=(K) = s(K)$. But so far, none of the seven knots has proved to be an example where the equilateral stick number differs from the stick number.

In [21], the authors obtained the general upper bound $s_=(K) \leq 2c(K) + 2$ for all nontrivial knots. Further, if K is non-alternating and prime, $s_=(K) \leq 2c(K) - 2$.

The lattice stick number was first investigated in [18]. There, the authors proved that $s_L(K) \geq 6b(K)$, where $b(K)$ is the bridge number, and $s_L(K) \geq 12$ for all nontrivial knots. Moreover, $s_L(K_1 \# K_2) \leq s_L(K_1) + s_L(K_2)$.

In [13] and [15], it was proved that $s_L(3_1) = 12$, $s_L(4_1) = 14$ and otherwise $s_L(K) \geq 15$. In [10], it was further proved that except for 3_1 and 4_1, $s_L(K) \geq 16$ and $s_L(5_1) = s_L(5_2) = 16$. In [4], it was shown that $s_L(8_{20}) = s_L(8_{21}) = s_L(9_{46}) = 18$. Moreover, the lattice stick number for certain torus knots was determined.

Theorem 84.2.1. *For $p \geq 2$, $s_L(T_{(p,p+1)}) = 6p$.*

In Figure 84.4, we see the stick conformations of the $(p, p+1)$-torus knots. The lattice stick numbers of certain satellites and compositions of these links can also be determined.

There are bounds in terms of the crossing number:

$$3\sqrt{c(K) + 1} + 3 \leq s_L(K) \leq 3c(K) + 2.$$

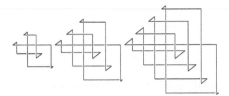

FIGURE 84.4: Minimal lattice stick conformations for the $(p, p + 1)$-torus knots.

The first bound comes from [8] and the second from [11]. In the case that the knot is prime and non-alternating, the second bound can be improved to $s_L(K) \leq 3c(K) - 4$. The same authors also proved that $s_L(K_1 \# K_2) \leq 2c(K_1) + 2c(K_2)$.

We now turn to the unit-length lattice stick number $e_L(K)$, which is clearly always as large as the lattice stick number. In [7], it was proved that $e_L(K) \geq 24$ and $e_L(K) = 24$ only for the trefoil knot. This was extended to show $e_L(4_1) = 30$ and $e_L(5_1) = 34$ in [29].

In [8], the authors proved that there exists a family of knots $\{K_n\}$ in the cubic lattice such that $e_L(K_n)/c(K_n)$ approaches 0 as $e_L(K_n)$ approaches infinity. Moreover for any knot in the cubic lattice, $c(K) \leq 3.2(e_L(K))^{4/3}$. In [16], it was proved that for all nontrivial 2-bridge knots, $e_L(K) \leq 8c(K)+2$. In [23], the authors prove that the cube number $C_L(K) = 2e_L(K)$ for all knots K. Also see [5, 22, 24] for results on stick numbers of knots in the hexagonal and other lattices.

Not much is known about the unit sphere stick number, however, if one treats the unit sphere as the Klein model of hyperbolic 3-space and then uses the fact that the relative positions of the edges in that model are preserved when you translate to the Poincaré ball model, one sees that the unit sphere stick number is equivalent to the least number of geodesics in $\mathbb{H}^3$ subsequently sharing ideal vertices, to construct the knot. This also proves that every knot and link can be represented as a set of sticks with vertices on the unit sphere, by drawing a projection on the boundary plane of the upper-half-space model of $\mathbb{H}^3$, and then using semi-circular geodesics sharing endpoints to mimic the projection and respect the crossings.

Upper bounds on the stick number and the equilateral stick number for spatial graphs (see Chapter 49) appear in [9].

84.3 Projection Stick Indices

There are two possible polygonal invariants one may want to investigate with regard to projections. There is the projection stick index $prs(K)$, which is the least number of sticks of any projection of a 3-dimensional stick conformation of the knot, each stick in the projection corresponding to a unique stick in the 3-dimensional realization. And there is the planar stick index $pls(K)$, which is the least number of sticks in the plane of a polygonal configuration such that when crossings are added in, the knot can be realized. Note that $pls(K) \leq prs(K)$. In many cases these are equal, but the expectation is that they can be quite different, although no explicit example is known.

Given that if we have a stick conformation of a knot in 3-space with $s(K)$ sticks, we can always project down one stick to see $s(K) - 1$ sticks. So

$$pls(K) \leq prs(K) \leq s(K) - 1.$$

For the trefoil knot T, these bounds are realized, since $pls(T) = prs(T) = 5 = s(K) - 1$. In the case of the 5_1 knot, as in Figure 84.2, $pls(K) = prs(K) = 5$ and $s(K) = 8$.

Note that $s(K) \leq 2prs(K)$, since a diagram realizing $prs(K)$ comes from projecting a stick conformation in 3-space, and there can be at most one stick in the 3-dimensional conformation that projects to each vertex in the planar projection. Eric Rawdon asked the question as to whether there are any examples when $s(K) = 2prs(K)$.

In [2], it is proved that $2b(K) < pls(K) \leq prs(K)$. In the case of certain torus knots, we know the planar stick index. For $p \geq 2$, $pls(T_{p,p+1}) = pls(T_{p,2p+1}) = 2p + 1$. This is also the projection stick indices.

As mentioned previously, for the $(p, 2p + 1)$-torus knots, the stick number is $4p$ and the projection stick number is $2p + 1$. So although this is within 1 of realizing $s(K) = 2prs(K)$, there has yet to be an example where the equation holds. Using results of [20], it is the same for the $(p, 2p + 2)$-torus knots.

On the other hand, if again, nT is the composition of n trefoils, $pls(nT) = 2n + 3$ from [3], whereas as mentioned previously, $s(nT) = 2n + 4$. Hence, $prs(nT) = 2n + 3$ and these are examples of knots where the minimum difference between $s(K)$ and $prs(K)$ is realized.

The *spherical stick number* of a knot K, denoted $ss(K)$, is the least number of great circle arcs making up a projection of the knot to the unit sphere. Somewhat surprisingly, it behaves quite differently from the planar stick number.

In [3] it was proved that for $p \geq 2$, $ss(T_{(p,p+1)}) = p + 1$.

Moreover, if T_L is the left-handed trefoil and T_R is the right-handed trefoil, then $ss[T_L \# T_R] \neq ss[T_R \# T_R] = 5$ (see Figure 84.5). Further, $ss[T_L \# T_L] = ss[T_L \# T_L \# T_R] = 5$. This is a very unusual characteristic for a naturally defined physical knot invariant: there exist nontrivial knots K_1 and K_2 such that $ss[K_1 \# K_2] = ss[K_1]$.

Note that by radially projecting the projection plane onto a projection sphere, we see $pls[K] \leq ss[K]$. Various other inequalities relating $ss[K]$ to other invariants appear in [3].

84.4   Bibliography

[1] C. Adams, B. Brennan, D. Greilsheimer, A. Woo, Stick numbers and composition of knots and links, *J. Knot Theory Ramifications*, 6 (2)(1997)149–161.

[2] C. Adams and T. Shayler, The projection stick index of knots, *J. Knot Thy. Ram.*, Vol. 18, Issue 7 (2009) 889-899.

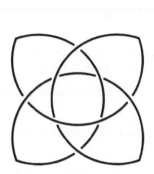

FIGURE 84.5: The least number of great circle arcs to construct the square knot $T_L \# T_R$ and the granny knot $T_R \# T_R$.

[3] C. Adams, D. Collins, K. Hawkins, C. Sia, R. Silversmith, B. Tshishiku, Planar and spherical stick indices of knots, *J. Knot Theory Ramifications* 20 (2011), no. 5, 721–739.

[4] C. Adams, M. Chu, T. Crawford, S.Jensen, K. Siegel, L. Zhang, Stick index of knots and links in the cubic lattice *J. Knot Theory Ramifications* 21 (2012), no. 5, 1250041, 16 pp.

[5] R. Bailey, H. Chaumont, M. Dennis, E. McMahon, J. McLoud-Mann, S. Melvin, G. Schuette, Simple hexagonal lattice stick numbers, *Involve*, Vol. 8, (2015) 503-512.

[6] J. Calvo, Jorge Alberto (2001), Geometric knot spaces and polygonal isotopy, *Jour. Knot Theory Ramifications*, 10 (2): 245–267

[7] Y. Diao, Minimal knotted polygons on the cubic lattice, *Jour. Knot Theory Ramifications*, 02 413, (1993).

[8] Y. Diao, C. Ernst, Total curvature, ropelength and crossing number of thick knots, *Math. Proc. Cambridge Philos. Soc.* 143 (2007), no. 1, 41–55.

[9] M. Lee, S. No, S. Oh, Stick number of spatial graphs, *Jour. of Knot Theory and Its Ramifications*, Vol. 26, No. 14, (2017)1750100(9 pages).

[10] Y. Huang and W. Yang, Lattice stick number of knots, *Jour. of Physics A: Math. Theor.* 50 (2017)50524 (21pp).

[11] K. Hong, S. No, and S. Oh, Upper bound on lattice stick number of knots, *Math. Proc. Cambridge Philos. Soc.* 155:1 (2013), 173–179.

[12] Y. Huh, S. No, and S. Oh, Stick Numbers of 2-bridge Knots and Links, *Proc. Amer. Math. Soc.* 139(11) (2011) 4143–4152.

[13] Y. Huh and S. Oh, Lattice stick numbers of small knots, *J. Knot Theory Ramifications* 14:7 (2005), 859–867.

[14] Y. Huh, S. Oh, An upper bound on stick number of knots, *J. Knot Theory Ramifications*, 20 (5): (2011) 741–747.

[15] Y. Huh and S. Oh, Knots with small lattice stick numbers, Jour. of Physics: Math,. and Theor., Vo. 43 No. 26 (2010).

[16] Y. Huh, K. Hong, H. Kim, S. No, S. Oh, Minimum lattice length and ropelength of 2-bridge knots and links, *Journal of Mathematical Physics* 55, 113503 (2014).

[17] E. Insko and R. Trapp, Supercoiled tangles and stick numbers of 2-bridge links, *J. Knot Theory Ramifications* 24, 1550029 (2015)

[18] E.J. Janse van Rensberg, SD Promislow, The curvature of lattice knots, *J. Knot Theory Ramifications* 8 (1999), no. 4, 463–490.

[19] G. T. Jin, Polygon indices and superbridge indices of torus knots and links, *J. of Knot Theory and its Ramifications*, 6 (2) (1997) 281–289

[20] M. Johnson, S. N. Mills, R. Trapp, Stick and Ramsey numbers of torus links, *J. Knot Theory Ramifications*, 22(7) (2013) 1350027.

[21] H. Kim, S. No, S. Oh Equilateral stick number of knots, *J. Knot Theory Ramifications*, 23 (2014) 1460008.

[22] C. Mann, J. McLoud-Mann, D. Milan, Stick numbers in the simple hexagonal lattice, *J. Knot Theory Ramifications*, Vol. 21, No. 14, (2012)15 pages.

[23] C. Mann, J. McLoud-Mann, On the relationship between minimal lattice knots and minimal cube knots, *J. Knot Theory Ramifications*, Vol. 14, No. 7, 841-851.

[24] C. Mann, B. McCarty, J. McLoud-Mann, R. Ranalli, N. Smith, *Minimal knotting numbers*, *J. Knot Thy. Ram.*, Vol. 18, No. 8, (2009)1159-1173.

[25] C. McCabe, An upper bound on edge numbers of 2-bridge knots and links, *J. Knot Theory Ramifications* 7 (1998), no. 6, 797–805.

[26] S. Negami, Ramsey theorems for knots, links and spatial graphs, *Trans. AMS*, 324 (2): (1991) 527–5

[27] R. Randell, An elementary invariant of knots, *J. Knot Theory Ramifications*. 3(1994) 279-286.

[28] E. Rawdon, R. Scharein Upper bounds for equilateral stick number, *Contemp. Math.* 304, Amer. Math. Soc., Providence, RI, 2002.

[29] R. Scharein, K. Ishihara, J. Arsuaga, Y. Diao, K. Shimokawa, M. Vazquez, Bounds for the minimum step number of knots in the simple cubic lattice *J. Phys. A* 42 (2009), no. 47, 475006, 24 pp.

Chapter 85

Random Knots

Kenneth Millett, University of California, Santa Barbara

85.1 Introduction

"Random knots" is more about a process than it is about a single knot in the sense that no single knot is "random." Rather, a large sample of elements of a collection of knots with a specified probability distribution is required in order to determine, with some measure of confidence, if the sample was randomly selected. In the case of knots, this is a rather more complex question than it might first appear. Traditionally, mathematical knots are understood as nice realizations of circles in Euclidean three-dimensional space where "nice" is intended to imply that certain pathologies have been avoided. Examples of such assumptions ensuring this are having a smooth parameterization of the curve with nowhere zero derivative, being defined by a piecewise linear function, or being an equilateral spatial polygon, as in Figure 85.1.

Perhaps the first interest in random knotting arose from the desire to express the potential spatial complexity of ring macromolecules occurring in chemistry. In 1961, Frisch and Wasserman [29] stated, "We guess that for a C_{60} chain, the probability that the formed ring would exist as a knot is about $10^{-3} - 10^{-2}$." They go on to say, "We conjecture that the probability that a cyclizing chain will produce a knot (moreover an optically active knot) approaches unity as the length of the chain increases without limit." At about the same time, Delbruck [19] considered cubic lattice models for simple macromolecular chains and asked: "What fraction of the permissible configurations of a chain of given length will contain a knot?" Inspired by these physical examples, one is led to expand the consideration to include knotting in open molecular chains or mathematical arcs, Figure 85.2.

FIGURE 85.1: A smooth trefoil and an equilateral hexagonal spatial trefoil [32].

It is quite remarkable that, without mastery of the theory of random knots in 3-space, it has been possible to prove these conjectures, specifically that the probability that a spatial chain or polygon contains a trefoil knot goes to one as the length goes to infinity [59, 53, 23, 22]. Using these same ideas, one can also prove that the probability of a trefoil slipknot, i.e. a knotted subchain contained in an unknotted chain, also goes to one as the length of the chain goes to infinity [48].

While natural and important, these and related questions proved to be very difficult to resolve. In the first instance, one is faced with a continuous variation of configurations in 3-space. In the second, a discrete family of configurations in the 3-dimensional cubic lattice, one ultimately encounters many of the same issues as the length increases due to the exponential growth in the size of the collection. Furthermore, ultimately, one must account for the physical properties of the macromolecules that are reflected in their geometry and excluded volume or thickness. As a consequence, the first efforts to explore these questions focused on mathematical models for self-avoiding walks, i.e. open chains, and polygons in three-dimensional Euclidean space. To illustrate the special facets of the spatial nature of the problem, we consider one of the classic Lewis Caroll Pillow Problems [11].

Three Points are taken at random on an infinite Plane. Find the chance of their being the vertices of an obtuse-angled Triangle.

Lewis Carroll's 58th Pillow Problem, 1884

That something is not quite right with Carroll's problem quickly becomes apparent when one is confronted with two versions of Carroll's method giving two equally compelling, but different answers. Indeed, it is only recently that new insights as to the natural structure on the space of triangles has provided the rigorously determined answer of $\frac{3}{2} - \frac{3ln2}{\pi}$ in Cantarella et al. [9] where one finds further references to the history of the problem. The critical issue stems from the lack of an appropriate probability measure on the continuous collection, or space, of triangles in the plane. Analogously, one is confronted with the same problem with respect to smooth or polygonal representatives of knots in 3-space. Delbruck avoids this facet of the problem by considering the finite discrete set of closed chains in the cubic lattice of a given length, up to lattice symmetry. Alas, the number of such chains grows so rapidly, with increasing number of edges, that one is

FIGURE 85.2: An open trefoil.

soon effectively confronted with insurmountable counting problems and must resort to statistical methods as is the case in the continuous three-dimensional space studies.

To begin close in spirit to the historical Frisch-Wasserman-Delbruck question, we will first discuss the efforts to understand "random polygonal knots" in 3-space. Next, we will discuss the efforts to discretize the question by studying polygonal chains in the cubic lattice. Following these, we will consider the efforts to make the problem even more discrete by employing various combinatorial representations of knots and recalling the earliest work to enumerate irreducible, or prime, knots. Finally, we will discuss outstanding issues arising from efforts to estimate average physical characteristics of macromolecules, problems related to representations of knots, and the challenges arising from the desire to rigorously study links and mathematical models of physical materials such as polymer gels.

85.2 Open and Closed Equilateral Chains

In order to determine the proportion of chains that are knotted, one must be able to quantify the likelihood that a specified collection of members of the population is observed. In the case of open or closed three-space chains, this requires an appropriate continuous probability density function with which one can determine the fraction of the members in a given subset compared to those of the entire set. Thus, one must require a density function for which the entire population has finite mass (in probability theory this is taken to be one). The continuum of n edge polygonal chains makes this difficult for closed chains, though easier for open chains as they are not subject to the closure constraint.

85.2.1 Spaces of Open and Closed Equilateral Chains

Open equilateral chains have been of interest since the 1800s when they arose implicitly in the analysis of Rayleigh [55] and, later, in the notion of a random walk in the work of Pearson [52]. Only interested in the distribution of ends of an n-step random walk, whereby one begins at the origin and takes a step of unit length in a direction selected without bias, the question of knotting of the trajectory did not arise. Note that each step of the random walk corresponds to the random choice of a unit vector in 3-space, i.e. a point on the 2-dimensional unit sphere. Thus, an n-step random walk corresponds to the selection of a point in the product of n unit spheres. To describe an open chain in 3-space, one requires that the union of edges corresponding to the step vectors be non-singular, i.e. non-intersecting, thereby removing a subset, called the discriminant, from this product and leaving an open 2n-dimensional subspace of full volume, Arm(n). The challenge of identifying knotting in an open chain was apparent to Frisch-Wasserman-Delbruck, resurfacing in the polymer studies of Edwards [25], and, later, in the protein analysis of Taylor [61]. Traditional mathematical knot theory requires closed chains in order to ensure that any knotted structure is preserved under spatial deformation [56]. Identifying the subspaces of Arm(n) consisting of chains of given knot types remains very difficult. Even now, one best employs the statistics of the probability distribution knot types arising from closures of the ends of the chain to "points at infinity" determined by the 2-sphere of directions centered at the center of mass of the chain [50] to identify the dominant knot type and thereby, assign a knot type to an open chain.

For closed chains, one needs only require that the sum of the n vectors is the 0 vector corresponding to ensure closure of the chain. Thus, the space of closed equilateral random chains is a 2n-6 dimensional subspace, Pol(n), given by polygons based at the origin modulo the action of rotation by SO(3). The identification of the subsets of given knot types is, even here, a challenging problem for which one has very limited geometric or topological information, even for small numbers of edges [5, 6, 32]. As a consequence, it has been necessary to employ statistical methods in order to estimate many of the averages of properties of open and closed polygonal chains or to quantify the extent of knotting among these open or closed polygonal chains. While one can easily randomly sample the space of open chains, i.e. random walks, in 3-space, it was only recently that symplectic geometric structures have been employed to determine average characteristics of closed chains in 3-space [10, 8] based on the fundamental construction of Hausmann and Knutson [33]. One takes any (n-3)-tuple of non-intersecting diagonals giving a triangulation of the standard regular n-gon, see Figure 85.4. Their 3-space lengths define the "action" coordinates of the spatial polygon which, together with the (n-3) folding angles at the common edge of the pairs of adjacent triangles that define the "angle" coordinates of the configuration. As shown in Figure 85.4, these determine the spatial polygon up to 3-space rigid orientation-preserving symmetries provided by SO(3). By uniformly sampling the polyhedron of allowable diagonal lengths, as constrained by the triangle inequalities, and uniformly sampling the angle coordinates, one is uniformly sampling Pol(n) and, thereby, randomly selecting a knot, i.e. a "random knot." Note that selecting a different set of diagonals defines a different set of action-angle coordinates and different symplectic parameterization of the configuration space.

Building on the ability to create random walks of a specified number of steps and knowing the associated endpoint probability distribution function, Diao et al. [21] have developed a method for randomly selecting closed polygons, even under confinement.

Thus, one is now able to rigorously randomly sample the space of equilateral polygons [8, 10, 32] using the "action-angle" coordinates by sampling the action polytope with the uniform measure and the angle coordinates, again with the uniform measure, or employing methods derived from knowledge of random walks [21]. There are, however, additional features that are important for applications to the biological and physical sciences. Most important among these is the thickness

FIGURE 85.3: A 100-step open chain starting at the origin (red dot).

of the conformation reflecting the physical volume associated to the macromolecule being studied. Employing an appropriate realization of the largest physical tube of radius "r" centered on the core polygon [51] to describe polygon's excluded volume, one wants to randomly sample the space of equilateral paths or closed polygons having thickness "r" or larger. While none of the previously discussed methods provide effective sampling methods when one adds this additional thickness constraint, Chapman and Plunkett have developed new methods inspired by the strategies employed for lattice knots, discussed in the next section, and have rigorously proved that these do provide a random sampling method [16, 54]. A central facet of their approach is to employ "spatial reflections," an example of lattice symmetry, to modify conformations of open or closed equilateral polygons in a manner that allows them to simultaneously simplify the conformation while ensuring that thickness is not reduced, Figure 85.5.

In the last 50 years, there have been quite a few efforts to randomly select n edge polygons that could serve as models of macromolecular structures in biophysics and polymer physics. Examples, discussed later, include the 1988 "hedgehog method" of Klenin et al. [40], the "polygonal fold" method of Millett [47], the "crankshaft" method of Alvardo et al. [4], and analogs of the more physical Kratky-Porod model [31, 41]. While these methods have been observed to produce collections with the expected statistics for the radius of gyration and are widely believed to randomly

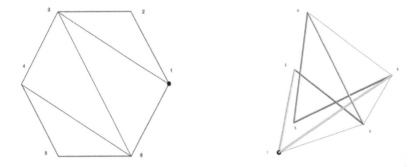

FIGURE 85.4: The action-angle coordinates realizing the symplectic representation of a spatial hexagonal trefoil [32].

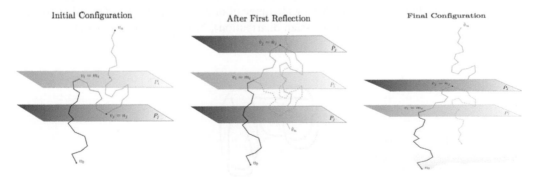

FIGURE 85.5: An example of spatial reflections employed to simplify equilateral configurations without reducing thickness [16, 54].

sample these configurations, there has not yet been a rigorous mathematical proof that they actually do so. In my opinion, this remains as one of the most important open conjectures in this area.

85.3 Identification of Knots in 3-Space

Implicit in the discussion of "Random Knots" is the challenge of identifying the knot type of a randomly selected spatial open or closed curve, either smooth or polygonal. As discussed earlier, this has been a goal from the very start of the mathematical study of knotting. A key in this effort is the fact that the topological type of every closed curve can be uniquely expressed as the connected sum, see Figure 85.6, of indecomposable, or prime, knots [2]. The identification of the topological knot type of a configuration is typically accomplished by the determination of one or more numbers or algebraic expressions, for example Vassiliev [62] invariants or the Alexander [3], Jones [37], or HOMFLY [28, 43] polynomials. These provide a sufficiently fine sieve to separate the knot types in the collection of conformations that the research project concerns as long as the expected minimal crossing number is not too large. Although, in the worst of cases, the calculation of the Jones or HOMFLY polynomials is computationally difficult, for the analysis of a collection of randomly generated configurations, these tools prove to be computationally effective and, most importantly, very efficient. In unusual cases, where one must achieve a much finer analysis, one must employ other methods such as representations of the fundamental group of the complement of the knot. Even for small composite knots, one can employ the fact that a knot polynomial of the composite is the product of the knot polynomials of its irreducible components or use the *knotfind* feature of the Hoste-Thistlethwaite *KnotScape*: https: //www.math.utk.edu/morwen/knotscape.html.

85.4 Some Classes of Knots in 3-Space

A good place to start a study of knots in 3-space is with equilateral polygons or with simple cubic lattice polygons as they closely resemble a wide range of other models, such as Gaussian polygons (in which the edge lengths satisfy a Gaussian distribution), ball-spring or rod models, or models based upon other spatial lattices.

FIGURE 85.6: Connected sum of prime knots.

85.4.1 Equilateral Polygonal Knots is 3-Space

Although, as discussed earlier, one has long known how to randomly generate open equilateral polygons, known as random walks, in three space, it has only been very recently that we have rigorously proven methods of randomly generating closed polygonal chains using the symplectic geometry structure or, if one constrains thickness, the Chapman-Plunkett reflection methods. Historically, this generation goal has been the subject of important efforts and, even though they have not yet been rigorously proven to be random, they are in such common use that it is important to be aware of them. The first of these was the *crankshaft* method applied to a *hedgehog* initial configuration to try to generate a "random" configuration [4, 40]. A hedgehog configuration is created by randomly selecting "n" unit vectors in 3-space and then randomly permuting them together with their negatives to create the edge vectors of a closed polygon, closed since the sum of the edge vectors is zero, see Figure 85.7. The crankshaft method consists of a large sequence of perturbations each of which is accomplished by randomly selecting two edge vectors and replacing them by the result of randomly rotating the triangle they describe about its base. The effect on the polygon resembles the motion of a crankshaft, thus its name, see Figure 85.8. The basic *polygonal fold* perturbation [47] is applied to a conformation by randomly selecting two vertices and, using the axis they determine, randomly replacing one of the determined arcs by its random rotation about the axis determined by these vertices, see Figure 85.8. In each of these methods, the basic perturbations are applied to a conformation until an independent conformation is attained. This is sampled, and the basic perturbations are then again applied in a procedure that continues until a sample of the desired size in the knot space is attained. While, to date, one does not have a rigorous proof that these methods provide a random sample of knot space, certain quantities of the collection of conformations such as the average radius of gyration or the average crossing number seem to have the theoretically predicted character consistent with being a random sample [4].

85.4.2 Knots in Lattices

Arcs and closed polygons in a lattice are a very important area of research due to the combinatorial richness of the structure. They provide a platform on which one can achieve rigorous proofs despite the rapid increase in complexity that occurs with increasing the number of edges. This is the reason

 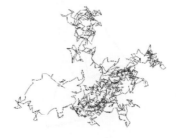

FIGURE 85.7: A hedgehog and associated hedgehog knot.

for the Monte Carlo efforts used to estimate structural characteristics. For example, in the case of the simple cubic lattice on which we focus here, one needs at least 24 edges [20] in order to achieve a closed path forming a trefoil knot and only a few less for an open trefoil (by essentially the same construction, Figure 85.9). Although one knows that there are finitely many self-avoiding polygons (SAPs) with n edges, determining exactly how many as a function of n is an exceedingly difficult task especially when one must determine, as well, exactly how many there are of each knot type. Thus, assuming each SAP has equal probability, one needs a method to randomly select SAPs of n edges in order to estimate the fraction of SAPs of a given knot type. For example there are $3,328$ lattice trefoils with exactly 24 edges [57] among the $512,705,615,350$ SAPs [17]. To randomly sample this finite set of n edge SAPs, one randomly picks two of a SAP's vertices and replaces one half of the SAP with the result of applying a lattice symmetry to this half when the union of the new segment and the remaining half is a SAP [42, 46]. One of the early results, based on the Keston pattern theorem [39], is the proof that the probability that a SAP contains a trefoil summand goes to one as the number of edges goes to infinity [53, 59]. This result was later extended to equilateral knots in 3-space [22]. These methods were then extended to give the same result for slipknots in the lattice and 3-space [49].

85.5   Random Knot Presentations

To overcome some of the geometric challenges we encounter in the study of continuous knot spaces, we can study knots via their projections using combinatorial or algebraic presentations of knots derived from an appropriate projection of the spatial knot to a 2-dimensional plane. In some approaches, the shadow of a projection has been chosen so as to describe a planar curve with helpful properties, i.e. no tangencies, only double point crossings (no triple or larger crossing points) or their polygonal equivalents in the case of polygonal knots in 3-space. One can encode the under/over crossing at a double point in the shadow by modifying the shadow to indicate which

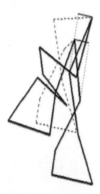   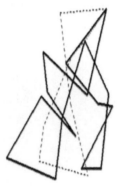

FIGURE 85.8: Similar stages of the crankshaft move, on right, and the polygonal fold move, on right [4].

strand goes over the other. See Figure 85.6 for an example of such presentations. The study of knots through the analysis of knot diagrams began at the beginning of the mathematical study of knotting and linking [30, 44, 60] and continues to be an attractive and powerful method of research.

85.5.1 Generic Knot Presentations

In Figure 85.10, we show the image of a projection of the 8_{20} knot determining a planar graph in which the crossings are viewed as vertices of the graph and, to its right, the traditional knot presentation in which the shadow is broken to indicate under- and over-crossing information. One combinatorial approach to the study of knots is to study the finite collection of such knot presentations with n crossing, up to planar or spherical regular deformation, positing that each is equally likely. In a recent exploration using *random knot diagrams*, Cantarella, Chapman and Mastin [7] randomly select a diagram from the finite collection of n crossing knot diagrams (they enumerate all such with 10 or fewer crossings) and determine the associated knot types. They show that roughly 78% of the approximately 1.6 billion diagrams present the unknot. Their data show, for example, that about 12.6% of the time, a random 8-crossing knot presentation is a trefoil knot. In some cases, one can prove results that are still unknown for the spatial knots. One example of this is Chapman's proof that almost all unknot diagrams contain a slipknot [14]. Further results of this spirit concerning knot diagrams are found in [12, 13, 15].

85.5.2 Gauss and Dowker-Thistlethwaite Code Presentation

Mathematicians have used symbolic knot codes to represent knot shadows since the time of Gauss [30]. A refined version of his code was developed and employed by Dowker-Thistlethwaite to enable the computer-assisted enumeration of irreducible, or prime, knots in 3-space [24]. The DT code is created by following the projected n-crossing curve and labeling each crossing in the order encountered but with a minus sign, if it is an even number crossing associated to an over crossing. Since each crossing has an odd and an even label, one lists the even labels in the order given by the odd entry, see Figure 85.11. Since there are $n!$ sequences of the even natural numbers $2, 4, ..2n$ and 2^n choices of signs for each, one can randomly generate a potential DT-code among the $2^n n!$

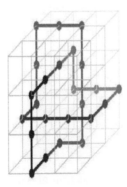

FIGURE 85.9: Cubic lattice containing a minimal trefoil [20].

choices and, of course, check to see if a given code is actually associated to an irreducible knot diagram. For $n = 8$, one must analyze $10, 321, 920$ possible codes. In general, this is a computationally difficult undertaking as the code may not actually represent an n-crossing knot so that the exact enumeration of all n crossing irreducible knots has only been achieved for small values of n [34, 35]. Nevertheless, one could estimate the probability that an n crossing presentation represents a specific knot type using a Monte Carlo method, even though the knot type identification problem may still be exceedingly difficult.

85.5.3 Braid Representation

Braids can also be used to represent knots, a strategy that has both topological and algebraic elements in its creation. A braid on n strands is a construction in which one has, in this description, n strands beginning at a vertical collection of n points on the left and proceeding monotonically to the right, crossing over and under adjacent strands, and ending at the corresponding n points on the right as shown in Figure 85.12. As a consequence, there are two parameters giving rise to the complexity of the representation: (1) the number of strands in the braid, called the *braid index n*, and (2) the number of crossings, called the *length* of the braid. In order to ensure a finite collection of configurations from which to randomly sample, one bounds the braid index, n, and the length of the braid, i.e. the number of crossings, m. This provides a bound of $(2(n-1))^m$ for the number of distinct configurations of uniform probability. For example, with $n = 3$ and $m = 8$, one has $65, 536$ configurations from which to randomly select. Some characteristics of random braids have been studied in [45, 38].

85.5.4 Arc or Grid Diagram Presentations and Codes

Still another representation of knots is described by arc or grid diagrams [18, 26, 36, 58] in which all the arcs are either vertical or horizontal. All the vertical arcs are above the horizontal arcs, all the horizontal arcs have distinct "y" coordinates, and all the vertical arcs have distinct "x" coordinates,

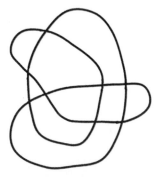

FIGURE 85.10: The shadow of a knot projection giving a planar graph in which crossings become vertices and the associated knot presentation with the shadow broken to indicate under/over crossings.

see Figure 85.13. One may choose to let the "x" and "y" coordinates be the integers $1, 2, ...k$. In this setting, an arc diagram can be encoded by a sequence of pairs of integers $1, 2, ..., k$, with each integer appearing exactly once as an initial integer and exactly once as a final integer, encoded by starting at the bottom and giving the initial and terminal point of the horizontal arc. Given all horizontal arcs, one connects the vertical over arcs between the corresponding horizontal arc termini lying in the same vertical position, as in Figure 85.13. The total number of such conformations for a k-by-k grid is $(k!)^2$.

85.5.5 Petal Diagram Representation and Codes

Yet another representation of knots is given by the petal diagram [1] in which all the crossings are vertically aligned giving a projection that resembles a flower with an odd number of petals, as in Figure 85.14. An odd number, n, of horizontal segments are placed with centers on an axis with equally spaced angles at levels given by a coding permutation. They are connected by curves realizing the given permutation of the n levels. Employing the number of petals as the measure of complexity of a representation, a random presentation is selected by taking a random permutation

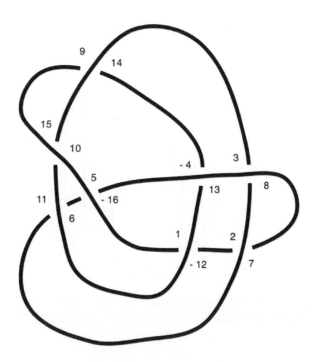

FIGURE 85.11: Generating a Dowker-Thistlethwaite code of an eight-crossing knot presentation [24] gives $-12\ 8\ -16\ 2\ 14\ 6\ -4\ 10$.

on that number of levels. This provides a method to select a random petal presentation that enables one to study the associated properties [26, 27] of such petal diagrams.

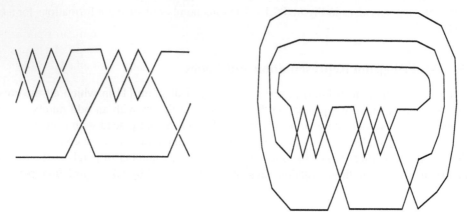

FIGURE 85.12: Generating the braid representation of the knot via the braid, index = 3 and length = 8, and its closure defining a knot (or link, depending on the braid).

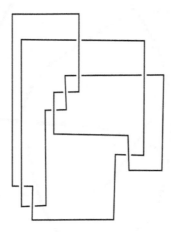

FIGURE 85.13: The arc representation of the knot corresponding to the code $(3, 8), (2, 4), (1, 3), (11, 9), (8, 10), (9, 5), (4, 6), (5, 7), (6, 11), (10, 2), (7, 1)$.

85.6 Some Concluding Thoughts

The effort to discover efficient algorithms that can be used to randomly select knots within populations of mathematical importance or in the natural sciences, such as biological macromolecules or polymer gels, has seen important progress in 3-space with the description and rigorous mathematical proof of randomness for the symplectic geometry method [10], the reflection methods [16, 54], and the lattice methods [46]. Nevertheless, many of the popular methods are believed to be random based solely on comparison of the statistics they generate with asymptotic theoretical measures. Rigorous proofs that they randomly select conformations are needed to provide a sound foundation supporting the conclusions found in a rather extensive literature. The methods required to provide such proofs would very likely lead to even stronger sampling methods and powerful new applications. In addition, the creation of new rigorously proven methods to randomly sample conformations with prescribed curvature, torsion, and thickness so as to better simulate macromolecules, e.g. worm-like chains, is also an important objective for research. Furthermore, in this direction, one has the goal of extending such rigorous methods to sample collections of chains that are appropriate mathematical models of, for example, polymer gels. Finally, more clearly understanding the relationships between the diverse mathematical models and the methods employed to select structures within them is an objective that would help test our knowledge across a wide range of biological and physical systems leading to stronger connection between the theoretical and experimental results.

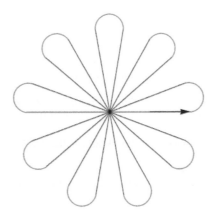

FIGURE 85.14: The petal representation of the knot 8_{20} corresponding to the code 1 3 5 8 2 7 4 9 6 [1].

85.7 Bibliography

[1] C. Adams, B. Crawford, M. Landry, A. Lin, M. Montee, S. Park, S. Venkatesh, and F. Yhee. Knot projections with a single multicrossing. *J. Knot Theory Ramifications*, 24(3):1550011, 2015.

[2] C. C. Adams. *The Knot Book*. Amer. Math. Soc., Providence, RI, 2004.

[3] J. W. Alexander. Topological invariants of knots and links. *Trans. Amer. Math. Soc.*, 30(2):275–306, 1928.

[4] S. Alvarado, J. A. Calvo, and K. C. Millett. The generation of random equilateral polygons. *J. Stat. Phys.*, 143(1):102–138, 2011.

[5] Jorge Alberto Calvo. The embedding space of hexagonal knots. *Topology Appl.*, 112(2):137–174, 2001.

[6] Jorge Alberto Calvo. Geometric knot spaces and polygonal isotopy. *J. Knot Theory Ramifications*, 10(2):245–267, 2001. Knots in Hellas '98, Vol. 2 (Delphi).

[7] J. Cantarella, H. Chapman, and M. Mastin. Knot probabilities in random diagrams. *J Phys, A: Math. Theor.*, 49:405001, 2016.

[8] J. Cantarella, B. Duplantier, C. Shonkwiler, and E. Uehara. A fast direct sampling algorithm for equilateral closed polygons. *J. Phys. A.*, 49(27):275202, 9, 2016.

[9] J. Cantarella, T. Needham, C. Shonkwiler, and G. Steward. Random triangles and polygons in the plane. arXiv:1702.01027:1–24, 2017.

[10] J. Cantarella and C. Shonkwiler. The symplectic geometry of closed equilateral random walks in 3-space. *Ann. Appl. Probab.*, 26(1):549–596, 2016.

[11] L. Carroll. *Curiosa Mathematica, Pillow Problems Thought Out during Wakeful Hours*. MacMillan, London, 1894.

[12] H. Chapman. Asymptotic laws for random knot diagrams. *J Phys, A: Math. Theor.*, 10:1751–8121, 2017.

[13] H. Chapman. On the structure and scarcity of alternating knots. arXiv:1804.09780 [math.GT], 2018.

[14] H. Chapman. Slipknotting in the knot diagram model. ArXiv 1803.07114, 2018.

[15] H. Chapman and A. Rechnitzer. A Markov chain sampler for plane curves. *ArXiv 1804.03311*, 1, 2018.

[16] K. Chapman. An ergodic algorithm for generating knots with a prescribed injectivity radius, 2019.

[17] N. Clisby, R. Liang, and G Slade. Self-avoiding walk enumeration via the lace expansion. *J. Phys. A: Math. Theor.*, 40(36):10973, 2007.

[18] Peter R. Cromwell. Embedding knots and links in an open book. I. Basic embedding knots and links in an open book. I. basic properties. *Topology and Its Applications*, 64(1):37–58, 1995.

[19] M. Delbrück. Knotting problems in biology. *Proc. Symp. Appl. Math.*, 14:55–58, 1962.

[20] Y. Diao. Minimal knotted polygons on the cubic lattice. *Journal of Knot Theory and Its Ramifications*, 02(04):413–425, 1993.

[21] Y. Diao, C. Ernst, A. Montemayor, and U. Ziegler. Generating equilateral random polygons in confinement. *Journal of Physics. A. Mathematical and Theoretical*, 44(40):1751–8113, 2011.

[22] Yuanan Diao. The knotting of equilateral polygons in $\mathbf{R}^3$. *J. Knot Theory Ramifications*, 4(2):189–196, 1995.

[23] Yuanan Diao, Nicholas Pippenger, and De Witt Sumners. On random knots. *J. Knot Theory Ramifications*, 3(3):419–429, 1994. Random knotting and linking (Vancouver, BC, 1993).

[24] C. H. Dowker and M. B. Thistlethwaite. On the classification of knots. *C. R. Math. Rep. Acad. Sci. Canada*, 4(2):129–131, 1982.

[25] S. F. Edwards. Statistical mechanics with topological constraints: I. *Proceedings of the Physical Society*, 91(3):513, 1967.

[26] C. Even-Zohar, J. Hass, N. Linial, and T. Nowik. Invariants of random knots and links. *Discrete and Computational Geometry*, 56(2):274–314, 2016.

[27] C. Even-Zohar, J. Hass, N. Linial, and T. Nowik. The distribution of knots in the Petaluma model. ArXiv, 1706,0657(v2):1–13, 2018.

[28] P. Freyd, D. Yetter, J. Hoste, W. B. R. Lickorish, K.C. Millett, and A. Ocneanu. A new polynomial invariant of knots and links. *Bull. Amer. Math. Soc. (N.S.)*, 12(2):239–246, 1985.

[29] H. L. Frisch and E. Wasserman. Chemical topology. *J. Am. Chem. Soc*, 18:3789–3795, 1961.

[30] K. F. Gauss. Zur mathematischen theorie der electrodynamischen wirkungen. *Werke vol 5 Konigl. Ges. Wiss. Gottingen*, page 605, 1877.

[31] A. Yu. Grosberg and A. R. Khokhlov. *Statistical Physics of Macromolecules*. American Institute of Physics, 1994.

[32] Kathleen Hake. *The Geometry of the Space of Knotted Polygons*. PhD thesis, University of California, Santa Barbara, 2018.

[33] J-C. Hausmann and A. Knutson. Polygon spaces and Grassmannians. *Enseign. Math.*, 63(4):173–198, 1997.

[34] J. Hoste, M. B. Thistlethwaite, and J. Weeks. The first 1,701,936 knots. *The Math. Intelligencer*, 20(4):33–48, 1998.

[35] Jim Hoste. The enumeration and classification of knots and links. In *Handbook of Knot Theory*, pages 209–232. Elsevier B. V., Amsterdam, 2005.

[36] Gyo Taek Jin and Wang Keun Park. Prime knots with arc index up to 11 and an upper bound of arc index for non-alternating knots. *Journal of Knot Theory and Its Ramifications*, 19(12):1655–1672, 2010.

[37] V. Jones. Hecke algebra representations of braid groups and link polynomials. *Ann. of Math.*, 126(2):335–388, 1987.

[38] Thomas Kerler and Yilong Wang. Random walk invariants of string links from R-matrices. *Algebraic & Geometric Topology*, 16(1):569–596, 2017.

[39] Harry Kesten. On the number of self-avoiding walks. *J. Mathematical Phys.*, 4:960–969, 1963.

[40] K. V. Klenin, A. V. Vologodskii, V. V. Anshelevich, A. M. Dykhne, and M. D. Frank-Kamenetskii. Effect of excluded volume on topological properties of circular DNA. *J. Biomol. Struct. Dyn.*, 5:1175–1185, 1988.

[41] O. Kratky and G. Porod. Röntgenuntersuchung gelöster fadenmoleküle. *Rev. Trav. Chim. Pays-Bas*, 68:1106–1123, 1949.

[42] M. Lal. 'Monte Carlo' computer simulation of chain molecules. i. *Molecular Physics*, 17(1):57–64, 1969.

[43] W. B. R. Lickorish and Kenneth C. Millett. A polynomial invariant of oriented links. *Topology*, 26(1):107–141, 1987.

[44] Johann B. Listing. *Vorstudien zur Topologie*. Vandenhoeck und Ruprecht, 1848.

[45] Jiming Ma. Components of random links. *Journal of Knot Theory and Its Ramifications*, 22(8):1350043, 11, 2013.

[46] N. Madras and A. D. Sokal. The pivot algorithm: A highly efficient monte carlo method for the self-avoiding walk. *J. Stat. Phys.*, 50(1-2):109–186, 1998.

[47] Kenneth C. Millett. Knotting of regular polygons in 3-space. *J. Knot Theory Ramifications*, 3(3):263–278, 1994.

[48] Kenneth C. Millett. Knots, slipknots, and ephemeral knots in random walks and equilateral polygons. *Journal of Knot Theory and its Ramifications*, 19(5):601–615, 2010.

[49] Kenneth C. Millett. Knots, slipknots, and ephemeral knots in random walks and equilateral polygons. *J. Knot Theory Ramifications*, 19(5):601–615, 2010.

[50] Kenneth C. Millett, Akos Dobay, and Andrzej Stasiak. Linear random knots and their scaling behavior. *Macromolecules*, 38(2):601–606, 2005.

[51] Kenneth C. Millett and Eric J. Rawdon. Energy, ropelength, and other physical aspects of equilateral knots. *J. Comput. Phys.*, 186(2):426–456, 2003.

[52] K. Pearson. The problem of the random walk. *Nature*, 72:294 EP–, 1905.

[53] Nicholas Pippenger. Knots in random walks. *Discrete Appl. Math.*, 25(3):273–278, 1989.

[54] L. Plunkett and K. Chapman. Off-lattice random walks with excluded volume: A new method of generation, proof of ergodicity and numerical results. *J Phys, A: Math. Theor.*, 49(13):135203, 28, 2016.

[55] L. Rayleigh. On the problems of random vibrations, and of random flights in one, two, or three dimensions. *Phil. Mag. S.*, 37(6):321–347, 1919.

[56] K. Reidemeister. *Knotentheorie*. Springer-Verlag, Berlin-New York, 1974.

[57] R Scharein, K Ishihara, J Arsuaga, Y Diao, K Shimokawa, and M Vazquez. Bounds for the minimum step number of knots in the simple cubic lattice. *J Phys, A: Math. Theor.*, 42(46):475006, 2009.

[58] Nancy Scherich. A simplification of grid equivalence. *Involve*, 8(5):721–734, 2015.

[59] D. W. Sumners and S. G. Whittington. Knots in self-avoiding walks. *J. Phys. A*, 21(7):1689–1694, 1988.

[60] P. G. Tait. *On knots I, II, III Scientific Papers*, volume I. Cambridge University Press, 1911.

[61] William R. Taylor. A deeply knotted protein and how it might fold. *Nature*, 406:916–919, 2000.

[62] Victor A. Vassiliev. Cohomology of knot spaces. Theory of singularities and its applications. *Adv. Soviet Math.*, 1:23–69, 1990.

Chapter 86

Open Knots

Julien Dorier, Swiss Institute of Bioinformatics

Dimos Goundaroulis, Baylor College of Medicine

Eric J. Rawdon, University of St. Thomas

Andrzej Stasiak, University of Lausanne, Center for Integrative Genomics

86.1 Introduction

The goal of this chapter is to describe different techniques used to measure knotting in open curves. Note that there is no "agreed upon" definition for describing knotting in open curves. As a result, we describe the context motivating each definition and then describe some advantages and disadvantages of the different approaches.

Traditional knot theory focuses on closed curves. For closed curves, ambient isotopy separates the essence of the knotting in the curve (i.e. the topology or knot type) from the geometry of the configurations. Unfortunately, all non-self-intersecting open curves are ambient isotopic. Therefore, a different paradigm is needed to extract information about knotting in open curves.

Measuring knotting in open curves seems like a reasonable thing to do mathematically. Furthermore, there are scientific motivations. In the 1980s, knot theory started to be used as a tool to understand biological systems involving objects (like DNA and proteins) that could be modeled (coarsely) as polygons. On the other end of the length-scale spectrum, objects such as solar flares and solar wind could be entangled. So while there are closed knotted curves in nature (or curves that behave like they are closed, e.g. when both ends are anchored), oftentimes one finds open curves. In fact, there are likely many more entangled open curves in nature than closed ones.

Most techniques to measure knotting in open curves require closing the curve and then rely on knot type classifications from traditional knot theory to describe the entanglement of the open curve. However, some new approaches utilize projections of an open curve (knotoids) or closing the curve using virtual crossings (see Chapter 36). It is important to note that the study of entanglement on open curves, to date, is more of a geometric exercise than a topological one. In particular, these techniques measure the entanglement of any given fixed open configuration, rather than studying classes of configurations with the same entanglement profile (i.e. knot type).

In the following sections, we explore different techniques for measuring knotting in an open curve configuration. We begin by motivating properties we might like to have in such a definition. Following that section, we present the different techniques grouped by class: single closure techniques, probabilistic closure techniques, projection techniques, and topological techniques.

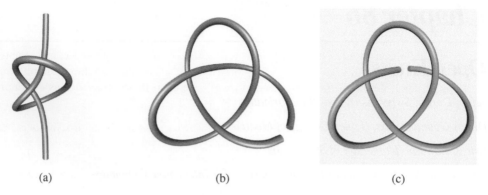

(a)   (b)   (c)

FIGURE 86.1: Three examples of open curves that many people would classify as being knotted. In (a) we show an overhand knot which, when closed in a simple fashion, becomes a trefoil knot, 3_1. In (b) and (c), small portions of a KnotPlot 3_1 configuration [21] have been removed. A reasonable classification scheme should call these knotted, although the positioning of the opening near the center in (c) may make us think that it is not as robustly knotted as (a) or (b).

86.2   Motivation

We might begin by asking what sorts of properties a measure of knotting in open curves should possess.

As a matter of everyday life, we see lots of entanglement: in headphone cables, garden hoses, climbing rigs, etc. If a person is given a piece of rope and told to tie a knot, the person typically has no problem creating something that most people would classify as being a knot. Thus, we might like the definition of open knotting to be consistent with the opinion of reasonable people, such as the authors of this chapter (this was termed the "reasonable-person test" in [15]). In particular, we are motivated by the three examples in Figure 86.1. Figure 86.1(a) is an example of what most people think of as a knot. Figures 86.1(b) and 86.1(c) are closed knots with small portions of the curve removed. These three examples all seem to be uncontroversially knotted, and the knotting appears to be trefoil-esque.

So, first, we might like the definition to assess these curves as being knotted, and to associate them with the trefoil in some fashion. We refer to this property as matching our intuition.

Second, we might like open knotting to converge to what we know about closed knots in some sense. For example, if we decrease the length of the missing pieces in Figures 86.1(b) and 86.1(c), we might hope for the open knotting classification to converge to the trefoil. Similarly, in Figure 86.1(a), as the tails get longer and longer we might hope for the open knotting classification to converge to the trefoil. We refer to this property as convergence.[1]

Third, the definition should be well-defined mathematically and able to be measured by a computer

---

[1]Millett called this property "continuity" [17].

(at least for polygons), without human interaction, in some reasonable amount of time. We refer to this property as computability.

Fourth, the definition should be continuous as a function of the curve (in some sense) and stable relative to small perturbations of the curve[2] In particular, the classification scheme should minimize the effect of small perturbations on the endpoints. We refer to these properties as continuity and stability.

There are some cases where the "correct" classification seems obvious, e.g. the examples in Figure 86.1. Another case occurs with full protein chains (see Chapter 90). Proteins have the pleasant property that the endpoints of protein chains typically lie on or very near the surface of the convex hull. In such a case (we also see this in Figure 86.1(a)), one can simply connect the endpoints via a circular arc lying outside of the convex hull to produce a closed knot, and that classification should satisfy any reasonable person.

King et al. [13] did the first search for protein slipknots, i.e. sub-arcs of proteins which are knotted, but for which the full protein chain is unknotted. When studying knotting in protein sub-arcs, we necessarily enter the uncomfortable situation where the endpoints are embedded within the entanglement, and the knotting is ambiguous. One could lament. However, perhaps this issue is more of an opportunity than a problem. Really, we should strive for a definition that is equally applicable to all open curves, not just the simplest cases.

86.3 Closure Techniques

The techniques described in this section are split into two classes: various methods that perform a single closure (Section 86.3.1) and various probabilistic methods that apply multiple closures (Section 86.3.2) to describe the entanglement of an open curve. The closure techniques classify the entanglement of open curves in terms of traditional knot types. Note that for many of the definitions below, we include an example image (Figures 86.4 and 86.5).

86.3.1 Single Closure Techniques

In this section, we review five single closure techniques. The underlying philosophy with these approaches is to select one closed configuration for each open configuration, and then describe the knot type of the open configuration as the knot type of the corresponding closed configuration.

In general, single closure methods have trouble with continuity and stability: in particular, small changes in the position of the endpoints can change the measured knotting immensely. Also, these methods tend to be undefined for some set of configurations (an issue of computability) and will generate classifications that do not match our intuition. Mathematicians tend to be disturbed by these fundamental issues. Physicists are more forgiving in this sense since these methods are quick to compute, and in many cases the classifications agree with the more involved methods. For example, when gathering data on large ensembles of random configurations, the statistics may be similar

---

[2]Millett called this property "robustness". [17].

between using single closure methods versus more involved methods, and the trade-off for speed can justify using the single closure methods.

86.3.1.1   Direct Closure

The easiest way to create a closed configuration from an open configuration is to add a line segment between the endpoints of the open configuration. See Figure 86.4(a). Of course, there are some configurations for which the additional segment will intersect the curve, but perhaps that set is small enough that one could find a way around that problem. The bigger issue with this approach is that the line segment can pierce the region defining the entanglement, which can cause the resulting classification to be in conflict with our intuition. For example, in Figure 86.1(a), the direct closure yields an unknot. Similarly, one can create examples where the direct closure would be a more complicated knot than one might classify by eye (see [2]).

86.3.1.2   Simplification

Many of the early schemes to measure knotting in proteins involved simplifying the open curve to the point where the knot type of the curve becomes clearer. In the protein world, this is called the KMT algorithm [14, 23], although the idea has deep mathematical roots, tracing back to at least [20]. Proteins are modeled, coarsely, as polygons, and the simplification sequentially removes edges as demonstrated in Figure 86.2. However, Millett et al. [16] showed that the order in which the moves are performed can result in different knot types after simple post-simplification closure, even when the endpoints are on the surface of the convex hull.

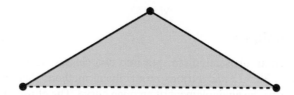

FIGURE 86.2: A configuration simplifying move from the KMT algorithm [14, 23]. To simplify an open polygon, one sequentially replaces the two adjacent edges with the dashed edge whenever the shaded triangle is not pierced by any other edge of the polygon. For closed polygons, such moves do not change the knot type. For open polygons, Millett et al. [16] showed that the order in which these moves are performed can affect the knot type classification.

To be fair, simplification is not a definition of knotting, per se, and the KMT algorithm has been employed primarily for proteins (whose endpoints generally are near the convex hull). The algorithm is effective in simplifying closed knots to make the computation of a knot type quicker (since knot type classifications can be exponential in the number of crossings).

86.3.1.3   Center of Mass Closure

In the direct closure method, the closure segment stays within the convex hull of a given open curve, thereby making it likely that the segment interacts in some fashion with the curve. The center of mass closure method is, at least in part, an attempt to minimize such interactions. Consider an open curve with endpoints v_1 and v_2 and center of mass c (note: for polygons, researchers typically use the center of mass of the vertex set instead of the edges). The closed polygon is then the union of the curve with two rays: one starting at v_1 with direction $v_1 - c$ and one starting at v_2 with direction $v_2 - c$. In other words, one extends rays from the endpoints in the direction away from the center of mass, and then the curve is closed at infinity. Of course, in practice, one only needs to add line segments along these rays of a sufficient length that one is guaranteed to be outside of the convex hull, and then find a way to connect the endpoints of the segments in a reasonable fashion. See Figure 86.4(b).

If the endpoints of the curve are on the surface of the convex hull, then this procedure is equivalent to connecting via a circular arc (mentioned earlier), and thus works roughly as desired. In particular, the method sees the trefoil for the images in Figure 86.1(a) and Figure 86.1(b), but not for Figure 86.1(c).

Beyond the continuity problem and the non-definability problem for some configurations, a main problem with this approach occurs when one or both endpoints are near the center of mass. In such a case, a small perturbation of an endpoint can result in huge differences in knot type (i.e. of arbitrarily large differences in crossing number).

86.3.1.4 Minimal Interference

The minimal interference method [25] is an effort to wed the best of the direct closure method and something akin to the center-of-mass method. In particular, if the endpoints are closer together than the sum of the minimal distances from the endpoints to the surface of the curve's convex hull, then one uses direct closure. Otherwise, one adds segments from each of the endpoints to the nearest point (which may not be unique, but usually is) on the convex hull and then connects the two points on the convex hull with a circular arc lying outside of the convex hull. See Figure 86.4(c).

This method seems quite reasonable. We have the typical single closure method problems of some-times not matching our intuition, continuity, stability, and having the quantity being undefined due to intersections between the added segments and the existing curve. Continuity and stability are the major issues. For example, one can construct examples where small perturbations lead to ar-bitrarily large changes in crossing number [2]. Still, intuitively, this method seems the best of the quantities presented to this point.

86.3.1.5 Nearest Neighbors between Open Curves and Closed Curves

There is a standard way to identify almost all equilateral open polygons (or arc-length parametrized open curves) with corresponding closed equilateral polygons (or arc-length parametrized curves) [12, 19] by viewing the closed polygons as a quotient space of the open polygons. Cantarella and Shonkwiler [6] point out that this is a natural closure algorithm, and Cantarella et al. [3] use a variant of this method to construct the closest closed equilateral polygon to a given open equilateral polygon. See Figure 86.4(d). The advantage of this approach is that there is a firm mathematical underpinning, so theorems are within reach. In particular, when the endpoints are relatively close, the local knotting structure is likely to be preserved. However, when the endpoints are far apart, the motion obtained via the geodesic between an open configuration and its corresponding closed configuration is likely to pass many edges through each other, leading to classifications that do not match our intuition. For example, if we trim the tails a bit on the configuration in Figure 86.1(a), the corresponding closed configuration is unknotted.

This technique is well defined and continuous on all but a set of polygons of measure zero. It satisfies our convergence and computability properties. However, again, one can construct examples where small perturbations lead to arbitrarily large changes in crossing number, leading to issues with stability. Overall, this technique has many attractive properties.

FIGURE 86.3: A polygonal arc which is a portion of the KnotPlot 6_2 knot configuration [21]. This is a base configuration used to demonstrate the different techniques in Figures 86.4, 86.5, 86.6, and 86.7.

86.3.2   Probabilistic Closure Techniques

Another overall approach is to think of the open curve as being an incomplete form, and then consider the knotting as a distribution of knot types that the curve could become. The main mathematical problem with approaches of this sort is that it is unclear what the theorems should be, or whether there even are good theorems. The main practical problem is that these approaches can take orders of magnitude longer to compute than the single closure methods, and one can only approximate the distributions.

On the other hand, these methods do generally satisfy all of desired properties mentioned earlier.

Overall, the idea here is to assign a distribution of objects to a given open curve. The main question is: what is the right set of objects?

86.3.2.1   Double Infinity Closure

If one were handed an entangled shoelace string and told to pull it tight, the person would have to choose in which direction to pull each end. Humans would generally choose to pull the ends of the string in fixed (but possibly different) directions as opposed to performing some complicated pattern. In the double infinity closure method, one extends rays from each endpoint in independently chosen directions and then closes at infinity [15]. There is an $S^2 \times S^2$ of such choices. Then the knotting of the open curve is the distribution of knot types obtained from this procedure. For any given open curve, there are pairs of directions which yield degenerate self-intersecting closed

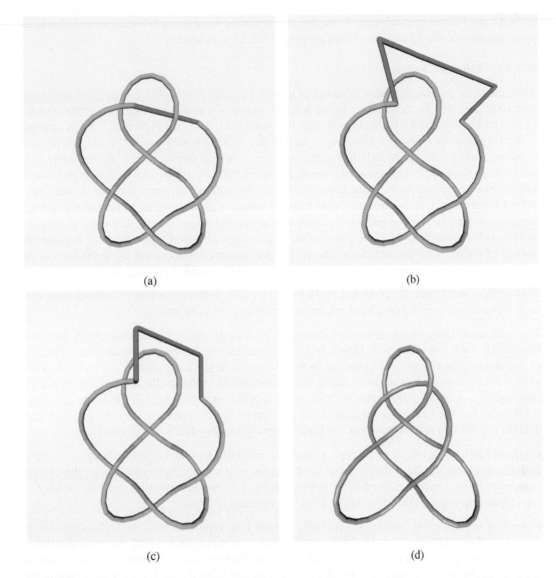

FIGURE 86.4: Examples of single closure techniques applied to the configuration in Figure 86.3. The single closure techniques (a) direct closure, (b) center of mass closure, (c) minimal interference, and (d) nearest closed neighbor applied to this configuration. For (c), one of the endpoints is on the convex hull and the other is very close to the convex hull. The teal arc in (c) is simply one example of a simple arc lying outside of the convex hull which connects the two endpoints.

curves, but the set of such pairs is of measure zero. Thus, this method is well defined. See Figure 86.5(a).

In the following section, we discuss a more commonly used probabilistic method, a sibling of this method called the infinity closure. Since the two methods share many advantages and disadvantages, we reserve our discussion of their properties for the next section.

86.3.2.2   Infinity Closure

In the previous section, the curve is closed at infinity by extending rays in independent directions from the endpoints. In the infinity closure method, one uses a common direction for the rays (so there is an S^2 worth of direction choices). See Figure 86.5(b). Also, see Figure 86.7 for an example of the map of knot types as a function of the ray directions. From a computational perspective, sampling from S^2 is significantly less strenuous than sampling from $S^2 \times S^2$. We typically have used 100 closures per open curve, so one would need $100^2 = 10000$ closures with the double infinity closure method to obtain the same density of information. Furthermore, there is precedence in knot theory for analyzing knotting over a single sphere of directions. In particular, the average crossing number and writhe are both computed by averaging over a sphere. The average crossing number is the average over S^2 of the number of crossings observed when the knot is projected in each of the S^2 different directions. The writhe is computed similarly, but using the sum of the values $+1$ for each right-handed crossing and -1 for each left-handed crossing in the projections.

This method was originally proposed by Millett et al. [18], although in lieu of parallel rays, they closed to a common point on a large sphere encompassing the open curve.

There are many things to like about this strategy: it yields a trefoil for a great majority of the directions in all three examples from Figure 86.1; the distributions change continuously as a function of the endpoint positions; no human interaction is needed in the computation; the method (while slower than the single closure methods) is straightforward to compute; the method sees a straight line segment as being 100% unknotted; the method provides a well-defined measurement of entanglement for open curves regardless of the position of the endpoints; and as the gap in Figure 86.1(b) and 86.1(c) approaches zero, the distribution approaches 100% trefoil knots.

On the negative side: one can only approximate the distribution of knot types (although one could, perhaps, compute the exact distributions for polygons with small numbers of edges); the computations are time-consuming; the definition seems difficult to use theoretically; and a distribution of knot types might not be the most desirable way to describe knotting.

Regarding the last point, researchers typically use the knot type seen with the highest percentage over the different closures to be the "knot type" of the open curve. Millett [17] observed that 996 of 1000 random equilateral open polygons with 300 edges had one knot type that appeared in at least 50% of the closure directions. This result suggests that assigning the predominate knot type in this fashion is generally quite reasonable. Of course, one can concoct open curves where two or more knot types occur in a near tie for the highest percentage, or where the direction sampling (with say 100 closure directions) is not sufficient to declare a clear predominate knot type. However, from Section 86.3.1 we know that assigning a single knot type to an open curve can produce awkward moments. At the least, the percentage of the predominate knot type provides a measure of confidence in the assignment. One could use a cut-off (such as 50%) in place of using the predominate knot type, or devise some other scheme, if one is weary of commitment.

86.3.2.3 Random Arc Closure

Instead of closing via rays heading to infinity, we could connect the endpoints via random arcs. In particular, for a given open polygonal arc, there is some fixed distance between the endpoints. One need only randomly sample from open polygons (or, conceivably, curves) with that endpoint distance. One could then use the distributions of knot types, or possibly the difference between that distribution and some other base distribution of knot types, to describe the knotting of the arc.

Cantarella's lab's plCurve library is able to generate random equilateral polygons with a given distance between the endpoints and a given number of edges, and is derived from theory in [4]. See Figure 86.5(c).

The main issue with this technique is how to choose the length, or lengths, of the connecting arcs. Perhaps there is a way around this problem, for example, by assigning a likely length or a distribution of likely lengths. If one chooses a fixed number of equilateral edges with which to close a given equilateral open chain, then one is in an honest-to-goodness knot space, one even with a metric. This underlying measure makes this technique attractive mathematically.

Beyond the issue of choosing the length of the connecting arcs, the random closures technique shares the same practical problems as the single and double infinity closures: namely, one can only approximate the distribution of knot types; the computations are time-consuming; the definition seems difficult to use theoretically; and a distribution of knot types might not be the most desirable way to describe knotting.

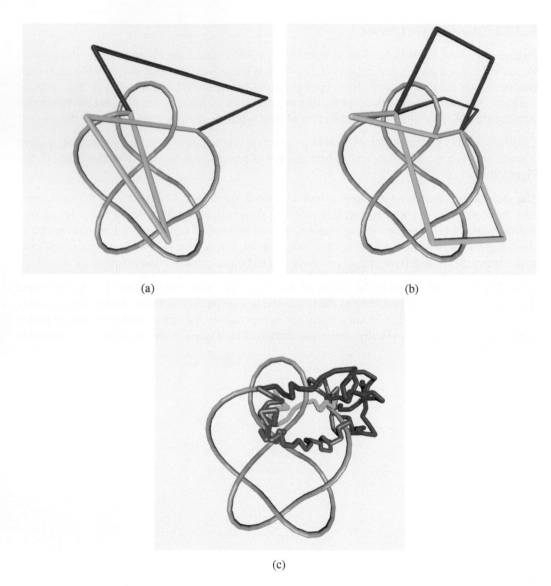

(a)   (b)

(c)

FIGURE 86.5: Examples of probabilistic closure techniques on the configuration from Figure 86.3. In (a), we show two examples of closures from the double infinity closure technique. While the polygons are officially closed at infinity, one can connect the endpoints of the line segments once both of the rays pierce the convex hull. In (b), we show four examples of parallel rays from the infinity closure technique. In (c), we show three connecting random arcs with different numbers of edges: 20 (green), 50 (blue), and 60 (purple).

86.4   Measuring Entanglement without Requiring a Closure

The introduction of the concept of knotoids by V. Turaev [26] brought into play new ways of measuring the entanglement of an open curve. In particular, one may leave the curve open and rely on the classification of its projections by their knotoid type to describe the entanglement. Knotoids are equivalence classes of open-ended diagrams that comprise an extension of classical knot theory (see Chapter 36 for an introduction to knotoids). Furthermore, knotoids generalize the notion of a long knot or 1-1 tangle, allowing the definition of an open knot type, at least at the diagrammatic level. In analogy to the techniques that search for a closure for the curve, it is the projections this time that are functions of the geometry of the curve. Here the underlying philosophy is to assign a knotoid type (or a probability distribution of knotoid types) to the open configuration.

For a given open curve, each projection direction (there are S^2 of these) determines a knotoid. Different choices of projection planes can change the placement of the endpoints which, in turn, can change the measured entanglement (i.e. knotoid type) of the curve dramatically, see Figure 86.6(a). Notice that the three projections shown are not isotopic to each other as knotoid diagrams. More precisely, knotoid (A) is a knotoid with 6 crossings, knotoid (B) is isotopic to a knotoid diagram with just two crossings, and knotoid (C) reduces to the trivial knotoid diagram which is just a single unknotted long segment. Choosing a single projection (which would yield a single knotoid) suffers from the same continuity and stability issues as the single closure techniques. Thus, we deduce that the knotoids approach is best applied as a probabilistic method.

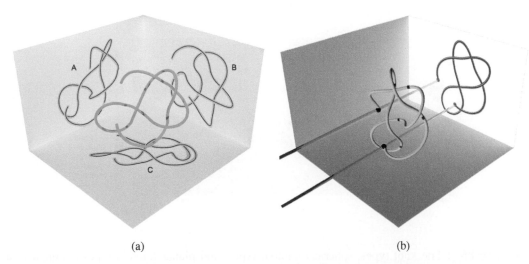

(a)   (b)

FIGURE 86.6: Knotoid projections from the open configuration of Figure 86.3. In (a), three knotoids are produced by projecting the configuration onto three different planes. In (b), a knotoid coming from a projection of the open configuration is shown together with the infinite lines that fix the endpoints region-wise.

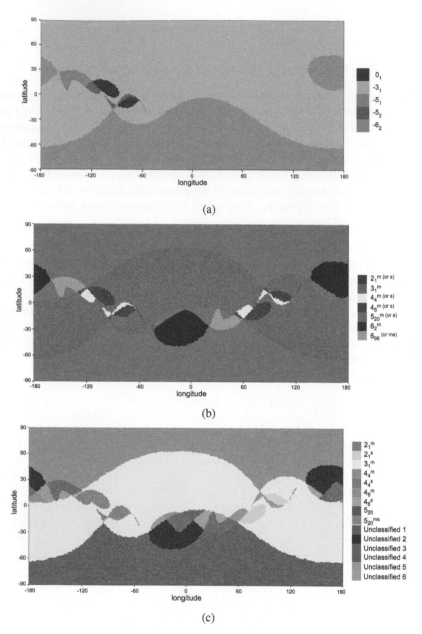

(a)

(b)

(c)

FIGURE 86.7: The knot types, spherical knotoid types, and planar knotoid types obtained as a function of the ray direction for the configuration from Figure 86.3. Each latitude/longitude pair on the unit sphere specifies a unique closure or projection direction. In (a), a sampling of different spherical coordinates yields a map of knot types (here, flattened from the sphere). The predominate knot type is the one with the largest area on the sphere, and this knot type is typically used as a proxy to describe the knotting of the open arc. In this case, the green region, corresponding to the -3_1 knot, has the largest area, so this configuration normally would be referred to as forming an open -3_1 knot. The maps in (b) for spherical and (c) for planar knotoids are similar. For (b) and (c) the latitude/longitude pair define a projection direction, which in turn defines a spherical or planar knotoid. The coloring is based on the spherical or planar knotoid types. For all three analyses, a sample of 40,000 directions was used.

Similar to the probabilistic closure method that produces a spectrum of knots, the knotoids approach assigns to a curve a spectrum of knotoids resulting from different directions of projections.

To compute knotoid types more efficiently, the knotoid projections can often be simplified using an algorithm similar to the KMT algorithm [14, 23]. To do so, one adds two parallel lines to the knotoid projection: these lines are perpendicular to the chosen projection plane and pass through one of the endpoints, see Figure 86.6(b). Then a triangle of the polygonal knotoid is simplified (as in Figure 86.2) only if it is not pierced by any other arc of the knotoid or by either of the two infinite lines. These KMT-like simplifications do not change the knotoid type and thus are able to give precise classifications even for sub-arcs where the endpoints lie inside the convex hull. On a side note, the introduction of the two infinite lines allows for the development of theorems, such as in [11] where the conditions are shown for when two such ensembles of curves and lines are isotopic with respect to the same plane of projection.

As in the case of the probabilistic closure methods, the knotoid approach classifies the configurations of Figure 86.1 as a trefoil for most of the projections, the distributions change continuously as a function of the endpoint positions, and, in general, it carries all the advantages and disadvantages of the infinity closure.

It is important to mention here that knotoids can be studied on any oriented surface like, for example, on S^2. In fact, there is a well-defined map between the set of all planar knotoids (knotoids in $\mathbb{R}^2$) and the set of all spherical knotoids (knotoids in S^2) which is induced by the inclusion of $\mathbb{R}^2$ in S^2 [11, 26]. Moreover, it can be shown that this map is surjective but not injective since one can find examples of knotoids that are nontrivial in $\mathbb{R}^2$ and trivial in S^2. It was shown recently that the entanglement analysis of subchains of an open curve using spherical knotoids provides a more detailed overview of the topology of a curve than using knot types via infinity closures [8]. This analysis can be further refined if one chooses to work with planar knotoids [10].

In Figure 86.7 one can see a comparison between the probabilistic methods using (single) infinity closures, spherical knotoids, and planar knotoids. Each small region in these maps corresponds to a region on the sphere (shown in spherical coordinates). The different knot/knotoid types are indicated by the different colors in the legend of each image. Each point corresponds to either a closure or a projection direction for the studied open curve. The notation for knotoids follows the classification in [9]. The exponents m, s, and ms correspond to the following involutive forms of the respective knotoids shown in the table of [9]: mirror reflected, symmetric, and mirror symmetric. Note that the projection map for the spherical case shows antipodal symmetry, while the one for the planar case does not. The knotoid types are computed using invariants. The Jones-type invariants for spherical knotoids available in the literature (like the Jones polynomial and the arrow polynomial) cannot distinguish between a knotoid and its mirror symmetric involution (nor between the mirror reflection of a knotoid and its symmetric involution). However, this issue may resolve itself in the future with a new invariant for spherical knotoids. The loop arrow polynomial [9, 10] is an invariant for planar knotoids that, in principle, can distinguish a planar knotoid from its involutions.

Returning to the comparison of projection maps between knots and knotoids, we see that new regions appear in the projection map as we transition from infinity closures to spherical knotoids and to planar knotoids. This behavior is good news and bad news. The knotoid methods provide additional information about the knotting in an open curve. However, since there are more regions with the knotoids than with the knot types using the infinity closure, the percentages associated with each knotoid type are smaller, which makes the assignment of a predominate knotoid type precarious.

We note here that a computational tool that analyzes open curves, using both knotoid approaches as well as the infinity closure approach, has been implemented [7]. The projection maps in Figure 86.7 were produced using this software.

An equivalent approach to spherical knotoids is the virtual closure technique [1] where the knotoid diagram is closed to a virtual knot. More precisely, the endpoints are joined by an arc that crosses the rest of the diagram using virtual crossings. The resulting virtual knot is independent of the choice of the closing arc. Since all virtual crossings occur along the closure arc, one may consider this arc as living in a single handle between the endpoints of the open arc and, thus, they do not interfere with the rest of the diagram. Therefore, the virtual closure technique is a probabilistic method equivalent to the spherical knotoids approach with one extra step, the addition of the virtual closure on a knotoid diagram.

86.5   Topological Techniques

Below we review two topological approaches to measure knotting in open curves.

86.5.1   Ball-Arc Pair

One topological approach is to use a ball-arc pair. In particular, a *ball-arc pair* is an embedded open curve (the arc) and a topological sphere (i.e. an object homeomorphic to a round sphere) that does not intersect the curve except at the two endpoints of the curve. One can analyze ball-arc pairs as topological objects. However, as we show below, the prescription of the ball to the open curve is equivalent to assigning a closing curve [2]. So while the ball-arc pair does allow one to analyze the knotting of an open curve topologically, that is because the geometry of the arc acts in conjunction with the choice of the topological sphere.

To illustrate this issue, we show a classic example from [22] in Figure 86.8. The topological spheres are shown as transparent tubes, the arcs are shown in red, and the intersection between the balls and the arcs are shown as black round spheres. Note that in (d), the topological sphere is the clear outside and the dark gray part "inside" (that looks like an overhand knot), while the arc is the short straight segment on the right. By deforming the sphere in (c), one can obtain (a), so the ball-arc pairs of (a) and (c) are equivalent. Similarly, the ball-arc pairs of (b) and (d) are equivalent. While the arcs of (a) and (d) are both straight, the ball-arc pairs of (a) and (d) are not equivalent, due to what might be considered an unusual choice of a topological sphere in (d). In (d), the union of the straight segment and any non-self-intersecting curve lying on the topological sphere that connects the two segment endpoints is a trefoil. Thus, choosing a topological sphere is equivalent to choosing a closing curve.

Certainly, it seems like one would want a straight line segment to be considered mainly unknotted, so this behavior limits the applicability of this approach. In the end, the ball-arc pair approach does allow one to prove theorems, but cannot independently describe the knotting of an open curve in a way that matches our intuition. To be fair, the ball-arc pair was not defined as a technique to classify knotting in a fixed open configuration.

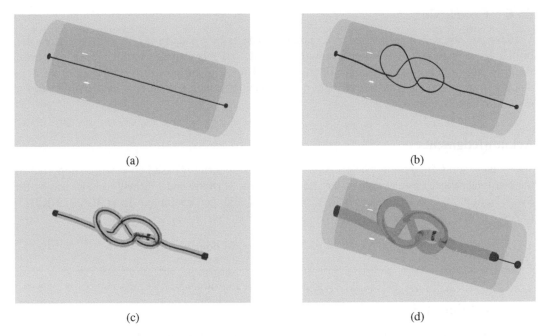

(a)

(b)

(c)

(d)

FIGURE 86.8: Four ball-arc pairs originally presented in [22]. The topological spheres are shown as transparent tubes, the arcs are shown in red, and the intersection between the balls and the arcs are shown as black round spheres. The ball-arc pairs from (a) and (c) are equivalent, as are the pairs in (b) and (d). However, the ball-arc pairs from (a) and (c) are not equivalent to the ball-arc pairs in (b) and (d).

86.5.2 Polygons with Fixed Edge Lengths

If we concentrate on open polygons, it is possible to use a topological approach. Consider the class of open polygons where the edge lengths are fixed (although they need not be all the same) and define equivalence in terms of ambient isotopies that keep the edge lengths fixed. In this situation, Cantarella and Johnston [5] showed that there are "stuck unknots." In other words, there exist open polygons that cannot be deformed into (non-self-intersecting) collinear segments via this restricted class of ambient isotopies. One could then define "knot types" as connected components of the given polygon space, i.e. polygons with a given array of edge lengths using ambient isotopies that fix the edge lengths. Again, this approach allows one to prove theorems. One might imagine that this technique could be useful for biological chains since chemical bonds are modeled as edges, and these lengths are usually fixed (or close to fixed). However, biological chains tend to have some electrostatic repulsion which keeps the edges from getting too close together. With this property, it is unlikely that the biological chains could be "stuck." This approach can be used for applications such as the movement of robot arms and motion planning (see Chapter 9 of [24]).

86.6   Conclusion

We have presented techniques used by researchers to describe the knotting of open curves, focusing primarily on open polygons. There are many options, and this chapter provides researchers with insights into which approach might be most useful for a given project. Furthermore, we hope that the discussions inspire researchers to develop new techniques and create new mathematics to understand knotting in open arcs.

Acknowledgments

We thank Jason Cantarella for helpful comments. This material is based upon work supported by the National Science Foundation under Grant Numbers 1418869 and 1720342 to EJR, and by the Swiss National Science Foundation [31003A_166684] and the Leverhulme Trust [RP2013-K-017] to AS.

86.7   Bibliography

[1] Keith Alexander, Alexander J. Taylor, and Mark R. Dennis. Proteins analysed as virtual knots. *Sci. Rep.*, 7:42300, 2017.

[2] Erin Brine-Doyle, Madeline Shogren, Emily Vecchia, and Eric J. Rawdon. Open knotting. In Philipp Reiter, Simon Blatt, and Armin Schikorra, editors, *New Directions in Geometric and Applied Knot Theory*. De Gruyter, 2018.

[3] Jason Cantarella, Kyle Chapman, Philipp Reiter, and Clayton Shonkwiler. Open and closed random walks with fixed edgelengths in $\mathbb{R}^d$. *J. Phys. A*, 51(43):434002, 2018.

[4] Jason Cantarella, Bertrand Duplantier, Clayton Shonkwiler, and Erica Uehara. A fast direct sampling algorithm for equilateral closed polygons. *J. Phys. A*, 49(27):275202, 2016.

[5] Jason Cantarella and Heather Johnston. Nontrivial embeddings of polygonal intervals and unknots in 3-space. *J. Knot Theory Ramifications*, 7(8):1027–1039, 1998.

[6] Jason Cantarella and Clayton Shonkwiler. Personal communication, 2018.

[7] Julien Dorier, Dimos Goundaroulis, Fabrizio Benedetti, and Andrzej Stasiak. Knoto-ID: A tool to study the entanglement of open protein chains using the concept of knotoids. *Bioinformatics*, 34(19):3402–3404, 2018.

[8] Dimos Goundaroulis, Julien Dorier, Fabrizio Benedetti, and Andrzej Stasiak. Studies of global and local entanglements of individual protein chains using the concept of knotoids. *Sci. Rep.*, 7:6309, 2017.

[9] Dimos Goundaroulis, Julien Dorier, and Andrzej Stasiak. A systematic classification of knotoids on the plane and on the sphere. arXiv:1902.07277, 2019.

[10] Dimos Goundaroulis, Neslihan Gügümcü, Sofia Lambropoulou, Julien Dorier, Andrzej Stasiak, and Louis Kauffman. Topological models for open-knotted protein chains using the concepts of knotoids and bonded knotoids. *Polymers*, 9(9):444, 2017.

[11] Neslihan Gügümcü and Louis H. Kauffman. New invariants of knotoids. *European J. Combin.*, 65:186–229, 2017.

[12] Michael Kapovich and John J. Millson. The symplectic geometry of polygons in Euclidean space. *J. Differential Geom.*, 44(3):479–513, 1996.

[13] Neil P. King, Eric O. Yeates, and Todd O. Yeates. Identification of rare slipknots in proteins and their implications for stability and folding. *J. Mol. Biol.*, 373:153–166, 2007.

[14] Kleanthes Koniaris and Murugappan Muthukumar. Self-entanglement in ring polymers. *J. Chem. Phys.*, 95(4):2873–2881, 1991.

[15] Marc L. Mansfield. Are there knots in proteins? *Nat. Struct. Biol.*, 1:213–214, 1994.

[16] Kenneth Millett, Akos Dobay, and Andrzej Stasiak. Linear random knots and their scaling behavior. *Macromolecules*, 38(2):601–606, 2005.

[17] Kenneth C. Millett. The length scale of 3-space knots, ephemeral knots, and slipknots in random walks. *Prog. Theor. Phys. Suppl.*, 191:182–191, 2011.

[18] Kenneth C. Millett, Akos Dobay, and Andrzej Stasiak. Linear random knots and their scaling behaviour. *Macromolecules*, 38:601–606, 2005.

[19] John J. Millson and Brett Zombro. A Kähler structure on the moduli space of isometric maps of a circle into Euclidean space. *Invent. Math.*, 123(1):35–59, 1996.

[20] Kurt Reidemeister. Elementare begründung der knotentheorie. *Abhandlungen aus dem Mathematischen Seminar der Universität Hamburg*, 5(1):24–32, 1927.

[21] Robert G. Scharein. KnotPlot. http://www.knotplot.com, 1998. Program for drawing, visualizing, manipulating, and energy minimizing knots.

[22] De Witt Sumners and Stuart G. Whittington. Knots in self-avoiding walks. *J. Phys. A*, 21(7):1689–1694, 1988.

[23] William R. Taylor. A deeply knotted protein and how it might fold. *Nature*, 406:916–919, 2000.

[24] Csaba D. Toth, Joseph O'Rourke, and Jacob E. Goodman, editors. *Handbook of Discrete and Computational Geometry*. CRC Press, third edition, 2017.

[25] Luca Tubiana, Enzo Orlandini, and Cristian Micheletti. Probing the entanglement and locating knots in ring polymers: A comparative study of different arc closure schemes. *Prog. Theor. Phys. Suppl.*, 191:192–204, 2011.

[26] Vladimir Turaev. Knotoids. *Osaka J. Math.*, 49(1):195–223, 2012.

Chapter 87

Random and Polygonal Spatial Graphs

Kenji Kozai, Rose-Hulman Institute of Technology

Random knots have been studied extensively as a way to model the behavior of long polymers such as DNA packed into confined spaces. Such polymers sometimes become tangled up, and a version of the Frisch-Wasserman-Delbruck conjecture says that the more densely packed the polymer is, the more likely it is to become knotted [8, 3]. If you have ever haphazardly stuffed earbuds with a long cable into your pockets, you have likely witnessed something along these lines when you discover the cord is a tangled mess when you remove it from your pocket.

Many authors have studied the knotting and linking behavior of polymers using a variety of mathematical models for the physical configuration of long chains [1, 2, 5, 4, 15, 16, 17, 19]. Much like there is a natural progression from studying knots to spatial graphs, there is a generalization of random knots to random spatial embeddings of graphs. One can think of a random spatial embedding of a graph as a way of modeling configurations of non-linear molecules or the more complicated secondary and tertiary structures of polymers, which often contain loops and other structures that can only be realized in three-dimensional space.

87.1   From Random Knots to Random Spatial Graphs

A random knot or random knot model is a choice of probability distribution on the space of all topological knot types. An example of one such model is the uniform random polygon model (URP). In this model, a random knot of length n is generated by the following procedure:

1. Pick n ordered points $p_1, p_2, \ldots, p_n$ from the unit cube $[0, 1]^3$ at random using the uniform distribution.

2. Connect p_i to p_{i+1} by a straight line segment between the two points.

Using this random knot model, one can calculate the mean writhe as a function of n, or the mean linking number of two knots of length n generated by the above procedure [1].

As in the random knot case, a *random spatial embedding* for a graph G is a probability distribution on the set of all spatial embeddings of G. A simple model of a random spatial graph embedding is *random linear embedding*.

Definition 87.1.1. A *random linear embedding* of a graph G is a probability distribution on the linear embeddings of G given by the following process:

1. Choose a point in $[0, 1]^3$ for each vertex of G using the uniform distribution on the cube.

2. Edges between two vertices are realized by the straight line segment between their associated points in $[0, 1]^3$.

As the probability of any four randomly chosen points being coplanar is 0, this will result in a spatial embedding of G with probability 1. The uniform random polygon model is a special case of a random linear embedding when the graph G is the cyclic graph C_n.

An advantage of this particular model is that it is relatively straightforward to obtain information about the linking number of two-component links in a random spatial embedding of a graph using known results and techniques from the URP model of knots. For example, a generalization of an argument from URPs shows that the mean squared linking number of a random linear embedding of an n cycle and m cycle is nmk, where k is a constant depending on the probability that two triangles are linked [1, 7].

The linking number of a two-component link is a rough measure of the topological complexity of the link. One way to measure the topological complexity of a spatial graph embedding is to find the linking number of all of the two-component links in the graph and sum those values to turn it into a single number. The value of this number averaged over all spatial embeddings is a reasonable first pass at finding the complexity of an "average" spatial embedding.

One quickly realizes that the mean value of the linking number will always be 0. This is because the linking number is signed, and if the orientation of one cycle is reversed, the sign of the linking number is changed. Therefore, for every configuration with a positive sum of linking numbers, there is a mirror configuration that has the opposite sign, so the mean value is 0. A solution to this problem is to instead consider the mean value of the sum of the squared linking numbers.

A careful enumeration of the disjoint pairs of cycles can then be used to find this quantity. Because of the linearity of expectations, the expected value of the sum is the sum of the expectations of the squared linking number of each of the individual two-component links, and the mean value of the squared linking number nmk can be applied to each of the two-component links. For the complete graphs K_n, it turns out that the mean sum of squared linking numbers is on the order of $n(n!)$. The complete graph on n vertices contains on the order of $(n-1)!$ disjoint pairs of cycles, so the mean squared linking number of the typical link in the graph is on the order of n^2.

87.2   Random Linear Embeddings of Small Graphs

It is also possible to deduce information about the probability of specific linear embeddings occurring for small graphs. For example, it is known that linear embeddings of K_6 contain either a single Hopf link or three Hopf links [9, 10, 14]. Moreover, in the first case, the embedding of K_6 is knotless, and in the latter case, the embedding of K_6 contains a trefoil knot as a 6-cycle.

Using a simulation to find a numerical value for the probability of two 3-cycles being linked, the

FIGURE 87.1: Mobius ladder configuration of $K_{3,3}$.

expected value of the sum of squared linking numbers for K_6 is found to be approximately 1.521. Since the sum of the squared linking numbers must either be 1 for the case with a single Hopf link or 3 for the case with three Hopf links, then $1.521 = p + 3(1 - p)$, where p is the probability that a random linear embedding of K_6 contains only one Hopf link. The value of p is therefore approximately 0.738. The topologically simpler configuration is much more likely to occur than the more complex configuration with three Hopf links and a trefoil knot. A similar computation for linear embeddings of $K_{3,3,1}$, which contain either 1, 2, 3, 4, or 5 non-trivial links [13], shows that the simplest configuration with only a single Hopf link dominates the other configurations, occurring over 58% of the time in random linear embeddings. These results indicate that at least for small graphs, the topologically simplest configuration seems to be the most likely configuration to occur.

Keeping in mind that one of the motivations for studying random graph embeddings is to model spatial configurations of non-linear molecules, even the complete graph K_6 is too large of a graph to expect to naturally occur linearly, as that would require six atoms to be pairwise bonded with each other. However, one interesting graph that does occur as a substructure of molecules is $K_{3,3}$ (some examples can be found in [6]). This subgraph has thus far been found to be in a specific spatial configuration known as a *Mobius ladder* (Figure 87.1). Since $K_{3,3}$ is a subgraph of K_6, probabilities for embeddings of K_6 determine the probability that a random linear embedding of $K_{3,3}$ is isotopic to a Mobius ladder.

The key step in determining whether an embedding of $K_{3,3}$ is isotopic to a Mobius ladder or not is to show that an embedding of $K_{3,3}$ is Mobius if and only if it is knotless. Once this fact is established, it suffices to study the completion of an embedding of $K_{3,3}$ to an embedding of K_6 by adding straight edges between vertices that do not already have an edge between them. As we have already seen, a random linear embedding of K_6 has a Hopf link with probability 0.738, so the probability that the K_6 contains a knot – which must necessarily be a 6-cycle – is $1 - 0.738 = 0.262$. However, in order for the knot to be in the original embedding of $K_{3,3}$, every edge of the knot in K_6 must be present in the $K_{3,3}$ subgraph. There are $\binom{6}{3} = 10$ distinct pairs of disjoint 3-cycles in K_6, and exactly one of those will be disjoint from a given 6-cycle. Hence, the probability that a linear embedding of $K_{3,3}$ contains a knot is 0.0262. Therefore, under the random linear embedding model, $K_{3,3}$ will nearly always be isotopic to a Mobius ladder, agreeing with known observations.

87.3 Stick Numbers and Polygonal Graphs

A *paneled embedding* of a graph is an embedding in which each cycle bounds a disk in the complement of the graph, and it is a slight generalization of a planar graph. The Mobius ladder configuration of $K_{3,3}$ in Figure 87.1 is paneled, and Figure 87.2 demonstrates a non-paneled embedding

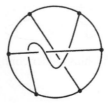

FIGURE 87.2: A non-paneled embedding of $K_{3,3}$.

of $K_{3,3}$ – any disk bounded by the cycle consisting of the bottom four vertices intersects the edge connecting the top right vertex to the bottom left vertex. It has been shown that $K_{3,3}$ has a unique paneled embedding [18], so in order to establish that an embedding of $K_{3,3}$ is isotopic to a Mobius ladder if and only if it is knotless, it suffices to show that the Mobius ladder embedding is paneled and that any knotless embedding of $K_{3,3}$ is paneled. The property of being paneled is closely related to the fundamental group of the complement of the graph.

Leveraging results in [18] and [11], it suffices to show that all of the proper subgraphs in a linear embedding of $K_{3,3}$ are paneled. Removing an edge from $K_{3,3}$ turns it into a graph with four vertices of degree 3 and two vertices of degree 2. Treating the two edges adjacent to a degree 2 vertex as a single piecewise linear edge made of two sticks, this can be viewed as an embedding of K_4 with seven sticks.

This leads to a connection to the concept of a stick number for graphs. A graph can be embedded in $\mathbb{R}^3$ in a manner so that the edges are piecewise linear, i.e. a polygonal path between the two vertices. Each linear piece in the polygonal path can be viewed as an individual stick, and the *stick number* of a graph is the minimum number of sticks necessary to construct a non-paneled embedding of the graph. For example, the stick number of a cyclic graph of length less than six is 6 since a non-trivial knot requires at least six sticks to construct.

In the graph case, it is possible to have knotless and linkless embeddings that are non-trivial in the sense that they are not paneled. Thus, it makes sense to distinguish between the stick number of a graph and the stick number of a knotless and linkless embedding of the graph.

Definition 87.3.1. The *stick number* of a graph G is the fewest number of linear segments required in a piecewise linear embedding of G that is not paneled.

Definition 87.3.2. The *stick number of a knotless and linkless embedding* of a graph G is the fewest number of linear segments required in a piecewise linear embedding of G that is not paneled and does not contain any non-trivial knots or links.

The well-known Kinoshita theta curve is an example of an embedding of a graph that is non-trivial but knotless. This theta curve is known to have a stick number of 8 [12]. An example of an eight-stick embedding can be found in Figure 87.3. Similarly, the stick number of a non-paneled knotless embedding of K_4 can be shown to be 9. This is one more than the number of sticks in a linear embedding of $K_{3,3}$ minus an edge, so a knotless embedding of K_4 within a linear embedding of $K_{3,3}$ is paneled, as desired.

The proofs of the stick numbers of these two graphs are essentially exhaustive searches. Since

FIGURE 87.3: The eight stick-realization of the Kinoshita theta curve.

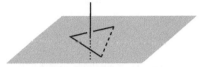

FIGURE 87.4: If the vertical edge does not pierce the two consecutive solid sticks in the plane, the two sticks can be replaced by the dotted stick without changing the isotopy class of the path.

we are looking for the minimum number of sticks required in a non-paneled embedding, it is necessary that any polygonal edge with two or more sticks is "pierced" by another edge of the graph. Otherwise, it is possible to reduce the stick number by replacing two adjacent sticks with a single one without changing the isotopy class of the graph embedding, as in Figure 87.4. This limits the possible point configurations of the vertices of the graph and the polygon edges, at which point one simply checks that if the number of sticks is less than the stick number, every embedding is either knotted or paneled.

87.4 Other Random Models and Invariants

The random linear embedding model, where edges are linear and vertices are chosen randomly from the unit cube, is only one mathematical model for random spatial embeddings of graphs. It has been studied because it is relatively simple to define, and work done by others on uniform random polygons can be adapted to the setting of graph embeddings. Other choices for random models are possible, such as choosing the vertices randomly from a sphere or using a Gaussian distribution. The choice to study linking numbers is also somewhat arbitrary and done because of the ease with which results from uniform random polygons can be used. It would be interesting to know if there are better measures for the complexity of a graph embedding that generalize other knot and link invariants besides the linking number.

The random linear embedding model has one glaring weakness if the purpose is to understand spatial configurations of molecules. The bond lengths between atoms in a molecule are essentially fixed, or at least highly restricted, but the random linear embedding model will have bond lengths that vary within the full range of distances between two points in the unit cube.

There exist random knot models that have greater restrictions on edge lengths, and it would be interesting to see if there are reasonable definitions of random graph embeddings that can generalize these models. In the graph setting, it is necessary to restrict all of the distances between a vertex and its edge-adjacent vertices, whereas in the case of a random knot, each vertex only has two vertices that are edge-adjacent to it. This makes it difficult to define the probability distribution on graph embeddings. In addition, the probability distribution on embeddings of subgraphs may not be the restriction of the distribution on the graph to the subgraph. This will likely make the analysis of spatial graph embeddings under any such model trickier.

87.5   Bibliography

[1] J. Arsuaga, T. Blackstone, Y. Diao, E. Karadayi, and M. Saito. Linking of uniform random polygons in confined spaces. *J. Phys. A*, 40(9):1925–1936, 2007.

[2] J. Arsuaga, B. Borgo, Y. Diao, and R. Scharein. The growth of the mean average crossing number of equilateral polygons in confinement. *J. Phys. A*, 42(46):465202, 9, 2009.

[3] M. Delbruck. Knotting problems in biology. *Proceedings of Symposia in Applied Mathematics*, 14:55–63, 1962.

[4] Y. Diao, C. Ernst, S. Saarinen, and U. Ziegler. Generating random walks and polygons with stiffness in confinement. *J. Phys. A*, 48(9):095202, 19, 2015.

[5] Y. Diao, N. Pippenger, and D.W. Sumners. On random knots. In *Random Knotting and Linking (Vancouver, BC, 1993)*, volume 7 of *Ser. Knots Everything*, pages 187–197. World Sci. Publ., River Edge, NJ, 1994.

[6] Erica Flapan and Gabriella Heller. Topological complexity in protein structures. *Mol. Based Math. Biol.*, 3:23–42, 2015.

[7] Erica Flapan and Kenji Kozai. Linking number and writhe in random linear embeddings of graphs. *J. Math. Chem.*, 54(5):1117–1133, 2016.

[8] H. L. Frisch and E. Wasserman. Chemical topology. *Journal of the American Chemical Society*, 83(18):3789–3795, 1961.

[9] C. Hughes. Linked triangle pairs in straight edge embeddings of K_6. *Pi Mu Epsilon Journal*, 12(4):213–218, 2006.

[10] Y. Huh and C. B. Jeon. Knots and links in linear embeddings of K_6. *J. Korean Math. Soc.*, 44(3):661–671, 2007.

[11] Youngsik Huh and Jung Hoon Lee. Linearly embedded graphs in 3-space with homotopically free exteriors. *Algebr. Geom. Topol.*, 15(2):1161–1173, 2015.

[12] Youngsik Huh and Seungsang Oh. Stick number of theta-curves. *Honam Math. J.*, 31(1):1–9, 2009.

[13] R. Naimi and E. Pavelescu. On the number of links in a linearly embedded $K_{3,3,1}$. *J. Knot Theory Ramifications*, 24(8):1550041, 2015.

[14] R. Nikkuni. A refinement of the Conway-Gordon theorems. *Topology Appl.*, 156(17):2782–2794, 2009.

[15] E. Panagiotou, K. C. Millett, and S. Lambropoulou. The linking number and the writhe of uniform random walks and polygons in confined spaces. *J. Phys. A*, 43(4):045208, 28, 2010.

[16] Nicholas Pippenger. Knots in random walks. *Discrete Appl. Math.*, 25(3):273–278, 1989.

[17] J. Portillo, Y. Diao, R. Scharein, J. Arsuaga, and M. Vazquez. On the mean and variance of the writhe of random polygons. *J. Phys. A*, 44(27):275004, 19, 2011.

[18] Neil Robertson, Paul Seymour, and Robin Thomas. Sachs' linkless embedding conjecture. *J. Combin. Theory Ser. B*, 64(2):185–227, 1995.

[19] K. Tsurusaki and T. Deguchi. Numerical analysis on topological entanglements of random polygons. In *Statistical Models, Yang-Baxter Equation and Related Topics, and Symmetry, Statistical Mechanical Models and Applications (Tianjin, 1995)*, pages 320–329. World Sci. Publ., River Edge, NJ, 1996.

Chapter 88

Folded Ribbon Knots in the Plane

Elizabeth Denne, Washington & Lee University

88.1 Introduction

We can create a ribbon knot in $\mathbb{R}^3$ by taking a long, rectangular piece of paper, tying a knot in it, and connecting the two ends. We then flatten the ribbon into the plane, origami style, with folds in the ribbon appearing only at the corners. In Figure 88.1 left and center, we show part of a trefoil knot and the corresponding folded ribbon trefoil. If we join the two ends of the ribbon and tighten, we get a "tight" folded ribbon trefoil shown on the right.

  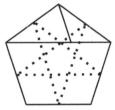

FIGURE 88.1: Creating a folded ribbon trefoil knot.

Such a *folded ribbon knot* was first modeled by L. Kauffman [29]. (He called them flat knotted ribbons.) Kauffman defined the *ribbonlength* (the length to width ratio) of folded ribbons knots, and asked to find the least length of ribbon needed for a knot type, given a choice of width. This is known as the *(folded) ribbonlength problem*. A minimal ribbonlength folded ribbon knot can be considered to be folded "tightly." This idea is neatly illustrated with the construction of a tight folded ribbon trefoil knot in Figure 88.1 right, which has a pentagon as its boundary shape. Kauffman [29] gave the conjectured minimal ribbonlength for this and a tight figure-eight knot.

Understanding folded ribbonlength reveals interesting relationships between geometry and topology, and there are natural connections between folded ribbons and other areas of mathematics and science. For example, the ribbonlength problem may also be thought of as a 2-dimensional analogue of the *ropelength problem*: that of finding the minimum amount of rope needed to tie a knot in a rope of unit diameter. (See for instance [8, 13, 24, 33, 37].) Folded ribbon knots arise naturally from considering (smooth) ribbon knots in space. These are used, for example, to model cyclic duplex DNA in molecular biology with the two boundaries corresponding to the two edges of the DNA ladder (see for instance [1, 7]). Folded ribbon knots have connections to other parts of knot theory as well. We will see later that grid diagrams of knots, and mosaic knots, can easily produce folded ribbon knots. Grid diagrams of knots have been extensively studied [36] and are used, for example, in the combinatorial formulation of knot Floer homology. Mosaic knots [35, 32] are used to model quantum knots which describe a physical quantum system.

In the past, there have been a number of recreational articles about tying knots in strips of paper. In particular, [1, 5, 15, 26, 39] all described the construction of a pentagon as the boundary shape of a folded ribbon trefoil knot. Some of these also found other regular n-gons; for example D.A. Johnson [26] gave the construction of a regular hexagon by folding a trivial 2-component link in a certain way. Constructing any regular polygons as the boundary of a folded ribbon knot appeared to be a harder problem. This was finally solved in the 1999 master's thesis of L. DeMaranville [16]. She described which (p, q) torus knots[1] can be easily converted to a folded ribbon knot, and showed how to build all regular n-gons for $n > 6$, by tying certain families of folded torus ribbon knots. While there is often more than one way to construct a particular n-gon, the $(p, 2)$ torus knots give all odd p-gons, and the $(q + 1, q)$ torus knots give all $2q$-gons (here $p, 2q > 6$). An interesting open question is to find what other shapes can be formed by torus links, and what can be said about their folded ribbonlengths.

88.2 Modeling Folded Ribbon Knots

We will assume that our readers are familiar with the definition of a knot, a link and a knot diagram. A formal definition of a folded ribbon knot can be found in [4, 17]. In this section, we provide enough details to give the reader the big picture. When we model a folded ribbon knot, we view the knot as a *polygonal knot diagram*. Figure 88.2 shows two different polygonal knot diagrams for the trefoil knot. A polygonal knot diagram has a finite number of *vertices* denoted by $v_1, ..., v_n$, and *edges* e_i defined by $e_1 = [v_1, v_2], ..., e_n = [v_n, v_1]$. If the knot diagram is oriented, then we assume that the labeling follows the orientation.

FIGURE 88.2: Polygonal knot diagrams of the trefoil knot with five and six edges.

We pause to note that we do not require our polygonal knot diagrams to be *regular*. For example, take a ribbon which is an annulus and then fold it flat with just two folds. The polygonal knot diagram is made of two edges and we understand that one edge always lies over the other so that the crossing information is consistent. We will refer to this diagram as the 2-stick unknot. Recall that the *stick index* of a knot K is defined to be the least number of line segments needed to construct a polygonal embedding of K in $\mathbb{R}^3$ (see [1]). We can define the *projection stick index* to

---

[1]Here, a (p, q) torus knot is assumed to wrap p times around the meridional direction and q times around the longitude direction of a torus.

be the minimum number of sticks needed for a polygonal knot diagram of K. We have just seen the unknot has projection stick index two, and regular stick index three. Together with undergraduate students, C. Adams [2, 3] showed that the projection stick index of the trefoil knot is five, while the regular stick index is six (illustrated in Figure 88.2). Indeed, we expect the projection stick index to be smaller than the stick index since the edges in the knot diagram are not rigid sticks in space, they have crossing information instead.

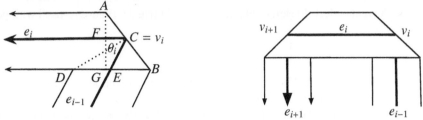

FIGURE 88.3: On the left, a close-up view of a ribbon fold. On the right, the construction of the ribbon centered on edge e_i. Figure re-used with permission from [17].

To construct a width w folded ribbon knot, we view a polygonal knot diagram K as the centerline of the ribbon. Then the fold lines of the ribbon are perpendicular to the angle bisectors at each of the knot diagram's vertices.

Definition 1. Given an oriented polygonal knot diagram K, we define the *fold angle* at vertex v_i to be the angle θ_i (where $0 \le \theta_i \le \pi$) between edges e_{i-1} and e_i. Then, the *oriented folded ribbon knot of width* w, denoted K_w, is constructed as follows:

1. First, construct the fold lines. At each vertex v_i of K, find the fold angle θ_i. If $\theta_i < \pi$, place a fold line of length $w/\cos(\frac{\theta_i}{2})$ centered at v_i perpendicular to the angle bisector of θ_i. If $\theta_i = \pi$, there is no fold line.

2. Second, add in the ribbon boundaries. For each edge e_i, join the ends of the fold lines at v_i and v_{i+1}. By construction, each boundary line is parallel to, and distance $w/2$ from K.

3. The ribbon inherits an orientation from K.

This construction is illustrated in Figure 88.3. On the left, the fold angle is $\theta_i = \angle ECF$, the angle bisector is DC, and the fold line is AB. Using the geometry of the figure we see that $\angle GAB = \theta_i/2$ in right triangle $\triangle AGB$. Thus $|AB| = w/\cos(\frac{\theta_i}{2})$ guarantees the ribbon width $|AG| = w$. We say that the fold angle is $\theta_i = \angle ECF$ is *positive*, since e_i is to the left of e_{i-1}. If e_i were to the right of e_{i-1}, then it would be *negative*.

Observe that near a fold line, there is a choice of which ribbon lies above the other. Thus a polygonal knot diagram with n vertices has 2^n possible folded ribbon knots depending on the choices made. There is an *overfold* at vertex v_i if the ribbon corresponding to segment e_i is over the ribbon of segment e_{i-1} (see Figure 88.4 right). Similarly, there is an *underfold* if the ribbon corresponding to e_i is under the ribbon of e_{i-1}.

Definition 2. The choice of overfold or underfold at each vertex of K_w is called the *folding information*, and is denoted by F.

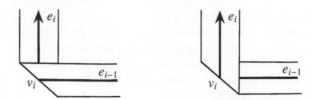

FIGURE 88.4: A right underfold (left) and a right overfold (right). Figure re-used with permission from [17].

In our construction of a folded ribbon knot, we start with a polygonal knot diagram then build the ribbon. Ribbons can be viewed as an immersion of an annulus or a Möbius strip into the plane, with singularities at the fold lines. (Recall that immersions are locally one-to-one maps so our 2-stick unknot satisfies this condition.) There appears to be no restriction placed on the width, and yet very wide ribbons might not be physically possible. For example, in Figure 88.3 (right), the ribbon width cannot be more than the length of e_i. To prevent this, we assume the fold lines are disjoint.

Definition 3. Given an oriented knot diagram K, we say the folded ribbon $K_{w,F}$ of width w and folding information F is *allowed* provided the following hold:

1. The ribbon has no singularities (is immersed), except at the fold lines which are assumed to be disjoint.

2. K_w has consistent crossing information, and moreover this agrees

   (a) with the folding information given by F, and

   (b) with the crossing information of the knot diagram K.

When a folded ribbon $K_{w,F}$ is allowed, the consistent crossing information means that a straight ribbon segment cannot "pierce" a fold, it either lies entirely above or below the fold, or lies between the two ribbons segments joined at the fold. From now on **we assume that our folded ribbon knots have an allowed width**. This is a reasonable assumption, since we can always construct folded ribbon knots for "small enough" widths.

Proposition 4 ([4])**.** Given any regular polygonal knot diagram K and folding information F, there is a constant $C > 0$ such that an allowed folded ribbon knot $K_{w,F}$ exists for all $w < C$.

88.3   Ribbonlength

Given a particular folded ribbon knot, it is very natural to wonder what is the least length of ribbon needed to tie it. More formally, we define a scale-invariant quantity, called *ribbonlength*, as follows.

Definition 5 ([29])**.** The *(folded) ribbonlength*, $\mathrm{Rib}(K_{w,F})$, of a folded ribbon knot $K_{w,F}$ is the quotient of the length of K to the width w:

$$\mathrm{Rib}(K_{w,F}) = \frac{\mathrm{Len}(K)}{w}.$$

When minimizing the ribbonlength of a folded ribbon knot, we have two choices. We can fix the width and minimize the length, or, we can fix the length and maximize the width. As mentioned above, Kauffman [29] gave the conjectured minimal ribbonlength for the trefoil and figure-eight knots. In 2008, B. Kennedy, T.W. Mattman, R. Raya and D. Tating [31] gave upper bounds on the ribbonlength of the $(p, 2)$, $(q + 1, q)$, and $(2q + 1, q)$ families of torus knots, using ideas in the master's theses of DeMaranville [16] and Kennedy [30]. They did not expect the bounds to be minimal, and in fact gave shorter versions of the $(5, 2)$ and $(7, 2)$ torus knots.

An interesting open question is to understand the relationship between the ribbonlength of a knot K and its crossing number[2] $\mathrm{Cr}(K)$. In particular, to find constants c_1, c_2, α, β such that

$$c_1 \cdot (\mathrm{Cr}(K))^{\alpha} \leq \mathrm{Rib}(K_w) \leq c_2 \cdot (\mathrm{Cr}(K))^{\beta}. \tag{88.1}$$

Y. Diao and R. Kusner [22] conjecture that $\alpha = 1/2$ and $\beta = 1$ in Equation 88.1. Kennedy *et al.* [31] made a first pass at the bounds in Equation 88.1 by using the fact that the crossing number of a (p, q) torus knot is $\min\{p(q - 1), q(p - 1)\}$. This allowed them to show the ribbonlength's upper bound is quadratic in the crossing number for the $(p, 2)$ torus knots, and linear for the $(q + 1, q)$ and $(2q + 1, q)$ torus knots.

In 2017, G. Tian [38] used grid diagrams of knots to make further progress. A *grid diagram*, with grid number n, is an $n \times n$ square grid with n X's and n O's arranged so that each row and column contains exactly one X and one O. A grid diagram of the trefoil knot is given on the left in Figure 88.5. It turns out that every knot has a grid diagram associated to it [14, 36].

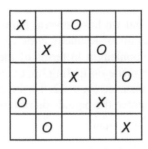 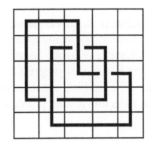

FIGURE 88.5: A grid diagram (left) and corresponding knot diagram (right) of a trefoil knot.

Now, any grid diagram gives a knot diagram in a standard way: connect O to X in each row, connect X to O in each column, and have the vertical line segments always cross over the horizontal ones. This process is illustrated on the right in Figure 88.5. From here, we can see that any grid diagram of a knot can be used to create a folded ribbon knot whose width is the sidelength of the squares in the grid.

---

[2]The crossing number of a knot is the minimum number of crossings in any regular knot diagram of the knot.

The *grid index*, $g(K)$, is the minimal grid number, and others have proved [6, 25] that the grid index is bounded above using the crossing number: $g(K) \leq \text{Cr}(K) + 2$. Tian [38] argued that given a knot K represented by a $g(K) \times g(K)$ grid diagram, we can estimate the length of the folded ribbon knot. Each horizontal distance between the X and O is at most $g(K) - 1$, hence the sum of all horizontal distances is at most $g(K)(g(K) - 1)$. The same is true for vertical distances. Thus we obtain

$$\text{Rib}(K) \leq 2g(K)(g(K) - 1) \leq 2(\text{Cr}(K) + 1)(\text{Cr}(K) + 2) \leq 12(\text{Cr}(K))^2.$$

Tian then used a grid diagram of n-twist knots T_n, to show the ribbonlength of the corresponding folded ribbon T_n knot is $2(4n + 8)$. Since the crossing number of T_n is $n + 2$, we find $\text{Rib}(T_n) \leq 8\text{Cr}(T_n)$. A similar argument for (p, q) torus knots shows the ribbonlength is also bounded above by $8\text{Cr}(K)$. Figure 88.6 shows the kind of grid diagrams for torus and twist knots used to obtain the bounds.

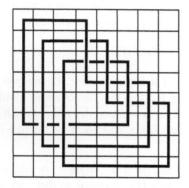 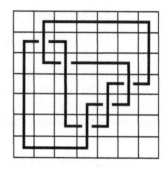

FIGURE 88.6: The grid diagram of the $(5, 3)$ torus knot on the left, and the 3-twist knot on the right.

Altogether, we know that $\beta \leq 2$ in Equation 88.1, and we have several families of knots where $\beta = 1$. Tian's results both improve and extend the results of Kennedy *et al.* [31] described above. So far, no one has proved any lower bounds on ribbonlength in terms of the crossing number. We end this section by noting that progress has been made on finding upper and lower bounds of ropelength in terms of the crossing number (see for instance [8, 9, 12, 13, 19, 20, 21]) . It may be possible to use ideas from this related problem to improve the bounds on ribbonlength.

88.4   Ribbon Equivalence and Ribbonlength

As we saw in Section 88.2, there is a choice of folding information at each vertex of the knot diagram. It turns out that the folding information affects the ribbonlength. To get a handle on these differences, we give three different definitions of ribbon equivalence (see also [4, 17]). We first begin by defining ribbon linking number.

The *linking number* is an invariant from knot theory (see for instance [1, 27, 34]) used to determine the degree to which components of a link are joined together. Given an oriented two-component link $L = A \cup B$, the linking number $\mathrm{Lk}(A, B)$ is defined to be one half the sum of $+1$ crossings and -1 crossings between A and B. (See Figure 88.7.)

FIGURE 88.7: The crossing on the left is labelled -1, the crossing on the right $+1$. Figure re-used with permission from [17].

Although we have described the construction of folded ribbon knots in $\mathbb{R}^2$, we can also consider the ribbons that these diagrams represent in $\mathbb{R}^3$. That is, as *framed knots*. The ribbon linking number was defined for these ribbons (see [10, 11, 27]), but it equally applies to our situation.

Definition 6. Given an oriented folded ribbon knot $K_{w,F}$, we define the *(folded) ribbon linking number* to be the linking number between the knot diagram and one boundary component of the ribbon. We denote this as $\mathrm{Lk}(K_{w,F})$, or $\mathrm{Lk}(K_w)$.

We are now ready to define three different kinds of ribbon equivalence, starting with the most restrictive.

Definition 7. (Link equivalence) Two oriented folded ribbon knots are *(ribbon) link equivalent* if they have equivalent knot diagrams with the same ribbon linking number.

For example, the left and center folded ribbon unknots in Figure 88.8 are link equivalent (with ribbon linking number 0), while the one on the right is not link equivalent to them (with ribbon linking number -2). This example shows that there can be different-looking folded ribbon knots with the same ribbon linking number.

Definition 8. (Topological equivalence) Two oriented folded ribbon knots are *topologically (ribbon) equivalent* if they have equivalent knot diagrams and, when considered as ribbons in $\mathbb{R}^3$, both ribbons are topologically equivalent to a Möbius strip or both ribbons are topologically equivalent to an annulus.

We say the knot diagrams are equivalent if the corresponding knots are equivalent (ambient isotopic). In the definition of topological equivalence note that the folding information is not considered, only the knot diagram and the topological type of the ribbon. For example, all of the 4-stick folded ribbon unknots in Figure 88.8 are topologically equivalent.

Definition 9. (Knot diagram equivalence) Two folded ribbon knots are *knot diagram equivalent* if they have equivalent knot diagrams.

This is the least restrictive kind of equivalence. For example, the 3-stick and 4-stick folded ribbon unknots in Figures 88.8 and 88.9 are knot diagram equivalent, but are not topologically equivalent or link equivalent.

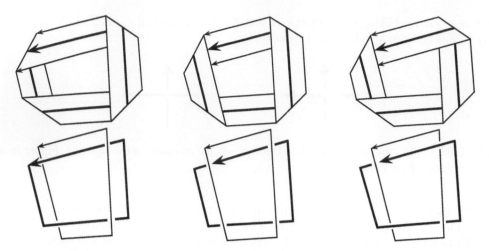

FIGURE 88.8: Three different 4-stick folded ribbon unknots (above) with corresponding link diagrams of one boundary component and the knot diagram (below). The left and center ribbon unknots have ribbon linking number 0, while the one on the right has ribbon linking number −2.

Remark 10. The *ribbonlength problem* asks us to minimize the ribbonlength of a folded ribbon knot, while staying in a fixed topological knot type. That is, with respect to knot diagram equivalence of folded ribbon knots. We can also ask to minimize the ribbonlength of folded ribbon knots with respect to topological and link equivalence.

We pause to remark that the previous work on ribbonlength [29, 31, 38] found upper bounds on the ribbonlength with respect to knot diagram equivalence. Together with undergraduate students [4, 17], we have made a first pass at finding bounds on ribbonlength with respect to topological and link equivalence. We start by considering unknots.

Any polygonal unknot diagram can be reduced to a 2-stick unknot, and the width of such a diagram can be made large as we like. Thus the minimum ribbonlength of any unknot with respect to knot diagram equivalence is 0. If we minimize ribbonlength with respect to topological equivalence, then we have already considered the topological annulus in the 2-stick unknot. The 3-stick unknot is a Möbius band, and we find upper bounds on the ribbonlength in the following proposition.

Proposition 11 ([17]). The minimum ribbonlength of an 3-stick folded ribbon unknot $K_{w,F}$ is less than or equal to

1. $3\sqrt{3}$ when the folds are all the same type (Figure 88.9 left),

2. $\sqrt{3}$ when one fold is of different type to the other two (Figure 88.9 center).

It turns out we can do even better and show that the equilateral triangle gives the ribbonlength minimizer in the first case.

Theorem 12 ([17]). The minimum ribbonlength for the 3-stick folded unknot is $3\sqrt{3}$ where all

FIGURE 88.9: On the left and the center, the two different kinds of folding information for 3-stick unknots. On the right, a 5-stick folded ribbon unknot. Figure re-used with permission from [17].

folds have the same folding information, and occurs when the knot diagram is an equilateral triangle.

Thus the ribbonlength of an unknot is less than or equal to $\sqrt{3}$ when the ribbon is topologically equivalent to a Möbius strip, and is 0 when the ribbon is topologically equivalent to an annulus. When ribbon linking number is taken into consideration the situation is more complex. For example, for 3-stick unknots with ribbon linking number ± 3 the minimum ribbonlength is equal to $3\sqrt{3}$, and is less than or equal to $\sqrt{3}$ for 3-stick unknots with ribbon linking number ± 1. What about ribbonlength for other values of ribbon linking numbers ($\pm 2, \pm 3$, etc.)? Assume that all folds of an n-stick folded ribbon unknot are the same. Then for odd $n \geq 4$ the ribbon linking number is n, while for even $n \geq 4$ the ribbon linking number is $n/2$. In either case we can get an upper bound on the ribbonlength of n-stick unknots by considering the case where the knot diagrams are regular n-gons (as in Figure 88.9 right).

Proposition 13 ([17]). The ribbonlength of an n-stick folded ribbon unknot (for $n \geq 4$) is less than or equal to $n \cot(\frac{\pi}{n})$.

Thus Proposition 13 gives a reasonable upper bound on ribbonlength with respect to link equivalence for n-stick unknots with n odd sides and ribbon linking number n. Just how far can we improve our ribbonlength bounds (with respect for link equivalence) for the ribbon unknots with any ribbon linking number?

Moving now from unknots to nontrivial knots, it is natural to wonder about the relationship between the ribbonlength of a knot, the number of edges in the knot diagram, and the projection stick index of the knot. Recall that Kennedy *et al.* [31], found that there was a smaller ribbonlength for the $(5, 2)$ and $(7, 2)$ torus knots, simply by adding two more edges to the knot diagram and rearranging. The knot diagram for the $(5, 2)$ torus knot with smaller ribbonlength is shown on the left in Figure 88.10. In [17], we showed that the two folded ribbon $(5, 2)$ torus knots corresponding to the knot diagrams in Figure 88.10 are not ribbon link equivalent. We found a sequence of Reidemeister moves connecting the two knot diagrams, and we showed the corresponding folded ribbons differed by a full twist (regardless of the starting folding information). This example shows two interesting things. Firstly, the minimal ribbonlength of a knot (with respect to knot diagram equivalence) does not necessarily occur when the knot diagram has its number of edges equal to the projection stick index. Secondly, the candidate for minimal ribbonlength is *not* ribbon link equivalent to the folded ribbon knot whose diagram has its number of edges equal to the projection stick index. There is much to think about here!

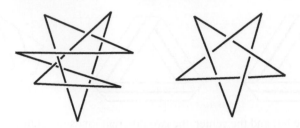

FIGURE 88.10: Two different polygonal diagrams of the $(5, 2)$ torus knot. The one on the left has a smaller ribbonlength than the one on the right, despite having 2 more edges. Figure re-used with permission from [17].

In summary, we have seen that folded ribbon knots are both interesting to study in their own right, and have many connections to other parts of knot theory. We close by listing just some of the open questions that have come up in this survey of folded ribbon knots.

- Prove that the ribbonlength of the trefoil knot is minimized by the configuration with pentagonal boundary given in Figure 88.1 (right).

- Improve the upper and lower bounds relating ribbonlength to the crossing number given in Equation 88.1.

- Is a minimal ribbonlength folded ribbon knot always given by a knot diagram that has more edges than the projection stick index?

- Is a minimal ribbonlength folded ribbon knot ever ribbon link equivalent to a folded ribbon knot whose diagram has its number of edges equal to the projection stick index?

- What are the ribbon linking numbers generated by a knot diagram when its number of edges equal to the projection stick index? How is this related to minimum ribbonlength?

- Minimize the ribbonlength of nontrivial knots with respect to link equivalence.

Acknowledgments

I wish to thank Jason Cantarella for many helpful math conversations, and Aaron Abrams for his discussions on the definition of a folded ribbon. I also thank John M. Sullivan and Nancy Wrinkle for discussions on smooth ribbon knots immersed in the plane [18].

I have worked with undergraduate students over many years on the folded ribbonlength problem. Special thanks go to former Smith College students: Shivani Aryal, Eleanor Conley, Shorena Kalandarishvili, Emily Meehan and Rebecca Terry. Special thanks also go to former Washington & Lee Students: Mary Kamp, and Catherine (Xichen) Zhu.

88.5 Bibliography

[1] C. Adams, 1994: *The Knot Book: An Elementary Introduction to the Mathematical Theory of Knots,* New York, W.H. Freeman and Company.

[2] C. Adams, T. Shayler, 2009: The projection stick index of knots. *J. Knot Theory Ramifications* **18** no. 7, 889–899.

[3] C. Adams, D. Collins, K. Hawkins, C. Sia, R. Silversmith, B. Tshishiku, 2011: Planar and spherical stick indices of knots. *J. Knot Theory Ramifications* **20** no. 5, 721–739.

[4] E. Denne, J.C. Harden, T. Larsen, E. Meehan, 2020: Ribbonlength of families of folded ribbon knots. Preprint: https://arxiv.org/abs/2010.04188.

[5] C.W. Ashley, 1944: *The Ashley Book of Knots.* New York Doubleday.

[6] Y. Bae, C-Y. Park, 2000: An upper bound of arc index of links. *Math. Proc. Cambridge Philos. Soc.* **129**, Issue 3 Nov., 491–500.

[7] D. Buck, E. Flapan (editors), 2009: *Applications of knot theory. Proceedings of Symposia in Applied Mathematics* Vol. 66, American Mathematical Society.

[8] G. Buck, J. Simon, 1999: Thickness and crossing number of knots. *Topology Appl.* **91** no. 3, 245–257.

[9] G. Buck, J. Simon, 2007: Total curvature and packing of knots. *Topology Appl.* **154**, 192–204.

[10] G. Călugăreanu, 1959: L'integrale de Gauss et l'analyse des noeds trideminsionnels. *Revue de Mathématiques Pures et Appliquées* **4**, 5–20.

[11] G. Călugăreanu, 1961: Sur les classes d'isotope des neuds tridimensional et leur invariants. *Czechsolovak Math. J.* **11**, 588–625.

[12] J. Cantarella, X.W.C. Faber, C.A. Mullikin, 2004: Upper bounds for ropelength as a function of crossing number. *Topology Appl.* **135**, 253-264.

[13] J. Cantarella, R.B. Kusner, J.M. Sullivan, 2002: On the minimum ropelength of knots and links. *Invent. Math.* **150**, 257–286

[14] P.R. Cromwell, 1995: Embedding knots and links in an open book I: Basic properties. *Topology Appl.* **64**, 37–58.

[15] H.M. Cundy, A.P. Rollet, 1952: *Mathematical Models.* Oxford, Clarendon Press.

[16] L. DeMaranville, 1999: *Construction of polygons by tying knots with ribbons.* Master's thesis, CSU Chico, Chico CA.

[17] E. Denne, M. Kamp, R. Terry, X. Zhu, 2017: Ribbonlength of folded ribbon unknots in the plane. *Knots, Links, Spatial Graphs and Algebraic Invariants.* Contemporary Mathematics, Vol. 689, Providence RI, Amer. Math. Soc., 37–51.

[18] E. Denne, J. M. Sullivan, N. Wrinkle, Ribbonlength for knot diagrams. In preparation.

[19] Y. Diao, C. Ernst, V. Kavuluru, U. Ziegler, 2006: Numerical upper bounds on ropelengths of large physical knots. *J. Phys. A: Math. Gen.* **39**, 4829–4843.

[20] Y. Diao, C. Ernst, A. Por, U. Ziegler, 2009: The ropelengths of knots are almost linear in terms of their crossing numbers. Preprint, arXiv: 0912.3282v1.

[21] Y. Diao, C. Ernst, Xingxing Yu, 2004: Hamiltonian knot projections and lengths of thick knots. *Topology Appl.* **136** no. 3, 7–36.

[22] Y. Diao, R. Kusner. Private communication. July 2018.

[23] M. H. Eggar, 2000: On White's formula. *J. Knot Theory Ramifications* **9** no. 5, 611–615.

[24] O. Gonzalez, J.H. Maddocks, 1999: Global curvature, thickness, and the ideal shapes of knots. *Proc. Nat. Acad. Sci.* (USA) **96**, 4769–4773

[25] K. Hong, H. Kim, S. Oh, S. No, 2014: Minimum lattice length and ropelength of knots, *J. Knot Theory Ramifications* **23** no. 7, 1460009 [10 pages].

[26] D.A. Johnson, 1957: *Paper Folding for the Mathematics Class*. Washington D.C, National Council of Teachers of Mathematics.

[27] L.H. Kauffman, 1987: *On Knots*. Princeton University Press, Princeton NJ, 498 pp.

[28] L.H. Kauffman, 1990: An Invariant of Regular Isotopy. *Trans Amer. Math. Soc.* **318**, no. 2, 417–471.

[29] L.H. Kauffman, 2005: Minimal flat knotted ribbons. *Physical and numerical models in knot theory,* Singapore, World Sci. Publ., Vol. 36 Ser. Knots Everything, 495–506.

[30] B. Kennedy, 2007: *Ribbonlength of torus knots*. Master's Thesis, CSU Chico CA.

[31] B. Kennedy, T.W. Mattman, R. Raya, D. Tating, 2008: Ribbonlength of torus knots. *J. Knot Theory Ramifications* **17** no. 1, 13–23.

[32] T. Kuriya, O. Shehab, 2014: The Lomonaco-Kauffman conjecture. *J. Knot Theory Ramifications*, **23** no. 1, 1450003 [20 pages].

[33] R.A. Litherland, J. Simon, O. Durumeric, E. Rawdon, 1999: Thickness of knots. *Topology* **91**, 233–244.

[34] C. Livingston, 1993: *Knot Theory*. Carus Mathematical Monographs Vol. 24, Mathematical Association of America, Washington DC, 240 pp.

[35] S.J. Lomonaco, L. Kauffman, 2008: Quantum knots and mosaics, *Quantum Inf. Process.*, **7**, 85–115.

[36] P.S. Ozsváth, A.I. Stipsicz, Z. Szabó, 2015: *Grid Homology for Knots and Links*. Mathematical Surveys and Monographs, Vol. 208, American Mathematical Society, Providence RI.

[37] P. Pieranski, 1998: In search of ideal knots. In A. Stasiak, V. Katritch and L. Kauffman, editors, *Ideal Knots*, World Scientific Publ., 20–41.

[38] G.M. Tian, 2017: *Linear Upper Bound on the Ribbonlength of Torus Knots and Twist Knots*. Preprint, June 2017: https: //math.mit.edu/research/highschool/rsi/documents/2017Tian.pdf.

[39] D. Wells, 1991: *The Penguin Dictionary of Curious and Interesting Geometry.* New York, Penguin Books.

[40] J. H. White, 1969: Self-linking and Gauss integral in higher dimensions, *Amer. J. of Math.* **91**, 693–728.

Part XV

Knots and Science

Chapter 89

DNA Knots and Links

Isabel K. Darcy, University of Iowa

89.1 Introduction

Imagine that you have a very long knotted rope that forms a closed loop. Suppose you could cut the rope anywhere you choose, and pass another segment through this break before resealing the break. What is the likelihood you could unknot this rope blindfolded? This is what the protein topoisomerase does when acting on circular DNA. Topoisomerase is a protein that can change the topology of DNA by cutting the DNA, allowing another segment of DNA to pass through the cut before resealing the break. Topoisomerase is extremely good at unknotting DNA. Why this may be the case and other topological issues involving DNA will be discussed below.

The structure of DNA will be introduced in section 89.2. Topological issues involving double-stranded DNA and linking number will be discussed in Section 89.3. How DNA knots are created and how they can be used to analyse protein action will be the topic of Section 89.4. DNA links will be briefly discussed in Section 89.5, followed by concluding remarks in Section 89.6.

89.2 DNA Structure

DNA can be thought of as a twisted ladder [1] as shown in Fig. 89.1a. This double helical model was proposed in 1953 [22, 47, 52, 21]. According to Franklin and Gosling [22], while the helical structure could not be proven in 1953, the "helical structure" is "highly probable" where "one repeating unit contains ten nucleotides on each of two . . . co-axial molecules," the "period is 34 Å," "phosphate groups lie on the outside of the structural unit, on a helix of diameter about 20 Å," and "the sugar and base groups must accordingly be turned inwards towards the helical axis." Others proposed various helical structures for DNA (e.g., [25, 26]) including a triple helix [29]. DNA can form a variety of specialized structures including triple and quadruple helices [38], but we will focus on the standard double helical DNA structure.

The rungs on the double-stranded DNA (dsDNA) ladder consist of base pairs where adenine (A) normally pairs with thymine (T) and guanine (G) normally pairs with cytosine (C) [47]. This Watson-Crick base pairing allows for the storage and replication of DNA as well as the transcription of DNA into RNA, which is then translated into protein [47, 48]. Linear DNA will also circularize if it has sticky ends: complementary single-stranded segments of DNA at each end (Fig. 89.1b).

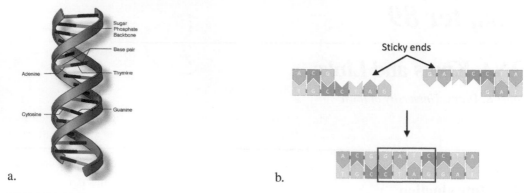

FIGURE 89.1: (**a**) Structure of DNA is helical where the rungs of the ladder are formed by the base pairs A-T and C-G (figure courtesy: National Human Genome Research Institute. https://www.genome.gov/glossary/resources/double_helix.ppt). (**b**) If a DNA segment has a single-stranded region at the end, it is said to have a sticky end. For example, the top right segment has an overhang consisting of bases G, A, T, C. These are complementary to the single-stranded end of the other DNA segment. Thus these two segments can stick together via base pairing as shown on the bottom. If the two sticky ends belong to a single linear DNA molecule, then when the two ends join, a circular DNA molecule is formed (figure modified from Simon Caulton's https://commons.wikimedia.org/wiki/File:BamHI.png).

89.3 DNA Replication, Linking Number, and Supercoiling

The structure of DNA both causes and prevents topological problems. With respect to the latter, DNA does not normally form a Möbius band. If a linear piece of DNA has sticky ends, it can circularize. However, the two strands of dsDNA are chemically oriented in an anti-parallel fashion. Since the strands are anti-parallel, when the sticky ends join, a twisted annulus is formed instead of a Möbius band. The two edges of the annulus represent the two strands of the DNA. One can "replicate" a twisted annulus by cutting it along its centerline (axial curve). Note one obtains two twisted annuli that are linked together. However, if one cuts a Möbius band along its centerline, one obtains a single twisted annulus that is twice as long as the original Möbius band. Thus if DNA were to form a Möbius band, this would cause difficulties for DNA replication when trying to separate the two joined copies of DNA as a cell divides into two daughter cells.

The twisted annulus conformation of DNA, however, is also a topological problem. The two strands of dsDNA must separate in order for DNA replication and transcription to occur. But the two strands of circular dsDNA are linked. While human DNA is linear, it is attached to a scaffold at many places. Thus knotting and linking can also be a problem for linear DNA.

The linking of two oriented closed circles can be measured using the topological invariant linking number (Lk). To calculate the linking number, one first projects the link onto a 2-dimensional plane. Each crossing is given a sign dependent on how the link is oriented as shown in Fig. 89.2a. Lk is one half the sum of the signed crossings between the two curves. Note self-crossings are not included in this sum. See Fig. 89.2b, c and d for examples.

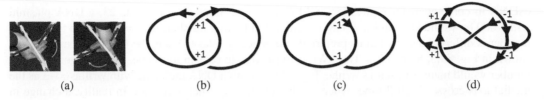

FIGURE 89.2: (**a**) Signed crossings. The yellow and red arrows indicate the global orientation of a knot or link. One can determine the sign of the crossing by rotating the yellow arrow on the over strand so that its head aligns with the red arrowhead on the under strand. The counter-clockwise rotation indicated in the left figure means this is a +1 crossing. The clockwise rotation indicated in the right figure means this is a −1 crossing. Alternatively to show a crossing is positive one can place their right hand around the over strand as shown on the left with fingres pointing in the direction of the arrow on the under strand, then the thumb points out in the direction of the over strand. Similarly, one can show a crossing is negative using one's left hand as shown on the right. (**b**) Linking number depends on how the circles are oriented. The linking number for the two circles with orientation shown is $\frac{1}{2}(1 + 1) = +1$. (**c**) If one changes the orientation of one of the circles in (b), one obtains a link with Lk = -1. (**d**) This link has Lk = 0 despite the fact that these two circles cannot be separated.

Although the two strands of dsDNA are chemically oriented in an anti-parallel fashion, for mathematical convenience, the strands are given a parallel orientation when calculating linking number between these two strands. A small circular DNA plasmid with 4000 base pairs with about 10 base pairs per turn can have a linking number around 400. These links must be removed when DNA replicates. Thus there were mathematical objections to the helical structure of DNA. In fact alternative non-helical models were proposed in the 1970s [32, 36]. But also in the 1970s, proteins that can change the topology of DNA were discovered [45, 7, 27, 19]. These proteins were named topoisomerase and are responsible for keeping DNA unknotted, unlinked, and nicely organized. Note there are a number of other proteins that are involved in DNA replication, transcription, and organization, but a discussion of those is beyond the scope of this chapter. In Section 89.4, we will discuss topoisomerases and another class of proteins, recombinases, which can create DNA knots and links.

Linking number is the sum of two geometric invariants, twist (Tw) and writhe (Wr) [49, 12, 30, 23, 24]. If DNA is modeled by a twisted annulus, twist measures how many times one strand of DNA twists around the centerline (axial curve) of this annulus. Writhe is the average of the sum of the signed crossing numbers of this centerline averaged over all planar projections. The formula Lk = Tw + Wr can be found in many molecular biology books [4, 51]. For a nice introduction to this formula with many examples, see [50].

The above calculation of Lk = 400 for a circular DNA plasmid with 4000 base pairs was oversimplified. In reality, the energetically preferred number of base pairs per turn depends on the solution in which the DNA resides (for example, how much salt is dissolved in the liquid solution of the test tube or cell containing the DNA). Under standard conditions DNA prefers to twist about 10.5 base pairs per turn. Thus an energetically relaxed circular DNA plasmid with 4000 bp would have an Lk = 381 (4000/10.5 rounded to the nearest integer). However, in normal biological conditions, DNA is usually underwound by about 6. This makes it easier for the two strands of dsDNA to come apart for replication and transcription. Since 6% of 381 is 23, a DNA plasmid with 4000 bp would have an Lk = 358 or a change in linking number, ΔLk = 23. When the linking number is different than that in its preferred energetically relaxed state, the DNA will supercoil as shown in Fig. 89.3. Since Lk = Tw + Wr, if twist were to remain the same, this change in linking number would manifest itself as writhe. Fig. 89.3b shows a DNA molecule with writhe = −3 as the medial axis crosses itself 3 times where each crossing has a negative sign. In reality, a change in linking number will result in some change in both twist and writhe.

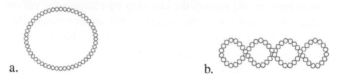

a.   b.

FIGURE 89.3: (**a**) Relaxed DNA (from https://commons.wikimedia.org/wiki/File:Relaxed_DNA_-molecule.jpg). (**b**) supercoiled DNA (from https://en.wikipedia.org/wiki/File: Negative.jpg.)

89.4   DNA Knots

Suppose you need to pack a long extension cord into a stuffed suitcase. If the extension cord contains local knots and is simply thrown into the suitcase in a disorganized manner, it will take up more room. It is easier to pack the suitcase and to work with the extension cord if it does not contain local knots and if it is nicely organized (for example, coiled nicely like a spool of thread). Similarly, for DNA to fit nicely into a cell and for it to be accessible for transcription, it must be kept nicely organized. Knotted DNA can increase the rate of genetic mutations [16].

Topoisomerase is one of the proteins needed for keeping DNA organized, including unknotted and unlinked. Some anti-cancer and some anti-bacterial drugs target topoisomerase since interfering with the unlinking of replicating DNA can lead to cell death. There are several different types of topoisomerase. Some types will only cut a single strand of dsDNA while others will cut both strands of dsDNA. We will focus on the latter. In this case topoisomerase action on dsDNA can be modeled by a crossing change (Fig. 89.4a).

Protein-bound DNA can be modeled by tangles [18]. A tangle consists of strings embedded inside a 3-dimensional ball. The protein-complex is modeled by the 3-dimensional ball. The DNA bound by protein is modeled by the strings embedded within the ball. For example, topoisomerase bound to DNA can be modeled by a ball containing a crossing. Thus, in Figs 89.4a, b, c, the green circles containing a single crossing represents the protein topoisomerase.

A typical system of tangle equations is shown in Fig. 89.4d. Circular DNA bound by a protein complex can be partitioned into two tangles. The tangle B in Fig. 89.4d represents the protein-bound DNA. The tangle U_f represents the remaining DNA unbound by protein. The circular DNA before protein action is called the substrate. After protein action, the product is modeled by changing the tangle B to the tangle E. To model topoisomerase action, changing the tangle B to the tangle E corresponds to a crossing change (Fig. 89.4a). One solution to the system of tangle equations in Fig.

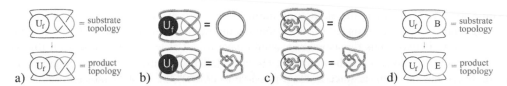

FIGURE 89.4: **(a)** Tangle equation modeling topoisomerase. The topoisomerase binding to a DNA crossing is modeled by the green circle containing a single crossing. The tangle U_f represents the DNA not bound by protein. **(b)** Tangle equation modeling topoisomerase where the DNA substrate topology (starting conformation) is unknotted and the product after topoisomerase action is the four crossing figure 8 knot 4_1. **(c)** A solution to the tangle equation shown in (b). **(d)** Generic tangle model for modeling protein action. The protein action is represented by replacing the tangle B with the tangle E while the unbound DNA remains unchanged and is modeled by the tangle U_f.

89.4b is shown in Fig. 89.4c. There are actually two solutions to the system of tangle equations in Fig. 89.4b. The TopoICE-X software [13] within KnotPlot [35] can be used to solve such systems of equations.

While this tangle model may seem overly simplified and not quite correct, it has been extremely useful. A protein-complex is generally not a ball and protein bound DNA normally wraps around the protein-complex as opposed to being embedded within a ball. However, this model gives an excellent starting point for biologists to model protein action on DNA. It is difficult to model such reactions if one does not know how to draw the DNA. One can solve tangle equations to determine possible topologies for the unbound DNA as well as protein-bound DNA (when not known), after which one can push the protein-bound DNA onto the boundary of the protein and use other techniques to obtain a more accurate model. For example, if you push the two DNA segments forming a crossing onto the boundary of the ball, one can obtain a "hooked juxtaposition": the DNA crossing has the geometry of a hook as shown in Fig. 89.5a. Some topoisomerases will unknot and unlink DNA well below thermodynamic equilibrium [33]. There are many hypotheses regarding why topoisomerase is so efficient at unknotting DNA [2, 53, 31, 40, 44, 42, 43]. In one model, the topoisomerase only performs crossing changes when the DNA crossing has the geometry of a hook [6].

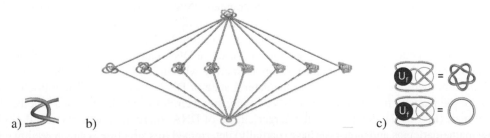

FIGURE 89.5: **(a)** Hooked juxtaposition. **(b)** Possible reaction pathways from the knot 5_1 to the unknot. **(c)** A tangle equation which has no solution as it is not possible to convert the knot 5_1 to the unknot via a single crossing change.

The minimum number of times topoisomerase needs to act to unknot a particular knot is related to a well-known knot invariant: the unknotting number of a knot, which is the minimum number of crossing changes needed to convert this knot into the unknot. For example, the unknotting number of the knot 5_1 is two. Thus topoisomerase would have to act at least twice to unknot this knot. Possible reaction pathways are shown in Fig. 89.5b [13]. This also means that the tangle equation in Fig. 89.5c has no solution since it is not possible to convert the knot 5_1 into the unknot via a single crossing change.

Other proteins can also create DNA knots and links. A recombinase is a protein that will cut two segments of DNA and exchange the ends before resealing the break. Site-specific recombinases bind specific DNA sequences. These sequences are modeled by the arrows in Fig. 89.6. If the two binding sites are oriented as shown in Fig. 89.6a, then these sequences are said to be directly repeated and recombination at these sites will change the number of components, for example converting a one-component knot into a two-component link. If the two binding sites are oriented as shown in Fig. 89.6b, then these sequences are said to be inversely repeated and recombination at these sites will preserve the number of components. Recombination can result in gene rearrangement. Viruses use recombination to insert their genome into a host's genome. Recombinases are also used in gene therapy. Different recombinases have different mechanisms (e.g. [18, 11, 14, 39]). See [41] for a nice 3D model of Xer recombination. The software TopoICE-R [15] in KnotPlot and the software TangleSolve [34] can be used to solve tangle equations modeling recombination.

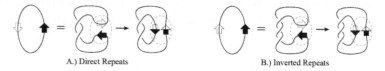

A.) Direct Repeats   B.) Inverted Repeats

FIGURE 89.6: Recombination.

89.5   DNA Catenanes

The linking of the two single strands of double-stranded DNA was discussed in Section 89.3. Two circular dsDNA molecules can also be linked. To differentiate the former from the latter, biologists use the term *catenane* for the latter. Recombinases can form both DNA knots and catenanes. As mentioned in Section 89.3, DNA replication can result in DNA catenanes which are unlinked by topoisomerase. While in most organisms, topoisomerase will unlink DNA below thermodynamic equilibrium, there are organisms where the natural state of DNA is catenated. The mitochondrial DNA of the parasite trypanosomes consists of many small circles of DNA catenated together like chain mail (Fig. 89.7) along with a few larger circles of DNA intertwined with the small circles. How mathematicians and biologists have (partially) determined this structure is an interesting topological problem [10, 17, 28]. How this DNA replicates is another interesting topological problem [37, 9]. Since trypanosomes infect tsetse flies, which infect humans and other mammals with potentially fatal diseases such as sleeping sickness and Chagas disease, there is much interest in understanding these organisms.

FIGURE 89.7: **(a)** Chain mail model of mini-circle DNA. **(b)** Electron microscopy of a small portion of mini-circle DNA extracted from the mitochondrial DNA of the parasite trypanosomes. Note that in both figures, one circle is linked to three other circles. Figures are from [10] (reprinted with permission from Elsevier).

89.6 Conclusion

The structure of DNA causes some interesting topological problems involving knotting, linking and tangling. Knots, links, and tangles have been used to probe the structure of DNA as well as how proteins act on DNA. They have also been used to predict experimental results, determine possible reaction pathways, and to probe how DNA is packed in cells. For more on DNA topology, please see [8, 3, 46, 5, 20].

89.7 Bibliography

[1] DNA from the beginning: An animated primer of 75 experiments that made modern genetics. http://www.{DNA}ftb.org/19/; accessed 13-August-2018.

[2] A.D. Bates and A. Maxwell. DNA topology: topoisomerases keep it simple. *Curr Biol.*, 7(12):R778–81, 1997.

[3] A.D. Bates and A. Maxwell. *DNA Topology*. Oxford Bioscience. OUP Oxford, 2005.

[4] Jeremy M. (Jeremy Mark) Berg, Tymoczko JL, and Stryer L. *Biochemistry*. W.H. Freeman, fifth edition, 2002.

[5] Dorothy Buck. DNA topology. In *Applications of Knot Theory*, volume 66 of *Proc. Sympos. Appl. Math.*, pages 47–79. Amer. Math. Soc., Providence, RI, 2009.

[6] G.R. Buck and E.L. Zechiedrich. DNA disentangling by type-2 topoisomerases. *J Mol Biol.*, 340(5):933–939, 2004.

[7] James C. Wang. Reflections on an accidental discovery. In *DNA Topoisomerases in Cancer Therapy: Present and Future*, pages 1–13. Springer, 2003.

[8] C.R. Calladine, H. Drew, B. Luisi, and A. Travers. *Understanding DNA: The Molecule and How It Works*. Elsevier Science, third edition.. edition, 2004.

[9] Junghuei Chen, Paul T. Englund, and Nicholas R. Cozzarelli. Changes in network topology during the replication of kinetoplast DNA. *EMBO Journal*, 14(24):6339–6347, 1995.

[10] Junghuei Chen, Carol A Rauch, James H White, Paul T Englund, and Nicholas R Cozzarelli. The topology of the kinetoplast DNA network. *Cell*, 80(1):61–69, 1995.

[11] N.J. Crisona, R.L. Weinberg, B.J. Peter, D.W. Sumners, and N.R. Cozzarelli. The topological mechanism of phage lambda integrase. *J Mol Biol.*, 289(4):747–75, 1999.

[12] George C˘ alugăreanu. L'intégrale de Gauss et l'analyse des nœuds tridimensionnels. *Rev. Math. Pures Appl.*, 4:5–20, 1959.

[13] I. K. Darcy, R. G. Scharein, and A. Stasiak. 3d visualization software to analyze topological outcomes of topoisomerase reactions. *Nucleic Acids Research*, 36(11):3515–3521, 2008.

[14] Isabel K. Darcy. Biological distances on DNA knots and links: Applications to XER recombination. *J. Knot Theory Ramifications*, 10(2):269–294, 2001. Knots in Hellas '98, Vol. 2 (Delphi).

[15] Isabel K. Darcy and Robert G. Scharein. TopoICE-R: 3D visualization modeling the topology of DNA recombination. *Bioinformatics*, 22(14):1790–1791, 2006.

[16] Richard W. Deibler, Jennifer K. Mann, De Witt L. Sumners, and Lynn Zechiedrich. Hin-mediated DNA knotting and recombining promote replicon dysfunction and mutation. *BMC Molecular Biology*, 8(1):44, May 2007.

[17] Y. Diao, K. Hinson, Y. Sun, and J. Arsuaga. The effect of volume exclusion on the formation of DNA minicircle networks: implications to kinetoplast DNA. *J. Phys. A*, 48(43):435202, 13, 2015.

[18] C. Ernst and D. W. Sumners. A calculus for rational tangles: Applications to DNA recombination. *Math. Proc. Cambridge Philos. Soc.*, 108(3):489–515, 1990.

[19] Leroy F. Liu, Clare Liu, and Bruce Alberts. T4 dna topoisomerase: A new ATP-dependent enzyme essential for initiation of t4 bacteriophage dna replication. *Nature*, 281:456–61, 1979.

[20] Erica Flapan. *Knots, Molecules, and the Universe: An Introduction to Topology*. American Mathematical Society, Providence, RI, 2016. With contributions by Maia Averett, Lance Bryant, Shea Burns, Jason Callahan, Jorge Calvo, Marion Moore Campisi, David Clark, Vesta Coufal, Elizabeth Denne, Berit Givens, McKenzie Lamb, Emille Davie Lawrence, Lew Ludwig, Cornelia Van Cott, Leonard Van Wyk, Robin Wilson, Helen Wong and Andrea Young.

[21] Rosalind E. Franklin and R. G. Gosling. Evidence for 2-chain helix in crystalline structure of sodium deoxyribonucleate. *Nature*, 172(4369), July 1953.

[22] Rosalind E. Franklin and R. G. Gosling. Molecular configuration in sodium thymonucleate. *Nature*, 171(4356), April 1953.

[23] F. Brock Fuller. The writhing number of a space curve. *Proc. Nat. Acad. Sci. U.S.A.*, 68:815–819, 1971.

[24] F. Brock Fuller. Decomposition of the linking number of a closed ribbon: a problem from molecular biology. *Proc. Nat. Acad. Sci. U.S.A.*, 75(8):3557–3561, 1978.

[25] Sven Furberg. *An X-ray study of some nucleosides and nucleotides*. PhD thesis, University of London, 1949.

[26] Sven Furberg. On the structure of nucleic acids. *Acta Chemica Scandinavica*, 6:634–640, 1952.

[27] M. Gellert, K. Mizuuchi, M. H. O'Dea, and H. A. Nash. DNA gyrase: An enzyme that introduces superhelical turns into DNA. *Proceedings of the National Academy of Sciences*, 73(11):3872–3876, 1976.

[28] Davide Michieletto, Davide Marenduzzo, and Enzo Orlandini. Is the kinetoplast DNA a percolating network of linked rings at its critical point? *Physical Biology*, 12(3):036001, 2015.

[29] Linus Pauling and Robert B Corey. A proposed structure for the nucleic acids. *Proceedings of the National Academy of Sciences*, 39(2):84–97, 1953.

[30] William F. Pohl. The self-linking number of a closed space curve. *J. Math. Mech.*, 17:975–985, 1967/1968.

[31] J Roca. Varying levels of positive and negative supercoiling differently affect the efficiency with which topoisomerase II catenates and decatenates DNA. *J Mol Biol.*, 305(3):441–50, 2001.

[32] G. A. Rodley, R. S. Scobie, R. H. T. Bates, and R. M. Lewitt. A possible conformation for double-stranded polynucleotides. *Proceedings of the National Academy of Sciences of the United States of America*, 73(9):2959–2963, September 1976.

[33] V.V. Rybenkov, C. Ullsperger, A.V. Vologodskii, and N.R. Cozzarelli. Simplification of DNA topology below equilibrium values by type II topoisomerases. *Science*, 277:690–693, 1997.

[34] Y. Saka and M. Vazquez. TangleSolve: Topological analysis of site-specific recombination. *Bioinformatics*, 18:1011–1012, 2002.

[35] Robert G. Scharein. *Interactive Topological Drawing*. PhD thesis, Department of Computer Science, The University of British Columbia, 1998.

[36] V. Sesisekharan, N. Pattabiraman, and Goutam Gupta. Some implications of an alternative structure for DNA. *Proceedings of the National Academy of Sciences of the United States of America*, 75(9):4092–4096, September 1978.

[37] T A Shapiro and P T Englund. The structure and replication of kinetoplast DNA. *Annual Review of Microbiology*, 49(1):117–143, 1995. PMID: 8561456.

[38] Richard R. Sinden. DNA structure and function, 1994.

[39] Robert Stolz, Masaaki Yoshida, Reuben Brasher, Michelle Flanner, Kai Ishihara, David J. Sherratt, Koya Shimokawa, and Mariel Vazquez. Pathways of DNA unlinking: A story of stepwise simplification. *Scientific Reports*, 7:12420, 2017.

[40] S Trigueros, J Salceda, I Bermudez, X Fernandez, and J. Roca. Asymmetric removal of supercoils suggests how topoisomerase II simplifies DNA topology. *J Mol Biol.*, 335(3):723–31, 2004.

[41] Mariel Vazquez, Sean D. Colloms, and De Witt Sumners. Tangle analysis of Xer recombination reveals only three solutions, all consistent with a single three-dimensional topological pathway. *Journal of Molecular Biology*, 346(2):493–504, 2005.

[42] Alexander Vologodskii. Disentangling dna molecules. *Physics of Life Reviews*, 18:118 – 134, 2016.

[43] Alexander Vologodskii. Type ii topoisomerases: Experimental studies, theoretical models, and logic: Reply to comments on disentangling dna molecules. *Physics of Life Reviews*, 18:165 – 167, 2016.

[44] AV Vologodskii, W Zhang, VV Rybenkov, AA Podtelezhnikov, D Subramanian, JD Griffith, and NR. Cozzarelli. Mechanism of topology simplification by type II DNA topoisomerases. *Proc Natl Acad Sci*, 98(6):3045–9, 2001.

[45] James C. Wang. Interaction between DNA and an Escherichia coli protein ω. *Journal of Molecular Biology*, 55(3):523 – IN16, 1971.

[46] J.C. Wang. *Untangling the Double Helix: DNA Entanglement and the Action of the DNA Topoisomerases*. Cold Spring Harbor Laboratory Press, 2009.

[47] J. D. Watson and F. H. C. Crick. Molecular structure of nucleic acids: A structure for deoxyribose nucleic acid. *Nature*, 171(4356), April 1953.

[48] James D. Watson. Genetical implications of the structure of deoxyribonucleic acid, 1953.

[49] James H. White. Self-linking and the Gauss integral in higher dimensions. *Amer. J. Math.*, 91:693–728, 1969.

[50] James H. White. An introduction to the geometry and topology of DNA structure. In *Mathematical Methods for DNA Sequences*, pages 225–253. CRC, Boca Raton, FL, 1989.

[51] Wikibooks. Principles of biochemistry — wikibooks, the free textbook project, 2011. https://en.wikibooks.org/wiki/Pinciples_of_Biochemistry/Nucleic_acid_I:_DNA_and_its_nucleotides; accessed 13-August-2018.

[52] M. H. F. Wilkins, A. R. Stokes, and H. R. Wilson. Molecular structure of nucleic acids: Molecular structure of deoxypentose nucleic acids. *Nature*, 171(4356), April 1953.

[53] J Yan, MO Magnasco, and JF Marko. A kinetic proofreading mechanism for disentanglement of DNA by topoisomerases. *Nature*, 401(6756):932–935, 1999.

Chapter 90

Protein Knots, Links and Non-Planar Graphs

Helen Wong, Claremont McKenna College

Proteins are large polypeptide chains that perform the many necessary actions inside a cell to support life. Each protein has its own highly specialized function within the cell. This is encoded in both the sequence of amino acids which makes up the backbone of the protein chain, and in the intricate way the various sections of the backbone fold into a more compact three-dimensional structure. Once folded, a protein's three-dimensional structure determines how it brings together and tightly binds specific molecules of a particular chemical reaction. Misfolded proteins are generally unstable and may lead to a class of diseases called proteinopathies, which include Alzheimer's and Parkinson's diseases. Thus, biologists are very interested in how proteins fold, and more particularly, in how certain proteins fold into non-planar conformations.

The Protein Data Bank (PDB) [1] is an online repository of 3-dimensional coordinates describing the relative location of atoms in a protein molecule. Because it is impossible to observe a protein inside a cell, researchers gather clues about how a protein folds and functions based on these snapshots of the protein, taken either in crystal form or in solution. In particular, it allows detection of knotting or linking in the protein backbone, or of other non-planar features when auxiliary covalent bonds and non-protein cofactor molecules are included. There is strong evidence that topologically complex features incur extra stability and other functional advantages to the proteins, and the latest survey shows about 6% of the PDB with some form of topological complexity [7].

90.1   Knots in the Protein Backbone

The backbones of all proteins are topologically trivial, equivalent to a linear segment with two endpoints. However, proteins are not usually flexible enough to undo any knotting in its backbone. Indeed, most knotting in proteins is deeply embedded, far away from the endpoints, and the topological complexity itself confers thermodynamic, conformational stability to the whole molecule. Thus knotting is effectively trapped, and closed topological knots are a reasonable model for proteins.

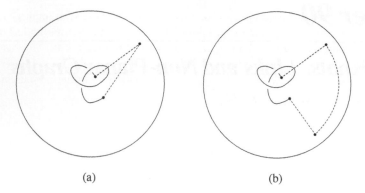

(a)   (b)

FIGURE 90.1: An open chain is closed by connecting the endpoints with one or two points on the sphere-at-infinity.

90.1.1 Closure and Knot Identification Methods

The protein backbone, which is an open chain, needs to be converted into a closed mathematical knot, before its topological knot type can be identified. The simplest method is direct closure, which bridges the two termini of the protein chain with a straight line segment. However, this obviously has serious drawbacks. For example, the added bridge may create or negate crossings in the protein knot, and when the knot is shallow, small changes in the coordinates of the endpoints can affect the knot type detected. usually chosen to avoid the rest of the protein as much as possible.

To avoid shallow knots, certain algorithms [13] simulate pulling the ends of the chain to obtain a tighter knot deep in the chain, from which a direct closure would unambiguously capture the knot. Algorithms of this sort tend to be highly sensitive to initial conditions, e.g. related to the direction in which one pulls the ends. Alternatively, the endpoints could be extended to a *sphere-at-infinity*, which is a sphere centered around the protein and has very large radius relative to the protein. Either the two ends are joined by two line segments to a single point on the sphere-at-infinity, as in Figure 90.1a; or, each end is extended by line segments to two points on the sphere-at-infinity, which are then joined by another arc on the sphere-at-infinity, as in Figure 90.1b. Further steps could be applied to avoid dependence on the initial choice of point(s) on the sphere-at-infinity. For example, one method locates a minimally interference closure [24]. In another, probabilistic approach, knot types of several hundred randomly chosen closures are computed, and the most commonly found knot type is the one assigned to the protein [18]. There is considerable debate about which method of knot closure is the most robust.

Once it is closed, identifying the knot type can also be difficult, because the typical protein's 3-dimensional structure involves many atoms and have large numbers of crossings. Simplification procedures can be used to reduce the protein structure [22, 13, 15], and sometimes visual identification of the knot type is possible. A more reliable method uses computer programs to calculate the Alexander and/or HOMFLY-PT polynomials. Although they're not perfect invariants, in practice the polynomial invariants suffice to identify the simple knots which biologists expect to be found in proteins. Usually, the Alexander polynomial is computed to identify the knot, and when a chiral knot is identified, then the HOMFLY-PT is used to further distinguish between the two mirror forms of the knot.

We also mention that recent research uses the theory of *knotoids* [25] to identify knotting in proteins. Knotoid diagrams are like the usual knot diagrams, except that the underlying curve is an interval with endpoints, and the Reidemeister moves for knotoids are like the usual Reidemeister moves for knots, except that they must occur away from the endpoints. Many knot invariants generalize to knotoids, and these invariants can be used to identify knotoid types in proteins [10].

90.1.2 Knotting in Subchains

Even more information can be extracted by computing the knot type of all possible subchains of the protein backbone. This information, called the protein's *knot fingerprint*, is available at the online KnotProt server [12, 21], and can often reveal rich insight into the specific conformations of proteins in the PDB. For example, one can determine depth of knotting by considering the relative length of the *knot core*, which is the shortest subchain of the protein backbone which has a non-trivial knot type. In addition, this gives information about which of the endpoints is implicated in the folding, as the one closest to the knot core is usually assumed to be the more active one in threading events.

The knot fingerprint further detects *slipknots* [14, 21], whose overall knot type is trivial but contains subchains of non-trivial knot type. For example, Figure 90.2 shows a protein chain whose overall knot type is trivial. Because a piece of the protein is folded back on itself (which can be thought of as a Reidemeister II move), cutting off a piece of the protein at the top yields a subchain with a non-trivial 4_1 knot type. It is believed that slipknots of this type are formed to facilitate threading during protein folding, though they may or may not remain in the final protein structure [20]. However, analysis of slipknotted proteins show that more complicated configurations, not due to simple Reidemeister II moves, can occur as well and may confer stability to the protein [21].

FIGURE 90.2: A slipknot, where the chain is trivial but contains a 4_1 subchain.

90.1.3 Knotting in the Protein Backbone

As of a 2017 survey of the PDB [11], nineteen kinds of proteins have been found with 3_1 knot type, two with 4_1, one with 5_2, and one with 6_1. The few number of knots, as well as the strong negative correlation between the crossing number of a knot and the number of proteins of that type, are seen to confirm biological intuition that knots are associated with high entropic costs and low likelihood.

All of the currently known protein knots are *twist knots*. Taylor [23] posited that all the proteins form using only one threading. Specifically, a section of the protein first folds into a twisted loop, which is then threaded through by one of the ends of the protein, as in Figure 90.3. Taylor thus asserts that only twist knots, which have unknotting number one, will ever be found in protein knots. In particular, Taylor uses this reasoning to contrast the two five-crossing knots: the 5_2 twist knot has been found in proteins, whereas the 5_1 torus knot, which has unknotting number two, has not.

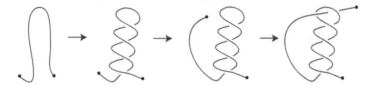

FIGURE 90.3: The 6_1 twist knot is formed using Taylor's method, with one threading.

Although Taylor's model is generally accepted for simple knots, it is unclear how accurate it is for the more complex 5_2 and 6_1 knots that have been found in proteins. While they can be formed from threading a loop with three or four half-twists, respectively, biologists argue that the torsional stress from the twisting may discourage this folding pathway. Instead, some have also suggested that the knots in proteins may result from multiple threadings [2].

90.2   Complex Topology from Including Disulfide and Cofactor Bonds

Proteins are subject to different kinds of strong bonds, which stabilize the 3-dimensional structure and enhance functionality. One important type is a disulfide bond, which occurs between two sulfur atoms in the protein backbone. Another is a cofactor bond, which occurs between one or more atoms on the protein backbone and a non-protein helper-molecule or metal ion. To include such information, the protein is modeled by a topological graph consisting of the (open-ended) protein backbone together with extra edges representing the disulfide or cofactor bonds.

These topological graphs built from disulfide and cofactor bond data often have subgraphs which are knotted, linked, or otherwise non-planar. In fact, the first knots and links in proteins were discovered by Liang and Mislow in metalloproteins in this way [16, 17]. They pointed out, for example, that the human copper lactoferrin protein has one subgraph which is a trefoil knot, and another subgraph which is a Hopf link. The presence of these knotted subgraphs was used to explain why lactoferrin is both non-planar and chiral.

Since then, numerous other topologically non-planar examples have been found [9], including proteins which are encircled by non-protein porphyrin rings to form key link rings and proteins which are attached to metallic clusters to form a non-planar graph like a Mobius ladder, which is depicted in Figure 90.4. Interestingly, many of the early examples involve metalloproteins, and the non-planar subgraphs require both disulfide and cofactor bonds. In particular, if either the disulfide or cofactor bonds are not included, then no such conclusions can be made based on the topology of the backbone alone. Thus, any topological analysis for them necessarily depends on whether disulfide bonds, cofactor bonds, or both are included in the graph.

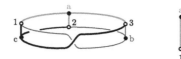

FIGURE 90.4: The Mobius ladder on the left is topologically equivalent to a $K_{3,3}$ graph on the right, and hence the Mobius ladder non-planar. Certain proteins can contain a Mobius ladder as subgraph.

Many proteins exhibit complex topology, even when auxiliary molecules and cofactor bonds are disregarded. Using only disulfide bonds between atoms of the protein backbone, subgraphs with Hopf links and Solomon's links have been identified [8]. By Kuratowski's theorem, a graph is non-planar if and only if it contains as a subgraph a complete subgraph on 5 vertices, K_5, or a complete bipartite graph on two sets of 3 vertices, $K_{3,3}$. Flapan and Heller [9] show many proteins are non-planar in this way. For example, a family of plant cyclotides contain a so-called cyclic cystine 'knot' motif, which form Mobius bands where the three rungs of the ladder represent disulfide bonds [4].

A recently discovered motif is that of a *lasso*, see Figure 90.5, which occurs when an endpoint of the protein pierces through a loop formed when two nearby points on the protein backbone are closed up by a disulfide bond [19]. When the loop closed by the disulfide bond is relatively small, the endpoint is effectively trapped. Although a protein with a lasso has structures which are visibly non-planar, this is not detected by existing topological models. For example, no subgraph is knotted or linked, and every subgraph is topologically equivalent to a planar one. How to capture their non-planarity using topological tools, and their subsequent classification, remains an active area of research [6].

FIGURE 90.5: A lasso occurs when an endpoint of the protein pierces through a covalent loop, formed by a disulfide bond (in thick grey).

Proteins can often form multi-chain complexes, where the different components are held in proximity by strong disulfide bonds. The online server LinkProt [3, 5] shows that many different topologies can occur. For example, the two components of a two-protein complex can form a Hopf link, Solomon link, or be a disjoint union of two trefoil knots. The two components are usually held together by disulfide bonds, which confer added structural stability. The largest examples of a protein link may be capsids, which form shells to surround and protect bacterial viruses. For example, the capsid for the HK97 consists of 72 proteins, cross-linked to make a chain mail shell in the shape of an icosahedron [26].

90.3 Bibliography

[1] Helen M Berman, John Westbrook, Zukang Feng, Gary Gilliland, T N Bhat, Helge Weissig, Ilya N Shindyalov, and Philip E Bourne. The protein data bank. *Nucleic Acids Research*, 28(1):235–242, 2000.

[2] Daniel Bölinger, Joanna I Sułkowska, Hsiao-Ping Hsu, Leonid A Mirny, Mehran Kardar, José N Onuchic, and Peter Virnau. A Stevedore's protein knot. *PLOS Comput. Biol.*, 6(4):e1000731, 2010.

[3] Daniel R Boutz, Duilio Cascio, Julian Whitelegge, L Jeanne Perry, and Todd O Yeates. Discovery of a thermophilic protein complex stabilized by topologically interlinked chains. *Journal of Molecular Biology*, 368(5):1332–1344, 2007.

[4] David J Craik, Norelle L Daly, Trudy Bond, and Clement Waine. Plant cyclotides: A unique family of cyclic and knotted proteins that defines the cyclic cystine knot structural motif. *Journal of molecular biology*, 294(5):1327–1336, 1999.

[5] Pawel Dabrowski-Tumanski, Aleksandra I Jarmolinska, Wanda Niemyska, Eric J Rawdon, Kenneth C Millett, and Joanna I Sulkowska. LinkProt: A database collecting information about biological links. *Nucleic Acids Research*, 45(Database issue):D243–D249, 2017.

[6] Pawel Dabrowski-Tumanski, Wanda Niemyska, Pawel Pasznik, and Joanna I Sulkowska. Lassoprot: server to analyze biopolymers with lassos. *Nucleic Acids Research*, 44(W1):W383–W389, 2016.

[7] Pawel Dabrowski-Tumanski and Joanna I Sulkowska. To tie or not to tie? that is the question. *Polymers*, 9(9):454, 2017.

[8] Pawel Dabrowski-Tumanski and Joanna I Sulkowska. Topological knots and links in proteins. *Proceedings of the National Academy of Sciences*, 114(13):3415–3420, 2017.

[9] Erica Flapan and Gabriella Heller. Topological complexity in protein structures. *Molecular Based Mathematical Biology*, 3(1), 2015.

[10] Dimos Goundaroulis, Neslihan Gügümcü, Sofia Lambropoulou, Julien Dorier, Andrzej Stasiak, and Louis Kauffman. Topological models for open-knotted protein chains using the concepts of knotoids and bonded knotoids. *Polymers*, 9(9):444, 2017.

[11] Sophie E Jackson, Antonio Suma, and Cristian Micheletti. How to fold intricately: Using theory and experiments to unravel the properties of knotted proteins. *Curr. Opin. Struct. Biol.*, 42:6–14, 2017.

[12] M Jamroz, W Niemyska, E J Rawdon, A Stasiak, K C Millett, P Sułkowski, and J I Sułkowska. KnotProt: A database of proteins with knots and slipknots. *Nucleic Acids Res.*, 43(D1):D306–D314, 2015.

[13] Firas Khatib, Matthew T Weirauch, and Carol A Rohl. Rapid knot detection and application to protein structure prediction. *Bioinformatics*, 22(14):e252–e259, 2006.

[14] Neil P King, Eric O Yeates, and Todd O Yeates. Identification of rare slipknots in proteins and their implications for stability and folding. *J. Mol. Biol.*, 373:153–166, 2007.

[15] Kleanthes Koniaris and Murugappan Muthukumar. Self-entanglement in ring polymers. *The Journal of Chemical Physics*, 95(4):2873–2881, 1991.

[16] Chengzhi Liang and Kurt Mislow. A left-right classification of topologically chiral knots. *J. Math. Chem.*, 15(1):35–62, 1994.

[17] Chengzhi Liang and Kurt Mislow. Topological features of protein structures: Knots and links. *Journal of the American Chemical Society*, 117(15):4201–4213, 1995.

[18] Kenneth C Millett and Benjamin M Sheldon. Tying down open knots: A statistical method for identifying open knots with applications to proteins. In *Physical and Numerical Models in Knot Theory: Including Applications to the Life Sciences*, pages 203–217. World Scientific, 2005.

[19] Wanda Niemyska, Pawel Dabrowski-Tumanski, Michal Kadlof, Ellinor Haglund, Piotr Sułkowski, and Joanna I Sulkowska. Complex lasso: new entangled motifs in proteins. *Scientific reports*, 6:36895, 2016.

[20] Joanna I Sułkowska, Jeffrey K Noel, César A Ramírez-Sarmiento, Eric J Rawdon, Kenneth C Millett, and José N Onuchic. Knotting pathways in proteins. *Biochem. Soc. Trans.*, 41:523–527, 2013.

[21] Joanna I Sułkowska, Eric J Rawdon, Kenneth C Millett, José N Onuchic, and Andrzej Stasiak. Conservation of complex knotting and slipknotting patterns in proteins. *Proc. Natl. Acad. Sci. U.S.A.*, 109(26):E1715–E1723, 2012.

[22] William R Taylor. A deeply knotted protein structure and how it might fold. *Nature*, 406:916–919, 2000.

[23] William R Taylor. Protein knots and fold complexity: some new twists. *Comp. Biol. Chem.*, 31(3):151–162, 2007.

[24] Luca Tubiana, Enzo Orlandini, and Cristian Micheletti. Probing the entanglement and locating knots in ring polymers: A comparative study of different arc closure schemes. *Progress of Theoretical Physics Supplement*, 191:192–204, 2011.

[25] Vladimir Turaev et al. Knotoids. *Osaka Journal of Mathematics*, 49(1):195–223, 2012.

[26] William R Wikoff, Lars Liljas, Robert L Duda, Hiro Tsuruta, Roger W Hendrix, and John E Johnson. Topologically linked protein rings in the bacteriophage HK97 capsid. *Science*, 289(5487):2129–2133, 2000.

Chapter 91

Synthetic Molecular Knots and Links

Erica Flapan, Pomona College

Knots and links have been observed in biologically based molecules like DNA and proteins. Such molecules are long and flexible, so it is not surprising that they can become entangled. However, *synthetic molecules*, which are those made in a laboratory, are normally quite small and rigid. As a result, most are topologically planar. Starting in the early 20th century, chemists proposed knots and links, as well as closed ladders in the form of a Möbius strip as goals for synthesis. The motivation for such goals is in part to stimulate the development of new techniques of synthesis, and in part because such molecules can have unforeseen applications. For example, knotted, linked, and entangled molecules exhibit greater elasticity than simple molecules, and hence are being used in the development of artificial materials which have the flexibility of rubber. Also, because of the capability of individual components of a link to rotate and in some cases turn over, molecular links have been proposed as potential switches in microscopic transistors [24] and as the basis for molecular machines [13, 17].

For example, we can see on the left side of Figure 91.1 that turning over the red component can be utilized as an on-off switch; while on the right side of the figure, the movement of the grey circle from the red, to the green, to the blue region, and then back to the red region can be the basis for a rotary motor.

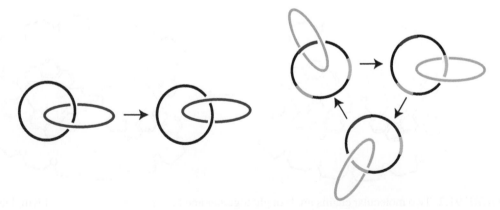

FIGURE 91.1: On the left, turning over the red circle is the basis for an on-off switch; and on the right, the circular motion of the grey circle is the basis for a motor.

Knotted and linked molecules occur naturally in biopolymers like DNA, RNA, and proteins. However, synthetically creating such molecules is not easy. Molecular knots and links were proposed as targets of synthesis as early as 1912, but the first successful synthesis of a molecular link occurred in 1961 [15] and the first synthesis of a molecular knot only occurred in 1989 [11].

In particular, Jean-Pierre Sauvage and co-workers [12] pioneered the use of a metal template to synthesize links as well as the first molecular knot. This approach involves bringing together intertwined chains and holding them in place using a metal template as a scaffolding. The ends of the chains are then joined to create closed curves, after which the metal template is removed leaving a molecular knot or link.

Figure 91.2 illustrates how this approach was used to synthesize a trefoil knot. We start with two molecular chains, one drawn in blue and the other drawn in red in the top left of the figure. On the top right we see that they are made to twist together and held in place by copper molecules, represented by the atomic symbol "Cu." Then the ends of the red and blue chains are joined with long chains of carbons and oxygens as illustrated on the bottom left. Finally, the copper is removed, and we obtain the trefoil knot illustrated on the bottom right of the figure. More details on the use of metal templates and other techniques to synthesize molecular knots and links are provided in a number of excellent review articles [4, 6, 14, 16, 18, 20].

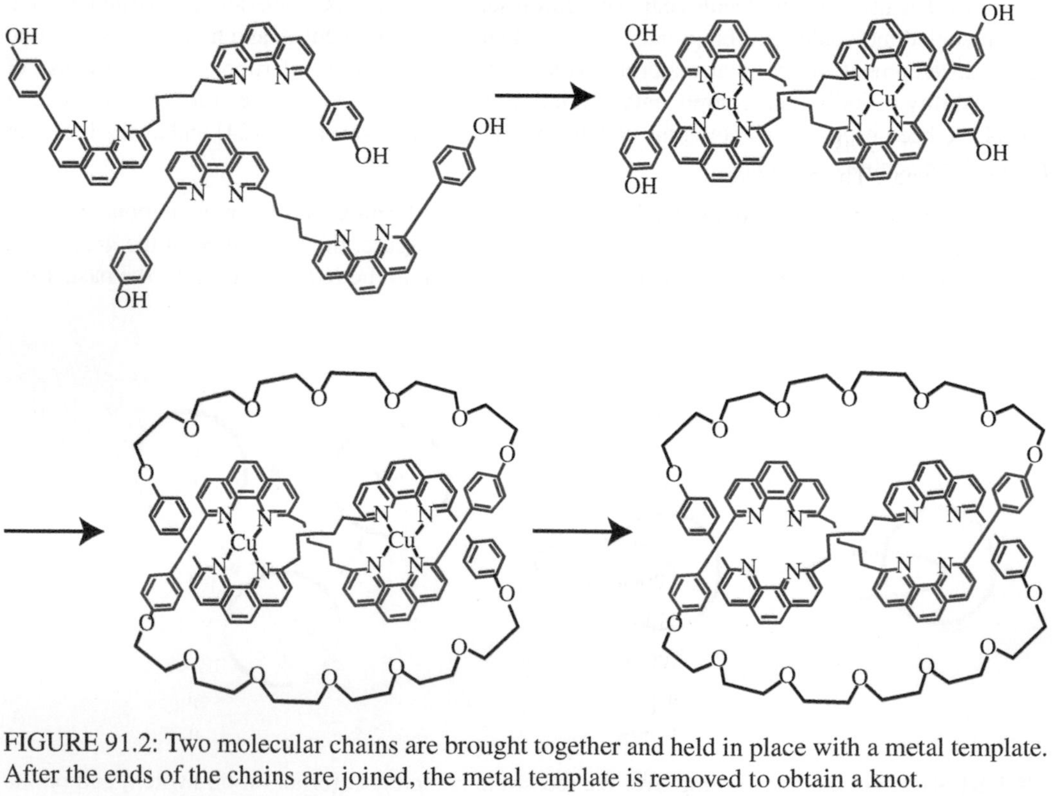

FIGURE 91.2: Two molecular chains are brought together and held in place with a metal template. After the ends of the chains are joined, the metal template is removed to obtain a knot.

Building on the successful synthesis of left-handed and right-handed trefoil knots, Sauvage and co-workers were able to synthesize both a granny knot and a square knot [8]. However, in spite of having only four crossings, it has been surprisingly difficult to create a synthetic molecular 4_1 knot. While the trefoil knot was synthesized in 1989, the 4_1 was only synthesized in 2008 [7]. Furthermore, the synthetic 4_1 knot is part of a metal structure which cannot be removed from the knot. This is in contrast with metal templates which can be removed as in Figure 91.2. More recently, a molecular 4_1 knot was obtained as a product of molecular *self-assembly* [22]. This is a novel technique in which molecular building blocks are designed so that they will come together into the desired form without any control from external sources. It is worth noting that both of these molecular 4_1 knots have the symmetric form illustrated on the left in Figure 91.3, rather than the usual form with only four crossings.

4_1 5_1

FIGURE 91.3: The form of a molecular 4_1 knot, and the five twisted strands as they are held in place to create a 5_1 knot.

The family of $(2, n)$ torus knots are obtained by twisting two strands together n times and then joining the ends. The trefoil is the $(2, 3)$ torus knot and the 5_1 knot is the $(2, 5)$ torus knot. In 2012, Ayme and co-workers [5] synthesized a 5_1 knot by using metal ions to hold five separate arcs in a twisted form as illustrated on the right in Figure 91.3 where the metal ions are indicated by grey disks.

More recently, in 2017, Danon and co-workers synthesized an 8_{19} knot [10]. Because 8_{19} is a $(3, 4)$ torus knot rather than a $(2, n)$ torus knot, it required a more complex strategy for synthesis. This was done by creating four semi-circular pieces, braiding them together, holding them in place with metal ions, joining the arcs in four places (as illustrated in Figure 91.4), and finally removing the metal ions. This is illustrated in Figure 91.4 where the metal ions are grey disks. The molecule consists of 192 atoms, and was awarded the Guinness World Record for being the tightest knot ever tied.

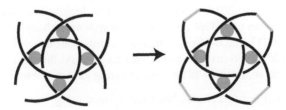

FIGURE 91.4: The synthesis of a 8_{19} knot.

Chemists have also sought to synthesize composite knots. However, until 2018, the only such molecular knots were granny knots (i.e., the connected sum of two identical trefoils) and square knots (i.e., the connected sum of two trefoils which are mirror images). Furthermore, in synthesizing the granny knots and square knots, the chemists were unable to control which mirror form of the trefoil resulted. Thus they ended up with a mixture of granny knots of both mirror forms mixed together with square knots [8]. By contrast, in 2018, Zhang and co-workers synthesized composite knots consisting of three identical trefoil knots [26]. In fact, their strategy enabled them to also synthesize a 3-component link. In both cases, they started with six arcs braided together and held in place with metal ions as illustrated in the center in Figure 91.5. By connecting the ends of the arcs in one way they get the sum of three trefoil knots as illustrated on the right and by connecting the arcs another way they get a 9_7^3 link as illustrated on the left where each ring is of a different color.

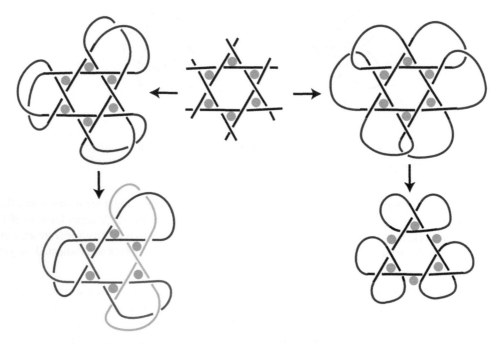

FIGURE 91.5: The synthesis of a knot and link with nine crossings.

The link illustrated in Figure 91.5 is one of the most complex molecular links to be synthesized to date. Prior to its synthesis in 2018, a number of other molecular links had been synthesized. In particular, molecular versions of a Hopf link [12], a 3-component chain [23], a 4-component chain [2], a 5-component chain [1], and 6-component and 7-component networks of rings [3] have been produced this way. In addition, Sauvage and co-workers have synthesized links with 3, 4, 5, and 6 rings all linked to a single central ring as keys on a keyring [23]. See Figure 91.6.

Several other complicated links have also been synthesized using metal templates. This includes the Borromean rings [9, 21], a link of three rings which are pairwise linked [25], a $(2, 4)$ torus link (also known as a Solomon Link) [19], and a $(2, 6)$ torus link (also known as a Star of David Link). We illustrate these links in Figure 91.7. Note that the $(2, 4)$ and $(2, 6)$ torus links were synthesized with twisting strategies analogous to those used to synthesize the $(2, n)$ torus knots 3_1 and 5_1.

chain link

keyring

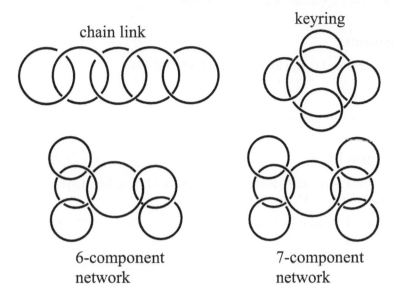

6-component
network

7-component
network

FIGURE 91.6: Chains, keyring links, and 6-component and 7-component networks of rings have
been synthesized.

FIGURE 91.7: These links have also been synthesized: the Borromean rings, three rings that are
pairwise linked, the Solomon link, and the Star of David link.

Developing strategies to synthesize molecular knots and links is an active area of organic chemistry. As a result, there are likely to be more and more synthetic knots and links in the future. Also, as techniques for synthesizing knots and links become more widely known, we can expect more diverse and innovative applications for such molecules.

Acknowledgments

The author was supported in part by NSF Grant DMS-1607744.

91.1   Bibliography

[1] D. Amabilino, P. Ashton, A. Reder, N. Spencer, and J. Stoddart. Olympiadane. *Angewandte Chimie Int. Ed. Engl.*, 33:1286–1290, 1994.

[2] D. Amabilino, P. Ashton, A. Reder, N. Spencer, and J. Stoddart. The two-step self-assembly of [4]- and [5]catenanes. *Angewandte Chimie Int. Ed. Engl.*, 33:433–437, 1994.

[3] D. Amabilino, P. Ashton, A. Reder, N. Spencer, and J. Stoddart. Oligocatenanes made to order. *J. Am. Chem. Soc.*, 120:4295–4307, 1998.

[4] J.-F. Ayme, J. Beves, C. Campbell, and D. Leigh. Template synthesis of molecular knots. *Chem. Soc. Rev.*, 42:1700–1712, 2013.

[5] J.-F. Ayme, J. Beves, D. Leigh, R. McBurney, K. Rissanen, and D. Schultz. A synthetic molecular pentafoil knot. *Nature Chemistry*, 4:15–20, 2012.

[6] J. Beves. Metal template synthesis of molecular knots and links. *Chimia*, 6:170–173, 2012.

[7] F. Bitsch, C. Dietrich-Buchecker, A. Khemiss, J.-P. Sauvage, and A. Van Dorsselaer. Multiring interlocked systems: structure elucidated by electrospray mass spectorometry. *J. Am. Chem. Soc.*, 113:4023–4025, 1991.

[8] R. Carina, C. Dietrich-Buchecker, and J.-P. Sauvage. Molecular composite knots. *J. Am. Chem. Soc.*, 118:9100–9116, 1996.

[9] K. Chichak, S. Cantrill, A. Pease, S. Chiu, G. Cave, J. Atwood, and J. Stoddart. Molecular Borromean rings. *Science*, 28:1308–1312, 2004.

[10] J. Danon, A. Krüger, D. Leigh, J.-F. Lemonnier, Vitorica-Yrezabal I. Stephens, A., and S. Woltering. Braiding a molecular knot with eight crossings. *Science*, 335:159–162, 2017.

[11] C. Dietrich-Buchecker and J.-P. Sauvage. A synthetic molecular trefoil knot. *Angew. Chem. Int. Ed. Engl.*, 28:189–192, 1989.

[12] C. Dietrich-Buchecker, J.-P. Sauvage, and J.-P. Kintzinger. Une nouvelle famille de molecules: Les metallo-catenanes. *Tetrahedron Letters*, 24:15095–5098, 1983.

[13] S. Durot, F. Reviriego, and J.-P. Sauvage. Copper-complexed catenanes and rotaxanes in motion: 15 years of molecular machines. *Dalton Transactions*, 39:10557–10570, 2010.

[14] S. Fielden, D. Leigh, and S. Woltering. Molecular knots. *Angew. Chem. Int. Ed. Engl.*, 56:11166–11194, 2017.

[15] H. Frisch and E. Wasserman. Chemical topology. *J. Am. Chem. Soc.*, 83:13789–3795, 1961.

[16] G. Gil-Ramírez, D. Leigh, and A. Stephens. Catenanes: Fifty years of molecular links. *Angew. Chem. Int. Ed. Engl.*, 54:6110–6150, 2015.

[17] E. Kay, D. Leigh, and F. Zerbetto. Synthetic molecular motors and mechanical machines. *Angew. Chem. Int. Ed. Engl.*, 46:72–191, 2007.

[18] O. Lukin and F. Vögtle. Knotting and threading of molecules: Chemistry and chirality of molecular knots and their assemblies. *Angew. Chem. Int. Ed. Engl.*, 44:1456–1477, 2005.

[19] J.-F. Nierengarten, C. Dietrich-Buchecker, and J.-P. Sauvage. Synthesis of a doubly interlocked [2]-catenane. *J. Am. Chem. Soc.*, 116:375–376, 1994.

[20] Z. Niu and H. Gibson. Polycatenanes. *Chem. Review*, 109:6024–6046, 2009.

[21] C. Pentecost, N. Tangchaivang, S. Cantrill, K. Chichak, A. Peters, and J. F. Stoddart. Making molecular borromean rings. a gram-scale synthetic procedure for the undergraduate organic lab. *Journal of Chemical Education*, 84:1855–1859, 2007.

[22] N. Ponnuswamy, F. Cougnon, G. Pantos, and J. Sanders. Homochiral and meso figure eight knots and a solomon link. *J. Am. Chem. Soc.*, 136:8243–8251, 2014.

[23] J.-P. Sauvage and J. Weiss. Synthesis of dicopper(i) [3]catenanes: Multiring interlocked coordinating systems. *J. Am. Chem. Soc.*, 107:6108–6110, 1985.

[24] F. Vögtle, W. Müller, U. Müller, M. Bauer, and K. Rissanen. Photoswitchable catenanes. *J. Angew. Chem. Int. Ed. Engl.*, 32:1295–1297, 1993.

[25] C. Wood, T. Ronson, Holstein-J. Belenguer, A., and J. Nitschke. Two-stage directed self-assembly of a cyclic [3]catenane. *Nature Chemistry*, 7:354–358, 2015.

[26] L. Zhang, A. Stephens, A. Nussbaumer, J.-F. Lemonnier, P. Jurček, I. Vitorica-Yrezabal, and D. Leigh. Stereoselective synthesis of a composite knot with nine crossings. *Nature Chemistry*, 10:1083–1088, 2018.

Index

Printed and bound by CPI Group (UK) Ltd, Croydon, CR0 4YY
29/10/2024
01780642-0001